S0-ACV-255

COMMON FOURIER TRANSFORM PAIRS

$1, \quad -\infty < t < \infty \leftrightarrow 2\pi\delta(\omega)$

$-0.5 + u(t) \leftrightarrow \dfrac{1}{j\omega}$

$u(t) \leftrightarrow \pi\delta(\omega) + \dfrac{1}{j\omega}$

$\delta(t) \leftrightarrow 1$

$\delta(t - c) \leftrightarrow e^{-j\omega c}, \quad c$ any real number

$e^{-bt}u(t) \leftrightarrow \dfrac{1}{j\omega + b}, \quad b > 0$

$e^{j\omega_0 t} \leftrightarrow 2\pi\delta(\omega - \omega_0), \omega_0$ any real number

$p_\tau(t) \leftrightarrow \tau \operatorname{sinc} \dfrac{\tau\omega}{2\pi}$

$\tau \operatorname{sinc} \dfrac{\tau t}{2\pi} \leftrightarrow 2\pi p_\tau(\omega)$

$\left(1 - \dfrac{2|t|}{\tau}\right)p_\tau(t) \leftrightarrow \dfrac{\tau}{2}\operatorname{sinc}^2\left(\dfrac{\tau\omega}{4\pi}\right)$

$\dfrac{\tau}{2}\operatorname{sinc}^2\left(\dfrac{\tau t}{4\pi}\right) \leftrightarrow 2\pi\left(1 - \dfrac{2|\omega|}{\tau}\right)p_\tau(\omega)$

$\cos \omega_0 t \leftrightarrow \pi[\delta(\omega + \omega_0) + \delta(\omega - \omega_0)]$

$\cos(\omega_0 t + \theta) \leftrightarrow \pi[e^{-j\theta}\delta(\omega + \omega_0) + e^{j\theta}\delta(\omega - \omega_0)]$

$\sin \omega_0 t \leftrightarrow j\pi[\delta(\omega + \omega_0) - \delta(\omega - \omega_0)]$

$\sin(\omega_0 t + \theta) \leftrightarrow j\pi[e^{-j\theta}\delta(\omega + \omega_0) - e^{j\theta}\delta(\omega - \omega_0)]$

Fundamentals of Signals and Systems

Using the Web and MATLAB®

Second Edition

EDWARD W. KAMEN
BONNIE S. HECK

School of Electrical and Computer Engineering
Georgia Institute of Technology

Prentice Hall
Upper Saddle River, New Jersey 07458

Library of Congress Cataloging-in-Publication Data

Kamen, Edward W.
Fundamentals of signals and systems Using the Web and MATLAB, 2/E / Edward W. Kamen,
Bonnie S. Heck.
p. cm.
Includes bibliographical references and index.
ISBN 0-13-017293-6
1. Signal processing—Digital techniques. 2. System analysis.
3. MATLAB. I. Heck, Bonnie S. II. Title.
TK5102.9.K35 1999
621.581′34—dc20 99-336
CIP

Publisher: **TOM ROBBINS**
Vice-president and editorial director: **MARCIA HORTON**
Executive managing editor: **VINCE O'BRIEN**
Managing editor: **DAVID A. GEORGE**
Cover designer: **BRUCE KENSELAAR**
Marketing manager: **DANNY HOYT**
Manufacturing manager: **TRUDY PISCIOTTI**
Manufacturing buyer: **BETH STURLA**
Editorial assistant: **JESSE POWERS**
Production/editorial supervision: **PREPARE, INC.**
Vice-president of production and manufacturing: **DAVID W. RICCARDI**

© 2000, 1997 by Prentice-Hall, Inc.
Upper Saddle River, New Jersey 07458

All rights reserved. No part of this book may be reproduced, in any form or by any means, without permission in writing from the publisher.

The author and publisher of this book have used their best efforts in preparing this book. These efforts include the development, research, and testing of the theories to determine their effectiveness. The author and publisher make no warranty of any kind, expressed or implied, with regard to the documentation contained in this book.

MATLAB® is a registered trademark of the MathWorks, Inc.

Printed in the United States of America

10 9

ISBN 0-13-017293-6

Prentice-Hall International (UK) Limited, *London*
Prentice-Hall of Australia Pty. Limited, *Sydney*
Prentice-Hall Canada Inc., *Toronto*
Prentice-Hall Hispanoamericana, S.A., *Mexico*
Prentice-Hall of India Private Limited, *New Delhi*
Prentice-Hall of Japan, Inc., *Tokyo*
Pearson Education Asia Pte. Ltd., *Singapore*
Editora Prentice-Hall do Brasil, Ltda., *Rio de Janeiro*

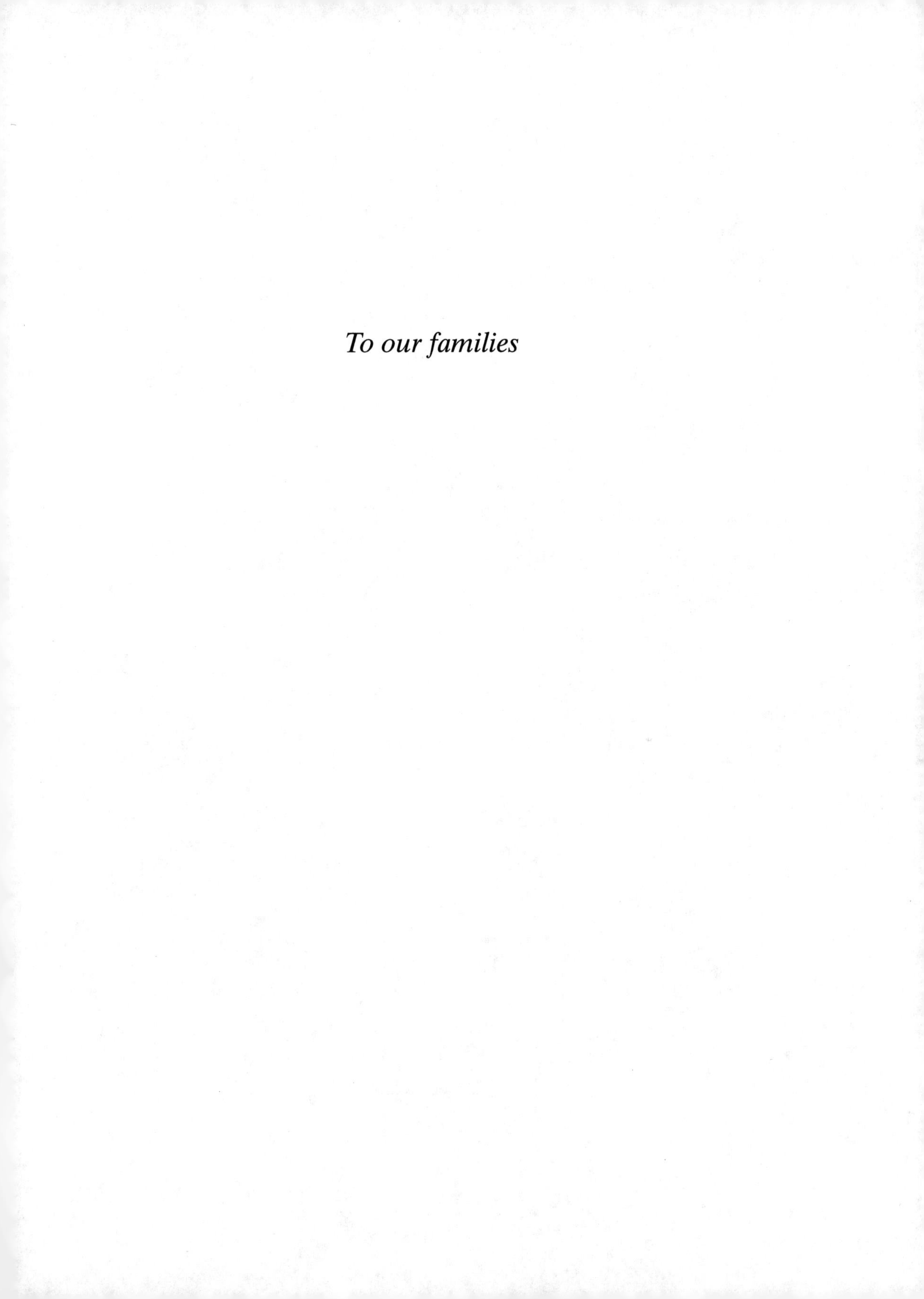

To our families

Contents

Preface

This book contains an introductory, yet comprehensive, treatment of continuous-time and discrete-time signals and systems with demos on the Web and MATLAB examples integrated throughout the text. The second edition contains modifications of the material in the first edition to improve the presentation, additional illustrative examples and homework problems, a new chapter on communication systems, and the use of numerous on-line demos that illustrate the concepts and techniques presented in the book. The demos are available at the Web site (http://users.ece.gatech.edu/~bonnie/book) that accompanies this text. The chapter on communications covers both analog and digital modulation with an emphasis on the digital case, including phase-shift keying, frequency-shift keying, quadrature amplitude modulation, and on-off keying. The background required for reading the book consists of the usual freshman/sophomore courses in calculus and elementary differential equations. It is also helpful, but not necessary, to have had some exposure to physics. The book is also intended to be used for self-study. Both authors have been teaching the material in the book to electrical engineering juniors for many years, and Bonnie Heck has been actively involved in the use of the Web for enhancing education in the fields of signals, systems, and controls.

A key feature of the text is the use of the on-line demos on the Web. In many of these demos, students may change various values to see what the result is. For example, the frequencies comprising a signal can be changed with the resulting effect on the signal displayed, and the parameters of a system's frequency response function (or transfer function) can be changed with the effect on system perfor-

Mass–Spring–Damper System

mance displayed. In some of the demos, the students can hear the sounds that correspond to the signals being considered. There is also a demo on a mass–spring–damper system that provides an animation of the output response resulting from the application of various inputs. Via this demo, students can actually see the characteristics of the response to an impulsive input, step input, and sinusoidal input. The reference to a demo in the text is given by an icon in the left-hand margin, as illustrated here.

Another key feature of the book is the use of MATLAB (Version 5.0 or higher) to generate computer implementations of the techniques for signal and system analysis and design covered in this work. Along with the on-line demos, the MATLAB implementations provide the reader with the opportunity to verify that the theory does work and allow the reader to experiment with the application of the techniques studied. Use of the various MATLAB commands is illustrated in numerous examples throughout the text. This includes a discussion in Chapter 1 on the use of MATLAB to plot signals. A large number of the homework problems also involve MATLAB. All of the MATLAB programs and M-files that are used in the examples are available from the Web site that accompanies the text.

The book includes a wide range of examples and problems on different areas in engineering including electrical circuits, mechanical systems, and biological systems. Other features of the book are a parallel treatment of continuous-time and discrete-time signals and systems, three chapters on communication systems, digital filtering, and feedback control that prepare students for senior electives in these topics, and a chapter on the state-space formulation of systems.

The book begins with the time-domain aspects of signals and systems in Chapters 1 through 3. This includes the basic properties of signals and systems, the input/output differential equation model, the input/output difference equation model, and the convolution models. Chapter 4 begins the treatment of signals and systems from the frequency-domain standpoint. Starting with signals that are a sum of sinusoids, the presentation then goes into the Fourier series representation of periodic signals and on to the Fourier transform of aperiodic signals. Then, the analysis of continuous-time systems via the Fourier transform is considered in Chapter 5, along with the application to sampling and signal reconstruction. In Chapter 6, the Fourier transform is applied to the study of the transmission and reception of signals in a communication system. In Chapter 7, the Fourier analysis of discrete-time signals and systems is pursued, including a brief treatment of the fast Fourier transform (FFT).

The second half of the book begins in Chapter 8 with the development of the Laplace transform and the transfer function representation for linear time-invariant continuous-time systems. In Chapter 9, the transfer function representation is used to study the response of a system to particular types of inputs, such as the unit-step function and sinusoidal functions. This leads to the notion of the frequency response function first considered in Chapter 5. The transfer function framework is then applied to the problem of control in Chapter 10. In Chapter 11, the z-transform and the transfer function representation of linear time-invariant discrete-time systems is considered, and in Chapter 12, this framework is applied to the design of digital filters and controllers. Finally, in Chapter 13 the fundamentals of the state description of continuous-time and discrete-time systems are presented.

The book can be used in a two-quarter or two-semester course sequence with Chapters 1 through 7 (or parts of these chapters) covered in one course and Chapters 8 through 13 (or parts thereof) covered in the other. By selecting appropriate sections and chapters from the book, an instructor can cover the continuous-time case in the first course and the discrete-time case in the second course. It is possible to cover the main portion of the material in this book in a one-semester course. A schedule for a one-semester course consisting of forty-two 50-minute lectures is given in the following table:

SCHEDULE OF COVERAGE FOR A ONE-SEMESTER COURSE

Text coverage	Number of 50-minute lectures
Chapter 1: All sections	4
Chapter 2: 2.1–2.3	4
Chapter 3: 3.1–3.4	4
Chapter 4: All sections	5
Chapter 5: All sections	5
Chapter 7: 7.1,7.2,7.4	4
Chapter 8: 8.1–8.5	6
Chapter 9: 9.1,9.4,9.5	4
Chapter 11: All sections	6
	42

The authors wish to thank our colleague Jim McClellan for his thorough review and numerous constructive comments on various drafts of the previous edition of the book. We would also like to thank Gary E. Ford, University of California, Davis, for his comments on the previous edition, and Tom Robbins for his comments on both the previous edition and this edition of the book. We are particularly indebted to Steve McLaughlin and John Barry in our department at Georgia Tech who provided many helpful comments on the material in Chapter 6. The second-named author wishes to thank her former students, John Finney and James Moan, who wrote preliminary versions of the MATLAB tutorial that is available on the web site, and Darren Garner, James Ho, Jason Meeks, Johnny Wang, and Brian Wilson for their efforts in generating the demos that are on the web site.

E.W.K
B.S.H

◆

CHAPTER 1

Fundamental Concepts

1.1 SIGNALS AND SYSTEMS

The concepts of signals and systems arise in virtually all areas of technology, ranging from appliances or devices found in homes to very sophisticated engineering innovations. In fact, it can be argued that much of the development of high technology is a result of advancements in the theory and techniques of signals and systems. This section begins with a brief introduction to the concepts of signals and systems and the closely related field of signal processing. In Sections 1.2 and 1.3 the focus is on various fundamental aspects of continuous-time and discrete-time signals, and in Section 1.4 a number of specific examples of systems are given. Then in Section 1.5, the basic system properties of causality, linearity, finite dimensionality, and time invariance are defined.

Signals

A signal $x(t)$ is a *real-valued,* or *scalar-valued,* function of the time variable t. The term *real valued* means that for any fixed value of the time variable t, the value of the signal at time t is a real number. Common examples of signals are voltage or current waveforms in an electrical circuit, audio signals such as speech or music waveforms, bioelectric signals such as an electrocardiogram (ECG) or an electroencephalogram (EEG), forces or torques in a mechanical system, flow rates of liquids or gases in a chemical process, and so on.

Given a signal $x(t)$ that is very complicated, it is often not possible to determine a mathematical function that is exactly equal to $x(t)$. An example is a speech signal, such as the 50-millisecond (ms) segment of speech shown in Figure 1.1. The segment of speech shown in Figure 1.1 is the "sh"-to-"u" transition in the utterance of the word "should." Due to their complexity, signals such as speech waveforms are usually not specified in function form. Instead, they may be given by a set of sample values. For example, if $x(t)$ denotes the speech signal in Figure 1.1, the signal can be represented by the set of sample values

$$\{x(t_0), x(t_1), x(t_2), x(t_3), \ldots, x(t_N)\}$$

where $x(t_i)$ is the value of the signal at time t_i, $i = 0, 1, 2, \ldots, N$, and $N + 1$ is the number of sample points. This type of signal representation can be generated by sampling the speech signal. Sampling is discussed briefly in Section 1.3 and then is studied in depth in later chapters.

In addition to the representation of a signal in function form or by a set of sample values, signals can also be characterized in terms of their "frequency content" or "frequency spectrum." The representation of signals in terms of the frequency spectrum is accomplished by using the Fourier transform, which is studied in Chapters 4 to 7.

Signal Processing

A very important component of technology is signal processing, that is, the processing of signals for various purposes, such as the extraction of the information carried in a signal. Determining the information contained in a signal may not be a simple problem; in particular, knowing the functional form or sample values of a signal does

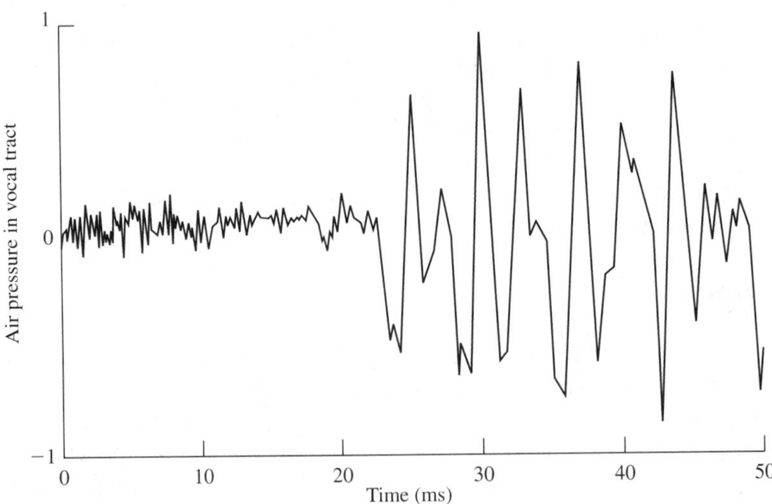

Figure 1.1 Segment of speech.

not directly reveal (in general) the information carried in a signal. An interesting example is the extraction of information carried in a speech signal. For example, it is a nontrivial matter to develop a speech processing scheme that is capable of identifying the person who is speaking from a segment of speech. Of course, to be able to identify the speaker correctly, the speech processor must have stored in its memory the "speech patterns" of a collection of people, one of whom is the speaker. The question here is what is an appropriate speech pattern. In other words, exactly what is it about one's voice that distinguishes it from that of others? One way to answer this is to consider the characterization of speech signals in terms of their frequency spectrum. The concept of the frequency spectrum of a signal is studied in Chapters 4 to 7.

The extraction of information from signals is also of great importance in the medical field in the processing of bioelectric signals. For instance, an important problem is determining the health of a person's heart from the information contained in a collection of ECG signals taken from surface electrodes placed on the person. A specific objective is to be able to detect if there is any heart damage that may be a result of coronary artery disease or from a prolonged state of hypertension. A trained physician may be able to detect heart disease by "reading" ECG signals, but due to the complexity of these signals, it is not likely that a "human processor" will be able to extract all the information contained in these signals. This is a problem area where signal processing techniques can be applied, and in fact, progress has been made on developing automated processing schemes for bioelectric signals.

Another important problem in signal processing is the reconstruction of signals that have been corrupted by spurious signals or noise. For example, suppose that a sensor provides measurements $m(t)$ of a signal $x(t)$ with the measurements given by $m(t) = x(t) + n(t)$, where $n(t)$ is measurement noise or is a term resulting from distortion arising in the sensor. This type of situation arises in target tracking, where $x(t)$ may be the distance (e.g., range) from some target to a radar antenna, and $m(t)$ is the measurement of range provided by the radar. Since the energy reflected from a target is so small, radar measurements of a target's position are always very noisy and usually are embedded in the "background noise." The problem of determining a "good estimate" of a signal $x(t)$ from measurements $m(t) = x(t) + n(t)$ is referred to as an "estimation" or "filtering" problem. In some cases, the estimation of $x(t)$ can be solved very effectively by considering the frequency spectra of $x(t)$ and the corrupting signal $n(t)$.

It is important to note that estimators and filters, and other types of signal processors, can in fact be viewed as systems, and thus system-theoretic concepts and techniques can be applied to the analysis and design of signal processors. So the field of systems can be motivated in part by applications to signal processing.

Systems

A system is an interconnection of components (e.g., devices or processes) with terminals or access ports through which matter, energy, or information can be applied or extracted. As illustrated in Figure 1.2, a common way of viewing a system is in terms of a "black box" with input and output terminals. In the figure, $x_1(t), x_2(t), \ldots, x_p(t)$ are the signals applied to the p input terminals of the system and $y_1(t), y_2(t), \ldots, y_q(t)$ are the resulting responses appearing at the q output terminals of the system. In general, p is not equal to q; in other words, the number of input terminals may not equal the number of output terminals. When $p = q = 1$, the system is a single-input single-output system. It should be noted that the input and output terminals shown in Figure 1.2 do not include "ground" connections. For example, if the $x_i(t)$ and the $y_i(t)$ are voltages relative to ground, the ground is not viewed as a terminal.

In this book the emphasis is on single-input single-output systems. The multi-input multi-output case is considered in Chapter 13 when the state model is presented.

A system may be subjected to different types of inputs. Common examples are control inputs, reference inputs, and disturbance inputs such as noise. Certain types of input signals, such as disturbance inputs, may not be directly measurable. In contrast, the output signals of a system are usually assumed to be measurable using sensing devices.

Some common examples of systems are listed below:

1. An electrical circuit with inputs equal to driving voltages and/or currents and with outputs equal to voltages and/or currents at various points in the circuit.
2. A communications system with inputs equal to the signals to be transmitted and with outputs equal to the received signals.
3. A biological system such as the human heart with inputs equal to the electrical stimuli applied to the heart muscle and with output equal to the flow rate of blood through the heart.
4. A robotic manipulator with inputs equal to the torques applied to the links of the robot and with output equal to the position of the end effector (hand).
5. An oil refinery with input equal to the flow rate of oil and with output equal to the flow rate of gasoline.
6. A manufacturing system with inputs equal to the flow rate of raw materials and with output equal to the rate of production of the finished product.

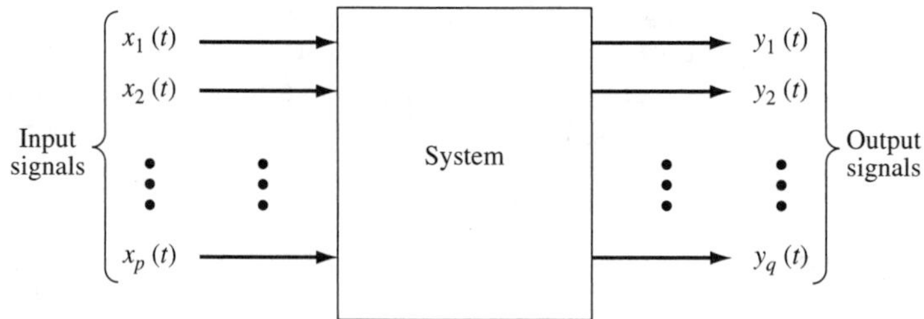

Figure 1.2 System with p inputs and q outputs.

To undertake an in-depth study of a system, such as one of the examples mentioned above, it is very useful to have a *mathematical model* of the system. A mathematical model consists of a collection of equations describing the relationships between the signals appearing in the system. It is important to note that if a system is specified by a digital computer or analog computer simulation, a mathematical model can be generated by writing down the equations corresponding to the signal-flow diagram of the simulation.

A mathematical model of a system is usually an idealized representation of the system. In other words, many actual (physical) systems cannot be described exactly by a mathematical model. For instance, this is the case for all five of the systems mentioned above. However, a sufficiently accurate mathematical model can often be generated so that system behavior and properties can be studied in terms of the model. Mathematical models are also very useful in the design of new systems having various desirable operating characteristics, for example, in the design of "controllers" whose purpose is to modify system behavior to meet some performance objectives. Thus mathematical models are used extensively in both system analysis and system design.

If a model of a system is to be useful, it must be tractable, and thus an effort should always be made to construct the simplest possible model of the system under study. But the model must also be sufficiently accurate, which means that all primary characteristics (all first-order effects) must be included in the model. Usually, the more characteristics that are put into a model, the more complicated the model is, so there is a trade-off between simplicity of the model and accuracy of the model.

There are two basic types of mathematical models: input/output representations describing the relationship between the input and output signals of a system, and the state or internal model describing the relationship among the input, state, and output signals of a system. Input/output representations are studied in the first 12 chapters; the state model is considered in Chapter 13.

Four types of input/output representations are studied in this book:

1. The input/output differential equation or difference equation
2. The convolution model
3. The Fourier transform representation
4. The transfer function representation

As will be shown, the Fourier transform representation can be viewed as a special case of the transfer function representation. Hence there are only three fundamentally different types of input/output representations that will be studied in this book.

The first two representations listed above and the state model are referred to as *time-domain models* since these representations are given in terms of functions of time. The last two of the representations listed above are referred to as *frequency-domain models* since they are specified in terms of functions of a complex variable that is interpreted as a frequency variable. Both time-domain and frequency-domain models are used in system analysis and design. These different types of models are often used together to maximize understanding of the behavior of the system under study.

1.2 CONTINUOUS-TIME SIGNALS

A signal $x(t)$ is said to be a *continuous-time signal* or an *analog signal* when the time variable t takes its values from the set of real numbers. Some common examples of continuous-time signals were mentioned in Section 1.1. In the first part of this section, some simple examples of continuous-time signals that can be expressed in function form are given. Then various fundamental aspects of continuous-time signals are pursued.

Step and Ramp Functions

Two simple examples of continuous-time signals are the unit-step function $u(t)$ and the unit-ramp function $r(t)$. These functions are plotted in Figure 1.3.

The *unit-step function* $u(t)$ is defined mathematically by

$$u(t) = \begin{cases} 1, & t \geq 0 \\ 0, & t < 0 \end{cases}$$

Here *unit step* means that the amplitude of $u(t)$ is equal to 1 for all $t \geq 0$. [Note that $u(0) = 1$; in some textbooks, $u(0)$ is defined to be zero.] If K is an arbitrary nonzero real number, $Ku(t)$ is the step function with amplitude K for $t \geq 0$.

For any continuous-time signal $x(t)$, the product $x(t)u(t)$ is equal to $x(t)$ for $t \geq 0$ and is equal to zero for $t < 0$. Thus multiplication of a signal $x(t)$ with $u(t)$ eliminates any nonzero values of $x(t)$ for $t < 0$.

The *unit-ramp function* $r(t)$ is defined mathematically by

$$r(t) = \begin{cases} t, & t \geq 0 \\ 0, & t < 0 \end{cases}$$

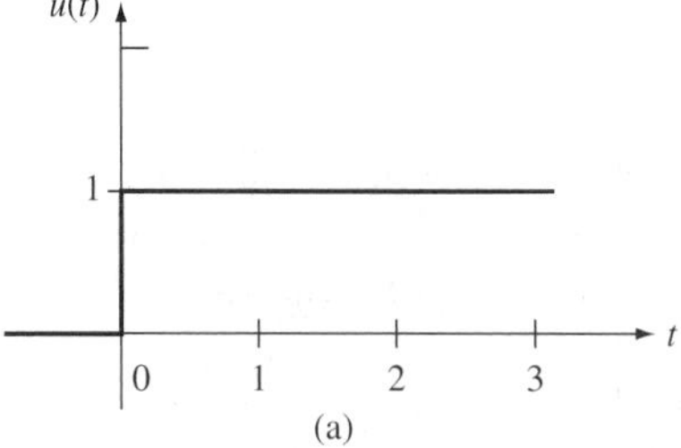

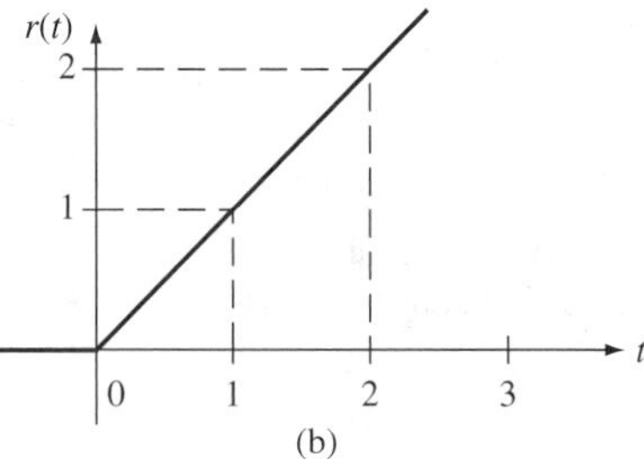

Figure 1.3 (a) Unit-step and (b) unit-ramp functions.

Note that for $t \geq 0$, the slope of $r(t)$ is 1. Thus $r(t)$ has "unit slope," which is the reason $r(t)$ is called the unit-ramp function. If K is an arbitrary nonzero scalar (real number), the ramp function $Kr(t)$ has slope K for $t \geq 0$.

The unit-ramp function $r(t)$ is equal to the integral of the unit-step function $u(t)$; that is,

$$r(t) = \int_{-\infty}^{t} u(\lambda)\, d\lambda$$

Conversely, the first derivative of $r(t)$ with respect to t is equal to $u(t)$, except at $t = 0$, where the derivative of $r(t)$ is not defined.

The Impulse

The *unit impulse* $\delta(t)$, also called the *delta function* or the *Dirac distribution*, is defined by

$$\delta(t) = 0, \qquad t \neq 0$$

$$\int_{-\varepsilon}^{\varepsilon} \delta(\lambda)\, d\lambda = 1 \qquad \text{for any real number } \varepsilon > 0$$

The first condition states that $\delta(t)$ is zero for all nonzero values of t, while the second condition states that the area under the impulse is 1, so $\delta(t)$ has unit area.

It is important to point out that the value $\delta(0)$ of $\delta(t)$ at $t = 0$ is not defined; in particular, $\delta(0)$ is not equal to infinity. The impulse $\delta(t)$ can be approximated by a pulse centered at the origin with amplitude A and time duration $1/A$, where A is a very large positive number. The pulse interpretation of $\delta(t)$ is displayed in Figure 1.4.

For any real number K, $K\delta(t)$ is the impulse with area K. It is defined by

$$K\delta(t) = 0, \qquad t \neq 0$$

$$\int_{-\varepsilon}^{\varepsilon} K\delta(\lambda)\, d\lambda = K \qquad \text{for any real number } \varepsilon > 0$$

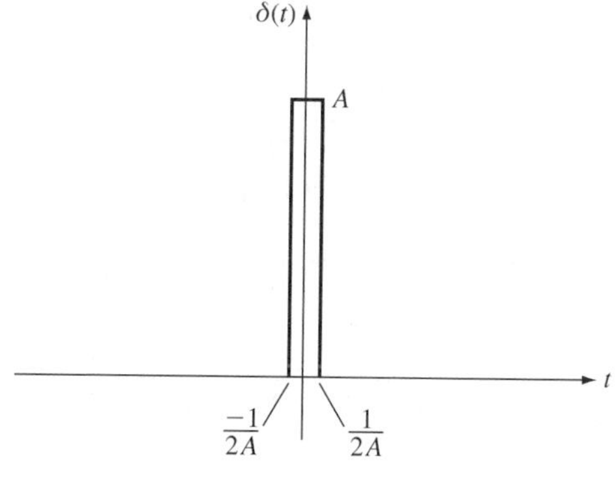

Figure 1.4 Pulse interpretation of $\delta(t)$.

The graphical representation of $K\delta(t)$ is shown in Figure 1.5. The notation "(K)" in the figure refers to the area of the impulse $K\delta(t)$.

The unit-step function $u(t)$ is equal to the integral of the unit impulse $\delta(t)$; more precisely,

$$u(t) = \int_{-\infty}^{t} \delta(\lambda)\, d\lambda, \qquad \text{all } t \text{ except } t = 0$$

To verify this relationship, first note that for $t < 0$,

$$\int_{-\infty}^{t} \delta(\lambda)\, d\lambda = 0, \qquad \text{since } \delta(t) = 0 \text{ for all } t < 0$$

For $t > 0$,

$$\int_{-\infty}^{t} \delta(\lambda)\, d\lambda = \int_{-t}^{t} \delta(\lambda)\, d\lambda = 1, \qquad \text{since } \int_{-\varepsilon}^{\varepsilon} \delta(\lambda)\, d\lambda = 1 \text{ for any } \varepsilon > 0$$

Periodic Signals

Let T be a fixed positive real number. A continuous-time signal $x(t)$ is said to be periodic with period T if

$$x(t + T) = x(t) \text{ for all } t, \qquad -\infty < t < \infty \tag{1.1}$$

Signals and Sounds

Note that if $x(t)$ is periodic with period T, it is also periodic with period qT, where q is any positive integer. The *fundamental period* is the smallest positive number T for which (1.1) holds.

An example of a periodic signal is the sinusoid

$$x(t) = A\cos(\omega t + \theta), \qquad -\infty < t < \infty \tag{1.2}$$

Here A is the amplitude, ω the frequency in radians per second (rad/sec), and θ the phase in radians. The frequency f in hertz (Hz) (or cycles per second) is $f = \omega/2\pi$.

To see that the sinusoid given by (1.2) is periodic, note that for any value of the time variable t,

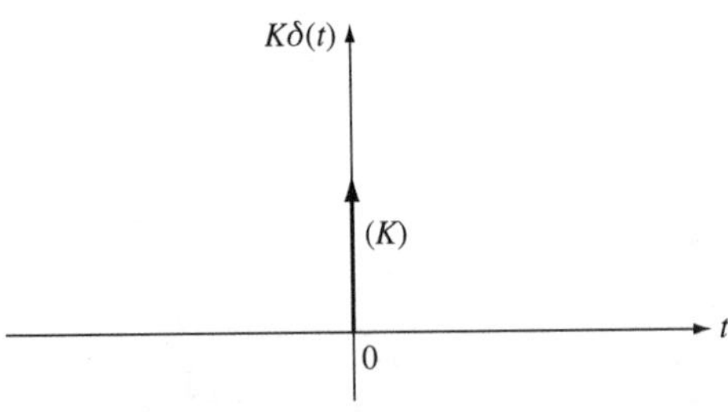

Figure 1.5 Graphical representation of the impulse $K\delta(t)$.

$$A\cos\left[\omega\left(t+\frac{2\pi}{\omega}\right)+\theta\right] = A\cos(\omega t + 2\pi + \theta) = A\cos(\omega t + \theta)$$

Thus the sinusoid is periodic with period $2\pi/\omega$, and in fact, $2\pi/\omega$ is the fundamental period. The sinusoid $x(t) = A\cos(\omega t + \theta)$ is plotted in Figure 1.6 for the case when $-\pi/2 < \theta < 0$. Note that if $\theta = -\pi/2$, then

$$A\cos(\omega t + \theta) = A\sin\omega t$$

An important question for signal analysis is whether or not the sum of two periodic signals is periodic. Suppose that $x_1(t)$ and $x_2(t)$ are periodic signals with fundamental periods T_1 and T_2, respectively. Then is the sum $x_1(t) + x_2(t)$ periodic; that is, is there a positive number T such that

$$x_1(t+T) + x_2(t+T) = x_1(t) + x_2(t) \qquad \text{for all } t \tag{1.3}$$

It turns out that (1.3) is satisfied if and only if the ratio T_1/T_2 can be written as the ratio q/r of two integers q and r. This can be shown by noting that if $T_1/T_2 = q/r$, then $rT_1 = qT_2$, and since r and q are integers, $x_1(t)$ and $x_2(t)$ are periodic with period rT_1. Thus the expression (1.3) follows with $T = rT_1$. In addition, if r and q are coprime (i.e., r and q have no common integer factors other than 1), then $T = rT_1$ is the fundamental period of the sum $x_1(t) + x_2(t)$.

Example 1.1 *Sum of Periodic Signals*

Periodicity of Sums of Sinusoids

Let $x_1(t) = \cos(\pi t/2)$ and $x_2(t) = \cos(\pi t/3)$. Then $x_1(t)$ and $x_2(t)$ are periodic with fundamental periods $T_1 = 4$ and $T_2 = 6$, respectively. Now

$$\frac{T_1}{T_2} = \frac{4}{6} = \frac{2}{3}$$

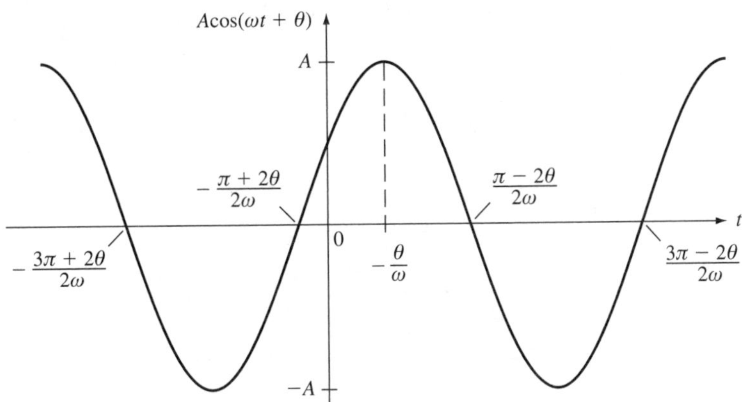

Figure 1.6 Sinusoid $A\cos(\omega t + \theta)$ with $-\pi/2 < \theta < 0$.

Then with $q = 2$ and $r = 3$, it follows that the sum $x_1(t) + x_2(t)$ is periodic with fundamental period $rT_1 = (3)(4) = 12$ seconds.

Time-Shifted Signals

Given a continuous-time signal $x(t)$, it is often necessary to consider a *time-shifted* version of $x(t)$: If t_1 is a positive real number, the signal $x(t - t_1)$ is $x(t)$ shifted to the right by t_1 seconds and $x(t + t_1)$ is $x(t)$ shifted to the left by t_1 seconds. For instance, if $x(t)$ is the unit-step function $u(t)$ and $t_1 = 2$, $u(t - t_1)$ is the 2-second right shift of $u(t)$ and $u(t + t_1)$ is the 2-second left shift of $u(t)$. These shifted signals are plotted in Figure 1.7. To verify that $u(t - 2)$ is given by the plot in Figure 1.7a, evaluate $u(t - 2)$ for various values of t. For example, $u(t - 2) = u(-2) = 0$ when $t = 0$, $u(t - 2) = u(-1) = 0$ when $t = 1$, $u(t - 2) = u(0) = 1$ when $t = 2$, and so on.

For any fixed positive or negative real number t_1, the time shift $K\delta(t - t_1)$ of the impulse $K\delta(t)$ is equal to the impulse with area K located at the point $t = t_1$; in other words,

$$K\delta(t - t_1) = 0, \qquad t \neq t_1$$

$$\int_{t_1-\varepsilon}^{t_1+\varepsilon} K\delta(\lambda - t_1)\, d\lambda = K, \qquad \text{any } \varepsilon > 0$$

The time-shifted unit impulse $\delta(t - t_1)$ is useful in defining the *sifting property* of the impulse given by

$$\int_{t_1-\varepsilon}^{t_1+\varepsilon} f(\lambda)\delta(\lambda - t_1)\, d\lambda = f(t_1), \qquad \text{for any } \varepsilon > 0$$

where $f(t)$ is any real-valued function that is continuous at $t = t_1$. (Continuity of a function is defined below.) To prove the sifting property, first note that since $\delta(\lambda - t_1) = 0$ for all $\lambda \neq t_1$, it follows that

$$f(\lambda)\delta(\lambda - t_1) = f(t_1)\delta(\lambda - t_1)$$

Thus

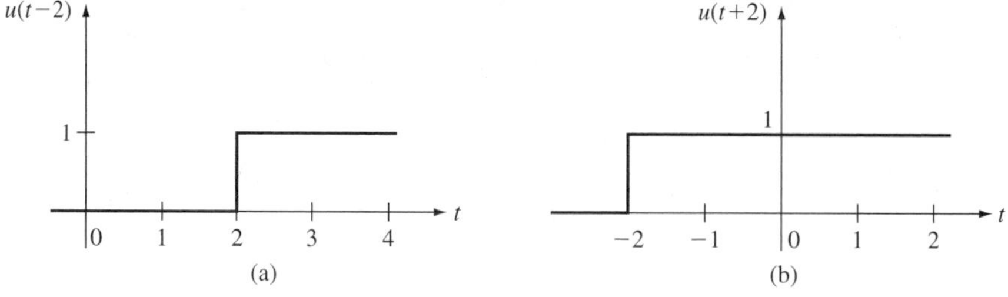

Figure 1.7 Two-second shifts of $u(t)$: (a) right shift; (b) left shift.

$$\int_{t_1-\varepsilon}^{t_1+\varepsilon} f(\lambda)\delta(\lambda - t_1)\, d\lambda = f(t_1)\int_{t_1-\varepsilon}^{t_1+\varepsilon} \delta(\lambda - t_1)\, d\lambda$$
$$= f(t_1)$$

which proves the sifting property.

Continuous and Piecewise-Continuous Signals

A continuous-time signal $x(t)$ is said to be *discontinuous* at a fixed point t_1 if $x(t_1^-) \neq x(t_1^+)$, where $t_1 - t_1^-$ and $t_1^+ - t_1$ are infinitesimal positive numbers. Roughly speaking, a signal $x(t)$ is discontinuous at a point t_1 if the value of $x(t)$ "jumps" as t goes through the point t_1.

A signal $x(t)$ is *continuous* at the point t_1 if $x(t_1^-) = x(t_1) = x(t_1^+)$. If a signal $x(t)$ is continuous at all points t, $x(t)$ is said to be a *continuous signal.* The reader should note that the term *continuous* is used in two different ways; that is, there is the notion of a continuous-time signal and there is the notion of a continuous-time signal that is continuous (as a function of t). This dual use of *continuous* should be clear from the context.

Many continuous-time signals of interest in engineering are continuous. Examples are the ramp function $Kr(t)$ and the sinusoid $A\cos(\omega t + \theta)$. Another example of a continuous signal is the triangular pulse function displayed in Figure 1.8. As indicated in the figure, the triangular pulse is equal to $(2t/\tau) + 1$ for $-\tau/2 \leq t \leq 0$ and is equal to $(-2t/\tau) + 1$ for $0 \leq t \leq \tau/2$.

There are also many continuous-time signals of interest in engineering that are not continuous at all points t. An example is the step function $Ku(t)$, which is discontinuous at the point $t = 0$ (assuming that $K \neq 0$). Another example of a signal that is not continuous everywhere is the rectangular pulse function $p_\tau(t)$, defined by

$$p_\tau(t) = \begin{cases} 1, & \dfrac{-\tau}{2} \leq t < \dfrac{\tau}{2} \\ 0, & t < \dfrac{-\tau}{2},\ t \geq \dfrac{\tau}{2} \end{cases}$$

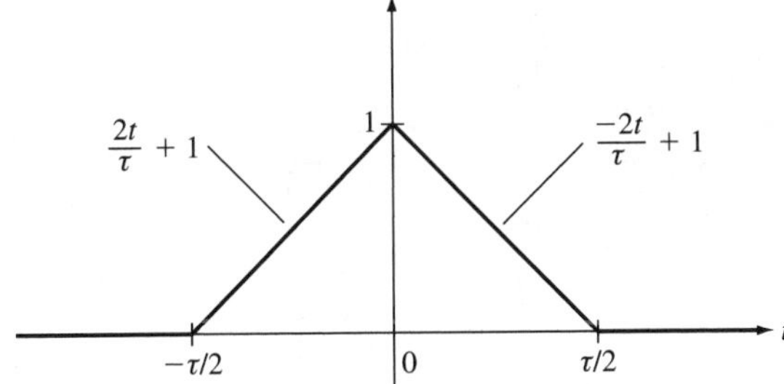

Figure 1.8 Triangular pulse function.

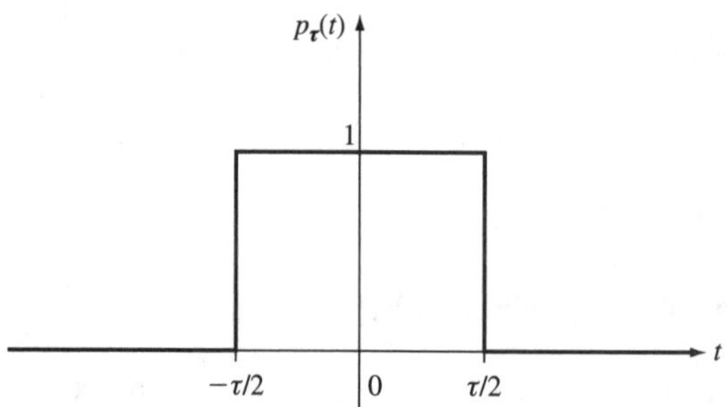

Figure 1.9 Rectangular pulse function.

Here τ is a fixed positive number equal to the time duration of the pulse. The rectangular pulse function $p_\tau(t)$ is displayed in Figure 1.9. It is obvious from Figure 1.9 that $p_\tau(t)$ is continuous at all t except $t = -\tau/2$ and $t = \tau/2$.

Note that $p_\tau(t)$ can be expressed in the form

$$p_\tau(t) = u\left(t + \frac{\tau}{2}\right) - u\left(t - \frac{\tau}{2}\right)$$

Note also that the triangular pulse function shown in Figure 1.8 is equal to $(1 - 2|t|/\tau)p_\tau(t)$, where $|t|$ is the absolute value of t defined by $|t| = t$ when $t > 0$, $|t| = -t$ when $t < 0$.

A continuous-time signal $x(t)$ is said to be *piecewise continuous* if it is continuous at all t except at a finite or countably infinite collection of points t_i, $i = 1, 2, 3, \ldots$. Examples of piecewise-continuous functions are the step function $Ku(t)$ and the rectangular pulse function $p_\tau(t)$. Another example of a piecewise-continuous signal is the pulse train shown in Figure 1.10. This signal is continuous at all t except at $t = 0, \pm 1, \pm 2, \ldots$. Note that the pulse train is a periodic signal with fundamental period equal to 2.

Derivative of a Continuous-Time Signal

A continuous-time signal $x(t)$ is said to be *differentiable* at a fixed point t_1 if

$$\frac{x(t_1 + h) - x(t_1)}{h}$$

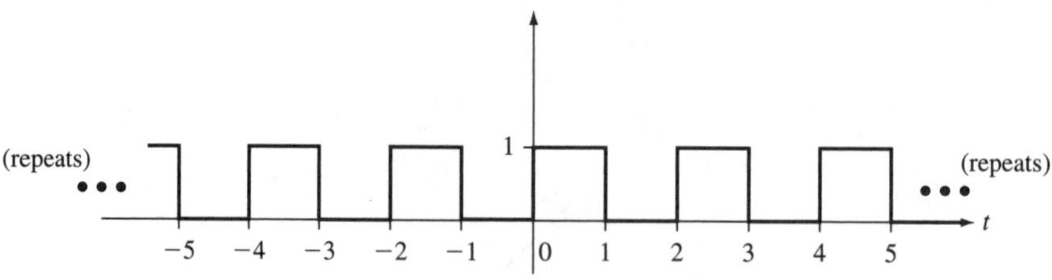

Figure 1.10 Signal that is discontinuous at $t = 0, \pm 1, \pm 2, \ldots$.

has a limit as $h \to 0$, independent of whether h approaches zero from above ($h > 0$) or from below ($h < 0$). If the limit exists, $x(t)$ has a *derivative* at the point t_1 defined by

$$\left.\frac{dx(t)}{dt}\right|_{t=t_1} = \lim_{h\to 0} \frac{x(t_1 + h) - x(t_1)}{h}$$

This definition of the derivative of $x(t)$ is sometimes called the *ordinary derivative* of $x(t)$.

To be differentiable at a point t_1, it is necessary (but not sufficient in general) that the signal $x(t)$ be continuous at t_1. Hence continuous-time signals that are not continuous at all points cannot be differentiable at all points. In particular, piecewise-continuous signals are not differentiable at all points. However, piecewise-continuous signals may have a derivative in the generalized sense. Suppose that $x(t)$ is differentiable at all t except $t = t_1$. Then the *generalized derivative* of $x(t)$ is defined to be

$$\frac{dx(t)}{dt} + [x(t_1^+) - x(t_1^-)]\delta(t - t_1)$$

where $dx(t)/dt$ is the ordinary derivative of $x(t)$ at all t except $t = t_1$, and $\delta(t - t_1)$ is the unit impulse concentrated at the point $t = t_1$. Thus the generalized derivative of a signal at a point of discontinuity t_1 is equal to an impulse located at t_1 and with area equal to the amount the function "jumps" at the point t_1.

Example 1.2 *Generalized Derivative of Step Function*

Let $x(t)$ be the step function $Ku(t)$. The ordinary derivative of $Ku(t)$ is equal to zero at all t except at $t = 0$. Therefore, the generalized derivative of $Ku(t)$ is equal to

$$K[u(0^+) - u(0^-)]\delta(t - 0) = K\delta(t)$$

For $K = 1$, it follows that the generalized derivative of the unit-step function $u(t)$ is equal to the unit impulse $\delta(t)$.

Example 1.3 *Generalized Derivative of Piecewise-Continuous Signal*

Consider the piecewise-continuous signal $x(t)$ defined by

$$x(t) = \begin{cases} 2t + 1, & 0 \le t < 1 \\ 1, & 1 \le t \le 2 \\ -t + 3, & 2 \le t \le 3 \\ 0, & \text{all other } t \end{cases}$$

The signal $x(t)$ is plotted in Figure 1.11. From the plot it is clear that $x(t)$ is continuous at all t except at $t = 0, 1$. The ordinary derivative of $x(t)$ at all t (except $t = 0, 1, 2, 3$) is equal to

$$2[u(t) - u(t - 1)] - [u(t - 2) - u(t - 3)]$$

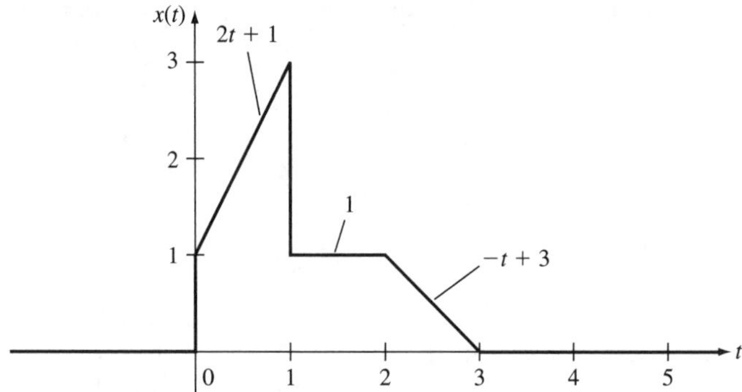

Figure 1.11 Signal in Example 1.3.

The generalized derivative of $x(t)$ is equal to

$$2[u(t) - u(t-1)] - [u(t-2) - u(t-3)] + [x(0^+) - x(0^-)]\delta(t) + [x(1^+) - x(1^-)]\delta(t-1)$$

which simplifies to

$$2[u(t) - u(t-1)] - [u(t-2) - u(t-3)] + \delta(t) - 2\delta(t-1)$$

The generalized derivative of $x(t)$ is displayed in Figure 1.12. Note that even though the ordinary derivative does not exist at $t = 2$ and $t = 3$, no impulses appear in the generalized derivative at these two time points since $x(2^-) = x(2^+)$ and $x(3^-) = x(3^+)$.

Signals Defined Interval by Interval

Continuous-time signals are often defined interval by interval. For example, suppose that $x(t)$ is given by

$$x(t) = \begin{cases} x_1(t), & t_1 \le t < t_2 \\ x_2(t), & t_2 \le t < t_3 \\ x_3(t), & t \ge t_3 \end{cases}$$

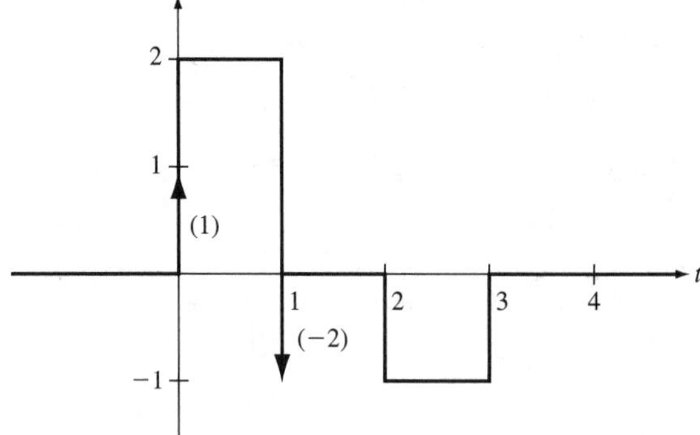

Figure 1.12 Generalized derivative of the signal in Example 1.3.

where $x_1(t)$, $x_2(t)$, and $x_3(t)$ are arbitrary functions of t. Any such signal can be expressed analytically in terms of the unit-step function $u(t)$ and time shifts of $u(t)$. For example, for the signal above,

$$x(t) = x_1(t)[u(t - t_1) - u(t - t_2)] + x_2(t)[u(t - t_2) - u(t - t_3)] + x_3(t)u(t - t_3), \quad t \geq t_1 \tag{1.4}$$

Rearranging terms in (1.4) yields

$$x(t) = x_1(t)u(t - t_1) + [x_2(t) - x_1(t)]u(t - t_2) + [x_3(t) - x_2(t)]u(t - t_3), \quad t \geq t_1 \tag{1.5}$$

From (1.5) it follows that $x(t)$ can be expressed in the form

$$x(t) = f_1(t)u(t - t_1) + f_2(t)u(t - t_2) + f_3(t)u(t - t_3), \quad t \geq t_1 \tag{1.6}$$

where

$$f_1(t) = x_1(t)$$

$$f_2(t) = x_2(t) - x_1(t)$$

$$f_3(t) = x_3(t) - x_2(t)$$

Conversely, suppose that $x(t)$ is given in the form (1.6) for some functions $f_1(t)$, $f_2(t), f_3(t)$. Then

$$x(t) = \begin{cases} f_1(t), & t_1 \leq t < t_2 \\ f_1(t) + f_2(t), & t_2 \leq t < t_3 \\ f_1(t) + f_2(t) + f_3(t), & t \geq t_3 \end{cases} \tag{1.7}$$

The signal $x(t)$ given by (1.7) can also be written in the form

$$x(t) = f_1(t)[u(t - t_1) - u(t - t_2)] + [f_1(t) + f_2(t)] \times [u(t - t_2) - u(t - t_3)] + [f_1(t) + f_2(t) + f_3(t)]u(t - t_3), \quad t \geq t_1$$

Example 1.4 *Analytical Form of Signal*

For the signal $x(t)$ in Example 1.3,

$$x(t) = (2t + 1)[u(t) - u(t - 1)] + (1)[u(t - 1) - u(t - 2)] + (-t + 3)[u(t - 2) - u(t - 3)]$$

Writing $x(t)$ in the form (1.6) yields

$$x(t) = (2t + 1)u(t) - 2tu(t - 1) + (-t + 2)u(t - 2) + (t - 3)u(t - 3)$$

Using MATLAB for Continuous-Time Signals

A continuous-time signal $x(t)$ given in analytical form can be defined and displayed using the software MATLAB. Since MATLAB is used throughout this book, the reader should become familiar with the basic commands and is invited to review the short tutorial that is available from the Web site that accompanies this text. To illustrate its use, consider the signal $x(t)$ given by

$$x(t) = e^{-0.1t} \sin \tfrac{2}{3}t$$

A plot of $x(t)$ versus t for a range of values of t can be generated by using the MATLAB software. For example, for t ranging from 0 to 30 seconds with 0.1-second increments, the MATLAB commands for generating $x(t)$ are

```
t = 0:0.1:30;
x = exp(-.1*t).*sin(2/3*t);
axis([0 30 -1 1]);
plot(t,x)
grid
ylabel('x(t)')
xlabel('Time (sec)')
```

In this program, the time values for which x is to be plotted are stored as elements in the vector `t`. Each of the expressions `exp(-.1*t)` and `sin(2/3*t)` creates a vector with elements equal to the expression evaluated at the corresponding time values. The resulting vectors must be multiplied element by element to define the vector `x`. Then `x` is plotted versus `t`. The `axis` command is used to overwrite the default values (usually, the default is acceptable and this command is not needed). Note that use of the `axis` command varies with the version of MATLAB being used.

The resulting plot of $x(t)$ is shown in Figure 1.13. Note that the MATLAB-generated plot is in box form, with the axes labeled as shown. The format of the plot

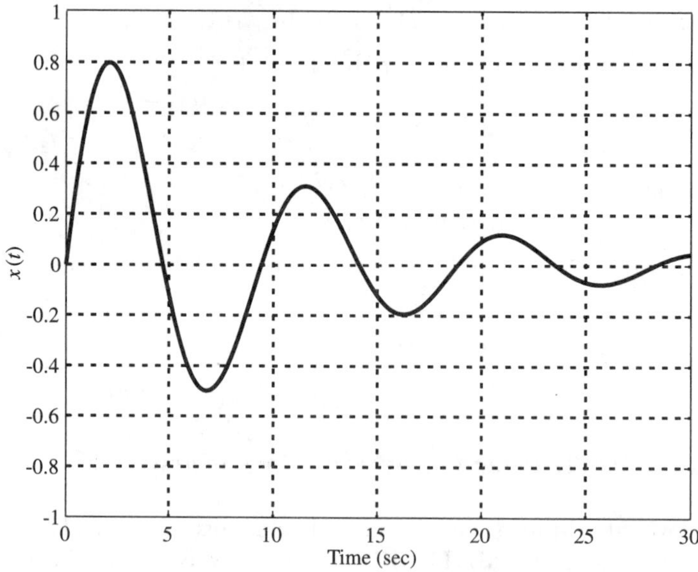

Figure 1.13 MATLAB plot of the signal $x(t) = e^{-0.1t} \sin \frac{2}{3} t$.

differs from those given previously. In this book, MATLAB-generated plots will always be in box form, whereas plots not generated by MATLAB will be given in the form used previously (as in Figure 1.11).

It is important to note that in generating a MATLAB plot of a continuous-time signal, the increment in the time step must be chosen to be sufficiently small to yield a smooth plot. If the increment is chosen to be too large (for a given signal), then when the values of the signal are connected by straight lines (in the computer generation of the plot), the resulting plot will look jagged. To see this effect, the reader is invited to rerun the program above using a time increment of 1 second to plot $x(t) = e^{-0.1t} \sin 2/3\, t$. For the plots in this book it was found that using 200 to 400 points per plot resulted in a small enough time increment. See Problem 1.2 for more information on selecting the time increment.

The program given above is stored as an "M-file" called `fig 1_13.m` that is available from the Web site. All MATLAB M-files used in this book are included with a title that matches the figure number or the example number; for example, `ex2_3.m` is the M-file containing the commands for Example 2.3.

1.3 DISCRETE-TIME SIGNALS

The time variable t is said to be a *discrete-time variable* if t takes on only the *discrete values* $t = t_n$ for some range of integer values of n. For example, t could take on the integer values $t = 0, 1, 2, \ldots$; that is, $t = t_n = n$ for $n = 0, 1, 2, \ldots$. A *discrete-time signal* is a signal that is a function of the discrete-time variable t_n; in other words, a discrete-time signal has values (is defined) only at the discrete-time points $t = t_n$, where n takes on only integer values. Discrete-time signals arise in many areas of business, economics, science, and engineering. In applications to business or economics, the discrete-time variable t_n may be the day, month, quarter, or year of a specified period of time. In Section 1.4 an example is given where the discrete-time variable is the month of a loan-repayment period, and in Section 2.3 an example is given where t_n is the nth day during a manufacturing operation.

In this book a discrete-time signal defined at the time points $t = t_n$ will be denoted by $x[n]$. Note that in the notation "$x[n]$," the integer variable n corresponds to the time instants t_n. Also note that brackets are used to denote a discrete-time signal $x[n]$, in contrast to a continuous-time signal $x(t)$, which is denoted using parentheses. A plot of a discrete-time signal $x[n]$ will always be given in terms of the values of $x[n]$ versus the integer time variable n. The values $x[n]$ will be indicated on the plot by closed circles with vertical lines connecting the circles to the time axis. This results in a *stem plot*, which is the standard way of displaying a discrete-time signal. For example, suppose that the discrete-time signal $x[n]$ is given by

$$x[0] = 1, x[1] = 2, x[2] = 1, x[3] = 0, x[4] = -1$$

with $x[n] = 0$ for all other n. Then the stem plot of $x[n]$ is shown in Figure 1.14. A plot of this signal can be generated using the MATLAB commands

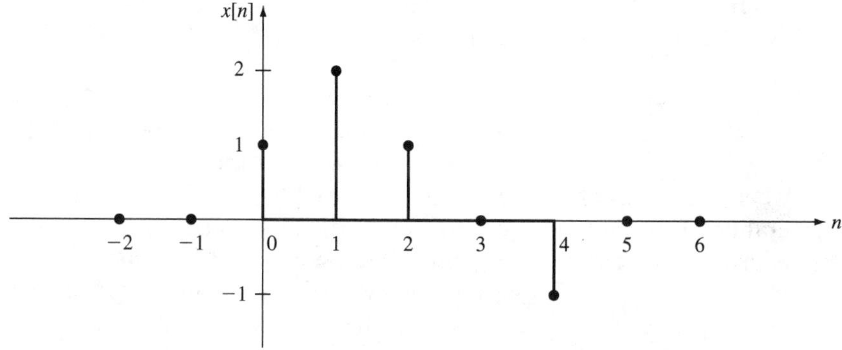

Figure 1.14 Plot of a discrete-time signal.

```
n = -2:6;
x = [0 0 1 2 1 0 -1 0 0];
stem(n,x);
xlabel('n')
ylabel('x[n]')
```

The MATLAB-generated plot of $x[n]$ is shown in Figure 1.15. Again note that the MATLAB plot is in box form, in contrast to the format of the plot given in Figure 1.14. As in the continuous-time case, MATLAB plots are always displayed in box form. Plots of discrete-time signals not generated by MATLAB will be given in the form shown in Figure 1.14. It should be noted that the MATLAB `stem` command produces

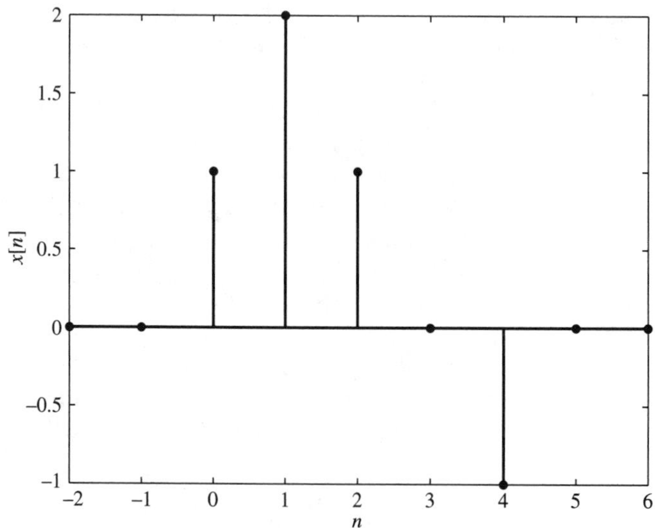

Figure 1.15 MATLAB plot of $x[n]$.

a plot that has "open circles". To match with the convention of "closed circles", the `stem` command was modified as discussed in the MATLAB tutorial on the Web site. All MATLAB stem plots in this text were generated using the modified command.

Sampling

One of the most common ways in which discrete-time signals arise is in sampling continuous-time signals: As illustrated in Figure 1.16, suppose that a continuous-time signal $x(t)$ is applied to an electronic switch that is closed briefly every T seconds. If the amount of time during which the switch is closed is much smaller than T, the output of the switch can be viewed as a discrete-time signal that is a function of the discrete-time points $t_n = nT$, where $n = \ldots, -2, -1, 0, 1, 2, \ldots$. The resulting discrete-time signal is called the *sampled version* of the original continuous-time signal $x(t)$ and T is called the *sampling interval.* Since the time duration T between adjacent sampling instants $t_n = nT$ and $t_{n+1} = (n + 1)T$ is equal to a constant, the sampling process under consideration here is called *uniform sampling. Nonuniform sampling* is sometimes utilized in applications but is not considered in this book.

To be consistent with the notation introduced above for discrete-time signals, the discrete-time signal resulting from the uniform sampling operation illustrated in Figure 1.16 will be denoted by $x[n]$. Note that in this case, the integer variable n denotes the time instant nT. By definition of the sampling process, the value of $x[n]$ for any integer value of n is given by

$$x[n] = x(t)\big|_{t=nT} = x(nT)$$

A large class of discrete-time signals can be generated by sampling continuous-time signals. For instance, if the continuous-time signal $x(t)$ displayed in Figure 1.13 is sampled with $T = 1$, the result is the discrete-time signal $x[n]$ plotted in Figure 1.17. This plot can be obtained by running the program that generated Figure 1.13 where the time increment is 1 second and the `plot(t,x)` command is replaced with `stem(t,x)`.

Step and Ramp Functions

Two simple examples of discrete-time signals are the discrete-time unit-step function $u[n]$ and the discrete-time unit-ramp function $r[n]$, which are defined by

$$u[n] = \begin{cases} 1, & n = 0, 1, \ldots \\ 0, & n = -1, -2, \ldots \end{cases}$$

$$r[n] = \begin{cases} n, & n = 0, 1, \ldots \\ 0, & n = -1, -2, \ldots \end{cases}$$

These two discrete-time signals are plotted in Figure 1.18.

The discrete-time step function $u[n]$ can be obtained by sampling the continuous-time step function $u(t)$. If the unit-ramp function $r(t) = tu(t)$ is sampled, the result is the discrete-time ramp function $r[n]$, given by

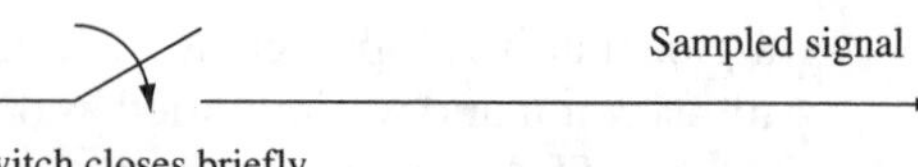

Figure 1.16 Sampling process.

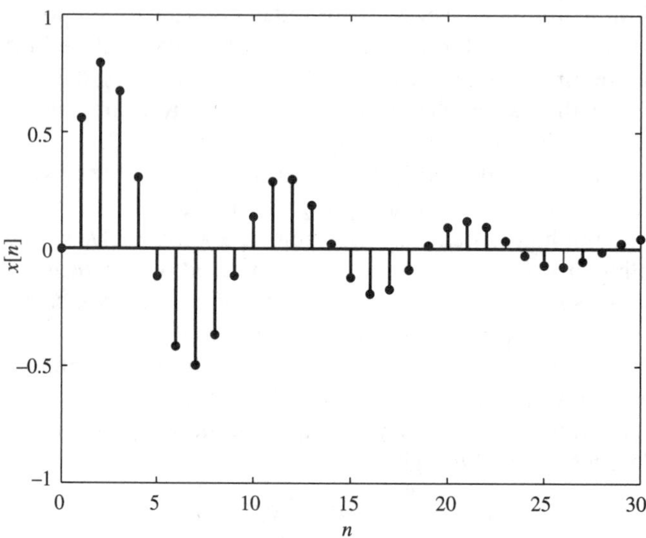

Figure 1.17 Sampled continuous-time signal.

$$r[n] = r(t)\big|_{t=nT} = r(nT)$$

The discrete-time signal $r[n]$ is plotted in Figure 1.19. Note that although the discrete-time signals in Figures 1.18b and 1.19 are given by the same notation $r[n]$, these two signals are not the same unless the sampling interval T is equal to 1. To distinguish between these two signals, the one plotted in Figure 1.19 could be denoted by $r_T[n]$, but the standard convention (which is followed here) is not to show the dependence on T in the notation for the sampled signal.

Unit Pulse

It should first be noted that there is no sampled version of the unit impulse $\delta(t)$ since $\delta(0)$ is not defined. However, there is a discrete-time signal that is the discrete-time counterpart of the unit impulse. This is the *unit-pulse function* $\delta[n]$, defined by

$$\delta[n] = \begin{cases} 1, & n = 0 \\ 0, & n \neq 0 \end{cases}$$

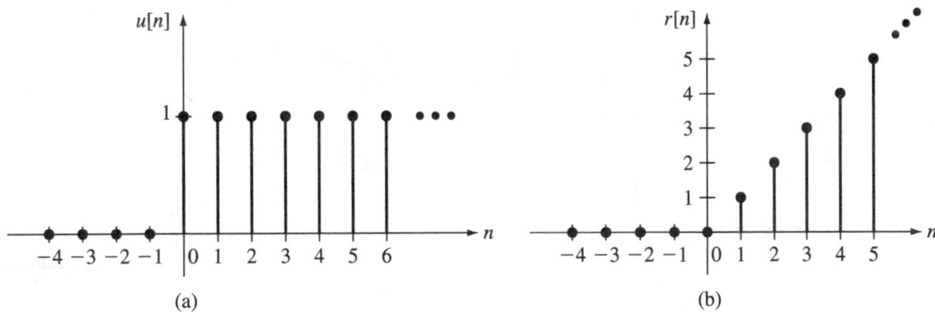

Figure 1.18 (a) Discrete-time unit step and (b) unit-ramp functions.

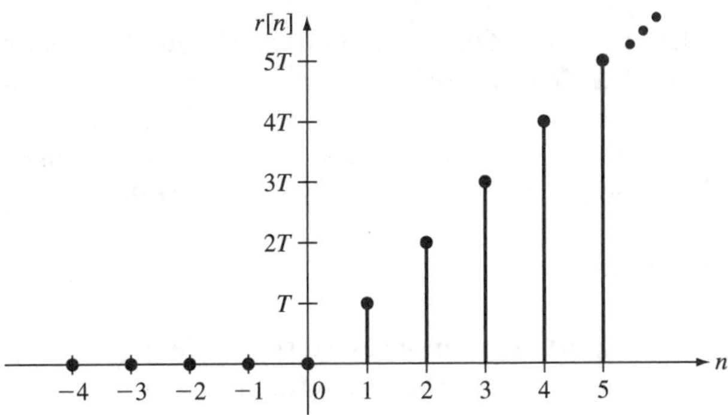

Figure 1.19 Discrete-time ramp function.

The unit-pulse function is plotted in Figure 1.20. It should be stressed that $\delta[n]$ is not a sampled version of the unit impulse $\delta(t)$.

Periodic Discrete-Time Signals

A discrete-time signal $x[n]$ is periodic if there exists a positive integer r such that

$$x[n + r] = x[n] \qquad \text{for all integers } n$$

Hence $x[n]$ is periodic if and only if there is a positive integer r such that $x[n]$ repeats itself every r time instants, where r is called the *period*. The fundamental period is the smallest value of r for which the signal repeats. For example, let us examine the periodicity of a discrete-time sinusoid given by

$$x[n] = A \cos(\Omega n + \theta)$$

The signal is periodic if

$$A \cos[\Omega(n + r) + \theta] = A \cos(\Omega n + \theta)$$

Recall that the cosine function repeats every 2π radians, so that

$$A \cos(\Omega n + \theta) = A \cos(\Omega n + 2\pi q + \theta)$$

for all integers q. Therefore, the signal $A \cos(\Omega n + \theta)$ is periodic if and only if there exists a positive integer r such that $\Omega r = 2\pi q$ for some integer q, or equivalently,

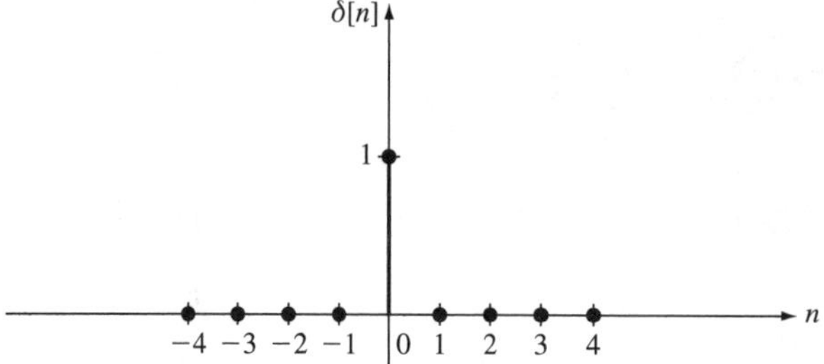

Figure 1.20 Unit-pulse function.

that the discrete-time frequency Ω is such that $\Omega = 2\pi q/r$ for some positive integers q and r.

The discrete-time sinusoid $x[n] = A\cos(\Omega n + \theta)$ is plotted in Figure 1.21 for two different values of Ω. For the case when $\Omega = \pi/3$ and $\theta = 0$, which is plotted in Figure 1.21a, the corresponding period is found to be $r = 2\pi q/\Omega = 6q$ and the fundamental period is 6. The case when $\Omega = 1$ and $\theta = 0$ is plotted in Figure 1.21b. Note that in this case the envelope of the signal is periodic but the signal itself is not periodic.

Discrete-Time Rectangular Pulse

Let L be a positive odd integer. An important example of a discrete-time signal is the *discrete-time rectangular pulse function* $p_L[n]$ of length L defined by

(a)

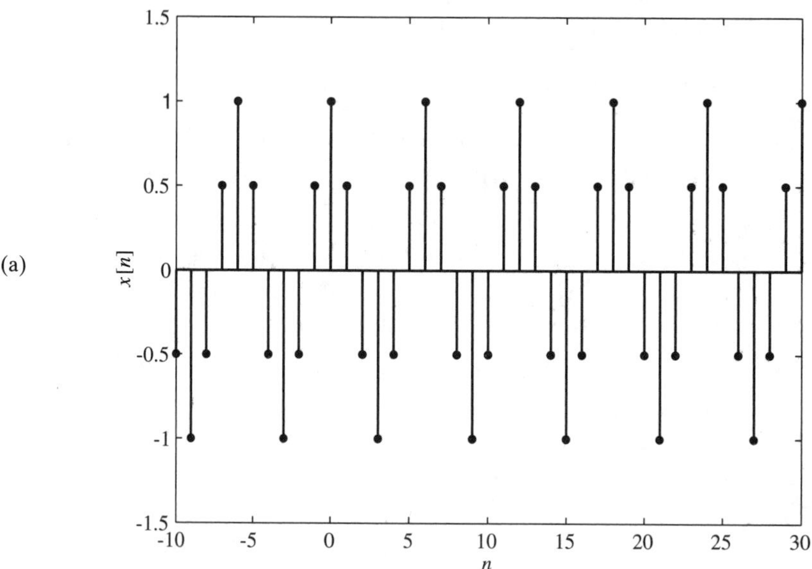

Figure 1.21 Discrete-time sinusoids with $\theta = 0$ and (a) $\Omega = \pi/3$ and (b) $\Omega = 1$.

(b)

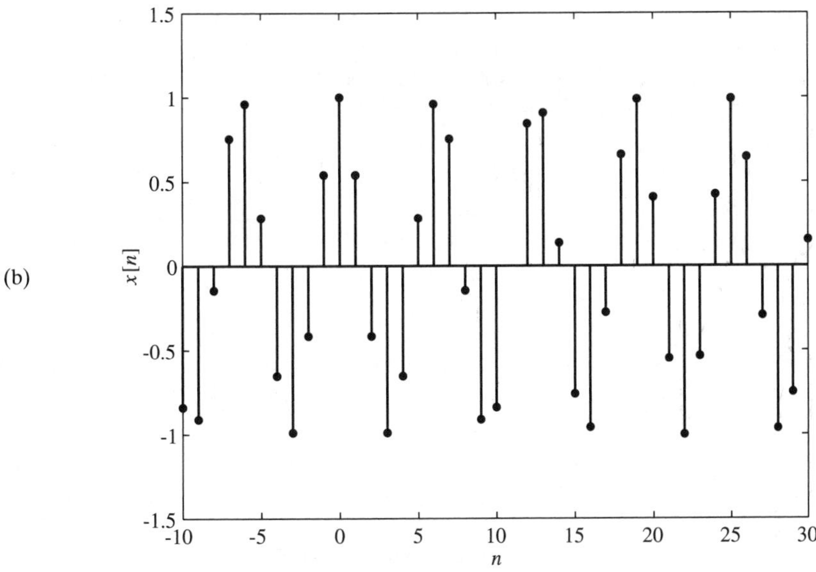

Figure 1.21 *(continued)*

$$p_L[n] = \begin{cases} 1, & n = -(L-1)/2, \ldots, -1, 0, 1, \ldots, (L-1)/2 \\ 0, & \text{all other } n \end{cases}$$

The discrete-time rectangular pulse is displayed in Figure 1.22.

Digital Signals

Let $\{a_1, a_2, \ldots, a_N\}$ be a set of N real numbers. A *digital signal* $x[n]$ is a discrete-time signal whose values belong to the finite set $\{a_1, a_2, \ldots, a_N\}$; that is, at each time instant

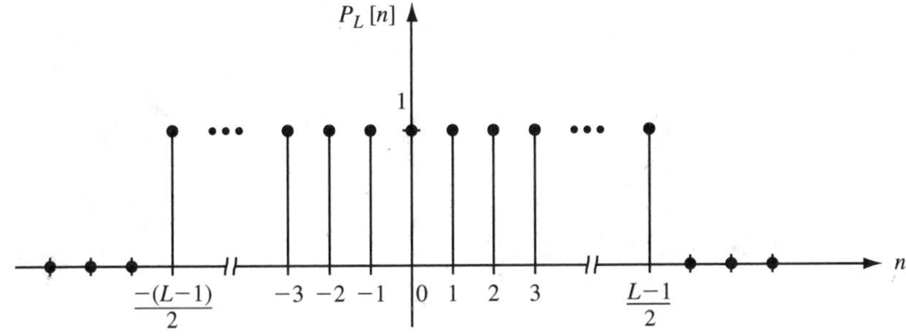

Figure 1.22 Discrete-time rectangular pulse.

$t_n, x(t_n) = x[n] = a_i$ for some i, where $1 \leq i \leq N$. So a digital signal can have only a finite number of different values.

A sampled continuous-time signal is not necessarily a digital signal. For example, the sampled unit-ramp function $r[n]$ shown in Figure 1.19 is not a digital signal since $r[n]$ takes on an infinite range of values when $n = \ldots, -2, -1, 0, 1, 2, \ldots$.

A *binary signal* is a digital signal whose values are equal to 1 or 0; that is, $x[n] = 0$ or 1 for $n = \ldots, -2, -1, 0, 1, 2, \ldots$. The sampled unit-step function and the unit-pulse function are both examples of binary signals.

Time-Shifted Signals

Given a discrete-time signal $x[n]$ and a positive integer q, the discrete-time signal $x[n - q]$ is the q-step right shift of $x[n]$ and $x[n + q]$ is the q-step left shift of $x[n]$. For example, $p_3[n - 2]$ is the two-step right shift of the discrete-time rectangular pulse $p_3[n]$, and $p_3[n + 2]$ is the two-step left shift of $p_3[n]$. The shifted signals are plotted in Figure 1.23.

Discrete-Time Signals Defined Interval by Interval

As in the case of continuous-time signals, discrete-time signals are sometimes defined interval by interval. For instance, $x[n]$ may be specified by

$$x[n] = \begin{cases} x_1[n], & n_1 \leq n < n_2 \\ x_2[n], & n_2 \leq n < n_3 \\ x_3[n], & n \geq n_3 \end{cases}$$

where $x_1[n]$, $x_2[n]$, and $x_3[n]$ are discrete-time signals and n_1, n_2, and n_3 are integers with $n_1 < n_2 < n_3$. With $u[n]$ equal to the discrete-time unit-step function, $x[n]$ can be written in the form

$$x[n] = x_1[n](u[n - n_1] - u[n - n_2]) + x_2[n](u[n - n_2] - u[n - n_3]) \\ + x_3[n]u[n - n_3], \qquad n > n_1 \tag{1.8}$$

Combining terms in (1.8) gives

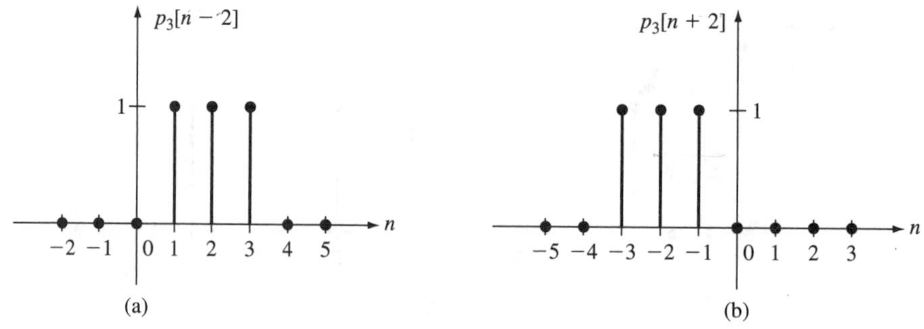

Figure 1.23 Two-step shifts of $p_3[n]$: (a) right shift; (b) left shift.

$$x[n] = x_1[n]u[n - n_1] + (x_2[n] - x_1[n])u[n - n_2] \\ + (x_3[n] - x_2[n])u[n - n_3], \qquad n > n_1 \tag{1.9}$$

Note that the expressions (1.8) and (1.9) are simply discrete-time versions of the corresponding expressions in the continuous-time case [see equations (1.4) and (1.5)].

1.4 EXAMPLES OF SYSTEMS

To provide some concreteness to the concept of a system, in this section five specific examples of a system are given. The first four examples are *continuous-time systems,* for which the time variable t takes its values from the set of real numbers. The fifth example is a discrete-time system, for which the time variable t takes on only the discrete values t_n, where again the index n is integer valued. Following the notation introduced in Section 1.3, the input and output signals of a discrete-time system will be denoted by $x[n]$ and $y[n]$, respectively.

To keep the discussion at an introductory level, in this section the presentation is restricted to simple examples of systems. More complicated examples are considered in subsequent chapters.

RC Circuit

Consider the *RC* circuit shown in Figure 1.24. The *RC* circuit can be viewed as a single-input single-output continuous-time system with input $x(t)$ equal to the current $i(t)$ into the parallel connection and with output $y(t)$ equal to the voltage $v_C(t)$ across the capacitor. By Kirchhoff's current law (see Section 2.2),

$$i_C(t) + i_R(t) = i(t) \tag{1.10}$$

where $i_C(t)$ is the current in the capacitor and $i_R(t)$ is the current in the resistor. Now

$$i_C(t) = C\frac{dv_C(t)}{dt} = C\frac{dy(t)}{dt} \tag{1.11}$$

and

$$i_R(t) = \frac{1}{R}v_C(t) = \frac{1}{R}y(t) \tag{1.12}$$

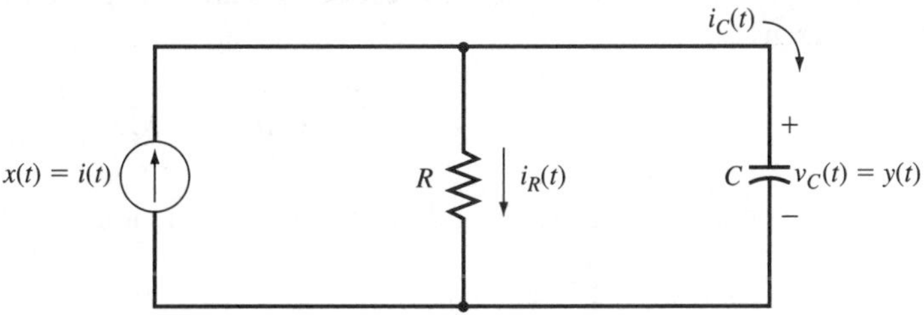

Figure 1.24 *RC* circuit.

Inserting (1.11) and (1.12) into (1.10) yields the following linear differential equation:

$$C\frac{dy(t)}{dt} + \frac{1}{R}y(t) = i(t) = x(t) \tag{1.13}$$

The differential equation (1.13) is called the *input/output differential equation* of the circuit. It provides an implicit relationship between the input $x(t)$ and the output $y(t)$. An explicit expression for $y(t)$ in terms of $x(t)$ can be generated by solving the input/output differential equation (1.13) for an arbitrary input $x(t)$ applied for $t \geq t_0$. The steps are as follows.

First, it is easy to see that if $x(t) = 0$ for all $t \geq t_0$, the solution $y(t)$ to (1.13) is given by

$$y(t) = e^{-(1/RC)(t-t_0)}y(t_0), \qquad t \geq t_0 \tag{1.14}$$

where $y(t_0)$ is the initial voltage on the capacitor at time t_0. To show that (1.14) is the solution to (1.13) when $x(t) = 0$ for $t \geq t_0$, simply verify that (1.13) is satisfied when the expression (1.14) for $y(t)$ is inserted into (1.13). The reader is invited to check this.

Now suppose that the input $x(t)$ is not necessarily zero for $t \geq t_0$. The solution to (1.13) can be computed by first multiplying both sides of (1.13) by an *integrating factor* equal to $(1/C)e^{(1/RC)t}$. This gives

$$e^{(1/RC)t}\left[\frac{dy(t)}{dt} + \frac{1}{RC}y(t)\right] = \frac{1}{C}e^{(1/RC)t}x(t) \tag{1.15}$$

The left-hand side of (1.15) is equal to the derivative with respect to t of the function $e^{(1/RC)t}\,y(t)$, and thus

$$\frac{d}{dt}[e^{(1/RC)t}y(t)] = \frac{1}{C}e^{(1/RC)t}x(t)$$

This equation is in the form

$$\frac{dv(t)}{dt} = q(t) \tag{1.16}$$

where

$$v(t) = e^{(1/RC)t}y(t) \tag{1.17}$$

$$q(t) = \frac{1}{C}e^{(1/RC)t}x(t) \tag{1.18}$$

Integrating both sides of (1.16) with respect to t and using the *fundamental theorem of calculus* gives

$$v(t) = v(t_0) + \int_{t_0}^{t} q(\lambda)\,d\lambda, \qquad t \geq t_0 \tag{1.19}$$

Inserting the expressions (1.17) and (1.18) into (1.19) yields

$$e^{(1/RC)t}y(t) = e^{(1/RC)t_0}y(t_0) + \int_{t_0}^{t}\frac{1}{C}e^{(1/RC)\lambda}x(\lambda)\,d\lambda, \qquad t \geq t_0 \tag{1.20}$$

Finally, multiplying both sides of (1.20) by $e^{-(1/RC)t}$ gives

$$y(t) = e^{-(1/RC)(t-t_0)}y(t_0) + \int_{t_0}^{t} \frac{1}{C} e^{-(1/RC)(t-\lambda)}x(\lambda)\, d\lambda, \qquad t \geq t_0 \tag{1.21}$$

The expression (1.21) for $y(t)$ is the complete output response of the RC circuit resulting from initial voltage $y(t_0)$ and arbitrary input $x(t) = i(t)$ applied for $t \geq t_0$.

It should be noted that if the input $x(t)$ contains an impulse $K\delta(t - t_0)$, it is necessary to take the initial condition at time t_0^-, where $t_0 - t_0^-$ is an infinitesimal positive number. The reason for this is that an impulsive input can instantaneously change the value of the output. If the initial condition is taken at time t_0^-, the expression (1.21) for the output becomes

$$y(t) = e^{-(1/RC)(t-t_0)}y(t_0^-) + \int_{t_0^-}^{t} \frac{1}{C} e^{-(1/RC)(t-\lambda)}x(\lambda)\, d\lambda, \qquad t \geq t_0$$

From (1.21), it is seen that the output response $y(t)$ consists of two terms: The first term, $e^{-(1/RC)(t-t_0)}y(t_0)$, is the part of the output response due to the initial condition $y(t_0)$ at time t_0, and the second term,

$$\int_{t_0}^{t} \frac{1}{C} e^{-(1/RC)(t-\lambda)}x(\lambda)\, d\lambda$$

is the part of the output response resulting from the application of the input current $x(t)$ for $t \geq t_0$.

Now since the capacitor is the only component in the RC circuit that can store energy, there is no *initial energy* in the circuit at time $t = t_0$ if and only if $y(t_0) = 0$. As an alternative terminology, it is also said that the RC circuit is in the *zero state* (or is *at rest*) at time $t = t_0$ if and only if $y(t_0) = 0$.

If the RC circuit is in the zero state at time $t = t_0$ [i.e., $y(t_0) = 0$], then from (1.21) the output response reduces to

$$y(t) = \int_{t_0}^{t} \frac{1}{C} e^{-(1/RC)(t-\lambda)}x(\lambda)\, d\lambda, \qquad t \geq t_0 \tag{1.22}$$

The expression (1.22) specifies the *input/output relationship* when the RC circuit is in the zero state (at rest) before the application of the input.

Note that by the input/output relationship (1.22), the output $y(t)$ at time t depends on the input $x(\lambda)$ for $t_0 \leq \lambda \leq t$. Hence the value of the output at time t depends in general on the values of the input signal over the time interval from the initial time t_0 to the present time t.

From (1.22), the output response resulting from the application of any input current $x(t)$ can be computed. For example, suppose that the initial time t_0 is taken to be zero and the input $x(t)$ is the unit-step function $u(t)$. Then from (1.22),

$$\begin{aligned} y(t) &= \int_{0}^{t} \frac{1}{C} e^{-(1/RC)(t-\lambda)}\, d\lambda \\ &= Re^{-(1/RC)(t-\lambda)}\Big|_{\lambda=0}^{\lambda=t} \\ &= R[1 - e^{-(1/RC)t}], \qquad t \geq 0 \end{aligned} \tag{1.23}$$

The output response $y(t)$ given by (1.23) is called the *step response* since $y(t)$ is the output when the input is the unit-step function $u(t)$ with the system at rest prior to the application of $u(t)$. If at $t = 0$ a constant current source of amplitude 1 is switched on [so that $x(t) = u(t)$], the resulting voltage across the capacitor would be given by (1.23). For the case $R = 1$ and $C = 1$, the step response is as plotted in Figure 1.25.

Car on a Level Surface

Consider an automobile on a horizontal surface, as illustrated in Figure 1.26. As indicated, the output $y(t)$ is the position of the car at time t relative to some reference, and the input $x(t)$ is the drive or braking force applied to the car at time t. It follows from Newton's second law of motion that $y(t)$ and $x(t)$ are related by the following second-order linear differential equation:

$$M\frac{d^2y(t)}{dt^2} + k_f\frac{dy(t)}{dt} = x(t) \tag{1.24}$$

where M is the mass of the car and k_f is the coefficient representing frictional losses. Note that k_f will change if there is a significant change in the road surface, for example, in going from a paved to an unpaved surface.

As in the RC circuit example, to compute an explicit expression for the output $y(t)$, it is necessary to solve the input/output differential equation (1.24). To accomplish this, first let $v(t) = dy(t)/dt$, so that $v(t)$ is the velocity of the car at time t. Rewriting (1.24) in terms of $v(t)$ gives

$$M\frac{dv(t)}{dt} + k_f v(t) = x(t) \tag{1.25}$$

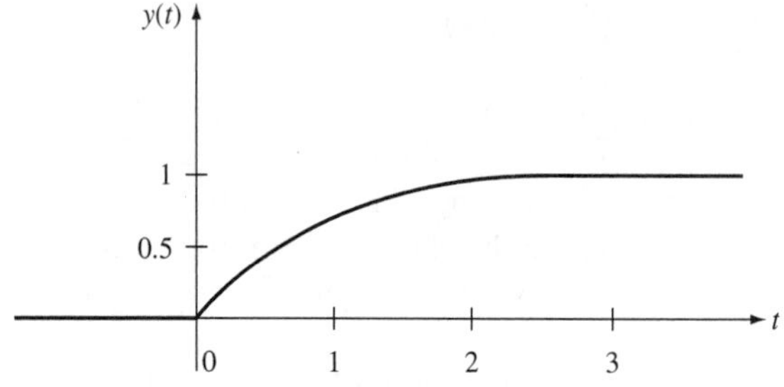

Figure 1.25 Step response of RC circuit when $R = C = 1$.

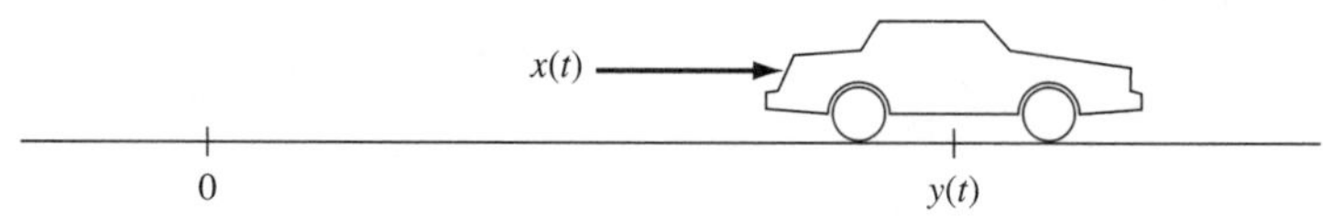

Figure 1.26 Car with drive or braking force $x(t)$.

The first-order differential equation (1.25) is called the *velocity model* of the car. It follows from the results in the *RC*-circuit example that the complete solution to (1.25) is given by

$$v(t) = e^{-(k_f/M)(t-t_0)}v(t_0) + \int_{t_0}^{t} \frac{1}{M} e^{-(k_f/M)(t-\lambda)}x(\lambda)\, d\lambda, \qquad t \geq t_0 \tag{1.26}$$

where $v(t_0)$ is the initial velocity of the car at time t_0 and $x(t)$ is the input force applied for $t \geq t_0$.

Now since $v(t) = dy(t)/dt$, integrating both sides of this equation gives

$$y(t) = y(t_0) + \int_{t_0}^{t} v(\tau)\, d\tau \tag{1.27}$$

Setting $x(t) = 0$ for $t \geq t_0$ and inserting (1.26) into (1.27) yields

$$y(t) = y(t_0) + \int_{t_0}^{t} e^{-(k_f/M)(\tau-t_0)}v(t_0)\, d\tau$$

$$y(t) = y(t_0) + \frac{M}{k_f}v(t_0)[1 - e^{-(k_f/M)(t-t_0)}], \qquad t \geq t_0 \tag{1.28}$$

The expression (1.28) is the output response resulting from initial position $y(t_0)$ and initial velocity $v(t_0)$ with no input applied for $t \geq t_0$. The expression (1.28) is valid as long as $k_f \neq 0$.

From (1.28), it is clear that if the initial velocity $v(t_0)$ is zero, then $y(t) = y(t_0)$ for all $t \geq t_0$. In other words, if the car does not have any initial velocity and no input is applied, it remains at its initial position $y(t_0)$. Letting $t \to \infty$ in (1.28) reveals that the car stops at the position

$$y(\infty) = y(t_0) + \frac{M}{k_f}v(t_0) \tag{1.29}$$

Note that it takes an infinite amount of time before the car stops, whereas an actual car would stop in a finite amount of time. This discrepancy is a result of the fact that the model of the car is an idealized model. Despite this, the model is sufficiently accurate to be useful in studying the "dynamics" of the car.

Also note that by measuring the stopping position $y(\infty)$, (1.29) can be solved for k_f. Hence it is possible to identify (i.e., determine) the friction coefficient k_f in the system representation (1.24).

Now the objective is to compute the output response $y(t)$ resulting from an arbitrary input $x(t)$ with the system at rest at time t_0. In this example, "at rest" means that the initial position $y(t_0)$ and initial velocity $v(t_0)$ are both zero. Setting $y(t_0) = 0$ and $v(t_0) = 0$ and inserting (1.26) into (1.27) gives

$$y(t) = \int_{t_0}^{t} \left[\int_{t_0}^{\tau} \frac{1}{M} e^{-(k_f/M)(\tau-\lambda)}x(\lambda)\, d\lambda \right] d\tau \tag{1.30}$$

This expression for $y(t)$ simplifies to

$$y(t) = \int_{t_0}^{t} \frac{1}{k_f}[1 - e^{-(k_f/M)(t-\lambda)}]x(\lambda)\,d\lambda, \qquad t \geq t_0 \tag{1.31}$$

The verification that (1.30) does reduce to (1.31) is left to the reader (see Problem 1.15).

The expression (1.31) is the output response resulting from the application of an input force $x(t)$ with the system at rest at initial time t_0 [i.e., $y(t_0) = 0$ and $v(t_0) = 0$]. From the input/output relationship (1.31) it is possible to compute the output $y(t)$ resulting from the application of any input force $x(t)$ applied to the car. For example, if $t_0 = 0$ and $x(t)$ is the unit-step function $u(t)$, the response is

$$\begin{aligned} y(t) &= \int_0^t \frac{1}{k_f}[1 - e^{-(k_f/M)(t-\lambda)}]\,d\lambda \\ &= \frac{1}{k_f}\left[\lambda - \frac{M}{k_f}e^{-(k_f/M)(t-\lambda)}\right]_{\lambda=0}^{\lambda=t} \\ &= \frac{1}{k_f}\left[t - \frac{M}{k_f} + \frac{M}{k_f}e^{-(k_f/M)t}\right], \qquad t \geq 0 \end{aligned} \tag{1.32}$$

The output $y(t)$ given by (1.32) is the step response of the car. Note that $y(t) \to \infty$ as $t \to \infty$; in other words, the application of a constant positive force will move the car to infinity. With the normalized mass $M = 1$ and with $k_f = 0.1$, the step response is as plotted in Figure 1.27. If both sides of the expression (1.32) for the step response are differentiated, the result is that the velocity $v(t)$ of the car in response to a step input is given by

$$v(t) = \frac{dy(t)}{dt} = \frac{1}{k_f}[1 - e^{-(k_f/M)t}], \qquad t \geq 0 \tag{1.33}$$

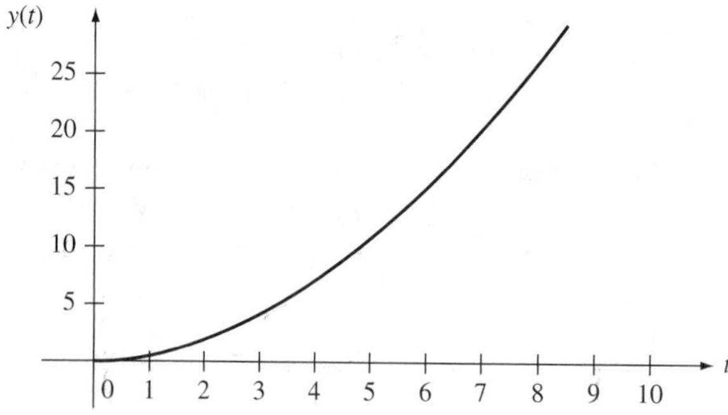

Figure 1.27 Step response of car with $M = 1$ and $k_f = 0.1$.

From (1.33), it is clear that $v(t) \to 1/k_f$ as $t \to \infty$. This result says that the velocity of the car will approach the "steady-state value" $1/k_f$ if a constant force of amplitude 1 is applied to the car.

Mass–Spring–Damper System

The simplest model for many vibratory systems is the mass–spring–damper system shown schematically in Figure 1.28. The mass–spring–damper system is an accurate representation of many actual structures or devices; examples include an accelerometer (a device for measuring acceleration), a seismometer (a device for measuring the vibration of the earth), and a vibration absorber (a mounting device that is used to absorb vibration of equipment). Other systems, such as a machine tool or a compressor on a resilient mount, can be modeled as a mass–spring–damper system for simplified analysis. This system, while crude, demonstrates most of the phenomena associated with vibratory systems, and, as such, it is the fundamental building block for the study of vibration.

Physically, the mass M is supported by a spring with stiffness constant K and a damper with damping constant D. An external force $x(t)$ is applied to the mass and causes the mass to move upward or downward with displacement $y(t)$, measured with respect to an equilibrium value. (That is, $y(t) = 0$ when no external force is applied.) When the mass is above its equilibrium value, $y(t) > 0$, and when the mass is below its equilibrium value, $y(t) < 0$. The movement of the mass is resisted by the spring (if the mass is moving downward, it compresses the spring, which then acts to push upward on the mass). The damper acts to dissipate energy by converting mechanical energy to thermal energy, which leaves the system in the form of heat. For example, a shock absorber in a car contains a damper.

The input/output differential equation for the mass–spring–damper system is given by

$$M\frac{d^2y(t)}{dt^2} + D\frac{dy(t)}{dt} + Ky(t) = x(t)$$

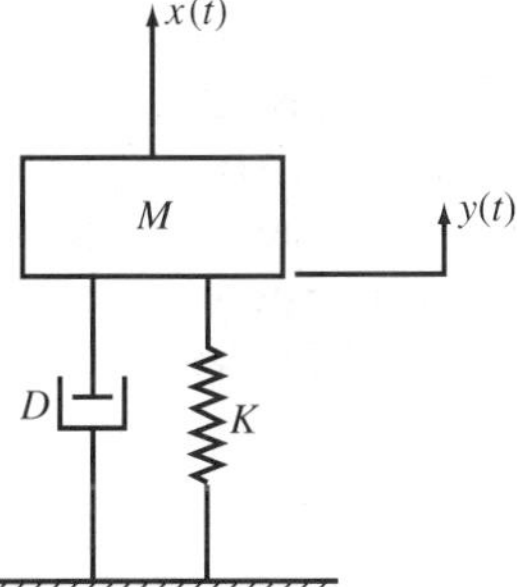

Figure 1.28 Schematic diagram of a mass–spring–damper system.

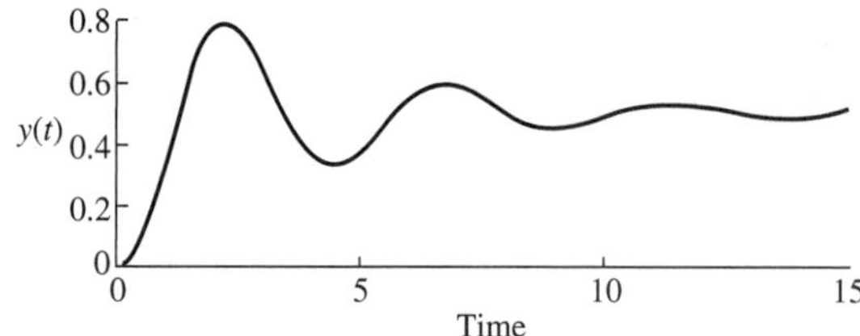

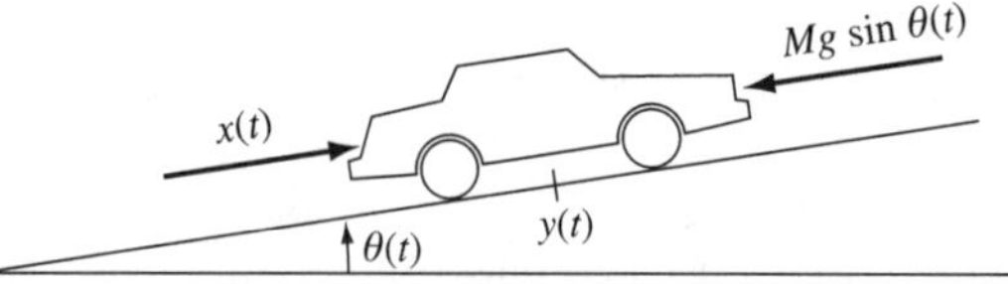

Figure 1.29 Response of the mass–spring–damper system to a unit step input with $M = 1$, $K = 2$, and $D = 0.5$.

Mass Spring Damper System

A demonstration of the mass–spring–damper system is available on-line at the Web site that accompanies this text. The demo allows the user to select different inputs for $x(t)$, such as a step function or a sinusoid, and view the resulting animated response of the system as the mass moves in response to the input. The user can choose different values of M, D, and K to view their effects on the system response. For many combinations of values for M, D, and K, the response $y(t)$ to a step input, $x(t) = u(t)$, is a decaying oscillation that settles to a constant (or steady-state) value, as seen in Figure 1.29. The oscillation is due to the transfer of energy between kinetic energy (proportional to the velocity squared of the mass) and the potential energy (energy stored in the spring as it compresses or stretches). The decay of the oscillation is due to the dissipation of energy that occurs in the damper.

A detailed discussion of vibrations is not the objective of this example or the on-line demo. However, the mass–spring–damper is a system whose response can be visualized readily via animation. A series RLC circuit is governed by the same general equation and responds in the same manner as this system, but the response cannot be visualized easily via animation. Therefore, the mass–spring–damper system and the accompanying on-line demo will be used throughout this text to demonstrate basic system input/output concepts.

Simple Pendulum

Consider a pendulum of length L and mass M as illustrated in Figure 1.30. Here the input $x(t)$ is the force applied to the mass M tangential to the direction of motion of the mass, and $Mg \sin \theta(t)$ is the force due to gravity tangential to the motion. The output $y(t)$ is defined to be the angle $\theta(t)$ between the pendulum and the vertical position.

From the laws of mechanics (see Section 2.2), the input and output are related by the following second-order differential equation:

$$I\frac{d^2\theta(t)}{dt^2} + MgL \sin \theta(t) = Lx(t) \tag{1.34}$$

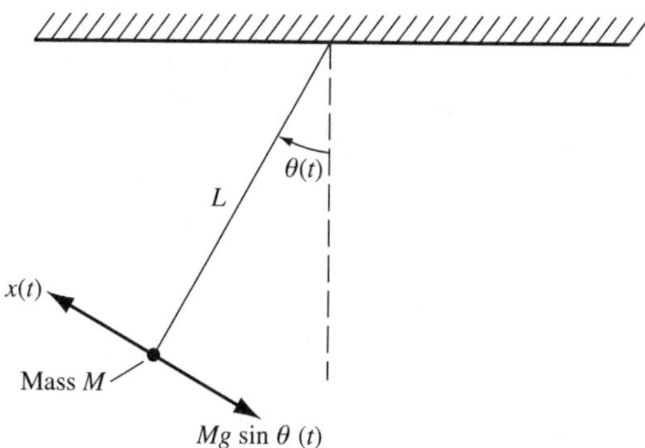

Figure 1.30 Simple pendulum.

where g is the gravity constant and I is the moment of inertia given by $I = M(L^2)$. As a result of the term $\sin \theta(t)$, the input/output differential equation (1.34) is a nonlinear differential equation. Due to the nonlinearity, it is not possible to derive an explicit expression for $y(t)$ in terms of $x(t)$ as was the case in the preceding two examples. However, $y(t)$ can still be computed (at least approximately) by using numerical techniques for solving nonlinear differential equations. The solution of nonlinear differential equations is considered in Section 2.5.

It is interesting to observe that if the magnitude $|\theta(t)|$ of the angle $\theta(t)$ is small, so that $\sin \theta(t)$ is approximately equal to $\theta(t)$, the nonlinear differential equation (1.34) can be approximated by the linear differential equation

$$I\frac{d^2\theta(t)}{dt^2} + MgL\theta(t) = Lx(t) \tag{1.35}$$

The representation (1.35) is called a *small-signal model* since it is a good approximation to the given system if $|\theta(t)|$ is small. It is possible to derive an explicit expression for the output $\theta(t)$ of the small-signal model (see Problem 2.6).

Example of a Discrete-Time System

The repayment of a bank loan can be modeled as a discrete-time system in the following manner: With $n = 0, 1, 2, \ldots$, the input $x[n]$ is the amount of the loan payment in the nth month, and the output $y[n]$ is the balance of the loan after the nth month. Here n is the time index that denotes the month, and the input $x[n]$ and output $y[n]$ are discrete-time signals that are functions of n. The initial condition $y[0]$ is the amount of the loan. Usually, the loan payments $x[n]$ are constant; that is, $x[n] = c$, $n = 1, 2, 3, \ldots$, where c is a constant. In this example, $x[n]$ is allowed to vary from month to month (i.e., the loan payments may not be equal).

The repayment of the loan is described by the difference equation

$$y[n] - \left(1 + \frac{I}{12}\right)y[n-1] = -x[n], \qquad n = 0, 1, 2, \ldots \tag{1.36}$$

where I is the yearly interest rate in decimal form. For example, if the yearly interest rate were 10 percent, I would be equal to 0.1. The term $(I/12)y[n-1]$ in (1.36) is the interest on the loan in the nth month. Equation (1.36) is a first-order linear *difference equation.* It is the input/output difference equation of the system consisting of the loan-repayment process. It is interesting to observe that by allowing the interest rate I to be a variable function of n, the representation (1.36) describes a variable-rate loan for which the interest rate may change from month to month.

The output $y[n]$ can be computed by solving (1.36) recursively as follows. First, rewrite (1.36) in the form

$$y[n] = \left(1 + \frac{I}{12}\right)y[n-1] - x[n] \tag{1.37}$$

Here it is assumed that I is constant (independent of n). Now inserting $n = 1$ in (1.37) yields

$$y[1] = \left(1 + \frac{I}{12}\right)y[0] - x[1] \tag{1.38}$$

Inserting $n = 2$ into (1.37) gives

$$y[2] = \left(1 + \frac{I}{12}\right)y[1] - x[2] \tag{1.39}$$

Taking $n = 3$ in (1.37) gives

$$y[3] = \left(1 + \frac{I}{12}\right)y[2] - x[3] \tag{1.40}$$

By continuing in this manner, $y[n]$ can be computed for any finite range of integer values of n. From (1.38)–(1.40), it is seen that the next value of the output is computed from the present value of the output plus an input term. This is why the process is called a *recursion.* In this example, the recursion is a first-order recursion.

A MATLAB program for carrying out the recursion defined by (1.37) is given in Figure 1.31. The inputs to the program are the loan amount, the interest rate, and the monthly payment. The statement "y = [ ];" is used to initialize y as a vector with no elements. The elements of y are then computed recursively to be the loan balance at the end of the nth month where the index of the vector corresponds to month n. Note that elements in vectors are denoted in MATLAB with parentheses. The program continues in a loop until the loan balance is negative, which means that the loan is paid off.

As an example, the MATLAB program was run with $y[0] = \$6000$, interest rate equal to 12 percent, and monthly payment equal to \$200 (so that $I = 0.12$ and $c = 200$). The resulting loan balance $y[n]$ is shown in Table 1.1. When the monthly payment is \$300, the loan balance $y[n]$ is as displayed in Table 1.2. Note that in the

```
% Loan Balance program
% Program computes loan balance y[n]
y0 = input ('Amount of loan ');
I = input ('Yearly Interest rate ');
c = input ('Monthly loan payment ');      % x[n] = c
y = [];         % defines y as an empty vector
y(1) = (1 + (I/12))*y0 - c;
for n=2:360
   y(n) = (1 + (I/12))*y(n-1) - c;
   if y(n) < 0, break, end
end
```

Figure 1.31 MATLAB program for computing loan balance.

TABLE 1.1 LOAN BALANCE WITH $200 MONTHLY PAYMENTS

n	$y[n]$	n	$y[n]$
1	$5859.99	19	$3086.47
2	5718.59	20	2917.33
3	5575.78	21	2746.51
4	5431.54	22	2573.97
5	5285.85	23	2399.71
6	5138.71	24	2223.71
7	4990.1	25	2045.95
8	4840	26	1866.41
9	4688.4	27	1685.07
10	4535.29	28	1501.92
11	4380.64	29	1316.94
12	4224.44	30	1130.11
13	4066.69	31	941.41
14	3907.36	32	750.83
15	3746.43	33	558.33
16	3583.89	34	363.92
17	3419.73	35	167.56
18	3253.93	36	−30.77

TABLE 1.2 LOAN BALANCE WITH $300 MONTHLY PAYMENTS

n	$y[n]$	n	$y[n]$
1	$5759.99	13	$2685.76
2	5517.59	14	2412.61
3	5272.77	15	2136.74
4	5025.5	16	1858.11
5	4775.75	17	1576.69
6	4523.51	18	1292.46
7	4268.75	19	1005.38
8	4011.43	20	715.43
9	3751.55	21	422.59
10	3489.06	22	126.81
11	3223.95	23	−171.92
12	2956.19		

first case, it takes 36 months to pay off the loan, whereas in the latter case, the loan is paid off in 23 months. In taking out a loan, the number of months in the payoff period is usually specified and then the monthly payment is determined. It is possible to solve for the monthly payment using the representation (1.36) [or (1.37)], but this is not pursued here (see Problem 1.20).

An expression for the loan balance $y[n]$ can be derived in terms of the loan amount $y[0]$ and the loan payments $x[n]$: Inserting the expression (1.38) for $y[1]$ into (1.39) yields

$$\begin{aligned} y[2] &= \left(1 + \frac{I}{12}\right)\left[\left(1 + \frac{I}{12}\right)y[0] - x[1]\right] - x[2] \\ &= \left(1 + \frac{I}{12}\right)^2 y[0] - \left(1 + \frac{I}{12}\right)x[1] - x[2] \end{aligned} \tag{1.41}$$

Inserting (1.41) into (1.40) gives

$$y[3] = \left(1 + \frac{I}{12}\right)^3 y[0] - \left(1 + \frac{I}{12}\right)^2 x[1] - \left(1 + \frac{I}{12}\right)x[2] - x[3] \tag{1.42}$$

From the pattern in (1.38), (1.41), and (1.42), it is seen that for an arbitrary integer value $n \geq 1$,

$$y[n] = \left(1 + \frac{I}{12}\right)^n y[0] - \sum_{i=1}^{n}\left(1 + \frac{I}{12}\right)^{n-i} x[i], \qquad n = 1, 2, 3, \ldots \tag{1.43}$$

The expression (1.43) for $y[n]$ is the loan balance for $n \geq 1$ starting from loan amount $y[0]$ and with loan payments $x[n], n \geq 1$. It follows from (1.43) that if the loan payments $x[n]$ are not sufficiently large, the loan balance $y[n]$ will grow as n increases. In particular, if $x[n] = 0$ for all $n \geq 1$, then

$$y[n] = \left(1 + \frac{I}{12}\right)^n y[0], \qquad n \geq 1$$

Clearly, $y[n] \to \infty$ as $n \to \infty$, since $1 + (I/12) > 1$ when $I > 0$.

1.5 BASIC SYSTEM PROPERTIES

The extent to which a system can be studied using analytical techniques depends on the properties of the system. Two of the most fundamental properties are linearity and time invariance. It will be seen in this book that there exists an extensive analytical theory for the study of systems possessing the properties of linearity and time invariance. These two properties and other basic properties are defined in this section.

Throughout this section the focus is on single-input single-output systems with input $x(t)$ and output $y(t)$. It is assumed that $y(t)$ is the output response of the system resulting from input $x(t)$ with no initial energy in the system prior to the application of the input. In the following development, the time variable t may take on real values or only the discrete values $t = t_n$; that is, the system may be continuous time or discrete time.

Causality

A system is said to be *causal* or *nonanticipatory* if for any time t_1, the output response $y(t_1)$ at time t_1 resulting from input $x(t)$ does not depend on values of the input $x(t)$ for $t > t_1$. Thus in a causal system it is not possible to get an output before an input is applied to the system (assuming no initial energy). A system is said to be *noncausal* or *anticipatory* if it is not causal. Although all systems that arise in nature are causal (or appear to be causal), there are applications in engineering where noncausal systems arise. An example is the off-line processing (or batch processing) of data. This will be discussed in a later chapter.

Example 1.5 *Ideal Predictor*

Consider the continuous-time system given by the input/output relationship

$$y(t) = x(t + 1)$$

This system is noncausal since the value $y(t)$ of the output at time t depends on the value $x(t + 1)$ of the input at time $t + 1$. Noncausality can also be seen by considering the response of the system to a 1-second input pulse shown in Figure 1.32a. From the relationship $y(t) = x(t + 1)$, it can be seen that the output $y(t)$ resulting from the input pulse is the pulse shown in Figure 1.32b. Since the output pulse appears before the input pulse is applied, the system is noncausal. The system with the input/output relationship $y(t) = x(t + 1)$ is called an *ideal predictor.* Most physicists would argue that ideal predictors do not exist, at least not in this universe.

Example 1.6 *Ideal Time Delay*

Consider the system with input/output relationship

$$y(t) = x(t - 1)$$

This system is causal since the value of the output at time t depends only on the value of the input at time $t - 1$. If the pulse shown in Figure 1.33a is applied to this system, the output is the pulse shown in Figure 1.33b. From Figure 1.33 it is clear that the system delays the input pulse by 1 second. In fact, the system delays all inputs by 1 second; in other words, the system is an *ideal time delay.* There are a number of techniques for generating

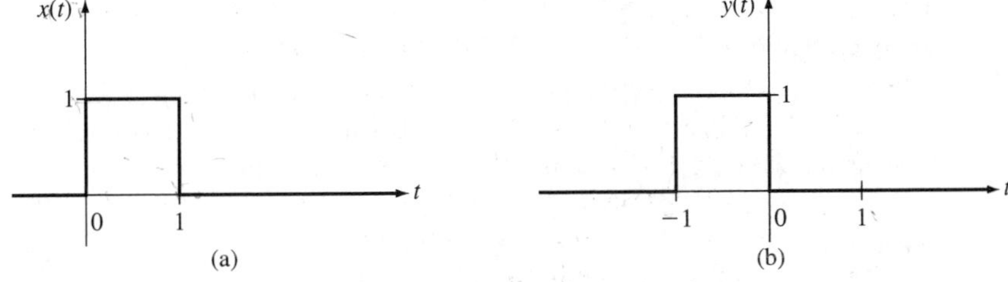

Figure 1.32 (a) Input and (b) output pulse of system in Example 1.5.

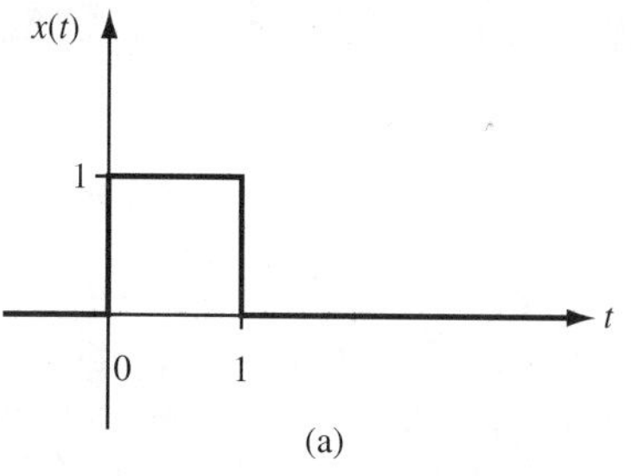

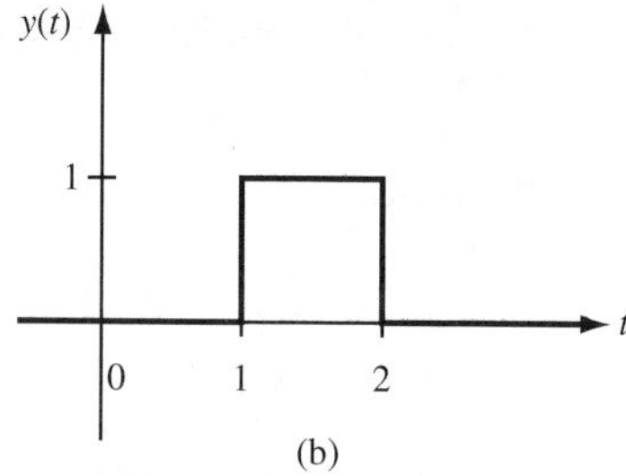

Figure 1.33 (a) Input and (b) output pulse of system in Example 1.6.

time delays. For instance, the time delay between the record and playback heads of a tape recorder can be used to create time delays with delays on the order of milliseconds.

Example 1.7 *RC Circuit*

Consider the *RC* circuit that was studied in Section 1.4. If the initial time t_0 is taken to be zero, by (1.22) the input/output relationship for the circuit is

$$y(t) = \int_0^t \frac{1}{C} e^{-(1/RC)(t-\lambda)} x(\lambda)\, d\lambda, \qquad t \geq 0 \tag{1.44}$$

Now suppose that $x(t) = 0$ for all $t < t_1$, where t_1 is an arbitrary positive real number. Then $x(\lambda) = 0$ for all $\lambda < t_1$ and the integral in (1.44) is zero when $t < t_1$. Hence $y(t) = 0$ for all $t < t_1$, so the *RC* circuit is causal.

Memoryless systems and systems with memory. A causal system is *memoryless* or *static* if for any time t_1, the value of the output at time t_1 depends only on the value of the input at time t_1.

Example 1.8 *Ideal Amplifier/Attenuator*

Suppose that $y(t) = Kx(t)$, where K is a fixed real number. At any time t_1, $y(t_1) = Kx(t_1)$, and thus $y(t_1)$ depends only on the value of the input at time t_1. Hence the system is memoryless. Since an ideal amplifier or attenuator can be represented by the input/output relationship $y(t) = Kx(t)$, it is obvious that these devices are memoryless.

A causal system that is not memoryless is said to have *memory.* A system has memory if the output at time t_1 depends in general on the past values of the input $x(t)$ for some range of values of t up to $t = t_1$.

Example 1.9 *RC Circuit*

Again consider the *RC* circuit with the input/output relationship (1.44). From (1.44) it can be seen that the output $y(t)$ of the *RC* circuit at time t depends on the values of the input $x(\lambda)$ for $0 \leq \lambda \leq t$. Thus the circuit does have memory.

Linearity

A system is said to be *additive* if for any two inputs $x_1(t)$ and $x_2(t)$, the response to the sum of inputs $x_1(t) + x_2(t)$ is equal to the sum of the responses to the inputs, assuming no initial energy before the application of the inputs. More precisely, if $y_1(t)$ is the response to input $x_1(t)$ and $y_2(t)$ is the response to input $x_2(t)$, the response to $x_1(t) + x_2(t)$ is equal to $y_1(t) + y_2(t)$. A system is said to be *homogeneous* if for any input $x(t)$ and any real scalar a, the response to the input $ax(t)$ is equal to a times the response to $x(t)$, again assuming no initial energy before the application of the input.

A system is linear if it is both additive and homogeneous; that is, for any inputs $x_1(t), x_2(t)$, and any scalars a_1, a_2, the response to the input $a_1x_1(t) + a_2x_2(t)$ is equal to a_1 times the response to input $x_1(t)$ plus a_2 times the response to input $x_2(t)$. So if $y_1(t)$ is the response to $x_1(t)$ and $y_2(t)$ is the response to $x_2(t)$, the response to $a_1x_1(t) + a_2x_2(t)$ is equal to $a_1y_1(t) + a_2y_2(t)$, again assuming no initial energy in the system before the application of the inputs. A system that is not linear is said to be *nonlinear.*

Linearity is an extremely important property. If a system is linear, it is possible to apply the vast collection of existing results on linear operations in the study of system behavior and structure. In contrast, the analytical theory of nonlinear systems is very limited in scope. In practice, a given nonlinear system is often approximated by a linear system so that analytical techniques for linear systems can then be applied. A widely used approximation method is based on linearization, which is considered in Section 2.5.

A very common type of nonlinear system is a circuit containing diodes, as shown in the following example.

Example 1.10 *Circuit with Diode*

Consider the circuit with the ideal diode shown in Figure 1.34. Here the output $y(t)$ is the voltage across the resistor with resistance R_2. The ideal diode is a short circuit when the voltage $x(t)$ is positive and it is an open circuit when $x(t)$ is negative. Thus the input/output relationship of the circuit is given by

$$y(t) = \begin{cases} \dfrac{R_2}{R_1 + R_2} x(t) & \text{when } x(t) \geq 0 \\ 0 & \text{when } x(t) \leq 0 \end{cases} \tag{1.45}$$

Now suppose that the input $x(t)$ is the unit-step function $u(t)$. Then, from (1.45), the resulting response is

$$y(t) = \frac{R_2}{R_1 + R_2} u(t) \tag{1.46}$$

If the unit-step input is multiplied by the scalar -1, so that the input is $-u(t)$, by (1.45) the resulting response is zero for all $t \geq 0$. But this is not equal to -1 times the response

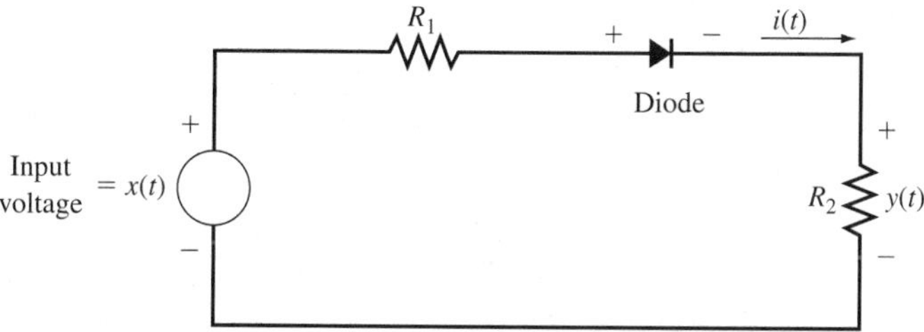

Figure 1.34 Resistive circuit with an ideal diode.

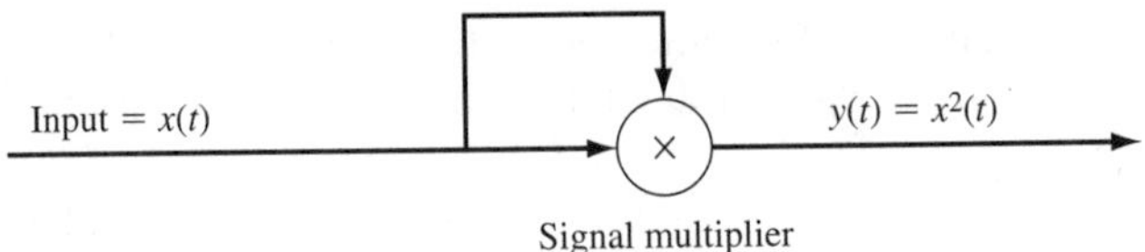

Figure 1.35 Realization of $y(t) = x^2(t)$.

to $u(t)$ given by (1.46). Hence the system is not homogeneous, and thus it is not linear. It is also easy to see that the circuit is not additive.

Nonlinearity may also be a result of the presence of signal multipliers. This is illustrated by the following example.

Example 1.11 *Square-Law Device*

Consider the continuous-time system with the input/output relationship

$$y(t) = x^2(t) \tag{1.47}$$

This system can be realized using a signal multiplier as shown in Figure 1.35. The signal multiplier in Figure 1.35 can be built (approximately) using operational amplifiers and diodes.

The system defined by (1.47) is sometimes called a *square-law device.* Note that the system is memoryless. Given a scalar a and an input $x(t)$, by (1.47) the response to $ax(t)$ is $a^2x^2(t)$. But a times the response to $x(t)$ is equal to $ax^2(t)$, which is not equal to $a^2x^2(t)$ in general. Thus the system is not homogeneous, and the system is not linear.

Another way in which nonlinearity arises is in systems containing devices that go into "saturation" when signal levels become too large, as in the following example.

Example 1.12 *Amplifier*

Consider an ideal amplifier with the input/output relationship $y(t) = Kx(t)$, where K is a fixed positive real number. A plot of the output $y(t)$ versus the input $x(t)$ is given in Figure 1.36. The ideal amplifier is clearly linear, but this is not the case for an actual (nonideal) amplifier, since the output $y(t)$ will not equal $Kx(t)$ for arbitrarily large input

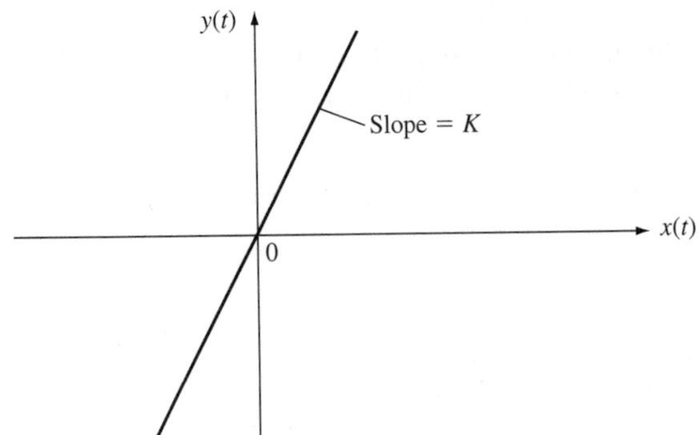

Figure 1.36 Output versus input in an ideal amplifier.

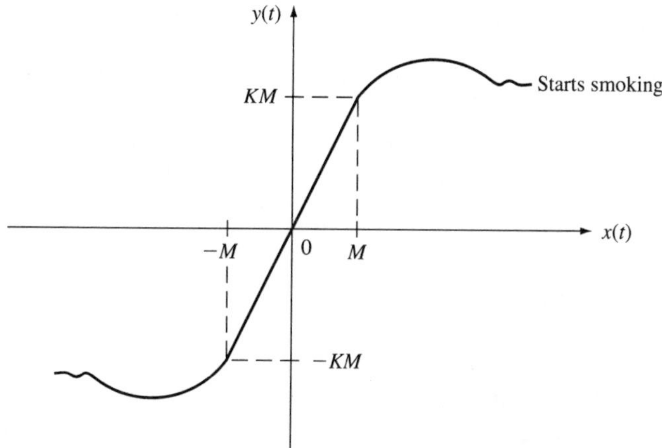

Figure 1.37 Output versus input in a nonideal amplifier.

signals. In a nonideal amplifier, the output versus input characteristics may be as shown in Figure 1.37. From the figure it is clear that $y(t) = Kx(t)$ only when the magnitude $|x(t)|$ of the input is less than M. The nonideal amplifier is not homogeneous since the response to $ax(t)$ is not equal to a times the response to $x(t)$ unless $|x(t)| < M$ and $|ax(t)| < M$. The nonideal amplifier can be viewed as a linear system only if it can be guaranteed that the magnitude of the input applied to the amplifier will never exceed M.

Although nonlinear systems are very common, many systems arising in practice can be modeled as linear systems. A large class of linear systems is defined in the next example.

Example 1.13 *Linear Time-Invariant Systems*

Consider a single-input single-output continuous-time system with the input/output relationship

$$y(t) = \int_{t_0}^{t} h(t - \lambda)x(\lambda)\, d\lambda \qquad t \geq t_0 \tag{1.48}$$

where $h(t)$ is an arbitrary real-valued function of t with $h(t) = 0$ for all $t < 0$. In Chapter 3 it will be seen that $h(t)$ is the impulse response of the system defined by (1.48). Note that if $h(t) = (1/C)\exp[(-1/RC)t]$ for $t \geq 0$, then (1.48) is the input/output relationship of the RC circuit considered in Section 1.4, or if

$$h(t) = \frac{1}{k_f}[1 - e^{-(k_f/M)t}] \qquad \text{for } t \geq 0$$

(1.48) is the input/output relationship of the automobile considered in Section 1.4. Now suppose that the input is $ax(t) + bv(t)$, where $x(t)$, $v(t)$, a, and b are arbitrary. By (1.48) the resulting output response is

$$y(t) = \int_{t_0}^{t} h(t - \lambda)[ax(\lambda) + bv(\lambda)]\, d\lambda$$

Using the linearity property of integration gives

$$y(t) = a\int_{t_0}^{t} h(t-\lambda)x(\lambda)\,d\lambda + b\int_{t_0}^{t} h(t-\lambda)v(\lambda)\,d\lambda$$

Hence $y(t)$ is equal to a times the response to $x(t)$ plus b times the response to $v(t)$, which proves that the system is linear. Since the input/output relationship of the RC circuit and the automobile can be expressed in the form (1.48), it follows that both of these systems are linear.

Time Invariance

Given a real number t_1 and a continuous-time signal $x(t)$, recall that $x(t - t_1)$ is equal to $x(t)$ shifted to the right by t_1 seconds if $t_1 > 0$, and that $x(t - t_1)$ is equal to $x(t)$ shifted to the left by t_1 seconds if $t_1 < 0$. Now consider a continuous-time system with input $x(t)$ and output $y(t)$. The system is said to be *time invariant* or *constant* if for any input $x(t)$ and any t_1, the response to the shifted input $x(t - t_1)$ is equal to $y(t - t_1)$, where $y(t)$ is the response to $x(t)$ with zero initial energy. Therefore, in a time-invariant system the response to a left or right shift of the input $x(t)$ is equal to a corresponding shift in the response $y(t)$ to $x(t)$ (assuming no initial energy). In a time-invariant system, there are no changes in the system structure as a function of time t. A system is *time varying* or *time variant* if it is not time invariant.

Example 1.14 *Amplifier with Time-Varying Gain*

Suppose that $y(t) = tx(t)$. It is easy to see that this system is memoryless and linear. Now for any t_1,

$$y(t - t_1) = (t - t_1)x(t - t_1)$$

But the response to input $x(t - t_1)$ is $tx(t - t_1)$, which does not equal $(t - t_1)x(t - t_1)$ in general. Hence $y(t - t_1)$ is not equal to the t_1-second shift of the response to $x(t)$, and thus the system is time varying. Note that this system can be viewed as an ideal amplifier with time-varying gain t.

Example 1.15 *First-Order System*

Consider the continuous-time system given by the input/output differential equation

$$\dot{y}(t) + a(t)y(t) = bx(t) \tag{1.49}$$

where $\dot{y}(t)$ is the derivative of $y(t)$. In (1.49), the coefficient b is a constant; however, the coefficient $a(t)$ may be time varying. Now let $\gamma(t)$ denote the output response of the system when the input $x(t)$ is equal to some signal $v(t)$ with zero initial energy in the system. Then $\gamma(t)$ and $v(t)$ satisfy the differential equation (1.49); that is,

$$\dot{\gamma}(t) + a(t)\gamma(t) = bv(t) \tag{1.50}$$

Replacing t by $t - t_1$ in (1.50) gives

$$\dot{\gamma}(t - t_1) + a(t - t_1)\gamma(t - t_1) = bv(t - t_1) \tag{1.51}$$

Now let $\beta(t)$ denote the response of the system to the shifted input $v(t - t_1)$. By definition of the input/output relationship (1.49), the response $\beta(t)$ resulting from input $v(t - t_1)$ is given by

$$\dot{\beta}(t) + a(t)\beta(t) = bv(t - t_1) \tag{1.52}$$

Comparing (1.51) and (1.52) yields the conclusion that $\beta(t)$ is always equal to $\gamma(t - t_1)$ if and only if $a(t) = a(t - t_1)$. For this to be true for any t_1, the coefficient $a(t)$ must be constant; that is, $a(t) = a$ for all t. Thus the system defined by (1.49) is time invariant if and only if the coefficient $a(t)$ is constant.

A discrete-time system is time invariant if for any integer q and any input $x[n]$, the response to the q-step shift $x[n - q]$ of the input $x[n]$ is equal to $y[n - q]$, where $y[n]$ is the response to $x[n]$ assuming no initial energy. Hence, as is the case for a continuous-time system, a discrete-time system is time invariant if the response to a shift of the input $x[n]$ is equal to a corresponding shift of the response $y[n]$ to $x[n]$ (assuming no initial energy).

Finite Dimensionality

Given a continuous-time system with input $x(t)$ and output $y(t)$, for any nonnegative integer i, let $y^{(i)}(t)$ and $x^{(i)}(t)$ denote the ith derivative of the output $y(t)$ and input $x(t)$. [When $i = 0$, $y^{(i)}(t) = y(t)$ and $x^{(i)}(t) = x(t)$.] The system is said to be *finite dimensional* or *lumped* if for some positive integer N the Nth derivative of the output at time t is equal to a function of $y^{(i)}(t)$ and $x^{(i)}(t)$ at time t for $0 \le i \le N - 1$. The Nth derivative of the output at time t may also depend on the ith derivative of the input at time t for $i \ge N$. In mathematical terms, the system is finite dimensional if for some positive integer N and nonnegative integer M, $y^{(N)}(t)$ can be written in the form

$$y^{(N)}(t) = f(y(t), y^{(1)}(t), \ldots, y^{(N-1)}(t), x(t), x^{(1)}(t), \ldots, x^{(M)}(t), t) \tag{1.53}$$

where f is a function of the variables $y(t), y^{(1)}(t), \ldots, y^{(N-1)}(t), x(t), x^{(1)}(t), \ldots, x^{(M)}(t)$ and time t. The integer N is called the *order* of the differential equation (1.53). The integer N is also referred to as the *order* or *dimension* of the system with the input/output relationship (1.53).

The input/output relationship (1.53) is called the *input/output differential equation* of the system. Hence a system has an input/output differential equation representation if and only if the system is finite dimensional. Recall that the RC circuit, the automobile, and the simple pendulum considered in Section 1.4 have input/output differential equation representations, and thus all three of these systems are finite dimensional. In addition, the RC circuit has order one ($N = 1$), while both the automobile and the pendulum have order two ($N = 2$).

A continuous-time system with memory is *infinite dimensional* if it is not finite dimensional. Thus a system with memory is infinite dimensional if and only if it is not possible to express the Nth derivative of the output in the form (1.53) for some positive integer N.

Example 1.16 *System with Time Delay*

Consider the continuous-time system given by

$$\frac{dy(t)}{dt} + ay(t - 1) = x(t) \tag{1.54}$$

where a is an arbitrary nonzero constant. Due to the delay term $y(t - 1)$ in (1.54), it is not possible to write (1.54) in the form (1.53) for any positive integer N. Thus the system is infinite dimensional. The system defined by (1.54) is an example of a *system with time delays.* Systems with time delays are always infinite dimensional. The input/output relationship (1.54) of the system is called a *delay differential equation.* Delay differential equations are much more complicated than ordinary differential equations. In particular, it is seldom possible to express solutions of a delay differential equation in analytical form.

Now suppose that the given system is a discrete-time system with input $x[n]$ and output $y[n]$. The system is *finite dimensional* if for some positive integer N and nonnegative integer M, $y[n]$ can be written in the form

$$y[n] = f(y[n-1], y[n-2], \ldots, y[n-N], x[n], x[n-1], \ldots, x[n-M], n) \quad (1.55)$$

The integer N is called the order or dimension of the system, and the input/output relationship (1.55) is called the *input/output difference equation.* So a discrete-time system has an input/output difference equation representation if it is finite dimensional. Recall that the loan-repayment system studied in Section 1.4 has an input/output difference equation representation, and hence this system is finite dimensional (with order $N = 1$).

Note that for the discrete-time system defined by the input/output difference equation (1.55), the output $y[n]$ depends only on previous values of the output and the current and previous values of the input $x[n]$. Hence $y[n]$ does not depend on future values of $x[n]$, and therefore the system is causal. The representation (1.55) can be generalized to the noncausal case by allowing $y[n]$ to depend on $x[n + i]$ for $i \geq 1$, but this will not be considered here.

A causal discrete-time system that cannot be expressed in the form (1.55) is *infinite dimensional.* An example of an infinite-dimensional discrete-time system is given below.

Example 1.17 *Infinite Dimensionality*

Consider the discrete-time system given by the input/output relationship

$$y[n] = \sum_{i=0}^{n-1} \frac{1}{n-i} x[i], \qquad n \geq 1$$

This system is clearly causal and linear. (Is it time invariant?) It is not at all obvious, but it turns out that $y[n]$ cannot be expressed in the form (1.55) for any positive integer N, and thus the system is infinite dimensional. For this example, infinite dimensionality can be shown by using *proof by contradiction;* that is, assume that (1.55) is valid for some N, and then show that a contradiction results. The interested reader is invited to carry out the details of the proof.

Linear finite-dimensional systems. Suppose that the system under study is a finite-dimensional continuous-time system with the input/output differential equation

$$y^{(N)}(t) = f(y(t), y^{(1)}(t), \ldots, y^{(N-1)}(t), x(t), x^{(1)}(t), \ldots, x^{(M)}(t), t) \quad (1.56)$$

The system defined by (1.56) can be shown to be linear if and only if (1.56) is a linear differential equation, which is the case if and only if there exist scalar functions $a_0(t), a_1(t), \ldots, a_{N-1}(t), b_0(t), b_1(t), \ldots, b_M(t)$, such that

$$f(y(t), \ldots, y^{(N-1)}(t), x(t), \ldots, x^{(M)}(t), t) = -\sum_{i=0}^{N-1} a_i(t)y^{(i)}(t) + \sum_{i=0}^{M} b_i(t)x^{(i)}(t)$$

Thus the system defined by (1.56) is linear if and only if the Nth derivative of the output can be expressed in the form

$$y^{(N)}(t) = -\sum_{i=0}^{N-1} a_i(t)y^{(i)}(t) + \sum_{i=0}^{M} b_i(t)x^{(i)}(t) \tag{1.57}$$

The input/output differential equations of the RC circuit and the automobile studied in Section 1.4 are both linear, and thus these systems are linear. (Recall that linearity of the RC circuit and the automobile was shown to be the case using the result in Example 1.13.) However, the input/output differential equation (1.34) of the pendulum is nonlinear, and therefore the pendulum is a nonlinear system.

Now consider a finite-dimensional discrete-time system with the input/output difference equation

$$y[n] = f(y[n-1], y[n-2], \ldots, y[n-N], x[n], x[n-1], \ldots, x[n-M], n) \tag{1.58}$$

It can be shown that the system defined by (1.58) is linear if and only if (1.58) is a linear difference equation, which is the case if and only if (1.58) can be written in the form

$$y[n] = -\sum_{i=1}^{N} a_i(n)y[n-i] + \sum_{i=0}^{M} b_i(n)x[n-i] \tag{1.59}$$

Recall that the loan-repayment system has an input/output difference equation of the form (1.59), and thus this system is linear.

Linear time-invariant finite-dimensional systems. Consider a linear finite-dimensional continuous-time system specified by the input/output differential equation (1.57). The system can be shown to be time invariant (see Example 1.15) if and only if the coefficients of the differential equation are constant (independent of t); that is,

$$a_i(t) = a_i \text{ and } b_i(t) = b_i \qquad \text{for all } i \text{ and all real numbers } t$$

where the a_i and the b_i are constants.

The input/output differential equations of the RC circuit, automobile, and pendulum studied in Section 1.4 have constant coefficients, and thus all three of these systems are time invariant. If the system is a linear finite-dimensional discrete-time system with the input/output difference equation (1.59), it can be shown that the system is time invariant if and only if the coefficients of the difference equation are constant; that is,

$$a_i(n) = a_i \quad \text{and} \quad b_i(n) = b_i \qquad \text{for all } i \text{ and integers } n$$

Example 1.18 *Linear Time-Invariance*

The input/output difference equation (1.36) of the loan-repayment system has constant coefficients if the interest rate I is constant, and thus in this case the system is time invariant. If the interest rate varies as a function of n (a variable-rate loan), one of the coefficients of the input/output difference equation is time varying, and thus the system is time varying.

Example 1.19 *System Properties*

Consider a system defined by

$$y(t) = \int_{t-1}^{t} x(\lambda)d\lambda$$

To check for causality, examine the response when $x(t) = 0$, for $t < t_0$:

$$y(t) = \int_{t-1}^{t} 0d\lambda = 0, \quad t < t_0$$

thus, the system is causal. Also, $y(t)$ depends in general on $x(\lambda)$ for $t - 1 \leq \lambda \leq t$, so the system has memory. Let $y_1(t)$ be the response to $x_1(t)$ and $y_2(t)$ be the response to $x_2(t)$. Now, consider the response $y(t)$ to $x(t) = a_1x_1(t) + a_2x_2(t)$:

$$\begin{aligned} y(t) &= \int_{t-1}^{t} [a_1x_1(\lambda) + a_2x_2(\lambda)]d\lambda \\ &= a_1\int_{t-1}^{t} x_1(\lambda)d\lambda + a_2\int_{t-1}^{t} x_2(\lambda)d\lambda \\ &= a_1y_1(t) + a_2y_2(t) \end{aligned}$$

Therefore, the system is linear. Now consider the response $\tilde{y}(t)$ to $x(t - t_1)$:

$$\tilde{y}(t) = \int_{t-1}^{t} x(\lambda - t_1)d\lambda$$

Let $\tau = \lambda - t_1$, then

$$\tilde{y}(t) = \int_{t-t_1-1}^{t-t_1} x(\tau)\,d\tau$$

Since this is equal to $y(t - t_1)$, the system is time-invariant.

Example 1.20 *System Properties*

Consider a system defined by

$$y(t) = \int_{t-1}^{t} tx(\lambda)\,d\lambda$$

By the same arguments used in Example 1.19, the system is causal, has memory, and is linear. Now, consider time invariance. The delayed response is given by

$$y(t - t_1) = \int_{t-t_1-1}^{t-t_1} (t - t_1)x(\tau)\,d\tau$$

Let $\tilde{y}(t)$ denote the response to $x(t - t_1)$:

$$\tilde{y}(t) = \int_{t-1}^{t} tx(\lambda - t_1)\,d\lambda$$

Let $t = \lambda - t_1$. Then,

$$\tilde{y}(t) = \int_{t-t_1-1}^{t-t_1} tx(\tau)\,d\tau$$

Since this is not equal to $y(t - t_1)$, the system is not time invariant.

The properties of linearity and time invariance make a system much easier to analyze. In particular, if system outputs $y_1(t)$ and $y_2(t)$ are known for inputs $x_1(t)$ and $x_2(t)$, respectively, then the outputs for any input of the form $x(t) = a_1x_1(t - t_1) + a_2x_2(t - t_2)$

can be calculated easily. Furthermore, since derivatives and integrals are linear operations, the following holds for linear time-invariant systems:

If $y(t)$ is the response to $x(t)$, then $dy(t)/dt$ is the response to $dx(t)/dt$. Also,

$$\int_0^t y(\lambda)\,d\lambda \text{ is the response to } \int_0^t x(\lambda)\,d\lambda.$$

Example 1.21 *Linear Time-Invariant System Responses*

Suppose that a linear time-invariant system has the input $x(t) = \sin 4t$ and responds with an output of $y(t) = 0.5\sin(4t - \pi/6)$. Then, the response to the input $x(t) = 4\sin 4t + 2\sin(4(t-2))$ is $y(t) = 2\sin(4t - \pi/6) + \sin(4(t-2) - \pi/6)$. The response to the input $x(t) = 4\cos 4t$ is $y(t) = 2\cos(4t - \pi/6)$.

PROBLEMS

1.1. Consider the continuous-time signals displayed in Figure P1.1.

(a) Show that each of these signals is equal to a sum of rectangular pulses $p_\tau(t)$ and/or triangular pulses $(1 - 2|t|/\tau)p_\tau(t)$.

(b) Use MATLAB to plot the signals in Figure P1.1.

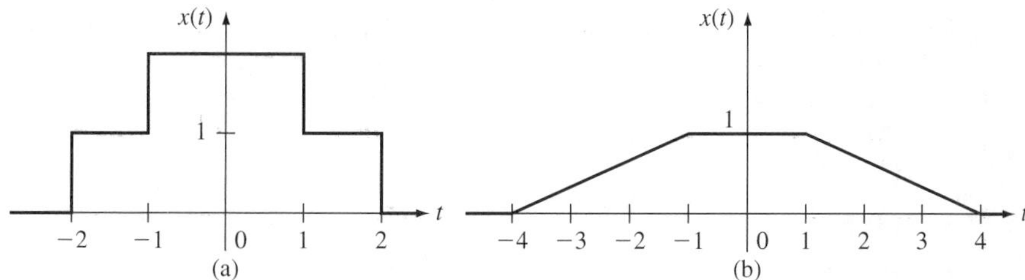

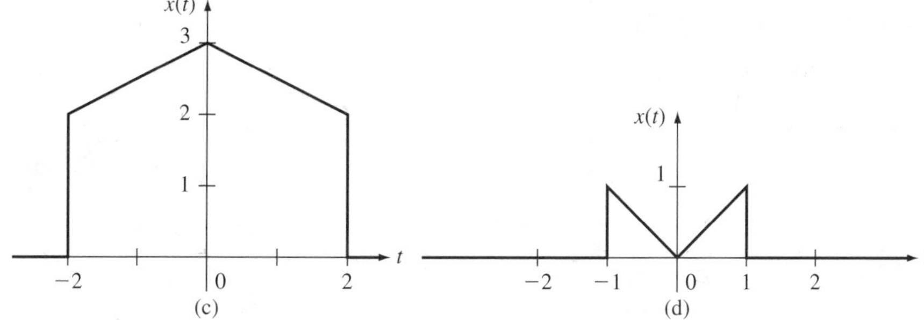

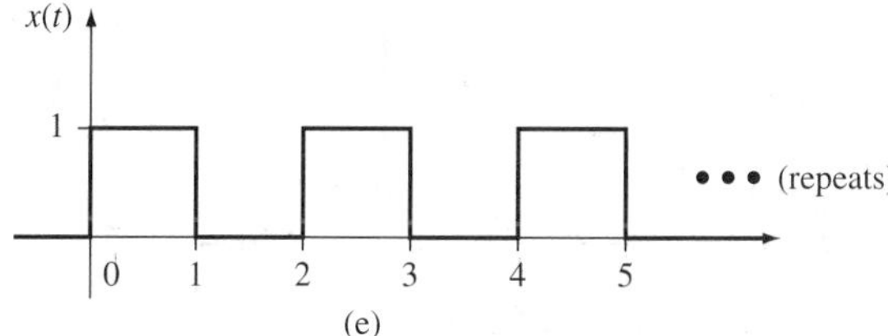

Figure P1.1

1.2. Obtaining a computer-generated plot of a continuous-time signal requires some care in choosing the time increment Δt(the spacing between points). As mentioned in Section 1.2, too large of an increment will cause a jagged plot. Moreover, a very large time increment may introduce a phenomena known as *aliasing,* which distorts the information given in the signal. (Aliasing is covered in greater detail in Chapter 6.) To avoid aliasing when defining a computer-generated sinusoid such as $x(t) = \cos(\omega t + b)$, choose $\Delta t \leq \pi/\omega$. A rule of thumb for defining a decaying sinusoid such as $x(t) = e^{-at}\cos\omega t$ is choose $\Delta t \leq \pi/(4\sqrt{a^2 + \omega^2})$. (Choosing even smaller values of Δt creates smoother plots.)

(a) Compute the maximum time increment for plotting $x(t) = \sin \pi t$ in MATLAB. Verify your result by plotting $x(t)$ for $t = 0$ to $t = 20$ sec with the following time increments: $\Delta t = 0.1$ sec, $\Delta t = 0.5$ sec, $\Delta t = 0.9$ sec, $\Delta t = 1.5$ sec. Note the apparent change in frequency in the plot due to the aliasing effect for $\Delta t = 1.5$. How do you expect your plot to appear when $\Delta t = 2$ sec? Verify your result.

(b) Compute the maximum time increment for plotting $x(t) = e^{-0.1t}\cos \pi t$. Verify your result by plotting $x(t)$ for $t = 0$ to $t = 20$ sec with $\Delta t = 0.1, 0.5, 1.5,$ and 2 sec.

(c) Compute the maximum time increment for plotting $x(t) = e^{-t}\cos(\pi t/4)$. Verify your result by plotting $x(t)$ for $t = 0$ to $t = 10$ sec with $\Delta t = 0.1, 1, 2,$ and 3 sec.

The problem with aliasing exists not only with plotting but with all digital processing of continuous-time signals. A computer program that emulates a continuous-time system needs to have an input signal defined so that the signal has very little aliasing.

1.3. Use MATLAB to plot the following functions for $-1 \leq t \leq 5$ sec. Label your axes appropriately.

(a) $u(t)$

(b) $r(t)$

(c) $x(t) = 3e^{-2t}u(t)$

(d) $x(t) = \begin{cases} e^{-t}, & 0 \leq t \leq 2 \\ 0, & \text{otherwise} \end{cases}$

(e) $x(t) = \sin 2t + 2\cos(3t - 0.2)$

(f) $x(t) = p_2(t)$, where $p_\tau(t)$ is the rectangular pulse defined in Figure 1.9

(g) $x(t)$ as shown in Figure P1.3a

(h) $x(t)$ as shown in Figure 1.11

(i) $x(t) = e^{2t}\sin 3t\, u(t)$

(j) $x(t) = e^{-2t}\sin 3t\, u(t)$

(k) $x(t)$ as shown in Figure P1.3b

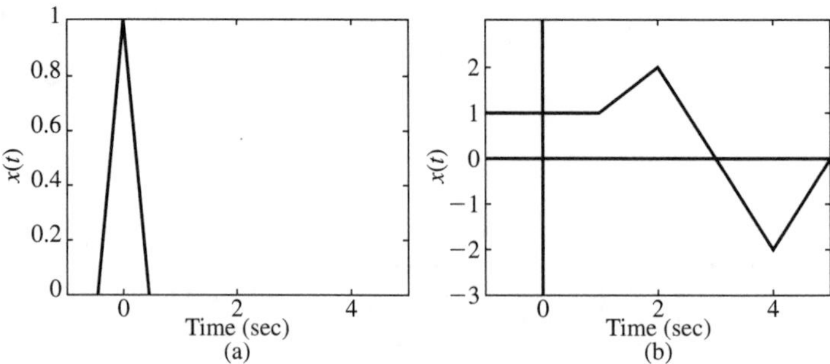

Figure P1.3

1.4. Sketch the continuous-time signals in parts (a) to (d).

(a) $x(t) = u(t+1) - 2u(t-1) + u(t-3)$

(b) $x(t) = (t+1)u(t-1) - tu(t) - u(t-2)$

(c) $x(t) = e^{-t}u(t) + e^{-t}[\exp(2t-4) - 1]u(t-2) - e^{t-4}u(t-4)$

(d) $x(t) = \cos t\left[u\left(t+\frac{\pi}{2}\right) - 2u(t-\pi)\right] + (\cos t)u\left(t - \frac{3\pi}{2}\right)$

(e) Use MATLAB to plot the signals defined in parts (a) to (d).

1.5. For each of the signals in Problem 1.4, give the generalized derivative in analytical form, and then sketch the generalized derivative.

1.6. Express each of the signals in Problem 1.4 in the form

$$x(t) = x_1(t)[u(t-t_1) - u(t-t_2)] + x_2(t)[u(t-t_2) - u(t-t_3)] + \cdots$$

Give the signals $x_1(t), x_2(t), \ldots$ in the simplest possible analytical form.

1.7. Consider the continuous-time signals shown in Figure P1.7.

(a) Express each signal in the form

$$x(t) = f_1(t)u(t-t_1) + f_2(t)u(t-t_2) + \cdots$$

Give the signals $f_1(t), f_2(t), \ldots$ in the simplest possible analytical form.

(b) Use MATLAB to plot the signals shown in Figure P1.7.

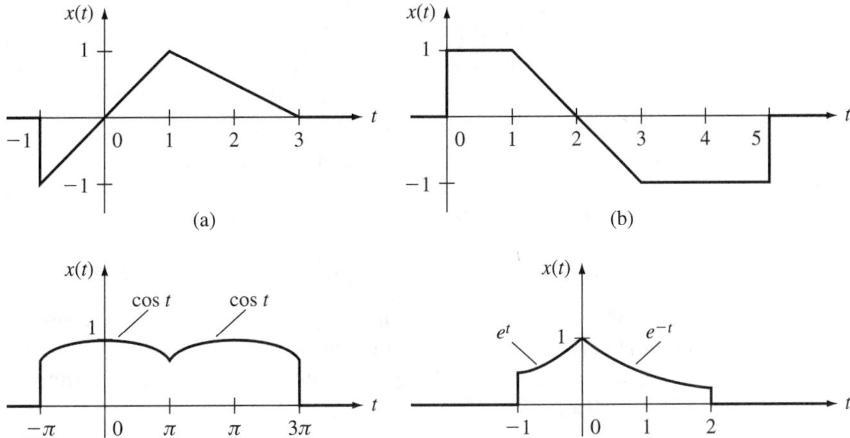

Figure P1.7

1.8. For each of the signals in Problem 1.7, give the generalized derivative in analytical form, and then sketch the generalized derivative.

1.9. Given a continuous-time signal $x(t)$ and a constant c, consider the signal $x(t)u(t - c)$.

(a) Show that there exists a signal $v(t)$ such that

$$x(t)u(t - c) = v(t - c)u(t - c)$$

Express $v(t)$ in terms of $x(t)$.

(b) Determine the simplest possible analytical form for $v(t)$ when:

(i) $x(t) = e^{-2t}$ and $c = 3$

(ii) $x(t) = t^2 - t + 1$ and $c = 2$

(iii) $x(t) = \sin 2t$ and $c = \pi/4$

1.10. Plot the following discrete-time signals using MATLAB for $-5 \le n \le 15$ using the `stem` command if available, otherwise, use `plot(n,x,'o')`, which places circles at the desired points. Label your axes appropriately.

(a) $x[n] = u[n]$

(b) $x[n] = r[n]$

(c) $x[n] = (0.8)^n u[n]$

(d) $x[n] = (-0.8)^n u[n]$

(e) $x[n] = \sin(\pi n/4)$

(f) $x[n] = \sin(\pi n/2)$

(g) $x[n] = (0.9)^n[\sin(\pi n/4) + \cos(\pi n/4)]$

(h) $x[n] = 2^n u[n]$

(i) $x[n] = \begin{cases} 1 & -4 \le n \le 4 \\ 0 & \text{otherwise} \end{cases}$

1.11. Sketch the following discrete-time signals.

(a) $x[n] = u[n] - 2u[n - 1] + u[n - 4]$

(b) $x[n] = (n + 2)u[n + 2] - 2u[n] - nu[n - 4]$

(c) $x[n] = \delta[n + 1] - \delta[n] + u[n + 1] - u[n - 2]$

(d) $x[n] = e^{0.2n}u[n + 1] + u[n] - 2e^{0.1n}u[n - 3] - (1 - e^{0.1n})^2 u[n - 5]$

(e) Use MATLAB to plot the signals defined in parts (a) to (d).

1.12. Express each of the discrete-time signals in Problem 1.11 in the form

$$x[n] = x_1[n](u[n - n_1] - u[n - n_2]) + x_2[n](u[n - n_2] - u[n - n_3]) + \cdots$$

Give the signals $x_1[n], x_2[n], \ldots$ in the simplest possible analytical form.

1.13. Use an analytical method to determine if the signals below are periodic; if so, find the fundamental period. Use MATLAB to plot each signal and verify your prediction of periodicity. Use a small enough time increment for the continuous-time signals to make your plot smooth (see Problem 1.2).

(a) $x(t) = \cos 2\pi t + \sin 5\pi t$

(b) $x(t) = \cos 2\pi t + \sin 2t$

(c) $x(t) = \cos 2\pi t + \sin 6t$

(d) $x[n] = \cos 2n$

(e) $x[n] = \cos(2\pi n/3)$

(f) $x[n] = \cos(\pi n^2/2)$ (*Hint:* to get n^2 when n is stored in a vector, type `n.^2` instead of `n*n`)

1.14. Find an expression for each of the signals $x(t)$ plotted in Figure P1.14.

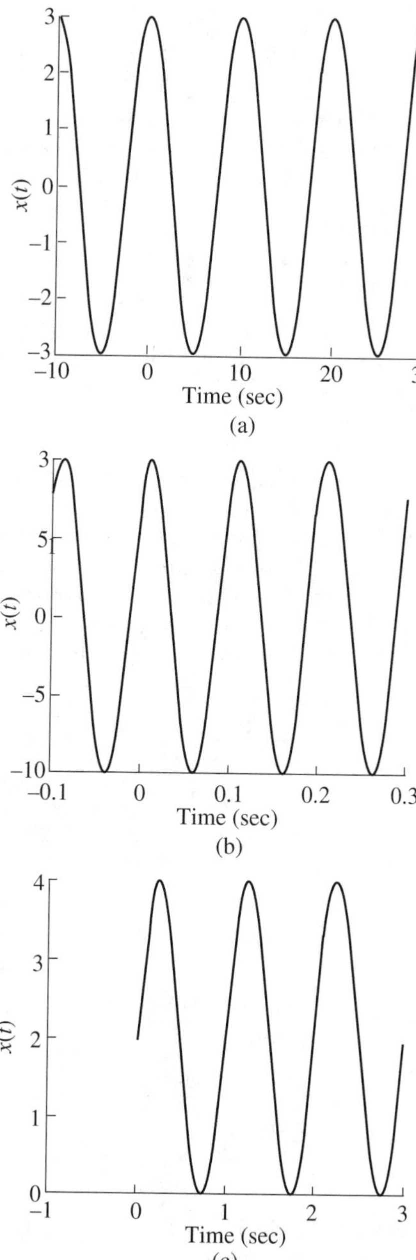

Figure P1.14

1.15. Prove the expression given in Equation (1.30) reduces to that in (1.31). *Hint:* Use the integration property that

$$\int_a^b \int_c^\tau f_1(\tau)f_2(\lambda)\, d\lambda\, d\tau = \int_c^{c+b-a} \int_\lambda^b f_1(\tau)f_2(\lambda)\, d\tau\, d\lambda$$

1.16. Consider the RC circuit in Figure 1.24. With $R = 1000\ \Omega$ and $C = 100\ \mu\text{F}$, calculate the responses to the inputs given below. For parts (a) and (b), assume that the system is in the zero state prior to the input. In each case, sketch the response $y(t)$.
(a) $x(t) = u(t)$
(b) $x(t) = \delta(t)$
(c) $x(t) = u(t),\ y(0) = -1$

1.17. Consider the car moving on a level surface shown in Figure 1.26. With $M = 1200$ kg and $k_f = 250$ kg/s, calculate the responses to the inputs given below. Assume that the system is in the zero state prior to the input. In each case, sketch the position of the car $y(t)$ and the velocity $v(t)$
(a) $x(t) = 2500u(t)$
(b) $x(t) = 2500u(t) - 2500u(t-2)$

1.18. A tank filled with water is shown in Figure P1.18. Here, $h(t)$ is the water level at time t in meters (m), $x(t)$ is the flow rate (m^3/s) of the water coming into the tank, and $y(t)$ is the flow rate (m^3/s) of the water coming out of the tank. Let $v(t)$ represent the volume of water in the tank (m^3) at time t. Write an expression for $dv(t)/dt$. Let R be the valve resistance, $y(t) = Rh(t)$. Assume that the tank is cylindrical with the area of the base equal to A (m^2). Then, $v(t) = Ah(t)$. Now, using the above relationships along with the expression derived for $dv(t)/dt$, write an input/output differential equation of the system.

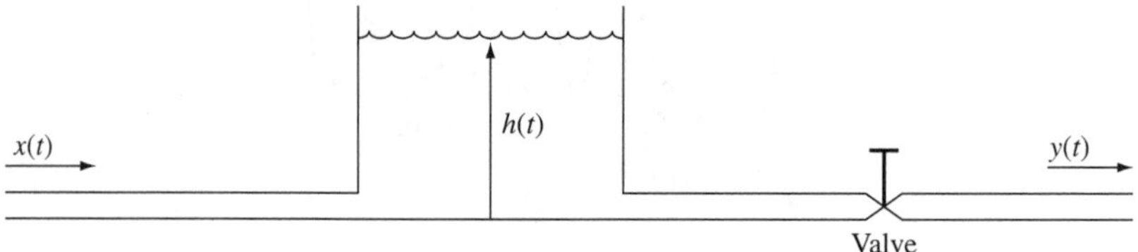

Figure P1.18

1.19. As discussed in McClamroch [1980], the ingestion and metabolism of a drug in a human is described by the equations

$$\frac{dq(t)}{dt} = -k_1 q(t) + x(t)$$

$$\frac{dy(t)}{dt} = k_1 q(t) - k_2 y(t)$$

where the input $x(t)$ is the ingestion rate of the drug, the output $y(t)$ is the mass of the drug in the bloodstream, and $q(t)$ is the mass of the drug in the gastrointestinal tract. The constants k_1 and k_2 are metabolism rates that satisfy the inequality $k_1 > k_2 > 0$. The constant k_2 characterizes the excretory process of the individual. Determine the second-order input/output differential equation of the system.

1.20. Again consider the loan-balance system with the input/output difference equation

$$y[n+1] - \left(1 + \frac{I}{12}\right) y[n] = -x[n+1]$$

Recall that $y[0]$ is the amount of the loan, $y[n]$ is the loan balance at the end of the nth month, $x[n]$ is the loan payment in the nth month, and I is the interest rate in decimal form. It is assumed that the monthly payments $x[n]$ for $n \geq 1$ are equal to a constant c. Suppose that the number of months in the repayment period is N. Derive an expression for the monthly payments c in terms of $y[0]$, N, and I. *Hint:* Use the relationship

$$\sum_{i=1}^{N} a^{N-i} = \frac{1 - a^N}{1 - a}, \qquad a \neq 1$$

1.21. A savings account in a bank with interest acruing quarterly can be modeled by the input/output difference equation

$$y[n+1] - \left(1 + \frac{I}{4}\right) y[n] = x[n+1]$$

where $y[n]$ is the amount in the account at the end of the nth quarter, $x[n]$ is the amount deposited in the nth quarter, and I is the yearly interest rate in decimal form.

(a) Suppose that $I = 10\%$. Compute $y[n]$ for $n = 1, 2, 3, 4$ when $y[0] = 1000$ and $x[n] = 1000$ for $n \geq 1$.

(b) Suppose that $x[n] = c$ for $n \geq 1$ and $y[0] = 0$. Given an integer N, suppose that it is desired to have an amount $y[N]$ in the savings account at the end of the Nth quarter. Derive an expression for N in terms of $y[n]$, c, and I.

(c) Suppose that an IRA (individual retirement account) is set up with $y[0] = 2000$, $I = 10\%$, and $x[n] = \$500, n \geq 1$ (n = quarter). How many years will it take to amass \$1,000,000 in the account?

(d) Modify the loan balance program given in Figure 1.31 to compute the savings amount. Repeat parts (a) and (c) using your new MATLAB program.

1.22. For each difference equation given below, write a MATLAB M-file that performs a recursion to solve for $y[n]$ for $0 \leq n \leq 10$, and plot y versus n on a stem plot.

(a) $y[n] = 0.5y[n-1] + u[n-1]; y[-1] = 0$

(b) $y[n] = 2y[n-1]; y[-1] = 1$

(c) $y[n] = 0.5y[n-1] + 0.1y[n-2] + u[n-1]; y[-2] = 1, y[-1] = 0$

(d) $y[n] = 0.1y[n-1] + 0.5y[n-2] + (0.5)^n; y[-1] = y[-2] = 0$

1.23. A machine is repetitively processing parts as illustrated in Figure P1.23. Here $x[n]$ for $n \geq 0$ is the time at which the nth part arrives at the machine for processing and $y[n]$ is the time at which the machine completes the processing of the nth part. The machine cannot process the $(n+1)$th part until it has processed the nth part and the $(n+1)$th part has arrived at the machine for processing. The processing of each part requires S seconds.

(a) Derive an expression for $y[n+1]$ as a function of $y[n]$, $x[n+1]$, and S.

(b) Based on your result in part (a), is the machine system linear or nonlinear, time invariant or time varying? Justify your answer.

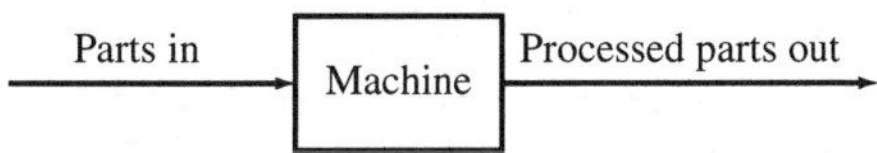

Figure P1.23

1.24. Determine whether the following continuous-time systems are causal or noncausal, have memory or are memoryless. Justify your answers. In the following parts, $x(t)$ is an arbitrary input and $y(t)$ is the zero-state response to $x(t)$.

(a) $y(t) = x(t) + 1$

(b) $y(t) = |x(t)| = \begin{cases} x(t) & \text{when } x(t) \geq 0 \\ -x(t) & \text{when } x(t) < 0 \end{cases}$

(c) $\dfrac{dy(t)}{dt} = y(t)x(t)$

(d) $\dfrac{dy}{dt} = 3y(t) + 2x(t)$

(e) $y(t) = \sin x(t)$

(f) For any $x(t)$, $y(t) = \begin{cases} 1, & t \geq 0 \\ 0, & t < 0 \end{cases}$

(g) $y(t) = e^{-t}x(t)$

(h) $y(t) = \begin{cases} x(t) & \text{when } |x(t)| \leq 10 \\ 10 & \text{when } |x(t)| > 10 \end{cases}$

(i) $y(t) = \displaystyle\int_{-\infty}^{t} (t - \lambda)x(\lambda)\, d\lambda$

(j) $\dfrac{dy}{dt} = 3ty(t) + 2x(t)$

1.25. For each of the systems in Problem 1.24, determine whether the system is linear or nonlinear. Justify your answers.

1.26. For each of the systems in Problem 1.24, determine whether the system is time invariant or time varying. Justify your answers.

1.27. A continuous-time system is said to have a *dead zone* if the output response $y(t)$ is zero for any input $x(t)$ with $|x(t)| < A$ where A is a constant called the *threshold.* An example is a dc motor that is unable to supply any torque until the input voltage exceeds a threshold value. Show that any system with a dead zone is nonlinear.

1.28. Determine whether the circuit with the ideal diode in Figure P1.28 is causal or noncausal, linear or nonlinear, time invariant or time varying. Justify your answers.

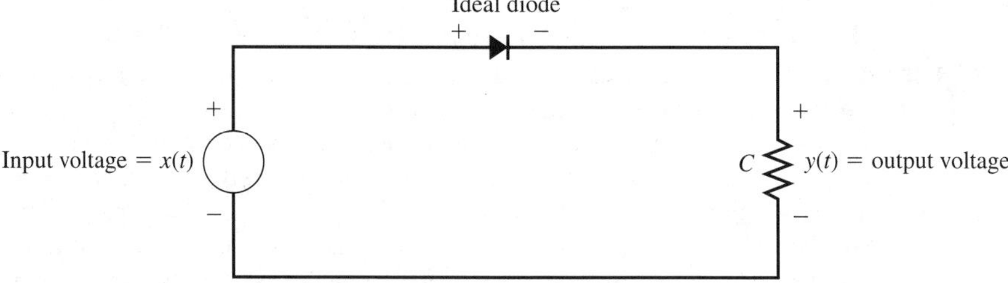

Figure P1.28

1.29. An automobile on an inclined surface is illustrated in Figure P1.29. The velocity model of the car is given by the input/output differential equation

$$M\frac{dv(t)}{dt} + k_f v(t) = x(t) - Mg \sin \theta(t)$$

Here $x(t)$ is the drive or braking force applied to the car, $Mg \sin \theta(t)$ is the force on the car due to gravity, g is the gravity constant, $\theta(t)$ is the angle of the inclined surface, and $v(t) = dy(t)/dt$ is the velocity of the car.

In all of the following parts, the output of the system is the velocity $v(t)$ of the car.

(a) With $\theta(t) = 0$ for all t and with the input equal to $x(t)$, determine whether the car system is linear and time invariant. Justify your answers.

(b) Now suppose that $\theta(t) \neq 0$. With the input equal to $x(t)$, determine whether the system is linear and time invariant. Justify your answers.

(c) The system can be made into a two-input system by taking the first input to be $x(t)$ and the second input to be $\sin \theta(t)$. With this definition of the inputs, is the system linear and time invariant? Justify your answers.

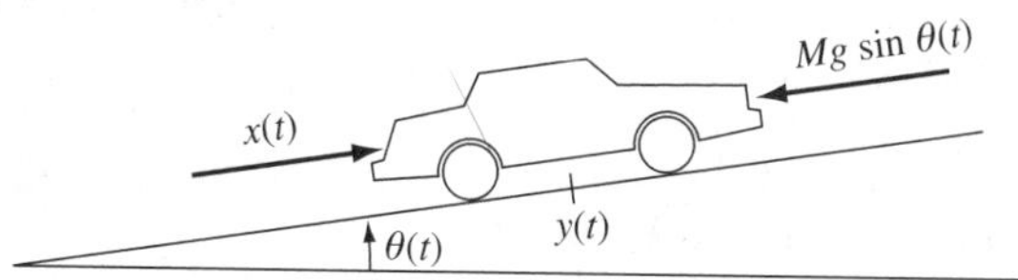

Figure P1.29

1.30. As shown in Figure P1.30, a concentrated dye is mixed together with water in a tank. The process is modeled by the differential equation

$$\frac{dy(t)}{dt} = \frac{1}{V}[Dq(t) - (q(t) + w)y(t)]$$

where $y(t)$ is the dye concentration in the mixing tank, V the volume of the mixing tank, D the dye concentration in the dye storage tank, $q(t)$ the flow rate of the dye into the mixing tank, and w the flow rate of water into the mixing tank. It is assumed that w is a constant. With $y(t)$ defined to be the output of the system and $q(t)$ the input to the system, determine if the system is causal or noncausal, linear or nonlinear, time invariant or time varying. Justify your answers.

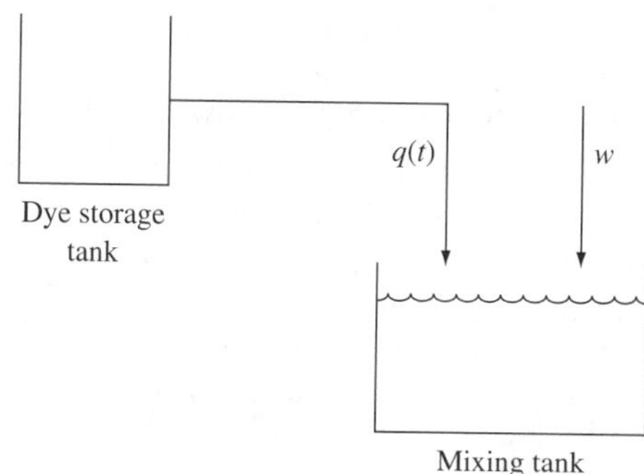

Figure P1.30

1.31. A linear time-invariant system responds to the following inputs with the corresponding outputs:

If $x(t) = u(t)$, then $y(t) = (1 - e^{-2t})u(t)$

If $x(t) = \cos(2t)$, then $y(t) = 0.707 \cos(2t - \pi/4)$

Find $y(t)$ for the following inputs:
(a) $x(t) = 2u(t) - 2u(t-1)$
(b) $x(t) = 4\cos(2(t-2))$
(c) $x(t) = 5u(t) + 10\cos(2t)$
(d) $x(t) = \delta(t)$
(e) $x(t) = tu(t)$

1.32. To understand better the concept of linearity in discrete-time systems, write a MATLAB M-file that performs the recursion to compute the response $y[n]$ for $0 \le n \le 10$ of the following system to the various inputs below. Assume that the initial condition is zero.

$$y[n] = 0.5y[n-1] + x[n]$$

(a) Compute and plot the response $y_1[n]$ to the input $x[n] = u[n]$.
(b) Compute and plot the response $y_2[n]$ to the input $x[n] = 2u[n]$. Compare this response to that obtained in part (a).
(c) Compute and plot the response $y_3[n]$ to the input $x[n] = \sin(\pi n/4)u[n]$.
(d) Compute and plot the response $y_4[n]$ to the input $x[n] = 2u[n] + \sin(\pi n/4)u[n]$. Compare this response $2y_1[n] + y_3[n]$.
(e) Repeat parts (a) to (d) with the initial condition of $y[-1] = -1$. Does linearity still hold?
(f) Repeat parts (a) to (d) with the system defined by

$$y[n] = 0.1y^2[n-1] + x[n]$$

Is this system linear? Justify your answer.

1.33. To understand better the concept of time-invariance in discrete-time systems, write a MATLAB M-file that performs recursion to compute the response $y[n]$ for $0 \le n \le 20$ of the following system to the various inputs below. Assume that the initial condition is zero.

$$y[n] = 0.5y[n-1] + x[n]$$

(a) Compute and plot the response $y_1[n]$ to the input $x[n] = u[n]$.
(b) Compute and plot the response $y_2[n]$ to the input $x[n] = u[n-2]$. Compare this response to that obtained in part (a).
(c) Compute and plot the response $y_3[n]$ to the input $x[n] = \sin(\pi n/4)u[n]$.
(d) Compute and plot the response $y_4[n]$ to the input $x[n] = \sin(\pi[n-4]/4)u[n-4]$. Compare this response to $y_3[n-4]$.
(e) Repeat parts (a) to (d) with the following system:

$$y[n] = 0.8^n y[n-1] + x[n]$$

Is this system time invariant? Justify your answer.

1.34. Determine whether the following discrete-time systems are causal or noncausal, have memory or are memoryless, are linear or nonlinear, are time invariant or time varying. Justify your answers. In the following parts, $x[n]$ is an arbitrary input, the system is initially at rest, and $y[n]$ is the response to $x[n]$.
(a) $y[n] = x[n] + x[n-1]$
(b) $y[n] = x[n] + x[n+1]$
(c) $y[n+1] = y[n]x[n]$
(d) $y[n] = u[n]x[n]$
(e) $y[n+1] + ny[n] = x[n]$
(f) $y[n] = \sum_{i=-\infty}^{n} (0.5)^{n-i}x[i]$
(g) $y[n] = \sum_{i=-\infty}^{n} (0.5)^{n}x[i]$

◆

CHAPTER 2

Systems Defined by Differential or Difference Equations

This chapter deals with the study of finite-dimensional continuous-time and discrete-time systems specified in terms of input/output differential equations or input/output difference equations. Linear input/output differential equations with constant coefficients are considered first. Then in Section 2.2 it is shown how such equations arise in the modeling of electrical circuits and mechanical systems. In Section 2.3 the presentation focuses on discrete-time systems specified by a linear constant-coefficient input/output difference equation that can be solved very easily using recursion. The recursion process is implemented using a MATLAB program that yields a software realization of the discrete-time system under consideration. Then in Section 2.4 it is shown that continuous-time systems can be studied using discrete-time representations by "discretizing in time" the input/output differential equation. Section 2.5 deals with systems defined by time-varying or nonlinear input/output equations.

2.1 LINEAR INPUT/OUTPUT DIFFERENTIAL EQUATIONS WITH CONSTANT COEFFICIENTS

Consider the single-input single-output continuous-time system given by the input/output differential equation

$$y^{(N)}(t) + \sum_{i=0}^{N-1} a_i y^{(i)}(t) = \sum_{i=0}^{M} b_i x^{(i)}(t) \tag{2.1}$$

where $y^{(i)}(t)$ is the ith derivative of the output $y(t)$ and $x^{(i)}(t)$ is the ith derivative of the input $x(t)$. Here it is assumed that the coefficients $a_0\, a_1, \ldots, a_{N-1}$ and $b_0, b_1, \ldots,$ b_M are real constants.

Recall from Chapter 1 that since (2.1) is a linear differential equation with constant coefficients, the system defined by (2.1) is linear, time invariant, and finite dimensional. The positive integer N in (2.1) is the order or dimension of the system. The class of linear time-invariant continuous-time systems that can be described by a differential equation of the form (2.1) is very large. Examples involving electrical circuits and mechanical systems are given in the next section.

Initial Conditions

To solve (2.1) for $t > 0$, it is necessary to specify the N initial conditions.

$$y(0), y^{(1)}(0), \ldots, y^{(N-1)}(0) \tag{2.2}$$

where $y^{(i)}(t)$ is the ith derivative of $y(t)$. To solve (2.1) for $t > 0$, the initial conditions could be taken at time $t = 0^-$, where 0^- is an infinitesimal negative number. In this case, the N initial conditions are

$$y(0^-), y^{(1)}(0^-), \ldots, y^{(N-1)}(0^-) \tag{2.3}$$

If the Mth derivative of the input $x(t)$ contains an impulse $k\delta(t)$ or a derivative of an impulse, to solve (2.1) for $t > 0$ it is necessary to take the initial time to be 0^-. The reason for this is that if the term

$$b_M \frac{d^M x(t)}{dt^M}$$

in (2.1) contains an impulse at $t = 0$, the values of the output $y(t)$ and its derivatives up to order $N - 1$ may change instantaneously at time $t = 0$. So initial conditions must be taken just prior to time $t = 0$.

Using solution techniques for ordinary linear differential equations, (2.1) can be solved for the output response $y(t)$ resulting from the initial conditions (2.2) or (2.3) and an input $x(t)$. In Chapter 8, it is shown that (2.1) can be solved using the Laplace transform.

First-Order Case

In the first-order case ($N = 1$), it is possible to express the solution to (2.1) in a useful general form: Suppose that the system is given by the first-order differential equation

$$\frac{dy(t)}{dt} + ay(t) = bx(t) \tag{2.4}$$

where a and b are arbitrary constants. Equation (2.4) can be solved using the integrating-factor method as illustrated in the RC circuit example in Chapter 1. The result is that the output response $y(t)$ resulting from initial condition $y(0)$ and input $x(t)$ applied for $t \geq 0$ is given by

$$y(t) = e^{-at}y(0) + \int_0^t e^{-a(t-\lambda)}bx(\lambda)\, d\lambda, \qquad t \geq 0 \tag{2.5}$$

If the initial time is taken to be 0^-, the output response is given by

$$y(t) = e^{-at}y(0^-) + \int_{0^-}^t e^{-a(t-\lambda)}bx(\lambda)\, d\lambda, \qquad t \geq 0 \tag{2.6}$$

Note that if $a = 1/RC$ and $b = 1/C$, (2.5) or (2.6) gives the output response of the RC circuit studied in Chapter 1.

The first-order case given by (2.4) can be generalized by adding a term involving the derivative of the input: Suppose that the system is given by the input/output differential equation

$$\frac{dy(t)}{dt} + ay(t) = b_1\frac{dx(t)}{dt} + b_0x(t) \tag{2.7}$$

Equation (2.7) is still a first-order differential equation, but its solution is more complicated than the solution of (2.4) due to the $dx(t)/dt$ term. It is possible to solve (2.7) without having to differentiate the input $x(t)$. This will be shown by first reducing (2.7) to an equation of the form (2.4). Let

$$q(t) = y(t) - b_1x(t) \tag{2.8}$$

so that

$$y(t) = q(t) + b_1x(t) \tag{2.9}$$

Differentiating both sides of (2.8) yields

$$\frac{dq(t)}{dt} = \frac{dy(t)}{dt} - b_1\frac{dx(t)}{dt}$$

Using the expression (2.7) for $dy(t)/dt$ gives

$$\frac{dq(t)}{dt} = -ay(t) + b_1\frac{dx(t)}{dt} + b_0x(t) - b_1\frac{dx(t)}{dt}$$

$$= -ay(t) + b_0x(t) \tag{2.10}$$

Inserting the expression (2.9) for $y(t)$ into (2.10) results in

$$\frac{dq(t)}{dt} = -a(q(t) + b_1x(t)) + b_0x(t)$$

$$= -aq(t) + (b_0 - ab_1)x(t) \tag{2.11}$$

Now (2.11) is the form of (2.4), so $q(t)$ can be solved using (2.5). This gives

$$q(t) = e^{-at}q(0) + \int_0^t e^{-a(t-\lambda)}(b_0 - ab_1)x(\lambda)\,d\lambda, \qquad t \geq 0 \tag{2.12}$$

Then, inserting the expression (2.12) for $q(t)$ into (2.9) and using the relationship $q(0) = y(0) - b_1x(0)$ yields

$$\begin{aligned} y(t) &= e^{-at}[y(0) - b_1x(0)] \\ &\quad + \int_0^t e^{-a(t-\lambda)}(b_0 - ab_1)x(\lambda)\,d\lambda + b_1x(t), \qquad t \geq 0 \end{aligned} \tag{2.13}$$

The expression (2.13) for $y(t)$ is the complete solution of (2.7) with initial condition $y(0)$ and with input $x(t)$ applied for $t \geq 0$.

If the derivative of the input $x(t)$ contains an impulse $k\delta(t)$, it is necessary to take the initial time to be 0^-, in which case the output response is

$$y(t) = e^{-at}[y(0^-) - b_1x(0^-)] + \int_{0^-}^t e^{-a(t-\lambda)}(b_0 - ab_1)x(\lambda)\,d\lambda + b_1x(t), \qquad t \geq 0$$

If $x(0^-) = 0$, the output response $y(t)$ is

$$y(t) = e^{-at}y(0^-) + \int_{0^-}^t e^{-a(t-\lambda)}(b_0 - ab_1)x(\lambda)\,d\lambda + b_1x(t), \qquad t \geq 0 \tag{2.14}$$

Note that if the input $x(t)$ is the unit-step function $u(t)$, the derivative of $x(t)$ is the unit impulse $\delta(t)$, and (2.14) must be used to compute the output response. Evaluating (2.14) with $x(t) = u(t)$ gives

$$\begin{aligned} y(t) &= e^{-at}y(0^-) + \int_{0^-}^t e^{-a(t-\lambda)}(b_0 - ab_1)\,d\lambda + b_1, \qquad t \geq 0 \\ &= e^{-at}y(0^-) + \left(\frac{b_0}{a} - b_1\right)[1 - e^{-at}] + b_1, \qquad t \geq 0 \\ &= \left[y(0^-) - \frac{b_0}{a} + b_1\right]e^{-at} + \frac{b_0}{a}, \qquad t \geq 0 \end{aligned} \tag{2.15}$$

From (2.15) it is seen that there is an instantaneous change of amount b_1 in the value of the output $y(t)$ in going from 0^- to 0^+ seconds. This is due to the $b_1[dx(t)/dt]$ term in the input/output differential equation (2.7) of the system.

Solution Of Second-Order and Higher-Order Equations

The form of the solution (2.5) in the first-order case can be generalized to the Nth-order case by working in terms of matrices. The matrix-equation formulation is constructed by defining a state model for the system. The details are carried out in Chapter 13. Also, as noted above, second-order and higher-order differential equations can be solved using the Laplace transform, which is studied in Chapter 8.

2.2 SYSTEM MODELING

In many applications the input/output differential equation of a system can be determined by applying the laws of physics. Examples given in Chapter 1 included the RC circuit, the car on a level surface, and the simple pendulum. This section deals with the process of determining the input/output differential equation for a class of electrical circuits and mechanical systems. The development begins with electrical circuits consisting of an interconnection of resistors, capacitors, and inductors.

Electrical Circuits

Schematic representations of a resistor, capacitor, and inductor are shown in Figure 2.1. With respect to the voltage $v(t)$ and the current $i(t)$ defined in Figure 2.1, the voltage–current relationship for the resistor is

$$v(t) = Ri(t) \tag{2.16}$$

for the capacitor it is

$$\frac{dv(t)}{dt} = \frac{1}{C}i(t) \quad \text{or} \quad v(t) = \frac{1}{C}\int_{-\infty}^{t} i(\lambda)\, d\lambda \tag{2.17}$$

and for the inductor it is

$$v(t) = L\frac{di(t)}{dt} \quad \text{or} \quad i(t) = \frac{1}{L}\int_{-\infty}^{t} v(\lambda)\, d\lambda \tag{2.18}$$

Now consider the process of determining the input/output differential equation of an electrical circuit consisting of an interconnection of resistors, capacitors, and inductors. The input $x(t)$ to the circuit is a voltage or current driving source and the output $y(t)$ is a voltage or current at some point in the circuit. The input/output differential equation of the circuit can be determined using Kirchhoff's voltage and current laws. The voltage law states that at any fixed time the sum of the voltages around a closed loop in the circuit must be equal to zero. The current law states that at any fixed time the sum of the currents entering a node (a junction of circuit elements) must equal the sum of the currents leaving the node.

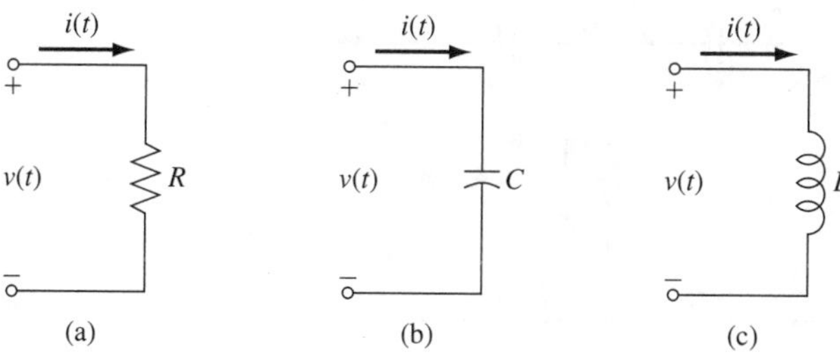

Figure 2.1 Basic circuit elements: (a) resistor; (b) capacitor; (c) inductor.

Via the voltage–current relationships (2.16)–(2.18) and Kirchhoff's voltage and current laws, node and/or loop equations can be written for the circuit, which can then be combined to yield the input/output differential equation. The process is illustrated by the following example. For an in-depth treatment on the writing of circuit equations, see Van Valkenburg [1974], Hayt and Kemmerly [1986], or Nilsson [1986].

Example 2.1 *Bridged-T Circuit*

Consider the bridged-T circuit shown in Figure 2.2. The input $x(t)$ is the voltage applied to the circuit and the output $y(t)$ is the sum of the voltages across the resistor with resistance R_1 and the capacitor with capacitance C_2. With $i_1(t)$ equal to the current in the capacitor with capacitance C_1 and $i_2(t)$ equal to the current in the other capacitor, by Kirchhoff's current law the current in the resistor with resistance R_1 is equal to $i_1(t) - i_2(t)$. Using Kirchhoff's voltage law yields the following loop (or mesh) equations:

$$\frac{1}{C_1}\int i_1(\lambda)\,d\lambda + R_1[i_1(t) - i_2(t)] = x(t) \tag{2.19}$$

$$\frac{1}{C_1}\int i_1(\lambda)\,d\lambda + \frac{1}{C_2}\int i_2(\lambda)\,d\lambda + R_2 i_2(t) = 0 \tag{2.20}$$

In addition, by definition of the output $y(t)$,

$$y(t) = x(t) + R_2 i_2(t) \tag{2.21}$$

Since the currents $i_1(t)$ and $i_2(t)$ do not appear in the input/output differential equation, it is necessary to combine (2.19), (2.20), and (2.21) into a single equation with $i_1(t)$ and $i_2(t)$ eliminated. First, to simplify the development the integrals in (2.19) and (2.20) will be eliminated by differentiating both sides of these equations. This gives

$$\frac{1}{C_1} i_1(t) + R_1\left[\frac{di_1(t)}{dt} - \frac{di_2(t)}{dt}\right] = \frac{dx(t)}{dt} \tag{2.22}$$

$$\frac{1}{C_1} i_1(t) + \frac{1}{C_2} i_2(t) + R_2\frac{di_2(t)}{dt} = 0 \tag{2.23}$$

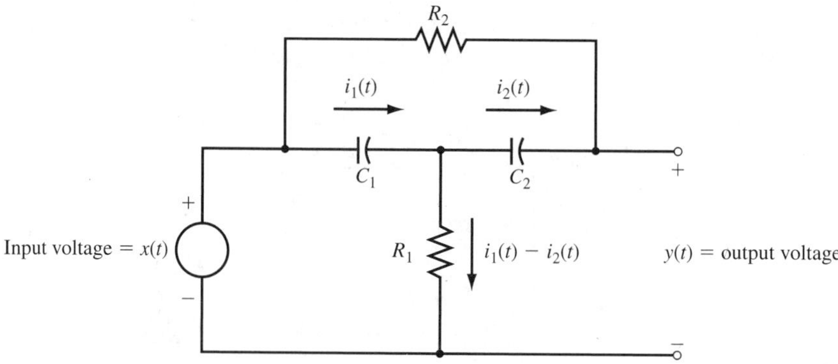

Figure 2.2 Bridged-T circuit.

Rewriting (2.23) results in

$$i_1(t) = -\frac{C_1}{C_2} i_2(t) - R_2 C_1 \frac{di_2(t)}{dt} \tag{2.24}$$

Inserting (2.24) into (2.22) gives

$$-\frac{1}{C_2} i_2(t) - R_2 \frac{di_2(t)}{dt} + R_1 \left[-\frac{C_1}{C_2} \frac{di_2(t)}{dt} - R_2 C_1 \frac{d^2 i_2(t)}{dt^2} - \frac{di_2(t)}{dt} \right] = \frac{dx(t)}{dt}$$

$$-\frac{1}{C_2} i_2(t) - \left(R_2 + \frac{R_1 C_1}{C_2} + R_1 \right) \frac{di_2(t)}{dt} - R_1 R_2 C_1 \frac{d^2 i_2(t)}{dt^2} = \frac{dx(t)}{dt} \tag{2.25}$$

Now by (2.21)

$$i_2(t) = \frac{1}{R_2} [y(t) - x(t)] \tag{2.26}$$

Inserting (2.26) into (2.25) yields

$$-\frac{1}{R_2 C_2} [y(t) - x(t)] - \frac{1}{R_2} \left(R_2 + \frac{R_1 C_1}{C_2} + R_1 \right) \left[\frac{dy(t)}{dt} - \frac{dx(t)}{dt} \right]$$

$$- R_1 C_1 \left[\frac{d^2 y(t)}{dt^2} - \frac{d^2 x(t)}{dt^2} \right] = \frac{dx(t)}{dt}$$

Finally, rearranging terms results in the input/output differential equation

$$R_1 C_1 \frac{d^2 y(t)}{dt^2} + \left(1 + \frac{R_1 C_1}{R_2 C_2} + \frac{R_1}{R_2} \right) \frac{dy(t)}{dt} + \frac{1}{R_2 C_2} y(t)$$

$$= R_1 C_1 \frac{d^2 x(t)}{dt^2} + \left(\frac{R_1 C_1}{R_2 C_2} + \frac{R_1}{R_2} \right) \frac{dx(t)}{dt} + \frac{1}{R_2 C_2} x(t)$$

Note that the order of this input/output differential equation is equal to two.

Mechanical Systems

Motion in a mechanical system can always be resolved into translational and rotational components. Translational motion is considered first.

In linear translational systems there are three fundamental types of forces that resist motion. They are inertia force of a moving body, damping force due to viscous friction, and spring force. By Newton's second law of motion, the inertia force $x(t)$ of a moving body is equal to its mass M times its acceleration, that is,

$$x(t) = M \frac{d^2 y(t)}{dt^2} \tag{2.27}$$

where $y(t)$ is the position of the body at time t.

The damping force $x(t)$ due to viscous friction is proportional to velocity, so that

$$x(t) = k_d \frac{dy(t)}{dt} \tag{2.28}$$

where k_d is the damping constant. Viscous friction is often represented by a dashpot consisting of an oil-filled cylinder and piston. A schematic representation of the dashpot is shown in Figure 2.3.

The restoring force $x(t)$ of a spring is proportional to the amount $y(t)$ it is stretched, that is,

$$x(t) = k_s y(t) \tag{2.29}$$

where k_s is a constant representing the stiffness of the spring. The schematic representation of a spring is shown in Figure 2.4.

The input/output differential equation of a translational mechanical system can be determined by applying D'Alembert's principle, which is a slight variation of Newton's second law of motion. By D'Alembert's principle, at any fixed time the sum of all the external forces applied to a body in a given direction and all the forces resisting motion in that direction must be equal to zero. D'Alembert's principle is the mechanical analog of Kirchhoff's laws in circuit analysis. The application of D'Alembert's principle is illustrated in the following example, which is taken from McClamroch [1980].

Example 2.2 *Automobile Suspension System*

Consider the automobile suspension system shown schematically in Figure 2.5. The mass M_1 is equal to the effective mass of the wheel and the mass M_2 is equal to one-fourth of the mass of the automobile frame. The constant k_s is the stiffness of the automobile suspension spring, k_t is the stiffness of the tire, and k_d is the damping constant of the shock absorber. The input $x(t)$ to the system is the ground level and the output $y(t)$ is the vertical position of the automobile frame relative to some equilibrium position. The vertical position of the wheel relative to the equilibrium position is $q(t)$. It is assumed that the bottom of the tire remains in contact with the road surface at all times. The gravitational forces acting on the frame and wheel are neglected. Using (2.27)–(2.29) and D'Alembert's principle applied to each mass in the system yields

$$M_1 \frac{d^2q(t)}{dt^2} + k_t[q(t) - x(t)] = k_s[y(t) - q(t)] + k_d\left[\frac{dy(t)}{dt} - \frac{dq(t)}{dt}\right] \tag{2.30}$$

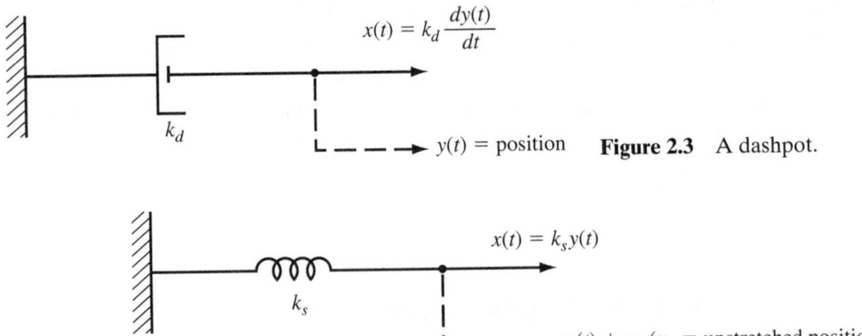

Figure 2.3 A dashpot.

Figure 2.4 A spring.

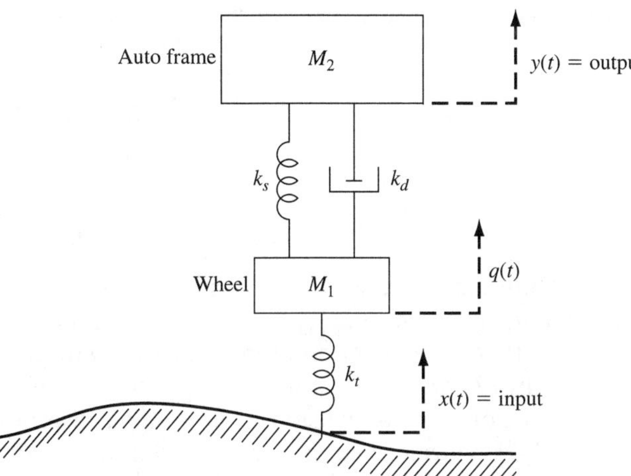

Figure 2.5 Automobile suspension system.

$$M_2\frac{d^2y(t)}{dt^2} + k_s[y(t) - q(t)] + k_d\left[\frac{dy(t)}{dt} - \frac{dq(t)}{dt}\right] = 0 \tag{2.31}$$

The input/output differential equation can be determined by combining (2.30) and (2.31) to eliminate $q(t)$. This will result in a fourth-order input/output differential equation. The derivation of the input/output equation is somewhat complicated. This is left to those readers who enjoy a challenge (see Section 8.5).

Rotational Mechanical Systems. In analogy with the three types of forces resisting translational motion, there are three types of forces resisting rotational motion. They are inertia torque given by

$$x(t) = I\frac{d^2\theta(t)}{dt^2} \tag{2.32}$$

damping torque given by

$$x(t) = k_d\frac{d\theta(t)}{dt} \tag{2.33}$$

and spring torque given by

$$x(t) = k_s\theta(t) \tag{2.34}$$

In (2.32)–(2.34), $\theta(t)$ is the angular position at time t, I is the moment of inertia, and k_d and k_s are the rotational damping and stiffness constants, respectively.

For rotational systems, D'Alembert's principle states that at any fixed time the sum of all external torques applied to a body about any axis and all torques resisting motion about that axis must be equal to zero. The input/output differential equation of a rotational system can be derived using (2.32)–(2.34) and D'Alembert's principle. The process is very similar to the steps carried out for the translational system in Example 2.2.

2.3 LINEAR INPUT/OUTPUT DIFFERENCE EQUATIONS WITH CONSTANT COEFFICIENTS

Now consider the single-input single-output discrete-time system defined by the input/output difference equation

$$y[n] + \sum_{i=1}^{N} a_i y[n - i] = \sum_{i=0}^{M} b_i x[n - i] \tag{2.35}$$

where n is the integer-valued discrete-time index, $x[n]$ is the input, and $y[n]$ is the output. Here it is assumed that the coefficients $a_1, a_2, \ldots, a_N$ and $b_0, b_1, \ldots, b_M$ are constants.

Recall from Chapter 1 that since (2.35) is a linear difference equation with constant coefficients, the system defined by (2.35) is linear, time invariant, and finite dimensional. The integer N in (2.35) is the order or dimension of the system. Also, any discrete-time system of the form (2.35) is causal since the output $y[n]$ at time n depends only on previous values of the output and the current and previous values of the input $x[n]$.

Solution by Recursion

Unlike linear input/output differential equations, linear input/output difference equations can be solved by a direct numerical procedure. More precisely, the output $y[n]$ for some finite range of integer values of n can be computed recursively as follows. First, rewrite (2.35) in the form

$$y[n] = -\sum_{i=1}^{N} a_i y[n - i] + \sum_{i=0}^{M} b_i x[n - i] \tag{2.36}$$

Then setting $n = 0$ in (2.36) gives

$$y[0] = -a_1 y[-1] - a_2 y[-2] - \cdots - a_N y[-N] + b_0 x[0] + b_1 x[-1] + \cdots + b_M x[-M]$$

Thus the output $y[0]$ at time 0 is a linear combination of $y[-1], y[-2], \ldots, y[-N]$ and $x[0], x[-1], \ldots, x[-M]$.

Setting $n = 1$ in (2.36) gives

$$y[1] = -a_1 y[0] - a_2 y[-1] - \cdots - a_N y[-N + 1] + b_0 x[1] + b_1 x[0] + \cdots + b_M x[-M + 1]$$

So $y[1]$ is a linear combination of $y[0], y[1], \ldots, y[-N + 1]$ and $x[1]x[0], \ldots, x[-M + 1]$.

If this process is continued, it is clear that the next value of the output is a linear combination of the N past values of the output and $M + 1$ values of the input. At each step of the computation, it is necessary to store only the N past values of the output (plus, of course, the input values). This process is called an

*N*th-*order recursion.* Here the term *recursion* refers to the property that the next value of the output is computed from the N previous values of the output (plus the input values). The discrete-time system defined by (2.35) [or (2.36)] is sometimes called a *recursive discrete-time system* or a *recursive digital filter* since its output can be computed recursively. Here it is assumed that at least one of the coefficients a_i in (2.35) is nonzero. If all the a_i are zero, the input/output difference equation (2.35) reduces to

$$y[n] = \sum_{i=0}^{M} b_i x[n - i]$$

In this case the output at any fixed time point depends only on values of the input $x[n]$, and thus the output is not computed recursively. Such systems are said to be *nonrecursive.*

It is important to note that the recursive structure described above is a result of finite dimensionality. If the system is infinite dimensional, the output response $y[n]$ cannot be computed recursively; that is, the next value of $y[n]$ cannot be computed from a finite linear combination of the previous values $y[n - \text{i}]$ (where $i = 1, 2, \ldots, N$ for some N). For example, as noted in Example 1.17, the discrete-time system with the input/output relationship

$$y[n] = \sum_{i=0}^{n-1} \frac{1}{n - i} x[i], \qquad n \geq 1$$

is infinite dimensional. Thus in this case $y[n]$ cannot be expressed in the form (2.35) for any N, and therefore $y[n]$ cannot be computed recursively.

From the expression (2.36) for $y[n]$, it follows that if $M = N$ the computation of $y[n]$ for each integer value of n requires (in general) $2N$ additions and $2N + 1$ multiplications. So the "computational complexity" of the Nth-order recursion is directly proportional to the order N of the recursion. In particular, note that the number of computations required to compute $y[n]$ does not depend on n.

Finally, from (2.35) or (2.36) it is clear that the computation of the output response $y[n]$ for $n \geq 0$ requires that the N initial conditions $y[-N], y[-N + 1], \ldots, y[-1]$ must be specified. In addition, if the input $x[n]$ is not zero for $n < 0$, the evaluation of (2.35) or (2.36) also requires the M initial input values $x[-M], x[-M + 1], \ldots, x[-1]$.

Example 2.3 *Second-Order System*

Consider the discrete-time system given by the second-order input/output difference equation

$$y[n] - 1.5y[n - 1] + y[n - 2] = 2x[n - 2] \tag{2.37}$$

Writing (2.37) in the form (2.36) results in the input/output equation

$$y[n] = 1.5y[n - 1] - y[n - 2] + 2x[n - 2] \tag{2.38}$$

Now suppose that the input $x[n]$ is the discrete-time unit-step function $u[n]$ and that the initial output values are $y[-2] = 2$ and $y[-1] = 1$. Then setting $n = 0$ in (2.38) gives

$$y[0] = 1.5y[-1] - y[-2] + 2x[-2]$$

$$y[0] = (1.5)(1) - 2 + (2)(0) = -0.5$$

Setting $n = 1$ in (2.38) gives

$$y[1] = 1.5y[0] - y[-1] + 2x[-1]$$

$$y[1] = (1.5)(-0.5) - 1 + 2(0) = -1.75$$

Continuing the process yields

$$\begin{aligned} y[2] &= (1.5)y[1] - y[0] + 2x[0] \\ &= (1.5)(-1.75) - 0.5 + (2)(1) = -0.125 \\ y[3] &= (1.5)y[2] - y[1] + 2x[1] \\ &= (1.5)(-0.125) + 1.75 + (2)(1) = 3.5625 \end{aligned}$$

and so on.

In solving (2.35) or (2.36) recursively, the process of computing the output $y[n]$ can begin at any time point desired. In the development above, the first value of the output that was computed was $y[0]$. If the first desired value is the output $y[q]$ at time q, the recursion process should be started by setting $n = q$ in (2.36). In this case, the initial values of the output that are required are $y[q - N]$, $y[q - N + 1], \ldots, y[q - 1]$.

The Nth-order difference equation (2.35) can be solved by using the MATLAB M-file `recur` that is available from the Web site. An abbreviated version of the program that contains all of the important steps is given in Figure 2.6. To use the `recur` M-file, the user must input the system coefficients (i.e., the a_i and b_i), the initial values of $y[n]$ and $x[n]$, the desired range of n, and the input $x[n]$. The program first initializes the solution vector `y` by augmenting the initial conditions with zeros. This predefines the length of `y` that makes the loop more efficient. The input vector `x` is also augmented by the initial conditions of the input. The summations given in (2.35) are computed by multiplying vectors; for example, the summation on the left-hand side of (2.35) can be written as

$$\sum_{i=1}^{N} a_i y[n - i] = [a_N \quad a_{N-1} \quad \cdots \quad a_1] \begin{bmatrix} y[n - N] \\ y[n - N + 1] \\ \vdots \\ y[n - 1] \end{bmatrix}$$

Each summation could also be evaluated by using an inner loop in the program; however, MATLAB does not process loops very efficiently, so loops are avoided whenever possible. In the last line of the program, the initial conditions are stripped off the vector y with the result being a vector y that contains the values of $y[n]$ for the time

```
function y = recur(a,b,n,x,x0,y0);
N = length(a); y = [y0 zeros(1,length(n))];
M = length(b)-1; x = [x0 x];
a1 = a(length(a):-1:1);          % reverses the elements in a
b1 = b(length(b):-1:1);          % reverses the elements in b
for i=N+1:N+length(n),
  y(i) = -a1*y(i-N:i-1)' + b1*x(i-N:i-N+M)';
end
y = y(N+1:N+length(n));
```

Figure 2.6 MATLAB program `recur`.

indices defined in the vector n. It should be noted that there are other programs in the MATLAB toolboxes that solve the equation in (2.35); however, these programs utilize concepts that have not yet been introduced.

The following commands demonstrate how `recur` is used to compute the output response for the system in Example 2.3.

```
a=[-1.5 1]; b=[0 0 2];
y0 = [2 1]; x0 = [0 0];
n=0:20;
x = ones(1,length(n));
y = recur (a,b,n,x,x0,y0);
stem(n,y)          % produces a "stem plot"
xlabel('n')
ylabel('y[n]')
```

The M-file computes the response y for $n = 0, 1, \ldots, 20$ and then plots y versus n with labels on the axes. The resulting output response is given in Figure 2.7. In future examples, the plotting commands will not be shown except when the variables to be plotted are not obvious.

Complete Solution

By solving (2.35) or (2.36) recursively, it is possible to generate an expression for the complete solution $y[n]$ resulting from initial conditions and the application of the input $x[n]$. The process is illustrated by considering the first-order linear difference equation

$$y[n] = -ay[n-1] + bx[n], \qquad n = 1, 2, \ldots \tag{2.39}$$

with the initial condition $y[0]$. First, setting $n = 1$, $n = 2$, and $n = 3$ in (2.39) gives

$$y[1] = -ay[0] + bx[1] \tag{2.40}$$

$$y[2] = -ay[1] + bx[2] \tag{2.41}$$

$$y[3] = -ay[2] + bx[3] \tag{2.42}$$

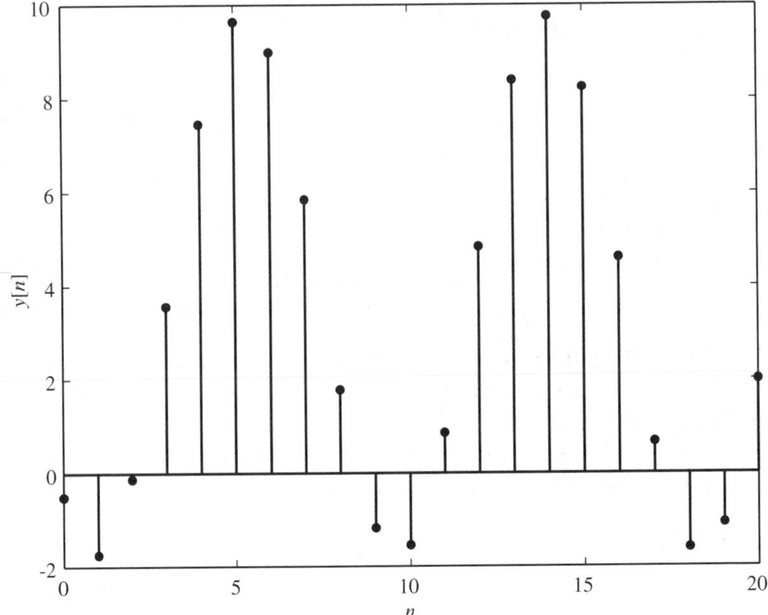

Figure 2.7 Plot of output response in Example 2.3.

Inserting the expression (2.40) for $y[1]$ into (2.41) gives

$$y[2] = -a(-ay[0] + bx[1]) + bx[2]$$

$$y[2] = a^2y[0] - abx[1] + bx[2] \tag{2.43}$$

Inserting the expression (2.43) for $y[2]$ into (2.42) yields

$$y[3] = -a(a^2y[0] - abx[1] + bx[2]) + bx[3]$$

$$y[3] = -a^3y[0] + a^2bx[1] - abx[2] + bx[3] \tag{2.44}$$

From the pattern in (2.40), (2.43), and (2.44), it can be seen that for $n \geq 1$,

$$y[n] = (-a)^n y[0] + \sum_{i=1}^{n} (-a)^{n-i} bx[i] \tag{2.45}$$

Equation (2.45) gives the complete output response $y[n]$ for $n \geq 1$ resulting from initial condition $y[0]$ and the input $x[n]$ applied for $n \geq 1$. This solution is the discrete-time counterpart to the solution (2.5) of the general first-order input/output differential equation (2.4).

Example 2.4 *Inventory Level*

Consider a manufacturer that produces a specific product. Let $y[n]$ denote the number of the product in inventory at the end of the nth day, let $p[n]$ denote the number of the product whose manufacturing is completed during the nth day, and let $d[n]$ denote the number of the product that is delivered (to customers) during the nth day. Then the number $y[n]$ of the product in the inventory at the end of the nth day must be equal to $y[n-1]$ plus the difference between $p[n]$ and $d[n]$. In mathematical terms,

$$y[n] = y[n-1] + p[n] - d[n] \qquad \text{for } n = 1, 2, \ldots$$

Here $y[0]$ is the initial number of the product in inventory.

Now with $x[n]$ defined to be the difference $p[n] - d[n]$, this equation is in the form (2.39) with $a = -1$ and $b = 1$. Hence, from (2.45), the solution is

$$y[n] = y[0] + \sum_{i=1}^{n} x[i]$$

$$y[n] = y[0] + \sum_{i=1}^{n} (p[i] - d[i])$$

One of the objectives in manufacturing is to keep the level of the inventory fairly constant; in particular, depletion of the inventory should obviously be avoided; otherwise, there will be a delay in delivery of the product to customers. From the expression for $y[n]$ above, it is seen that $y[n]$ can be kept constant by setting $p[n] = d[n]$. In other words, the number of the product whose manufacturing is completed during the nth day should be equal to the number of the product delivered during the nth day. However, it is not possible to set $p[n] = d[n]$ since a product cannot be manufactured "instantaneously" and $d[n]$ depends on customer orders and is not known in advance. If the manufacture of the product requires less than 1 day, it is possible to set

$$p[n] = d[n-1]$$

that is, the number $p[n]$ of the product whose manufacturing is to be completed during the nth day is set equal to the number $d[n-1]$ of deliveries during the preceding day. Inserting $p[n] = d[n-1]$ into the expression above for $y[n]$ yields

$$y[n] = y[0] + \sum_{i=1}^{n} (d[i-1] - d[i])$$

From this it is clear that the inventory will never be depleted if the initial inventory is sufficiently large to handle the variations in the number of deliveries from day to day. More precisely, depletion of the inventory will not occur if

$$y[0] > -\sum_{i=1}^{n} (d[i-1] - d[i])$$

This is an interesting result, for it tells the manufacturer how much of the product should be kept in stock to avoid delays in delivery due to inventory depletion.

Expressions for the solution of second- and higher-order linear constant-coefficient difference equations can be derived in terms of matrix equations. As in the continuous-time case, the matrix formulation is constructed by defining a state model for the given system. The reader is referred to Chapter 13. In Chapter 11 it is shown that solutions to Nth-order difference equations can be computed by applying the z-transform.

Example 2.5 *Raising Hamsters*

Consider the problem of raising hamsters starting from a single adult male/female pair. It is assumed that two male/female pairs are born every three months to each pair of adults. It is also assumed that male/female pairs remain permanently paired and that the hamsters do not die during the period under consideration here. A newborn male and female hamster can begin to produce offspring in 6 months. Now let $y[n]$ denote the number of male/female pairs at the end of the nth 3-month period. It is assumed that $y[0] = 1$, so that $y[1] = 3$. For $n \geq 1$, $y[n] - y[n-1]$ is the number of pairs born at the end of the nth period. This must be equal to $2y[n-2]$ (rather than $2y[n-1]$) since newborn hamsters cannot have offspring for 6 months. Therefore, the process is described by the difference equation

$$y[n] - y[n-1] = 2y[n-2]$$

Note that this process does not have an input. Solving the difference equation with $y[0] = 1$ and $y[1] = 3$ results in the output values $y[2] = 5$, $y[3] = 11$, $y[4] = 21$, $y[5] = 43, \ldots$. The output response $y[n]$ can also be expressed in the analytical form

$$y[n] = \tfrac{1}{3}[4(2^n) - (-1)^n], \qquad n \geq 2$$

Solutions of this form can be generated by using the z-transform techniques discussed in Chapter 11. To verify that the solution is correct, insert it into the difference equation to see that the difference equation is satisfied. The reader is invited to check this. The solution can also be verified numerically by comparing the analytical solution to the solution obtained from the MATLAB M-file `recur`. Since there is no input, the b_i coefficients in the difference equation (2.35) are all zero. The corresponding commands are

```
a = [-1 -2]; b = [0 0 0];
n = 2:10;
x = zeros(1,length(n));              % creates a zero vector
x0 = [0 0]; y0 = [1 3];
y1 = recur(a,b,n,x,x0,y0);          % solution via recursion
y2 = (4*(2).^n - (-1).^n)/3;        % analytical solution
% n is a vector, so the operation .^ is used to compute
%   the exponent separately for each element
```

The reader is invited to plot the results to see what happens to the hamster population in 30 months. It is important to note that the rapid growth in the hamster population (as seen in the plot) is a result of the assumptions in the modeling process, that is, that no hamsters die during the process and that the food supply is large enough to sustain the same reproduction rate.

2.4 DISCRETIZATION IN TIME OF DIFFERENTIAL EQUATIONS

As an application of the difference equation framework, in this section it is shown that a linear constant-coefficient input/output differential equation can be discretized in time, resulting in a difference equation that can then be solved by recursion. This discretization in time actually yields a discrete-time representation of the continuous-time system defined by the given input/output differential equation. The development begins with the first-order case.

First-Order Case

Consider the linear time-invariant continuous-time system with the first-order input/output differential equation

$$\frac{dy(t)}{dt} = -ay(t) + bx(t) \tag{2.46}$$

where a and b are constants. Equation (2.46) can be discretized in time by setting $t = nT$, where T is a fixed positive number and n takes on integer values only. This results in the equation

$$\left.\frac{dy(t)}{dt}\right|_{t=nT} = -ay(nT) + bx(nT) \tag{2.47}$$

Now the derivative in (2.47) can be approximated by

$$\left.\frac{dy(t)}{dt}\right|_{t=nT} = \frac{y(nT + T) - y(nT)}{T} \tag{2.48}$$

If T is suitably small and $y(t)$ is continuous, the approximation (2.48) to the derivative $dy(t)/dt$ will be accurate. This approximation is called the *Euler approximation* of the derivative.

Inserting the approximation (2.48) into (2.47) gives

$$\frac{y(nT + T) - y(nT)}{T} = -ay(nT) + bx(nT) \tag{2.49}$$

To be consistent with the notation that is being used for discrete-time signals, the input signal $x(nT)$ and the output signal $y(nT)$ will be denoted by $x[n]$ and $y[n]$, respectively; that is,

$$x[n] = x(t)|_{t=nT} \quad \text{and} \quad y[n] = y(t)|_{t=nT}$$

In terms of this notation, (2.49) becomes

$$\frac{y[n + 1] - y[n]}{T} = -ay[n] + bx[n] \tag{2.50}$$

Finally, multiplying both sides of (2.50) by T and replacing n by $n - 1$ results in a discrete-time approximation to (2.46) given by the first-order input/output difference equation

$$y[n] - y[n - 1] = -aTy[n - 1] + bTx[n - 1]$$

or

$$y[n] = (1 - aT)y[n - 1] + bTx[n - 1] \tag{2.51}$$

The difference equation (2.51) is called the *Euler approximation* of the given input/output differential equation (2.46) since it is based on the Euler approximation of the derivative.

The discrete values $y[n] = y(nT)$ of the solution $y(t)$ to (2.46) can be computed by solving the difference equation (2.51). The solution of (2.51) with initial condition $y[0]$ and with $x[n] = 0$ for all n is given by

$$y[n] = (1 - aT)^n y[0], \qquad n = 0, 1, 2, \ldots \tag{2.52}$$

To verify that (2.52) is the solution, insert the expression (2.52) for $y[n]$ into (2.51) with $x[n] = 0$. This gives

$$\begin{aligned}(1 - aT)^n y[0] &= (1 - aT)(1 - aT)^{n-1} y[0] \\ &= (1 - aT)^n y[0]\end{aligned}$$

Hence (2.51) is satisfied, which shows that (2.52) is the solution.

The expression (2.52) for $y[n]$ gives approximate values of the solution $y(t)$ to (2.46) at the times $t = nT$ with arbitrary initial condition $y[0]$ and with zero input. To compare (2.52) with the exact values of $y(t)$ for $t = nT$, first recall that from the results in Section 2.1, the exact solution $y(t)$ to (2.46) with initial condition $y(0)$ and with zero input is given by

$$y(t) = e^{-at} y(0), \qquad t \geq 0 \tag{2.53}$$

Setting $t = nT$ in (2.53) gives the following exact expression for $y[n]$:

$$y[n] = e^{-anT} y[0], \qquad n = 0, 1, 2, \ldots \tag{2.54}$$

Now since

$$e^{ab} = (e^a)^b \qquad \text{for any real numbers } a \text{ and } b$$

(2.54) can be written in the form

$$y[n] = (e^{-aT})^n y[0], \qquad n = 0, 1, 2, \ldots \tag{2.55}$$

Further, inserting the expansion

$$e^{-aT} = 1 - aT + \frac{a^2T^2}{2} - \frac{a^3T^3}{6} + \cdots$$

for the exponential into (2.55) results in the following exact expression for the values of $y(t)$ at the times $t = nT$:

$$y[n] = \left(1 - aT + \frac{a^2T^2}{2} - \frac{a^3T^3}{6} + \cdots\right)^n y[0], \qquad n = 0, 1, 2, \ldots \tag{2.56}$$

Comparing (2.52) and (2.56) shows that (2.52) is an accurate approximation if $1 - aT$ is a good approximation to the exponential $\exp(-aT)$. This will be the case if the magnitude of aT is much less than 1, in which case the magnitudes of the powers of aT will be much smaller than the quantity $1 - aT$.

Example 2.6 *RC Circuit*

Consider the RC circuit given in Figure 2.8. As shown in Section 1.4 [see (1.13)], the circuit has the input/output differential equation

$$C\frac{dy(t)}{dt} + \frac{1}{R}y(t) = x(t) \tag{2.57}$$

where $x(t)$ is the current applied to the circuit and $y(t)$ is the voltage across the capacitor. Writing (2.57) in the form (2.46) reveals that in this case, $a = 1/RC$ and $b = 1/C$. Hence the discrete-time representation (2.51) for the RC circuit is given by

$$y[n] = \left(1 - \frac{T}{RC}\right)y[n-1] + \frac{T}{C}x[n-1] \tag{2.58}$$

The difference equation (2.58) can be solved recursively to yield approximate values $y[n]$ of the voltage on the capacitor resulting from initial voltage $y[0]$ and input current $x(t)$ applied for $t \geq 0$. The recursion can be carried out using the MATLAB M-file `recur`, where the coefficients are identified by comparing (2.58) to (2.36). This yields $a_1 = -(1 - T/RC)$, $b_0 = 0$, and $b_1 = T/C$. The commands for the case when $R = C = 1$, $y[0] = 0, x(t) = 1, t \geq 0$, and $T = 0.2$ are found to be

```
R = 1; C = 1; T = .2;
a =-(1-T/R/C); b = [0 T/C];
y0 = 0; x0 = 1;
n = 1:40;
x = ones(1,length(n));
y1 = recur(a,b,n,x,x0,y0);                % approximate solution
```

The computation of the exact solution was carried out in Section 1.4 [see (1.23)] and is given by

$$y(t) = 1 - e^{-t}, \qquad t \geq 0$$

The MATLAB commands used to compute the exact y for $t = 0$ to $t = 8$ and the commands used to plot both solutions are

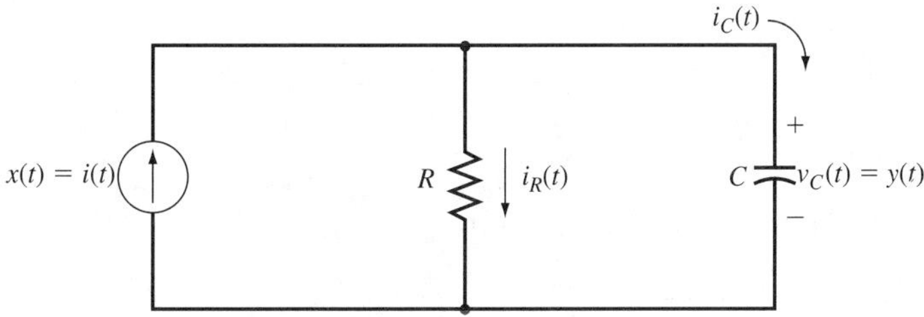

Figure 2.8 *RC* circuit in Example 2.6.

```
t = 0:0.04:8;
y2 = 1 - exp(-t);            % exact solution
% augment the initial condition onto the vector
y1 = [y0 y1];
n = 0:40;                           % redefines n accordingly
plot(n*T,y1,'o',t,y2,'-');
```

A plot of the resulting output (the step response) for the approximation is displayed in Figure 2.9 along with the exact step response. Since `y1` is an approximation to a continuous-time signal, the plot is not displayed using the stem form (which is used for discrete-time signals).

Note that since the value of $aT = T/RC = 0.2$ is small compared to 1, the approximate step response is close to that of the exact response. A better approximation can be obtained by taking a smaller value for T and then using the MATLAB program above. The reader is invited to try this.

Second-Order Case

The discretization technique for first-order differential equations described above can be generalized to second- and higher-order differential equations. In the second-order case the following approximations can be used:

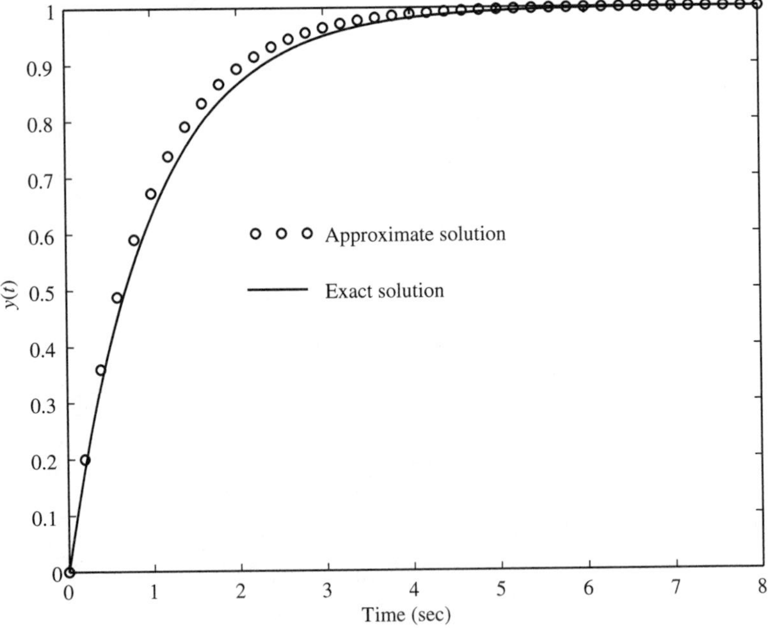

Figure 2.9 Exact and approximate step response in Example 2.6.

$$\left.\frac{dy(t)}{dt}\right|_{t=nT} = \frac{y(nT + T) - y(nT)}{T} \tag{2.59}$$

$$\left.\frac{d^2y(t)}{dt^2}\right|_{t=nT} = \frac{dy(t)/dt|_{t=nT+T} - dy(t)/dt|_{t=nT}}{T} \tag{2.60}$$

Combining (2.59) and (2.60) yields the following approximation to the second derivative:

$$\left.\frac{d^2y(t)}{dt^2}\right|_{t=nT} = \frac{y(nT + 2T) - 2y(nT + T) + y(nT)}{T^2} \tag{2.61}$$

The approximation (2.61) is the Euler approximation to the second derivative.

Now consider a linear time-invariant continuous-time system with the second-order input/output differential equation

$$\frac{d^2y(t)}{dt^2} + a_1\frac{dy(t)}{dt} + a_0y(t) = b_1\frac{dx(t)}{dt} + b_0x(t) \tag{2.62}$$

Setting $t = nT$ in (2.62) and using the approximations (2.59) and (2.61) results in the following time discretization of (2.62):

$$\frac{y[n + 2] - 2y[n + 1] + y[n]}{T^2} + a_1\frac{y[n + 1] - y[n]}{T} + a_0y[n] = b_1\frac{x[n + 1] - x[n]}{T} + b_0x[n] \tag{2.63}$$

Replacing n by $n - 2$ in (2.63) and multiplying both sides of (2.63) by T^2 yields the difference equation

$$y[n] + (a_1T - 2)y[n - 1] + (1 - a_1T + a_0T^2)y[n - 2] = b_1Tx[n - 1] + (b_0T^2 - b_1T)x[n - 2] \tag{2.64}$$

Equation (2.64) is the discrete-time approximation to the second-order input/output differential equation (2.62).

The discrete values $y(nT)$ of the solution $y(t)$ to (2.62) can be computed by solving the difference equation (2.64). To solve (2.64), the recursion will be started at $n = 2$ so that the initial values $y[0] = y(0)$ and $y[1] = y(T)$ are required. The initial value $y(T)$ can be generated by using the approximation

$$\dot{y}(0) = \frac{y(T) - y(0)}{T} \tag{2.65}$$

where $\dot{y}(t)$ denotes the derivative of $y(t)$. Solving (2.65) for $y(T)$ gives

$$y[1] = y(T) = y(0) + T\dot{y}(0) \tag{2.66}$$

With the initial values $y[0]$ and $y[1]$, the second-order difference equation (2.64) can be solved using the MATLAB M-file `recur`. The process is illustrated by the following example.

Example 2.7 *Series RLC Circuit*

Consider the series *RLC* circuit shown in Figure 2.10. As indicated, the input $x(t)$ is the voltage applied to the circuit and the output $y(t)$ is the voltage $v_C(t)$ across the capacitor. Using Kirchhoff's voltage law (see Section 2.2) results in the following input/output differential equation for the circuit:

$$LC\frac{d^2v_C(t)}{dt^2} + RC\frac{dv_C(t)}{dt} + v_C(t) = x(t) \tag{2.67}$$

Equation (2.67) is a second-order differential equation that can be written in the form (2.62) with

$$a_1 = \frac{RC}{LC} = \frac{R}{L}, \qquad a_0 = \frac{1}{LC}, \qquad b_1 = 0, \qquad b_0 = \frac{1}{LC} \tag{2.68}$$

Inserting (2.68) into the discretized equation (2.64) yields

$$v_C[n] + \left(\frac{RT}{L} - 2\right)v_C[n-1] + \left(1 - \frac{RT}{L} + \frac{T^2}{LC}\right)v_C[n-2] = \frac{T^2}{LC}x[n-2] \tag{2.69}$$

Equation (2.69) is the difference equation approximation of the *RLC* circuit. The voltage $v_C(t)$ across the capacitor will be computed using the discretization (2.69) in the case when $R = 2$, $L = C = 1$, $v_C(0) = 1$, $\dot{v}_C(0) = -1$, and $x(t) = (\sin t)u(t)$. To solve the difference equation (2.69) for $n \geq 2$, the initial conditions are $x[0] = \sin 0 = 0$, $x[1] = \sin T$, $v_C[0] = 1$, and from (2.66),

$$v_C[1] = v_C(T) = v_C(0) + T\dot{v}_C(0) = 1 - T$$

Now the difference equation (2.69) can be solved for $n \geq 2$ using the MATLAB M-file `recur`. To distinguish between the differential equation coefficients and the difference equation coefficients, let c_i's replace a_i's and d_i's replace b_i's in the general form of a difference equation given in (2.35). The corresponding coefficients are found from (2.69) as $c_1 = (RT/L - 2)$, $c_2 = (1 - RT/L + T^2/LC)$, $d_0 = d_1 = 0$, $d_2 = T^2/LC$. For the values given above, and with $T = 0.2$, the MATLAB commands are

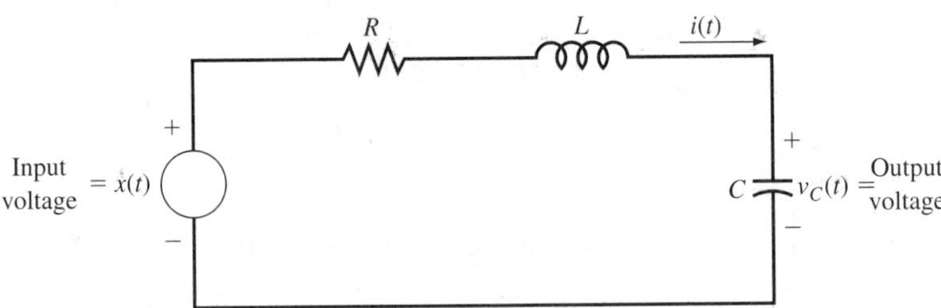

Figure 2.10 Series RLC circuit.

```
T = 0.2; R = 2;L=1;C=1;
vc0 = [1 1-T]; x0 = [0 sin(T)];
n = 2:40;
c = [R*T/L-2 1-R*T/L+T^2/L/C];
d = [0 0 T^2/L/C];
x = sin(T*n);
vc1 = recur(c,d,n,x,x0,vc0);
% augment the initial conditions onto vc1 for plotting
vc1 = [vc0 vc1];
```

The output from the MATLAB program is plotted in Figure 2.11.

The exact solution to the input/output differential equation (2.67) can be computed using solution techniques for differential equations (or the Laplace transform method given in Chapter 8). The result is that the exact solution is

$$v_C(t) = 0.5(3 + t)e^{-t} - \cos t, \qquad t \geq 0$$

The exact solution $v_C(t)$ can be evaluated using the MATLAB commands

```
t = 0:.04:8;
vc2 = 0.5*((3+t).*exp(-t)-cos(t));
```

The resulting response values are also plotted in Figure 2.11. From the plots it is seen that there is a significant error in the approximation. To obtain a better approximation, the discretization interval T can be decreased. In fact, as $T \to 0$, the approximation should approach the true response values. To check this, the MATLAB software was used

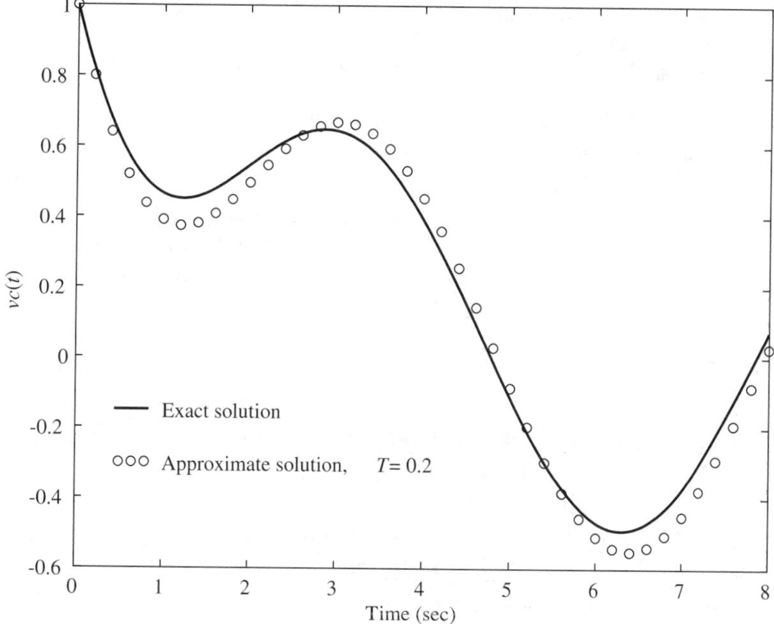

Figure 2.11 Exact and approximate output responses in Example 2.7.

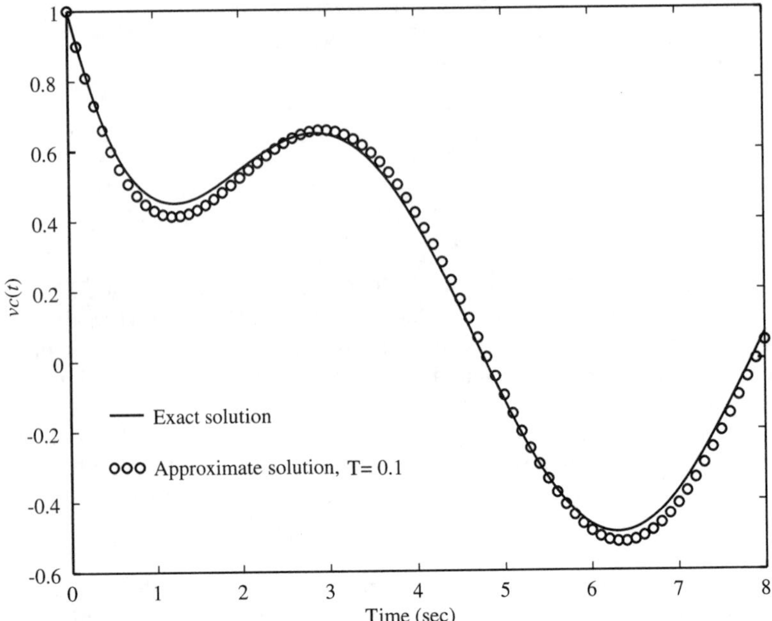

Figure 2.12 Exact and approximate output responses in Example 2.7.

again to compute the approximate and exact responses for $T = 0.1$, with the results shown in Figure 2.12. Clearly, the approximation with $T = 0.1$ shown in Figure 2.12 is closer to the exact solution than the approximation with $T = 0.2$ shown in Figure 2.11.

For continuous-time systems given by an input/output differential equation, there are a number of discretization techniques that are much more accurate (for a given value of T) than the technique above based on the Euler approximation of derivatives. One of these is based on an approximation to the integral expression of the solution to the input/output differential equation [e.g., see (2.5)]. The details of this discretization technique are delayed until Chapter 13, when the state-space formulation is presented.

2.5 SYSTEMS DEFINED BY TIME-VARYING OR NONLINEAR EQUATIONS

In applications, the system under study may be given by a linear input/output differential equation with time-varying coefficients or by a nonlinear input/output differential equation. These two possibilities are considered briefly in this section, beginning with the case of time-varying coefficients. A corresponding development for discrete-time systems is also presented.

Linear Input/Output Differential Equations with Time-Varying Coefficients

The class of linear time-invariant systems given by the input/output differential equation (2.1) can be extended to include time-varying systems by allowing the coefficients of (2.1) to vary with time. For example, consider the first-order case given by

$$\frac{dy(t)}{dt} + a(t)y(t) = b(t)x(t) \tag{2.70}$$

In (2.70), the coefficient $a(t)$ and/or the coefficient $b(t)$ may vary with t. If both $a(t)$ and $b(t)$ are constant (i.e., do not vary with t), (2.70) is a first-order differential equation with constant coefficients that was studied in detail in Section 2.1.

Systems given by an input/output differential equation with time-varying coefficients arise in many applications. Examples include circuits with time-varying components. For instance, resistors may have a time-varying resistance due to heating effects. Both capacitors and inductors may be time varying. A capacitor with a time-varying capacitance is illustrated in Figure 2.13.

The time variance $C(t)$ of the capacitance is a result of changing the position of the dielectric. To determine the current–voltage relationship in this case, first observe that the charge $q(t)$ on the capacitor plates is given by

$$q(t) = C(t)v_c(t) \tag{2.71}$$

where $v_c(t)$ is the voltage across the capacitor. The current $i_c(t)$ into the capacitor is equal to the derivative of the charge $q(t)$. Thus, taking the derivative of both sides of (2.71) yields

$$i_c(t) = C(t)\frac{dv_c(t)}{dt} + \dot{C}(t)v_c(t) \tag{2.72}$$

where $\dot{C}(t) = dC(t)/dt$. Note that if $C(t)$ is constant, so that $\dot{C}(t) = 0$, (2.72) reduces to the current–voltage relationship of the time-invariant capacitor.

Now the RC circuit studied in Chapter 1 shall be reconsidered with the capacitor given by the relationship (2.72). The circuit is shown in Figure 2.14. From Kirchhoff's current law

$$i_c(t) + \frac{1}{R}v_c(t) = i(t) \tag{2.73}$$

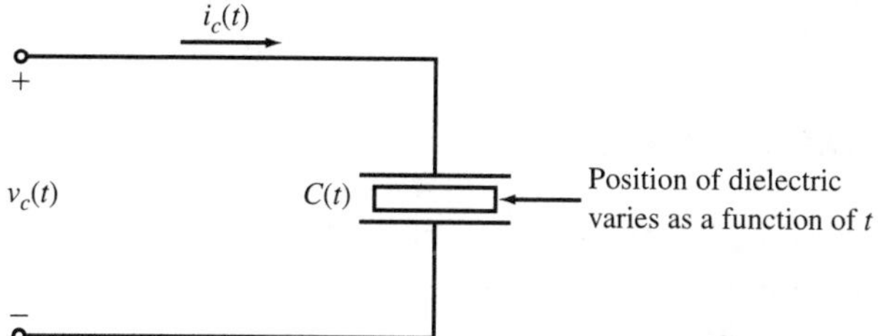

Figure 2.13 Time-varying capacitance.

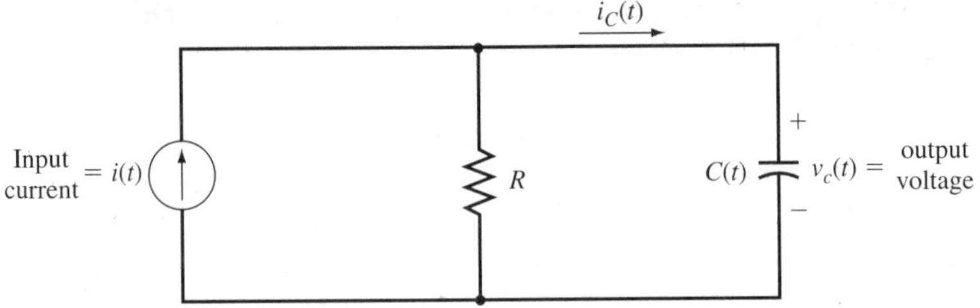

Figure 2.14 RC circuit with time-varying capacitor.

Inserting the expression (2.72) for $i_c(t)$ into (2.73) yields

$$C(t)\frac{dv_c(t)}{dt} + \left[\frac{1}{R} + \dot{C}(t)\right]v_c(t) = i(t) \tag{2.74}$$

If $C(t) > 0$ for all t, both sides of (2.74) can be divided by $C(t)$. This gives

$$\frac{dv_c(t)}{dt} + \frac{(1/R) + \dot{C}(t)}{C(t)}v_c(t) = \frac{1}{C(t)}i(t) \tag{2.75}$$

Equation (2.75) is the input/output differential equation of the RC circuit with time-varying capacitor. Note that (2.75) is a first-order differential equation with both coefficients equal to functions of t [assuming that $C(t)$ varies with t].

Again consider the first-order input/output differential equation given by (2.70). Via the integrating-factor method, it follows that the solution $y(t)$ to (2.70) with initial condition $y(0)$ and input $x(t)$ is given by

$$y(t) = \exp\left[\int_0^t -a(\tau)\,d\tau\right]y(0) + \int_0^t \exp\left[\int_\lambda^t -a(\tau)\,d\tau\right]b(\lambda)x(\lambda)\,d\lambda, \quad t \geq 0 \tag{2.76}$$

When both coefficients $a(t)$ and $b(t)$ are constant, (2.76) reduces to the expression (2.5) for the solution of the first-order differential equation (2.4) with constant coefficients. To see this, note that if the coefficient $a(t)$ is equal to a constant a, then

$$\int_\lambda^t -a(\tau)\,d\tau = \int_\lambda^t -a\,d\tau = -a\Big]_{\tau=\lambda}^{\tau=t} = -a(t-\lambda) \tag{2.77}$$

Inserting (2.77) into (2.76) and setting $b(t)$ equal to a constant b results in (2.5).

Unlike the constant-coefficient case, second- and higher-order differential equations with time varying coefficients cannot be solved exactly in general. In such cases, solutions must be obtained using numerical-solution techniques.

Nonlinear Input/Output Differential Equations

Nonlinear systems are often specified by a nonlinear input/output differential equation. For example, consider the nonlinear time-invariant system given by the first-order input/output differential equation

$$\frac{dy(t)}{dt} = f(y(t), x(t)), \qquad t > 0 \tag{2.78}$$

Here $f(y(t), x(t))$ is a function of the output $y(t)$ at time t and the input $x(t)$ at time t.

Recall from Chapter 1 that the system defined by (2.78) is linear if and only if (2.78) is a linear differential equation. The differential equation (2.78) is linear if and only if there exist constants a and b such that

$$f(y(t), x(t)) = -ay(t) + bx(t) \tag{2.79}$$

In this case, the input/output differential equation (2.78) becomes

$$\frac{dy(t)}{dt} = -ay(t) + bx(t)$$

This equation was studied in Section 2.1.

It is now assumed that the system defined by (2.78) is nonlinear so that the function f cannot be expressed in the form (2.79). In general, due to the nonlinearity it is not possible to derive an analytical expression for the output response $y(t)$ resulting from an initial condition $y(0)$ and an input $x(t)$. However, it may be possible to approximate the system by a linear model that is valid for some range of operating values. One procedure for doing this is based on an approximation of the system behavior with respect to some *nominal operation.* This technique is called *linearization* and is carried out as follows.

Let $y_{\text{nom}}(t)$ denote the solution to (2.78) when the input $x(t)$ is equal to a given function $x_{\text{nom}}(t)$, called the *nominal input function,* and where the initial condition $y_{\text{nom}}(0)$ is some given real number y_0, called the *nominal initial condition.* In other words,

$$\frac{dy_{\text{nom}}(t)}{dt} = f(y_{\text{nom}}(t), x_{\text{nom}}(t)), \qquad t > 0 \tag{2.80}$$

with initial condition $y_{\text{nom}}(0) = y_0$.

The function $y_{\text{nom}}(t)$ is called the *nominal output response* resulting from the nominal input function $x_{\text{nom}}(t)$ and nominal initial condition y_0. The nominal response $y_{\text{nom}}(t)$ is often computed using a numerical solution technique for nonlinear differential equations. The function $y_{\text{nom}}(t)$ must be computed before the linearization procedure can be applied.

Now the idea is to linearize the system with respect to the nominal behavior given by the nominal functions $x_{\text{nom}}(t)$ and $y_{\text{nom}}(t)$. Suppose that the new input $x(t)$ is given by

$$x(t) = x_{\text{nom}}(t) + \Delta x(t) \tag{2.81}$$

and the new initial condition is

$$y(0) = y_0 + \Delta y_0 \tag{2.82}$$

In (2.81) and (2.82) it is assumed that the magnitudes of $\Delta x(t)$ and Δy_0 are small in comparison with the magnitudes of $x_{\text{nom}}(t)$ and $y_{\text{nom}}(0)$. The input $\Delta x(t)$ and initial condition Δy_0 are referred to as *perturbations* in the nominal input $x_{\text{nom}}(t)$ and nominal initial condition y_0, respectively.

The output response $y(t)$ resulting from the input $x(t) = x_{\text{nom}}(t) + \Delta x(t)$ and initial condition $y(0) = y_0 + \Delta y_0$ can be written in the form

$$y(t) = y_{\text{nom}}(t) + \Delta y(t) \tag{2.83}$$

In (2.83), $\Delta y(t)$ is the perturbation in the nominal output response $y_{\text{nom}}(t)$ resulting from the perturbations $\Delta x(t)$ and Δy_0 in the nominal input and initial condition. The output $y(t)$ is the solution to (2.78) with input $x(t)$ given by (2.81) and with initial condition $y(0)$ given by (2.82). Hence

$$\frac{dy(t)}{dt} = \frac{dy_{\text{nom}}(t)}{dt} + \frac{d\Delta y(t)}{dt} = f(y_{\text{nom}}(t) + \Delta y(t), x_{\text{nom}}(t) + \Delta x(t)) \tag{2.84}$$

Under certain conditions on the function f, the term

$$f(y_{\text{nom}}(t) + \Delta y(t), x_{\text{nom}}(t) + \Delta x(t))$$

can be explained into a Taylor series about the nominal functions $y_{\text{nom}}(t)$ and $x_{\text{nom}}(t)$. The result is

$$\begin{aligned} f(y_{\text{nom}}(t) &+ \Delta y(t), x_{\text{nom}}(t) + \Delta x(t)) \\ &= f(y_{\text{nom}}(t), x_{\text{nom}}(t)) + c(t)\,\Delta y(t) + d(t)\,\Delta x(t) \\ &\quad + \text{higher-order terms involving } (\Delta y(t))^2, \Delta y(t)\,\Delta x(t), (\Delta x(t))^2, \text{etc.} \end{aligned} \tag{2.85}$$

In (2.85), $c(t)$ and $d(t)$ are the partial derivatives of f with respect to x and y, with the partial derivatives evaluated at the nominal functions. In mathematical terms,

$$c(t) = \left.\frac{\partial f}{\partial y}\right|_{\substack{y(t) = y_{\text{nom}}(t) \\ x(t) = x_{\text{nom}}(t)}} \tag{2.86}$$

$$d(t) = \left.\frac{\partial f}{\partial x}\right|_{\substack{y(t) = y_{\text{nom}}(t) \\ x(t) = x_{\text{nom}}(t)}} \tag{2.87}$$

If the magnitudes of $\Delta y(t)$ and $\Delta x(t)$ are suitably small, the higher-order terms in (2.85) can be neglected, which gives

$$\begin{aligned} f(y_{\text{nom}}(t) + \Delta y(t), x_{\text{nom}}(t) + \Delta x(t)) &= f(y_{\text{nom}}(t), x_{\text{nom}}(t)) \\ &\quad + c(t)\,\Delta y(t) + d(t)\,\Delta x(t) \end{aligned} \tag{2.88}$$

Now inserting (2.88) into (2.84) gives

$$\frac{dy_{\text{nom}}(t)}{dt} + \frac{d\,\Delta y(t)}{dt} = f(y_{\text{nom}}(t), x_{\text{nom}}(t)) + c(t)\,\Delta y(t) + d(t)\,\Delta x(t) \tag{2.89}$$

and inserting (2.80) into (2.89) yields

$$\frac{d\,\Delta y(t)}{dt} = c(t)\,\Delta y(t) + d(t)\,\Delta x(t) \tag{2.90}$$

The linear differential equation (2.90) is called the *linearized equation* with respect to the nominal functions $y_{\text{nom}}(t)$ and $x_{\text{nom}}(t)$. The representation (2.90) describes the perturbation in system behavior due to the perturbation $\Delta x(t)$ in the nominal input $x_{nom}(t)$ and the perturbation Δy_0 in the nominal initial condition $y_{nom}(0)$. The input/output differential equation (2.90) is an accurate model of system behavior as long as the magnitudes of $\Delta y(t)$ and $\Delta x(t)$ are small in comparison with the magnitudes of $x_{nom}(t)$ and $y_{nom}(t)$.

Since the differential equation (2.90) is linear, it can be studied using linear techniques. This is illustrated in the following example.

Example 2.8 *High-Speed Vehicle*

Consider a high-speed vehicle on a horizontal surface. The vehicle is described by the velocity model

$$\frac{dv(t)}{dt} + \frac{k_f}{M}v(t) + \frac{k_d}{M}v^2(t) = \frac{1}{M}x(t) \tag{2.91}$$

In (2.91) $v(t)$ is the velocity of the vehicle at time $t, x(t)$ is the drive or braking force at the time t, M is the mass of the vehicle, k_f is the frictional loss coefficient, and k_d is the drag coefficient due to air resistance. Note that (2.91) is the same as the velocity model of the car in Section 1.4, except that (2.91) contains an additional term due to air resistance. Also note that the air resistance term depends on the square of velocity, and thus (2.91) is a nonlinear differential equation. The system can be linearized with respect to the nominal operation consisting of the vehicle moving at a constant velocity v_0; that is

$$v_{nom}(t) = v_0 \qquad \text{for all } t \geq 0 \tag{2.92}$$

The nominal output given by (2.92) is produced by applying the nominal input function

$$x_{nom}(t) = (k_f + k_d v_0)v_0, \qquad t \geq 0 \tag{2.93}$$

with the nominal initial condition

$$v_{nom}(0) = v_0$$

To verify this, simply check that the differential equation (2.91) is satisfied with $v_{nom}(t)$ and $x_{nom}(t)$ given by (2.92) and (2.93), respectively. Now in this example,

$$f(v(t), x(t)) = -\frac{k_f}{M}v(t) - \frac{k_d}{M}v^2(t) + \frac{1}{M}x(t)$$

Then

$$\frac{\partial f}{\partial v} = -\frac{k_f}{M} - 2\frac{k_d}{M}v(t) \tag{2.94}$$

$$\frac{\partial f}{\partial x} = \frac{1}{M} \tag{2.95}$$

Evaluating the partial derivatives (2.94) and (2.95) at the nominal functions given by (2.92) and (2.93) and using (2.86) and (2.87) gives

$$c(t) = -\frac{1}{M}(k_f + 2k_d v_0)$$

$$d(t) = \frac{1}{M}$$

Note that in this example, both $c(t)$ and $d(t)$ are constant. The linearized equation is then

$$\frac{d\,\Delta v(t)}{dt} = -\frac{1}{M}(k_f + 2k_d v_0)\,\Delta v(t) + \frac{1}{M}\Delta x(t) \tag{2.96}$$

The differential equation (2.96) describes the perturbed behavior of the vehicle resulting from some small perturbation $\Delta x(t)$ in the nominal input $x_{\text{nom}}(t)$ given by (2.93) and/or some small perturbation Δv_0 in the nominal velocity v_0. Since (2.96) is a first-order linear differential equation with constant coefficients, it can be solved using the general form derived in Section 2.1. For example, suppose that $\Delta x(t)$ is a small step change in the drive force applied to the vehicle; that is,

$$\Delta x(t) = \varepsilon u(t)$$

where $|\varepsilon|$ is a small number. With $\Delta v_0 = 0$, the resulting solution $\Delta v(t)$ to the linearized equation (2.96) is

$$\begin{aligned}\Delta v(t) &= \int_0^t \exp\left[\frac{-1}{M}(k_f + 2k_d v_0)(t-\lambda)\right]\frac{1}{M}\varepsilon\, d\lambda\\ &= \frac{\varepsilon}{k_f + 2k_d v_0}\exp\left[\frac{-1}{M}(k_f + 2k_d v_0)(t-\lambda)\right]\Bigg|_{\lambda=0}^{\lambda=t}\\ &= \frac{\varepsilon}{k_f + 2k_d v_0}\left\{1 - \exp\left[\frac{-1}{M}(k_f + 2k_d v_0)t\right]\right\}, \quad t \ge 0\end{aligned} \tag{2.97}$$

The expression (2.97) is the change in the velocity from the nominal v_0 resulting from the perturbation $\varepsilon u(t)$ in the force applied to the vehicle. The total velocity $v(t)$ of the vehicle resulting from the force $x(t) = x_{\text{nom}}(t) + \varepsilon u(t)$ is given by

$$\begin{aligned}v(t) &= v_0 + \Delta v(t)\\ &= v_0 + \frac{\varepsilon}{k_f + 2k_d v_0}\left\{1 - \exp\left[\frac{-1}{M}(k_f + 2k_d v_0)t\right]\right\}, \quad t \ge 0\end{aligned}$$

Time-Varying and Nonlinear Discrete-Time Systems

The linear constant-coefficient input/output difference equation (2.35) studied in Section 2.3 can be generalized to include the time-varying case by allowing the coefficients to be functions of the discrete-time index n. In the first-order case, the time-varying version of the input/output difference equation has the general form

$$y[n] + a(n)y[n-1] = b_0(n)x[n] + b_1(n)x[n-1] \tag{2.98}$$

In Equation (2.98) the coefficients $a(n)$, $b_0(n)$, and $b_1(n)$ are in general functions of the integer variable n. The discrete-time system defined by (2.98) is time invariant if and only if the coefficients $a(n), b_0(n)$, and $b_1(n)$ are constant (i.e., do not vary with n).

Example 2.9 *Loan-Repayment Process*

Again consider the loan-repayment process defined in Section 1.4. It is now assumed that the loan is a variable-rate loan, so that the annual interest rate $I(n)$ in the nth month varies as a function of n, where n is the month of the repayment period. The input/output difference equation of the loan-repayment process is

$$y[n] - \left(1 + \frac{I(n)}{12}\right)y[n-1] = -x[n] \tag{2.99}$$

where $x[n]$ is the loan payment in the nth month and $y[n]$ is the loan balance at the end of the nth month. Clearly, the difference equation (2.99) is a special case of (2.98), where

$$a(n) = -1 - \frac{I(n)}{12}$$

$$b_0(n) = -1$$

$$b_1(n) = 0$$

Nonlinear input/output difference equations. Nonlinear discrete-time systems may be modeled by a nonlinear input/output difference equation. In the first-order case, a nonlinear input/output difference equation has the general form

$$y[n] = f(y[n-1], x[n], x[n-1]) \tag{2.100}$$

where $f(y[n-1], x[n], x[n-1])$ is a function of the output at time n and the input at times n and $n-1$.

The first-order input/output difference equation (2.100) can be solved by recursion for any finite range of integer values of n. If an explicit expression is given for the function f in the input/output difference equation, the recursion process can be programmed as in the linear case.

Example 2.10 *Square Root Operation*

Given a positive real number a, the square root of a can be computed from the nonlinear difference equation

$$y[n] = 0.5\left(y[n-1] + \frac{a}{y[n-1]}\right)$$

For any initial value $y[0] > 0$, the solution $y[n]$ to this equation converges to $\sqrt{a}$. For example, setting $a = 2$ and $y[0] = a = 2$ gives

$$y[1] = 0.5\left(2 + \frac{2}{2}\right) = \frac{3}{2} = 1.5$$

$$y[2] = 0.5\left(\frac{3}{2} + \frac{2}{3}(2)\right) = \frac{17}{12} = 1.4166667$$

$$y[3] = 0.5\left(\frac{17}{12} + \frac{12}{17}(2)\right) = 1.4142157$$

Clearly, $y[n]$ is converging to $\sqrt{a} = 1.4142136$. The MATLAB commands for computing the first five values of $y[n]$ are

```
y0=2; a=2;
y=zeros(1,5);
y(1) = 0.5*(y0 + a/y0);
for n=2:5
  y(n) = 0.5*(y(n-1) + a/y(n-1));
end
```

PROBLEMS

2.1. Solve the following differential equations:

(a) $\frac{dy(t)}{dt} - 2y(t) = x(t); \quad x(t) = u(t), \quad y(0) = 1$

(b) $\frac{dy(t)}{dt} + 2y(t) = x(t); \quad x(t) = u(t), \quad y(0) = 4$

(c) $\frac{dy(t)}{dt} + 2y(t) = x(t); \quad x(t) = \sin(4t)u(t), \quad y(0) = 0$

(d) $\frac{dy(t)}{dt} + 10y(t) = 2x(t); \quad x(t) = e^{4t}u(t), \quad y(0) = 0$

(e) $\frac{dy(t)}{dt} - 2y(t) = \frac{dx(t)}{dt} - 2x(t); \quad x(0) = u(t), \quad y(0) = 1$

2.2. Consider the automobile on an inclined surface described in Problem 1.29 and shown in Figure P1.29. The differential equation is given by

$$M\frac{dv(t)}{dt} + k_f v(t) = x(t) - Mg\sin\theta(t)$$

(a) Compute the response of the velocity model if $x(t) = u(t)$, $v(0) = 0$, and $\theta(t) = u(t)$.

(b) Compute the response of the velocity model if $x(t) = u(t-5)$, $v(0) = 0$, and $\theta(t) = u(t-5)$.

(c) Compute the response of the velocity model if $x(t) = u(t)$, $v(0) = v_0$ and

$$\theta(t) = \begin{cases} \theta & \text{for } 0 \le t \le 10 \\ 0 & \text{otherwise} \end{cases}$$

where θ is a constant.

(d) Compute the step response of the velocity model if θ(t) is equal to $tu(t) - (t-10)u(t-10)$.

2.3. In Problem 1.19 the ingestion and metabolism of a drug in a human was modeled by the equations

$$\frac{dq(t)}{dt} = -k_1 q(t) + x(t)$$

$$\frac{dy(t)}{dt} = k_1 q(t) - k_2 y(t)$$

where $x(t)$ is the ingestion rate of the drug, the output $y(t)$ is the mass of the drug in the bloodstream, and $q(t)$ is the mass of the drug in the gastrointestinal tract.

(a) Now suppose that $k_1 = 0.05$, $k_2 = 0.02$, $y(0) = 0$, $q(0) = 10$ milligrams (mg), and $x(t) = 0$ for $t \ge 0$. This corresponds to a person having recently ingested a drug that resulted in a 10-mg amount in the gastrointestinal tract at time $t = 0$. Solve for $y(t)$ for $t > 0$ and plot the response. [Hint: First compute the solution $q(t)$ of the first differential equation. Using your expression for $q(t)$, solve for $y(t)$ from the second differential equation.]

(b) For your result in part (a), at what time is the mass of the drug in the bloodstream at a peak level? Compute the peak value of the drug concentration in the bloodstream.

2.4. Consider the mixing of a concentrated dye with water as described in Problem P1.30. Recall that the system is given by the differential equation

$$\frac{dy(t)}{dt} = \frac{1}{V}[Dq(t) - (q(t) + w))y(t)]$$

where $y(t)$ is the dye concentration in the mixing tank, V the volume of the mixing tank, D the dye concentration in the dye storage tank, $q(t)$ the flow rate of the dye into the mixing tank, and w the flow rate of water into the mixing tank. It is assumed that w is a constant.

(a) Assuming that $y(0) = 0$ and $q(t)$ is equal to a constant q for all t, derive an expression for the output $y(t)$ for $t \geq 0$.

(b) Assuming that $y(0) = y_0$, determine $q(t)$ so that $y(t) = y_0$ for all $t \geq 0$.

2.5. A top view of a single human eye is illustrated in Figure P2.5. The input $x(t)$ is the angular position $\theta_T(t)$ of the target and the output $y(t)$ is the angular position $\theta_E(t)$ of the eye, with both angular positions defined relative to the resting position. It is assumed that the only possible movement is rotation about the center of the eye in the horizontal plane. The single-eye system illustrated in Figure P2.5 is equivalent to a two-eye system with the target located at an infinite distance. The idealized rapid-eye-movement model (called the ideal saccade model) is given by the equations

$$T_c \frac{d\theta_E(t)}{dt} + \theta_E(t) = R(t)$$

$$R(t) = T_c \frac{d\theta_T(t - d)}{dt} + \theta_T(t - d)$$

In these equations $R(t)$ is the firing rate of action potentials in the nerve to the eye muscle, T_c is the viscoelastic restoring force of the eyeball, and d is the time delay through the central nervous system. A typical value of d is 200 ms. Now suppose that the target suddenly moves from the resting position to position A at time $t = 0$; that is, $\theta_T(t) = Au(t)$, where $u(t)$ is the unit-step function.

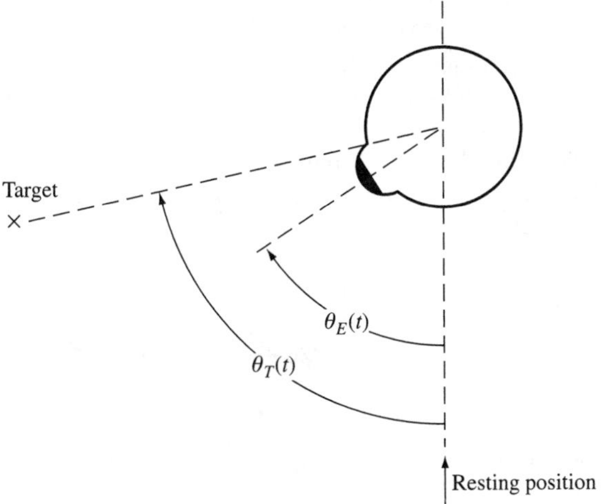

Figure P2.5

(a) Compute $R(t)$ for $t > 0$ assuming that $\theta_E(0) = 0$.
(b) Compute $\theta_E(t)$ for $t > 0$ assuming that $\theta_E(0) = 0$.
(c) Based on your result in part (b), does the eye "lock onto" the new target position? Explain.
(d) Repeat parts (a), (b), and (c) for the case when the target begins to move at time $t = 0$ with constant velocity V; that is, $\theta_T(t) = Vtu(t)$.

2.6. Consider the simple pendulum given by the small-signal model

$$\frac{d^2\theta(t)}{dt^2} + \frac{MgL}{I}\theta(t) = \frac{L}{I}x(t)$$

Recall from Chapter 1 that $\theta(t)$ is the angle of the pendulum from the vertical reference and $x(t)$ is the force applied to the pendulum tangential to the motion. Let $q_1(t)$ be the solution to

$$\frac{dq_1(t)}{dt} + j\sqrt{\frac{MgL}{I}}\, q_1(t) = \frac{L}{I}x(t)$$

where $j = \sqrt{-1}$. Let $q_2(t)$ be the solution to

$$\frac{dq_2(t)}{dt} - j\sqrt{\frac{MgL}{I}}\, q_2(t) = q_1(t)$$

In all of the following parts, take the initial time t_0 to be zero and assume that the system is initially at rest.

(a) Show that

$$\frac{d^2q_2(t)}{dt^2} + \frac{MgL}{I}q_2(t) = \frac{L}{I}x(t)$$

(b) Determine an integral expression that relates the input $x(t)$ to the solution $q_1(t)$ of the first first-order differential equation.
(c) Determine an integral expression that relates $q_1(t)$ to the solution $q_2(t)$ of the second first-order differential equation.
(d) Combining parts (b) and (c), determine an integral expression that relates the input $x(t)$ to the output $\theta(t)$.
(e) Using your result in part (d), compute the step response of the pendulum.

2.7. Consider the circuit in Figure P2.7. As shown, a 1-V voltage source is connected in series with a switch that has been closed for a very long time and then is opened at time $t = 0$. The input voltage $x(t)$ is assumed to be zero for $t < 0$.
(a) Determine the initial condition $y(0)$.
(b) Determine the differential equation relating the input $x(t)$ and the output $y(t)$ for $t \geq 0$.
(c) Compute $y(t)$ for $t > 0$ when $x(t) = 0$ for all t.

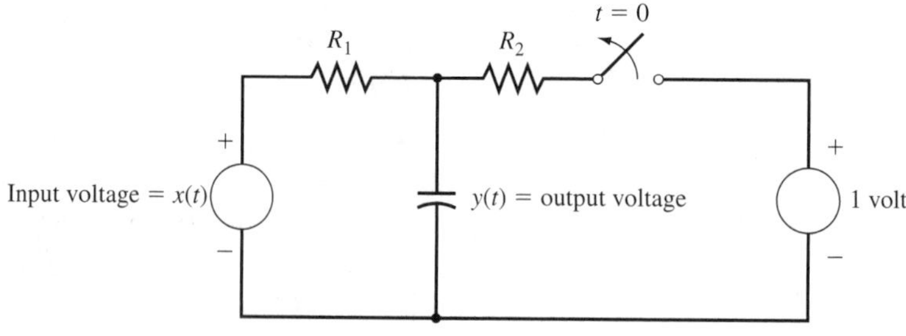

Figure P2.7

2.8. Consider the RL circuit shown in Figure P2.8.

(a) Determine the input/output differential equation of the circuit.

(b) Compute the step response of the RL circuit.

(c) Compute $v_L(t)$ for all $t > 0$ when $v_L(0) = v_0$ and $x(t) = 1$ for $t \geq 0$.

(d) Compute $v_L(t)$ for all $t > 0$ when $v_L(0) = 0$ and $x(t) = 1 + (R/L)t$ for $t \geq 0$.

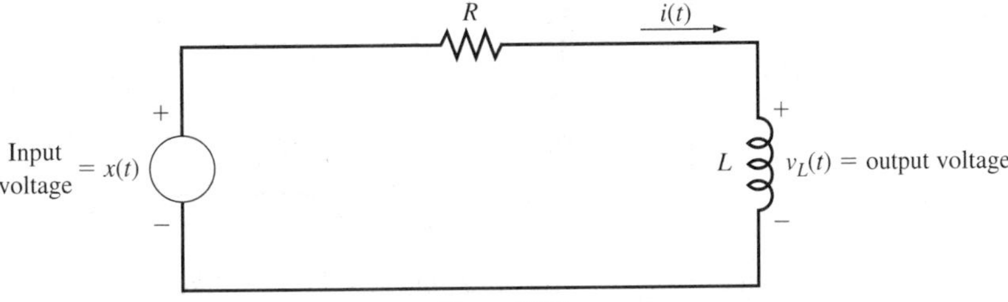

Figure P2.8

2.9. For the *RLC* circuit in Figure P2.9, find the input/output differential equation.

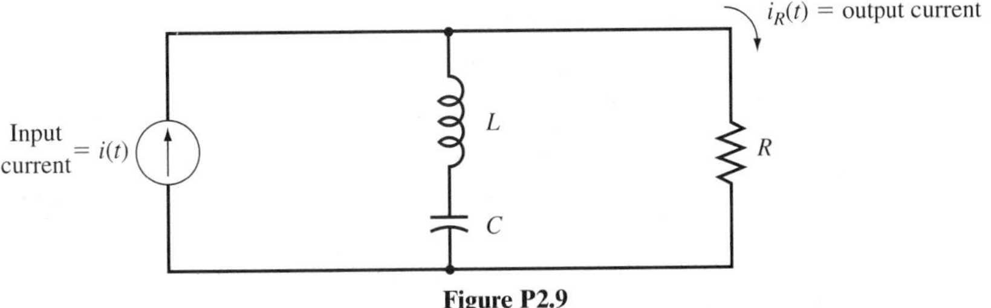

Figure P2.9

2.10. Find the input/output differential equations for the *RC* circuits in Figure P2.10.

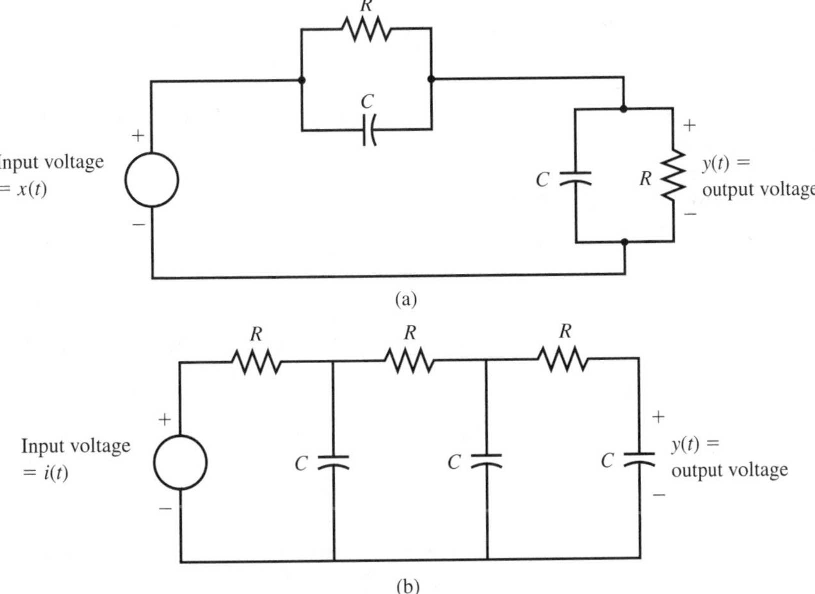

Figure P2.10

2.11. Find the input/output differential equation relating the input $i(t)$ to the output $v_C(t)$ for the circuit in Figure P2.11.

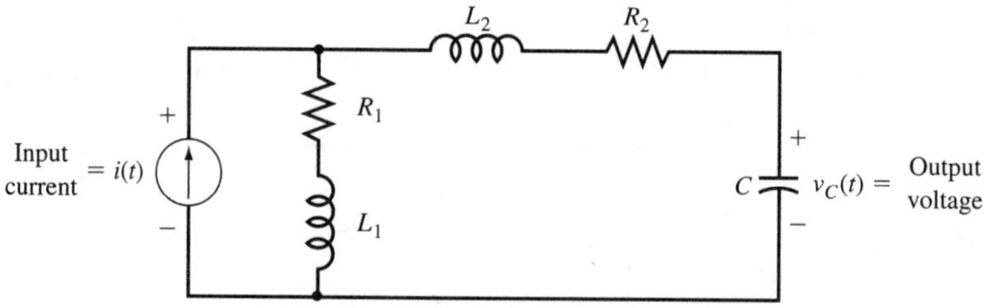

Figure P2.11

2.12. For the circuit in Figure P2.12, determine the input/output differential equation when:

(a) The output $y(t) = i_R(t)$
(b) The output $y(t) = i_L(t)$
(c) The output $y(t) = v_L(t)$
(d) The output $y(t) = v_C(t)$

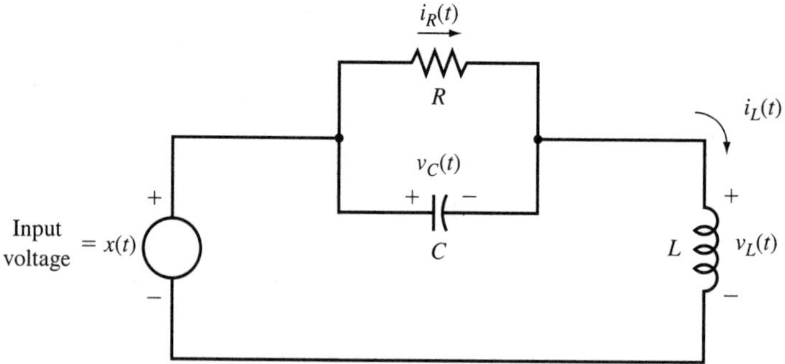

Figure P2.12

2.13. For the series RLC circuit in Figure P2.13, find the input/output differential equation when:

(a) $y(t) = v_R(t)$
(b) $y(t) = i(t)$
(c) $y(t) = v_L(t)$
(d) $y(t) = v_C(t)$

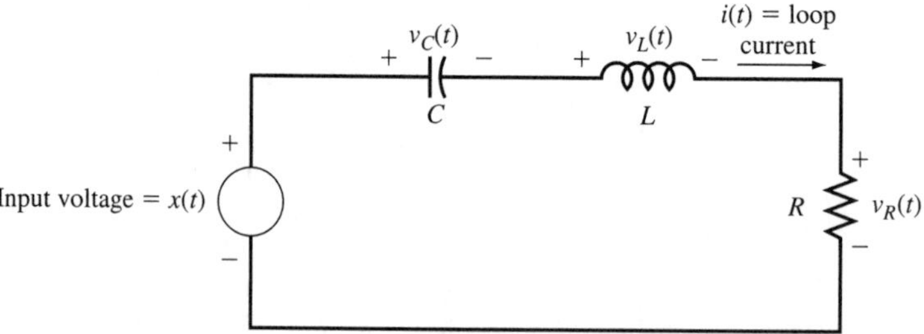

Figure P2.13

2.14. A mass M sits on top of a vibration absorber as illustrated in Figure P2.14 As shown in Figure P2.14, a force $x(t)$ (e.g., a vibrational force) is applied to the mass M, whose base is located at position $y(t)$. Derive the input/output differential equation of the system.

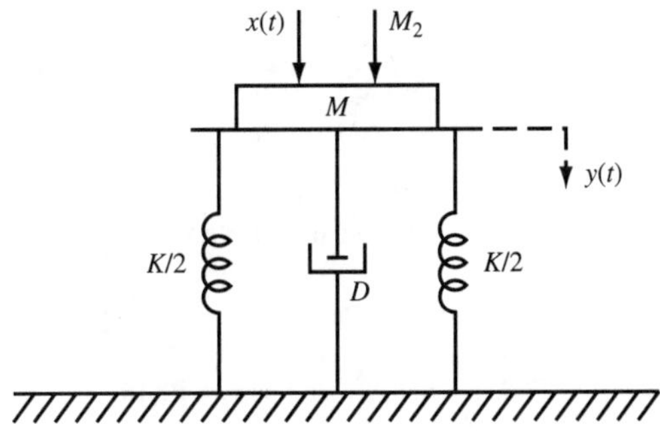

Figure P2.14

2.15. Consider the system consisting of two masses and three springs shown in Figure P2.15. The masses are on wheels that are assumed to be frictionless. The input $x(t)$ to the system is the force $x(t)$ applied to the first mass. The position of the first mass is $q(t)$ and the position of the second mass is the output $y(t)$, where both $q(t)$ and $y(t)$ are defined with respect to some equilibrium position. Determine the input/output differential equation of the system.

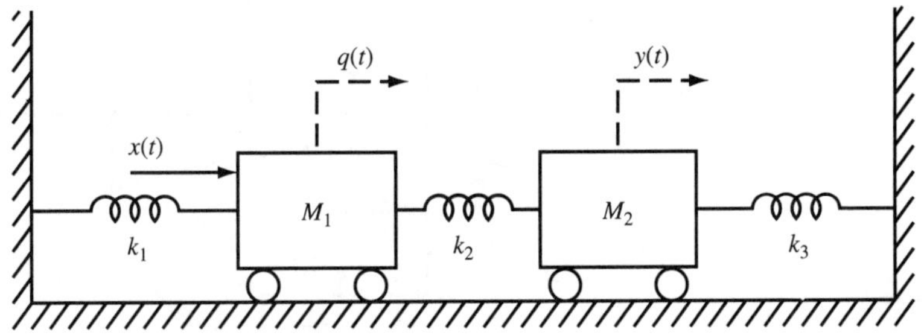

Figure P2.15

2.16. Two pendulums are connected together by a long spring as shown in Figure P2.16. Each pendulum has mass M and length L, the angular position of the pendulum on the left is the output $y(t)$, and the angular position of the one on the right is $\theta(t)$. The input to the system is the force $x(t)$ applied to the mass on the right, with the force tangential to the motion of the mass. The distance between the two pendulums when they are in the vertical (resting) position is denoted by d_0. *Assuming that the angles $y(t)$ and $\theta(t)$ are small for all t*, derive the input/output differential equation of the system. (Recall that if $|\theta|$ is small, then $\sin\theta \approx \theta$.)

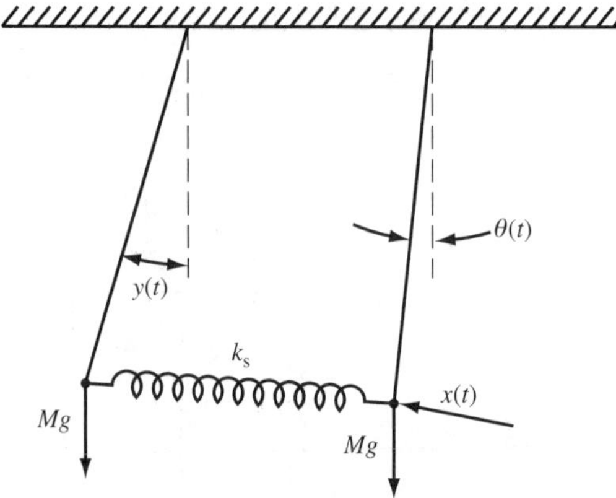

Figure P2.16

2.17. Two systems are connected together as shown in Figure P2.17. System 1 has the input/output differential equation

$$\frac{dy(t)}{dt} = y(t) + v(t)$$

and system 2 has the input/output differential equation

$$\frac{dw(t)}{dt} = 2w(t) + y(t)$$

Determine the differential equation relating the system input $x(t)$ and the system output $y(t)$.

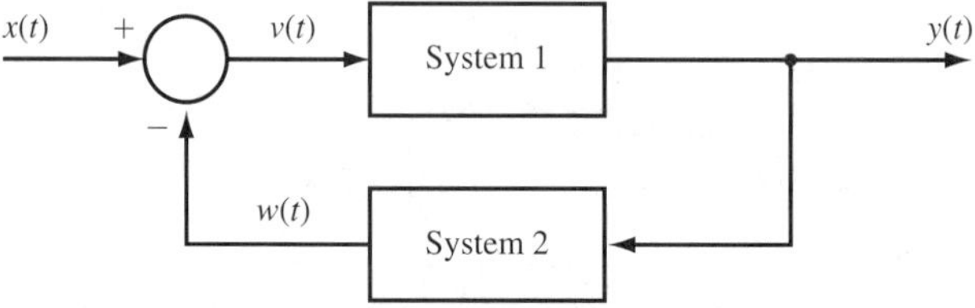

Figure P2.17

2.18. For each of the following difference equations
(i) $y[n + 1] + 1.5y[n] = x[n]$
(ii) $y[n + 1] + 0.8y[n] = x[n]$
(iii) $y[n + 1] - 0.8y[n] = x[n]$
use the method of recursion to solve the following problems.
(a) Compute $y[n]$ for $n = 0, 1, 2$, when $x[n] = 0$ for all n and $y[-1] = 2$.
(b) Compute $y[n]$ for $n = 0, 1, 2$, when $x[n] = u[n]$ and $y[-1] = 0$.
(c) Compute $y[n]$ for $n = 0, 1, 2$, when $x[n] = u[n]$ and $y[-1] = 2$.

2.19. For the difference equations given in Problem 2.18:
(a) Find a closed-form solution for $y[n]$ when $x[n] = 0$ for all n and $y[0] = 2$.
(b) Find a closed-form solution for $y[n]$ when $x[n] = u[n]$ and $y[0] = 0$.

(c) Find a closed-form solution for $y[n]$ when $x[n] = u[n]$ and $y[0] = 2$.

(d) Use the M-file `recur` to solve the difference equations for the cases defined in parts (a) to (c). Plot the corresponding answers from parts (a) to (c) along with those found from MATLAB.

2.20. For the difference equations given below, solve for the sequence $y[n]$ using the program `recur` for $0 \leq n \leq 10$ and plot y versue n on a stem plot.

(a) $y[n] = 0.5y[n-1] + u[n-1]; y[-1] = 0$

(b) $y[n] = 2y[n-1]; y[-1] = 1$

(c) $y[n] = 0.5y[n-1] + 0.1y[n-2] + u[n-1]; y[-2] = 1, y[-1] = 0$

(d) $y[n] = 0.5y[n-2] + 0.1y[n-1] + (0.5)^n u[n]; y[-1] = y[-2] = 0$

2.21. A discrete-time system is given by the input/output difference equation

$$y[n+2] + 1.5y[n+1] + 0.5y[n] = x[n+2] - x[n]$$

(a) Compute $y[n]$ for $n = 0, 1, 2, 3$ when $y[-2] = -1, y[-1] = 2$, and $x[n] = 0$ for all n.

(b) Compute $y[n]$ for $n = 0, 1, 2, 3$ when $y[-2] = y[-1] = 0$, and $x[n] = 1$ for $n \geq -2$.

(c) Compute $y[n]$ for $n = 0, 1, 2, 3$ when $y[-2] = -1, y[-1] = 2$, and $x[n] = 1$ for $n \geq -2$.

(d) Compute $y[n]$ for $n = 0, 1, 2, 3$ when $y[-2] = 2, y[-1] = 3$, and $x[n] = \sin(\pi n/2)$ for $n \geq 0$.

(e) Compute $y[n]$ for $n = 0, 1, 2, 3$ when $y[-2] = -2$, $y[-1] = 4$, and $x[n] = (0.5)^{n-1}u[n-1]$ for all n.

2.22. Given two real numbers a and b, the ratio a/b can be computed by recursively solving the difference equation

$$y[n+1] + (b-1)y[n] = a, \qquad n \geq 0$$

(a) Derive an analytical expression for the solution $y[n]$ assuming an arbitrary initial condition $y[0]$.

(b) Under what conditions does $y[n]$ converge to a/b as $n \to \infty$?

(c) Compute $y[10]$ when $y[0] = 1, a = 1$, and $b = 1.5$.

2.23. Consider the discrete-time signal $x[n]$ where $x[n] = 0$ for $n \leq 0$ and $x[1] = x[2] = 1$, $x[3] = 2, x[4] = 3, x[5] = 5, x[6] = 8, x[7] = 13$, and so on. This signal is called the *Fibonacci sequence.* Show that there are constants a_0 and a_1 such that the Fibonacci sequence $x[n]$ is the solution to the difference equation

$$x[n+2] + a_1 x[n+1] + a_0 x[n] = \delta[n+1]$$

with initial data $x[-2] = x[-1] = 0$. Here $\delta[n+1]$ is the unit pulse located at $n = -1$ (i.e., $\delta[n+1] = 0$ for all n except $n = -1$, and $\delta[0] = 1$). Determine a_0 and a_1.

2.24. When the unit pulse $\delta[n]$ is applied to a particular linear time-invariant discrete-time system which is initially at rest, the resulting response is $y[n] = (-1)^n, n = 0, 1, 2, \ldots$. Compute the input/output difference equation of the system. (*Hint:* observe that $y[n+2] = y[n]$ for $n = 0, 1, 2, \ldots$.)

2.25. Two discrete-time systems are connected in cascade as shown in Figure P2.25. System 1 has the input/output difference equation

$$w[n+1] = x[n]$$

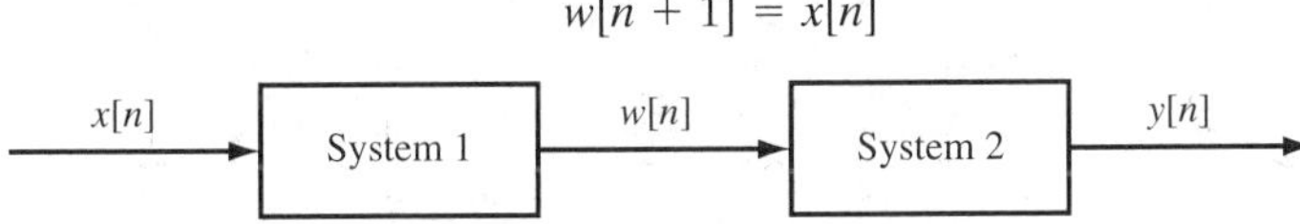

Figure P2.25

and system 2 has the input/output difference equation

$$y[n+1] + 2y[n] = w[n]$$

(a) Determine the input/output difference equation of the overall system.
(b) Compute $y[n]$ for all $n > 0$ when $y[0] = 1$, $w[0] = 2$, and $x[n] = 0$ for $n \geq 0$.
(c) Derive an analytical expression for the output response $y[n]$ when $y[0] = w[0] = 0$ and $x[n] = A$ for $n \geq 0$.

2.26. Consider the discrete-time model for a savings account given by the difference equation

$$y[n+1] - \left(1 + \frac{I[n]}{4}\right)y[n] = x[n+1]$$

Here interest accrues quarterly, n is the quarter, $y[n]$ the amount in the account at the end of the nth quarter, $x[n]$ the amount deposited in the nth quarter, and $I[n]$ the yearly interest rate in decimal form in the nth quarter.

(a) Suppose that $I[n] = 0.1 + 0.01n(u[n]-\mathrm{u}[\mathrm{n}-5])$, so the interest rate is increasing by 1% each of the first five quarters. Compute $y[n]$ for n=1, 2, 3, 4 when $y[0] = 1000$ and $x[n] = 1000$ for $n \geq 1$.
(b) Repeat part (a) with $I[n] = 0.1 - 0.01n(u[n] - \mathrm{u}[\mathrm{n} - 5])$, so that now the interest rate is decreasing by 1% each of the first five quarters.
(c) Write an M-file that computes $y[n]$ using recursion and plot the response for $0 \leq n \leq 20$.

2.27. Consider the *RL* circuit shown in Figure P2.27.

(a) Compute the output voltage $y(t)$ for all $t > 0$ when $y(0) = 0$ and $i(t) = u(t) - \mathrm{u}(\mathrm{t} - 1)$, where $u(t)$ is the step function. Express $y(t)$ in analytical form.
(b) Using Euler's approximation of derivatives with T arbitrary and input $x(t)$ arbitrary, derive a difference equation model for the *RL* circuit.
(c) Using your answer to part (b) and the M-file `recur` with $T = 0.1$, and $i(t) = u(t) - u(t-1)$, plot the approximation to $y(t)$ for $t = 0$ to $t = 5$ seconds. Take $y(-\mathrm{T}) = 0$. Compare your results with the exact solution plotted from the answer obtained in part (a).

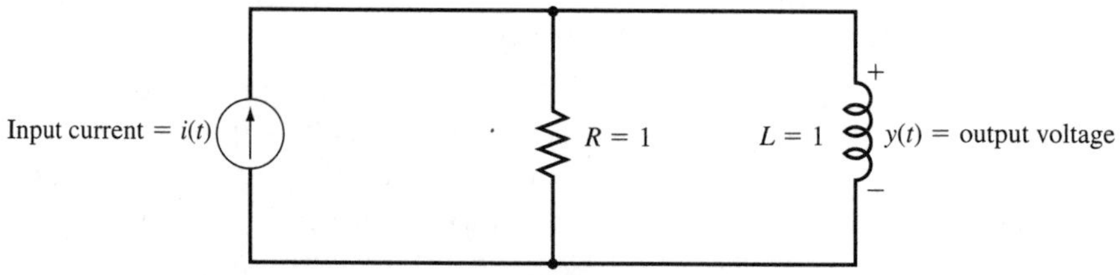

Figure P2.27

2.28. Consider the following differential equation:

$$\frac{d^2y(t)}{dt^2} + \frac{dy(t)}{dt} + 4.25y(t) = 0, \qquad y(0) = 2, \quad \dot{y}(0) = 1$$

(a) Show that the solution is given by $y(t) = e^{-0.5t}(\sin 2t + 2\cos 2t)$.
(b) Using Euler's approximation of derivatives with T arbitrary and input $x(t)$ arbitrary, derive a difference equation model.
(c) Using the answer in part (b) and the M-file `recur` with $T = 0.1$ sec, compute the approximation to $y(t)$.

(d) Repeat part (c) for $T = 0.05$ second.

(e) Plot the reponses obtained in parts (a), (c), and (d) for $0 \le t \le 10$ and compare the results.

2.29. Consider the following differential equation:

$$\frac{d^2y(t)}{dt^2} + 3\frac{dy(t)}{dt} + 2y(t) = 0, \qquad y(0) = 1, \quad \dot{y}(0) = 0$$

(a) Show that the solution is given by $y(t) = -e^{-2t} + 2e^{-t}$.

(b) Using Euler's approximation of derivatives with T arbitrary and input $x(t)$ arbitrary, derive a difference equation model.

(c) Using the answer in part (b) and the M-file `recur` with $T = 0.4$ second, compute the approximation to $y(t)$.

(d) Repeat part (c) for $T = 0.1$ second.

(e) Plot the responses obtained in parts (a), (c) and (d) for $0 \le t \le 10$ and compare the results.

2.30. Consider the following differential equation:

$$\frac{d^2y(t)}{dt^2} + 2\frac{dy(t)}{dt} + y(t) = 0, \qquad y(0) = 2, \quad \dot{y}(0) = -1$$

(a) Show that the solution is given by $y(t) = 2e^{-t} + te^{-t}$.

(b) Using Euler's approximation of derivatives with T arbitrary and input $x(t)$ arbitrary, derive a difference equation model.

(c) Using the answer in part (b) and the M-file `recur` with $T = 0.4$ second, compute the approximation to $y(t)$.

(d) Repeat part (c) for $T = 0.1$ second.

(e) Plot the responses obtained in parts (a), (c) and (d) for $0 \le t \le 10$ and compare the results.

2.31. A continuous-time system is given by the input/output differential equation

$$\frac{d^2y(t)}{dt^2} + a\frac{dy(t)}{dt} = bx(t)$$

Recall that this is the input/output differential equation of a car on a level surface as derived in Chapter 1 if $a = k_f/M$ and $b = 1/M$.

(a) Determine a discretization of the system by using the Euler approximation. Let the discretization interval T be arbitrary.

(b) With $a = 0.1, b = 1$, and $T = 1$, use the discretization in part (a) and the M-file `recur` to compute an approximation of $y(t)$ for $0 \le t \le 50$ when $y(0) = 0$ and $\dot{y}(0) = 0$, and $x(t) = \sin(0.1\pi t)$ for $t \ge 0$.

(c) Compare your results in part (b) with the exact values of the response found from evaluating equation (1.31).

(d) Repeat parts (b) and (c) with $T = 0.5$ second.

2.32. In Problems 1.19 and 2.3 the ingestion and metabolism of a drug in a human was modeled by the equations

$$\frac{dq(t)}{dt} = -k_1q(t) + x(t)$$

$$\frac{dy(t)}{dt} = k_1q(t) - k_2y(t)$$

(a) By first discretizing each of the two equations above using Euler's approximation, compute a discretization relating $x(t)$ and $y(t)$. Let the discretization interval T be arbitrary.

(b) Using your result in part (a) and the M-file `recur`, compute an approximation for $y(t)$ for $1 \le t \le 60$ when $k_1 = 0.05, k_2 = 0.02, T = 1, y(0) = 0, q(0) = 10$, and $x(t) = 0$ for $t \ge 0$. Compare your results with those obtained in part (a) of Problem 2.3.

2.33. Consider the pendulum given by the input/output differential equation

$$I\frac{d^2\theta(t)}{dt^2} + MgL \sin \theta(t) = Lx(t)$$

(a) Discretize the system by using the Euler approximation. Let T be arbitrary. (Note: do not approximate the system by the small-signal model.)

(b) Let $I = M = L = 1$, $g = 9.8$. Using the discretization in part (a) with $T = 0.04$, compute $\theta(0.04n)$ for $n = 1, 2, 3, \ldots, 50$ when $\theta(0) = \pi/2$ (radians), $\dot{\theta}(0) = 0$, and $x(t) = 0$ for all $t \ge 0$.

(c) Repeat part (b) with $T = 0.02$. Are the results significantly different?

(d) Now consider the small-signal model given by

$$\frac{d^2\theta(t)}{dt^2} + 9.8\theta(t) = x(t)$$

When $\theta(0) = \pi/2, \dot{\theta}(0) = 0$, and $x(t) = 0$ for $t \ge 0$, the exact solution to the small-signal model is $\theta(t) = (\pi/2)\cos 3.13t$. Compare these values with the values of $\theta(0.02n)$ obtained in part (c). (The values should be plotted.) What do you conclude from this? In particular, is the small-signal model accurate when $\theta(t)$ is large, as is the case here?

Mass Spring Damper System

2.34. Consider the mass–spring–damper system described in Chapter 1 and in the on-line demo. The differential equation for the system is given by

$$M\frac{d^2y(t)}{dt^2} + D\frac{dy(t)}{dt} + Ky(t) = x(t)$$

(a) Discretize the system by using Euler approximation. Let T be arbitrary.

(b) Write a MATLAB program to simulate this system.

(c) Using the program developed in part (b), simulate the unit step response (that is, the response $y(t)$ when $x(t) = u(t)$) for $M = 10$, $D = 1$, and $K = 1$. Do this for three cases (i) $T = 1$ s, (ii) $T = 10$ s, and (iii) $T = 100$ s. Simulate the response long enough so that $y(t)$ appears to reach a constant steady-state value.

(d) Plot your approximations for $y(t)$ versus time (defined at $t = nT$) for all three values of T on the same axes. Clearly identify each plot according to its value of T.

(e) Compute the response for an input of $x(t) = 10\sin(0.2\pi t)$ when $T = 1$ s, and plot $y(t)$. Determine the amplitude of the resulting sinusoid.

(f) Use the on-line demo to check your results in parts (d) and (c). The on-line demo uses Runge-Kutta integration, which is similar to, but more accurate than, the Euler integration method used in part (a). The "Show Input/Output Summary" button can be used to view the results.

2.35. Consider the connection shown in Figure P2.35. System 1 is a continuous-time system with the input/output differential equation $\dot{y}_1(t) + y_1(t) = x(t)$, and system 2 is a discrete-time system with the input/output difference equation $y[n + 1] + 3y[n] = 2y_1[n]$. Compute the exact output values $y[1], y[2], y[3]$ when

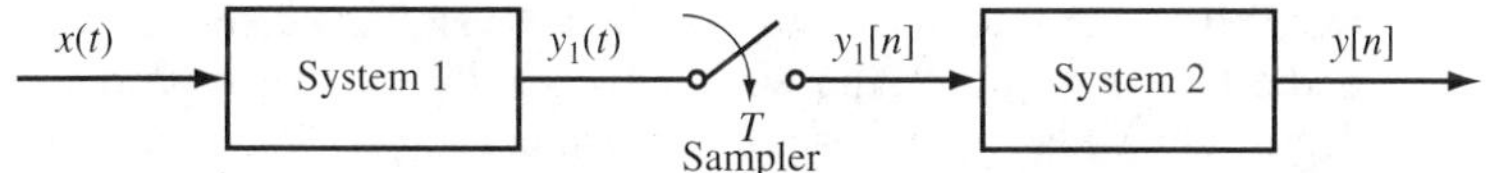

Figure P2.35

(a) $x(t) = 0$ for all $t \geq 0$ and $y_1(0) = 1, y(0) = 1$.
(b) $x(t) = 1$ for all $t \geq 0$ and $y_1(0) = 1, y(0) = 1$.
(c) $x(t) = e^{-t}$ for all $t \geq 0$ and $y_1(0) = -2, y(0) = -1$.

2.36. Consider the automobile with time-varying mass $M(t) = 1 - 0.05t[u(t) - u(t - 10)] - 0.5u(t - 10)$. The velocity model of the car is

$$M(t)\frac{dv(t)}{dt} + k_f v(t) = x(t)$$

where $v(t)$ is the velocity of the car, $x(t)$ is the drive or braking force, and k_f is the coefficient of friction. Suppose that $k_f = 0.1$. Compute the velocity $v(t)$ for all $t > 0$ when $v(0) = 0$ and $x(t) = 1$ for all $t \geq 0$. Your expression for $v(t)$ should be completely evaluated (i.e., all integrals should be evaluated).

2.37. Consider the nonlinear continuous-time system given by the input/output differential equation

$$\frac{dy(t)}{dt} + y^2(t) + ay(t) = bx(t)$$

where a and b are arbitrary constants.

(a) Suppose that $x(t) = 0$ for $t \geq 0$, and let $v(t) = 1/y(t)$. Show that $v(t)$ satisfies the differential equation

$$\frac{dv(t)}{dt} - av(t) = 1, \qquad t \geq 0$$

(b) With $v(0)$ equal to an arbitrary constant, derive an expression for the solution $v(t)$ to the differential equation $\dot{v}(t) - av(t) = 1$.

(c) Using your result in part (b), derive an expression for the solution $y(t)$ to the differential equation

$$\frac{dy(t)}{dt} + y^2(t) + ay(t) = 0$$

with initial condition $y(0)$.

(d) We continue to assume that $x(t) = 0$ for all $t \geq 0$. Determine all values of $y(0)$ such that $y(t) = \infty$ for some finite value of t (i.e., $0 < t < \infty$). Such values of t are called *finite escape times.*

2.38. In McClamroch [1980] the fish population in a fishery is modeled by the differential equation

$$\frac{dy(t)}{dt} = ky(t)[A - y(t)] - x(t)$$

where the output $y(t)$ is the total weight of fish that can be harvested and the input $x(t)$ is the harvesting rate. The term $kAy(t)$ is the rate of reproduction and $ky^2(t)$ is the death rate, where k and A are positive constants. Thus the reproduction rate is proportional to the number of fish, while the death rate is proportional to the square of the number of fish.

(a) Determine the nominal input $x_{nom}(t)$ that results in the nominal output $y_{nom}(t) = A$ for $t > 0$ with nominal initial condition $y_{nom}(0) = A$. What does this nominal behavior correspond to in terms of the dynamics of the fish population? Explain.

(b) Compute the linearized equation with respect to the nominal functions defined in part (a).

(c) Using your result in part (b), compute the approximate output $y(t)$ when $y(0) = A$ and $x(t) = C$ for $t \geq 0$, where C is a positive constant (C = harvesting rate).

(d) For your answer in part (c), how large can the harvesting rate C be without depleting the fish population, that is, so that $y(t)$ does not converge to zero as $t \to \infty$?

2.39. Consider the simple pendulum that is shown in Figure 1.30 and described by equation (1.34).

(a) Compute the linearized equation with respect to the nominal functions $x_{nom}(t) = 0$ for all $t \geq 0$, $\theta_{nom}(t) = \pi$ radians for all $t \geq 0$.

(b) Using your result in part (a), determine the approximate response $\theta(t)$ when $\theta(0) = \pi + \alpha$, where $|\alpha|$ is a small number. We are assuming that $\dot{\theta}(0) = 0$ and $x(t) = 0$ for all $t \geq 0$.

(c) Using your result in part (a), compute the approximate response $\theta(t)$ when $\theta(0) = \pi$ and $\dot{\theta}(0) = \beta$, where $|\beta|$ is a small number. We are still assuming that $x(t) = 0$ for $t \geq 0$.

(d) Compare your results in parts (b) and (c) with the corresponding results in the cases $\theta_{nom}(t) = 0$ for $t \geq 0$ and $\theta_{nom}(t) = \pi/2$ for $t \geq 0$.

2.40. A continuous-time system is specified by the input/output differential equation

$$\frac{dy(t)}{dt} = -y^2(t) + x(t)$$

(a) Compute the linearized equation with respect to the nominal behavior $x_{nom}(t) = 0$ for all $t \geq 0$, and $y_{nom}(0) = 0$.

(b) Compute the nominal output response $y_{nom}(t)$ when $x_{nom}(t) = 0$ for all $t \geq 1$ and $y_{nom}(1) = 1$.

(c) Compute the linearized equation with respect to the nominal functions in part (b).

2.41. Consider the following nonlinear difference equations:

(i) $y[n+2] = y[n+1]x[n+1] + y[n] + x[n]$

(ii) $y[n+2] = y[n+1]y[n] + x[n+1]$

(a) For each of the systems, compute the output response $y[n]$ for $n = 3, 4, 5$ when $y[1] = -1, y[2] = 3$, and $x[n] = \tan(\pi n/5)(u[n] - u[n-5])$.

(b) Write an M-file that computes $y[n]$ using recursion and plot the response for $1 \leq n \leq 20$.

CHAPTER 3

Convolution Representation

In Chapter 2 time-domain models of continuous-time and discrete-time systems were given in terms of input/output differential equations and input/output difference equations. In this chapter another type of time-domain model is considered based on the convolution operation. In Section 3.1 the convolution representation of linear time-invariant discrete-time systems is studied, and then in Section 3.2 the evaluation of the convolution operation for discrete-time signals is considered. The convolution representation of linear time-invariant continuous-time systems and the convolution of continuous-time signals are studied in Sections 3.3 and 3.4. In Section 3.5 it is shown that the continuous-time convolution representation can be discretized in time, which results in a discrete-time simulation of the given continuous-time system. Although the convolution representation does not apply to time-varying systems, in Section 3.6 it is shown that there is an input/output relationship for linear time-varying systems that reduces to the convolution relationship in the time-invariant case.

3.1 CONVOLUTION REPRESENTATION OF LINEAR TIME-INVARIANT DISCRETE-TIME SYSTEMS

Consider a single-input single-output discrete-time system with input $x[n]$ and output $y[n]$. Throughout this section it is assumed that $y[n]$ is the output response resulting from input $x[n]$ with no initial energy in the system prior to the application of

$x[n]$. It is also assumed that the system under consideration is causal, linear, and time-invariant, but the system is not necessarily finite dimensional.

Unit-Pulse Response

Let $h[n]$ denote the output response of the system when the input $x[n]$ is equal to the unit pulse $\delta[n]$ with no initial energy in the system at time $n = 0$. (Recall that $\delta[0] = 1$ and $\delta[n] = 0$ for all $n \neq 0$.) The response $h[n]$ is called the *unit-pulse response* of the system. Note that since $\delta[n] = 0$ for $n = -1, -2, \ldots$, by causality the unit-pulse response $h[n]$ must be zero for all integers $n < 0$ (since in a causal system there can be no response before an input is applied). An example of the form of the unit-pulse response $h[n]$ is illustrated in Figure 3.1.

It is sometimes possible to determine the unit-pulse response $h[n]$ experimentally by applying the unit pulse $\delta[n]$ to the system and measuring the resulting output response. In this experiment the system must not have any initial energy prior to the application of the unit pulse. The unit-pulse response may also be computable from some known mathematical representation of the system. For example, suppose that the system is given by the input/output difference equation (2.35) studied in Section 2.3. Then $h[n]$ can be calculated by solving (2.35) with $x[n] = \delta[n]$ and with the initial conditions $y[i] = 0$ for $i = -1, -2, \ldots, -N$.

Example 3.1 *First-Order System*

Consider the finite-dimensional discrete-time system given by the input/output difference equation

$$y[n] + ay[n-1] = bx[n] \tag{3.1}$$

where a and b are constants. The unit-pulse response for this system can be computed by solving (3.1) with initial condition $y[-1] = 0$ and with input $x[n] = \delta[n]$. From the discussion in Section 2.3, the solution to (3.1) can be expressed in the form

$$y[n] = \sum_{i=0}^{n} (-a)^{n-i} b\delta[i], \qquad n = 0, 1, 2, \ldots \tag{3.2}$$

But since $\delta[n] = 0$ for all $n \neq 0$ and $\delta[0] = 1$, (3.2) reduces to

$$y[n] = (-a)^n b, \qquad n = 0, 1, 2, \ldots$$

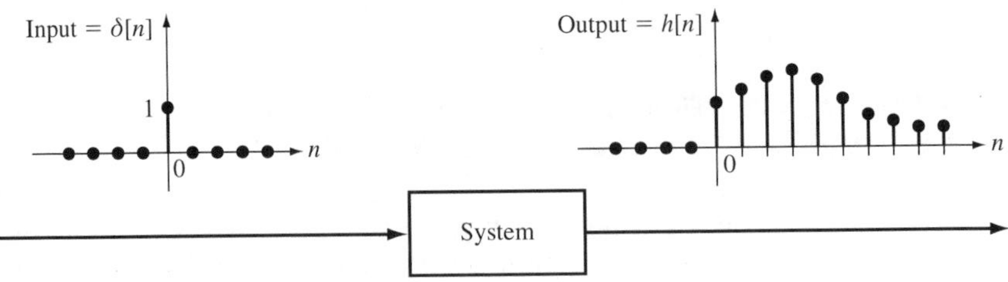

Figure 3.1 Generation of the unit-pulse response.

Hence the unit-pulse response $h[n]$ for this system is given by

$$h[n] = \begin{cases} (-a)^n b, & n = 0, 1, 2, \ldots \\ 0, & n = -1, -2, -3, \ldots \end{cases}$$

Returning to the general case, consider a causal linear time-invariant discrete-time system with input $x[n]$, output $y[n]$, and unit-pulse response $h[n]$. Since the system is time invariant, for any positive integer i the output response resulting from the i-step right shift of the unit pulse $\delta[n]$ must be equal to the i-step right shift of $h[n]$; that is, the response to $\delta[n - i]$ must be equal to $h[n - i]$. An example of the form of $h[n - i]$ is illustrated in Figure 3.2.

Now let $x[n]$ be an arbitrary input with $x[n] = 0$ for $n = -1, -2, \ldots$. To compute the resulting output response (assuming zero initial energy), first note that since $\delta[n - i] = 0$ when $i \neq n$ and $\delta[n - i] = 1$ when $i = n$, $x[n]$ can be expressed in the form

$$x[n] = x[0]\delta[n] + x[1]\delta[n-1] + x[2]\delta[n-2] + \cdots$$

$$x[n] = \sum_{i=0}^{\infty} x[i]\delta[n - i], \qquad n = 0, 1, 2, \ldots \tag{3.3}$$

Then to compute the output response resulting from the input $x[n]$, the idea is first to compute the output response, denoted by $y_i[n]$, due to the ith term $x[i]\delta[n - i]$ comprising $x[n]$. Since the system is linear and thus is homogeneous, and the response to $\delta[n - i]$ is $h[n - i]$, the response $y_i[n]$ to $x[i]\delta[n - i]$ is given by

$$y_i[n] = x[i]h[n - i] \tag{3.4}$$

By the additivity property, the response to the sum given by (3.3) must be equal to the sum of the $y_i[n]$ defined by (3.4). Thus the response to $x[n]$ is

$$y[n] = \sum_{i=0}^{\infty} x[i]h[n - i], \qquad n \geq 0 \tag{3.5}$$

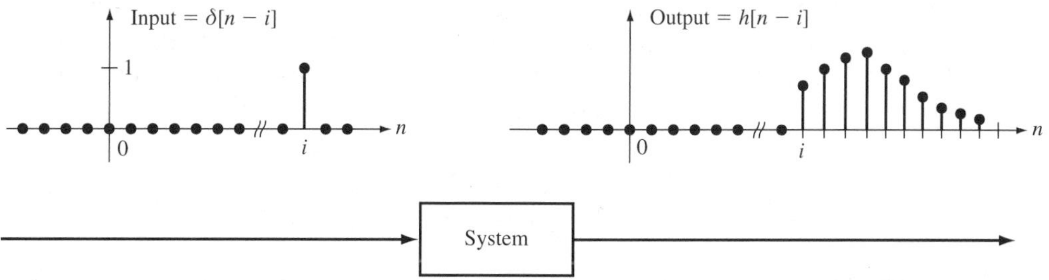

Figure 3.2 The shifted unit-pulse response $h[n - i]$.

The operation defined by the expression

$$\sum_{i=0}^{\infty} x[i]h[n-i]$$

is called the *convolution* of $x[n]$ and $h[n]$, and is denoted by the symbol "$*$"; that is,

$$x[n] * h[n] = \sum_{i=0}^{\infty} x[i]h[n-i]$$

Rewriting (3.5) in terms of the convolution symbol gives

$$y[n] = x[n] * h[n], \qquad n \geq 0 \tag{3.6}$$

By (3.6), the output response $y[n]$ resulting from input $x[n]$ with no initial energy at time $n = 0$ is equal to the convolution of the unit-pulse response $h[n]$ with the input $x[n]$. Equation (3.6) [or (3.5)] is called the *convolution representation* of the system. This is a time domain model since the components of (3.6) are functions of the discrete-time index n.

An interesting consequence of the convolution representation (3.6) is the result that the system is determined completely by the unit-pulse response $h[n]$. In particular, if $h[n]$ is known, the output response resulting from any input $x[n]$ can be computed by evaluating (3.6). The evaluation of the convolution operation and the basic properties of this operation are studied in the next section.

3.2 CONVOLUTION OF DISCRETE-TIME SIGNALS

In Section 3.1 the convolution of an input $x[n]$ and the unit-pulse response $h[n]$ was defined, with both $x[n]$ and $h[n]$ equal to zero for all $n < 0$. In this section the convolution operation is defined for arbitrary discrete-time signals $x[n]$ and $v[n]$ that are not necessarily zero for $n < 0$.

Given two discrete-time signals $x[n]$ and $v[n]$, the convolution of $x[n]$ and $v[n]$ is defined by

$$x[n] * v[n] = \sum_{i=-\infty}^{\infty} x[i]v[n-i] \tag{3.7}$$

The summation on the right-hand side of (3.7) is called the *convolution sum.* If $x[n]$ and $v[n]$ are zero for all integers $n < 0$, then $x[i] = 0$ for all integers $i < 0$ and $v[n-i] = 0$ for all integers $n - i < 0$ (or $n < i$). Thus the summation on i in (3.7) may be taken from $i = 0$ to $i = n$, and the convolution operation is given by

$$x[n] * v[n] = \begin{cases} 0, & n = -1, -2, \ldots \\ \displaystyle\sum_{i=0}^{n} x[i]v[n-i], & n = 0, 1, 2, \ldots \end{cases} \tag{3.8}$$

The computation of the convolution (3.7) or (3.8) can be carried out by first changing the discrete-time index n to i in the signals $x[n]$ and $v[n]$. The resulting signals $x[i]$ and $v[i]$ are then functions of the discrete-time index i. The next step is to determine $v[n-i]$ and then form the product $x[i]v[n-i]$. The signal $v[n-i]$ is a *folded* and *shifted* version of the signal $v[i]$. More precisely, $v[-i]$ is $v[i]$ folded about the vertical axis, and $v[n-i]$ is $v[-i]$ shifted by n steps. If $n > 0$, $v[n-i]$ is a n-step right shift of $v[-i]$. If the signal $v[n]$ is given in analytical form, $v[n-i]$ can be formed by simply replacing n by $n-i$ in the expression for $v[n]$. Once the product $x[i]v[n-i]$ is formed, the value of the convolution $x[n] * v[n]$ at the point n is computed by summing the values of $x[i]v[n-i]$ as i ranges over the set of integers. The computation is facilitated by using graphical representations of the signals. The process is illustrated by the following example.

Example 3.2 *Convolution of Rectangular Pulses*

Suppose that $x[n]$ and $v[n]$ are equal to the rectangular pulse $p[n]$ defined by

$$p[n] = \begin{cases} 1, & 0 \le n \le 9 \\ 0, & \text{all other } n \end{cases}$$

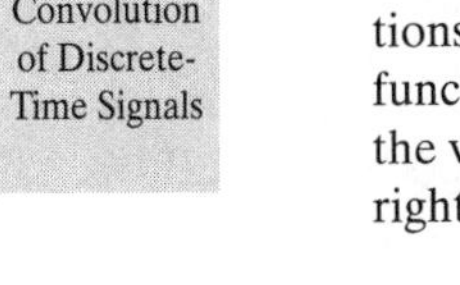

The pulse $p[n]$ is plotted in Figure 3.3. Since both $x[n]$ and $v[n]$ are zero for $n < 0$, the convolution $x[n] * v[n]$ is given by (3.8). To evaluate (3.8), the graphical representations of $x[i]$ and $v[n-i]$ as functions of i are first considered: The plot of $v[-i]$ as a function of i is given in Figure 3.4. Note that $v[-i]$ is equal to the pulse $p[i]$ folded about the vertical axis. Now for any positive value of n, $v[n-i]$ is equal to $v[-i]$ shifted to the right by n steps. The plot of $v[n-i]$ as a function of i for $n \le 9$ is given in Figure 3.5b

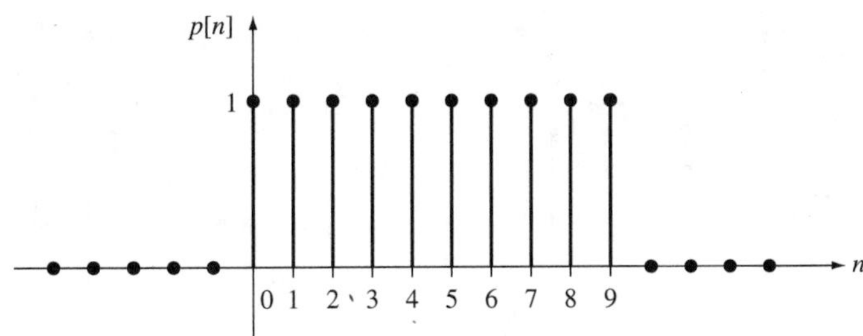

Figure 3.3 The pulse $p[n]$.

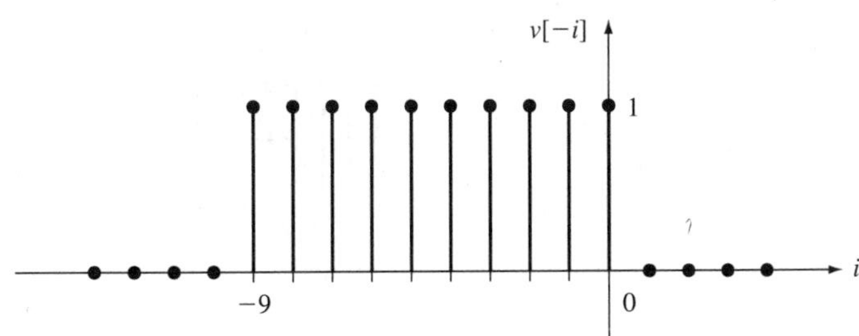

Figure 3.4 The folded pulse.

and the plot of $x[i]$ is given in Figure 3.5a. The product $x[i]v[n - i]$ is then formed by multiplying point-by-point the plots $x[i]$ and $v[n - i]$. For an arbitrary value of $n \leq 9$, the product $x[i]v[n - i]$ is illustrated in Figure 3.5c. The product is a rectangular pulse of amplitude one running from $i = 0$ to $i = n$, and thus the sum of the values of $x[i]v[n - i]$ is equal to $(n + 1)(1) = n + 1$. Hence

$$x[n] * v[n] = n + 1 \text{ for } 0 \leq n \leq 9$$

For an arbitrary value of n between 10 and 18, the computation of the product is illustrated in Figure 3.6. In this case the product is a rectangular pulse of amplitude one running from $i = n - 9$ to $i = 9$. Hence the sum of the values of $x[i]v[n - i]$ is equal to $[9 - (n - 9) + 1](1) = 19 - \text{n}$, and therefore

$$x[n] * v[n] = 19 - n \text{ for } 10 \leq n \leq 18$$

Finally, for $n > 18$, $x[i]$ and $v[n - i]$ do not overlap, so the product is zero, and hence

$$x[n] * v[n] = 0 \qquad \text{for } n > 18$$

A plot of the convolution $x[n] * v[n]$ is given in Figure 3.7. This result shows that the convolution of a rectangular pulse with itself is equal to a triangular pulse of twice the time duration of the rectangular pulse.

If the signals $x[n]$ and $v[n]$ are given by simple analytical expressions, the convolution $x[n] * v[n]$ may be computed analytically. This is illustrated by the following example.

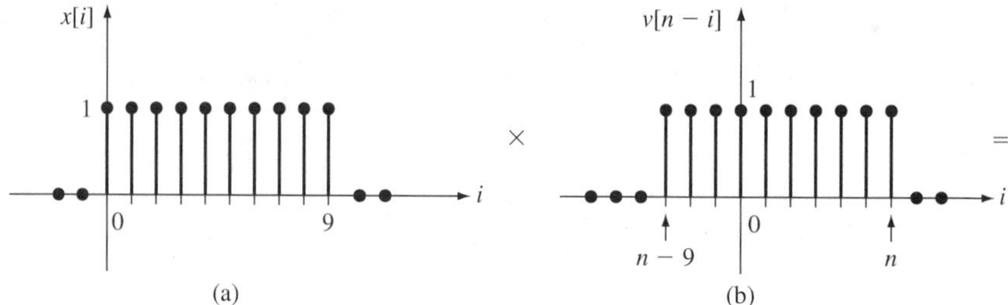

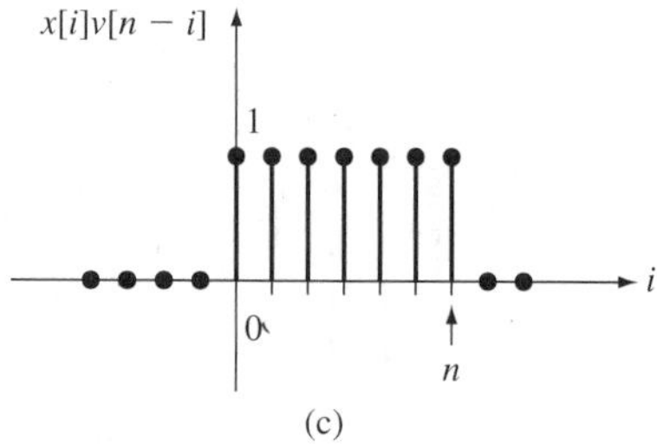

Figure 3.5 Plots of (a) $x[i]$; (b) $v[n - i]$; (c) $x[i]v[n - i]$ for $0 \leq n \leq 9$.

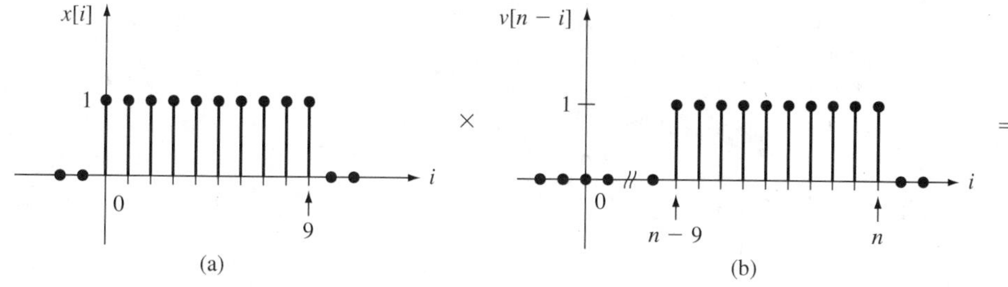

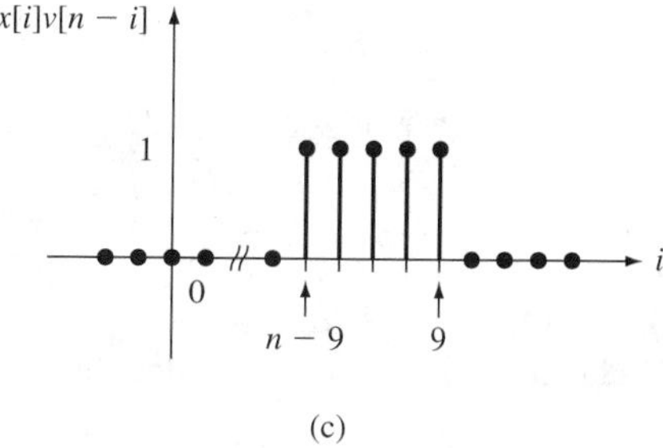

Figure 3.6 Plots of (a) $x[i]$; (b) $v[n-i]$; (c) $x[i]v[n-i]$ for $10 \leq n \leq 18$.

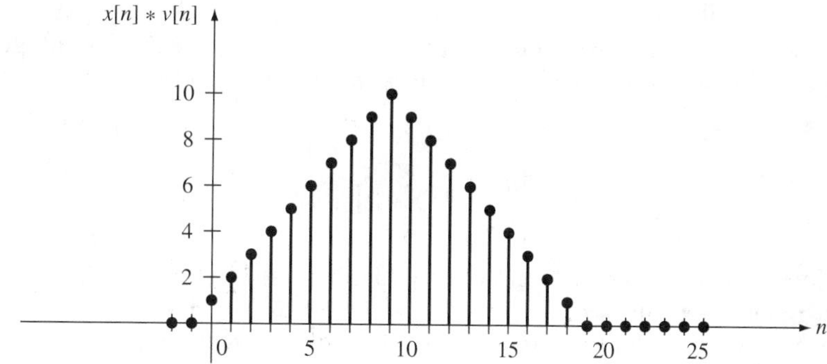

Figure 3.7 Plot of $x[n] * v[n]$.

Example 3.3 *Use of Analitical Form*

Suppose that $x[n] = a^n u[n]$ and $v[n] = b^n u[n]$, where $u[n]$ is the discrete-time unit-step function and a and b are fixed nonzero real numbers. Inserting $x[i] = a^i u[i]$ and $v[n-i] = b^{n-i} \mathrm{u}[n-i]$ into (3.8) gives

$$x[n] * v[n] = \sum_{i=0}^{n} a^i u[i] b^{n-i} u[n-i], \qquad n = 0, 1, 2, \ldots \tag{3.9}$$

Now $u[i] = 1$ and $u[n - i] = 1$ for all integer values of i ranging from $i = 0$ to $i = n$, and thus (3.9) reduces to

$$x[n] * v[n] = \sum_{i=0}^{n} a^i b^{n-i} = b^n \sum_{i=0}^{n} \left(\frac{a}{b}\right)^i, \qquad n = 0, 1, 2, \ldots \tag{3.10}$$

If $a = b$,

$$\sum_{i=0}^{n} \left(\frac{a}{b}\right)^i = n + 1$$

and

$$x[n] * v[n] = b^n(n + 1) = a^n(n + 1), \qquad n = 0, 1, 2, \ldots$$

If $a \neq b$,

$$\sum_{i=0}^{n} \left(\frac{a}{b}\right)^i = \frac{1 - (a/b)^{n+1}}{1 - a/b} \tag{3.11}$$

The relationship (3.11) can be verified by multiplying both sides of (3.11) by $1 - (a/b)$. Inserting (3.11) into (3.10) yields (assuming that $a \neq b$)

$$x[n] * v[n] = b^n \frac{1 - (a/b)^{n+1}}{1 - a/b} = \frac{b^{n+1} - a^{n+1}}{b - a}, \qquad n = 0, 1, 2, \ldots$$

It is easy to generalize (3.8) to the case when $x[n]$ and $v[n]$ are not necessarily zero for all integers $n < 0$. In particular, suppose that $x[n] = 0$ for all $n < N$ and $v[n] = 0$ for all $n < M$, where M and N are positive or negative integers. In this case the convolution operation (3.7) can be written in the form

$$x[n] * v[n] = \begin{cases} 0, & n < M + N \\ \displaystyle\sum_{i=N}^{n-M} x[i]v[n - i], & n \geq M + N \end{cases} \tag{3.12}$$

Note that the convolution sum in (3.12) is still finite, and thus the convolution $x[n] * v[n]$ exists.

The convolution operation (3.12) can be evaluated using an array which is defined as follows: Let the first row of the array consist of the values $x[N]$, $x[N + 1], \ldots$, and let the second row consist of the values $v[M]$, $v[M + 1], \ldots$. The third row of the array consists of the products of the first element in the second row and successive elements of the first row. The elements of the fourth row are the products of the second element of the second row and successive elements of the first row. The products in the third, fourth, ... rows are shifted one column to the right of the products in the previous row. This pattern is repeated as shown in the array below. Then the values of $y[n] = x[n] * v[n]$ are computed by summing the products in each column of the array starting with the third row. The values $y[n]$ of the convolution are displayed in the last row of the array as shown below:

$x[N]$	$x[N+1]$	$x[N+2]$	$x[N+3]$	$\cdots$
$v[M]$	$v[M+1]$	$v[M+2]$	$v[M+3]$	$\cdots$
$x[N]v[M]$	$x[N+1]v[M]$	$x[N+2]v[M]$	$x[N+3]v[M]$	$\cdots$
	$x[N]v[M+1]$	$x[N+1]v[M+1]$	$x[N+2]v[M+1]$	$\cdots$
		$x[N]v[M+2]$	$x[N+1]v[M+2]$	$\cdots$
			$\vdots$	$\vdots$
$y[M+N]$	$y[M+N+1]$	$y[M+N+2]$	$y[M+N+3]$	$\cdots$

The process is illustrated by the following example.

Example 3.4 *Array Method*

Suppose that $x[n] = 0$ for $n < -1, x[-1] = 1, x[0] = 2, x[1] = 3, x[2] = 4, x[3] = 5, \ldots,$ and $v[n] = 0$ for $n < -2$, $v[-2] = -1$, $v[-1] = 5$, $v[0] = 3$, $v[1] = -2$, $v[2] = 1, \ldots$. In this case, $N = -1$, $M = -2$, and the array is as follows:

1	2	3	4	5	$\cdots$
−1	5	3	−2	1	$\cdots$
−1	−2	−3	−4	−5	$\cdots$
	5	10	15	20	$\cdots$
		3	6	9	$\cdots$
			−2	−4	$\cdots$
				1	$\cdots$
−1	3	10	15	21	$\cdots$

The values $y[n]$ of the convolution $x[n] * v[n]$ are given by the last row; that is, $y[-3] = -1, y[-2] = 3, y[-1] = 10, y[0] = 15, y[1] = 21, \ldots,$ and $y[n] = 0$ for $n < -3$.

The convolution of two discrete-time signals can be carried out using the MATLAB M-file `conv`. To illustrate this, consider the convolution of the pulse $p[n]$ with itself, where $p[n]$ is defined in Example 3.2. The MATLAB commands for computing the convolution in this case are

```
p = [0 ones(1,10) zeros(1,5)]; % corresponds to n=-1 to n=14
x = p; v = p;
y = conv(x,v);
n = -2:25;
stem(n,y(1:length(n)))
```

The command `y = conv(x,v)` in this example results in a vector `y` that has length 32 with the first element corresponding to $y[-2]$. The values corresponding to $n = -2$ to $n = 25$ are then plotted to recover the graph in Figure 3.7. The reader is encouraged to read the comments in the MATLAB tutorial on the Web site.

9/15

Properties of the Convolution Operation

The convolution operation $x[n] * v[n]$ satisfies a number of useful properties that are given below.

Associativity. For any convolvable signals $x[n]$, $v[n]$, and $w[n]$,

$$x[n] * (v[n] * w[n]) = (x[n] * v[n]) * w[n]$$

so convolution satisfies the associativity property. The proof of associativity follows easily from the definition of convolution and is left to the reader.

Commutativity. The convolution operation $x[n] * v[n]$ is commutative; that is,

$$x[n] * v[n] = v[n] * x[n] \tag{3.13}$$

or

$$\sum_{i=-\infty}^{\infty} x[i]v[n-i] = \sum_{i=-\infty}^{\infty} v[i]x[n-i] \tag{3.14}$$

To prove (3.14), let $\bar{i} = n - i$ in the definition (3.7) of convolution. Then $i = n - \bar{i}$ and (3.7) becomes

$$x[n] * v[n] = \sum_{\bar{i}=-\infty}^{\infty} x[n-\bar{i}]v[\bar{i}] \tag{3.15}$$

Since the signals $x[n]$ and $v[n]$ are real valued and real numbers commute, (3.15) can be rewritten as

$$x[n] * v[n] = \sum_{\bar{i}=-\infty}^{\infty} v[\bar{i}]x[n-\bar{i}] \tag{3.16}$$

Finally, since the index $\bar{i}$ of the summation in (3.16) can be changed to i, the right-hand side of (3.16) is equal to the convolution $v[n] * x[n]$, which proves the commutativity property.

Distributivity with addition. The convolution operation is distributive with addition; that is, for any signals $x[n]$, $v[n]$, $w[n]$,

$$x[n] * (v[n] + w[n]) = x[n] * v[n] + x[n] * w[n] \tag{3.17}$$

The relationship (3.17) follows directly from the definition of convolution. The details are omitted.

Shift property. Given two signals $x[n]$ and $v[n]$ and a positive or negative integer q, consider the shifted signals $x[n-q]$ and $v[n-q]$. Recall that $x[n-q]$ and $v[n-q]$ are q-step right shifts (respectively, q-step left shifts) of $x[n]$ and $v[n]$ when

$q > 0$ (respectively, when $q < 0$). Now let $x_q[n] = x[n - q]$, $v_q[n] = v[n - q]$, and let $w[n] = x[n] * v[n]$. Then

$$w[n - q] = x_q[n] * v[n] = x[n] * v_q[n] \tag{3.18}$$

By (3.18), the q-step shift of the convolution $w[n] = x[n] * v[n]$ is equal to the q-step shift of $x[n]$ convolved with $v[n]$, which in turn is equal to $x[n]$ convolved with the q-step shift of $v[n]$.

To prove (3.18), from (3.7),

$$w[n - q] = \sum_{i=-\infty}^{\infty} x[i]v[n - q - i] \tag{3.19}$$

By definition of $v_q[n]$,

$$v[n - q - i] = v_q[n - i]$$

and thus (3.19) can be rewritten as

$$w[n - q] = \sum_{i=-\infty}^{\infty} x[i]v_q[n - i] = x[n] * v_q[n]$$

Hence the second part of (3.18) is proved. The first part of (3.18) is proved in a similar manner using the commutativity property. The details are omitted.

Convolution with the unit pulse. For any signal $x[n]$,

$$x[n] * \delta[n] = x[n] \tag{3.20}$$

where $\delta[n]$ is the unit pulse. By (3.20), the convolution of any discrete-time signal $x[n]$ with the unit pulse $\delta[n]$ reproduces $x[n]$. As a result of this property, the unit pulse $\delta[n]$ is said to be the *identity element* of the convolution operation. To prove (3.20), first note that by (3.7),

$$x[n] * \delta[n] = \sum_{i=-\infty}^{\infty} x[i]\delta[n - i] \tag{3.21}$$

Since $\delta[n - i] = 0$ when $n \neq i$, and $\delta[n - i] = 1$ when $n = i$, the summation in (3.21) reduces to $x[n]$. Hence (3.20) is verified.

Convolution with the shifted unit pulse. Given a positive or negative integer q, let $\delta_q[n]$ denoted the shifted unit pulse $\delta[n - q]$; that is, $\delta_q[n] = \delta[n - q]$. Then for any signal $x[n]$,

$$x[n] * \delta_q[n] = x[n - q] \tag{3.22}$$

The relationship (3.22) reveals that the convolution of $x[n]$ with the shifted unit pulse $\delta_q[n]$ is equivalent to shifting $x[n]$ by q steps. This characterization of the shift operation turns out to be very useful. The proof of (3.22) follows from (3.20) and the shift property (3.18).

Computation of System Output

Given a causal linear time-invariant discrete-time system with unit-pulse response $h[n]$, let $y[n]$ denote the output response resulting from input $x[n]$ with $x[n] = 0$ for $n < 0$ and with no initial energy in the system at time $n = 0$. As shown in Section 3.1, $y[n]$ is equal to the convolution of $x[n]$ and $h[n]$; that is,

$$y[n] = x[n] * h[n] = \sum_{i=0}^{\infty} x[i]h[n - i], \qquad n \geq 0 \tag{3.23}$$

Since $h[n] = 0$ for $n < 0$ (by causality), $h[n - i] = 0$ for $i > n$ and thus (3.23) reduces to

$$y[n] = \sum_{i=0}^{n} x[i]h[n - i], \qquad n \geq 0 \tag{3.24}$$

Also note that by commutativity of convolution, (3.24) can be rewritten in the form

$$y[n] = h[n] * x[n] = \sum_{i=0}^{n} h[i]x[n - i], \qquad n \geq 0 \tag{3.25}$$

From (3.24) or (3.25), it is seen that for each integer value of $n > 0$, the computation of $y[n]$ requires in general $n + 1$ multiplications and n additions. Thus the computational complexity of the input/output convolution relationship increases with n. For instance, to compute the output $y[n]$ at time index $n = 10{,}000$, in general a total of 10,001 multiplications and 10,000 additions will have to be performed in order to evaluate the summation in (3.24) or (3.25).

The dependency on n of the computational complexity of convolution is in contrast to the recursion process studied in Section 2.3, where the computational complexity depends only on the order N of the recursion. The difference in the complexity of the two computational procedures is due to finite dimensionality. That is, the recursion process requires and exploits finite dimensionality of the given discrete-time system, whereas the convolution procedure does not require finite dimensionality and does not make effective use of finite dimensionality when the given system has this property.

Since the output $y[n]$ is equal to the discrete-time convolution $x[n] * h[n]$ given by (3.23), the response $y[n]$ for any finite range of values of n can be computed using the MATLAB M-file `conv`. The procedure is illustrated by the following example.

Example 3.5 *Convolution of Sinusoids*

Suppose that the unit-pulse response $h[n]$ is equal to $\sin(0.5n)$ for $n \geq 0$, and the input $x[n]$ is equal to $\sin(0.2n)$ for $n \geq 0$. Plots of $h[n]$ and $x[n]$ are given in Figure 3.8. Now to compute the response $y[n]$ for $n = 0, 1, \ldots, 40$, use the commands

```
n=0:40;
x = sin(.2*n);
h = sin(.5*n);
y = conv(x,h);
stem(n,y(1:length(n)))
```

A MATLAB-generated plot of the response values is given in Figure 3.9.

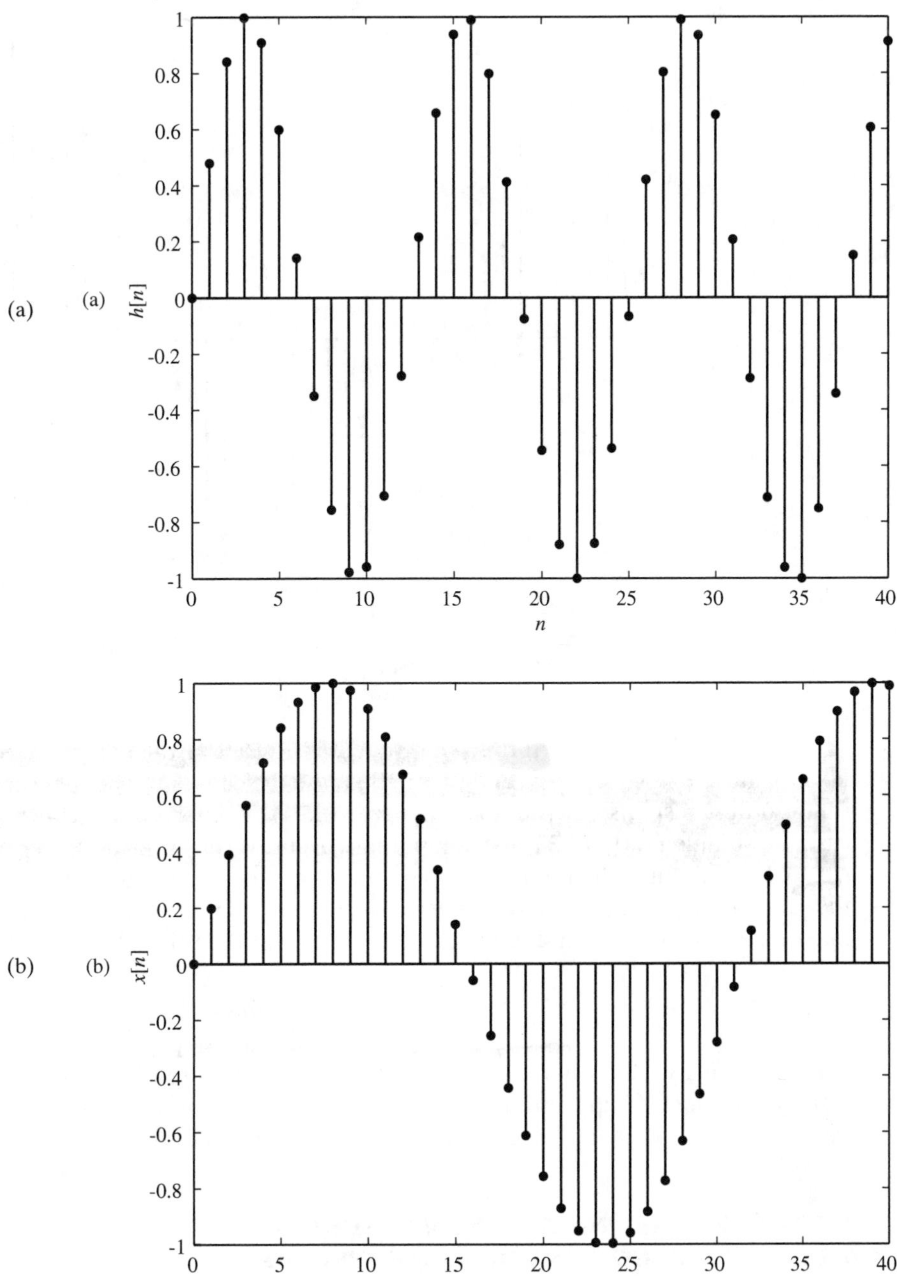

Figure 3.8 Plots of (a) $h[n]$ and (b) $x[n]$ in Example 3.5.

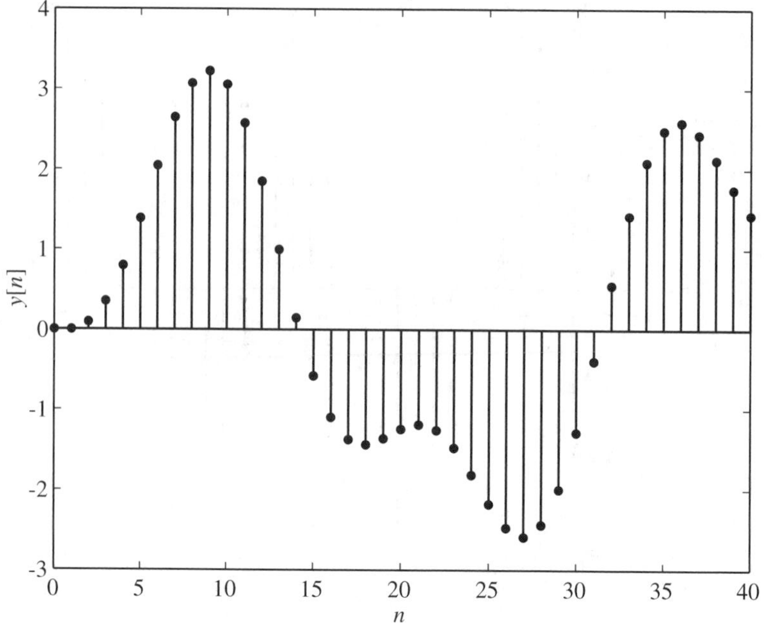

Figure 3.9 Plot of output response in Example 3.5.

Noncausal systems. If the given system is noncausal, the unit-pulse response $h[n]$ will not be zero for $n < 0$, and thus in this case the convolution expression (3.23) for the output response $y[n]$ does not reduce to (3.24). In other words, in the noncausal case the summation in (3.23) for computing $y[n]$ must run from $i = 0$ to $i = \infty$ (not $i = n$). In addition, if the input $x[n]$ is nonzero for values of n ranging from 0 to $-\infty$, the summation in (3.23) must start at $i = -\infty$. Hence the input/output convolution expression for a noncausal system (with $x[n] \neq 0$ for $n < 0$) is given by

$$y[n] = x[n] * h[n] = \sum_{i=-\infty}^{\infty} x[i]h[n - i] \tag{3.26}$$

It is interesting to note that although (3.26) is the input/output relationship for the system, in general (3.26) cannot be computed since it is a bi-infinite sum; that is, an infinite summation cannot be evaluated in a finite number of (computational) steps.

3.3 CONVOLUTION REPRESENTATION OF LINEAR TIME-INVARIANT CONTINUOUS-TIME SYSTEMS

Consider a single-input single-output causal linear time-invariant continuous-time system with input $x(t)$ and output $y(t)$. Throughout this section it is assumed that $y(t)$ is the response resulting from $x(t)$ with no initial energy in the system prior to the application of $x(t)$. As shown below, the input/output relationship of the system can be specified in terms of a convolution operation between the input $x(t)$ and the impulse response of the system. The development begins with the definition of the impulse response.

Impulse Response

Let $\delta(t)$ denote the unit impulse concentrated at $t = 0$. Recall that $\delta(t)$ is defined by

$$\delta(t) = 0 \qquad \text{for all } t \neq 0$$

$$\int_{-\varepsilon}^{\varepsilon} \delta(t)\, dt = 1 \qquad \text{for any } \varepsilon > 0$$

The *impulse response* $h(t)$ of a causal linear time-invariant continuous-time system is the output response when the input $x(t)$ is the unit impulse $\delta(t)$ with no initial energy in the system at time $t = 0^-$ [prior to the application of $\delta(t)$]. Since the system is assumed to be causal and $\delta(t) = 0$ for all $t < 0$, the impulse response $h(t)$ is zero for all $t < 0$. An example of the form of the impulse response $h(t)$ is illustrated in Figure 3.10.

The impulse response $h(t)$ could be determined experimentally by applying the unit impulse to the system and measuring the resulting output response. To carry out this experiment, it would be necessary to apply to the system a large-amplitude short-duration pulse [as an approximation to $\delta(t)$], but in practice it is usually not possible to apply such an input to a system. In Section 3.4 it is shown that the impulse response can be determined by measuring the output response resulting from the application of a unit-step input.

The impulse response $h(t)$ may be computable from some known mathematical representation of the given system. For instance, suppose that the system is modeled by the input/output differential equation (2.1) studied in Chapter 2. For such a system the impulse response $h(t)$ is equal to the solution to (2.1) with $x(t) = \delta(t)$ and with the initial conditions $y^{(i)}(0^-) = 0$ for $i = 0, 1, 2, \ldots, N - 1$.

Example 3.6 *Impulse Response of RC Circuit*

Consider the *RC* circuit shown in Figure 3.11. The input $x(t)$ is the current $i(t)$ into the parallel connection and the output $y(t)$ is the voltage $v_C(t)$ across the capacitor. In Section 1.4 it was shown that the input/output differential equation of the *RC* circuit is given by

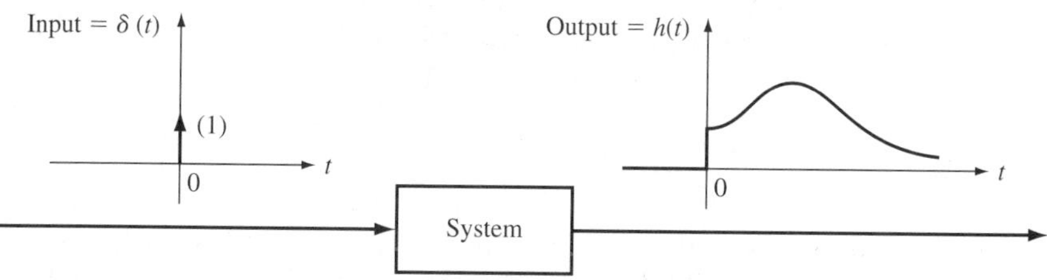

Figure 3.10 Generation of the impulse response.

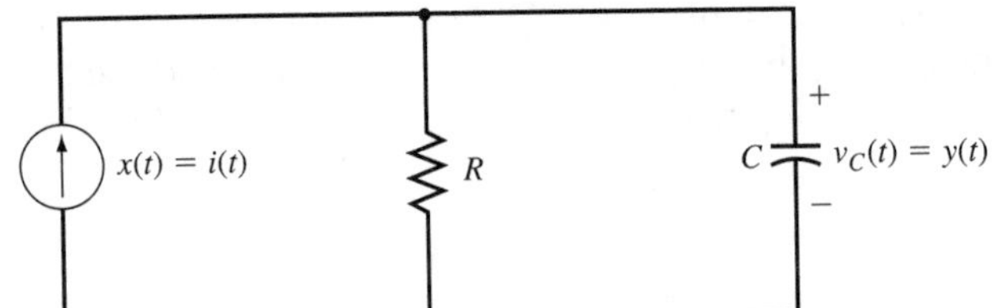

Figure 3.11 *RC* circuit.

$$C\frac{dy(t)}{dt} + \frac{1}{R}y(t) = x(t) \tag{3.27}$$

With initial condition $y(0^-) = 0$ and input $x(t)$ applied for $t \geq 0$, it follows from the results given in Section 1.4 that the solution to (3.27) is given by

$$y(t) = \int_{0^-}^{t} \frac{1}{C} e^{-(1/RC)(t-\lambda)} x(\lambda)\, d\lambda, \qquad t \geq 0 \tag{3.28}$$

Now the impulse response $h(t)$ for the *RC* circuit can be computed by setting $x(t) = \delta(t)$ in (3.28). This gives

$$y(t) = \int_{0^-}^{t} \frac{1}{C} e^{-(1/RC)(t-\lambda)} \delta(\lambda)\, d\lambda, \qquad t \geq 0 \tag{3.29}$$

By the shifting property of the impulse (see Section 1.2), (3.29) reduces to

$$y(t) = \frac{1}{C} e^{-(1/RC)t}, \qquad t \geq 0$$

Thus the impulse response $h(t)$ of the *RC* circuit is given by

$$h(t) = \begin{cases} 0, & t < 0 \\ \dfrac{1}{C} e^{-(1/RC)t}, & t \geq 0 \end{cases}$$

Consider a causal linear time-invariant continuous-time system with input $x(t)$, output $y(t)$, and impulse response $h(t)$. By time invariance of the system, for any positive number λ the output response resulting from the λ-second right shift of the unit impulse $\delta(t)$ must be equal to the λ-second right shift of the impulse response $h(t)$. Hence when $x(t) = \delta(t - \lambda)$, the resulting response is $h(t - \lambda)$. An example of the form of $h(t - \lambda)$ is illustrated in Figure 3.12.

Now let $x(t)$ be an arbitrary input with $x(t) = 0$ for all $t < 0$. By the sifting property of the impulse, $x(t)$ can be expressed in the form

$$x(t) = \int_{0^-}^{\infty} x(\lambda)\delta(t - \lambda)\, d\lambda, \qquad t \geq 0$$

Now for any fixed value of λ, let $y_\lambda(t)$ denote the output response resulting from the input $x(\lambda)\delta(t - \lambda)$ with no initial energy. Since the system is linear (and thus

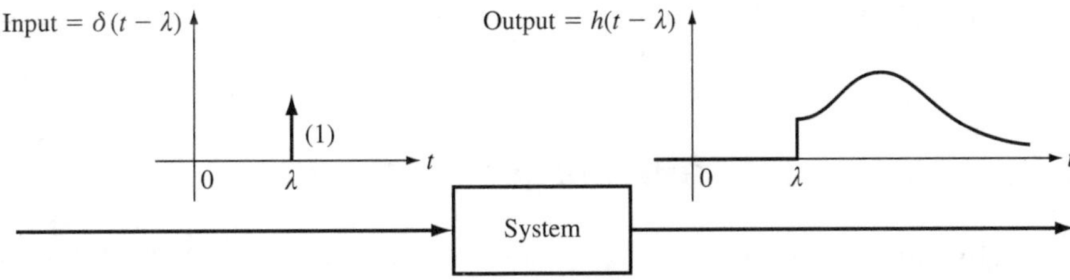

Figure 3.12 Time-shifted impulse response $h(t-\lambda)$.

is homogeneous), and $h(t-\lambda)$ is the response to $\delta(t-\lambda)$, the response $y_\lambda(t)$ to $x(\lambda)\delta(t-\lambda)$ is given by

$$y_\lambda(t) = x(\lambda)h(t-\lambda)$$

Then since $x(t)$ is the integral with respect to λ of $x(\lambda)\delta(t-\lambda)$, by the additivity property it follows that the response $y(t)$ to $x(t)$ is the integral with respect to λ of $y_\lambda(t)$. Therefore,

$$y(t) = \int_{0^-}^{\infty} x(\lambda)h(t-\lambda)\,d\lambda, \qquad t \geq 0 \tag{3.30}$$

If $x(\lambda)h(t-\lambda)$ does not contain an impulse concentrated at $\lambda = 0$, the lower limit of the integral in (3.30) can be taken to be zero. Hence the output response can be expressed in the form

$$y(t) = \int_{0}^{\infty} x(\lambda)h(t-\lambda)\,d\lambda, \qquad t \geq 0 \tag{3.31}$$

The operation on the right-hand side of (3.31) is the convolution of the input $x(t)$ and the impulse response $h(t)$. With the convolution operation denoted by $*$, the input/output relationship (3.31) becomes

$$y(t) = x(t) * h(t), \qquad t \geq 0 \tag{3.32}$$

The input/output relationship (3.32) [or (3.31)] is called the *convolution representation* of the system. Note that (3.31) is a natural continuous-time counterpart of the convolution representation (3.5) in the discrete-time case. Also note that the major difference between (3.5) and (3.31) is that the convolution sum in the discrete-time case becomes a convolution integral in the continuous-time case.

By the results above, the input/output relationship of a linear time-invariant continuous-time system is a convolution operation between the input $x(t)$ and the impulse response $h(t)$. One consequence of this relationship is that the system is completely determined by $h(t)$ in the sense that if $h(t)$ is known, the response to any input can be computed. Again, this corresponds to the situation in the discrete-time case where knowledge of the unit-pulse response $h[n]$ determines the system uniquely. The evaluation of the convolution operation in the continuous-time case is studied in the next section.

3.4 CONVOLUTION OF CONTINUOUS-TIME SIGNALS

Given two continuous-time signals $x(t)$ and $v(t)$, the convolution of $x(t)$ and $v(t)$ is defined by

$$x(t) * v(t) = \int_{-\infty}^{\infty} x(\lambda)v(t-\lambda)\,d\lambda \tag{3.33}$$

The integral in the right-hand side of (3.33) is called the *convolution integral*. If $x(t)$ and $v(t)$ are both zero for all $t < 0$, then $x(\lambda) = 0$ for all $\lambda < 0$ and $v(t - \lambda) = 0$ for all $t - \lambda < 0$ (or $t < \lambda$). In this case the integration in (3.33) may be taken from $\lambda = 0$ to $\lambda = t$, and the convolution operation is given by

$$x(t) * v(t) = \begin{cases} 0, & t < 0 \\ \displaystyle\int_0^t x(\lambda)v(t-\lambda)\,d\lambda, & t \geq 0 \end{cases} \tag{3.34}$$

The integral in (3.34) exists for all $t > 0$ if the functions $x(t)$ and $v(t)$ are absolutely integrable for all $t > 0$; that is,

$$\int_0^t |x(\lambda)|\,d\lambda < \infty \quad \text{and} \quad \int_0^t |v(\lambda)|\,d\lambda < \infty \qquad \text{for all } t > 0 \tag{3.35}$$

Most functions of interest in engineering satisfy (3.35) and therefore can be convolved.

To compute the convolution $x(t) * v(t)$, it is often useful to graph the functions in the integrand of the convolution integral. This can help to determine the integrand and integration limits of the convolution integral. The steps of this graphical aid to computing the convolution integral are listed below. Here it is assumed that both $x(t)$ and $v(t)$ are zero for all $t < 0$. If $x(t)$ and $v(t)$ are not zero for all $t < 0$, the shift property can be used to reduce the problem to the case when $x(t)$ and $v(t)$ are zero for all $t < 0$.

Step 1. Graph $x(\lambda)$ and $v(-\lambda)$ as functions of λ. The function $v(-\lambda)$ is equal to the function $v(\lambda)$ folded about the vertical axis.

Step 2. Let $[0,a]$ denote the set of all t such that $0 \leq t \leq a$, where a is a positive number. For t equal to an arbitrary point of the interval $[0,a]$, graph $v(t - \lambda)$ and the product $x(\lambda)v(t - \lambda)$ as functions of λ. Note that $v(t - \lambda)$ is equal to $v(-\lambda)$ shifted to the right by t seconds. The function $v(t - \lambda)$ is a folded and shifted version of $v(\lambda)$. The value of a in the interval $[0,a]$ is the largest value of a for which the product $x(\lambda)v(t - \lambda)$ has the same analytical form [or the plot of $x(\lambda)v(t - \lambda)$ has the same form] for all values of $t \in [0,a]$.

Step 3. Integrate the product $x(\lambda)v(t-\lambda)$ as a function of λ with the limits of integration from $\lambda = 0$ to $\lambda = t$. The result is the convolution $x(t) * v(t)$ for $t \in [0,a]$.

Step 4. For t equal to an arbitrary point of the interval $[a,b]$, graph $v(t-\lambda)$ and the product $x(\lambda)v(t-\lambda)$ as functions of λ. The value of b is the largest value of b for which the product $x(\lambda)v(t-\lambda)$ has the same analytical form for all values of $t \in [a,b]$.

Step 5. Integrate the product $x(\lambda)v(t-\lambda)$ as a function of λ with the limits of integration from $\lambda = 0$ to $\lambda = t$. The result is the convolution $x(t) * v(t)$ for $t \in [a,b]$. Repeat steps 4 and 5 as many times as necessary until $x(t) * v(t)$ is computed for all $t > 0$.

The procedure above is illustrated by the following two examples.

Convolution of Continuous-Time Signals

Example 3.7 *Convolution of Pulses*

Suppose that $x(t) = u(t) - 2u(t-1) + u(t-2)$, and that $v(t)$ is the pulse $v(t) = u(t) - u(t-1)$. The convolution of $x(t)$ and $v(t)$ is carried out via the steps given above.

Step 1. The functions $x(\lambda)$ and $v(-\lambda)$ are plotted in Figure 3.13.

Step 2. For $t \in [0,1]$, the plots of $x(\lambda)$, $v(t-\lambda)$ and the product $x(\lambda)v(t-\lambda)$ are given in Figure 3.14. For $t \in [1,2]$, the form of the product $x(\lambda)v(t-\lambda)$ changes as shown in Figure 3.15c. Thus the value of a is 1.

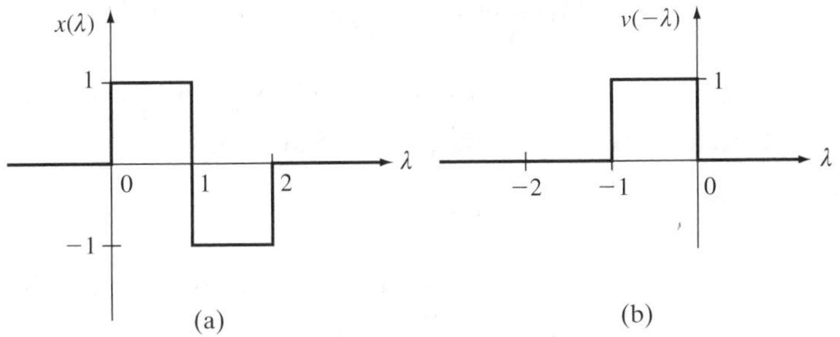

Figure 3.13 Plots of (a) $x(\lambda)$ and (b) $v(-\lambda)$.

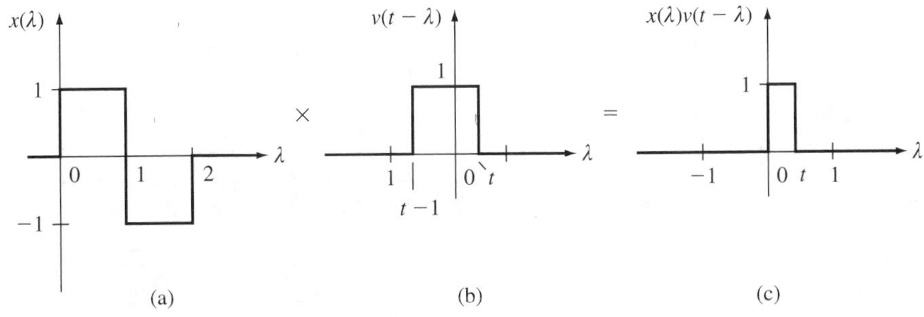

Figure 3.14 Plots of (a) $x(\lambda)$, (b) $v(t-\lambda)$, and (c) $x(\lambda)v(t-\lambda)$ for $t \in [0,1]$.

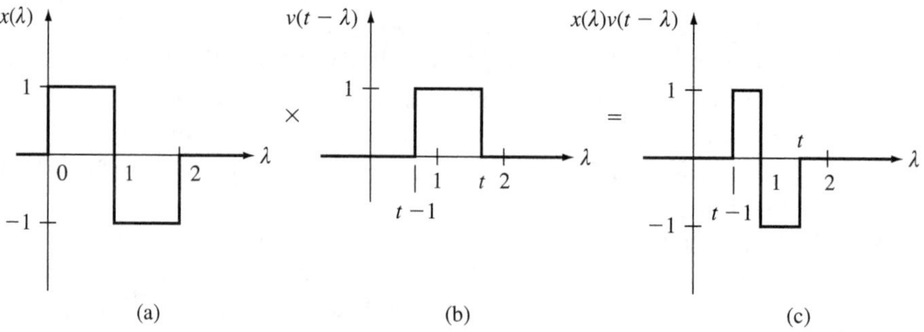

Figure 3.15 Plots of (a) $x(\lambda)$, (b) $v(t-\lambda)$, and (c) $x(\lambda)v(t-\lambda)$ for $t \in [1,2]$.

Step 3. Integrating the product $x(\lambda)v(t-\lambda)$ displayed in Figure 3.14c for $0 \le t \le 1$ gives

$$x(t) * v(t) = \int_0^t 1\, d\lambda = t$$

Step 4. For $t \in [2,3]$, the product $x(\lambda)v(t-\lambda)$ is plotted in Figure 3.16c. From Figures 3.15c and 3.16c, it is seen that the form of the product $x(\lambda)v(t-\lambda)$ again changes from the interval [1,2] to the interval [2,3]. Thus the value of b is 2.

Step 5. Integrating the product plotted in Figure 3.15c for $1 \le t \le 2$ yields

$$x(t) * v(t) = \int_{t-1}^{1} (1)\, d\lambda + \int_1^t (-1)\, d\lambda$$

$$= 1 - (t-1) + (-1)(t-1) = -2t + 3$$

Repeating step 5 with $t \in [2,3]$, from Figure 3.16c

$$x(t) * v(t) = \int_{t-1}^{2} (-1)\, d\lambda$$

$$= (-1)[2-(t-1)] = t-3, \qquad 2 \le t \le 3$$

Finally, for $t \ge 3$, the product $x(\lambda)v(t-\lambda)$ is zero since there is no overlap between $x(\lambda)$ and $v(t-\lambda)$ when $t \ge 3$. Hence

$$x(t) * v(t) = 0 \qquad \text{for all } t \ge 3$$

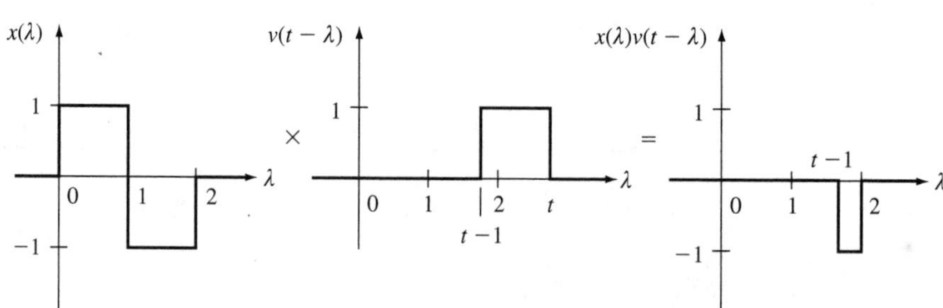

Figure 3.16 Plots of (a) $x(\lambda)$, (b) $v(t-\lambda)$, and (c) $x(\lambda)v(t-\lambda)$ for $t \in [2,3]$.

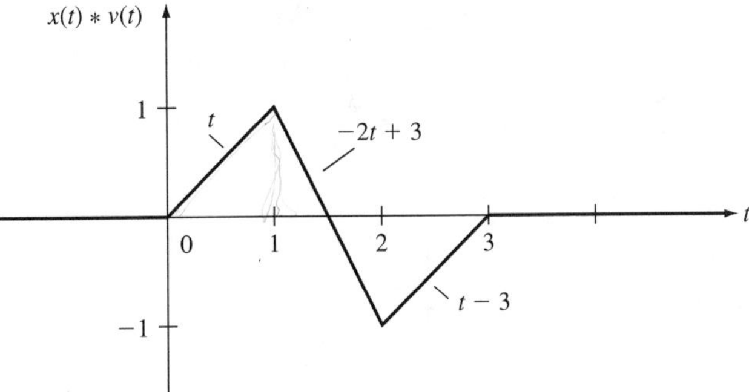

Figure 3.17 Sketch of $x(t) * v(t)$.

A sketch of the convolution $x(t) * v(t)$ is shown in Figure 3.17. The convolution $x(t) * v(t)$ can be expressed in analytical form as follows. From the plot in Figure 3.17

$$x(t) * v(t) = t[u(t) - u(t-1)] + (-2t+3)[u(t-1) - u(t-2)] + (t-3)[u(t-2) - u(t-3)]$$

$$= tu(t) + (-3t+3)u(t-1) + (3t-6)u(t-2) - (t-3)u(t-3)$$

Example 3.8 *Convolution of Exponential Segments*

Consider the signals $x(t)$ and $v(t)$ defined by

$$x(t) = \begin{cases} e^t, & 0 \le t < 1 \\ e^{2-t}, & 1 \le t < 2 \\ 0, & \text{all other } t \end{cases} \qquad v(t) = \begin{cases} e^{-t}, & 0 \le t \le 4 \\ 0, & \text{all other } t \end{cases}$$

The signals $x(t)$ and $v(t)$ are plotted in Figure 3.18. The functions $x(\lambda)$ and $v(-\lambda)$ are displayed in Figure 3.19. For $t \in [0,1]$, the functions $v(t-\lambda)$ and $x(\lambda)v(t-\lambda)$ are plotted in Figure 3.20, and for $t \in [1,2]$, these functions are plotted in Figure 3.21. Integrating the product $x(\lambda)v(t-\lambda)$ displayed in Figure 3.20b for $t \in [0,1]$ yields

$$x(t) * v(t) = \int_0^t e^{\lambda}e^{-(t-\lambda)}\,d\lambda = e^{-t}\int_0^t e^{2\lambda}\,d\lambda$$

$$= \frac{e^{-t}}{2}[e^{2\lambda}]_{\lambda=0}^{\lambda=t} = \frac{e^{-t}}{2}(e^{2t} - 1)$$

$$= \frac{1}{2}(e^t - e^{-t})$$

Integrating the product displayed in Figure 3.21b for $t \in [1,2]$ gives

$$x(t) * v(t) = \int_0^1 e^{\lambda}e^{-(t-\lambda)}\,d\lambda + \int_1^t e^{2-\lambda}e^{-(t-\lambda)}\,d\lambda$$

$$= \frac{e^{-t}}{2}[e^{2\lambda}]_{\lambda=0}^{\lambda=1} + e^2e^{-t}\int_1^t (1)\,d\lambda$$

$$= \frac{e^{-t}}{2}(e^2 - 1) + e^2e^{-t}(t-1)$$

$$= \left[\frac{e^2-1}{2} + e^2(t-1)\right]e^{-t}$$

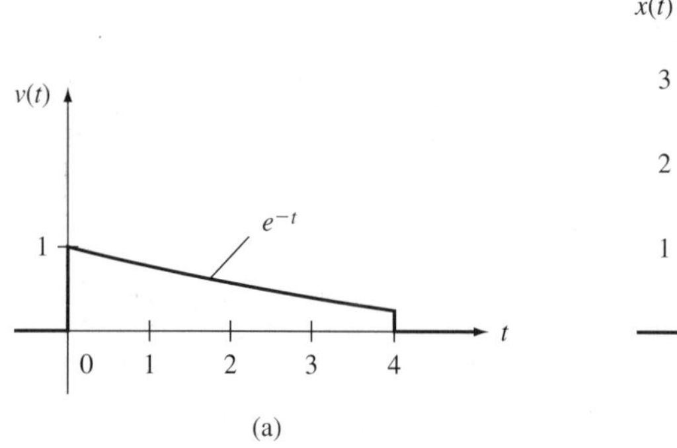

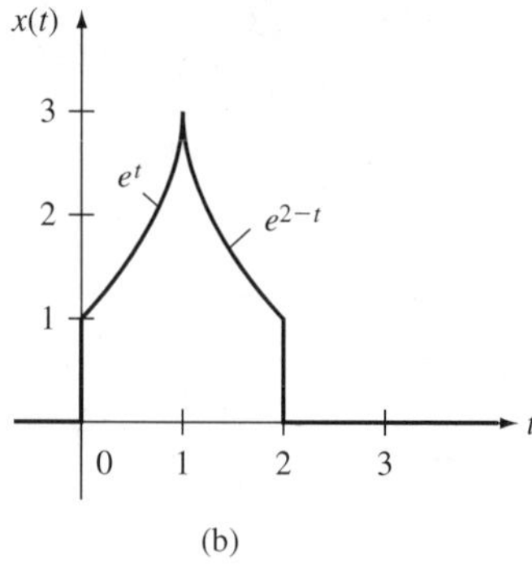

Figure 3.18 Plots of (a) $x(t)$ and (b) $v(t)$.

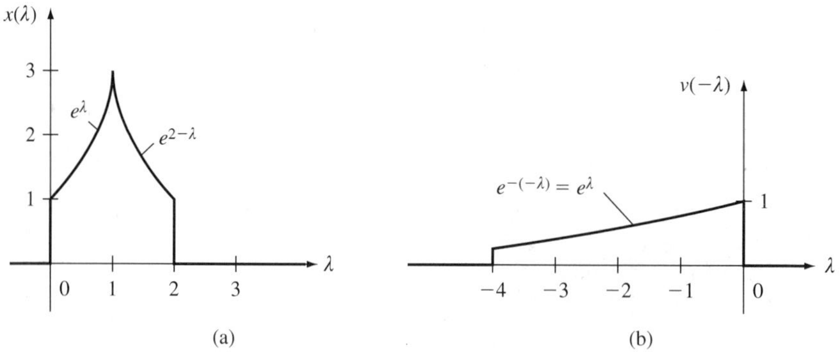

Figure 3.19 Functions (a) $x(\lambda)$ and (b) $v(-\lambda)$.

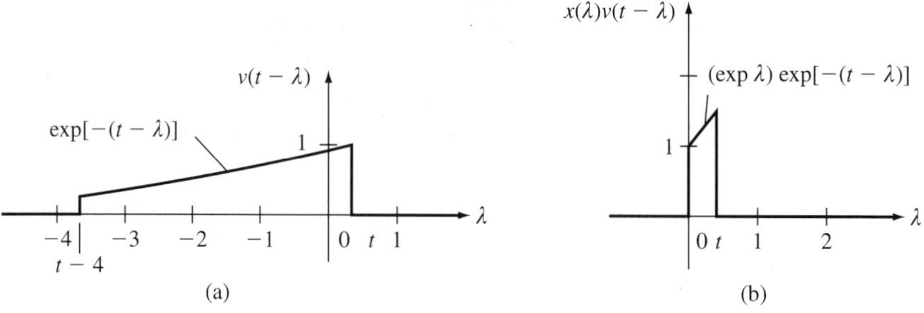

Figure 3.20 Functions (a) $v(t-\lambda)$ and (b) $x(\lambda)v(t-\lambda)$ for $t \in [0,1]$.

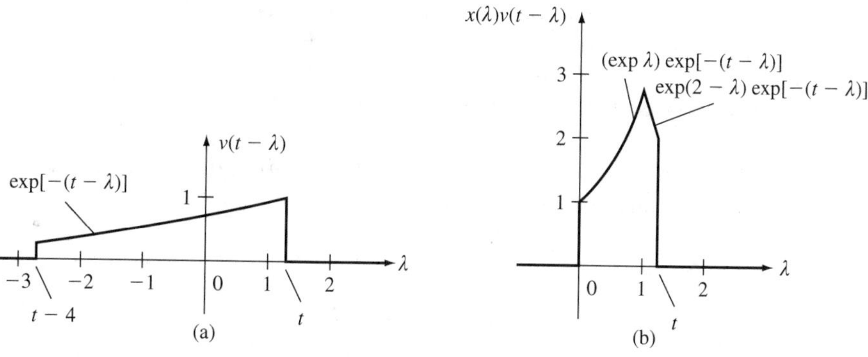

Figure 3.21 Functions (a) $v(t-\lambda)$ and (b) $x(\lambda)v(t-\lambda)$ for $t \in [1,2]$.

Continuing with the steps described above, for $t \in [2,4]$

$$x(t) * v(t) = \int_0^1 e^{\lambda}e^{-(t-\lambda)}\, d\lambda + \int_1^2 e^{2-\lambda}e^{-(t-\lambda)}\, d\lambda$$

$$= \frac{e^{-t}}{2}(e^2 - 1) + e^2e^{-t}\int_1^2 (1)\, d\lambda$$

$$x(t) * v(t) = \frac{e^{-t}}{2}(e^2 - 1) + e^2e^{-t}$$

$$= \left(\frac{e^2 - 1}{2} + e^2\right)e^{-t}$$

For $t \in [4,5]$,

$$x(t) * v(t) = \int_{t-4}^1 e^{\lambda}e^{-(t-\lambda)}\, d\lambda + \int_1^2 e^{2-\lambda}e^{-(t-\lambda)}\, d\lambda$$

$$= \frac{e^{-t}}{2}[e^{2\lambda}]_{\lambda=t-4}^{\lambda=1} + e^2e^{-t}\int_1^2 (1)\, d\lambda$$

$$= \frac{e^{-t}}{2}[e^2 - e^{2(t-4)}] + e^2e^{-t}$$

$$= \frac{1}{2}[3e^2 - e^{2(t-4)}]e^{-t}$$

For $t \in [5,6]$,

$$x(t) * v(t) = \int_{t-4}^2 e^{2-\lambda}e^{-(t-\lambda)}\, d\lambda$$

$$= e^2e^{-t}\int_{t-4}^2 (1)\, d\lambda$$

$$= e^2e^{-t}[2 - (t-4)]$$

$$= e^2(6 - t)e^{-t}$$

Finally, for $t \geq 6$, $x(t) * v(t) = 0$ since the functions $v(\lambda)$ and $x(t - \lambda)$ do not overlap when $t \geq 6$.

Properties of Convolution

As in the discrete-time case, the convolution operation satisfies a number of useful properties which are given below.

Associativity. For any convolvable signals $x(t)$, $v(t)$, and $w(t)$,

$$[x(t) * v(t)] * w(t) = x(t) * [v(t) * w(t)]$$

and thus the convolution operation is associative. The proof of associativity is omitted.

Commutativity. The convolution operation $x(t) * v(t)$ is commutative; that is,

$$x(t) * v(t) = v(t) * x(t) \tag{3.36}$$

By definition of convolution, (3.36) is equivalent to

$$\int_{-\infty}^{\infty} x(\lambda)v(t - \lambda)\, d\lambda = \int_{-\infty}^{\infty} v(\lambda)x(t - \lambda)\, d\lambda \tag{3.37}$$

The property (3.37) can be proved by considering the change of variable $\bar{t} = t - \lambda$ in the convolution integral in (3.33). The reader is invited to work out the details.

By (3.36) the convolution $x(t) * v(t)$ can be computed by folding and shifting either $x(t)$ or $v(t)$. When convolution is computed with the aid of graphical representations, it is best to fold and shift the simpler of the two functions $x(t)$ and $v(t)$.

Distributivity with addition. Convolution is distributive with addition; that is, for any convolvable signals $x(t)$, $v(t)$, $w(t)$,

$$x(t) * [v(t) + w(t)] = x(t) * v(t) + x(t) * w(t) \tag{3.38}$$

The proof of (3.38) is immediate from the definition of convolution.

Shift property. Given two continuous-time signals $x(t)$ and $v(t)$, and a positive or negative number c, let $x_c(t)$ and $v_c(t)$ denote the c-second shifts of $x(t)$ and $v(t)$ defined by

$$x_c(t) = x(t - c) \quad \text{and} \quad v_c(t) = v(t - c)$$

When $c > 0$, the signals $x_c(t)$ and $v_c(t)$ are c-second right shifts of $x(t)$ and $v(t)$, and $x_{-c}(t)$ and $v_{-c}(t)$ are c-second left shifts of $x(t)$ and $v(t)$. Now with $w(t) = x(t) * v(t)$, it is claimed that

$$w(t - c) = x_c(t) * v(t) = x(t) * v_c(t) \tag{3.39}$$

By (3.39), the c-second shift of the convolution $w(t) = x(t) * v(t)$ is equal to the c-second shift of $x(t)$ convolved with $v(t)$, which in turn is equal to $x(t)$ convolved with the c-second shift of $v(t)$. The proof of (3.39) follows directly from the definition of convolution. The details are omitted.

Derivative property. If the signal $x(t)$ has an ordinary first derivative $\dot{x}(t)$, the convolution $x(t) * v(t)$ has an ordinary first derivative and

$$\frac{d}{dt}[x(t) * v(t)] = \dot{x}(t) * v(t) \tag{3.40}$$

To prove (3.40), by definition of convolution

$$\begin{aligned}\frac{d}{dt}[x(t) * v(t)] = \frac{d}{dt}[v(t) * x(t)] &= \frac{d}{dt}\left[\int_{-\infty}^{\infty} v(\lambda)x(t-\lambda)\,d\lambda\right] \\ &= \int_{-\infty}^{\infty} v(\lambda)\left[\frac{d}{dt}x(t-\lambda)\right]d\lambda\end{aligned} \tag{3.41}$$

But

$$\frac{d}{dt}x(t-\lambda) = \dot{x}(t-\lambda)$$

and the right-hand side of (3.41) is equal to $v(t) * \dot{x}(t)$, which is equal to $\dot{x}(t) * v(t)$. Thus (3.40) is verified.

If both $x(t)$ and $v(t)$ are differentiable (i.e., have ordinary first derivatives), then using commutativity and applying the derivative property (3.40) twice yields the result that $x(t) * v(t)$ has an ordinary second derivative given by

$$\frac{d^2}{dt^2}[x(t) * v(t)] = \dot{x}(t) * \dot{v}(t)$$

This result shows that convolution is a *smoothing operation;* that is, the convolution $x(t) * v(t)$ is, in general, a smoother function of t than either $x(t)$ or $v(t)$. Here smoothness is measured in terms of the number of times a function can be differentiated in the ordinary sense.

Integration property. Let $x^{(-1)}(t)$ and $v^{(-1)}(t)$ denote the integrals of $x(t)$ and $v(t)$ defined by

$$x^{(-1)}(t) = \int_{-\infty}^{t} x(\lambda)\,d\lambda \quad \text{and} \quad v^{(-1)}(t) = \int_{-\infty}^{t} v(\lambda)\,d\lambda$$

Letting $(x * v)^{(-1)}$ denote the integral of $x * v$, it follows that

$$(x * v)^{(-1)} = x^{(-1)} * v = x * (v^{(-1)}) \tag{3.42}$$

The relationship (3.42) follows easily from the definition of convolution. The details are left to the reader.

Convolution with the unit impulse. Let $\delta(t)$ denote the unit impulse located at the origin. For any continuous-time signal $x(t)$, by definition of convolution

$$x(t) * \delta(t) = \delta(t) * x(t) = \int_{-\infty}^{\infty} \delta(\lambda)x(t-\lambda)\,d\lambda \tag{3.43}$$

Since $\delta(\lambda) = 0$ for all $\lambda \neq 0$, the integrand of the integral in (3.43) reduces to $\delta(\lambda)x(t)$, and thus

$$x(t) * \delta(t) = \int_{-\infty}^{\infty} \delta(\lambda)x(t)\,d\lambda = x(t)\int_{-\infty}^{\infty} \delta(\lambda)\,d\lambda$$

By definition of $\delta(t)$, it follows that

$$x(t) * \delta(t) = x(t) \tag{3.44}$$

By (3.44) it is seen that the convolution of any continuous-time signal $x(t)$ with the unit impulse $\delta(t)$ reproduces $x(t)$. Thus the unit impulse $\delta(t)$ is the identity element of the convolution operation.

Convolution with the shifted unit impulse. Given a positive or negative real number c, let $\delta_c(t)$ denote the unit impulse located at $t = c$; that is,

$$\delta_c(t) = \delta(t - c)$$

Given a signal $x(t)$, using (3.44) and the shift property (3.39) yields

$$x(t) * \delta_c(t) = x(t - c) \tag{3.45}$$

Hence convolving $x(t)$ with the shifted unit impulse $\delta_c(t)$ is equivalent to shifting $x(t)$ by c seconds.

Computation of System Output

Let $y(t)$ denote the output response of a causal linear time-invariant continuous-time system resulting from input $x(t)$ with $x(t) = 0$ for $t < 0$ and with no initial energy in the system at time $t = 0$. As shown in Section 3.3, $y(t)$ is equal to the convolution of the input $x(t)$ with the system's impulse response $h(t)$; that is

$$y(t) = x(t) * h(t) = \int_{0}^{t} x(\lambda)h(t-\lambda)\,d\lambda, \qquad t \geq 0 \tag{3.46}$$

Note that the upper limit of the integral in (3.46) is t [since $h(t) = 0$ for $t < 0$ by causality]. Also note that by commutativity of convolution, $y(t)$ can be expressed in the form

$$y(t) = h(t) * x(t) = \int_{0}^{t} h(\lambda)x(t-\lambda)\,d\lambda, \qquad t \geq 0 \tag{3.47}$$

Either (3.46) or (3.47) may be used to determine the response resulting from an input $x(t)$ with zero initial energy.

Example 3.9 *Finite-Time Integrator*

Given a fixed positive real number c, consider the system with impulse response

$$h(t) = u(t) - u(t - c)$$

Now if the system has zero initial energy at time $t = 0$, by the convolution representation (3.46), the output response $y(t)$ resulting from an arbitrary input $x(t)$ with $x(t) = 0$ for $t < 0$ is given by

$$y(t) = x(t) * h(t) = \int_0^t x(\lambda)h(t - \lambda)\, d\lambda, \qquad t \geq 0 \tag{3.48}$$

Since $h(t) = 1$ for $0 \leq t < c$ and $h(t) = 0$ for all other t,

$$h(t - \lambda) = \begin{cases} 1, & 0 \leq t - \lambda < c \\ 0, & \text{all other } t - \lambda \end{cases}$$

Now $0 \leq t - \lambda < c$ is equivalent to $-t \leq -\lambda < c - t$, which is equivalent to $t \geq \lambda > t - c$, and therefore

$$h(t - \lambda) = \begin{cases} 1, & t - c < \lambda \leq t \\ 0, & \text{all other } \lambda \end{cases} \tag{3.49}$$

Inserting (3.49) into (3.48) gives

$$y(t) = \int_{t-c}^t x(\lambda)\, d\lambda, \qquad t \geq c \tag{3.50}$$

The expression (3.50) for the output response $y(t)$ shows that at time t, the system processes an input $x(t)$ by integrating the input over the past c-second interval.

Representation in terms of the step response. Let $g(t)$ denote the output response of the system when the input $x(t)$ is the unit-step function $u(t)$ with no initial energy in the system at time $t = 0$. The response $g(t)$ is called the step response. From the convolution representation (3.47), it follows that

$$g(t) = h(t) * u(t) \tag{3.51}$$

Differentiating both sides of (3.51) and using the derivative property (3.40) of convolution gives

$$\dot{g}(t) = \dot{h}(t) * u(t) = h(t) * \dot{u}(t) \tag{3.52}$$

Since $\dot{u}(t) = \delta(t)$ and $h(t) = h(t) * \delta(t)$, from (3.52)

$$\dot{g}(t) = h(t) \tag{3.53}$$

Hence the impulse response $h(t)$ is equal to the derivative of the step response $g(t)$. In addition, integrating both sides of (3.53) yields

$$g(t) = \int_0^t h(\lambda)\, d\lambda$$

and thus the step response is equal to the integral of the impulse response.

Now let $x(t)$ be an arbitrary input with $x(t) = 0$ for $t < 0$. Inserting (3.53) into the convolution relationship (3.47) gives

$$y(t) = \dot{g}(t) * x(t) \tag{3.54}$$

Again using the derivative property of convolution,

$$\dot{g}(t) * x(t) = g(t) * \dot{x}(t)$$

and thus (3.54) can be written in the form

$$y(t) = g(t) * \dot{x}(t) \tag{3.55}$$

The relationships (3.54) and (3.55) are two additional convolution representations of the system, both of which are given in terms of the step response $g(t)$ of the system. Either (3.54) or (3.55) may be used to compute the output response resulting from an arbitrary input $x(t)$ with no initial energy in the system at time $t = 0$.

Noncausal systems. If the given system is noncausal, $h(t)$ will not be zero for $t < 0$, and in this case, the upper limit of the integral in the convolution expression (3.46) must be taken to be ∞ (not t). In addition, if $x(t)$ is nonzero for $t < 0$, the lower limit of the integral in (3.46) must be taken to be $-\infty$. Thus, for a noncausal system with input $x(t) \neq 0$ for $t < 0$, the input/output convolution relationship is given by

$$y(t) = x(t) * h(t) = \int_{-\infty}^{\infty} x(\lambda)h(t - \lambda)\, d\lambda \tag{3.56}$$

3.5 NUMERICAL CONVOLUTION

Consider the causal linear time-invariant continuous-time system given by the convolution relationship

$$y(t) = \int_{0}^{\infty} h(\lambda)x(t - \lambda)\, d\lambda \tag{3.57}$$

In this section a numerical procedure is given for computing the convolution integral in (3.57).

Given a fixed positive real number T, the convolution integral in (3.57) can be discretized in time by setting $t = nT$, where n is an integer variable. This gives

$$y(nT) = \int_{0}^{\infty} h(\lambda)x(nT - \lambda)d\lambda \tag{3.58}$$

The convolution integral in (3.58) can be evaluated by breaking the integral into a sum of integrals over T-second intervals; that is,

$$\begin{aligned} y(nT) = \int_{0}^{T} h(\lambda)x(nT - \lambda)\, d\lambda + \int_{T}^{2T} h(\lambda)x(nT - \lambda)\, d\lambda + \cdots \\ + \int_{iT}^{iT+T} h(\lambda)x(nT - \lambda)\, d\lambda + \cdots \end{aligned} \tag{3.59}$$

Rewriting (3.59) using the summation symbol yields

$$y(nT) = \sum_{i=0}^{\infty} \int_{iT}^{iT+T} h(\lambda)x(nT - \lambda)\, d\lambda \tag{3.60}$$

Now if T is taken to be suitably small, for each positive integer i, $h(\lambda)$ and $x(nT - \lambda)$ can be approximated on the interval $iT \le \lambda < iT + T$ by

$$h(\lambda) = h(iT), \qquad iT \le \lambda < iT + T \tag{3.61}$$

$$x(nT - \lambda) = x(nT - iT), \qquad iT \le \lambda < iT + T \tag{3.62}$$

Inserting the approximations (3.61) and (3.62) into (3.60) results in the following approximation of $y(nT)$:

$$y(nT) = \sum_{i=0}^{\infty} \int_{iT}^{iT+T} h(iT)x(nT - iT)\, d\lambda \tag{3.63}$$

Since $h(iT)x(nT - iT)$ is independent of the variable λ of integration, this product can be moved outside of the integrand of the integrals in (3.63). This gives

$$\begin{aligned} y(nT) &= \sum_{i=0}^{\infty} \left[\int_{iT}^{iT+T} (1)\, d\lambda \right] h(iT)x(nT - iT) \\ &= \sum_{i=0}^{\infty} [\lambda|_{\lambda=iT}^{\lambda=iT+T}] h(iT)x(nT - iT) \\ &= \sum_{i=0}^{\infty} Th(iT)x(nT - iT) \end{aligned} \tag{3.64}$$

The expression (3.64) is an approximation of the output response $y(nT)$ at the time points $t = nT$, where n takes on integer values. In general, the approximation will be more accurate the smaller T is.

Writing (3.64) in terms of the notation for discrete-time signals results in the input/output relationship

$$y[n] = \sum_{i=0}^{\infty} Th[i]x[n - i] \tag{3.65}$$

Note that (3.65) can be viewed as the convolution representation of a linear time-invariant discrete-time system with unit-pulse response $Th[n]$, where $h[n]$ is the sampled version of the impulse response $h(t)$ of the original continuous-time system. This discrete-time system can be viewed as a discrete-time simulation of the given continuous-time system. Thus it is possible to study any linear time-invariant continuous-time system with impulse response $h(t)$ in terms of the linear time-invariant discrete-time system with unit-pulse response $Th[n] = Th(nT)$.

The discrete-time representation (3.65) is not the same as the discrete-time simulation that was constructed in Section 2.4 by discretizing the input/output differential equation of the given continuous-time system. Hence the procedure above yields a second approach to the discretization in time of continuous-time systems.

Since the expression (3.65) for $y[n]$ is given in terms of a convolution sum, $y[n]$ can be computed simply by evaluating the sum. This yields a numerical procedure for computing the convolution integral in the continuous-time case. In particular, suppose that the input $x(t)$ applied to the continuous-time system is zero for all $t < 0$. Then $x[n] = 0$ for $n = -1, -2, \ldots$, and the approximation (3.65) for $y[n]$ becomes

$$y[n] = \sum_{i=0}^{n} Th[i]x[n-i], \qquad n = 0, 1, 2, \ldots \tag{3.66}$$

The evaluation of (3.66) can be carried out using the MATLAB M-file `conv`. This is illustrated by the following example.

Example 3.10 *Computation of a Car's Position*

Consider the car with the input/output differential equation (1.24). It follows directly from equation (1.31) in Section 1.4 that the impulse response of the car is given by

$$h(t) = \frac{1}{k_f}[1 - e^{-(k_f/M)t}], \qquad t \geq 0 \tag{3.67}$$

Let T be a fixed positive real number. Then the unit-pulse response of the discrete-time simulation of the car is

$$Th[n] = \frac{T}{k_f}[1 - e^{-(k_f T/M)n}], \qquad n = 0, 1, 2, \ldots$$

The response $y[n] = y(nT)$ of the car to an arbitrary input force $x(t)$ applied for $t \geq 0$ with zero initial position and zero initial velocity is then given (approximately) by

$$y[n] = \sum_{i=0}^{n} \frac{T}{k_f}[1 - e^{-(k_f T/M)i}]x[n-i], \qquad n \geq 0 \tag{3.68}$$

The MATLAB M-file `conv` was used to evaluate (3.68) for the case when $M = 1$, $k_f = 0.1$, $T = 2$, and with the input $x(t)$ equal to the force shown in Figure 3.22. The commands are

```
n=0:20; T = 2;      % since t = n*T, the end time is 40 sec.
kf = 0.1; m = 1;
h = (1-exp(-kf*n*T/m))/kf;
x = [ones(1,5) -ones(1,5) zeros(1,10)];
y = conv(T*h,x);
plot(n*T,y(1:length(n)), 'o');
```

The resulting values for $y[n]$ are plotted in Figure 3.23. Also plotted in Figure 3.23 are the exact values of $y[n]$ which were computed by first evaluating the convolution integral (3.57) with $h(t)$ given by (3.67) and with $x(t)$ equal to the force shown in Figure 3.22. The exact response is

$$y(t) = \begin{cases} 100(0.1t - 1 + e^{-0.1t}), & 0 \leq t < 10 \\ -100(0.1t - 3 + (2e - 1)e^{-0.1t}), & 10 \leq t < 20 \\ 100(1 - 2e + e^2)e^{-0.1t}, & t \geq 20 \end{cases}$$

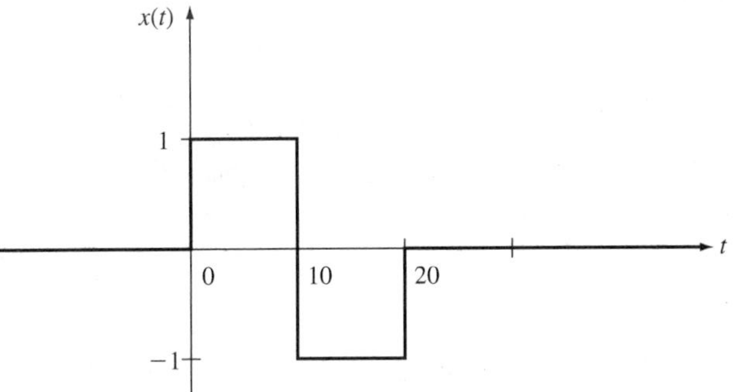

Figure 3.22 Force $x(t)$ applied to car in Example 3.10.

The MATLAB commands used to generate the exact solution, denoted by *ya*, are

```
t1 = 0:.1:9.9; % defines the segments of t
t2 = 10:.1:19.9;
t3 = 20:.1:40;
t = [t1,t2,t3];
ya = [100*(0.1*t1-1+exp(-0.1*t1)), . . .
    -100*(0.1*t2-3+(2*exp(1)-1)*exp(-0.1*t2)), . . .
    100*(1-2*exp(1)+exp(1)^2)*exp(-0.1*t3)];
```

From Figure 3.23 it can be seen that the approximate solution does differ significantly from the exact solution at each value of nT. To obtain a more accurate approximation,

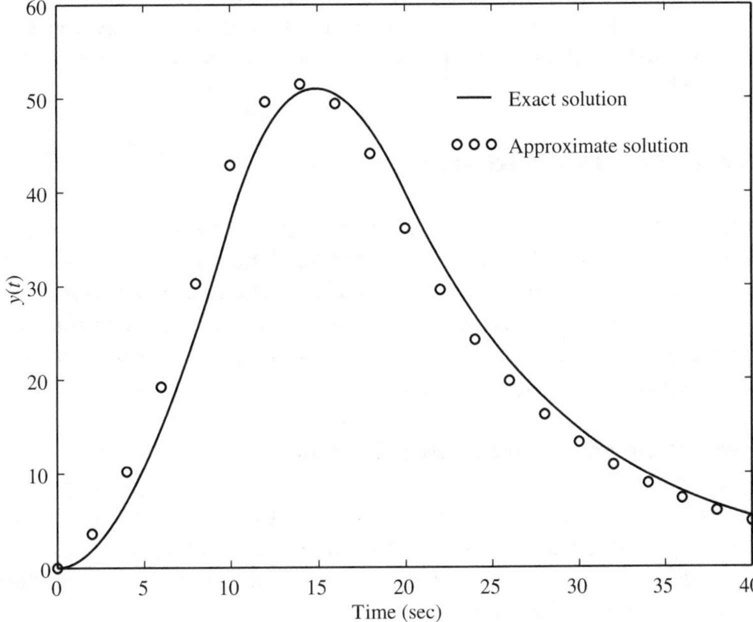

Figure 3.23 Exact and approximate output responses when $T = 2$.

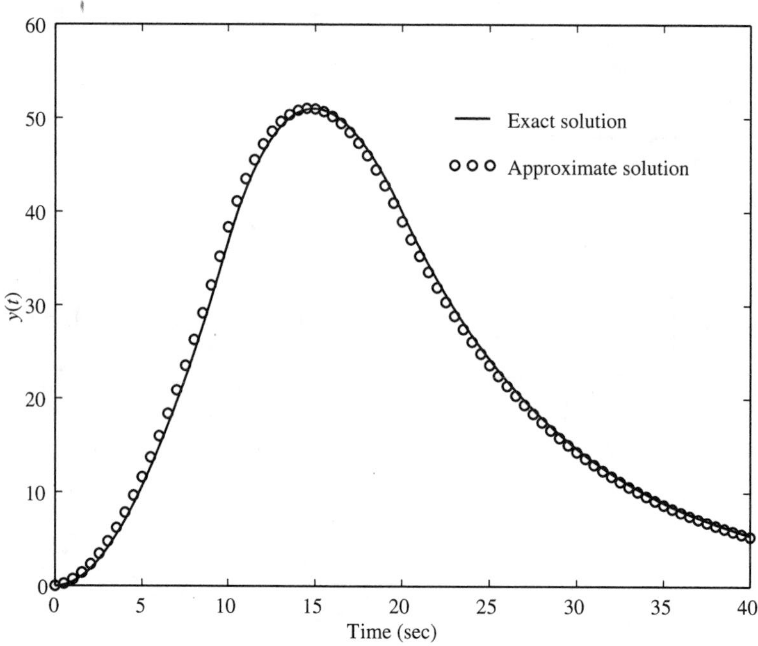

Figure 3.24 Exact and approximate output responses when $T = 0.5$.

a smaller discretization interval T can be chosen. For instance, if T is taken to be 0.5, the resulting approximate and exact solutions are as shown in Figure 3.24. Clearly, when $T = 0.5$ the approximation is closer to the exact solution than the approximation with $T = 2$.

3.6 LINEAR TIME-VARYING SYSTEMS

If the system under study is time varying, it is not possible to express the output response as the convolution of the input with the impulse response (or unit-pulse response). However, in the time-varying case there still is a useful expression for the output response that reduces to the convolution relationship in the time-invariant case. This input/output relationship is given below for linear time-varying systems. The discrete-time case is considered first.

Time-Varying Discrete-Time Systems

Given a causal linear discrete-time system, for each fixed integer i, let $h[n,i]$ denote the output response of the system when the input $x[n]$ is equal to the unit pulse $\delta[n - i]$ located at $n = i$, and with no initial energy in the system before the application of the pulse. Note that by causality,

$$h[n,i] = 0, \qquad n < i$$

Also note that

$$h[n,0] = h[n]$$

where $h[n]$ is the unit-pulse response defined in Section 3.1. The collection of responses $h[n,i]$ is the family of unit-pulse responses of the system.

Now let $x[n]$ be an arbitrary input with $x[n] = 0$ for $n = -1,-2,\ldots$. Then the output response $y[n]$ resulting from $x[n]$ with no initial energy at time index $n = 0$ is given by

$$y[n] = \sum_{i=0}^{n} x[i]h[n,i], \qquad n \geq 0 \tag{3.69}$$

Equation (3.69) is a generalization of the convolution relationship (3.24) for linear time-invariant discrete-time systems. Note that (3.69) and (3.24) are identical if and only if

$$h[n,i] = h[n-i] \qquad \text{for all } n \geq i \tag{3.70}$$

The condition (3.70) holds if and only if the given system is time invariant. Hence (3.69) reduces to the convolution relationship in the time-invariant case. The proof of (3.69) is similar to the proof of the convolution relationship (3.24) which was given in Section 3.1 (see Problem 3.35).

Example 3.11 *Response of Linear Time-Varying System*

Consider the linear time-varying discrete-time system given by the input/output difference equation

$$y[n] + a(n)y[n-1] = b(n)x[n-1] \tag{3.71}$$

In (3.71) the coefficients $a(n)$ and $b(n)$ are arbitrary functions of n. The unit-pulse responses $h[n,i]$ are the solutions of (3.71) with $x[n] = \delta[n-i]$ and with $y[i] = 0$. Solving (3.71) by recursion yields

$$h[n,i] = \begin{cases} 0, & n \leq i \\ b(i), & n = i+1 \\ (-1)^{n-i-1}a(n-1)a(n-2)\cdots a(i+1)b(i), & n \geq i+2 \end{cases} \tag{3.72}$$

Inserting (3.72) into (3.69) yields the input/output relationship

$$y[n] = \begin{cases} 0, & n = 0 \\ b(0)x[0], & n = 1 \\ \displaystyle\sum_{i=0}^{n-2} x[i](-1)^{n-i-1}a(n-1)a(n-2)\cdots a(i+1)b(i), & n \geq 2 \end{cases}$$

Continuous-Time Case

Now suppose that the system under consideration is a causal linear single-input single-output continuous-time system. For each fixed real number λ, let $h(t,\lambda)$ denote the output response resulting from the unit impulse $x(t) = \delta(t - \lambda)$ located at $t = \lambda$. The system is assumed to be at rest prior to the application of $\delta(t - \lambda)$. By causality

$$h(t,\lambda) = 0, \qquad t < \lambda$$

Note that

$$h(t,0) = h(t)$$

where $h(t)$ is the impulse response defined in Section 3.3. The collection of responses $h(t,\lambda)$ is the *family of impulse responses* of the system.

Now if $x(t)$ is an arbitrary input with $x(t) = 0$ for $t < 0$, the response resulting from $x(t)$ is

$$y(t) = \int_0^t x(\lambda)h(t,\lambda)\,d\lambda, \qquad t \geq 0 \tag{3.73}$$

The input/output relationship (3.73) is identical to the convolution relationship (3.46) if

$$h(t,\lambda) = h(t - \lambda) \qquad \text{for all } t \geq \lambda \tag{3.74}$$

The condition (3.74) is satisfied if and only if the given system is time invariant, and thus (3.73) reduces to the convolution expression in the time-invariant case.

Example 3.12 *Continuous-Time Case*

Consider the linear time-varying continuous-time system given by the input/output differential equation

$$\dot{y}(t) + a(t)y(t) = b(t)x(t) \tag{3.75}$$

where $a(t)$ and $b(t)$ are arbitrary functions of t. The impulse responses $h(t,\lambda)$ are the solutions of (3.75) with $x(t) = \delta(t - \lambda)$ and with $y(\lambda^-) = 0$. From the theory of differential equations with time-varying coefficients, it follows that

$$h(t,\lambda) = \exp\left[-\int_\lambda^t a(\beta)\,d\beta\right]b(\lambda) \tag{3.76}$$

Inserting (3.76) into (3.73) results in the input/output relationship

$$y(t) = \int_0^t x(\lambda)\exp\left[-\int_\lambda^t a(\beta)\,d\beta\right]b(\lambda)\,d\lambda, \qquad t \geq 0$$

Although it was possible to give $h(t,\lambda)$ in analytical form in Example 3.12, in general there is no closed-form analytical expression for the impulse responses in the time-varying case. In particular, this is usually the case for linear time-varying systems given by a second-order or higher-order input/output differential equation with time-varying coefficients.

PROBLEMS

3.1. Compute the unit-pulse response $h[n]$ for each of the following discrete-time systems (for $n = 0, 1, 2, 3$):

(a) $y[n + 1] + 2y[n] = x[n]$.

(b) $y[n + 2] - 2y[n + 1] + y[n] = x[n]$.

(c) $y[n+2] - 2y[n+1] + y[n] = x[n+1] + x[n]$.
(d) $y[n+2] + 1/2y[n+1] + 1/4y[n] = x[n+1] - x[n]$.
(e) $y[n+2] + 1/4y[n+1] - 3/8y[n+2] = 2x[n+2] - 3x[n]$.

3.2. Compute the unit-pulse response $h[n]$ for each of the following discrete-time systems (for all integers $n \geq 0$):
(a) $y[n+1] + 2y[n] = x[n]$.
(b) $y[n+1] - 1/2y[n] = x[n]$.
(c) $y[n+1] + 2y[n] = 2x[n+1] - 2x[n]$.
(d) $y[n+1] - 1/2y[n] = x[n+1] + 1/2x[n]$.

3.3. Let $r[n]$ denote the discrete-time unit-ramp function defined by $r[n] = v[n]$.
(a) Express $r[n]$ in terms of the unit-pulse function $\delta[n]$ and time shifts of $\delta[n]$.
(b) Let $h[n]$ represent the unit-pulse response of a linear time-invariant discrete-time system. Using your result in part (a), derive an expression for the response of the system to $r[n]$ in terms of $h[n]$.

3.4. For the discrete-time signals $x[n]$ and $v[n]$ shown in Figure P3.4:
(a) Compute the convolution $x[n] * v[n]$ for all $n \geq 0$. Sketch the results.
(b) Repeat part (a) using the M-file `conv`.

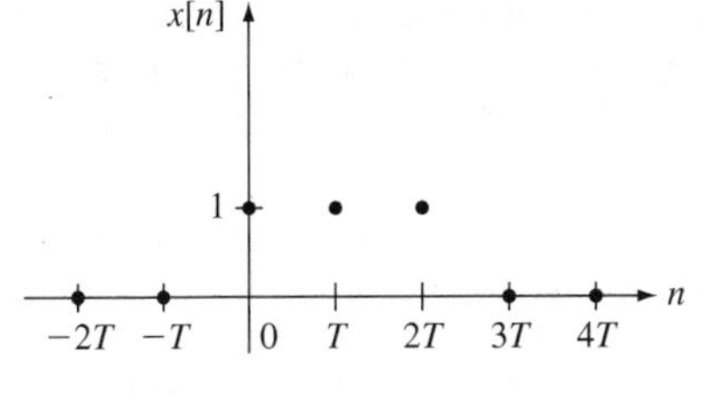

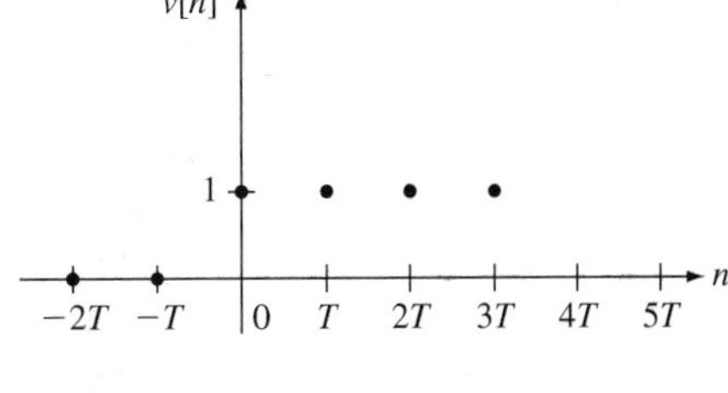

(a)

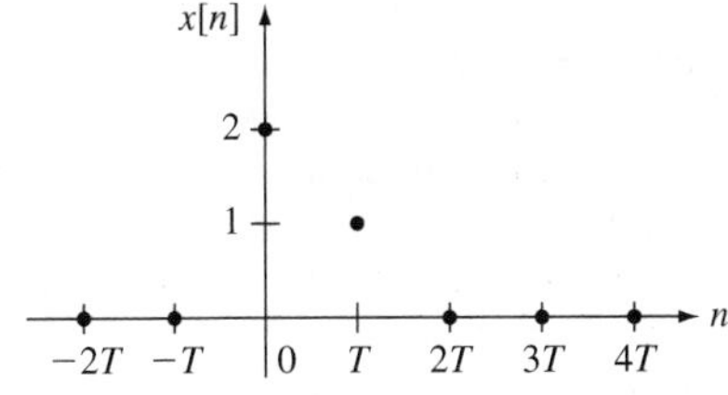

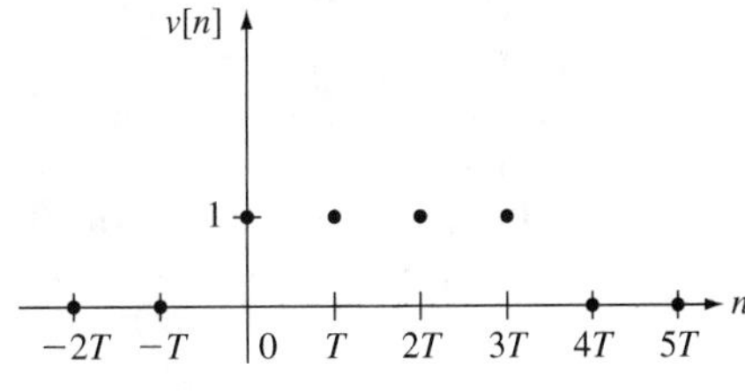

(b)

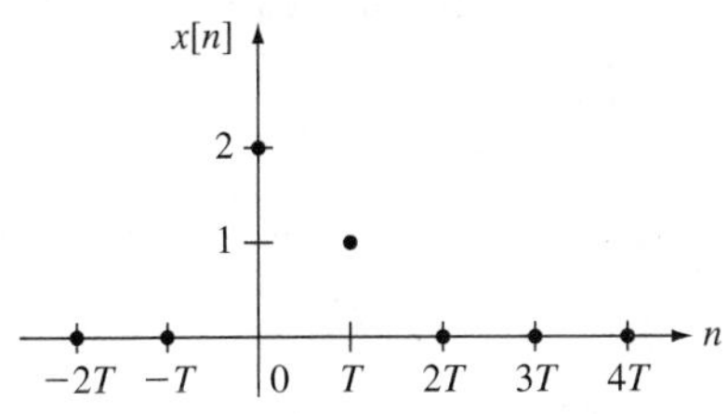

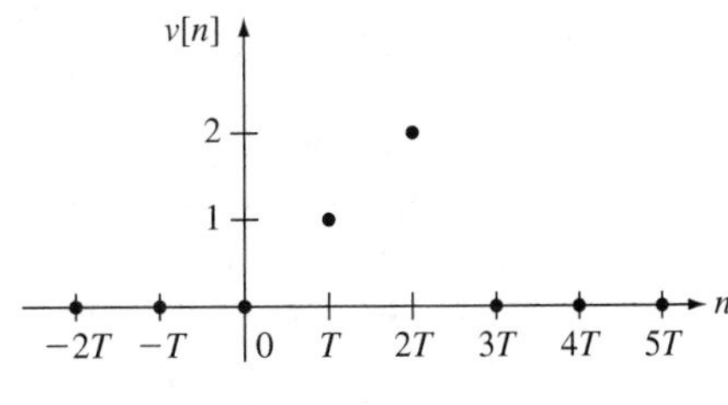

(c)

Figure P3.4

3.5. For the discrete-time signals $x[n]$ and $v[n]$ given in each of the following parts, compute the convolution $x[n] * v[n]$ for $n \geq 0$ [compute $x[n] * v[n]$ for $n \geq -2$ in part (b)].

(a) $x[0] = 1, x[1] = -2, x[3] = 3, x[n] = 0$ for all other integers n; $v[0] = 4, v[1] = -7, v[2] = 2, v[n] = 0$ for all other integers n.

(b) $x[n] = 1$ for $-1 \leq n \leq 2$ and $x[n] = 0$ for all other integers n; $v[-1] = 1$; $v[0] = 2, v[1] = 3, v[2] = 2, v[3] = 1$ and $v[n] = 0$ for all other integers n.

(c) $x[n] = 2^n$ for $n \leq 3$ and $x[n] = 0$ for $n \geq 4$; $v[0] = 2, v[1] = -3, v[2] = 0, v[3] = 6, v[n] = 0$ for all other integers n.

(d) $x[n] = 1/n$ for $2 \leq n \leq 5$ and $x[n] = 0$ for all other integers n; $v[2] = -2, v[3] = -5, v[n] = 0$ for all other integers n.

(e) $x[n] = u[n], v[n] = u[n]$ where $u[n]$ is the discrete-time step function.

(f) $x[n] = u[n], v[n] = \ln(n)$ for all integers $n \geq 1$ and $v[n] = 0$ for all integers $n < 1$.

(g) $x[n] = \delta[n] - \delta[n-2]$, where $\delta[n]$ is the unit pulse concentrated at $n = 0$; $v[n] = \cos(\pi n/3)$ for all integers $n \geq 0$, $v[n] = 0$ for all integers $n < 0$.

3.6. Convolve $v[n]$ with $x[n]$ for each of the cases below. Express your answer in closed form.

(a) $v[n] = 2^n u[n]$ and $x[n] = u[n]$

(b) $v[n] = (0.8)^n u[n]$ and $x[n] = u[n]$

(c) $v[n] = (0.8)^n u[n]$ and $x[n] = (0.5)^n u[n]$

(d) Use the M-file `conv` to compute the convolution in parts (a) to (c) for $0 \leq n \leq 20$; that is, define x and v for this range of n, compute the convolution and then save the values of $x[n] * v[n]$ only for this range of n. Plot the results using a `stem` plot. (See the comments in Problem 3.7 for more information regarding numerical convolution of infinite-duration signals.)

3.7. Care must be taken for using a computer to perform convolution on infinite-duration signals (i.e., signals that have nonzero values for an infinite number of points). Since you can store only a finite number of values for the signal, the numerical convolution returns an answer that is equivalent to the signal being zero outside the range of n defined for the stored points. In MATLAB, if $x[n]$ is defined for the range $0 \leq n \leq q$ and $v[n]$ is defined for the range $0 \leq n \leq r$, the result $y[n] = x[n] * v[n]$ will be defined over the range $0 \leq n \leq q + r$. However, the answer will only be correct for the range $0 \leq n \leq \min\{q,r\}$. As an example, consider the convolution of two step functions, $u[n] * u[n]$.

(a) Compute a closed-form expression for the actual convolution [see Problem 3.5(e)].

(b) Define a signal that is the truncated version of a step, $x[n] = u[n]$ for $n \leq q$ and $x[n] = 0$ for all other integers n. Compute $x[n] * x[n]$ for $q = 5$. Compare this result to that found in part (a) to see the effect of the truncation.

(c) Now, define a vector in MATLAB which is the truncated version of the signal; that is, x contains only the elements of $u[n]$ for $n \leq q$. Take $q = 5$. Compute the numerical convolution $x[n] * x[n]$ and plot the result for $0 \leq n \leq 2q$. Compare this result to the answers found in parts (a) and (b). For what range of n does the result accurately represent the convolution of the two step functions?

(d) Repeat parts (b) and (c) for $q = 10$.

3.8. Use the M-file `conv` to convolve the signals defined in Problem 3.5 and compare your answers to those found in Problem 3.5. Use the comments in Problems 3.6 and 3.7 when computing the convolutions for infinite-duration signals.

3.9. A linear time-invariant discrete-time system has the unit-pulse response

$$h[n] = (-1)^n \qquad \text{for } n \geq 0$$

(a) Compute the step response $g[n]$ for $n \geq 0$.

(b) Compute the output response $y[n]$ for $n \geq 0$ when the input is $x[n] = u[n] - u[n-5]$ with zero initial energy in the system prior to the application of the input.

(c) Plot the results obtained in parts (a) and (b).

3.10. A linear time-invariant discrete-time system has the step response

$$g[n] = \frac{2^n}{n+1}\left(\sin\frac{\pi n}{2}\right)u[n]$$

(a) Compute the unit-pulse response $h[n]$ for $n \geq 0$.

(b) Compute the output response $y[n]$ for $n \geq 0$ when $x[0] = 2, x[1] = -1, x[n] = 0$ for all other n. Assume that there is no energy in the system at time $n = 0$.

(c) Plot the results in parts (a) and (b).

3.11. Two linear time-invariant discrete-time systems are connected as shown in Figure P3.11. System 1 has unit-pulse response $h_1[n]$ and system 2 has unit-pulse response $h_2[n]$. Derive an expression for the unit-pulse response of the overall system. Express your answer in terms of $h_1[n]$ and $h_2[n]$.

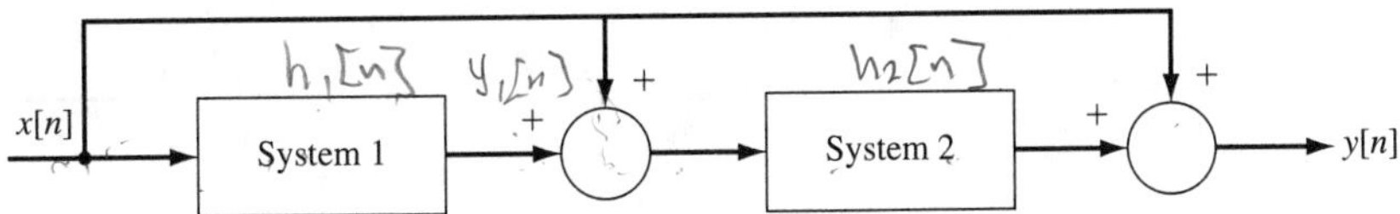

Figure P3.11

3.12. A discrete-time system has the following unit-pulse response:

$$h[n] = 0.3(0.7)^n u[n]$$

(a) Use conv to calculate the response of this system to $x[n] = u[n]$ and plot the response.

(b) Use conv to calculate the response of this system to $x[n] = \sin(n\pi/8)u[n]$ and plot the response.

(c) Use conv to calculate the response of this system to $x[n] = u[n] + \sin(n\pi/8)u[n]$ and plot the response.

(d) Find the first-order difference equation that describes this system where x is the input and y is the output.

(e) Using the result in part (d) and the M-file recur, calculate the response of the system to $x[n] = u[n]$ and compare to the answer obtained in part (a).

3.13. A discrete-time system has the following unit-pulse response:

$$h[n] = ((0.5)^n - (0.25)^n)u[n]$$

(a) Use conv to calculate the response of this system to $x[n] = u[n]$ and plot the response.

(b) Use conv to calculate the response of this system to $x[n] = \sin(n\pi/4)u[n]$ and plot the response.

(c) Use conv to calculate the response of this system to $x[n] = u[n] + \sin(n\pi/4)u[n]$ and plot the response.

(d) Show that the following difference equation has the unit-pulse response given above.

$$y[n+2] - 0.75y[n+1] + 0.125y[n] = 0.25x[n+1]$$

(e) Using the difference equation in part (d) and the M-file `recur`, calculate the response of the system to $x[n] = u[n]$ and compare to the answer obtained in part (a).

3.14. Compute the impulse response $h(t)$ for each of the following systems.

(a) $\dfrac{dy(t)}{dt} + 4y(t) = x(t)$ **(b)** $\dfrac{dy(t)}{dt} - 4y(t) = x(t)$

(c) $\dfrac{dy(t)}{dt} - y(t) = \dfrac{dx(t)}{dt} - 2x(t)$ **(d)** $\dfrac{dy(t)}{dt} + 4y(t) = x(t)$

3.15. Compute the impulse response $h(t)$ of the small-signal model of the pendulum given by the input/output differential equation

$$\frac{d^2\theta(t)}{dt^2} + \frac{MgL}{I}\theta(t) = \frac{L}{I}x(t)$$

Here M, g, L, and I are arbitrary positive numbers. (*Hint:* Use the results derived in Problem 2.6.)

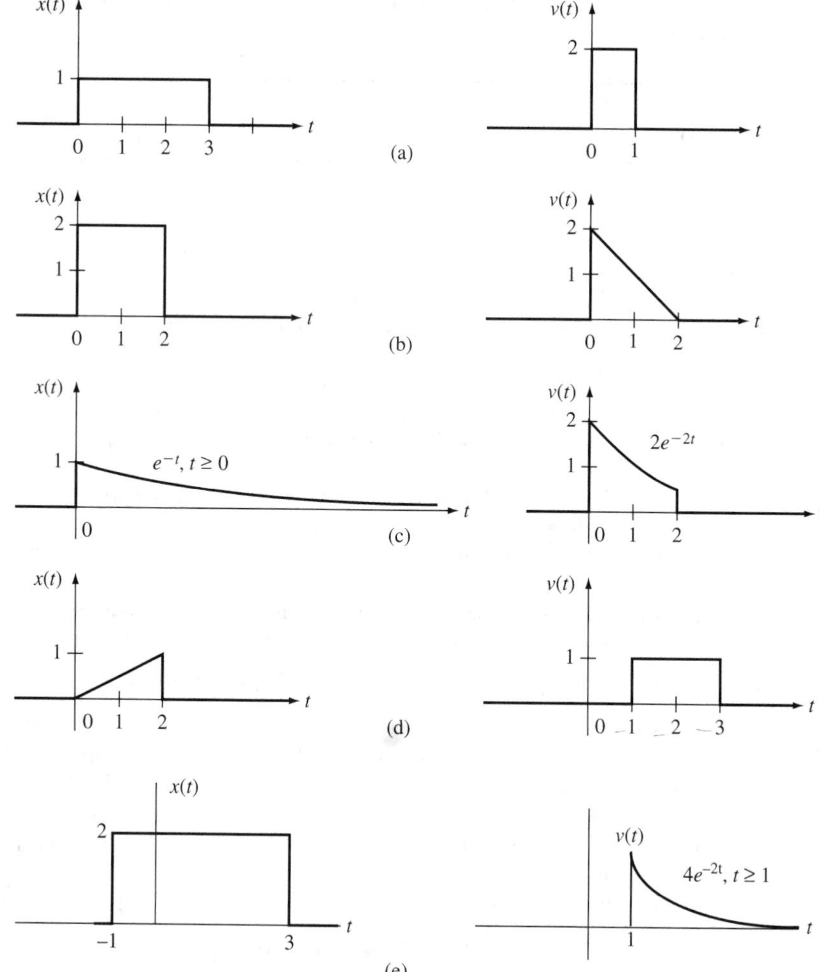

Figure P3.16

3.16. For the continuous-time signals $x(t)$ and $v(t)$ shown in Figure P3.16, compute the convolution $x(t) * v(t)$ for all $t \geq 0$ and plot your resulting signal.

3.17. Compute $x(t) * x(t)$ for $-\infty < t < \infty$ where $x(t) = (\sin \pi t)/(\pi t)$ for $-\infty < t < \infty$.

3.18. Compute $x(t) * v(t)$ for $-\infty < t < \infty$ where $x(t) = u(t) + u(t-1) - 2u(t-2)$ and $v(t) = 2u(t+1) - u(t) - u(t-1)$.

3.19. A continuous-time system has the input/output relationship

$$y(t) = \int_{-\infty}^{t} (t - \lambda + 2)x(\lambda)\, d\lambda$$

where it is assumed that the system has zero initial energy before the application of the input $x(t)$.

(a) Determine the impulse response $h(t)$ of the system.

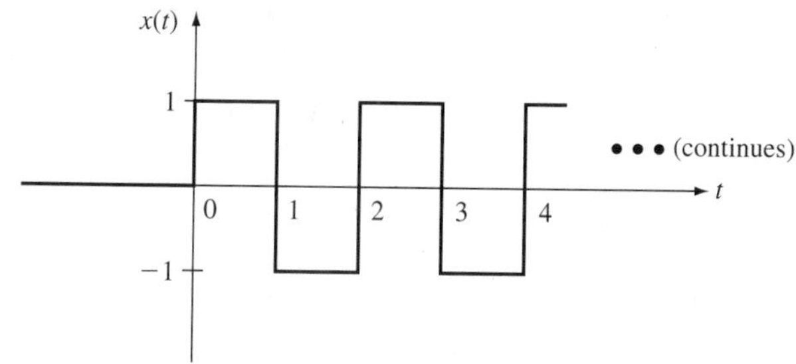

Figure P3.19

(b) Compute the output response $y(t)$ for $1 \leq t \leq 2$ *only* when the input $x(t)$ is as shown in Figure P3.19. It is assumed that the system has zero initial energy at time $t = 0$.

3.20. A linear time-invariant continuous-time system has impulse response

$$h(t) = e^{-t} + \sin t, \qquad t \geq 0$$

(a) Compute the step response $g(t)$ for all $t \geq 0$.

(b) Compute the output response $y(t)$ for all $t \geq 0$ when the input is $u(t) - u(t-2)$ with no initial energy in the system at time $t = 0$.

3.21. A linear time-invariant continuous-time system has impulse response $h(t) = (\sin t)u(t-2)$. Compute the output response $y(t)$ for all $t \geq 0$ when $x(t) = u(t) - u(t-1)$ with no initial energy in the system.

3.22. A linear time-invariant continuous-time system has impulse response $h(t)$ displayed in Figure P3.22a. Find the output response $y(t)$ for $4 \leq t \leq 6$ only resulting from the input shown in Figure 3.22b. Assume that there is no initial energy.

3.23. A linear time-invariant continuous-time system has impulse response $h(t)$ shown in Figure P3.22a. The input $x(t) = 2e^{-t}[u(t) - u(t-1)]$ is applied to the system with no initial energy at time $t = 0$. Compute the resulting output response $y(t)$ for $0 \leq t \leq 2$ only.

3.24. A linear time-invariant continuous-time system has impulse response $h(t)$ given by

$$h(t) = \sin t[u(t) - u(t-\pi)] + \sin t[u(t-2\pi) - u(t-3\pi)]$$

The input

$$x(t) = 4[u(t) - u(t-\pi)] - [u(t-\pi) - u(t-2\pi)] + t[u(t-3\pi) - u(t-4\pi)]$$

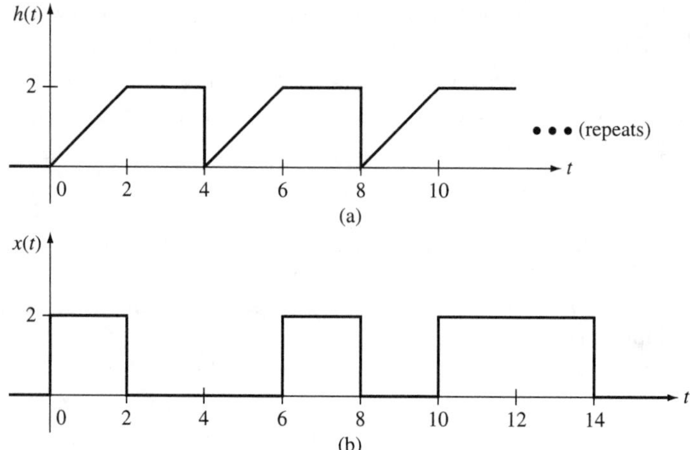

Figure P3.22

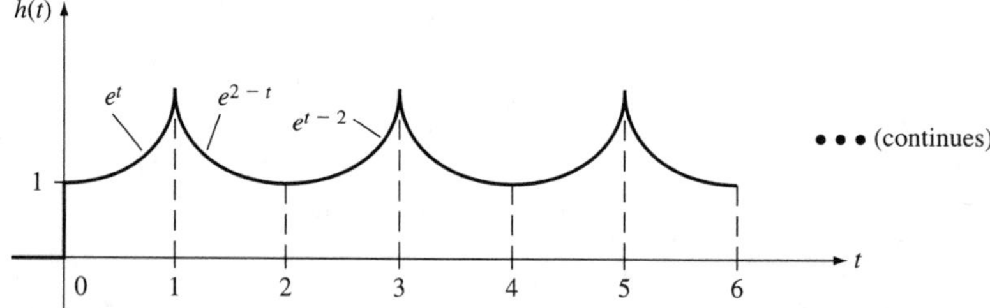

Figure P3.23

is applied to the system with no initial energy. By folding and shifting $x(t)$, compute the resulting output response $y(t)$ for $2\pi \leq t \leq 3\pi$.

3.25. Consider the cascade connection shown in Figure P3.25. System 1 is given by the input/output differential equation

$$\frac{dw(t)}{dt} = x(t)$$

It is known that system 2 is linear and time invariant and that when $w(t) = u(t)$, the resulting output response $y(t)$ with zero initial energy is

$$y(t) = 0.5(1 - e^{-2t})u(t)$$

If the input $x(t)$ to the cascade connection is the signal

$$x(t) = \begin{cases} 1 - t, & 0 \leq t \leq 1, \\ 0, & \text{otherwise} \end{cases}$$

compute the resulting output response $y(t)$ assuming that there is zero initial energy in the system. *Use the convolution representation.*

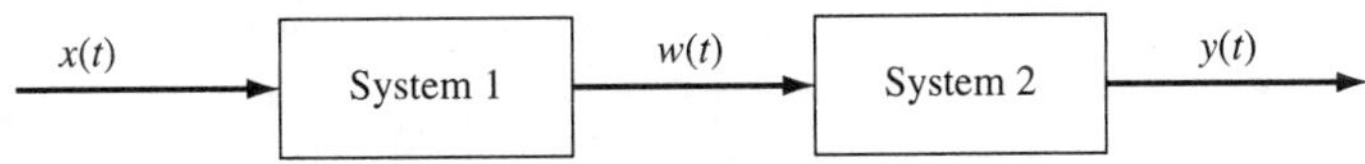

Figure P3.25

3.26. A linear time-invariant continuous-time system has impulse response $h(t)$. The output response of the system resulting from an input $x(t)$ is equal to $y(t)$, where again there is no initial energy in the system prior to the application of $x(t)$. Derive an expression for the response to the following inputs. Express your answer in terms of $y(t)$ and/or $x(t)$.

(a) $\dfrac{dx(t)}{dt}$

(b) $\dfrac{d^2x(t)}{dt^2}$

(c) $\displaystyle\int_{-\infty}^{t} x(\lambda)\, d\lambda$

(d) $x(t) * x(t)$

3.27. A linear time-invariant continuous-time system has step response $g(t)$. The system receives the input

$$x(t) = \sum_{i=0}^{N-1} c_i[u(t - t_i) - u(t - t_{i+1})]$$

where $t_0 < t_1 < \cdots < t_N$, N is a positive integer, and the c_i are constants. Express the resulting output in terms of $g(t)$ and time shifts of $g(t)$. Assume that there is no initial energy in the system prior to the application of the input.

3.28. Consider the linear time-invariant continuous-time system with the step response $g(t)$ displayed in Figure P3.28a. Using your result in Problem 3.27, compute the output response resulting from the input shown in Figure P3.28b. Assume that there is no initial energy.

3.29. For the system shown in Figure P3.29:

(a) Compute the impulse response $h(t)$.

(b) Compute the output response $y(t)$ resulting from input $x(t) = e^{-2t}u(t)$ with no initial energy in the system at time $t = 0$.

3.30. Use the numerical convolution method to compute an approximation to the continuous-time convolution of the signals given in Figure P3.16. Use a value of $T = 0.5$. Repeat for $T = 0.1$. Plot your results and compare your answers with those found in Problem 3.16.

3.31. Consider the series RLC circuit shown in Figure P3.31a. The circuit is equivalent to the cascade connection shown in Figure P3.31b, that is, the system in Figure P3.31b has the same input/output differential equation as the RLC circuit.

(a) Find the impulse responses of each of the subsystems in Figure P3.31b.

(b) Using your results in part (a), compute the impulse response of the RLC circuit.

(c) Suppose that $x(t) = \sin(t)u(t)$. Using the numerical convolution method, compute the values $y[n]$ of the resulting output response for $0 \le n \le 50$ and $T = 0.2$. Plot $y[n]$ versus nT to obtain the approximation for $y(t)$.

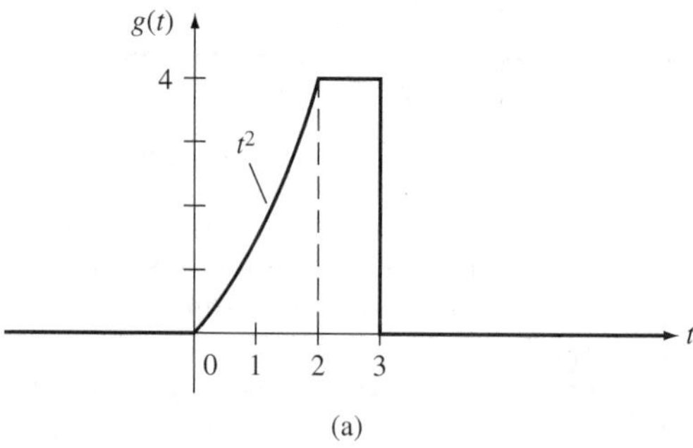

(a)

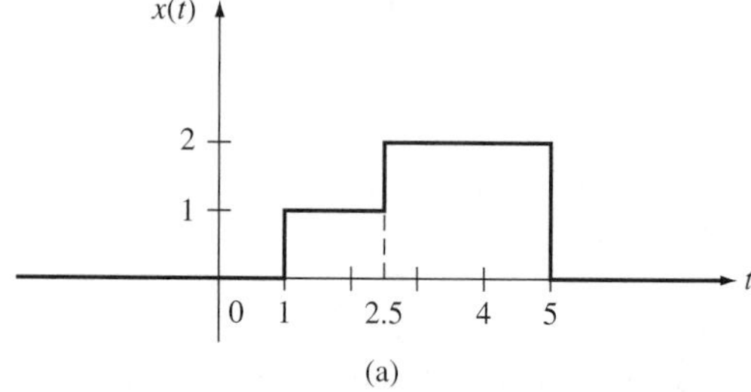

(a)

Figure P3.28

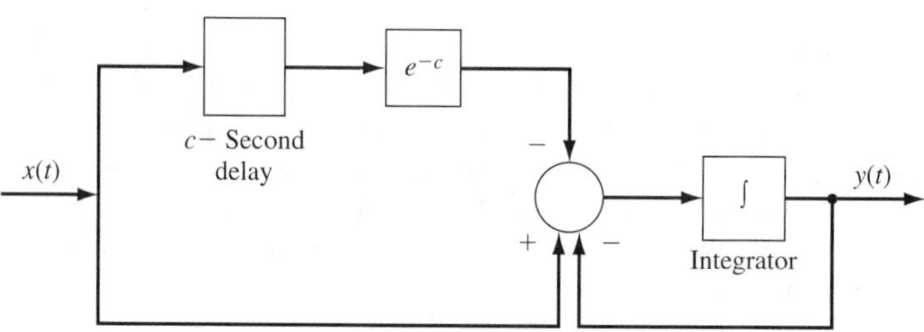

Figure P3.29

(d) Repeat part (c) with $T = 0.1$.

(e) Again suppose that $x(t) = \sin(t)u(t)$. Use the Euler approximation and the M-file `recur` to compute $y[n]$ for $0 \leq n \leq 50$ with $T = 0.2$. Take $y(0) = 0, \dot{y}(0) = 0$. Compare your results with those obtained in part (c). Which approximation scheme gives the better results?

(f) Repeat part (e) with $T = 0.1$.

3.32. Consider the top view of the single human eye shown in Figure P3.32. The input $x(t)$ is the angular position $\theta_T(t)$ of the target and the output $y(t)$ is the angular position $\theta_E(t)$ of the eye, with both angular positions defined relative to the resting position. In Problem 2.5 we defined the idealized rapid-eye-movement model. A more realistic model of the eye is given by the equations

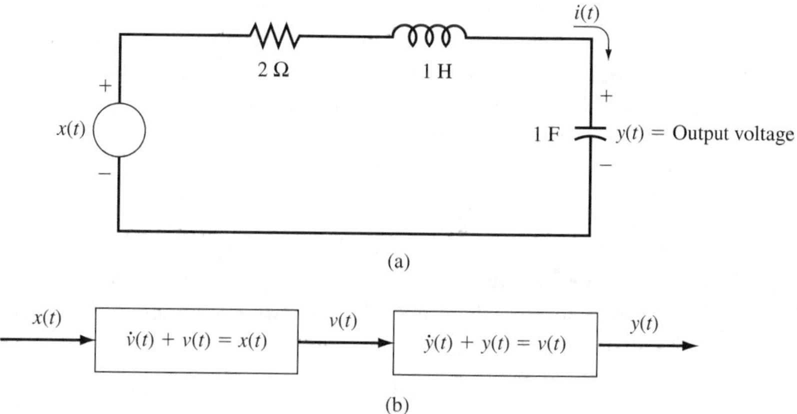

Figure P3.31

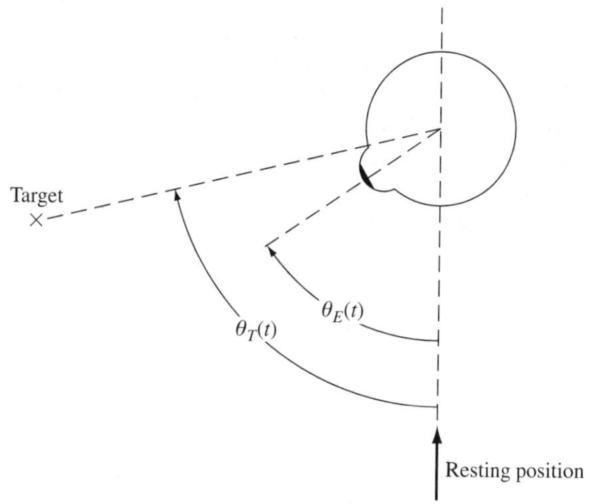

Figure P3.32

$$T_e \frac{d\theta_E(t)}{dt} + \theta_E(t) = R(t)$$

$$R(t) = b\theta_T(t - d) - b\theta_T(t - d - c) + \theta_T(t - d)$$

where $R(t)$ is the firing rate of action potentials in the nerve to the eye muscle, d is the time delay through the central nervous system, and T_e, b, and c are positive constants.

(a) *Using convolution,* derive an expression for $\theta_E(t)$ when the target suddenly moves from the resting position to position A at time $t = 0$; that is, $\theta_T(t) = Au(t)$. Assume that $\theta_E(0) = 0$.

(b) Using your result in part (a), show that there is a value of b for which $\theta_E(t) = A$ for all $t \geq d + c$; in other words, the eye locks onto the target at time $d + c$.

3.33. For the single-eye system in Problem 3.32, assume that $T_e = c = 0.1$, $d = 0.2$ and $\theta_T(t) = u(t)$. For the values of b given below, use the numerical convolution method with $T = 0.05$ to compute the approximate values $\theta_E(t)$ of the resulting output (eye position) for $0 \leq t \leq 2$.

(i) $b = 1$
(ii) $b = 0.2$
(iii) $b =$ values found in part (b) of Problem 3.32.

Does the eye lock onto the target for these values of b? Discuss your results.

3.34. Consider the automobile on a level surface with the time-varying mass $M(t)$ and with velocity model

$$\frac{dv(t)}{dt} + \frac{k_f}{M(t)} v(t) = \frac{1}{M(t)} x(t).$$

(a) Suppose that $M(t) = 1 - 0.05t[u(t) - u(t - 10)] - 0.5u(t - 10)$. Compute the impulse responses $h(t, \lambda)$ for all λ and all $t \geq \lambda$.

(b) Now suppose that the mass is piecewise constant; that is, $M(t) = M_i$ for $iT \leq t < iT + T$. Discretize the system using the integral approximation studied in Section 3.5. Take the discretization interval to be T.

(c) For the discretization computed in part (b), compute the unit-pulse responses $h[n,i]$, for all integers i and integers $k \geq i$.

3.35. Suppose a linear time-varying discrete-time system has the response $h[n,i]$ to a unit-pulse applied at $n = i$ (that is, $\delta[n - i]$). Use linearity to prove the expression in (3.69) which represents the relationship between an arbitrary input and the output of this system.

◆

CHAPTER 4

The Fourier Series and Fourier Transform

The fundamental notion of the frequency spectrum of a continuous-time signal is introduced in this chapter. As will be seen, the frequency spectrum displays the various sinusoidal components that comprise a given continuous-time signal. In general, the frequency spectrum is a complex-valued function of the frequency variable, and thus it is usually specified in terms of an amplitude spectrum and a phase spectrum.

The chapter begins with a study of signals that can be expressed as a sum of sinusoids, which includes periodic signals if an infinite number of terms is allowed in the sum. In the case of a periodic signal, the frequency spectrum can be generated by computing the *Fourier series.* The Fourier series is named after the French physicist Jean Baptiste Fourier (1768–1830), who was the first one to propose that periodic waveforms could be represented by a sum of sinusoids (or complex exponentials). It is interesting to note that in addition to his contributions to science and mathematics, Fourier was also very active in the politics of his time. For example, he played an important role in Napoleon's expeditions to Egypt during the late 1790s.

In the first section of the chapter, a frequency-domain analysis is given for continuous-time signals that can be expressed as a finite sum of sinusoids. This then leads to the Fourier series representation of periodic signals that is pursued in Section 4.2. In the last part of Section 4.2, the frequency spectrum of a periodic signal is defined in terms of the magnitudes and angles of the coefficients of the complex exponential terms comprising the Fourier series.

In the first part of Section 4.3 it is shown that it is also possible to generate a frequency domain representation of a nonperiodic signal. This representation is specified in terms of the Fourier transform, which is defined in Section 4.3. In contrast to a periodic signal, the amplitude and phase spectra of a nonperiodic signal consist of a continuum of frequencies. In Chapter 5 it will be seen that the characterization of periodic and nonperiodic input signals in terms of their frequency spectrum is very useful in determining how a linear time-invariant system processes inputs.

In Section 4.4 the properties of the Fourier transform are given. The last section of the chapter contains a brief treatment of the generalized Fourier transform.

4.1 REPRESENTATION OF SIGNALS IN TERMS OF FREQUENCY COMPONENTS

A fundamental concept in the study of signals is the notion of the *frequency content* of a signal. For a large class of signals, the frequency content can be generated by decomposing the signal into frequency components given by sinusoids. For example, consider the continuous-time signal $x(t)$ defined by the finite sum of sinusoids

$$x(t) = \sum_{k=1}^{N} A_k \cos(\omega_k t + \theta_k), \qquad -\infty < t < \infty \tag{4.1}$$

In (4.1), N is a positive integer, the A_k (which are assumed to be nonnegative) are the amplitudes of the sinusoids, the ω_k are the frequencies (in rad/sec) of the sinusoids, and the θ_k are the phases of the sinusoids.

In the case of the signal given by (4.1), the frequencies "present in the signal" are the frequencies $\omega_1, \omega_2, \ldots, \omega_N$ of the sinusoids comprising the signal, and the frequency components of the signal are the sinusoids $A_k \cos(\omega_k t + \theta_k)$ comprising the signal. It is important to observe that the signal given by (4.1) is characterized completely by the frequencies $\omega_1, \omega_2, \ldots, \omega_N$, the amplitudes $A_1, A_2, \ldots, A_N$, and the phases $\theta_1, \theta_2, \ldots, \theta_N$ in the representation given by (4.1).

The characteristics or "features" of a signal given by (4.1) can be studied in terms of the frequencies, amplitudes, and phases of the sinusoidal terms comprising the signal. In particular, the amplitudes $A_1, A_2, \ldots, A_N$ specify the relative weights of the frequency components comprising the signal, and these weights are a major factor in determining the "shape" of the signal. This is illustrated by the following example.

Example 4.1 *Sum of Sinusoids*

Sums of Sinusoids

Consider the continuous-time signal given by

$$x(t) = A_1 \cos t + A_2 \cos(4t + \pi/3) + A_3 \cos(8t + \pi/2), \qquad -\infty < t < \infty \tag{4.2}$$

This signal obviously has three frequency components with frequencies 1, 4, 8 rad/sec, amplitudes A_1, A_2, A_3, and phases $0, \pi/3, \pi/2$ rad. The goal here is to show that the shape of the signal depends on the relative magnitudes of the frequency components comprising the signal which are specified in terms of the amplitudes A_1, A_2, A_3. For this purpose, the following MATLAB commands were used to generate $x(t)$ for arbitrary values of A_1, A_2, and A_3:

```
t = 0:20/400:20;
w1 = 1; w2 = 4; w3 = 8;
A1 = input('Input the amplitude A1 for w1 = 1: ');
A2 = input('Input the amplitude A2 for w2 = 4: ');
A3 = input('Input the amplitude A3 for w3 = 8: ');
x = A1*cos(w1*t)+A2*cos(w2*t+pi/3)+A3*cos(w3*t+pi/2);
```

Using the commands above, MATLAB plots of $x(t)$ were generated for the three cases $A_1 = 0.5$, $A_2 = 1$, $A_3 = 0$; $A_1 = 1$, $A_2 = 0.5$, $A_3 = 0$; and $A_1 = 1$, $A_2 = 1$, $A_3 = 0$. The resulting plots are given in Figure 4.1. In all three of these cases, only the 1- and 4-rad/sec frequency components are present. In the first case, the 4-rad/sec component is twice as large as the 1-rad/sec component. The dominance of the 4-rad/sec component is obvious from Figure 4.1a. In the second case, the 1-rad/sec component

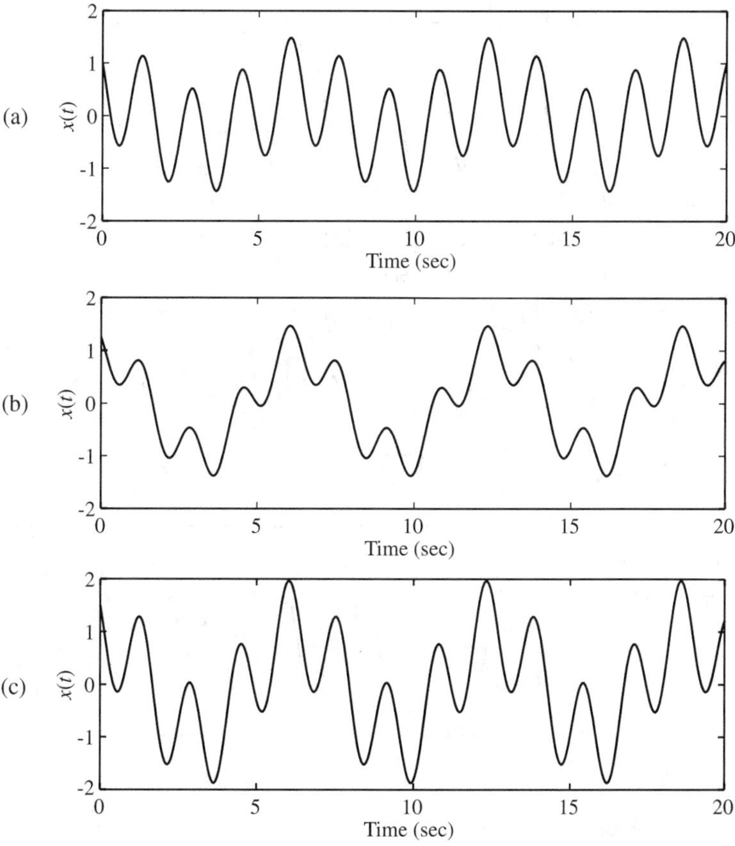

Figure 4.1 Plots of $x(t)$ for (a) $A_1 = 0.5, A_2 = 1, A_3 = 0$; (b) $A_1 = 1, A_2 = 0.5, A_3 = 0$; and (c) $A_1 = 1, A_2 = 1, A_3 = 0$.

dominates, which results in the signal shape shown in Figure 4.1b. In the third case, both frequency components have the same amplitude, which results in the waveform shown in Figure 4.1c.

The MATLAB program was then run again for the cases $A_1 = 0.5$, $A_2 = 1$, $A_3 = 0.5$; $A_1 = 1, A_2 = 0.5, A_3 = 0.5$; and $A_1 = 1, A_2 = 1, A_3 = 1$. In these three cases, all three frequency components are present, with the 4-rad/sec component dominating in the first case, the 1-rad/sec component dominating in the second case, and with all three components having the same amplitude in the third case. Figure 4.2 shows the plots of $x(t)$ for these three cases. For each of the plots in Figure 4.2, the reader should be able to distinguish all three of the frequency components comprising the signal.

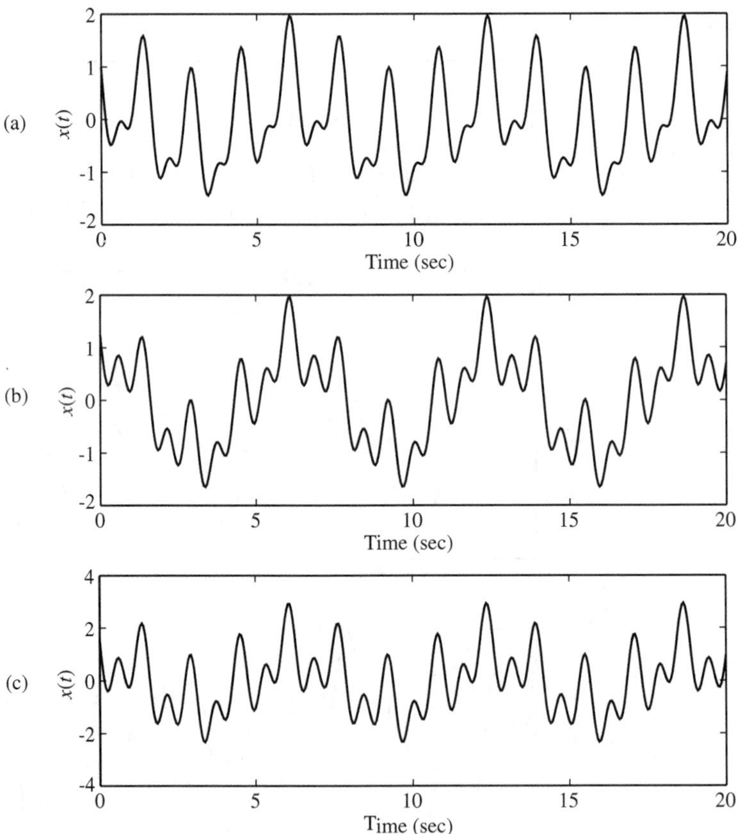

Figure 4.2 Plots of $x(t)$ for (a) $A_1 = 0.5$, $A_2 = 1$, $A_3 = 0.5$; (b) $A_1 = 1$, $A_2 = 0.5$,

Again consider the signal given by (4.1). With ω equal to the *frequency variable* (a real variable), the amplitudes A_k can be plotted versus ω. Since there are only a finite number of frequencies present in $x(t)$, the plot of A_k versus ω will consist of a finite number of points plotted at the frequencies ω_k present in $x(t)$. Usually, vertical lines are drawn connecting the values of the A_k with the points ω_k. The resulting plot is an example of a *line spectrum* and is called the *amplitude spectrum* of the signal $x(t)$. The amplitude spectrum shows the relative magnitudes of the various frequency components comprising the signal. For instance, consider the signal in Example 4.1 given by (4.2). For the various versions of the signal plotted in Figure 4.2, the amplitude spectrum is shown in Figure 4.3. Note the direct correspondence

Sums of Sinusoids

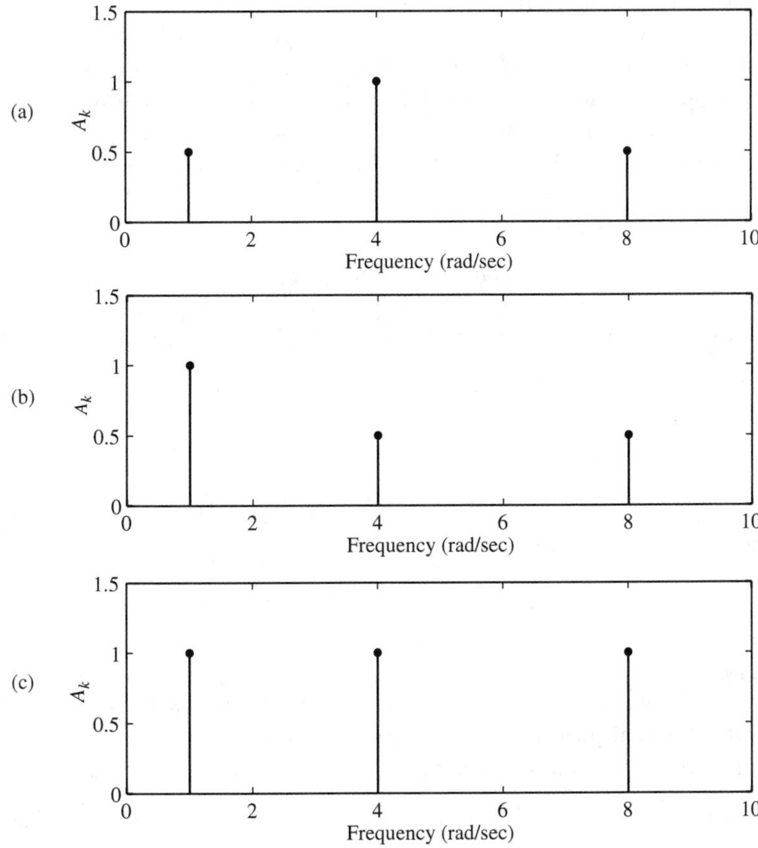

Figure 4.3 Amplitude spectra of the versions of $x(t)$ plotted in Figure 4.2.

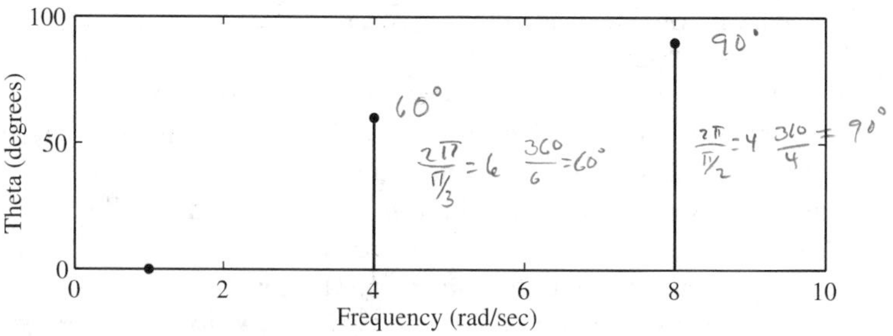

Figure 4.4 Phase spectrum of signal $x(t)$ defined by (4.2).

between the magnitudes of the spectral components shown in Figure 4.3 and the shape of the signals in Figure 4.2.

In addition to the amplitude spectrum, the signal defined by (4.1) also has a *phase spectrum,* which is a plot of the phase θ_k in degrees (or radians) versus the frequency variable ω. Again, in generating this plot vertical lines are drawn connecting the values of the θ_k with the frequency points ω_k, so the phase spectrum is also a line spectrum. For example, the phase spectrum of the signal given by (4.2) is plotted in Figure 4.4.

Complex Exponential Form

In signal and system analysis, a signal $x(t)$ of the form (4.1) is often expressed in terms of complex exponentials. To generate the complex exponential form, first consider the complex exponential $A_k e^{j(\omega_k t + \theta_k)}$. By Euler's formula (see Appendix A),

$$A_k e^{j(\omega_k t + \theta_k)} = A_k \cos(\omega_k t + \theta_k) + jA_k \sin(\omega_k t + \theta_k)$$

Hence,

$$A_k \cos(\omega_k t + \theta_k) = \mathrm{Re}[A_k e^{j(\omega_k t + \theta_k)}]$$

where $\mathrm{Re}[A_k e^{j(\omega_k t + \theta_k)}]$ is the real part of $A_k e^{j(\omega_k t + \theta_k)}$, and the signal $x(t)$ given by equation (4.1) can be written in the form

$$x(t) = \sum_{k=1}^{N} \mathrm{Re}[A_k e^{j(\omega_k t + \theta_k)}] \tag{4.3}$$

Thus, $x(t)$ can be expressed as a sum of the real parts of complex exponentials.

By considering complex exponentials with negative frequencies, we can remove the real-part operation in the preceding equation. This is accomplished by using the result that the real part of a complex number $s = a + jb$ is given by $a = \dfrac{s + \bar{s}}{2}$, where $\bar{s}$ is the complex conjugate of s. The complex conjugate of $A_k e^{j(\omega_k t + \theta_k)}$ is equal to $A_k e^{-j(\omega_k t + \theta_k)}$, and thus

$$\mathrm{Re}[A_k e^{j(\omega_k t + \theta_k)}] = \frac{A_k}{2} e^{j(\omega_k t + \theta_k)} + \frac{A_k}{2} e^{-j(\omega_k t + \theta_k)}$$

Therefore, (4.3) can be written in the form

$$x(t) = \sum_{k=1}^{N}\left[\frac{A_k}{2}e^{j(\omega_k t+\theta_k)} + \frac{A_k}{2}e^{-j(\omega_k t+\theta_k)}\right]$$

Then, defining

$$c_k = \frac{A_k}{2}e^{j\theta_k}, \qquad k = 1, 2, \ldots, N \tag{4.4}$$

and

$$c_{-k} = \frac{A_k}{2}e^{-j\theta_k}, \qquad k = 1, 2, \ldots, N \tag{4.5}$$

yields

$$x(t) = \sum_{k=1}^{N}[c_k e^{j\omega_k t} + c_{-k}e^{-j\omega_k t}]$$

$$x(t) = \sum_{k=1}^{N}[c_k e^{j\omega_k t} + c_{-k}e^{j(-\omega_k)t}]$$

$$x(t) = \sum_{k=1}^{N}c_k e^{j\omega_k t} + \sum_{k=1}^{N}c_{-k}e^{j(-\omega_k)t} \tag{4.6}$$

Note the appearance of frequency components at the negative frequencies $-\omega_k$. These frequencies do not have a meaning from a physical standpoint; rather, they are the result of the mathematical formulation (the complex exponential form).

Now defining $\omega_k = -\omega_k$ for $k = -1, -2, \ldots, -N$ and replacing the summation index k by $-k$ in the second term on the right-hand side of equation (4.6) gives

$$x(t) = \sum_{k=1}^{N} c_k e^{j\omega_k t} + \sum_{k=-N}^{-1} c_k e^{j\omega_k t}$$

$$x(t) = \sum_{\substack{k=-N \\ k\neq 0}}^{N} c_k e^{j\omega_k t} \tag{4.7}$$

The expression (4.7) is the complex exponential form of the signal $x(t)$ given by (4.1). Note that by (4.4) and (4.5), the coefficients c_k of the complex form are in general complex numbers. To see this, use Euler's formula to express (4.4) and (4.5) in the form

$$c_k = \frac{A_k}{2}[\cos\theta_k + j\sin\theta_k], \qquad k = 1, 2, \ldots \tag{4.8}$$

$$c_{-k} = \frac{A_k}{2}[\cos\theta_k - j\sin\theta_k], \qquad k = 1, 2, \ldots \tag{4.9}$$

Hence c_k is a real number if and only if $\sin \theta_k = 0$, which is the case if and only if $\theta_k = n\pi$ for some integer n.

The decomposition of a signal $x(t)$ into the complex exponential form (4.7) turns out to be extremely useful in characterizing how a system operates on an input signal equal to $x(t)$ to produce an output signal. In particular, as will be seen in Chapter 5, the behavior of a system can be characterized in terms of what the system does to the complex exponentials comprising the input signal.

If the sum in (4.7) is allowed to be infinite (i.e., $N = \infty$), the class of signals that can be expressed in the form (4.7) includes periodic signals. The case of a periodic signal is developed in the next section.

Line Spectra

Again consider the sum $x(t)$ of sinusoids

$$x(t) = \sum_{k=1}^{N} A_k \cos(\omega_k t + \theta_k)$$

where N may be finite or infinite. As noted above, the frequency components comprising this signal may be displayed in terms of the amplitude and phase spectra given by plots of A_k and θ_k versus ω. This results in line spectra defined for nonnegative frequencies only. However, the line spectra for a signal $x(t)$ consisting of a sum of sinusoids are usually defined with respect to the complex exponential form (4.7). In this case, the amplitude spectrum is the plot of the magnitudes $|c_k|$ versus ω, and the phase spectrum is a plot of the angles $\angle c_k$ versus ω. This results in line spectra that are defined for both positive and negative frequencies. Again, it should be stressed that the negative frequencies are a result of the complex exponential form (consisting of a positive and a negative frequency component) and have no physical meaning.

From (4.4)–(4.5) it can be see that $|c_k| = |c_{-k}|$ for $k = 1, 2, \ldots$, and thus the amplitude spectrum is symmetrical about $\omega = 0$. That is, the values of the amplitude spectrum for positive frequencies are equal to the values of the amplitude spectrum for the corresponding negative frequencies. In other words, the amplitude spectrum is an even function of the frequency variable ω. It also follows from (4.4)–(4.5) that

$$\angle c_{-k} = -\angle c_k, \quad k = 1, 2, \ldots$$

which implies that the phase spectrum is an odd function of the frequency variable ω.

Example 4.2 *Line Spectra*

Consider the signal

$$x(t) = \cos t + 0.5\cos(4t + \pi/3) + \cos(8t + \pi/2)$$

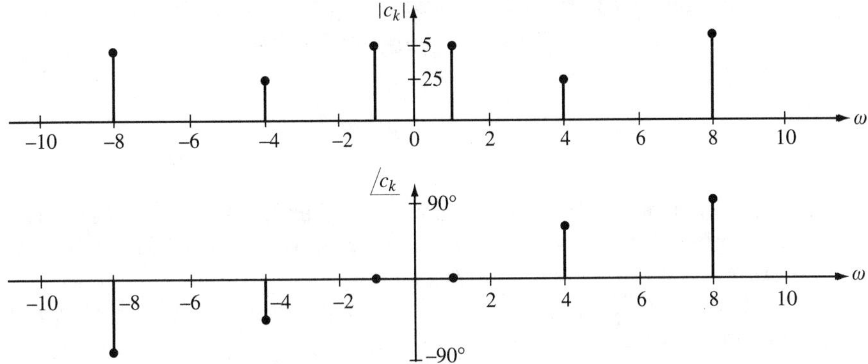

Figure 4.5 Line spectra for the signal in Example 4.2.

Using (4.4) and (4.5) gives

$$c_1 = \frac{1}{2} = 0.5,\ c_2 = \frac{0.5}{2}e^{j\pi/3} = 0.25\angle 60^\circ,\ c_3 = \frac{1}{2}e^{j\pi/2} = 0.5\angle 90^\circ$$

$$c_{-1} = \frac{1}{2} = 0.5,\ c_{-2} = \frac{0.5}{2}e^{-j\pi/3} = 0.25\angle -60^\circ,\ c_3 = \frac{1}{2}e^{-j\pi/2} = 0.5\angle -90^\circ$$

The amplitude and phase spectra are plotted in Figure 4.5.

4.2 FOURIER SERIES REPRESENTATION OF PERIODIC SIGNALS

Let T be a fixed positive real number. As first defined in Section 1.2, a continuous-time signal $x(t)$ is said to be periodic with period T if

$$x(t + T) = x(t) \qquad \text{for all } t, \quad -\infty < t < \infty \tag{4.10}$$

Recall that the fundamental period T is the smallest positive number for which (4.10) is satisfied. For example, the rectangular pulse train shown in Figure 4.6 is periodic with fundamental period $T = 2$.

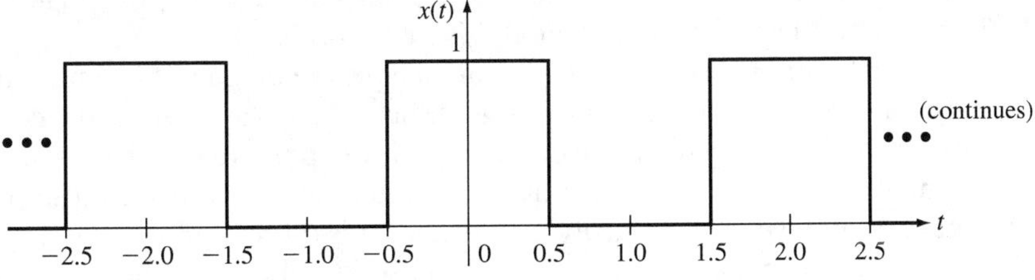

Figure 4.6 Periodic signal with fundamental period $T = 2$.

Let $x(t)$ be a periodic signal with fundamental period T. Then $x(t)$ can be expressed as a (in general infinite) sum of complex exponentials

$$x(t) = \sum_{k=-\infty}^{\infty} c_k e^{jk\omega_0 t}, \qquad -\infty < t < \infty \tag{4.11}$$

In the representation (4.11), c_0 is a real number, the c_k for $k \neq 0$ are in general complex numbers, and ω_0 is the *fundamental frequency* (in rad/sec) given by $\omega_0 = 2\pi/T$, where T is the fundamental period.

The coefficients c_k of the complex exponentials in (4.11) are computed using the formula

$$c_k = \frac{1}{T}\int_{-T/2}^{T/2} x(t)e^{-jk\omega_0 t}\, dt, \qquad k = 0, \pm 1, \pm 2, \ldots \tag{4.12}$$

It should be noted that the c_k given by (4.12) can be computed by integrating over any full period. For instance,

$$c_k = \frac{1}{T}\int_{0}^{T} x(t)e^{-jk\omega_0 t}\, dt, \qquad k = 0, \pm 1, \pm 2, \ldots$$

The term c_0 in (4.11) is the constant or dc component of $x(t)$ given by

$$c_0 = \frac{1}{T}\int_{-T/2}^{T/2} x(t)\, dt \tag{4.13}$$

The representation (4.11) is called the *Fourier series* of the periodic signal $x(t)$. Note that the frequencies that are present in $x(t)$ are integer multiples of the fundamental frequency ω_0. This is a key property of periodic signals. For a hint on how the Fourier series representation (4.11) is derived, see Problem 4.10.

It is worth noting that if a constant term c_0 is added to the sum in (4.7) and the sum is allowed to be infinite (i.e., $N = \infty$), the Fourier series (4.11) is a special case of (4.7) in that all the frequencies present in the signal are integer multiples of the fundamental frequency ω_0. To say this another way, with $N = \infty$ in (4.7) and with the addition of a constant term c_0, the class of signals that can be expressed in the form (4.7) includes the class of periodic signals given by (4.11).

The Fourier series representation of a periodic signal is a remarkable result. In particular, it shows that a periodic signal such as the waveform with "corners" in Figure 4.6 can be expressed as a sum of complex exponentials, which can be written as a sum of sinusoids. Since sinusoids are infinitely smooth functions (i.e., they have ordinary derivatives of arbitrarily high order), it is difficult to believe that signals with corners can be expressed as a sum of sinusoids. Of course, the key here is that the sum is an infinite sum. It is not surprising that Fourier had a difficult time convincing his peers (in this case the members of the French Academy of Science) that his theorem was true.

Fourier believed that any periodic signal could be expressed as a sum of complex exponentials (or sinusoids). However, this turned out not to be the case, although virtually all periodic signals arising in engineering do have a Fourier series representation. In particular, a periodic signal $x(t)$ has a Fourier series if it satisfies the *Dirichlet conditions* given by

1. $x(t)$ is absolutely integrable over any period; that is,

$$\int_a^{a+T} |x(t)|dt < \infty \qquad \text{for any } a$$

2. $x(t)$ has only a finite number of maxima and minima over any period.

3. $x(t)$ has only a finite number of discontinuities over any period.

Example 4.3 *Rectangular Pulse Train*

Consider the rectangular pulse train shown in Figure 4.6. This signal is periodic with fundamental period $T = 2$ and thus the fundamental frequency is $\omega_0 = 2\pi/2 = \pi$ rad/sec. The signal obviously satisfies the Dirichlet conditions, and thus it has a Fourier series representation. From (4.13), the constant component of $x(t)$ is

$$c_0 = \frac{1}{2}\int_{-1}^{1} x(t)\,dt = \frac{1}{2}\int_{-0.5}^{0.5} (1)\,dt = \frac{1}{2}$$

Evaluating (4.12) gives

$$\begin{aligned}
c_k &= \frac{1}{2}\int_{-1}^{1} x(t)e^{-jk\pi t}\,dt \\
&= \frac{1}{2}\int_{-0.5}^{0.5} e^{-jk\pi t}\,dt \\
&= -\frac{1}{j2k\pi} e^{-jk\pi t}\Big|_{t=-0.5}^{t=0.5} \\
&= -\frac{1}{j2k\pi}\left(-j\sin\frac{k\pi}{2} - j\sin\frac{k\pi}{2}\right) \\
&= \frac{1}{k\pi}\sin\frac{k\pi}{2}, \qquad k = \pm 1, \pm 2, \ldots \\
&= \begin{cases} 0, & k = \pm 2, \pm 4, \ldots \\ \dfrac{1}{k\pi}(-1)^{|(k-1)/2|}, & k = \pm 1, \pm 3, \ldots \end{cases}
\end{aligned}$$

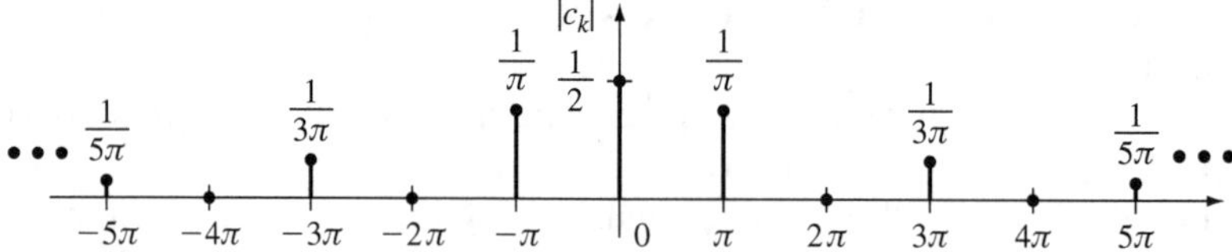

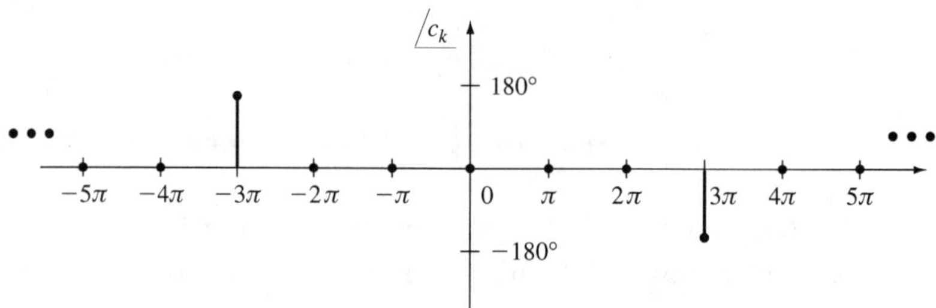

Figure 4.7 Line spectra for the rectangular pulse train.

Note that $c_k = 0$ for $k = \pm 2, \pm 4, \ldots$, and that the values for $k = \pm 1, \pm 3, \ldots$ are equal to $\pm\ 1/k\pi$. Then inserting the values for c_k into (4.11) results in the following Fourier series representation of the pulse train shown in Figure 4.6:

$$x(t) = \frac{1}{2} + \sum_{\substack{k=-\infty \\ k \text{ odd}}}^{\infty} \frac{1}{k\pi}(-1)^{|(k-1)/2|} e^{jk\pi t}, \qquad -\infty < t, \infty \tag{4.14}$$

To compute the amplitude and phase spectra for the rectangular pulse train, first note that

$$|c_k| = \begin{cases} 0, & k = 2, 4, \ldots \\ \dfrac{1}{k\pi}, & k = 1, 3, \ldots \end{cases}$$

$$\angle c_k = \begin{cases} 0, & k = 2, 4, \ldots \\ [(-1)^{(k-1)/2} - 1]\dfrac{\pi}{2}, & k = 1, 3, \ldots \end{cases}$$

The frequency spectra (amplitude and phase) are plotted in Figure 4.7.

Trigonometric Fourier Series

By using Euler's formula (see Problem 4.11), it is possible to write the Fourier series (4.11) in the following trigonometric form:

$$x(t) = c_0 + \sum_{k=1}^{\infty} 2|c_k| \cos(k\omega_0 t + \angle c_k), \qquad -\infty < t < \infty \tag{4.15}$$

where $|c_k|$ is the magnitude of c_k and $\angle c_k$ is the angle of c_k. The expression (4.15) is called the trigonometric Fourier series of $x(t)$. The constant c_0 is the *dc* or *constant component* of $x(t)$ and $2|c_k|\cos(k\omega_0 t + \angle c_k)$ is the kth *harmonic* of $x(t)$.

Example 4.4 *Trigonometric Form*

Consider the pulse train in Example 4.3 with the Fourier series (4.14). In this case

$$|c_k| = \begin{cases} 0, & k = 2, 4, \ldots \\ \dfrac{1}{k\pi}, & k = 1, 3, \ldots \end{cases}$$

and

$$\angle c_k = \begin{cases} 0, & k = 2, 4, \ldots \\ [(-1)^{(k-1)/2} - 1]\dfrac{\pi}{2}, & k = 1, 3, \ldots \end{cases}$$

and thus the trigonometric Fourier series is

$$x(t) = \frac{1}{2} + \sum_{\substack{k=1 \\ k \text{ odd}}}^{\infty} \frac{2}{k\pi}\cos\left(k\pi t + [(-1)^{(k-1)/2} - 1]\frac{\pi}{2}\right), \quad -\infty < t < \infty \tag{4.16}$$

Gibbs Phenomenon

Again consider the pulse train $x(t)$ with the trigonometric Fourier series representation (4.16). Given an odd positive integer N, let $x_N(t)$ denote the finite sum

$$x_N(t) = \frac{1}{2} + \sum_{\substack{k=1 \\ k \text{ odd}}}^{N} \frac{2}{k\pi}\cos\left(k\pi t + [(-1)^{(k-1)/2} - 1]\frac{\pi}{2}\right), \quad -\infty < t < \infty$$

Convergence of Fourier Series

By Fourier's theorem, $x_N(t)$ should converge to $x(t)$ as $N \to \infty$. In other words $|x_N(t) - x(t)|$ should be getting close to zero for all t as N is increased. Thus for a suitably large value of N, $x_N(t)$ should be a close approximation to $x(t)$. To see if this is the case, $x_N(t)$ can simply be plotted for various values of N. The MATLAB commands for generating $x_N(t)$ are

```
t = -3:6/1000:3;
N = input('Number of harmonics ');
c0 = 0.5;
w0 = pi;
xN = c0*ones(1,length(t));    % dc component
for k=1:2:N,                  % even harmonics are zero
  theta = ((-1)^((k-1)/2) - 1)*pi/2;
  xN = xN + 2/k/pi*cos(k*w0*t + theta);
end
```

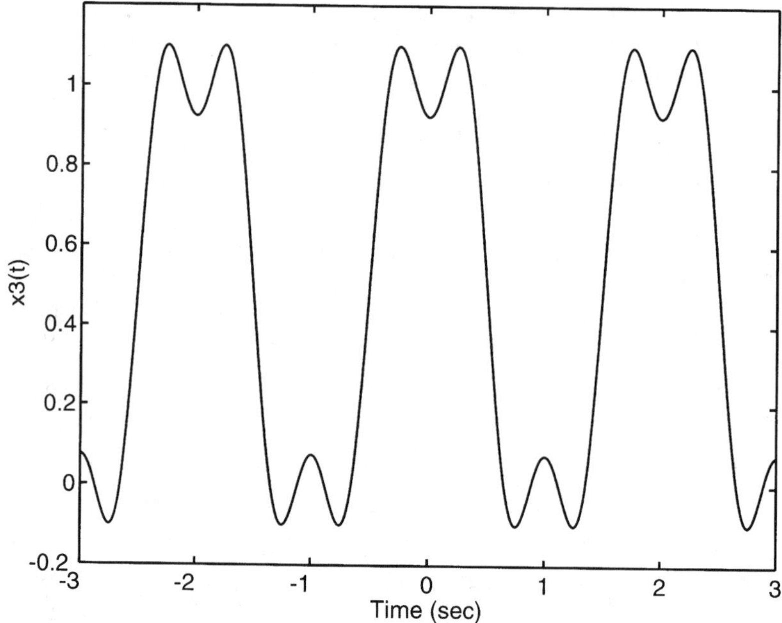

Figure 4.8 Plot of $x_N(t)$ when $N = 3$.

For the signal $x_N(t)$ given above, the even harmonics are zero and thus these terms are excluded in the loop to make the MATLAB program more efficient.

Now with $N = 3, x_N(t)$ becomes

$$x_3(t) = \frac{1}{2} + \frac{2}{\pi}\cos(\pi t) + \frac{2}{3\pi}\cos(3\pi t - \pi), \qquad -\infty < t < \infty$$

Setting $N = 3$ in the program above results in the plot of $x_3(t)$ shown in Figure 4.8. Note that even though $x_3(t)$ consists of the constant component and only two harmonics (the first and third), $x_3(t)$ does resemble the pulse train in Figure 4.6.

Increasing N to 9 produces the result shown in Figure 4.9. Comparing Figures 4.8 and 4.9 reveals that $x_9(t)$ is a much closer approximation to the pulse train $x(t)$ than $x_3(t)$. Of course, $x_9(t)$ contains the constant component and the first, third, fifth, seventh, and ninth harmonics of $x(t)$, and thus it would be expected to be a much closer approximation than $x_3(t)$.

Setting $N = 21$ produces the result in Figure 4.10. Except for the overshoot at the corners of the pulse, the waveform in Figure 4.10 is a much better approximation to $x(t)$ than $x_9(t)$. From a careful examination of the plot in Figure 4.10, it can be seen that the magnitude of the overshoot is approximately equal to 9%.

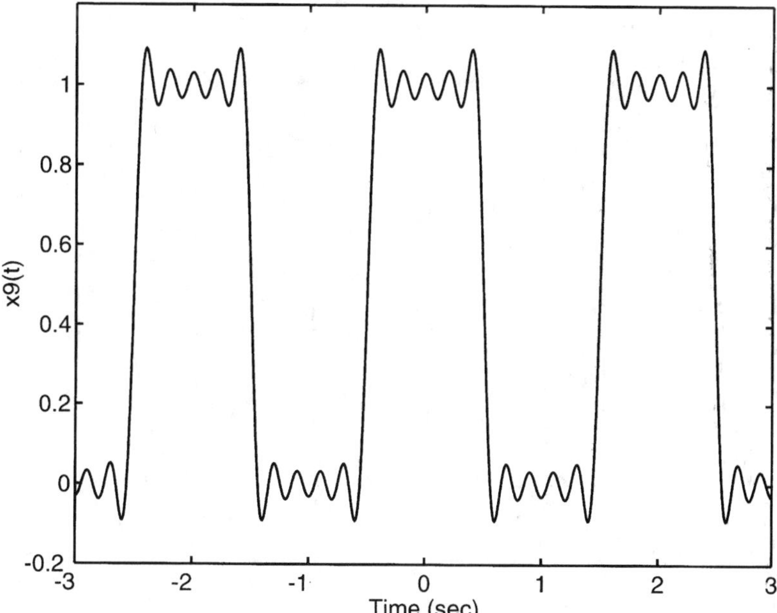

Figure 4.9 Approximation $x_9(t)$.

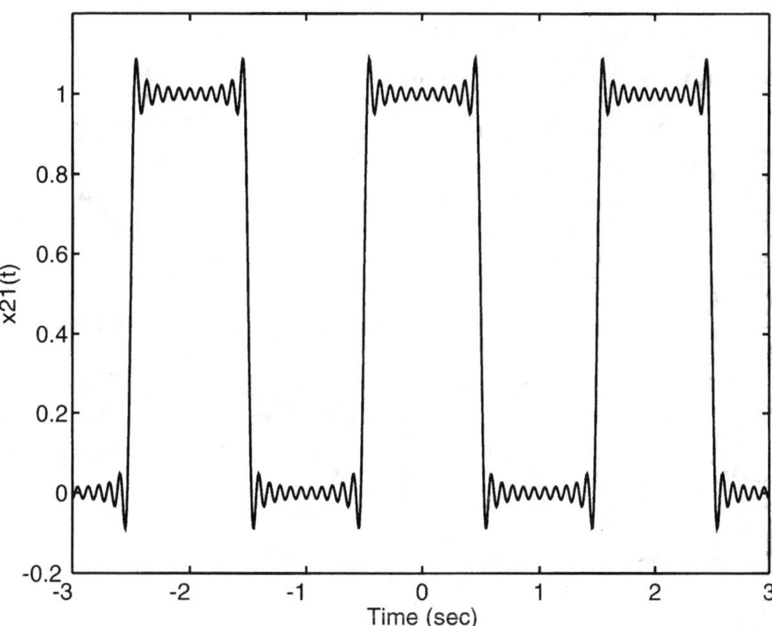

Figure 4.10 Approximation $x_{21}(t)$.

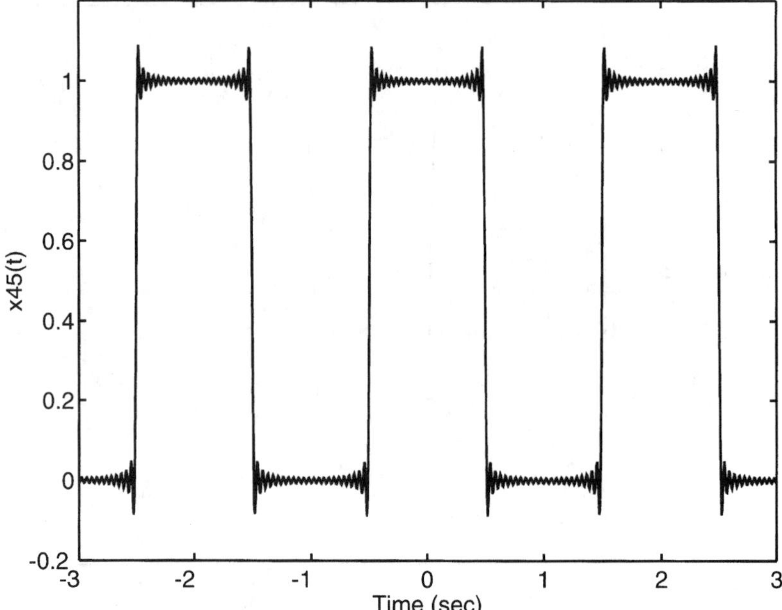

Figure 4.11 The signal $x_{45}(t)$.

Taking N = 45 yields the result displayed in Figure 4.11. Note that the 9% overshoot at the corners is still present. In fact, the 9% overshoot is present even in the limit as N approaches ∞. This characteristic was first discovered by Josiah Willard Gibbs (1839–1903), and thus the overshoot is referred to as the *Gibbs phenomenon*. Gibbs demonstrated the existence of the overshoot from mathematical properties rather than by direct computation.

Now let $x(t)$ be an arbitrary periodic signal. As a consequence of the Gibbs phenomenon, the Fourier series representation of $x(t)$ is not actually equal to the true value of $x(t)$ at any points where $x(t)$ is discontinuous. If $x(t)$ is discontinuous at $t = t_1$, the Fourier series representation is off by approximately 9% at t_1^- and t_1^+.

It should be noted that the finite sum $x_N(t)$ can also be calculated by truncating the exponential Fourier series directly:

$$x_N(t) = \sum_{k=-N}^{N} c_n e^{jk\omega_0 t}$$

The MATLAB commands for computing the truncated exponential Fourier series for the pulse train are

```
t = -3:6/1000:3;
N = input('Number of harmonics ');
c0 = 0.5;
w0 = pi;
xN = c0*ones(1,length(t));     % dc component
for k=1:N,
  ck = 1/k/pi*sin(k*pi/2);
  c_k = ck;
  xN = xN + ck*exp(j*k*w0*t) + c_k*exp(-j*k*w0*t);
end
```

The expression for c_k was calculated in Example 4.3, where it is seen by inspection that $c_{-k} = c_k$. Running the program above for $N = 3, 9, 21$, and 45 yields the same plots as those in Figures 4.8 to 4.11.

Parseval's Theorem

Let $x(t)$ be a periodic signal with period T. The average power P of the signal is defined by

$$P = \frac{1}{T}\int_{-T/2}^{T/2} x^2(t)\, dt \tag{4.17}$$

If $x(t)$ is the voltage across a 1-ohm resistor or the current in a 1-ohm resistor, the average power is given by (4.17). So the expression (4.17) is a generalization of the notion of average power to arbitrary signals.

Again let $x(t)$ be an arbitrary periodic signal with period T and consider the Fourier series of $x(t)$ given by (4.11). By Parseval's theorem, the average power P of the signal $x(t)$ is given by

$$P = \sum_{k=-\infty}^{\infty} |c_k|^2 \tag{4.18}$$

The relationship (4.18) is useful since it relates the average power of a periodic signal to the coefficients of the Fourier series of the signal. The proof of Parseval's theorem is beyond the scope of this book.

4.3 FOURIER TRANSFORM

A key feature of the Fourier series representation of periodic signals is the description of such signals in terms of the frequency content given by sinusoidal components. The question then arises as to whether or not nonperiodic signals, also called *aperiodic signals*, can be described in terms of frequency content. The answer is yes, and the analytical construct for doing this is the Fourier transform. As will be seen, the frequency components of nonperiodic signals are defined for all real values of the frequency variable ω, not just for discrete values of ω as in the case of a periodic signal. In other words, the spectra for a nonperiodic signal are not line spectra (unless the signal is equal to a sum of sinusoids).

In this section it is first shown that the frequency content of the rectangular pulse $x(t)$ shown in Figure 4.12a can be generated by first considering the Fourier series representation of a pulse train. This will then lead to the definition of the Fourier transform. The steps are as follows.

Given a fixed number $T > 1$, let $x_T(t)$ denote the pulse train with period T shown in Figure 4.12b. Note that the pulse $x(t)$ in Figure 4.12a is equal to the pulse train in the limit as $T \to \infty$; that is, in mathematical terms,

$$x(t) = \lim_{T \to \infty} x_T(t) \tag{4.19}$$

Now since $x_T(t)$ is periodic with fundamental period T, it has the complex exponential Fourier series

$$x_T(t) = \sum_{k=-\infty}^{\infty} c_k e^{jk\omega_0 t}, \qquad -\infty < t < \infty \tag{4.20}$$

where

$$c_k = \frac{1}{T}\int_{-T/2}^{T/2} x(t)e^{-jk\omega_0 t}\,dt, \qquad k = 0, \pm 1, \pm 2, \ldots \tag{4.21}$$

The goal here is to first determine what happens to the frequency components of $x_T(t)$ as $T \to \infty$; that is, when the pulse train $x_T(t)$ becomes the rectangular pulse $x(t)$ shown in Figure 4.12a. To investigate this, the coefficients c_k of the Fourier series (4.21) will be computed.

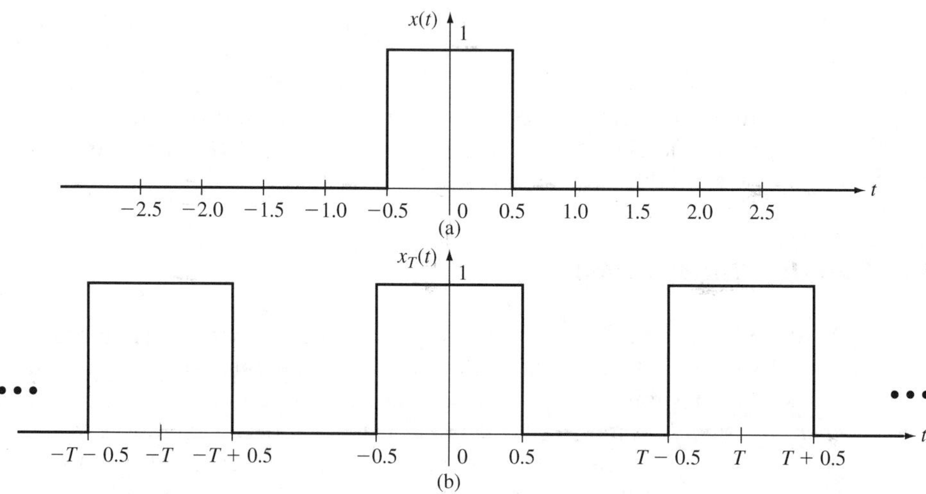

Figure 4.12 Plots of the (a) one-second rectangular pulse and (b) pulse train.

First, evaluating (4.21) for $k = 0$ gives

$$c_0 = \frac{1}{T}\int_{-T/2}^{T/2} x(t)\, dt = \frac{1}{T}\int_{-0.5}^{0.5} dt = \frac{1}{T}$$

Evaluating (4.21) for $k \neq 0$ gives

$$c_k = \frac{1}{T}\int_{-0.5}^{0.5} e^{-jk\omega_0 t}\, dt$$

$$= -\frac{1}{jk\omega_0 T}[e^{-jk\omega_0 t}]_{t=-0.5}^{t=0.5}$$

$$= -\frac{1}{jk\omega_0 T}[e^{-j(k\omega_0/2)} - e^{j(k\omega_0/2)}]$$

$$= -\frac{1}{jk\omega_0 T}\left(-j2\sin\frac{k\omega_0}{2}\right)$$

$$= \frac{2}{k\omega_0 T}\sin\frac{k\omega_0}{2}, \qquad k = \pm 1, \pm 2, \ldots \tag{4.22}$$

and since $\omega_0 = 2\pi/T$, (4.22) can be rewritten in the form

$$c_k = \frac{1}{k\pi}\sin\frac{k\omega_0}{2}, \qquad k = \pm 1, \pm 2, \ldots \tag{4.23}$$

Now the amplitude spectrum of $x_T(t)$ is the plot of $|c_k|$ versus $\omega = k\omega_0$. It turns out to be more appropriate to plot the amplitude spectrum scaled by T; that is, a plot of $T|c_k|$ versus $\omega = k\omega_0$ will be generated. The plots of $T|c_k|$ for $T = 2$, $T = 5$, and $T = 10$ are displayed in Figure 4.13.

From Figure 4.13 it can be seen that as T is increased, the "density" of the frequency components increases, whereas the envelope of the magnitudes of the scaled spectral components remains the same. In the limit as $T \to \infty$, the scaled frequency components converge into a continuum of frequency components whose magnitudes have the same shape as the envelope of the discrete spectra shown in Figure 4.13. To demonstrate this, it is first necessary to consider the *sinc function* defined by

$$\text{sinc}\,\lambda = \frac{\sin \pi\lambda}{\pi\lambda} \tag{4.24}$$

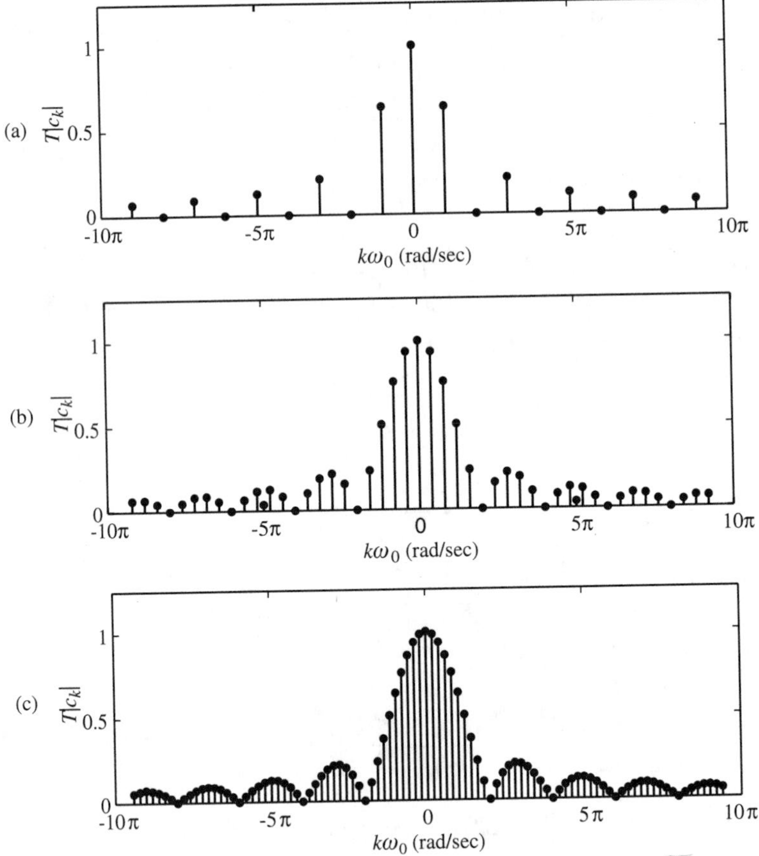

Figure 4.13 Plot of scaled spectrum of $x_T(t)$ for (a) $T = 2$, (b) $T = 5$, and (c) $T = 10$.

The sinc function is plotted in Figure 4.14. Note that sinc $0 = 1$. (This follows from l'Hôpital's rule.)

The coefficients c_k given by (4.23) can be written in terms of the sinc function as follows. First, by (4.24),

$$(\pi\lambda)[\text{sinc } \lambda] = \sin \pi\lambda$$

and thus setting $\lambda = k\omega_0/2\pi$ gives

$$\sin \frac{k\omega_0}{2} = \frac{k\omega_0}{2} \text{sinc } \frac{k\omega_0}{2\pi} \tag{4.25}$$

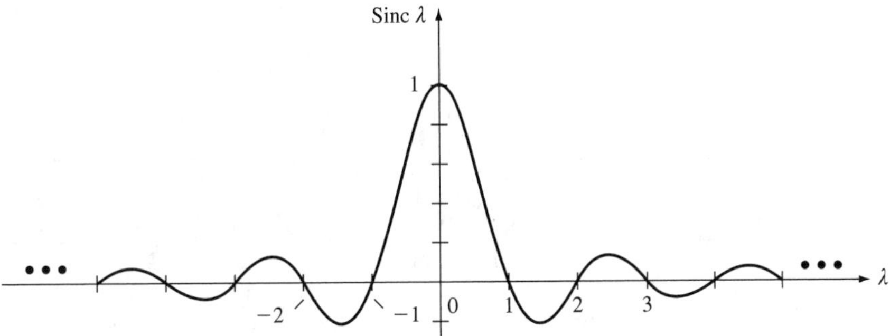

Figure 4.14 Plot of sinc λ.

Then inserting (4.25) into (4.23) yields

$$c_k = \frac{1}{k\pi}\frac{k\omega_0}{2}\operatorname{sinc}\frac{k\omega_0}{2\pi}$$
$$= \frac{\omega_0}{2\pi}\operatorname{sinc}\frac{k\omega_0}{2\pi}$$

But $\omega_0 = 2\pi/T$, and thus

$$c_k = \frac{1}{T}\operatorname{sinc}\frac{k\omega_0}{2\pi}, \qquad k = \pm 1, \pm 2, \ldots \tag{4.26}$$

and

$$Tc_k = \operatorname{sinc}\frac{k\omega_0}{2\pi}, \qquad k = \pm 1, \pm 2, \ldots \tag{4.27}$$

Finally, since $c_0 = 1/T$, $Tc_0 = 1$ and thus

$$Tc_0 = \operatorname{sinc} 0 \tag{4.28}$$

Therefore, (4.27) and (4.28) can be combined, which gives

$$Tc_k = \operatorname{sinc}\frac{k\omega_0}{2\pi}, \qquad k = 0, \pm 1, \pm 2, \ldots \tag{4.29}$$

Now as $T \to \infty$, the discrete frequency variable $k\omega_0$ becomes a real variable ω, and therefore taking the limit as $T \to \infty$ of both sides of (4.29) gives

$$\lim_{T\to\infty} Tc_k = \operatorname{sinc}\frac{\omega}{2\pi}, \qquad -\infty < \omega < \infty \tag{4.30}$$

and thus

$$\lim_{T\to\infty} T|c_k| = \left|\operatorname{sinc}\frac{\omega}{2\pi}\right|, \qquad -\infty < \omega < \infty \tag{4.31}$$

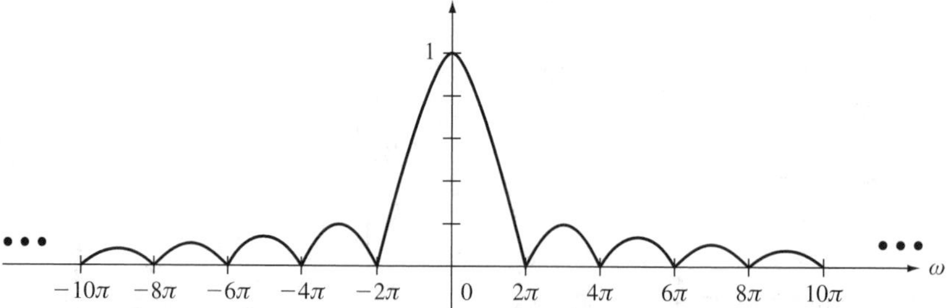

Figure 4.15 Amplitude spectrum of the rectangular pulse.

Equation (4.31) shows that in the limit as $T \to \infty$, the scaled amplitude spectrum does converge to a continuum of frequency components given mathematically by $|\text{sinc}(\omega/2\pi)|$, where the real variable ω (the frequency variable) represents the continuum of frequencies.

The amplitude spectrum of the rectangular pulse is defined to be the frequency function $|\text{sinc}(\omega/2\pi)|$ for $-\infty < \omega < \infty$, which is plotted in Figure 4.15. From the plot in Figure 4.15, it is interesting to note that most of the spectral content of the rectangular pulse is concentrated in the first "lobe," which runs from $\omega = 0$ to $\omega = 2\pi$ on the positive frequency range.

Now the Fourier transform [denoted by $X(\omega)$] of the rectangular pulse $x(t)$ is defined to be the limit of Tc_k as $T \to \infty$. Hence, by (4.29), the Fourier transform of the rectangular pulse is given by

$$X(\omega) = \text{sinc}\,\frac{\omega}{2\pi}, \qquad -\infty < \omega < \infty \tag{4.32}$$

The amplitude spectrum (see Figure 4.15) of the rectangular pulse is therefore equal to the magnitude $|X(\omega)|$ of the Fourier transform $X(\omega)$. The phase spectrum of the pulse is defined to be the angle function $\angle X(\omega)$. Since in this case $X(\omega)$ is real valued, for any given value of ω the phase is equal to either 0 or $\pm$ 180°. A plot of the phase spectrum of the pulse is given in Figure 4.16.

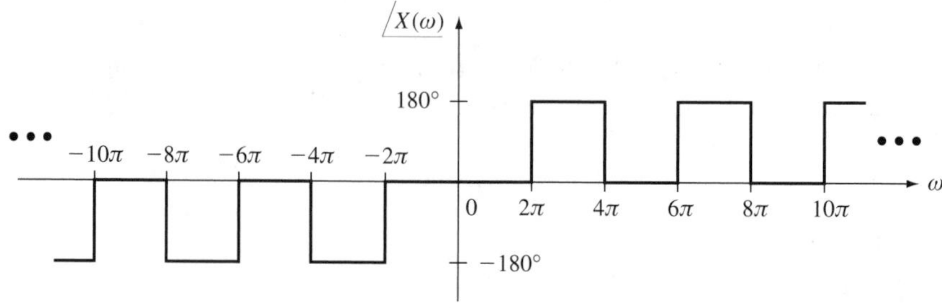

Figure 4.16 Phase spectrum of the rectangular pulse.

The Fourier transform $X(\omega)$ of the rectangular pulse $x(t)$ can be expressed in terms of $x(t)$ as follows. First, consider the expression for the coefficients c_k given by (4.21). Since $x(t) = 0$ for $t < -T/2$ and $t > T/2$, (4.21) can be rewritten as

$$c_k = \frac{1}{T}\int_{-\infty}^{\infty} x(t)e^{-jk\omega_0 t}\,dt, \qquad k = 0, \pm 1, \pm 2, \ldots$$

and thus

$$Tc_k = \int_{-\infty}^{\infty} x(t)e^{-jk\omega_0 t}\,dt, \qquad k = 0, \pm 1, \pm 2, \ldots \tag{4.33}$$

Now by definition of $X(\omega)$,

$$X(\omega) = \lim_{T\to\infty} Tc_k$$

and since $k\omega_0 \to \omega$ as $T \to \infty$, by (4.33),

$$X(\omega) = \int_{-\infty}^{\infty} x(t)e^{-j\omega t}\,dt, \qquad -\infty < \omega < \infty \tag{4.34}$$

Equation (4.34) shows that the Fourier transform $X(\omega)$ can be computed directly from the signal waveform $x(t)$ (in this case the rectangular pulse).

Conversely, it is possible to recompute the signal $x(t)$ from the transform $X(\omega)$ by using the relationship

$$x(t) = \frac{1}{2\pi}\int_{-\infty}^{\infty} X(\omega)e^{j\omega t}\,d\omega \tag{4.35}$$

The relationship (4.35) is referred to as the *inverse Fourier transform* since it shows how to reconstruct $x(t)$ from the transform $X(\omega)$. The relationship (4.35) can be verified by taking the limit as $T \to \infty$ of the Fourier series representation (4.20) of the periodic signal $x_T(t)$. The details of the derivation are omitted.

The Fourier Transform in the General Case

The definition (4.34) of the Fourier transform of the rectangular pulse can be extended to a large class of signals. In particular, given a signal $x(t)$, the Fourier transform $X(\omega)$ of $x(t)$ is defined to be the frequency function

$$X(\omega) = \int_{-\infty}^{\infty} x(t)e^{-j\omega t}\,dt, \qquad -\infty < \omega < \infty \tag{4.36}$$

where ω is the continuous frequency variable. In this book, the Fourier transform will always be denoted by an uppercase letter or symbol, whereas signals will usually be denoted by lowercase letters or symbols.

Note that due to the presence of the complex exponential $\exp(-j\omega t)$ in the integrand of the integral in (4.36), the values of $X(\omega)$ may be complex. Hence, in general, the Fourier transform $X(\omega)$ is a complex-valued function of the frequency variable ω, and thus in order to specify $X(\omega)$, in general it is necessary to display the magnitude function $|X(\omega)|$ and the angle function $\angle X(\omega)$. As in the case of the 1-second rectangular pulse considered above, the amplitude spectrum of a signal $x(t)$ is defined to be the magnitude function $|X(\omega)|$ of the Fourier transform $X(\omega)$, and the phase spectrum of $x(t)$ is defined to be the angle function $\angle X(\omega)$. As shown in the case of the 1-second rectangular pulse, the amplitude and phase spectra of a signal $x(t)$ are natural generalizations of the line spectra of periodic signals.

A signal $x(t)$ is said to have a Fourier transform in the ordinary sense if the integral in (4.36) converges (i.e., exists). The integral does converge if $x(t)$ is "well behaved" and if $x(t)$ is absolutely integrable, where the latter condition means that

$$\int_{-\infty}^{\infty} |x(t)|\, dt < \infty \tag{4.37}$$

Well behaved means that the signal has a finite number of discontinuities, maxima, and minima within any finite interval of time. Except for impulses, most signals of interest are well behaved. All actual signals (i.e., signals that can be physically generated) are well behaved and satisfy (4.37).

Since any well-behaved signal of finite duration in time is absolutely integrable, any such signal has a Fourier transform in the ordinary sense. An example of a signal that does not have a Fourier transform in the ordinary sense follows.

Example 4.5 *Constant Signal*

Consider the dc or constant signal

$$x(t) = 1, \qquad -\infty < t < \infty$$

Clearly, the constant signal is not an actual signal since no signal can be generated physically that is nonzero for all time. Nevertheless, the constant signal plays a very important role in the theory of signals and systems. The Fourier transform of the constant signal is

$$X(\omega) = \int_{-\infty}^{\infty} (1)e^{-j\omega t}\, dt \tag{4.38}$$

$$= \lim_{T\to\infty} \int_{-T/2}^{T/2} e^{-j\omega t}\, dt$$

$$= \lim_{T\to\infty} -\frac{1}{j\omega}\left[e^{-j\omega t}\right]_{t=-T/2}^{t=T/2}$$

$$= \lim_{T\to\infty} -\frac{1}{j\omega}\left[\exp\left(-\frac{j\omega T}{2}\right) - \exp\left(\frac{j\omega T}{2}\right)\right]$$

But $\exp(j\omega T/2)$ does not have a limit as $T \to \infty$, and thus the integral in (4.38) does not converge. Hence a constant signal does not have a Fourier transform in the ordinary sense. This can be seen by checking (4.37): The area under the constant signal is infinite, so the integral in (4.37) is not finite. Later, it will be shown that a constant signal has a Fourier transform in a generalized sense.

Example 4.6 *Exponential Signal*

Fourier Transform of Exponential Signal

Now consider the signal

$$x(t) = e^{-bt}u(t)$$

where b is a real constant and $u(t)$ is the unit-step function. Note that $x(t)$ is equal to $u(t)$ when $b = 0$. For an arbitrary value of b, the Fourier transform $X(\omega)$ of $x(t)$ is given by

$$X(\omega) = \int_{-\infty}^{\infty} e^{-bt}u(t)e^{-j\omega t}\,dt$$

and since $u(t) = 0$ for $t < 0$, $u(t) = 1$ for $t \geq 1$,

$$X(\omega) = \int_{0}^{\infty} e^{-bt}e^{-j\omega t}\,dt$$

$$= \int_{0}^{\infty} e^{-(b+j\omega)t}\,dt$$

Evaluating the integral gives

$$X(\omega) = -\frac{1}{b + j\omega}\left[e^{-(b+j\omega)t}\right]_{t=0}^{t=\infty}$$

The upper limit $t = \infty$ cannot be evaluated when $b \leq 0$, and thus for this range of values of b, $x(t)$ does not have an ordinary Fourier transform. Since $x(t) = u(t)$ when $b = 0$, it is seen that the unit-step function $u(t)$ does not have a Fourier transform in the ordinary sense. [But as shown in Section 4.6, $u(t)$ does have a generalized Fourier transform.]

When $b > 0$, $\exp(-bt) \to 0$ as $t \to \infty$, and thus

$$\lim_{t\to\infty} e^{-(b+j\omega)t} = \lim_{t\to\infty} e^{-bt}e^{-j\omega t} = 0$$

Hence for $b > 0$ $x(t)$ has a Fourier transform given by

$$X(\omega) = -\frac{1}{b + j\omega}(0 - 1) = \frac{1}{b + j\omega}$$

and the amplitude and phase spectra are given by

$$|X(\omega)| = \frac{1}{\sqrt{b^2 + \omega^2}}$$

$$\angle X(\omega) = -\tan^{-1}\frac{\omega}{b}$$

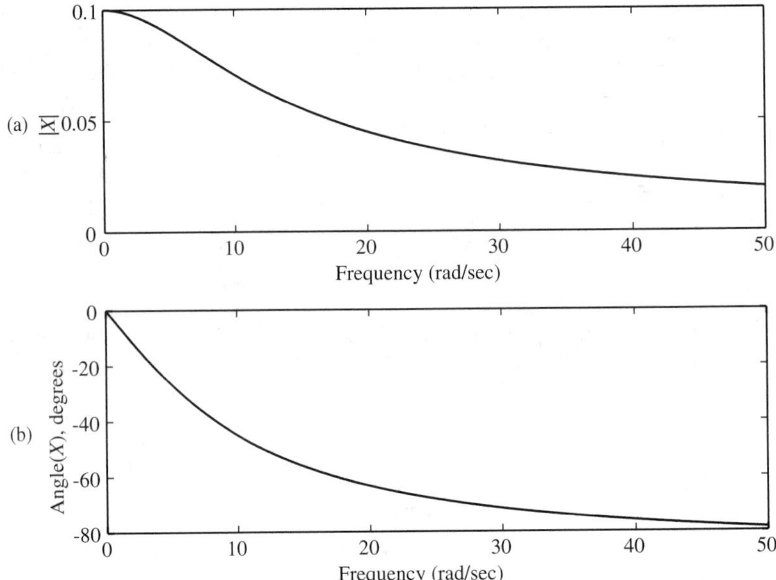

Figure 4.17 Plots of the (a) amplitude and (b) phase spectra of $x(t) = \exp(-10t)u(t)$.

Plots of the amplitude spectrum $|X(\omega)|$ and the phase spectrum $\underline{/X(\omega)}$ can be generated for the case $b = 10$ using the MATLAB commands given below.

```
w = 0:0.2:50;
b = 10;
X = 1./(b+j*w);
subplot(211),plot(w,abs(X));          % plot magnitude of X
subplot(212),plot(w,angle(X));        % plot angle of X
```

Note that the explicit expressions for $|X(\omega)|$ and $\underline{/X(\omega)}$ above are not needed to generate MATLAB plots of the amplitude and phase spectra. For the case $b = 10$, the MATLAB program above was run with the results displayed in Figure 4.17. From Figure 4.17a it is seen that most of the spectral content of the signal is concentrated in the low-frequency range with the amplitude spectrum decaying to zero as $\omega \to \infty$.

Rectangular and Polar Form of the Fourier Transform

Consider the signal $x(t)$ with Fourier transform

$$X(\omega) = \int_{-\infty}^{\infty} x(t)e^{-j\omega t}\,dt$$

As noted previously, $X(\omega)$ is a complex-valued function of the real variable ω. In other words, if a particular value of ω is inserted into $X(\omega)$, in general the result will be a complex number. Since complex numbers can be expressed in either rectangular or polar form, the Fourier transform $X(\omega)$ can be expressed in either rectangular or polar form. These forms are defined below.

Using Euler's formula, $X(\omega)$ can be written in the form

$$X(\omega) = \int_{-\infty}^{\infty} x(t)\cos \omega t \, dt - j\int_{-\infty}^{\infty} x(t)\sin \omega t \, dt$$

Now let $R(\omega)$ and $I(\omega)$ denote the real-valued functions of ω defined by

$$R(\omega) = \int_{-\infty}^{\infty} x(t)\cos \omega t \, dt$$

$$I(\omega) = -\int_{-\infty}^{\infty} x(t)\sin \omega t \, dt$$

Then the rectangular form of $X(\omega)$ is

$$X(\omega) = R(\omega) + jI(\omega) \tag{4.39}$$

The function $R(\omega)$ is the real part of $X(\omega)$ and the function $I(\omega)$ is the imaginary part of $X(\omega)$. Note that $R(\omega)$ and $I(\omega)$ could be computed first, and then $X(\omega)$ can be found using (4.39).

Now the polar form of the Fourier transform $X(\omega)$ is given by

$$X(\omega) = |X(\omega)| \exp[j \angle X(\omega)] \tag{4.40}$$

where $|X(\omega)|$ is the magnitude of $X(\omega)$ and $\angle X(\omega)$ is the angle of $X(\omega)$. It is possible to go from the rectangular form to the polar form by using the relationships

$$|X(\omega)| = \sqrt{R^2(\omega) + I^2(\omega)}$$

$$\angle X(\omega) = \tan^{-1} \frac{I(\omega)}{R(\omega)}$$

Note that if $x(t)$ is real valued, by (4.36)

$$X(-\omega) = \overline{X(\omega)} = \text{complex conjugate of } X(\omega)$$

Then taking the complex conjugate of the polar form (4.40) gives

$$X(-\omega) = |X(\omega)| \exp[-j \angle X(\omega)]$$

Thus

$$|X(-\omega)| = |X(\omega)|$$
$$\angle X(-\omega) = -\angle X(\omega)$$

Signals with Even or Odd Symmetry

Again suppose that $x(t)$ has Fourier transform $X(\omega)$ with $X(\omega)$ given in the rectangular form (4.39). As noted in Problem 4.12, a signal $x(t)$ is said to be even if $x(t) = x(-t)$, and the signal is said to be odd if $x(-t) = -x(t)$. If the signal $x(t)$ is even, it follows that the imaginary part $I(\omega)$ of the Fourier transform is zero and the real part $R(\omega)$ can be rewritten as

$$R(\omega) = 2\int_0^{\infty} x(t)\cos\omega t\,dt$$

Hence the Fourier transform of an even signal $x(t)$ is a real-valued function of ω given by

$$X(\omega) = 2\int_0^{\infty} x(t)\cos\omega t\,dt \tag{4.41}$$

If the signal $x(t)$ is odd, that is, $x(t) = -x(-t)$ for all $t > 0$, the Fourier transform of $x(t)$ is a purely imaginary function of ω given by

$$X(\omega) = -j2\int_0^{\infty} x(t)\sin\omega t\,dt \tag{4.42}$$

The expression (4.41) may be used to compute the Fourier transform of an even signal, and the expression (4.42) may be used to compute the Fourier transform of an odd signal.

Example 4.7 *Rectangular Pulse*

Given a fixed positive number τ, let $p_\tau(t)$ denote the rectangular pulse of duration τ seconds defined by

$$p_\tau(t) = \begin{cases} 1, & -\dfrac{\tau}{2} \le t < \dfrac{\tau}{2} \\ 0, & \text{all other } t \end{cases}$$

The rectangular pulse $p_\tau(t)$, which is plotted in Figure 4.18, is clearly an even signal, and thus (4.41) can be used to compute the Fourier transform. Setting $x(t) = p_\tau(t)$ in (4.41) yields

$$\begin{aligned} X(\omega) &= 2\int_0^{\tau/2} (1)\cos\omega t\,dt \\ &= \frac{2}{\omega}[\sin\omega t]_{t=0}^{t=\tau/2} \\ &= \frac{2}{\omega}\sin\frac{\omega\tau}{2} \end{aligned}$$

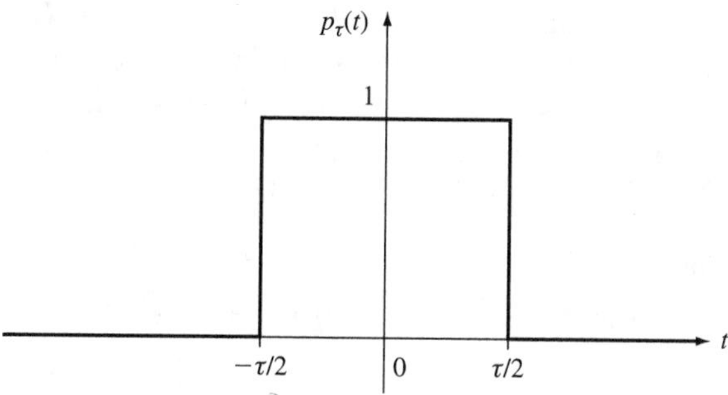

Figure 4.18 Rectangular pulse of duration τ seconds.

Expressing $X(\omega)$ in terms of the sinc function [defined by (4.24)] gives

$$X(\omega) = \tau \operatorname{sinc} \frac{\tau\omega}{2\pi}$$

Note that when $\tau = 1$, this result is consistent with the development given above for the case of the 1-second rectangular pulse [see (4.32)]. Also note that since the Fourier transform $X(\omega)$ in this example is real valued, $X(\omega)$ can be plotted versus ω. The result is displayed in Figure 4.19.

Bandlimited Signals

A signal $x(t)$ is said to be *bandlimited* if its Fourier transform $X(\omega)$ is zero for all $\omega > B$, where B is some positive number, called the *bandwidth* of the signal. If a signal $x(t)$ is bandlimited with bandwidth B, the signal does not contain any spectral components with frequency higher than B, which justifies the use of the term *bandlimited*. It turns out that any bandlimited signal must be of infinite duration in time; that is, bandlimited signals cannot be time limited. [A signal $x(t)$ is time limited if there exists a positive number T such that $x(t) = 0$ for all $t < -T$ and $t > T$.]

If a signal $x(t)$ is not bandlimited, it is said to have *infinite bandwidth* or an *infinite spectrum*. Since bandlimited signals cannot be time limited, time-limited signals cannot be bandlimited, and thus all time-limited signals have infinite bandwidth. In

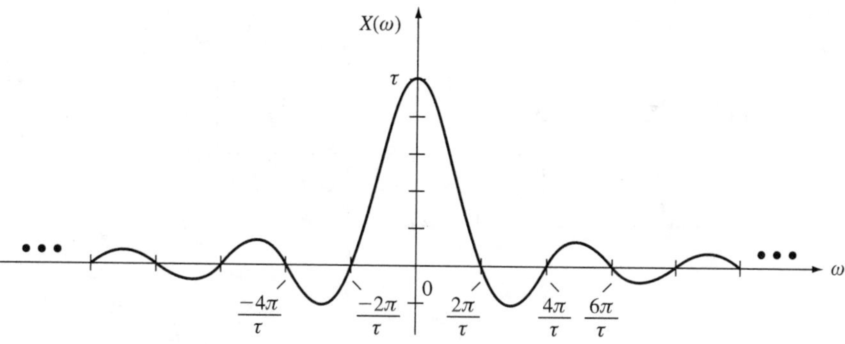

Figure 4.19 Fourier transform of the τ-second rectangular pulse.

addition, since all (physical) signals are time limited, any such signal must have an infinite bandwidth. However, for any well-behaved time-limited signal $x(t)$, it can be proved that the Fourier transform $X(\omega)$ converges to zero as $\omega \to \infty$. Therefore, for any time-limited signal arising in practice, it is always possible to assume that $|X(\omega)| \approx 0$ for all $\omega > B$, where B is chosen to be suitably large.

Example 4.8 *Frequency Spectrum*

Again consider the rectangular pulse function $x(t) = p_\tau(t)$. In Example 4.7 it was shown that the Fourier transform $X(\omega)$ is

$$X(\omega) = \tau \operatorname{sinc} \frac{\tau\omega}{2\pi}$$

The plots of the amplitude and phase spectra for this example are given in Figure 4.20. From Figure 4.20a, it is clear that the spectrum of the rectangular pulse is infinite; however, since the *sidelobes* shown in Figure 4.20a decrease in magnitude as the frequency ω is increased, it is clear that for any $c > 0$, there is a B (in general depending on c) such that $|X(\omega)| < c$ for all $\omega > B$. So if B is chosen to be sufficiently large, the rectangular pulse can be viewed as being "approximately bandlimited" with bandwidth B.

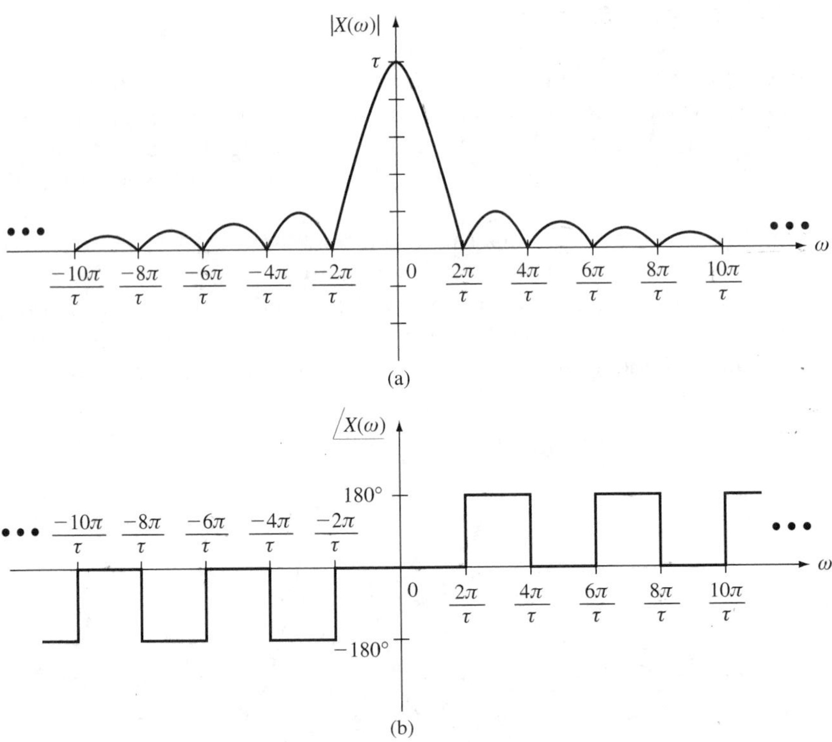

Figure 4.20 (a) Amplitude and (b) phase spectra of the rectangular pulse.

Note also that if the time duration τ of the rectangular pulse is made smaller, the amplitude spectrum "spreads out." This result shows that shorter-time-duration signals (e.g., a pulse with smaller time duration) have more spectral content at higher frequencies than that of longer-time-duration signals.

Inverse Fourier Transform

Given a signal $x(t)$ with Fourier transform $X(\omega)$, $x(t)$ can be recomputed from $X(\omega)$ by applying the inverse Fourier transform given by

$$x(t) = \frac{1}{2\pi}\int_{-\infty}^{\infty} X(\omega)e^{j\omega t}\, d\omega \tag{4.43}$$

To denote the fact that $X(\omega)$ is the Fourier transform of $x(t)$, or that $X(\omega)$ is the inverse Fourier transform of $x(t)$, the transform pair notation

$$x(t) \leftrightarrow X(\omega)$$

will sometimes be used. One of the most fundamental transform pairs in the Fourier theory is the pair

$$p_\tau(t) \leftrightarrow \tau \operatorname{sinc}\frac{\tau\omega}{2\pi} \tag{4.44}$$

The transform pair (4.44) follows from the results of Example 4.7. Note that by (4.44), a rectangular function in time corresponds to a sinc function in frequency, and conversely, a sinc function in frequency corresponds to a rectangular function in time.

It is sometimes possible to compute the Fourier transform or the inverse Fourier transform without having to evaluate the integrals in (4.36) and (4.43). In particular, it is possible to derive new transform pairs from a given transform pair [such as (4.44)] by using the properties of the Fourier transform. These properties are given in the next section.

4.4 PROPERTIES OF THE FOURIER TRANSFORM

The Fourier transform satisfies a number of properties that are useful in a wide range of applications. These properties are given in this section. In Chapters 5 and 6, some of these properties are applied to the study of modulation and sampling.

Linearity

The Fourier transform is a linear operation; that is, if $x(t) \leftrightarrow X(\omega)$ and $v(t) \leftrightarrow V(\omega)$, then for any real or complex scalars a, b,

$$ax(t) + bv(t) \leftrightarrow aX(\omega) + bV(\omega) \tag{4.45}$$

The property of linearity can be proved by computing the Fourier transform of $ax(t) + bv(t)$: By definition of the Fourier transform

$$ax(t) + bv(t) \leftrightarrow \int_{-\infty}^{\infty} [ax(t) + bv(t)]\, e^{-j\omega t}\, dt$$

By linearity of integration,

$$\int_{-\infty}^{\infty} [ax(t) + bv(t)]\, e^{-j\omega t}\, dt = a\int_{-\infty}^{\infty} x(t)e^{-j\omega t}\, dt + b\int_{-\infty}^{\infty} v(t)e^{-j\omega t}\, dt$$

and thus

$$ax(t) + bv(t) \leftrightarrow aX(\omega) + bV(\omega)$$

Example 4.9 *Sum of Rectangular Pulses*

Consider the signal shown in Figure 4.21. As illustrated in the figure, this signal is equal to a sum of two rectangular pulse functions. More precisely,

$$x(t) = p_4(t) + p_2(t)$$

Then by using linearity and the transform pair (4.44), it follows that the Fourier transform of $x(t)$ is

$$X(\omega) = 4\,\text{sinc}\,\frac{2\omega}{\pi} + 2\,\text{sinc}\,\frac{\omega}{\pi}$$

Left or Right Shift in Time

If $x(t) \leftrightarrow X(\omega)$, then for any positive or negative real number c,

$$x(t - c) \leftrightarrow X(\omega)e^{-j\omega c} \tag{4.46}$$

Note that if $c > 0$, $x(t - c)$ is a c-second right shift of $x(t)$, and if $c < 0$, $x(t - c)$ is a $(-c)$-second left shift of $x(t)$. Thus the transform pair (4.46) is valid for both left and right shifts of $x(t)$.

To verify the validity of the transform pair (4.46), first apply the definition of the Fourier transform to the shifted signal $x(t - c)$, which gives

$$x(t - c) \leftrightarrow \int_{-\infty}^{\infty} x(t - c)e^{-j\omega t}\, dt \tag{4.47}$$

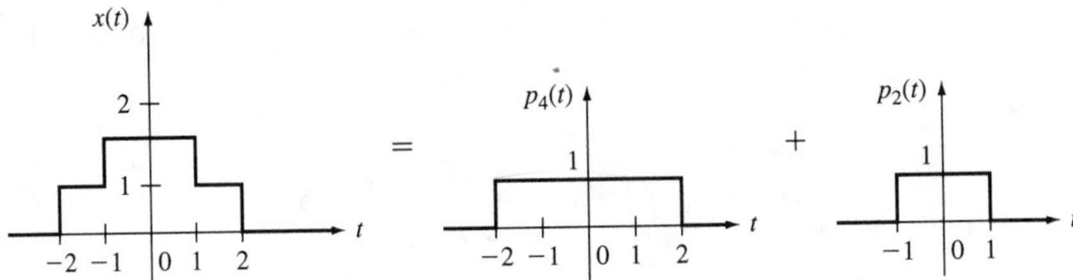

Figure 4.21 Signal in Example 4.9.

In the integral in (4.47), consider the change of variable $\bar{t} = t - c$. Then $t = \bar{t} + c$, $dt = d\bar{t}$, and (4.47) becomes

$$x(t - c) \leftrightarrow \int_{-\infty}^{\infty} x(\bar{t})e^{-j\omega(\bar{t}+c)}\, d\bar{t}$$

$$\leftrightarrow \left[\int_{-\infty}^{\infty} x(\bar{t})e^{-j\omega\bar{t}}\, d\bar{t}\right]e^{-j\omega c}$$

$$\leftrightarrow X(\omega)e^{-j\omega c}$$

Hence, (4.46) is obtained.

Example 4.10 *Right Shift of Pulse*

The signal $x(t)$ shown in Figure 4.22 is equal to a 1-second right shift of the rectangular pulse function $p_2(t)$; that is,

$$x(t) = p_2(t - 1)$$

The Fourier transform $X(\omega)$ of $x(t)$ can be computed using the time-shift property (4.46) and the transform pair (4.44). The result is

$$X(\omega) = 2\left(\operatorname{sinc}\frac{\omega}{\pi}\right)e^{-j\omega}$$

Note that since

$$\left|e^{-j\omega}\right| = 1 \qquad \text{for all } \omega$$

the amplitude spectrum $|X(\omega)|$ of $x(t) = p_2(t - 1)$ is the same as the amplitude spectrum of $p_2(t)$.

Time Scaling

If $x(t) \leftrightarrow X(\omega)$, for any positive real number a,

$$x(at) \leftrightarrow \frac{1}{a}X\left(\frac{\omega}{a}\right) \tag{4.48}$$

To prove (4.48), first apply the definition of the Fourier transform to $x(at)$, which gives

$$x(at) \leftrightarrow \int_{-\infty}^{\infty} x(at)e^{-j\omega t}\, dt \tag{4.49}$$

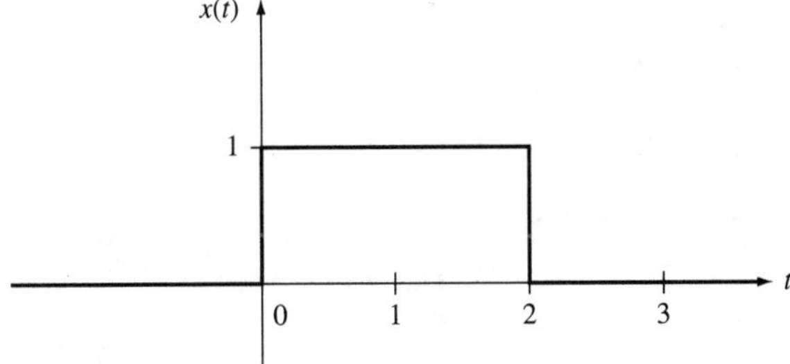

Figure 4.22 Signal in Example 4.10.

In the integral in (4.49) consider the change of variable $\bar{t} = at$. Then $t = \bar{t}/a$, $d\bar{t} = a(dt)$, and (4.49) becomes

$$x(at) \leftrightarrow \int_{-\infty}^{\infty} x(\bar{t}) \exp\left[-j\left(\frac{\omega}{a}\right)\bar{t}\right]\frac{1}{a}\,d\bar{t}$$

$$\leftrightarrow \frac{1}{a}\int_{-\infty}^{\infty} x(\bar{t}) \exp\left[-j\left(\frac{\omega}{a}\right)\bar{t}\right] d\bar{t}$$

$$\leftrightarrow \frac{1}{a}X\left(\frac{\omega}{a}\right)$$

Hence (4.48) is verified.

Given an arbitrary signal $x(t)$, if $a > 1, x(at)$ is a *time compression* of $x(t)$. For example, suppose that $x(t)$ is the 2-second rectangular pulse $p_2(t)$ and $a = 2$. The signals $p_2(t)$ and $p_2(2t)$ are displayed in Figure 4.23. Clearly, $p_2(2t)$ is a time compression of $p_2(t)$.

Now by (4.44), the Fourier transform of $p_2(t)$ is equal to 2 sinc(ω/π), and by (4.48) the Fourier transform of $p_2(2t)$ is equal to sinc($\omega/2\pi$). These transforms are displayed in Figure 4.24. As seen from this figure, the Fourier transform of $p_2(2t)$ is a *fre-*

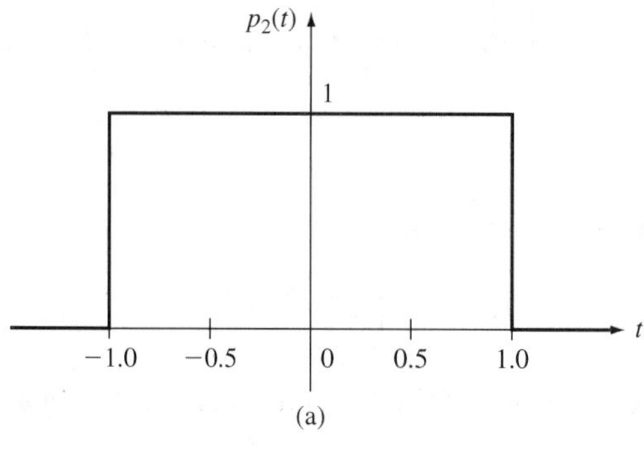

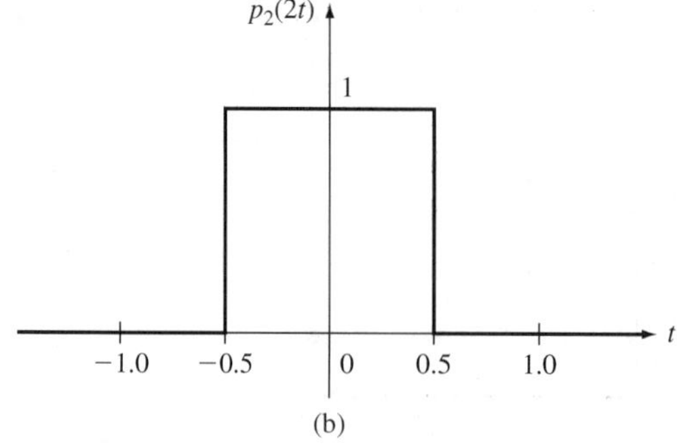

Figure 4.23 Signals (a) $p_2(t)$ and (b) $p_2(2t)$.

quency expansion of the Fourier transform of $p_2(t)$. Hence the shorter-duration pulse $p_2(2t)$ has a wider bandwidth than the longer-duration pulse $p_2(t)$.

For an arbitrary signal $x(t)$ with Fourier transform $X(\omega)$, if $a > 1$, $X(\omega/a)$ is a frequency expansion of $X(\omega)$. Thus, by (4.48) it is seen that a time compression of a signal $x(t)$ corresponds to a frequency expansion of the Fourier transform $X(\omega)$ of the signal. This again shows that shorter-time-duration signals have wider bandwidths than those of longer-time-duration signals.

Again let $x(t)$ be an arbitrary signal with Fourier transform $X(\omega)$. If $0 < a < 1$, $x(at)$ is a *time expansion* of $x(t)$ and $X(\omega/a)$ is a *frequency compression* of $X(\omega)$. In this case it follows from (4.48) that a time expansion of $x(t)$ corresponds to a frequency compression of $X(\omega)$. Thus longer-duration signals have smaller bandwidths.

Time Reversal

Given a signal $x(t)$, consider the time-reversed signal $x(-t)$. The signal $x(-t)$ is equal to $x(t)$ folded about the vertical axis. Now if $x(t) \leftrightarrow X(\omega)$, then

$$x(-t) \leftrightarrow X(-\omega) \tag{4.50}$$

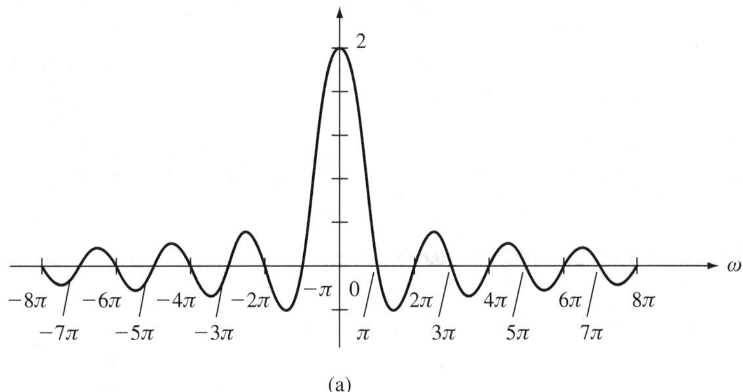

(a)

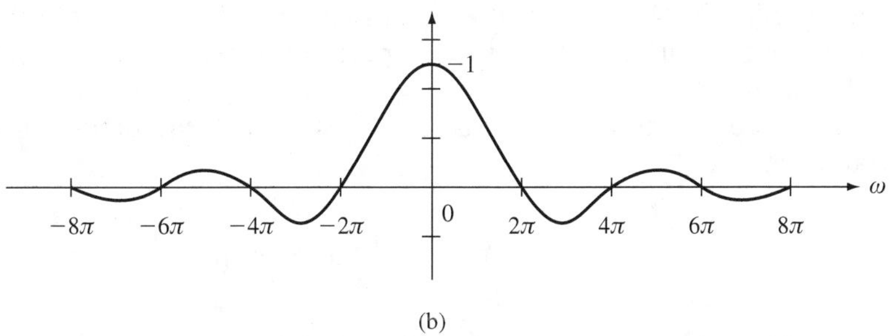

(b)

Figure 4.24 Fourier transforms of (a) $p_2(t)$ and (b) $p_2(2t)$.

To prove (4.50), simply replace t by $-t$ in the definition of the Fourier transform of $x(t)$.

If the signal $x(t)$ is real valued, from the definition (4.36) of the Fourier transform it follows that

$$X(-\omega) = \overline{X(\omega)}$$

where $\overline{X(\omega)}$ is the complex conjugate of $X(\omega)$. Hence the transform pair (4.50) can be rewritten as

$$x(-t) \leftrightarrow \overline{X(\omega)} \tag{4.51}$$

By (4.51), time reversal in the time domain corresponds to conjugation in the frequency domain.

Example 4.11 *Time-Reversed Exponential Signal*

Given a real number $b > 0$, consider the signal

$$x(t) = \begin{cases} 0, & t > 0 \\ e^{bt}, & t \leq 0 \end{cases}$$

Note that

$$x(-t) = e^{-bt}u(t)$$

and from the result in Example 4.6, the Fourier transform of $x(-t)$ is $1/(b + j\omega)$. Hence the Fourier transform of $x(t)$ is

$$X(\omega) = \overline{\frac{1}{b + j\omega}} = \frac{1}{b - j\omega}$$

Multiplication by a Power of t

If $x(t) \leftrightarrow X(\omega)$, for any positive integer n,

$$t^n x(t) \leftrightarrow (j)^n \frac{d^n}{d\omega^n} X(\omega) \tag{4.52}$$

Setting $n = 1$ in (4.52) yields the result that multiplication by t in the time domain corresponds to differentiation with respect to ω in the frequency domain (plus multiplication by j).

To prove (4.52) for the case $n = 1$, start with the definition of the Fourier transform:

$$X(\omega) = \int_{-\infty}^{\infty} x(t)e^{-j\omega t}\, dt \tag{4.53}$$

Differentiating both sides of (4.53) with respect to ω and multiplying by j yields

$$j\frac{dX(\omega)}{d\omega} = j\int_{-\infty}^{\infty} (-jt)x(t)e^{-j\omega t}\,dt$$

$$j\frac{dX(\omega)}{d\omega} = \int_{-\infty}^{\infty} tx(t)e^{-j\omega t}\,dt \tag{4.54}$$

The right-hand side of (4.54) is equal to the Fourier transform of $tx(t)$, and thus (4.52) is verified for the case $n = 1$. The proof for $n \geq 2$ follows by taking second- and higher-order derivatives of $X(\omega)$ with respect to ω. The details are omitted.

Example 4.12 *Product of t and a Pulse*

Let $x(t) = tp_2(t)$, which is plotted in Figure 4.25. The Fourier transform $X(\omega)$ of $x(t)$ can be computed using the property (4.52) and the transform pair (4.44). This yields

$$\begin{aligned} X(\omega) &= j\frac{d}{d\omega}\left(2\,\text{sinc}\,\frac{\omega}{\pi}\right) \\ &= j2\frac{d}{d\omega}\left(\frac{\sin\omega}{\omega}\right) \\ &= j2\frac{\omega\cos\omega - \sin\omega}{\omega^2} \end{aligned}$$

The amplitude spectrum $|X(\omega)|$ is plotted in Figure 4.26.

Multiplication by a Complex Exponential

If $x(t) \leftrightarrow X(\omega)$, then for any real number ω_0,

$$x(t)e^{j\omega_0 t} \leftrightarrow X(\omega - \omega_0) \tag{4.55}$$

So multiplication by a complex exponential in the time domain corresponds to a frequency shift in the frequency domain. The proof of (4.55) follows directly from the definition of the Fourier transform. The verification is left to the reader.

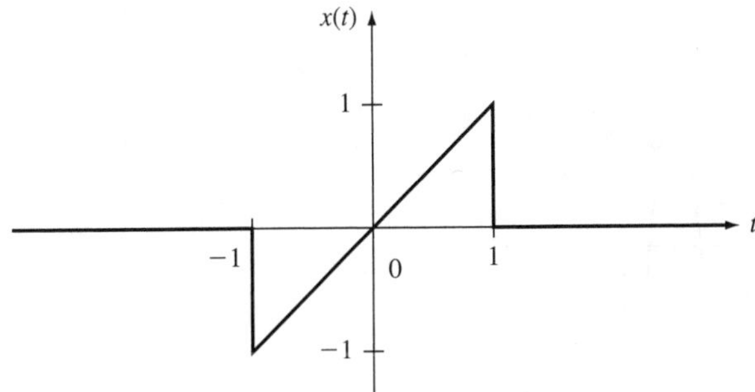

Figure 4.25 The signal $x(t) = tp_2(t)$.

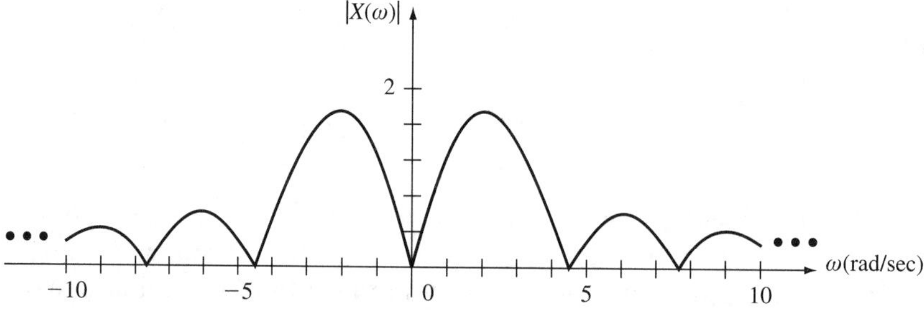

Figure 4.26 Amplitude spectrum of the signal in Figure 4.25.

Multiplication by a Sinusoid

If $x(t) \leftrightarrow X(\omega)$, then for any real number ω_0,

$$x(t) \sin \omega_0 t \leftrightarrow \frac{j}{2}[X(\omega + \omega_0) - X(\omega - \omega_0)] \tag{4.56}$$

$$x(t) \cos \omega_0 t \leftrightarrow \frac{1}{2}[X(\omega + \omega_0) + X(\omega - \omega_0)] \tag{4.57}$$

The proof of (4.56) and (4.57) follow from (4.55) and Euler's identity. The details are omitted.

As discussed in Chapter 6, the signals $x(t) \sin \omega_0 t$ and $x(t) \cos \omega_0 t$ can be viewed as amplitude-modulated signals. More precisely, in forming the signal $x(t) \sin \omega_0 t$, the carrier $\sin \omega_0 t$ is modulated by the signal $x(t)$. As a result of this characterization of $x(t) \sin \omega_0 t$ [and $x(t) \cos \omega_0 t$], the relationships (4.56) and (4.57) are called the *modulation theorems* of the Fourier transform. The relationships (4.56) and (4.57) show that modulation of a carrier by a signal $x(t)$ results in the frequency translations $X(\omega + \omega_0), X(\omega - \omega_0)$ of the Fourier transform $X(\omega)$.

Example 4.13 *Sinusoidal Burst*

Consider the signal $x(t) = p_\tau(t) \cos \omega_0 t$, which can be interpreted as a sinusoidal burst. For the case when $\tau = 0.5$ and $\omega_0 = 60$ rad/sec, the signal is plotted in Figure 4.27. By

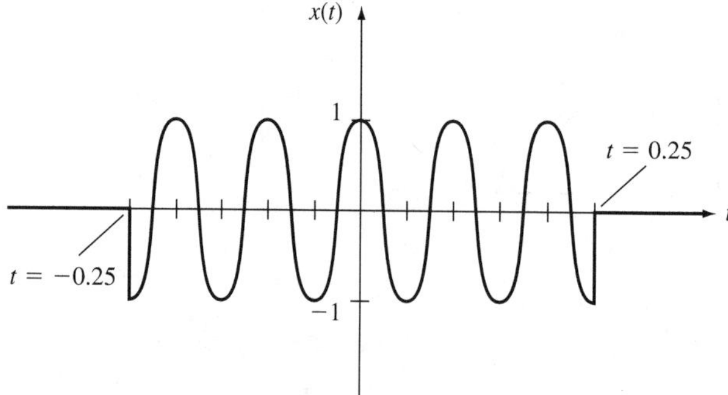

Figure 4.27 Sinusoidal burst.

the modulation property (4.57) and the transform pair (4.44), the Fourier transform of the sinusoidal burst is equal to

$$\frac{1}{2}\left[\tau \operatorname{sinc}\left(\frac{\tau(\omega+\omega_0)}{2\pi}\right) + \tau \operatorname{sinc}\left(\frac{\tau(\omega-\omega_0)}{2\pi}\right)\right]$$

For the case $\tau = 0.5$ and $\omega_0 = 60$ rad/sec, the transform of the sinusoidal burst is plotted in Figure 4.28.

Differentiation in the Time Domain

If $x(t) \leftrightarrow X(\omega)$, then for any positive integer n

$$\frac{d^n}{dt^n} x(t) \leftrightarrow (j\omega)^n X(\omega) \tag{4.58}$$

For the case $n = 1$, it follows from (4.58) that differentiation in the time domain corresponds to multiplication by $j\omega$ in the frequency domain. To prove (4.58) for the case $n = 1$, first observe that the Fourier transform of $dx(t)/dt$ is

$$\int_{-\infty}^{\infty} \frac{dx(t)}{dt} e^{-j\omega t}\, dt \tag{4.59}$$

The integral in (4.59) can be computed "by parts" as follows. With $v = e^{-j\omega t}$ and $w = x(t)$, $dv = -j\omega e^{-j\omega t}\, dt$ and $dw = [dx(t)/dt]\, dt$. Then

$$\int_{-\infty}^{\infty} \frac{dx(t)}{dt} e^{-j\omega t}\, dt = vw\Big|_{t=-\infty}^{t=\infty} - \int_{-\infty}^{\infty} w\, dv$$

$$= e^{-j\omega t} x(t)\Big|_{t=-\infty}^{t=\infty} - \int_{-\infty}^{\infty} x(t)(-j\omega)e^{-j\omega t}\, dt$$

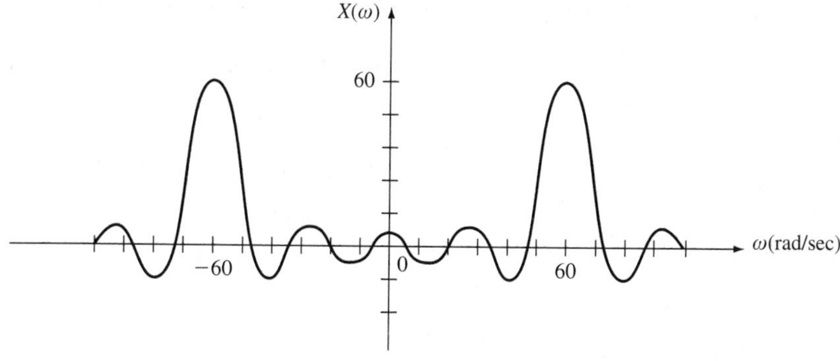

Figure 4.28 Fourier transform of the sinusoidal burst $x(t) = p_{0.5}(t) \cos 60t$.

Then if $x(t) \to 0$ as $t \to \pm\infty$,

$$\int_{-\infty}^{\infty} \frac{dx(t)}{dt} e^{-j\omega t}\, dt = (j\omega)X(\omega)$$

and thus (4.58) is valid for the case $n = 1$. The proof of (4.58) for $n \geq 2$ follows by repeated application of integration by parts.

Integration in the Time Domain

Given a signal $x(t)$, the integral of $x(t)$ is the function

$$\int_{-\infty}^{t} x(\lambda)\, d\lambda$$

Suppose that $x(t)$ has Fourier transform $X(\omega)$. In general, the integral of $x(t)$ does not have a Fourier transform in the ordinary sense, but it does have the generalized transform

$$\frac{1}{j\omega} X(\omega) + \pi X(0)\delta(\omega)$$

where $\delta(\omega)$ is the impulse function in the frequency domain. This results in the transform pair

$$\int_{-\infty}^{t} x(\lambda)\, d\lambda \leftrightarrow \frac{1}{j\omega} X(\omega) + \pi X(0)\delta(\omega) \tag{4.60}$$

Note that if the signal $x(t)$ has no dc component, e.g., $X(0) = 0$, then (4.60) reduces to

$$\int_{-\infty}^{t} x(\lambda)\, d\lambda \leftrightarrow \frac{1}{j\omega} X(\omega)$$

Hence, the second term on the right-hand side of the transform pair (4.60) is due to a possible dc component in $x(t)$.

Example 4.14 *Transform of a Triangular Pulse*

Consider the triangular pulse function $v(t)$ displayed in Figure 4.29. As first noted in Chapter 1, the triangular pulse can be expressed mathematically by

$$v(t) = \left(1 - \frac{2|t|}{\tau}\right) p_\tau(t)$$

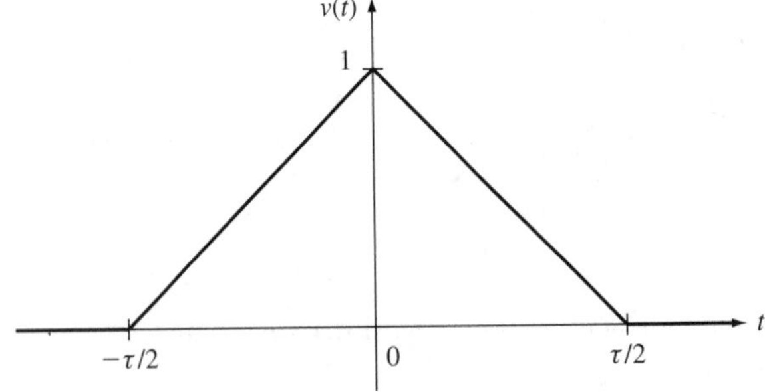

Figure 4.29 Triangular pulse.

where again $p_\tau(t)$ is the rectangular pulse of duration τ seconds. To compute the Fourier transform $V(\omega)$ of $v(t)$, the Fourier transform of the derivative of $v(t)$ will be computed first. Then using the integration property (4.60), it will be possible to determine $V(\omega)$.

The derivative of $v(t)$, which is denoted by $x(t)$, is shown in Figure 4.30. From the plot, it is clear that the derivative can be expressed mathematically as

$$x(t) = \frac{2}{\tau}p_{\tau/2}\left(t + \frac{\tau}{4}\right) - \frac{2}{\tau}p_{\tau/2}\left(t - \frac{\tau}{4}\right)$$

The Fourier transform $X(\omega)$ of $x(t)$ can be computed using the transform pair (4.44) and the shift property (4.46). This yields

$$\begin{aligned} X(\omega) &= \left(\operatorname{sinc}\frac{\tau\omega}{4\pi}\right)\left[\exp\left(\frac{j\tau\omega}{4}\right) - \exp\left(-\frac{j\tau\omega}{4}\right)\right] \\ &= \left(\operatorname{sinc}\frac{\tau\omega}{4\pi}\right)\left(j2\sin\frac{\tau\omega}{4}\right) \end{aligned}$$

Now since $v(t)$ is the integral of $x(t)$, by the integration property (4.60) the Fourier transform $V(\omega)$ of $v(t)$ is

$$\begin{aligned} V(\omega) &= \frac{1}{j\omega}\left(\operatorname{sinc}\frac{\tau\omega}{4\pi}\right)\left(j2\sin\frac{\tau\omega}{4}\right) + \pi X(0)\delta(\omega) \\ &= \frac{2}{\omega}\left[\frac{\sin(\tau\omega/4)}{\tau\omega/4}\right]\sin\frac{\tau\omega}{4} \\ &= \frac{\tau}{2}\frac{\sin^2(\tau\omega/4)}{(\tau\omega/4)^2} \\ &= \frac{\tau}{2}\operatorname{sinc}^2\frac{\tau\omega}{4\pi} \end{aligned}$$

Hence the end result is the transform pair

$$\left(1 - \frac{2|t|}{\tau}\right)p_\tau(t) \leftrightarrow \frac{\tau}{2}\operatorname{sinc}^2\frac{\tau\omega}{4\pi} \tag{4.61}$$

By (4.61), it is seen that the triangular pulse in the time domain corresponds to a sinc-squared function in the Fourier transform domain. In the case $\tau = 1$, the Fourier transform of the triangular pulse is plotted in Figure 4.31.

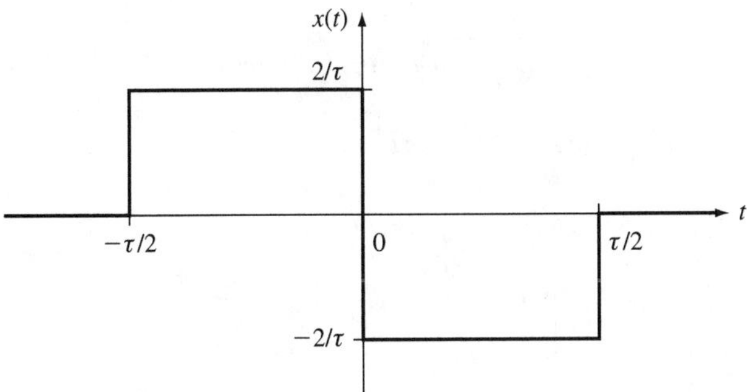

Figure 4.30 Derivative of the triangular pulse.

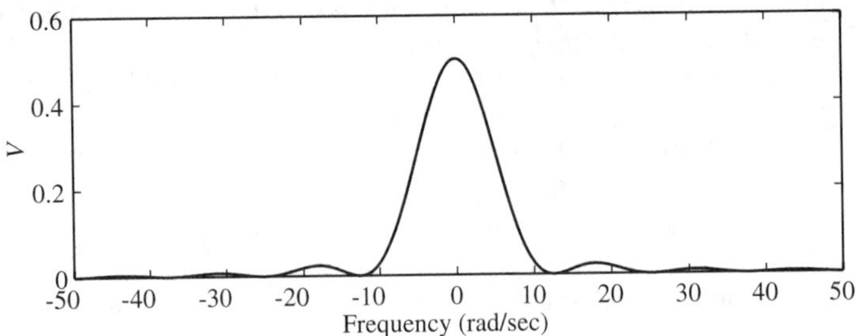

Figure 4.31 Fourier transform of the 1-second triangular pulse.

Convolution in the Time Domain

Given two signals $x(t)$ and $v(t)$ with Fourier transforms $X(\omega)$ and $V(\omega)$, the Fourier transform of the convolution $x(t) * v(t)$ is equal to the product $X(\omega)V(\omega)$, which results in the transform pair

$$x(t) * v(t) \leftrightarrow X(\omega)V(\omega) \tag{4.62}$$

Note that by (4.62), convolution in the time domain corresponds to multiplication in the frequency domain. In Chapter 5 it will be seen that this property is very useful in determining the relationship between the input and the output of a linear time-invariant continuous-time system.

To prove (4.62), first recall that by definition of convolution

$$x(t) * v(t) = \int_{-\infty}^{\infty} x(\lambda)v(t-\lambda)\,d\lambda$$

Hence the Fourier transform of $x(t) * v(t)$ is given by

$$\int_{-\infty}^{\infty}\left[\int_{-\infty}^{\infty} x(\lambda)v(t-\lambda)\,d\lambda\right]e^{-j\omega t}\,dt$$

This can be rewritten in the form

$$\int_{-\infty}^{\infty} x(\lambda)\left[\int_{-\infty}^{\infty} v(t-\lambda)e^{-j\omega t}\,dt\right]d\lambda$$

Using the change of variable $\bar{t} = t - \lambda$ in the second integral gives

$$\int_{-\infty}^{\infty} x(\lambda)\left[\int_{-\infty}^{\infty} v(\bar{t})e^{-j\omega(\bar{t}+\lambda)}\,d\bar{t}\right]d\lambda$$

This expression can be rewritten in the form

$$\left[\int_{-\infty}^{\infty} x(\lambda)e^{-j\omega\lambda}\,d\lambda\right]\left[\int_{-\infty}^{\infty} v(\bar{t})e^{-j\omega\bar{t}}\,d\bar{t}\right]$$

Clearly, the expression above is equal to $X(\omega)V(\omega)$, and thus (4.62) is verified.

Multiplication in the Time Domain

If $x(t) \leftrightarrow X(\omega)$ and $v(t) \leftrightarrow V(\omega)$, then

$$x(t)v(t) \leftrightarrow \frac{1}{2\pi}[X(\omega) * V(\omega)] = \frac{1}{2\pi}\int_{-\infty}^{\infty} X(\lambda)V(\omega - \lambda)\, d\lambda \tag{4.63}$$

From (4.63) it is seen that multiplication in the time domain corresponds to convolution in the Fourier transform domain. The proof of (4.63) follows from the definition of the Fourier transform and the manipulation of integrals. The details are omitted.

Parseval's Theorem

Again suppose that $x(t) \leftrightarrow X(\omega)$ and $v(t) \leftrightarrow V(\omega)$. Then

$$\int_{-\infty}^{\infty} x(t)v(t)\, dt = \frac{1}{2\pi}\int_{-\infty}^{\infty} \overline{X(\omega)}V(\omega)\, d\omega \tag{4.64}$$

where $\overline{X(\omega)}$ is the complex conjugate of $X(\omega)$. The relationship (4.64), which is called *Parseval's theorem,* follows directly from the transform pair (4.63). To see this, first note that the Fourier transform of the product $x(t)v(t)$ is equal to

$$\int_{-\infty}^{\infty} x(t)v(t)e^{-j\omega t}\, dt$$

But by the transform pair (4.63), the Fourier transform of $x(t)v(t)$ is equal to

$$\frac{1}{2\pi}\int_{-\infty}^{\infty} X(\omega - \lambda)V(\lambda)\, d\lambda$$

Thus

$$\int_{-\infty}^{\infty} x(t)v(t)e^{-j\omega t}\, dt = \frac{1}{2\pi}\int_{-\infty}^{\infty} X(\omega - \lambda)V(\lambda)\, d\lambda \tag{4.65}$$

The relationship (4.65) must hold for all real values of ω. Taking $\omega = 0$ gives

$$\int_{-\infty}^{\infty} x(t)v(t)\, dt = \frac{1}{2\pi}\int_{-\infty}^{\infty} X(-\lambda)V(\lambda)\, d\lambda \tag{4.66}$$

If $x(t)$ is real valued, $X(-\omega) = \overline{X(\omega)}$, and thus changing the variable of integration from λ to ω on the right-hand side of (4.66) results in (4.64).

Note that if $v(t) = x(t)$, Parseval's theorem becomes

$$\int_{-\infty}^{\infty} x^2(t)\, dt = \frac{1}{2\pi}\int_{-\infty}^{\infty} \overline{X(\omega)}X(\omega)\, d\omega \tag{4.67}$$

From the properties of complex numbers,

$$\overline{X(\omega)}X(\omega) = |X(\omega)|^2$$

and thus (4.67) can be written in the form

$$\int_{-\infty}^{\infty} x^2(t)\, dt = \frac{1}{2\pi}\int_{-\infty}^{\infty} |X(\omega)|^2\, d\omega \tag{4.68}$$

The left-hand side of (4.68) can be interpreted as the energy of the signal $x(t)$. Thus (4.68) relates the energy of the signal and the integral of the square of the magnitude of the Fourier transform of the signal.

Duality

Suppose that $x(t) \leftrightarrow X(\omega)$. A new continuous-time signal can be defined by setting $\omega = t$ in $X(\omega)$. This results in the continuous-time signal $X(t)$. The duality property states that the Fourier transform of $X(t)$ is equal to $2\pi x(-\omega)$; that is,

$$X(t) \leftrightarrow 2\pi x(-\omega) \tag{4.69}$$

In (4.69), $x(-\omega)$ is the frequency function constructed by setting $t = -\omega$ in the expression for $x(t)$.

For any given transform pair $x(t) \leftrightarrow X(\omega)$, by using duality the new transform pair (4.69) can be constructed. For example, applying the duality property to the pair (4.44) yields the transform pair

$$\tau \operatorname{sinc}\frac{\tau t}{2\pi} \leftrightarrow 2\pi p_\tau(-\omega) \tag{4.70}$$

Since $p_\tau(\omega)$ is an even function of $\omega, p_\tau(-\omega) = p_\tau(\omega)$ and (4.70) can be rewritten as

$$\tau \operatorname{sinc}\frac{\tau t}{2\pi} \leftrightarrow 2\pi p_\tau(\omega) \tag{4.71}$$

From (4.71) it is seen that a sinc function in time corresponds to a rectangular pulse function in frequency.

Applying the duality property to the transform pair (4.61) gives

$$\frac{\tau}{2}\operatorname{sinc}^2\frac{\tau t}{4\pi} \leftrightarrow 2\pi\left(1 - \frac{2|\omega|}{\tau}\right)p_\tau(\omega) \tag{4.72}$$

Thus the sinc-squared time function has a Fourier transform equal to the triangular pulse function in frequency.

The duality property is easy to prove: First, by definition of the Fourier transform

$$X(\omega) = \int_{-\infty}^{\infty} x(t)e^{-j\omega t}\, dt \tag{4.73}$$

Setting $\omega = t$ and $t = -\omega$ in (4.73) gives

$$X(t) = \int_{-\infty}^{\infty} x(-\omega)e^{j\omega t}\, d\omega$$

$$= \frac{1}{2\pi}\int_{-\infty}^{\infty} 2\pi x(-\omega)e^{j\omega t}\, d\omega$$

Thus $X(t)$ is the inverse Fourier transform of the frequency function $2\pi x(-\omega)$, which proves (4.69).

For the convenience of the reader, the properties of the Fourier transform are summarized in Table 4.1.

TABLE 4.1 PROPERTIES OF THE FOURIER TRANSFORM

Property	Transform Pair/Property
Linearity	$ax(t) + bv(t) \leftrightarrow aX(\omega) + bV(\omega)$
Right or left shift in time	$x(t - c) \leftrightarrow X(\omega)e^{-j\omega c}$
Time scaling	$x(at) \leftrightarrow \frac{1}{a}X\left(\frac{\omega}{a}\right) \quad a > 0$
Time reversal	$x(-t) \leftrightarrow X(-\omega) = \overline{X(\omega)}$
Multiplication by a power of t	$t^n x(t) \leftrightarrow j^n \frac{d^n}{d\omega^n} X(\omega) \quad n = 1, 2, \ldots$
Multiplication by a complex exponential	$x(t)e^{j\omega_0 t} \leftrightarrow X(\omega - \omega_0) \quad \omega_0$ real
Multiplication by sin $\omega_0 t$	$x(t) \sin \omega_0 t \leftrightarrow \frac{j}{2}[X(\omega + \omega_0) - X(\omega - \omega_0)]$
Multiplication by cos $\omega_0 t$	$x(t) \cos \omega_0 t \leftrightarrow \frac{1}{2}[X(\omega + \omega_0) + X(\omega - \omega_0)]$
Differentiation in the time domain	$\frac{d^n}{dt^n} x(t) \leftrightarrow (j\omega)^n X(\omega) \quad n = 1, 2, \ldots$
Integration	$\int_{-\infty}^{t} x(\lambda)\, d\lambda \leftrightarrow \frac{1}{j\omega} X(\omega) + \pi X(0)\delta(\omega)$
Convolution in the time domain	$x(t) * v(t) \leftrightarrow X(\omega)V(\omega)$
Multiplication in the time domain	$x(t)v(t) \leftrightarrow \frac{1}{2\pi} X(\omega) * V(\omega)$
Parseval's theorem	$\int_{-\infty}^{\infty} x(t)v(t)\, dt = \frac{1}{2\pi}\int_{-\infty}^{\infty} \overline{X(\omega)}V(\omega)\, d\omega$
Special case of Parseval's theorem	$\int_{-\infty}^{\infty} x^2(t)\, dt = \frac{1}{2\pi}\int_{-\infty}^{\infty} \lvert X(\omega)\rvert^2\, d\omega$
Duality	$X(t) \leftrightarrow 2\pi x(-\omega)$

4.5 GENERALIZED FOURIER TRANSFORM

In Example 4.6 it was shown that the unit-step function $u(t)$ does not have a Fourier transform in the ordinary sense. It is also easy to see that $\cos \omega_0 t$ and $\sin \omega_0 t$ do not have a Fourier transform in the ordinary sense. Since the step function and sinusoidal functions often arise in the study of signals and systems, it is very desirable to be able to define the Fourier transform of these signals. This can be done by defining the notion of the generalized Fourier transform, which is considered in this section.

First the Fourier transform of the unit impulse $\delta(t)$ will be computed. Recall that $\delta(t)$ is defined by

$$\begin{aligned} \delta(t) &= 0, \qquad t \neq 0 \\ \int_{-\varepsilon}^{\varepsilon} \delta(\lambda)\, d\lambda &= 1, \qquad \text{all } \varepsilon > 0 \end{aligned} \tag{4.74}$$

The Fourier transform of $\delta(t)$ is given by

$$\int_{-\infty}^{\infty} \delta(t) e^{-j\omega t}\, dt$$

Since $\delta(t) = 0$ for all $t \neq 0$,

$$\delta(t) e^{-j\omega t} = \delta(t)$$

and the Fourier transform integral reduces to

$$\int_{-\infty}^{\infty} \delta(t)\, dt$$

By (4.74) this integral is equal to 1, which results in the transform pair

$$\delta(t) \leftrightarrow 1 \tag{4.75}$$

This result shows that the frequency spectrum of $\delta(t)$ contains all frequencies with amplitude 1.

Now applying the duality property to (4.75) yields the transform pair

$$x(t) = 1, \qquad -\infty < t < \infty \leftrightarrow 2\pi\delta(\omega) \tag{4.76}$$

Hence the Fourier transform of a constant signal of amplitude 1 is equal to an impulse in frequency with area 2π. But from the results in Example 4.5, it was seen that the constant signal does not have a Fourier transform in the ordinary sense. The frequency function $2\pi\delta(\omega)$ is called the *generalized Fourier transform* of the constant signal $x(t) = 1, -\infty < t < \infty$.

Now consider the signal $x(t) = \cos \omega_0 t$, $-\infty < t < \infty$, where ω_0 is a fixed but arbitrary real number. Using the transform pair (4.76) and the modulation property reveals that $x(t)$ has the generalized Fourier transform

$$\pi[\delta(\omega + \omega_0) + \delta(\omega - \omega_0)]$$

Hence

$$\cos \omega_0 t \leftrightarrow \pi[\delta(\omega + \omega_0) + \delta(\omega - \omega_0)] \tag{4.77}$$

In a similar manner, it can be shown that $\sin \omega_0 t$ has the generalized transform

$$j\pi[\delta(\omega + \omega_0) - \delta(\omega - \omega_0)]$$

and thus

$$\sin \omega_0 t \leftrightarrow j\pi[\delta(\omega + \omega_0) - \delta(\omega - \omega_0)] \tag{4.78}$$

The plot of the Fourier transform of $\cos \omega_0 t$ is given in Figure 4.32. Note that the spectrum consists of two impulses located at $\pm\, \omega_0$ with each impulse having area π.

Fourier Transform of a Periodic Signal

Using the transform pair (4.76) and the property (4.55) results in the transform pair

$$e^{j\omega_0 t} \leftrightarrow 2\pi\delta(\omega - \omega_0) \tag{4.79}$$

The transform pair (4.79) can be used to compute the generalized Fourier transform of a periodic signal: Let $x(t)$ be periodic for $-\infty < t < \infty$ with period T. Then $x(t)$ has the complex exponential Fourier series

$$x(t) = \sum_{k=-\infty}^{\infty} c_k e^{jk\omega_0 t} \tag{4.80}$$

where $\omega_0 = 2\pi/T$. The Fourier transform of the right-hand side of (4.80) can be taken by using linearity and the transform pair (4.79). This gives

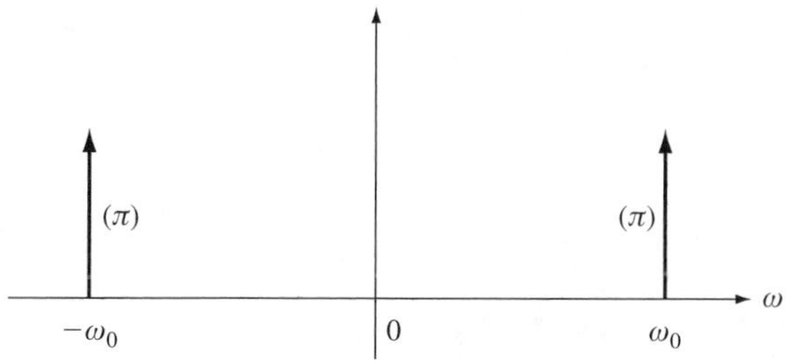

Figure 4.32 Fourier transform of $\cos \omega_0 t$.

$$X(\omega) = \sum_{k=-\infty}^{\infty} 2\pi c_k \delta(\omega - k\omega_0)$$

So the Fourier transform of a periodic signal is a train of impulse functions located at $\omega = k\omega_0, k = 0, \pm 1, \pm 2, \ldots$.

Transform of the Unit-Step Function

The (generalized) Fourier transform of the unit step $u(t)$ can be computed by using the integration property given by the transform pair (4.60). Since $u(t)$ is equal to the integral of the impulse $\delta(t)$ and the Fourier transform of $\delta(t)$ is the constant unit function, from (4.60), we see that the Fourier transform of $u(t)$ is given by

$$\frac{1}{j\omega}(1) + \pi(1)\delta(\omega) = \frac{1}{j\omega} + \pi\delta(\omega)$$

thus resulting in the transform pair

$$u(t) \leftrightarrow \frac{1}{j\omega} + \pi\delta(\omega) \tag{4.81}$$

In Table 4.2 a list of common Fourier transform pairs is given which includes the pairs that were derived in this chapter.

TABLE 4.2 COMMON FOURIER TRANSFORM PAIRS

$1, \quad -\infty < t < \infty \leftrightarrow 2\pi\delta(\omega)$
$-0.5 + u(t) \leftrightarrow \dfrac{1}{j\omega}$
$u(t) \leftrightarrow \pi\delta(\omega) + \dfrac{1}{j\omega}$
$\delta(t) \leftrightarrow 1$
$\delta(t - c) \leftrightarrow e^{-j\omega c}, \quad c$ any real number
$e^{-bt}u(t) \leftrightarrow \dfrac{1}{j\omega + b}, \quad b > 0$
$e^{j\omega_0 t} \leftrightarrow 2\pi\delta(\omega - \omega_0), \omega_0$ any real number
$p_\tau(t) \leftrightarrow \tau \operatorname{sinc} \dfrac{\tau\omega}{2\pi}$
$\tau \operatorname{sinc} \dfrac{\tau t}{2\pi} \leftrightarrow 2\pi p_\tau(\omega)$
$\left(1 - \dfrac{2\lvert t\rvert}{\tau}\right)p_\tau(t) \leftrightarrow \dfrac{\tau}{2}\operatorname{sinc}^2\left(\dfrac{\tau\omega}{4\pi}\right)$
$\dfrac{\tau}{2}\operatorname{sinc}^2\left(\dfrac{\tau t}{4\pi}\right) \leftrightarrow 2\pi\left(1 - \dfrac{2\lvert\omega\rvert}{\tau}\right)p_\tau(\omega)$
$\cos \omega_0 t \leftrightarrow \pi[\delta(\omega + \omega_0) + \delta(\omega - \omega_0)]$
$\cos(\omega_0 t + \theta) \leftrightarrow \pi[e^{-j\theta}\delta(\omega + \omega_0) + e^{j\theta}\delta(\omega - \omega_0)]$
$\sin \omega_0 t \leftrightarrow j\pi[\delta(\omega + \omega_0) - \delta(\omega - \omega_0)]$
$\sin(\omega_0 t + \theta) \leftrightarrow j\pi[e^{-j\theta}\delta(\omega + \omega_0) - e^{j\theta}\delta(\omega - \omega_0)]$

PROBLEMS

4.1. Express the following terms in polar notation.

(a) $e^{j\pi/4} + e^{-j\pi/8}$

(b) $(2 + 5j)e^{j10}$

(c) $1 + e^{j2} + e^{j4}$

(d) $1 + e^{j4}$

(e) $e^{j(\omega t + \pi/2)} + e^{j(\omega t - \pi/3)}$

4.2. Using complex notation, combine the expressions to form a single sinusoid for each of the following cases. (See Appendix A.)

(a) $2\cos(3t) - \cos(3t - \pi/4)$

(b) $\cos(10t + \pi/2) + 2\cos(10t - \pi/3)$

(c) $\cos(t) - \sin(t)$

(d) $10\cos(\pi t + \pi/3) + 8\cos(\pi t - \pi/3)$

4.3. Use MATLAB to plot the signals given in Problem 4.2, and verify the expression derived in Problem 4.2.

4.4. Each of the signals in Figure P4.4 is generated from a sum of sinusoids. Find the frequencies and the amplitudes of the sinusoids and draw the line spectrum (amplitude only) for each signal.

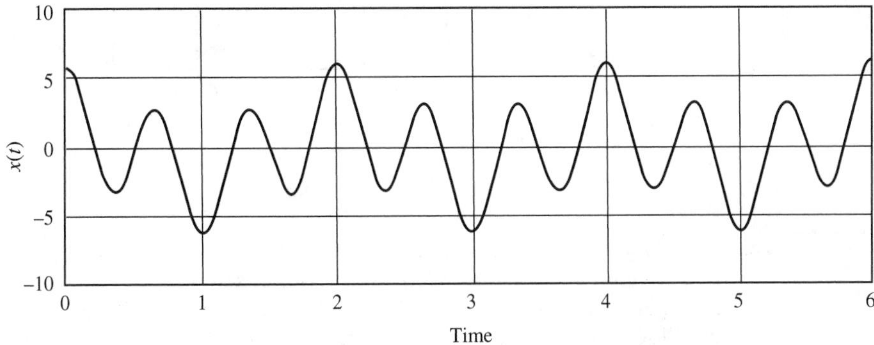

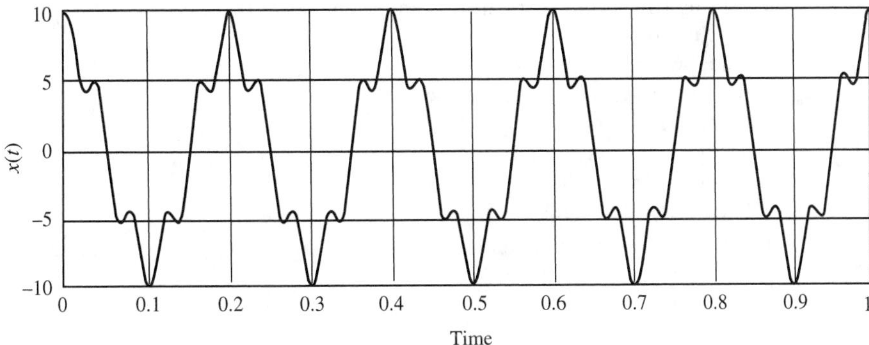

Figure P4.4

4.5. For each signal shown in Figure P4.5,

(a) Compute the complex exponential and trigonometric Fourier series.

(b) Compute and plot the truncated exponential series for $N = 3$, 10, and 30 using MATLAB when $T = 2$ and $a = 0.5$.

(c) Repeat part (b) using the truncated trigonometric series and compare your answer with part (b).

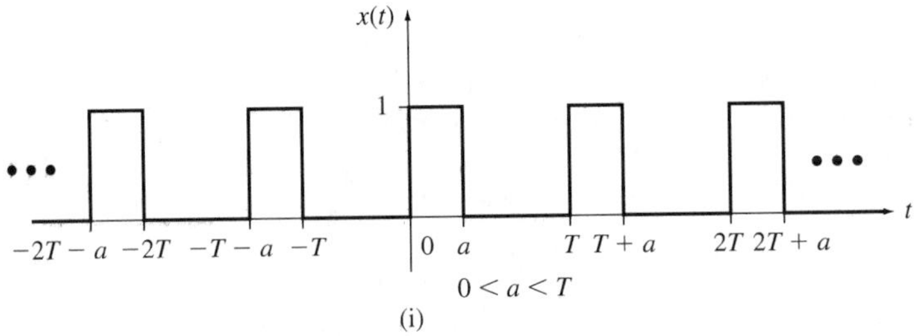

(i)

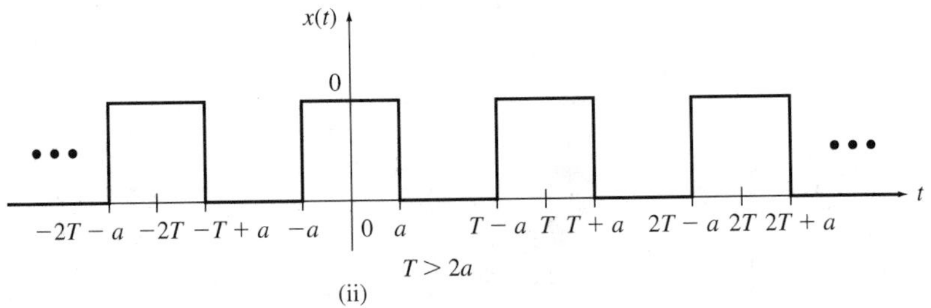

(ii)

Figure P4.5

4.6. For each of the periodic signals shown in Figure P4.6:

(i) Compute the complex exponential Fourier series.

(ii) Sketch the amplitude and phase spectra for $k = \pm 1, \pm 2, \pm 3, \pm 4, \pm 5$.

(iii) Plot the truncated complex exponential series for $N = 1, N = 5$, and $N = 30$.

4.7. For each of the following signals, compute the complex exponential Fourier series by using trigonometric identities, and then sketch the amplitude and phase spectra for all values of k.

(a) $x(t) = \cos(5t + \theta)$

(b) $x(t) = \sin t + \cos t$

(c) $x(t) = \sin^2 4t$

(d) $x(t) = \cos 2t \sin 3t$

(e) $x(t) = \cos^2 5t$

(f) $x(t) = \cos 3t + \cos 5t$

4.8. Determine the exponential Fourier series for the following periodic signals.

(a) $x(t) = \dfrac{\sin 2t + \sin 3t}{2 \sin t}$

(b) $x(t) = \sum_{k=-\infty}^{\infty} \delta(t - kT)$

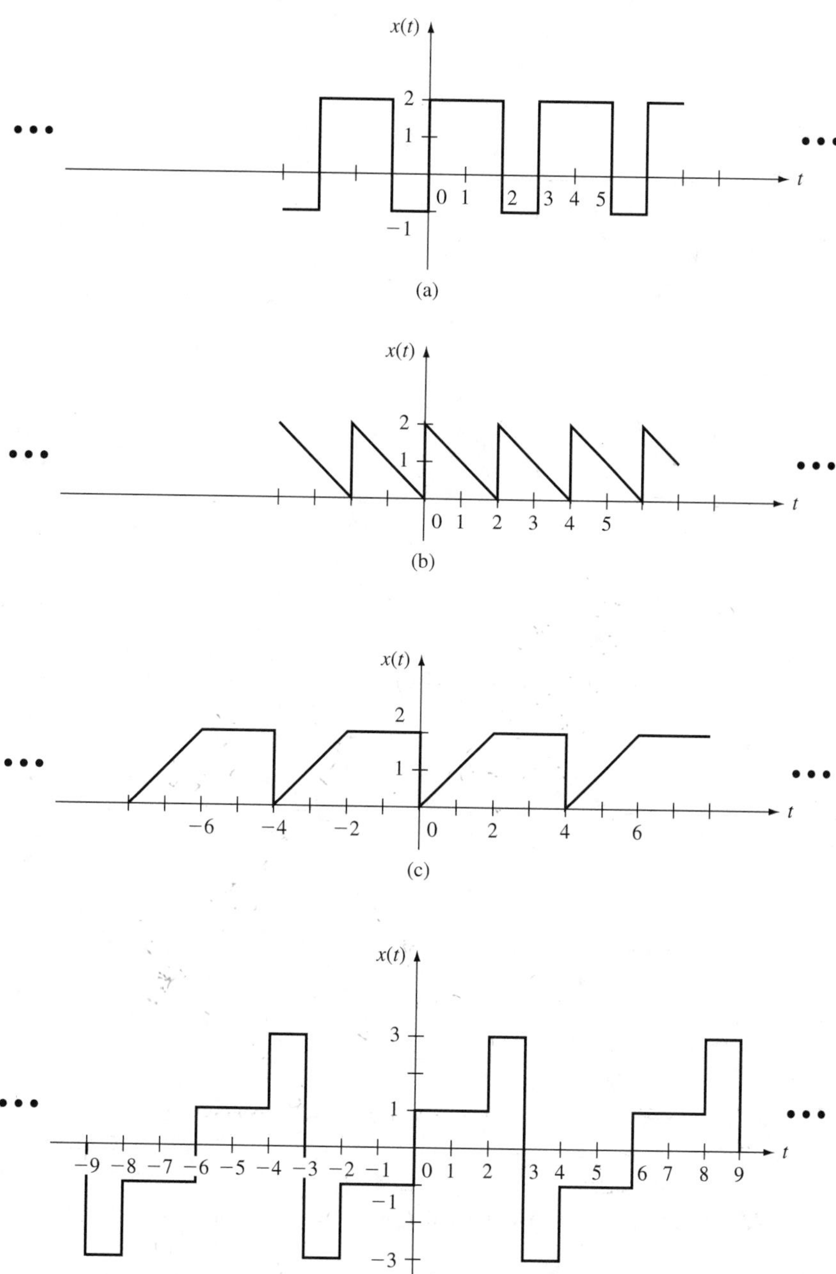
x(t)
2
1
−1
0 1 2 3 4 5
t
(a)
x(t)
2
1
0 1 2 3 4 5
t
(b)
x(t)
2
1
−6 −4 −2 0 2 4 6
t
(c)
x(t)
3
1
−1
−3
−9 −8 −7 −6 −5 −4 −3 −2 −1 0 1 2 3 4 5 6 7 8 9
t
(d)

Figure P4.6

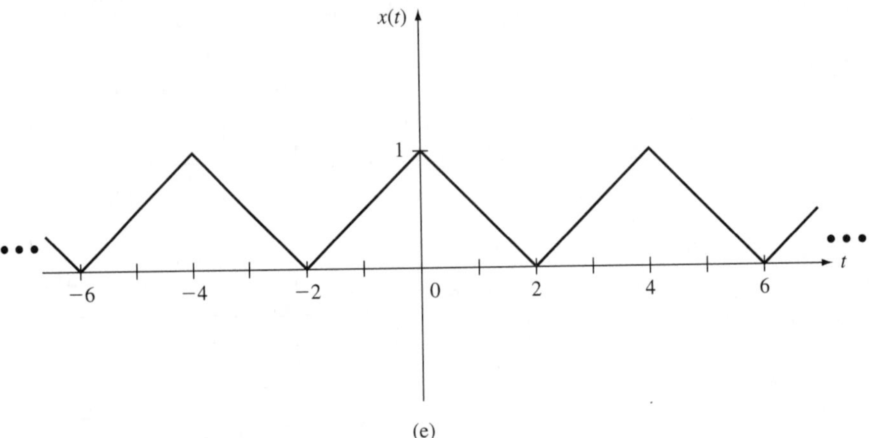

(e)

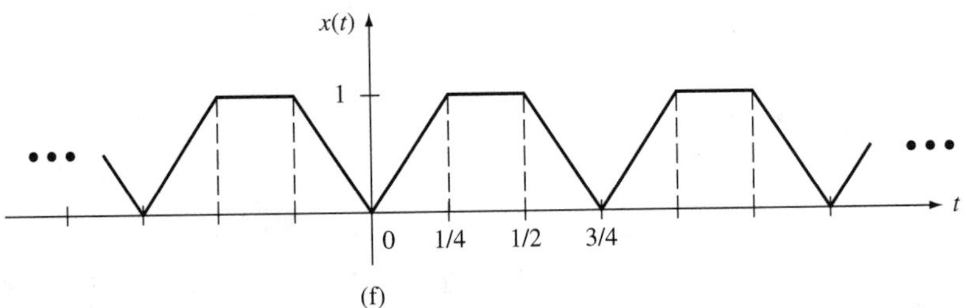

(f)

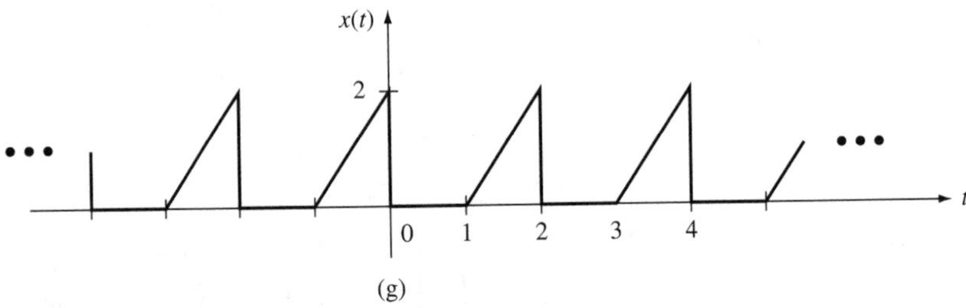

(g)

Figure P4.6 (continued)

4.9. A periodic signal with period T has Fourier coefficients c_k^x; that is,

$$x(t) = \sum_{k=-\infty}^{\infty} c_k^x \exp(jk\omega_0 t), \qquad \omega_0 = \frac{2\pi}{T}, \qquad -\infty < t < \infty$$

Compute the Fourier coefficients c_k^v for the periodic signal $v(t)$ where:

(a) $v(t) = x(t-1)$

(b) $v(t) = \dfrac{dx(t)}{dt}$

(c) $v(t) = x(t) \exp[j(2\pi/T)t]$

(d) $v(t) = x(t) \cos\left(\frac{2\pi}{T} t\right)$

4.10. The derivation of the Fourier series uses *orthogonal basis functions*, which are a set of functions of time, $\phi_k(t)$, such that the following holds over some specified time interval T:

$$\int_T \overline{\phi_k(t)}\phi_m(t)\, dt = 0$$

for all k and m such that $k \neq m$.

(a) Prove that $\phi_k(t) = e^{jk\omega_0 t}$ for $k = 0, \pm 1, \pm 2, \pm 3, \ldots$ are orthogonal basis functions over the time interval $T = 2\pi/\omega_0$.

(b) Suppose that $x(t)$ is periodic with period $T = 2\pi/\omega_0$. Approximate $x(t)$ by its Fourier series:

$$x(t) = \sum_{k=-\infty}^{\infty} c_k e^{jk\omega_0 t}$$

Using the notion of orthogonal basis functions, derive the expressions for c_0 and c_k given in (4.12) and (4.13). [*Hint:* To derive (4.12), multiply both sides of the Fourier series by $e^{-jk\omega_0 t}$ and integrate over T.]

4.11. One way of expressing the trigonometric Fourier series is

$$x(t) = a_0 + \sum_{k=1}^{\infty} a_k \cos k\omega_0 t + b_k \sin k\omega_0 t$$

(a) Using Euler's formula, derive an expression for the Fourier coefficients a_0, a_k, and b_k in terms of the Fourier coefficients c_0 and c_k of the complex exponential Fourier series. Also, give an explicit formula for a_0, a_k, and b_k in terms of integral expressions involving $x(t)$.

(b) Using the results in part (a) and trigonometric identities, derive the expression in (4.15).

4.12. The concept of odd and even functions can be used to simplify the calculation of the coefficients for the trigonometric Fourier series. Recall that a signal is defined as being *odd* if $x(t) = -x(-t)$ and *even* if $x(t) = x(-t)$.

(a) Determine if the signals in Figure P4.6 are odd, even, or neither.

(b) Consider two signals $x(t)$ and $z(t)$. Suppose that $x(t)$ is even and $z(t)$ is odd. Show that

$$\int_{-T/2}^{T/2} x(t)z(t)\, dt = 0$$

(c) Consider the trigonometric Fourier series given by

$$x(t) = a_0 + \sum_{k=1}^{\infty} a_k \cos k\omega_0 t + b_k \sin k\omega_0 t$$

Using the expressions derived for a_k and b_k in Problem 4.11, show that if $x(t)$ is even, then $b_k = 0$ for all k. Also show that if $x(t)$ is odd, then $a_k = 0$ for all k.

4.13. Using the results of Problem 4.12, calculate the trigonometric Fourier series for the signals given in Figures P4.6 (d) to (f).

4.14. Let

$$f(t) = \sum_{k=-\infty}^{\infty} f_k e^{jk\omega_0 t} \quad \text{and} \quad g(t) = \sum_{k=-\infty}^{\infty} g_k e^{jk\omega_0 t}$$

be the Fourier series expansions for $f(t)$ and $g(t)$. Is the following statement true: If $f_k = g_k$ for all k, then $f(t) = g(t)$ for all t? Justify your answer by giving a counterexample if the statement is false or by proving that the statement is true.

4.15. A continuous-time signal $x(t)$ has Fourier transform

$$X(\omega) = \frac{1}{j\omega + b}$$

where b is a constant. Determine the Fourier transform $V(\omega)$ of the following signals.

(a) $v(t) = x(5t - 4)$

(b) $v(t) = t^2 x(t)$

(c) $v(t) = x(t)e^{j2t}$

(d) $v(t) = x(t) \cos 4t$

(e) $v(t) = \dfrac{d^2x(t)}{dt^2}$

(f) $v(t) = x(t) * x(t)$

(g) $v(t) = x^2(t)$

(h) $v(t) = \dfrac{1}{jt - b}$

4.16. By first expressing $x(t)$ in terms of rectangular pulse functions and triangular pulse functions, compute the Fourier transform of the signals in Figure P4.16. Plot the magnitude and phase of the Fourier transform.

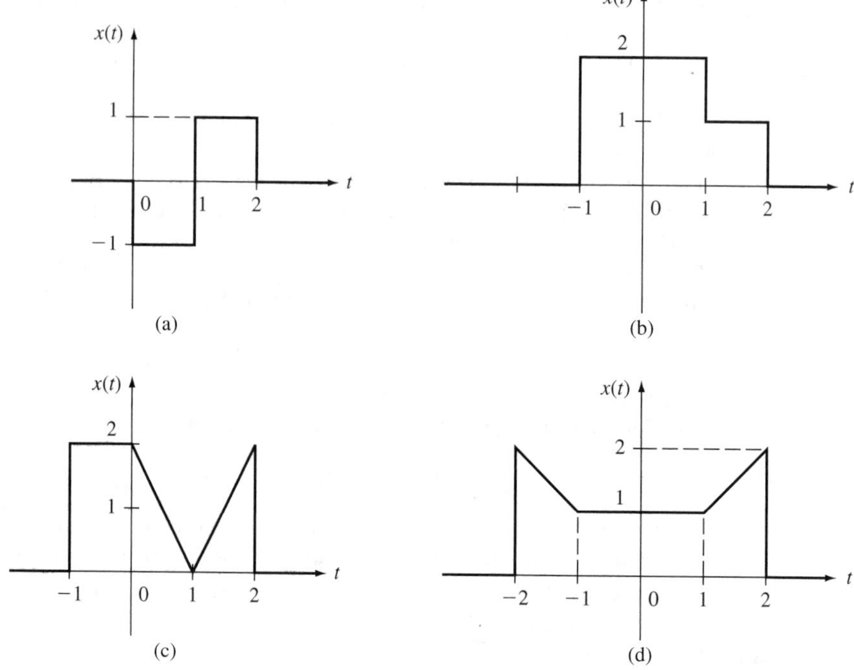

Figure P4.16

4.17. By using equations (4.33) and (4.34) and the Fourier transforms computed in Problem 4.16, determine the complex exponential Fourier series of the periodic signals in Figure P4.17.

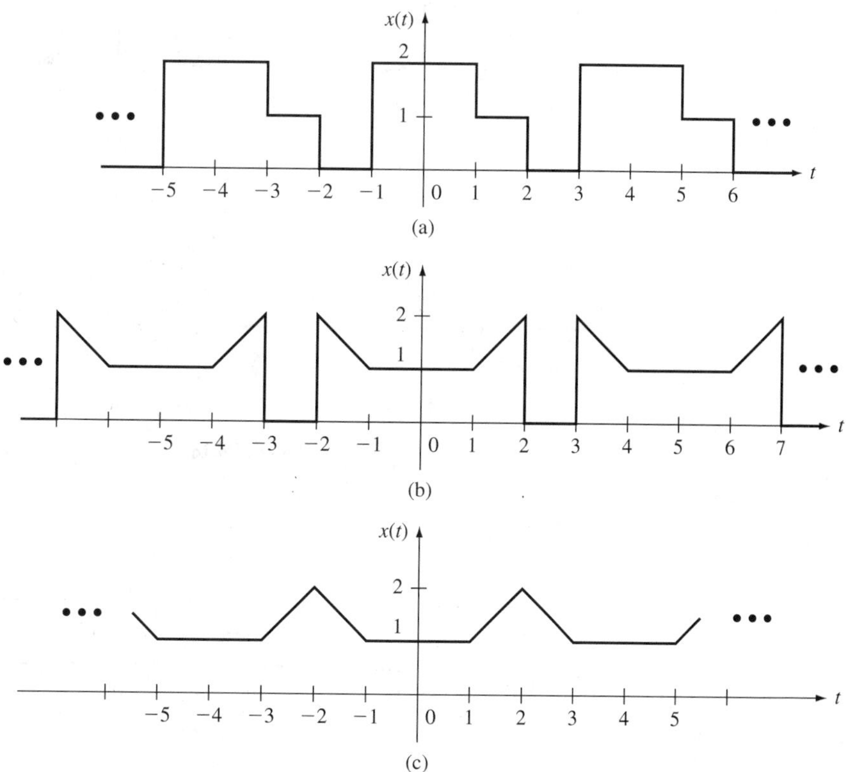

Figure P4.17

4.18. Compute the Fourier transform of the signals in Figure P4.18. Plot the magnitude and phase of the Fourier transform.

4.19. Compute the inverse Fourier transforms of the frequency functions $X(\omega)$ shown in Figure P4.19.

4.20. Compute the inverse Fourier transform of the following frequency functions.

(a) $X(\omega) = \cos 4\omega, -\infty < \omega < \infty$

(b) $X(\omega) = \sin^2 3\omega, -\infty < \omega < \infty$

(c) $X(\omega) = p_4(\omega) \cos \dfrac{\pi\omega}{2}$

(d) $X(\omega) = \dfrac{\sin(\omega/2)}{j\omega + 2} e^{-j\omega 2}, -\infty < \omega < \infty$

4.21. A signal $x(t)$ has Fourier transform

$$X(\omega) = \frac{1}{j}\left[\operatorname{sinc}\left(\frac{2\omega}{\pi} - \frac{1}{2}\right) - \operatorname{sinc}\left(\frac{2\omega}{\pi} + \frac{1}{2}\right)\right]$$

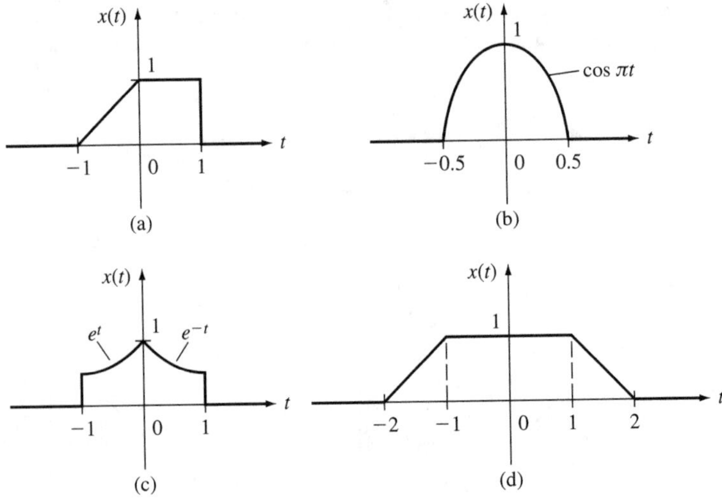

Figure P4.18

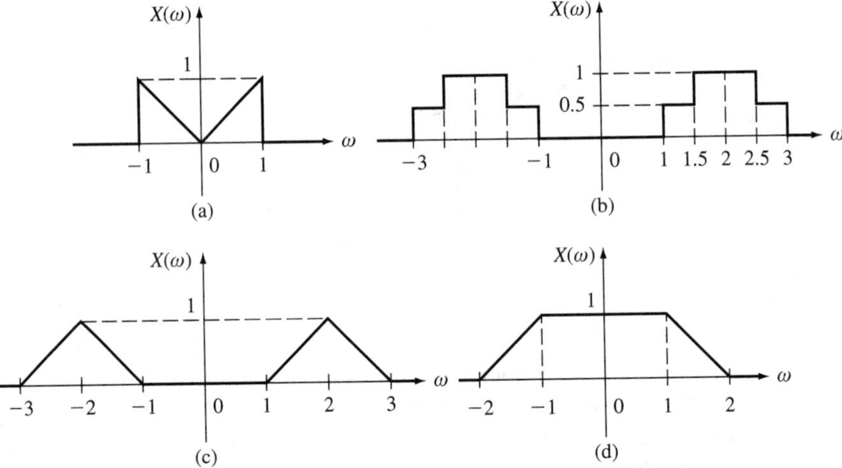

Figure P4.19

(a) Compute $x(t)$.
(b) Let $x_p(t)$ denote the periodic signal defined by

$$x_p(t) = \sum_{k=-\infty}^{\infty} x(t - 16k)$$

Compute the Fourier transform $X_p(\omega)$ of $x_p(t)$.

4.22. Compute the Fourier transform of the following signals.
(a) $x(t) = (e^{-t} \cos 4t)u(t)$

(b) $x(t) = te^{-1}u(t)$
(c) $x(t) = (\cos 4t)u(t)$
(d) $x(t) = e^{-|t|}, -\infty < t < \infty$
(e) $x(t) = e^{-t^2}, -\infty < t < \infty$

4.23. For the Fourier transforms $X(\omega)$ given in Figure P4.23, what characteristics does $x(t)$ have (i.e., real-valued, complex-valued, even, odd)? Calculate $x(0)$.

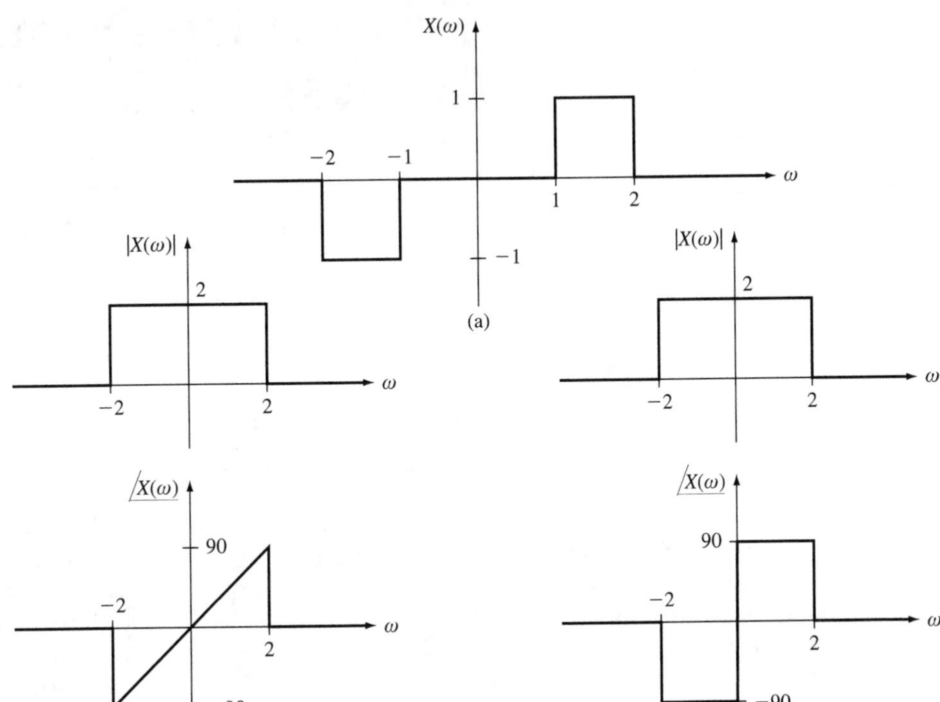

Figure P4.23

4.24. The Fourier transforms of $x(t)$ and $v(t)$ are defined below.

$$X(\omega) = \begin{cases} 2, & |\omega| < \pi \\ 0, & \text{otherwise} \end{cases}$$

$$V(\omega) = X(\omega - \omega_0) + X(\omega + \omega_0)$$

(a) Find $x(t)$ in closed form.
(b) Find $v(t)$ in closed form.

4.25. Compute the generalized Fourier transform of the following signals.
(a) $x(t) = 1/t, -\infty < t < \infty$
(b) $x(t) = 1 + 2e^{-j2\pi t} + 2e^{j2\pi t}, -\infty < t < \infty$
(c) $x(t) = 3\cos t + 2\sin 2t, -\infty < t < \infty$
(d) $x(t)$ as shown in Figure 4.6.

◆

CHAPTER 5

Frequency-Domain Analysis of Systems

In this chapter the Fourier series and Fourier transform are applied to the study of linear time-invariant continuous-time systems. The development begins in the next section with the fundamental result that the output response resulting from a sinusoidal input is also a sinusoid having the same frequency as the input, but which is amplitude scaled and phase shifted. This leads to the notion of the frequency response function of a system. In Section 5.2 it is shown that the response to a periodic input is also periodic and that the frequency spectrum of the response can be expressed in terms of the spectrum of the input and the frequency response function of the system. This result is then generalized in Section 5.3 to the case of a nonperiodic input, where it is shown that the Fourier transform of the output is the product of the Fourier transform of the input and the frequency response function. This frequency-domain description of system behavior gives a great deal of insight into how a system processes a given input to produce the resulting output. The theory is illustrated in Sections 5.4 and 5.5 when applications to ideal filtering and sampling are considered. Section 5.5 includes a proof of the famous sampling theorem. This very important result states that a bandlimited continuous-time signal can be completely reconstructed from a sampled version of the signal if the sampling frequency is suitably fast.

5.1 RESPONSE TO A SINUSOIDAL INPUT

Consider a linear time-invariant continuous-time system with impulse response $h(t)$. As discussed in Section 3.4 [see equation (3.56)], the output response $y(t)$ resulting from input $x(t)$ is given by the convolution relationship

$$y(t) = x(t) * h(t) = \int_{-\infty}^{\infty} x(\lambda)h(t-\lambda)\,d\lambda = \int_{-\infty}^{\infty} h(\lambda)x(t-\lambda)\,d\lambda \tag{5.1}$$

Recall that $y(t)$ is the output response with no initial energy in the system prior to the application of the input $x(t)$.

In this chapter it is not assumed that the system is necessarily causal, and thus the impulse response $h(t)$ may be nonzero for $t < 0$. Throughout this chapter it is assumed that the impulse response $h(t)$ is absolutely integrable; that is,

$$\int_{-\infty}^{\infty} |h(t)|\,dt < \infty \tag{5.2}$$

The condition (5.2) is a type of *stability condition* on the given system. A detailed development of the concept of system stability is left to a later chapter and thus is not pursued here. A key point to be made here is that the results in this chapter are *not valid in general* unless condition (5.2) is satisfied.

In this section the goal is to utilize the convolution representation (5.1) in order to derive an expression for the output response $y(t)$ when the input is the sinusoid

$$x(t) = A\cos(\omega_0 t + \theta), \qquad -\infty < t < \infty \tag{5.3}$$

Here the frequency ω_0 is assumed to be nonnegative and the phase θ is arbitrary.

It turns out that the derivation of an expression for the response to the sinusoidal input (5.3) is facilitated by first determining the response to the complex exponential input

$$x_c(t) = Ae^{j(\omega_0 t + \theta)}, \qquad -\infty < t < \infty \tag{5.4}$$

In (5.4), the subscript "c" stands for complex.

To derive an expression for the response $y_c(t)$ to $x_c(t)$, insert (5.4) into (5.1), which gives

$$\begin{aligned} y_c(t) &= \int_{-\infty}^{\infty} h(\lambda)Ae^{j(\omega_0(t-\lambda)+\theta)}\,d\lambda \\ &= \int_{-\infty}^{\infty} h(\lambda)e^{-j\omega_0\lambda}[Ae^{j(\omega_0 t+\theta)}\,d\lambda] \\ &= \left[\int_{-\infty}^{\infty} h(\lambda)e^{-j\omega_0\lambda}\,d\lambda\right][Ae^{j(\omega_0 t+\theta)}] \\ &= \left[\int_{-\infty}^{\infty} h(\lambda)e^{-j\omega_0\lambda}\,d\lambda\right]x_c(t) \end{aligned} \tag{5.5}$$

The response $y_c(t)$ can be expressed in terms of the function $H(\omega)$ defined by

$$H(\omega) = \int_{-\infty}^{\infty} h(\lambda)e^{-j\omega\lambda}\, d\lambda \tag{5.6}$$

where ω is a real variable. Note that $H(\omega)$ is the Fourier transform of the system's impulse response $h(t)$. (More will be said about this in the next section.) Note also that $H(\omega)$ exists as a result of the integrability condition (5.2).

Using the definition (5.6) of $H(\omega)$, the response $y_c(t)$ given by (5.5) can be written in the form

$$y_c(t) = H(\omega_0)x_c(t), \qquad -\infty < t < \infty \tag{5.7}$$

Inserting the expression (5.4) for $x_c(t)$ into (5.7) results in the following expression for $y_c(t)$:

$$y_c(t) = H(\omega_0)Ae^{j(\omega_0 t+\theta)}, \qquad -\infty < t < \infty \tag{5.8}$$

From (5.8) it is seen that the system processes the complex exponential input $x_c(t) = A\exp[j(\omega_0 t + \theta)]$ by scaling it by the quantity $H(\omega_0)$. Hence the response to a complex exponential input is also a complex exponential with the same frequency ω_0. This is a very fundamental result in the frequency-domain analysis of linear time-invariant systems.

The simple form for $y_c(t)$ given by (5.7) or (5.8) is a major reason for first considering the response to a complex exponential input. The computation of the response $y(t)$ to $x(t) = A\cos(\omega_0 t + \theta)$ can then be carried out by using this form. The derivation is as follows.

First apply Euler's formula to (5.4), which gives

$$x_c(t) = A\cos(\omega_0 t + \theta) + jA\sin(\omega_0 t + \theta) \tag{5.9}$$

Combining (5.3) and (5.9) yields

$$x_c(t) = x(t) + jA\sin(\omega_0 t + \theta) \tag{5.10}$$

By (5.10), it is seen that $x(t)$ is equal to the real part of $x_c(t)$. In mathematical terms,

$$x(t) = \mathrm{Re}[x_c(t)]$$

Now again let $y_c(t)$ denote the output response resulting from input $x_c(t)$. Inserting (5.10) into (5.1) gives

$$y_c(t) = \int_{-\infty}^{\infty} h(\lambda)[x(t-\lambda) + jA\sin(\omega_0(t-\lambda) + \theta)]\, d\lambda$$

and by linearity

$$y_c(t) = \int_{-\infty}^{\infty} h(\lambda)x(t-\lambda)\, d\lambda + j\int_{-\infty}^{\infty} h(\lambda)[A\sin(\omega_0(t-\lambda) + \theta)]\, d\lambda \tag{5.11}$$

Finally, taking the real part of both sides of (5.11) yields

$$\mathrm{Re}[y_c(t)] = \int_{-\infty}^{\infty} h(\lambda)x(t-\lambda)\,d\lambda$$

and thus

$$\mathrm{Re}[y_c(t)] = y(t) \tag{5.12}$$

where $y(t)$ is the response to the sinusoidal input (5.3). This is an expected result since $x(t) = \mathrm{Re}[x_c(t)]$ and the system is linear, and thus it should follow that $y(t) = \mathrm{Re}[y_c(t)]$. The derivation above shows that this is in fact the case.

By (5.12), the desired output $y(t)$ can be determined by taking the real part of the response $y_c(t)$ to the complex exponential input $x_c(t)$. This will be accomplished by first expressing $H(\omega_0)$ in the polar form

$$H(\omega_0) = |H(\omega_0)|\, e^{j\angle H(\omega_0)} \tag{5.13}$$

where $|H(\omega_0)|$ is the magnitude of $H(\omega_0)$ and $\angle H(\omega_0)$ is the angle of $H(\omega_0)$. Inserting (5.13) into (5.8) gives

$$\begin{aligned} y_c(t) &= [|H(\omega_0)|e^{j\angle H(\omega_0)}]\,Ae^{j(\omega_0 t+\theta)} \\ &= A|H(\omega_0)|e^{j(\omega_0 t+\theta+\angle H(\omega_0))} \end{aligned} \tag{5.14}$$

Then since

$$y(t) = \mathrm{Re}[y_c(t)] = \mathrm{Re}[H(\omega_0)x_c(t)]$$

taking the real part of both sides of (5.14) gives

$$y(t) = A|H(\omega_0)|\cos(\omega_0 t + \theta + \angle H(\omega_0)), \qquad -\infty < t < \infty \tag{5.15}$$

Hence the response resulting from the sinusoidal input $x(t) = A\cos(\omega_0 t + \theta)$ is also a sinusoid with the same frequency ω_0, but with the amplitude scaled by the factor $|H(\omega_0)|$ and with the phase shifted by amount $\angle H(\omega_0)$. This is quite a remarkable result, and in fact it forms the basis for the frequency-domain approach to the study of linear time-invariant systems.

Since the magnitude and phase of the sinusoidal response are given directly in terms of $|H(\omega)|$ and $\angle H(\omega)$, $|H(\omega)|$ is called the *magnitude function* of the system and $\angle H(\omega)$ is called the *phase function* of the system. In addition, since the response of the system to a sinusoid with frequency ω_0 can be determined directly from $H(\omega)$, $H(\omega)$ is often referred to as the *frequency response function* (or *system function*) of the system.

Example 5.1 *Response to Sinusoidal Inputs*

Response to Sinusoidal Inputs

Suppose that the frequency response function $H(\omega)$ is given by $|H(\omega)| = 1.5$ for $0 \le \omega \le 20$, $|H(\omega)| = 0$ for $\omega > 20$, and $\angle H(\omega) = -60°$ for all ω. Then if the input $x(t)$ is equal to $2\cos(10t + 90°) + 5\cos(25t + 120°)$ for $-\infty < t < \infty$, the response is given by $y(t) = 3\cos(10t + 30°)$ for $-\infty < t < \infty$.

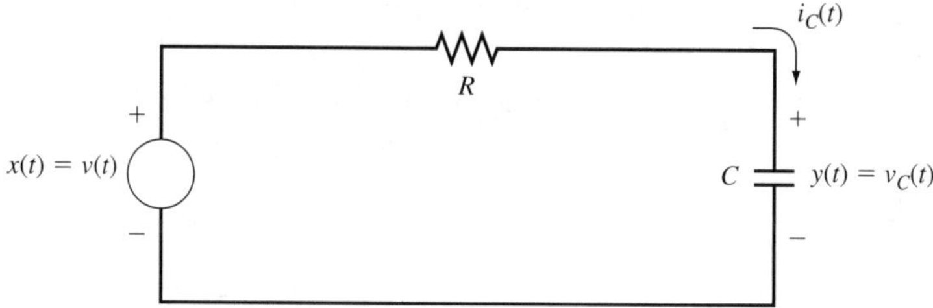

Figure 5.1 *RC* circuit in Example 5.2.

Example 5.2 *Frequency Analysis of an RC Circuit*

Consider the *RC* circuit shown in Figure 5.1. As is indicated in the figure, the input $x(t)$ is the voltage $v(t)$ applied to the circuit, and the output $y(t)$ is the voltage $v_C(t)$ across the capacitor. From basic circuit analysis (see Section 8.5), the complex impedance of the capacitor is equal to $1/Cs$, where $s = \sigma + j\omega$ is a complex variable. Then, if the input voltage is equal the complex exponential $x_c(t) = e^{st}$, by voltage division we see that the resulting output voltage $y_c(t)$ is given by

$$y_c(t) = \frac{1/Cs}{R + 1/Cs} e^{st} = \frac{1/RC}{s + 1/RC} e^{st} \tag{5.16}$$

Setting $s = j\omega_0$ in (5.16) and comparing the result with (5.8) yields

$$H(\omega_0) = \frac{1/RC}{j\omega_0 + 1/RC} \tag{5.17}$$

Replacing ω_0 in (5.17) by the real variable ω gives

$$H(\omega) = \frac{1/RC}{j\omega + 1/RC} \tag{5.18}$$

From (5.18), it is seen that the magnitude function $|H(\omega)|$ of the circuit is given by

$$|H(\omega)| = \frac{1/RC}{\sqrt{\omega^2 + (1/RC)^2}}$$

and the phase function $\angle H(\omega)$ is given by

$$\angle H(\omega) = -\tan^{-1} \omega RC$$

For any desired value of $1/RC$, the magnitude and phase functions can be computed using MATLAB. For instance, in the case when $1/RC = 1000$, the MATLAB commands for generating the magnitude and phase (angle) functions are

```
RC = 0.001;
w = 0:50:5000;
H = (1/RC)./(j*w+1/RC);
magH = abs(H);
angH = 180*angle(H)/pi;
```

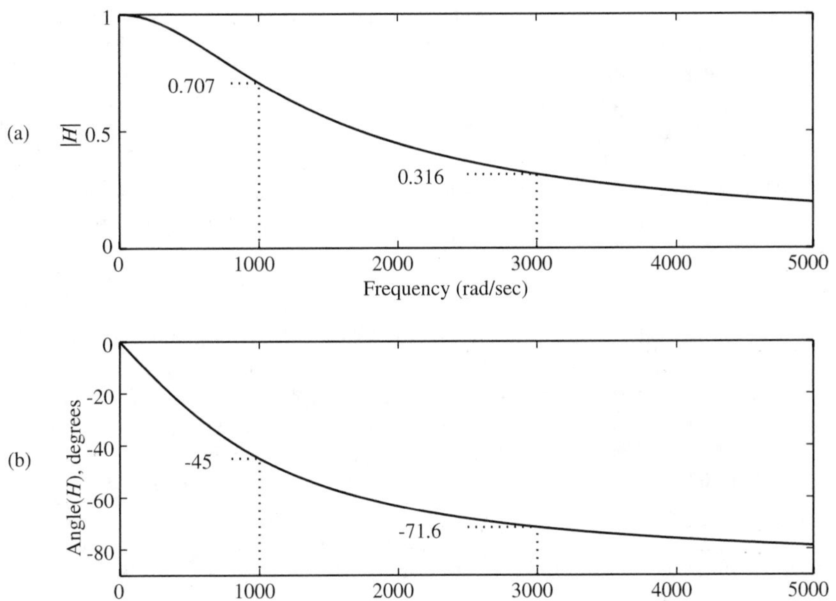

Figure 5.2 (a) Magnitude and (b) phase functions of the RC circuit in Example 5.2 for the case $1/RC = 1000$.

Using these commands and the plotting commands results in the plots of $|H(\omega)|$ and $\angle H(\omega)$ shown in Figure 5.2. From the figure, note that

$$|H(0)| = 1, \qquad \angle H(0) = 0 \tag{5.19}$$

$$|H(1000)| = \frac{1}{\sqrt{2}} = 0.707, \qquad \angle H(1000) = -45^\circ \tag{5.20}$$

$$|H(3000)| = 0.316, \qquad \angle H(3000) = -71.6^\circ \tag{5.21}$$

and

$$|H(\omega)| \to 0 \quad \text{as } \omega \to \infty, \qquad \angle H(\omega) \to -90^\circ \quad \text{as } \omega \to \infty \tag{5.22}$$

Now the output response $y(t)$ of the RC circuit resulting from a specific sinusoidal input $x(t) = A\cos(\omega_0 t + \theta)$ can be computed by inserting the appropriate values of $H(\omega_0)$ and $\angle H(\omega_0)$ into the expression (5.15). For example, suppose that $1/RC = 1000$. Then if $\omega_0 = 0$, using (5.19) the resulting output response is

$$y(t) = A(1)\cos(0t + \theta + 0) = A\cos\theta, \qquad -\infty < t < \infty \tag{5.23}$$

If $\omega_0 = 1000$ rad/sec, using (5.20) yields the response

$$y(t) = A(0.707)\cos(1000t + \theta - 45^\circ), \qquad -\infty < t < \infty \tag{5.24}$$

and if $\omega_0 = 3000$ rad/sec, using (5.21) gives

$$y(t) = A(0.316)\cos(3000t + \theta - 71.6^\circ), \qquad -\infty < t < \infty \tag{5.25}$$

Finally, it follows from (5.22) that the output response $y(t)$ goes to zero as $\omega_0 \to \infty$.

From (5.23), it is seen that when $\omega_0 = 0$, so that the input is the constant signal $x(t) = A \cos \theta$, the response is equal to the input. Hence the *RC* circuit passes a dc input without attenuation and without producing any phase shift. From (5.24), it is seen that when $\omega_0 = 1000$ rad/sec, the *RC* circuit attenuates the input sinusoid by a factor of 0.707, and phase shifts the input sinusoid by $-45°$, and by (5.25) it is seen that when $\omega_0 = 3000$ rad/sec, the attenuation factor is now 0.316 and the phase shift is $-71.6°$. Finally, as $\omega_0 \to \infty$, the magnitude of the output goes to zero while the phase shift goes to $-90°$.

The behavior of the *RC* circuit is summarized by noting that it passes low-frequency signals without any significant attenuation and without producing any significant phase shift. As the frequency increases, the attenuation and the phase shift become larger. Finally, as $\omega_0 \to \infty$, the *RC* circuit completely "blocks" the sinusoidal input. As a result of this behavior, the *RC* circuit is an example of a *lowpass filter,* that is, the circuit "passes" without much attenuation input sinusoids whose frequency ω_0 is less than 1000 rad/sec, and it significantly attenuates input sinusoids whose frequency ω_0 is much above 1000 rad/sec. As discussed later, the frequency range from 0 to 1000 rad/sec (in the case $1/RC = 1000$) is called the *bandwidth* of the *RC* circuit.

To further illustrate the lowpass filter characteristic of the *RC* circuit, now suppose that the input is the sum of two sinusoids

$$x(t) = A_1 \cos(\omega_0 t + \theta_1) + A_2 \cos(\omega_2 t + \theta_2)$$

Due to linearity of the *RC* circuit, the corresponding response $y(t)$ is the sum of the responses to the individual sinusoids

$$y(t) = A_1 |H(\omega_1)| \cos(\omega_1 t + \theta_1 + \angle H(\omega_1)) + A_2 |H(\omega_2)| \cos(\omega_2 t + \theta_2 + \angle H(\omega_2))$$

To demonstrate the effect of the lowpass filtering, the response to the input

$$x(t) = \cos 100t + \cos 3000t$$

will be calculated in the case when $1/RC = 1000$. The MATLAB commands used to generate $y(t)$ are

```
RC = 0.001;
t = -.1:.2/1000:.1;
w1 = 100; w2 = 3000;
Hw1 = (1/RC)/(j*w1+1/RC);
Hw2 = (1/RC)/(j*w2+1/RC);
x = cos(w1*t) + cos(w2*t);
y = abs(Hw1)*cos(w1*t+angle(Hw1))+...
    abs(Hw2)*cos(w2*t+angle(Hw2));
```

The plots for $x(t)$ and $y(t)$ are shown in Figure 5.3. Note that the amplitude of the low-frequency signal is approximately the same in both plots. However, the high-frequency component in $x(t)$ is very evident, but is much less significant in $y(t)$ due to the attenuation of high-frequency signals by the circuit.

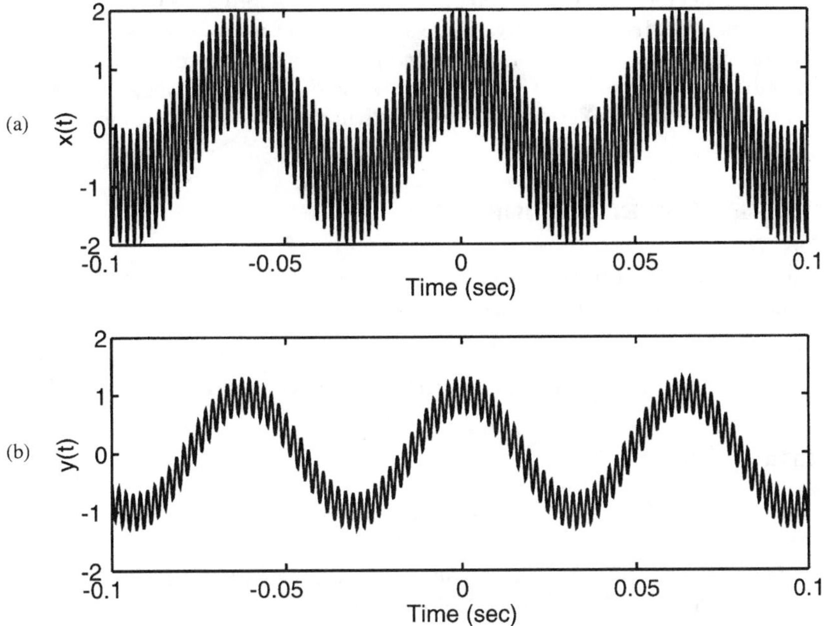

Figure 5.3 (a) Input and (b) output of RC circuit when $1/RC = 1000$.

Example 5.3 *Mass–Spring–Damper System*

Mass–Spring–Damper System

Consider the mass–spring–damper system that was defined in Chapter 1. (See Figure 1.28.) The input/output differential equation of the system is given by

$$M\frac{d^2y(t)}{dt^2} + D\frac{dy(t)}{dt} + Ky(t) = x(t)$$

where M is the mass, D is the damping constant, K is the stiffness constant, $x(t)$ is the force applied to the mass, and $y(t)$ is the displacement of the mass relative to the equilibrium position. With the input equal to the complex exponential $x_c(t) = e^{j\omega t}$, the output $y_c(t)$ has the form $y_c(t) = H(\omega)e^{j\omega t}$, where $H(\omega)$ is the frequency response function of the mass–spring–damper system. Inserting $x_c(t) = e^{j\omega t}$ and $y_c(t) = H(\omega)e^{j\omega t}$ into the input/output differential equation gives

$$M(j\omega)^2H(\omega)e^{j\omega t} + D(j\omega)H(\omega)e^{j\omega t} + KH(\omega)e^{j\omega t} = e^{j\omega t}$$

and solving for $H(\omega)$ gives

$$H(\omega) = \frac{1}{M(j\omega)^2 + D(j\omega) + K} = \frac{1}{K - M\omega^2 + jD\omega}$$

The magnitude $|H(\omega)|$ of the frequency response function $H(\omega)$ can be determined over a frequency range from some start frequency to some stop frequency by inputing a sine sweep (a sinusoid whose frequency is varied from the start frequency to the stop frequency). This can be carried out using the mass–spring–damper system on-line demo on the Web. Trying this for various values of M, D, and K will result in different shapes for the magnitude function $|H(\omega)|$. In fact, for various values of M, D, and K, the magnitude function $|H(\omega)|$ will have a peak at some positive value of ω. As discussed in Chapter 9, the peak is due to a resonance in the mass–spring–damper system.

5.2 RESPONSE TO PERIODIC INPUTS

Again consider the linear time-invariant continuous-time system given by the input/output convolution relationship

$$y(t) = x(t) * h(t) = \int_{-\infty}^{\infty} x(\lambda)h(t-\lambda)\,d\lambda \tag{5.26}$$

with the condition

$$\int_{-\infty}^{\infty} |h(t)|\,dt < \infty \tag{5.27}$$

As defined in Section 5.1, the frequency response function $H(\omega)$ of the system is equal to the Fourier transform of $h(t)$; that is,

$$H(\omega) = \int_{-\infty}^{\infty} h(t)e^{-j\omega t}\,dt$$

In the first part of this section, the output response resulting from a periodic signal is computed by extending the approach given in Section 5.1. Then using the Fourier transform theory, the computation of responses resulting from aperiodic inputs is considered.

Response to Periodic Inputs

Suppose that the input $x(t)$ to the system defined by (5.26) is periodic so that $x(t)$ has the complex exponential Fourier series

$$x(t) = \sum_{k=-\infty}^{\infty} c_k e^{jk\omega_0 t}, \qquad -\infty < t < \infty \tag{5.28}$$

It follows directly from the results in Section 5.1 that the output response resulting from the complex exponential input $c_k\exp(jk\omega_0 t)$ is equal to $H(k\omega_0)c_k\exp(jk\omega_0 t)$. Then by linearity, the response to the periodic input $x(t)$ is

$$y(t) = \sum_{k=-\infty}^{\infty} H(k\omega_0)c_k e^{jk\omega_0 t}, \qquad -\infty < t < \infty \tag{5.29}$$

Since the right-hand side of (5.29) is the complex exponential form of a Fourier series, it follows that the response $y(t)$ is periodic. In addition, since the fundamental frequency of $y(t)$ is ω_0, which is the fundamental frequency of the input $x(t)$, the period of $y(t)$ is equal to the period of $x(t)$. Hence the response to a periodic input with fundamental period T is periodic with fundamental period T.

Now let c_k^x denote the coefficients of the Fourier series for $x(t)$ and let c_k^y denote the coefficients of the Fourier series for the resulting output $y(t)$. From (5.29),

$$c_k^y = H(k\omega_0)c_k^x \tag{5.30}$$

Hence the Fourier coefficients of the response are equal to the product of the coefficients of the periodic input with the frequency response function $H(\omega)$ evaluated at $\omega = k\omega_0$. Taking the magnitude of both sides of (5.30) yields

$$|c_k^y| = |H(k\omega_0)|\,|c_k^x| \tag{5.31}$$

so the output amplitude spectrum is the product of the input amplitude spectrum and the system's magnitude function $|H(\omega)|$ with $\omega = k\omega_0$. Taking the angle of both sides of (5.30) gives

$$\angle c_k^y = \angle H(k\omega_0) + \angle c_k^x \tag{5.32}$$

so the output phase spectrum is the sum of the input phase spectrum and the system's phase function $\angle H(\omega)$ with $\omega = k\omega_0$.

The relationships (5.31) and (5.32) are very important. In particular, they describe how the system processes the various complex exponential components comprising the periodic input signal. From (5.31) it is possible to determine if the system will pass or will attenuate a given component of the input, and from (5.32) the phase shift the system will give to a particular component of the input can be determined.

Using (5.31) and (5.32), the Fourier series of the output can be computed directly from the coefficients of the input Fourier series. The process is illustrated by the following example.

Example 5.4 *Response to a Rectangular Pulse Train*

Response to Periodic Inputs

Again consider the *RC* circuit examined in Example 5.2 and shown in Figure 5.1. The objective is to determine the voltage $y(t)$ on the capacitor resulting from the rectangular pulse train $x(t)$ shown in Figure 5.4. From the results in Example 4.3, $x(t)$ has the complex exponential Fourier series

$$x(t) = \sum_{k=-\infty}^{\infty} c_k^x e^{jk\pi t}, \qquad -\infty < t < \infty$$

where $c_0^x = 0.5$, $c_{-k}^x = c_k^x$ for $k = 1,2,3, \ldots$, and

$$c_k^x = \begin{cases} \dfrac{1}{k\pi}, & k = 1, 5, 9 \ldots \\ -\dfrac{1}{k\pi}, & k = 3, 7, 11, \ldots \\ 0, & k = 2, 4, 6, \ldots \end{cases}$$

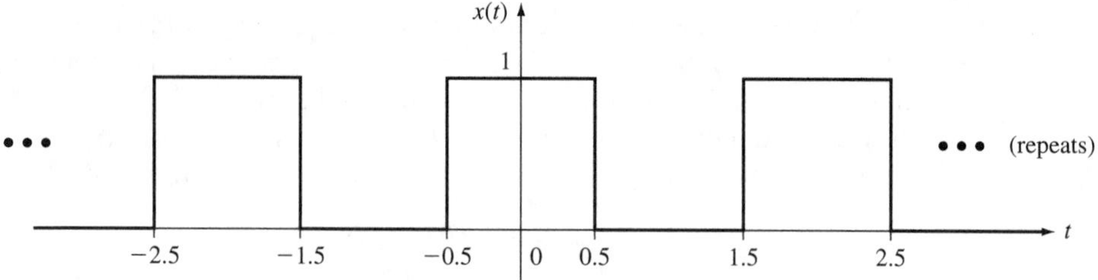

Figure 5.4 Periodic input signal in Example 5.4.

The amplitude spectrum (the plot of $|c_k^x|$ versus $\omega = k\omega_0$) of the rectangular pulse train is displayed in Figure 5.5. The plot shows that $x(t)$ has frequency components from dc all the way to infinite frequency, with the higher-frequency components having less significance. In particular, from the plot it is clear that most of the spectral content of $x(t)$ is contained in the frequency range 0 to 40 rad/sec. The "corners" comprising the rectangular pulse train are a result of the spectral lines in the limit as $k \to \infty$.

As computed in Example 5.1, the frequency response function of the RC circuit is

$$H(\omega) = \frac{1/RC}{j\omega + 1/RC}$$

Hence

$$H(k\omega_0) = H(k\pi) = \frac{1/RC}{jk\pi + 1/RC}$$

and

$$|H(k\pi)| = \frac{1/RC}{\sqrt{k^2\pi^2 + (1/RC)^2}}$$

$$\angle H(k\pi) = -\tan^{-1} k\pi RC$$

Note that the magnitude function $|H(k\pi)|$ rolls off as the integer k is increased. This is, of course, due to the lowpass characteristic of the RC circuit. As discussed in Example 5.1, in the case when $1/RC = 1000$, the 3-dB bandwidth of the RC circuit is 1000 rad/sec.

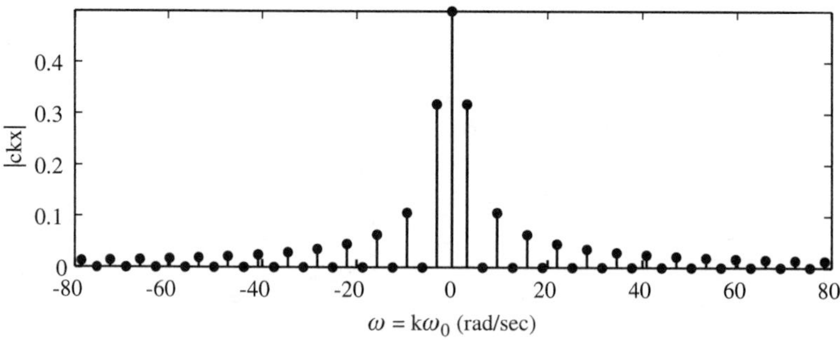

Figure 5.5 Amplitude spectrum of periodic input in Example 5.4.

For arbitrary positive values of R and C, the 3-dB bandwidth is equal to $1/RC$ rad/sec. Frequency components of the input below $1/RC$ are passed without significant attenuation, while frequency components above $1/RC$ are attenuated. Thus the larger the value of $1/RC$, the larger the bandwidth of the RC circuit is, resulting in higher-frequency components of the input being passed through the circuit.

Now inserting the expressions for c_k^x, $|H(k\pi)|$, and $\angle H(k\pi)$ into (5.31) and (5.32) yields the following expressions for the Fourier coefficients of the output:

$$c_0^y = H(0)c_0^x = 0.5 \tag{5.33}$$

$$|c_k^y| = \begin{cases} \dfrac{1}{k\pi}\dfrac{1/RC}{\sqrt{k^2\pi^2 + (1/RC)^2}}, & k \text{ odd} \\ 0, & k \text{ even} \end{cases} \tag{5.34}$$

$$\angle c_k^y = \begin{cases} -180^\circ - \tan^{-1} k\pi RC, & k = 3, 7, 11, \ldots \\ -\tan^{-1} k\pi RC, & \text{all other } k \end{cases} \tag{5.35}$$

The effect of the bandwidth of the circuit on the output can be seen by plotting the amplitude spectrum ($|c_k^y|$ versus $\omega = k\omega_0$) of the output for various values of $1/RC$. The output amplitude spectrum is displayed in Figure 5.6a to c for the values $1/RC = 1$, $1/RC = 10$, and $1/RC = 100$. Comparing Figures 5.5 and 5.6a reveals that when the 3-dB bandwidth of the RC circuit is 1 rad/sec (i.e., $1/RC = 1$), the circuit attenuates much of the spectral content of the rectangular pulse train. On the other hand, comparing Figures 5.5 and 5.6c shows that there is very little attenuation of the input spectral components when the 3-dB bandwidth is 100 rad/sec (i.e., $1/RC = 100$). It would therefore be expected that when $1/RC = 1$, the circuit will significantly distort the pulse train, whereas when $1/RC = 100$, there should not be much distortion. To verify this, the output response will be computed by first computing the Fourier series representation of the output.

From (5.29), the complex exponential Fourier series of the output is then given by

$$y(t) = 0.5 + \sum_{k=\pm1,\pm5,\pm9,\ldots} \frac{1}{k\pi}\frac{1/RC}{jk\pi + 1/RC} e^{jk\pi t} - \sum_{k=\pm3,\pm7,\pm11,\ldots} \frac{1}{k\pi}\frac{1/RC}{jk\pi + 1/RC} e^{jk\pi t} \tag{5.36}$$

Since the coefficients of the Fourier series (5.36) for $y(t)$ are getting very small as k increases, it is possible to determine the values of $y(t)$ by evaluating a suitable number of the terms comprising (5.36). The MATLAB commands used to obtain $y(t)$ for the case $1/RC = 1$ are

```
t = -3:6/800:3;
RC = 1; N = 50; w0 = pi;
c0 = 0.5; H0 = 1;
y = c0*H0*ones(1,length(t));
counter = 0;
for k=1:2:N,
  if counter == 0,
```

```
      ck = 1/k/pi;
      counter = 1;
   else
      ck = -1/k/pi;
      counter = 0;
   end
   H = (1/RC)/(j*k*w0 + 1/RC);
   y = y + ck*H*exp(j*k*w0*t) + ck*conj(H)*exp(-j*k*w0*t);
end
```

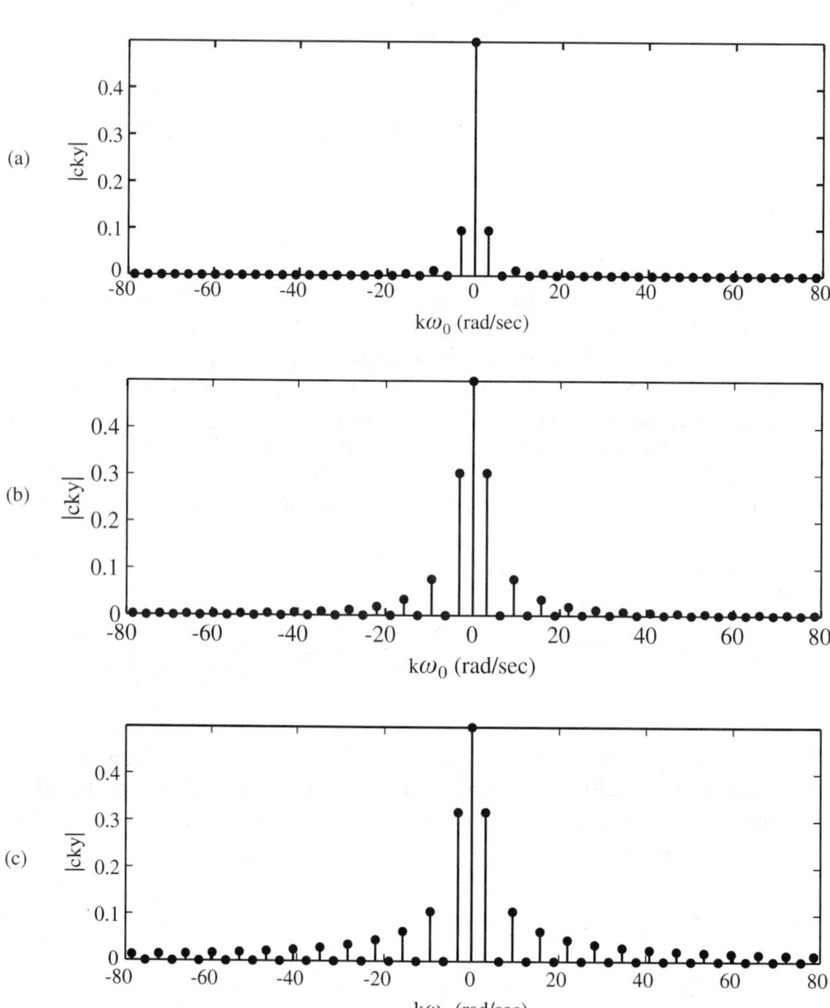

Figure 5.6 Amplitude spectrum of output when (a) $1/RC = 1$; (b) $1/RC = 10$; (c) $1/RC = 100$.

The value of $N = 50$ was chosen to be sufficiently large to achieve good accuracy in recovering the waveform of $y(t)$. Also note that c_k is real valued, so `conj(ck)` is not needed in defining $y(t)$.

The response $y(t)$ is displayed in Figure 5.7 for the values $1/RC = 1$, $1/RC = 10$, and $1/RC = 100$. From the figure it is seen that the response more closely resembles the input pulse train as the bandwidth of the RC circuit is increased from 1 rad/sec (Figure 5.7a) to 100 rad/sec (Figure 5.7c). Again, this result is expected since the circuit is passing more of the spectral content of the input pulse train as the bandwidth of the circuit is increased.

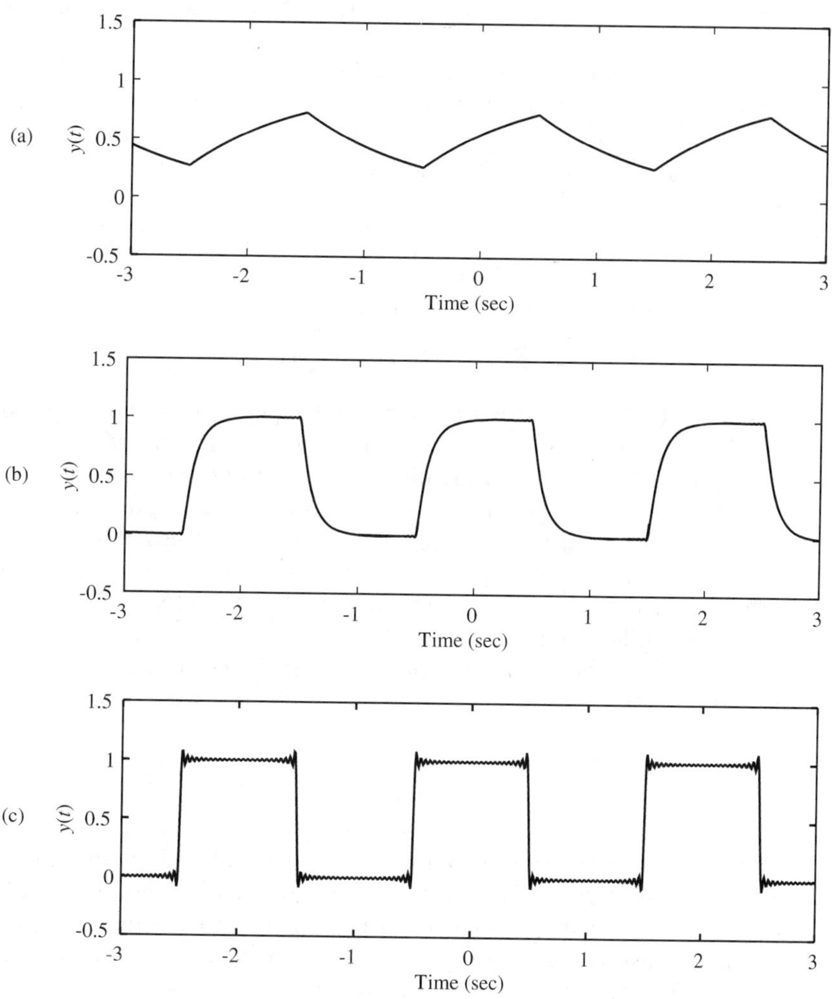

Figure 5.7 Plot of output when (a) $1/RC = 1$; (b) $1/RC = 10$; (c) $1/RC = 100$.

5.3 RESPONSE TO APERIODIC INPUTS

Again consider the system given by the convolution input/output relationship (5.26) with the condition (5.27). Taking the Fourier transform of both sides of (5.26) and using the convolution property of the transform results in the relationship

$$Y(\omega) = H(\omega)X(\omega) \tag{5.37}$$

where $Y(\omega)$ is the Fourier transform of the output, $X(\omega)$ is the Fourier transform of the input, and $H(\omega)$ is the system frequency function [i.e., $H(\omega)$ is the Fourier transform of $h(t)$]. Equation (5.37) is the Fourier transform domain (or the ω-domain) representation of the given system. This is a *frequency-domain representation* of the given system since the quantities in (5.37) are functions of the frequency variable ω.

From the ω-domain representation (5.37), it is seen that the frequency spectrum $Y(\omega)$ of the output is equal to the product of the system's frequency function $H(\omega)$ with the frequency spectrum $X(\omega)$ of the input. Taking the magnitude and angle of both sides of (5.37) shows that the amplitude spectrum $|Y(\omega)|$ of the output response $y(t)$ is given by

$$|Y(\omega)| = |H(\omega)|\,|X(\omega)| \tag{5.38}$$

and the phase spectrum $\angle Y(\omega)$ is given by

$$\angle Y(\omega) = \angle H(\omega) + \angle X(\omega) \tag{5.39}$$

Equation (5.38) shows that the amplitude spectrum of the output is equal to the product of the amplitude spectrum of the input with the system's magnitude function, and (5.39) shows that the phase spectrum of the output is equal to the sum of the phase spectrum of the input and the phase function of the system. As will be seen, these relationships provide a good deal of insight into how a system processes inputs. It is important to note that the relationships (5.37), (5.38), and (5.39) can be viewed as generalizations of the corresponding relationships (5.30), (5.31), and (5.32) in the case of periodic inputs.

Applying the inverse Fourier transform to both sides of the ω-domain representation (5.37) gives

$$y(t) = \frac{1}{2\pi}\int_{-\infty}^{\infty} H(\omega)X(\omega)e^{j\omega t}\, dt \tag{5.40}$$

The expression (5.40) is the output response resulting from input $x(t)$ with no initial energy in the system prior to the application of $x(t)$. The expression (5.40) can be used to compute $y(t)$. In some cases it is possible to determine $y(t)$ by working with Fourier transform pairs from a table rather than by computing the integral in (5.40). Examples are given in the next section.

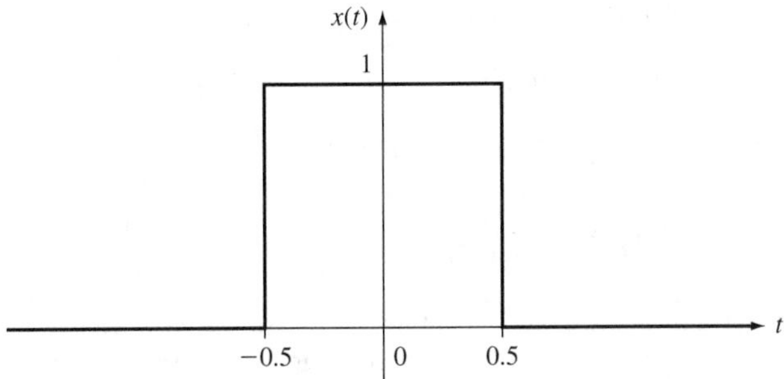

Figure 5.8 Input pulse in Example 5.5.

Example 5.5 *Response of RC Circuit to a Pulse*

Again consider the *RC* circuit shown in Figure 5.1. In this example the objective is to examine the response due to the rectangular pulse shown in Figure 5.8. The Fourier transform of $x(t)$ was found in Example 4.7 to be (setting $\tau = 1$)

$$X(\omega) = \operatorname{sinc}\frac{\omega}{2\pi} = \frac{\sin(\omega/2)}{\omega/2} = 2\,\frac{\sin(\omega/2)}{\omega}$$

The amplitude and phase spectra of the pulse are displayed in Figure 5.9.

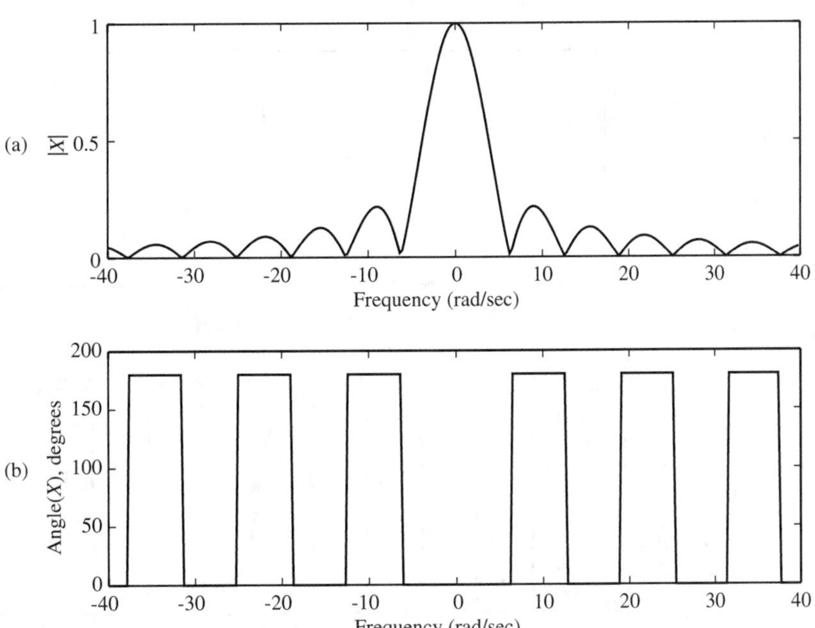

Figure 5.9 (a) Amplitude and (b) phase spectra of the input pulse.

From (5.37), the Fourier transform of the output is

$$Y(\omega) = X(\omega)H(\omega)$$

where, again,

$$H(\omega) = \frac{1/RC}{j\omega + 1/RC}$$

The amplitude and phase spectra of $y(t)$ are plotted in Figure 5.10 for the case $1/RC = 1$, and are plotted in Figure 5.11 for the case $1/RC = 10$. The amplitude spectrum $|Y(\omega)|$ was obtained by multiplying $|H(\omega)|$ by $|X(\omega)|$ at each frequency, and the phase $\angle Y(\omega)$ was obtained by adding $\angle H(\omega)$ to $\angle X(\omega)$. The MATLAB commands to obtain $|Y(\omega)|$ and $\angle Y(\omega)$ when $1/RC = 1$ are

```
RC = 1;
w = -40:.3:40;
X = 2*sin(w/2)./w;
H = (1/RC)./(j*w+1/RC);
Y = X.*H;
magY = abs(Y);
angY = 180*angle(Y)/pi;
```

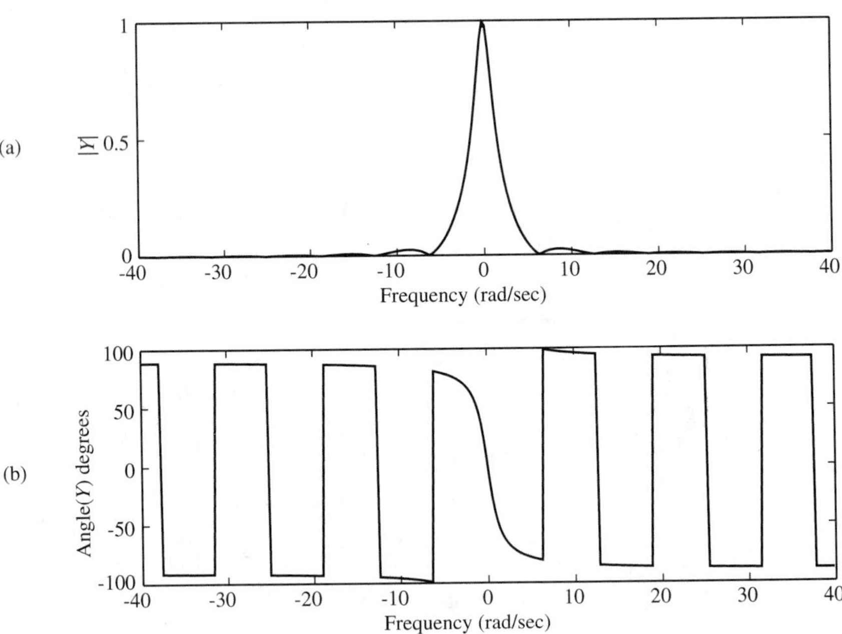

Figure 5.10 (a) Amplitude and (b) phase spectra of $y(t)$ when $1/RC = 1$.

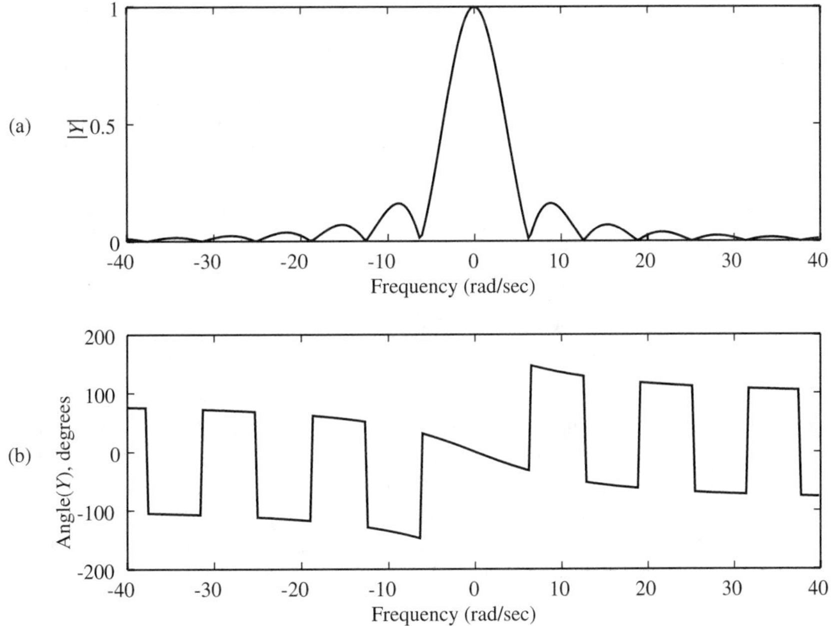

Figure 5.11 (a) Amplitude and (b) phase spectra of $y(t)$ when $1/RC = 10$.

As noted in Example 5.4, the larger the value of $1/RC$, the larger the 3-dB bandwidth of the RC circuit, which means that the sidelobes of $X(\omega)$ are attenuated less in passing through the filter. In particular, from Figures 5.9 and 5.10 it can be seen that when $1/RC = 1$, there is a significant amount of attenuation of the main lobe and sidelobes of $|Y(\omega)|$ in comparison to those of $|X(\omega)|$, which is a result of the 3-dB bandwidth of the circuit being set too low (in the case when $1/RC = 1$). As seen from Figure 5.11, increasing the bandwidth by setting $1/RC = 10$ results in much less attenuation of the sidelobes of $|Y(\omega)|$; that is, there is much less attenuation of the higher-frequency components of $x(t)$ in passing through the filter when the 3-dB bandwidth is 10 rad/sec.

The effect of the filtering can be examined in the time domain by computing $y(t)$ for various values of $1/RC$. The output $y(t)$ can be computed by taking the inverse Fourier transform of $H(\omega)X(\omega)$ as given by (5.40). But this computation is tedious to carry out for the present example. Instead, the response can be computed numerically by using the numerical convolution program given on the Web. In using this program, the time step T should be sufficiently small to achieve good accuracy. In this example, selecting $T \leq RC/10$ should be adequate. The computation of $y(t)$ using numerical convolution requires that the impulse response $h(t)$ of the circuit be computed. Since $h(t)$ is equal to the inverse Fourier transform of $H(\omega)$, it can be computed from the transform pairs in Table 4.2. The result is

$$h(t) = \begin{cases} \dfrac{1}{RC} e^{-(1/RC)t}, & t \geq 0 \\ 0, & t < 0 \end{cases}$$

The MATLAB commands used to compute $y(t)$ for $RC = 0.1$ are

```
T = .01; RC = 0.1;
nh = 0:400;      % defines indices for h, corresponds to
                 % t=0 to t=4
h = exp(-1/RC*nh*T)/RC;
nx = -60:100;    % defines indices for x, corresponds to
                 % t=.6 to t=1
x = [zeros(1,10) ones(1,101) zeros(1,50)];
y = conv(x,h*T);
ny = nx(1)+nh(1):nx(length(nx))+nh(length(nh));
plot(ny*T,y)
```

The time-domain responses for $1/RC = 1$ and for $1/RC = 10$ are shown in Figure 5.12. Note that in the higher-bandwidth case ($1/RC = 10$), when more of the main lobe and sidelobes of $X(\omega)$ are passed through the circuit, the output response looks more like the input pulse. However, even in this case the cutoff of the high-frequency components of $X(\omega)$ causes the corners of the rectangular pulse to be smoothed in the output response.

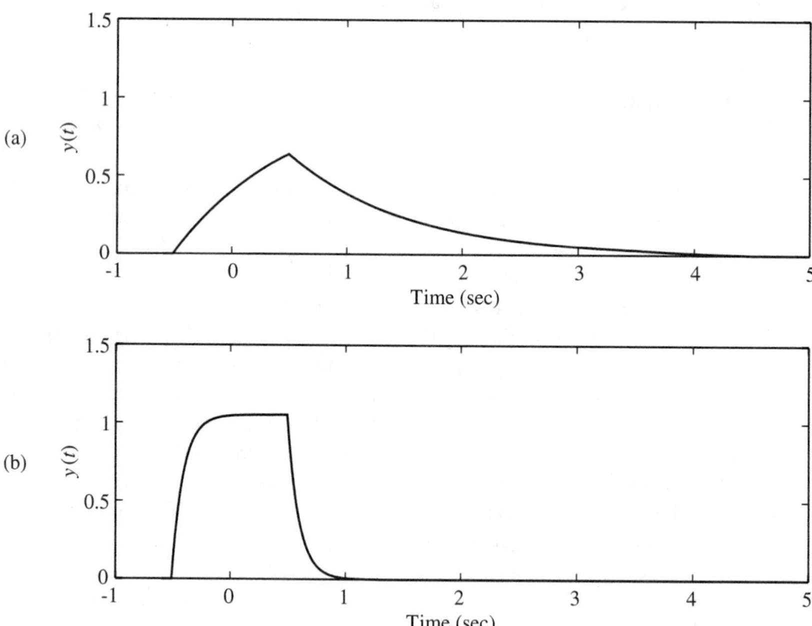

Figure 5.12 Output response when (a) $1/RC = 1$ and (b) $1/RC = 10$.

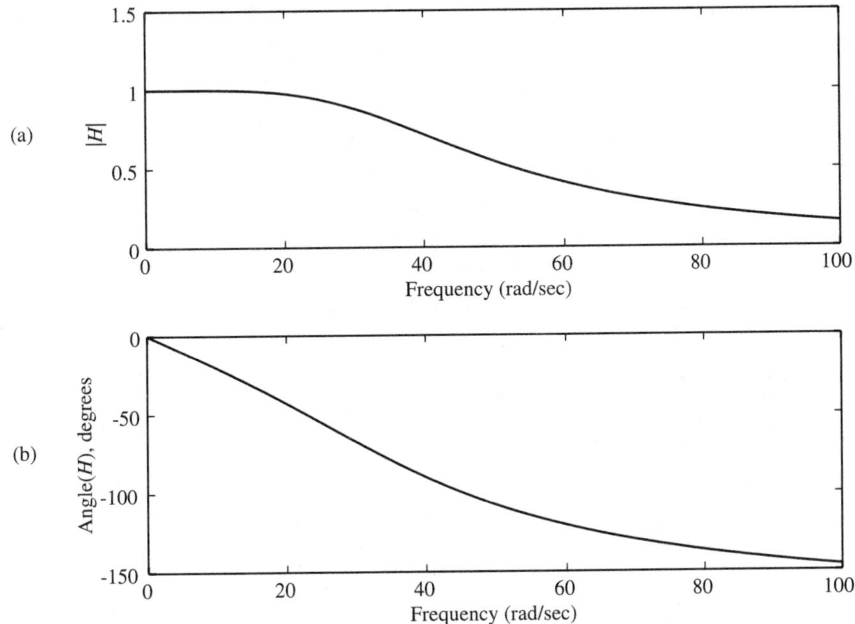

Figure 5.13 (a) Magnitude and (b) phase functions of system in Example 5.6.

Example 5.6 *Attention of High-Frequency Components*

Consider the system with the magnitude function $|H(\omega)|$ and phase function $\angle H(\omega)$ shown in Figure 5.13. The input $x(t)$ to the system is given in terms of its amplitude $|X(\omega)|$ and phase $\angle X(\omega)$ spectra shown in Figure 5.14. The Fourier transform $Y(\omega)$ of the output response to the input $x(t)$ is computed using the relationships (5.38) and (5.39). The resulting amplitude and phase spectra for $y(t)$ are plotted in Figure 5.15. Note that $X(\omega)$ has a continuum of frequencies with two peaks at $\omega = 10$ rad/sec and $\omega = 70$ rad/sec. Given the frequency response characteristic of the system, the effect of the system on the input is to attenuate the higher-frequency peak at $\omega = 70$ rad/sec. As a result, $y(t)$ should have a smoother appearance than $x(t)$, due to the lack of higher-frequency components. This is verified by comparing $x(t)$ and $y(t)$ as plotted in Figure 5.16. Here the MATLAB program that generates these plots uses advanced commands that will not be introduced until later chapters.

5.4 ANALYSIS OF IDEAL FILTERS

Given a linear time-invariant continuous-time system with frequency function $H(\omega)$, in Section 5.1 it was shown that the output response $y(t)$ resulting from the sinusoidal input

$$x(t) = A\cos(\omega_0 t + \theta), \qquad -\infty < t < \infty$$

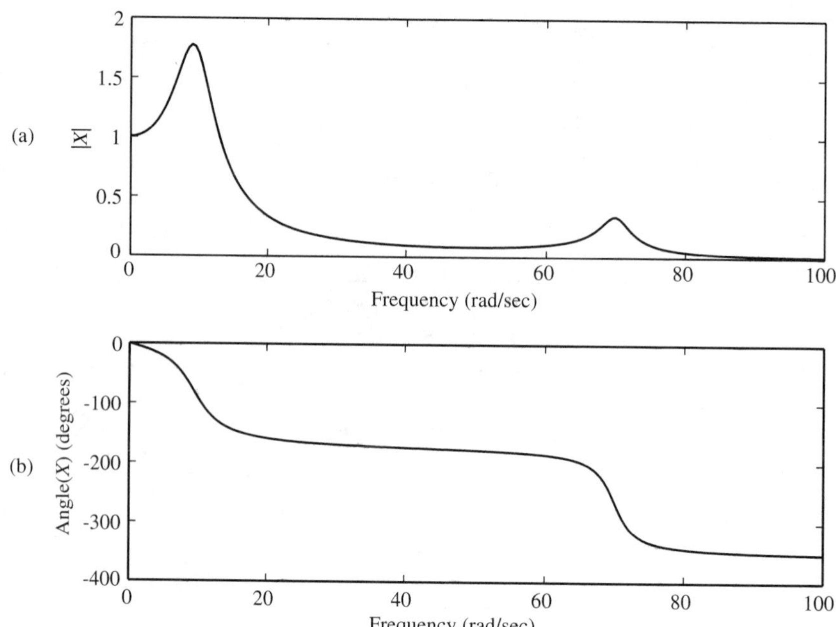

Figure 5.14 (a) Amplitude and (b) phase spectra of input in Example 5.6.

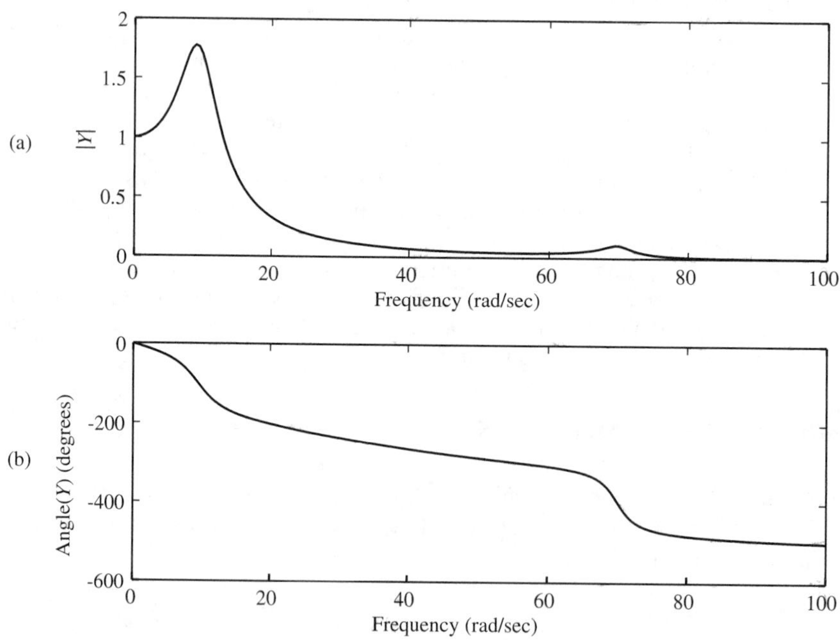

Figure 5.15 (a) Amplitude and (b) phase spectra of output in Example 5.6.

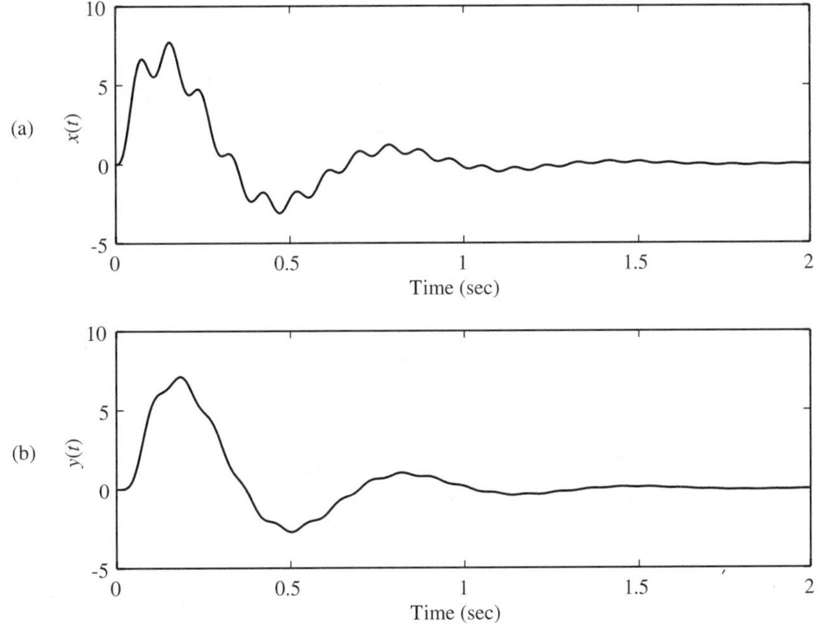

Figure 5.16 (a) Input and (b) resulting output in Example 5.6.

is given by

$$y(t) = A|H(\omega_0)| \cos(\omega_0 t + \theta + \angle H(\omega_0)), \qquad -\infty < t < \infty \tag{5.41}$$

From (5.41) it is clear that a sinusoid with a particular frequency ω_0 can be prevented from "going through" the system by selecting $H(\omega)$ so that $|H(\omega_0)|$ is zero or very small. The process of "rejecting" sinusoids having particular frequencies, or for a range of frequencies, is called *filtering* and a system that has this characteristic is called a *filter.* The concept of filtering was first considered in Section 5.2 in the specific context of an RC circuit. In this section, a general treatment of filtering is presented in the case of ideal filters. Nonideal filters will be studied in Chapter 9.

An ideal filter is a system that completely rejects sinusoidal inputs of the form $x(t) = A \cos \omega_0 t,\ -\infty < t < \infty$, for ω_0 in certain frequency ranges, and does not attenuate sinusoidal inputs whose frequencies are outside these ranges. There are four basic types of ideal filters: lowpass, highpass, bandpass, and bandstop. The magnitude functions of these four types of filters are displayed in Figure 5.17. Mathematical expressions for these magnitude functions are as follows:

$$\text{Ideal lowpass:} \quad |H(\omega)| = \begin{cases} 1, & -B \le \omega \le B \\ 0, & |\omega| > B \end{cases} \tag{5.42}$$

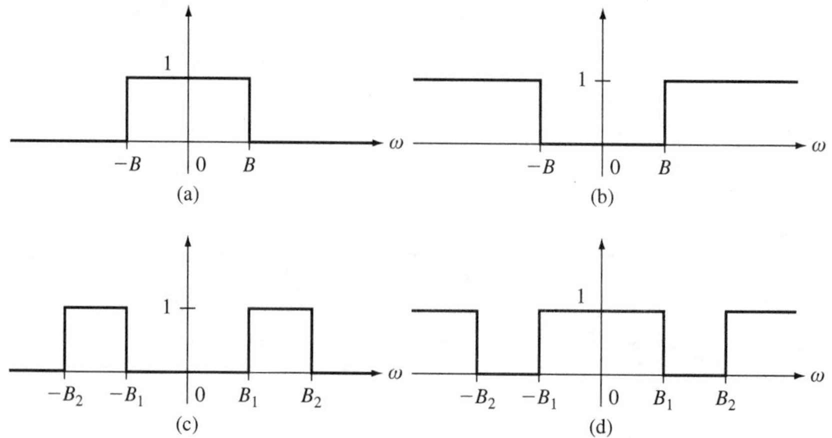

Figure 5.17 Magnitude functions of ideal filters: (a) lowpass; (b) highpass; (c) bandpass; (d) bandstop.

$$\text{Ideal highpass:} \quad |H(\omega)| = \begin{cases} 0, & -B < \omega < B \\ 1, & |\omega| \geq B \end{cases} \tag{5.43}$$

$$\text{Ideal bandpass:} \quad |H(\omega)| = \begin{cases} 1, & B_1 \leq |\omega| \leq B_2 \\ 0, & \text{all other } \omega \end{cases} \tag{5.44}$$

$$\text{Ideal bandstop:} \quad |H(\omega)| = \begin{cases} 0, & B_1 \leq |\omega| \leq B_2 \\ 1, & \text{all other } \omega \end{cases} \tag{5.45}$$

The *stopband* of an ideal filter is defined to be the set of all frequencies ω_0 for which the filter completely stops the sinusoidal input $x(t) = A \cos \omega_0 t, -\infty < t < \infty$. The *passband* of the filter is the set of all frequencies ω_0 for which the input $x(t)$ is passed without attenuation.

From (5.41) and (5.42), it is seen that the ideal lowpass filter passes with no attenuation sinusoidal inputs with frequencies ranging from $\omega = 0$ (rad/sec) to $\omega = B$ (rad/sec), while it completely stops sinusoidal inputs with frequencies above $\omega = B$. The filter is said to be *lowpass,* since it passes low-frequency sinusoids and stops high-frequency sinusoids. The frequency range $\omega = 0$ to $\omega = B$ is the passband of the filter and the range $\omega = B$ to $\omega = \infty$ is the stopband of the filter. The width B of the passband is defined to be the filter bandwidth.

From (5.41) and (5.43), it is seen that the highpass filter stops sinusoids with frequencies below B while it passes sinusoids with frequencies above B, hence the term *highpass.* The stopband of the highpass filter is the frequency range from $\omega = 0$ to $\omega = B$ and the passband is the frequency range from $\omega = B$ to $\omega = \infty$.

By (5.44) the passband of the bandpass filter is the frequency range from $\omega = B_1$ to $\omega = B_2$, while the stopband is the range from $\omega = 0$ to $\omega = B_1$ and the range from $\omega = B_2$ to $\omega = \infty$. The bandwidth of the bandpass filter is the width of the passband (i.e., $B_2 - B_1$). By (5.45) the stopband of the bandstop filter is the range from $\omega = B_1$ to $\omega = B_2$, while the passband is the range from $\omega = 0$ to $\omega = B_1$ and the range from $\omega = B_2$ to $\omega = \infty$.

More complicated examples of ideal filters can be constructed by cascading ideal lowpass, highpass, bandpass, and bandstop filters. For instance, by cascading bandpass filters with various values of B_1 and B_2 it is possible to construct an ideal *comb* filter, whose magnitude function is illustrated in Figure 5.18.

Phase Function

In the discussion above on ideal filters, nothing has been said regarding the phase of the filters. It turns out that to avoid phase distortion in the filtering process, a filter should have a linear phase characteristic over the passband of the filter. In other words, the phase function (in radians) should be of the form

$$\angle H(\omega) = -\omega t_d \qquad \text{for all } \omega \text{ in the filter passband} \tag{5.46}$$

where t_d is a fixed positive number. If ω_0 is in the passband of a linear phase filter, by (5.41) and (5.46) the response resulting from the input $x(t) = A \cos \omega_0 t$, $-\infty < t < \infty$, is given by

$$y(t) = A|H(\omega_0)| \cos(\omega_0 t - \omega_0 t_d), \qquad -\infty < t < \infty$$

$$y(t) = A|H(\omega_0)| \cos[\omega_0(t - t_d)], \qquad -\infty < t < \infty$$

Thus the linear phase characteristic results in a time delay of t_d seconds through the filter.

Note that if the input is

$$x(t) = A_0 \cos \omega_0 t + A_1 \cos \omega_1 t, \qquad -\infty < t < \infty \tag{5.47}$$

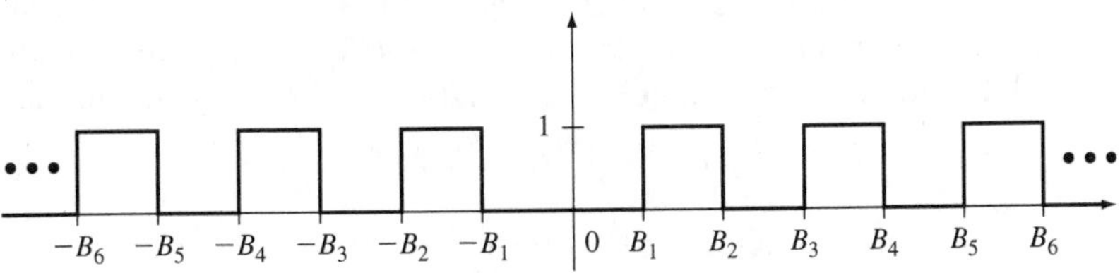

Figure 5.18 Magnitude function of an ideal comb filter.

where ω_0 and ω_1 are in the passband of the filter, by linearity the response is

$$y(t) = A_0|H(\omega_0)|\cos[\omega_0(t - t_d)] + A_1|H(\omega_1)|\cos[\omega_1(t - t_d)], \qquad -\infty < t < \infty$$

So again the output is a t_d-second time delay of the input; in particular, there is no distortion of the input. In contrast, if the phase function is not linear, there will be phase distortion in the filter output. To see this, suppose that the phase function $\angle H(\omega)$ is equal to some nonzero constant C. In this case the response to the input (5.47) is

$$y(t) = A_0|H(\omega_0)|\cos(\omega_0 t + C) + A_1|H(\omega_1)|\cos(\omega_1 t + C), \qquad -\infty < t < \infty$$

This output is not a time-delayed version of the input, so there is distortion in the filtering process. Therefore, for distortionless filtering the phase function of the filter should be as close to linear as possible over the passband of the filter.

Ideal Linear-Phase Lowpass Filter

Consider the ideal lowpass filter with the frequency function

$$H(\omega) = \begin{cases} e^{-j\omega t_d}, & -B \le \omega \le B \\ 0, & \omega < -B, \quad \omega > B \end{cases} \tag{5.48}$$

where t_d is a positive real number. Equation (5.48) is the polar-form representation of $H(\omega)$. From (5.48)

$$|H(\omega)| = \begin{cases} 1, & -B \le \omega \le B \\ 0, & \omega < -B, \quad \omega > B \end{cases}$$

and the phase in radians is

$$\angle H(\omega) = \begin{cases} -\omega t_d, & -B \le \omega \le B \\ 0, & \omega < -B, \quad \omega > B \end{cases}$$

The phase function $\angle H(\omega)$ of the filter is plotted in Figure 5.19. Note that over the frequency range 0 to B, the phase function of the system is linear with slope equal to $-t_d$.

The impulse response of the lowpass filter defined by (5.48) can be computed by taking the inverse Fourier transform of the frequency function $H(\omega)$. First, using the definition of the rectangular pulse, $H(\omega)$ can be expressed in the form

$$H(\omega) = p_{2B}(\omega)e^{-j\omega t_d}, \qquad -\infty < \omega < \infty \tag{5.49}$$

From Table 4.2 the following transform pair can be found:

$$\frac{\tau}{2\pi}\operatorname{sinc}\frac{\tau t}{2\pi} \leftrightarrow p_\tau(\omega) \tag{5.50}$$

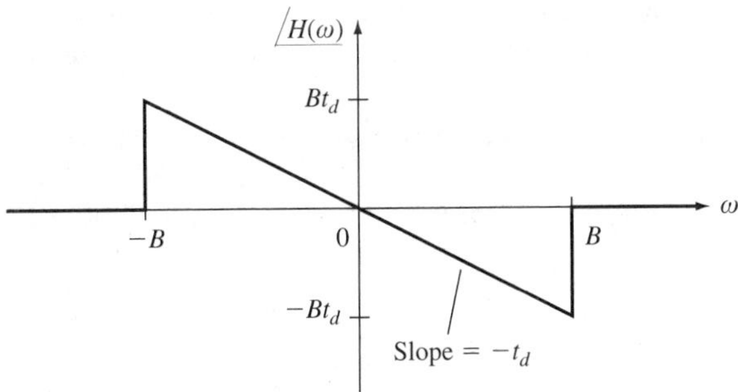

Figure 5.19 Phase function of ideal lowpass filter defined by (5.48).

Setting $\tau = 2B$ in (5.50) gives

$$\frac{B}{\pi}\,\text{sinc}\left(\frac{B}{\pi}t\right) \leftrightarrow p_{2B}(\omega) \tag{5.51}$$

Applying the time-shift property to the transform pair (5.51) gives

$$\frac{B}{\pi}\,\text{sinc}\left[\frac{B}{\pi}(t - t_d)\right] \leftrightarrow p_{2B}(\omega)e^{-j\omega t_d} \tag{5.52}$$

Since the right-hand side of the transform pair (5.52) is equal to $H(\omega)$, the impulse response of the ideal lowpass filter is

$$h(t) = \frac{B}{\pi}\,\text{sinc}\left[\frac{B}{\pi}(t - t_d)\right], \qquad -\infty < t < \infty \tag{5.53}$$

The impulse response $h(t)$ is plotted in Figure 5.20.

In Figure 5.20 it is clear that the impulse response $h(t)$ is not zero for $t < 0$, and thus the filter is noncausal. As a result, it is not possible to build an ideal lowpass

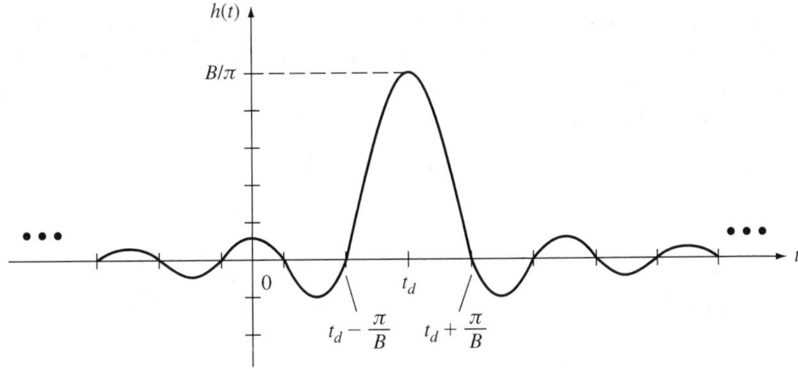

Figure 5.20 Impulse response of ideal linear-phase lowpass filter.

filter. In fact, any ideal filter is noncausal and thus cannot be realized. For "real-time" filtering, it is necessary to consider causal filters, which are studied in Section 9.7.

Response to nonsinusoidal inputs. Again consider the ideal lowpass filter defined by (5.48). The ω-domain representation $Y(\omega) = H(\omega)X(\omega)$ will be utilized to compute the output response resulting from a nonsinusoidal input $x(t)$. In particular, suppose that

$$x(t) = \operatorname{sinc}\frac{t}{\pi}, \qquad -\infty < t < \infty$$

Then from Table 4.2, the input spectrum is

$$X(\omega) = \pi p_2(\omega)$$

Since

$$|X(\omega)| = 0 \qquad \text{for } \omega > 1$$

the bandwidth of the signal $x(t) = \operatorname{sinc} t/\pi$ is equal to 1.

Now using (5.48),

$$Y(\omega) = H(\omega)X(\omega) = p_{2B}(\omega)e^{-j\omega t_d}\pi p_2(\omega)$$

If $2B > 2$ or $B > 1$,

$$p_{2B}(\omega)p_2(\omega) = p_2(\omega)$$

and thus the output spectrum is

$$Y(\omega) = \pi p_2(\omega)e^{-j\omega t_d}$$

Using the time-shift property gives

$$y(t) = \operatorname{sinc}\frac{t - t_d}{\pi} = x(t - t_d), \qquad -\infty < t < \infty$$

This result shows that if the bandwidth B of the filter is wide enough to pass all the frequency components of the input signal, the filter simply passes the input with a t_d-second time delay.

If $2B \leq 2$ or $B \leq 1$,

$$p_{2B}(\omega)p_2(\omega) = p_{2B}(\omega)$$

and thus

$$Y(\omega) = \pi p_{2B}(\omega)e^{-j\omega t_d} = \pi H(\omega)$$

Therefore,

$$y(t) = \pi h(t) = B \operatorname{sinc}\left[\frac{B}{\pi}(t - t_d)\right], \qquad -\infty < t < \infty \tag{5.54}$$

In this case, the bandwidth B of the filter is not wide enough to pass all the frequency components of the input, and thus the output is not a time-delayed version of the input. In fact, by (5.54) the response to $x(t) = \text{sinc}(t/\pi)$ is equal to a scalar multiple of the impulse response when $B < 1$. This is a very interesting result since it implies that the response to the input $x(t) = \pi\delta(t)$ is the same as the response to the input $x(t) = \text{sinc}(t/\pi)$.

Now suppose that

$$x(t) = \left(\text{sinc}\,\frac{t}{\pi}\right)(\cos 2t), \qquad -\infty < t < \infty$$

Then by the modulation property of the Fourier transform,

$$X(\omega) = \frac{\pi}{2}\,[p_2(\omega + 2) + p_2(\omega - 2)]$$

The amplitude spectrum $|X(\omega)|$ is plotted in Figure 5.21. In this case $|X(\omega)| = 0$ for $\omega > 3$, and thus the bandwidth of $x(t)$ is equal to 3. The transform of the resulting output is

$$Y(\omega) = p_{2B}(\omega)e^{-j\omega t_d}\,\frac{\pi}{2}\,[p_2(\omega + 2) + p_2(\omega - 2)]$$

If $B > 3$, the filter again passes all frequency components of the input, and thus

$$y(t) = x(t - t_d) = \left[\text{sinc}\,\frac{t - t_d}{\pi}\right]\cos\,[2(t - t_d)]$$

If $B < 1$, the filter does not pass any of the frequency components of the input $x(t) = [\text{sinc}\,(t/\pi)](\cos 2t)$, and thus

$$Y(\omega) = 0$$

which implies that

$$y(t) = 0, \qquad -\infty < t < \infty$$

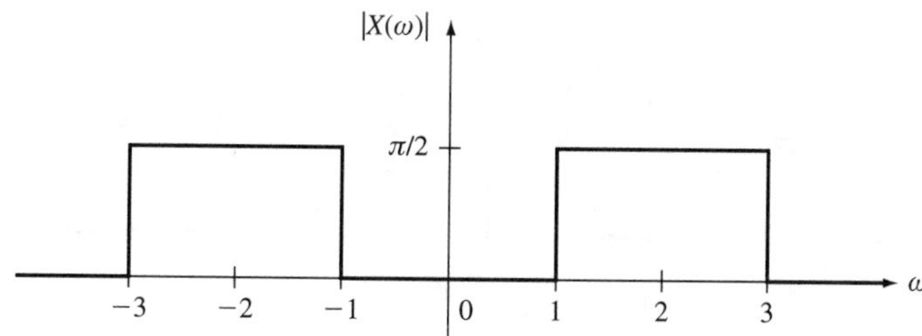

Figure 5.21 Amplitude spectrum of the signal $[\text{sinc}(t/\pi)](\cos 2t)$.

If $1 < B < 3$, the filter passes the frequency components in the range from 1 to B. In this case, the product

$$p_{2B}(\omega)[p_2(\omega + 2) + p_2(\omega - 2)]$$

is computed in Figure 5.22. From Figure 5.22

$$p_{2B}(\omega)[p_2(\omega + 2) + p_2(\omega - 2)] = p_{B-1}\left(\omega + \frac{B+1}{2}\right) + p_{B-1}\left(\omega - \frac{B+1}{2}\right)$$

Let $V(\omega) = p_{B-1}(\omega)$. Then

$$Y(\omega) = \frac{\pi}{2}\left[V\left(\omega + \frac{B+1}{2}\right) + V\left(\omega - \frac{B+1}{2}\right)\right]e^{-j\omega t_d}$$

Using the modulation and shift properties yields

$$y(t) = \pi v(t - t_d)\cos\left[\frac{B+1}{2}(t - t_d)\right], \qquad -\infty < t < \infty$$

But

$$v(t) = \frac{B-1}{2\pi}\operatorname{sinc}\left(\frac{B-1}{2\pi}t\right), \qquad -\infty < t < \infty$$

and thus

$$y(t) = \frac{B-1}{2}\operatorname{sinc}\left[\frac{B-1}{2\pi}(t - t_d)\right]\cos\left[\frac{B+1}{2}(t - t_d)\right] \qquad -\infty < t < \infty$$

Clearly, $y(t)$ is not a time-delayed version of the input $x(t) = [\operatorname{sinc}(t/\pi)](\cos 2t)$. So "cutting off" some of the frequency components of the input results in a significant distortion of the input.

Now let $x(t)$ be an arbitrary input with Fourier transform $X(\omega)$. It is assumed that $|X(\omega)| = 0$ for all $\omega > \Omega$; that is, Ω is the bandwidth of the input sig-

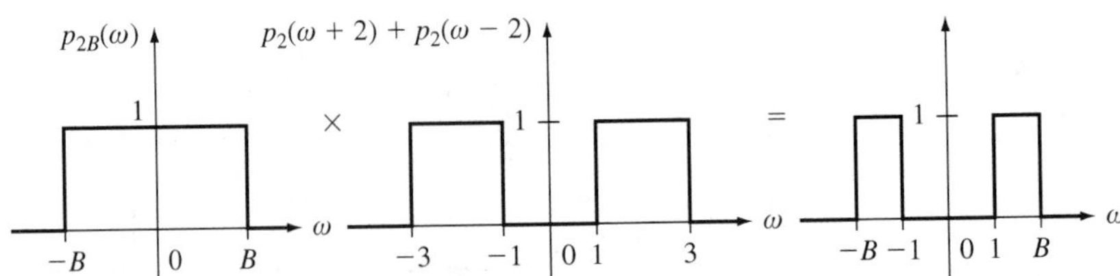

Figure 5.22 Computation of $p_{2B}(\omega)[p_2(\omega + 2) + p_2(\omega - 2)]$.

nal $x(t)$. If $\Omega < B$, the lowpass filter passes all the frequency components of the input; that is,

$$Y(\omega) = X(\omega)e^{-j\omega t_d}$$

Hence

$$y(t) = x(t - t_d)$$

If $\Omega > B$, the filter does not pass all the frequency components of the input, and thus the filter output $y(t)$ will be a distorted version of the input.

Ideal Linear-Phase Bandpass Filters

The analysis given above can be extended to the other types of ideal filters mentioned in the first part of this section. For example, the frequency function of an ideal bandpass filter is given by

$$H(\omega) = \begin{cases} e^{-j\omega t_d}, & B_1 \le |\omega| \le B_2 \\ 0, & \text{all other } \omega \end{cases}$$

where t_d, B_1, and B_2 are positive real numbers. The magnitude function $|H(\omega)|$ is plotted in Figure 5.17c and the phase function $\angle H(\omega)$ (in radians) is plotted in Figure 5.23. Since the passband of the filter is from B_1 to B_2, for any input signal $x(t)$ whose frequency components are contained in the region from B_1 to B_2, the filter will pass the signal with no distortion, although there will be a time delay of t_d seconds.

5.5 SAMPLING

An important operation in various application domains, such as communications and controls, is the sampling of a continuous-time signal $x(t)$. In uniform sampling (the case of interest here), the sample values of $x(t)$ are equal to the values $x(nT)$, where n is the integer index, $n = 0, \pm 1, \pm 2, \ldots$, and T is the sampling interval. To simplify

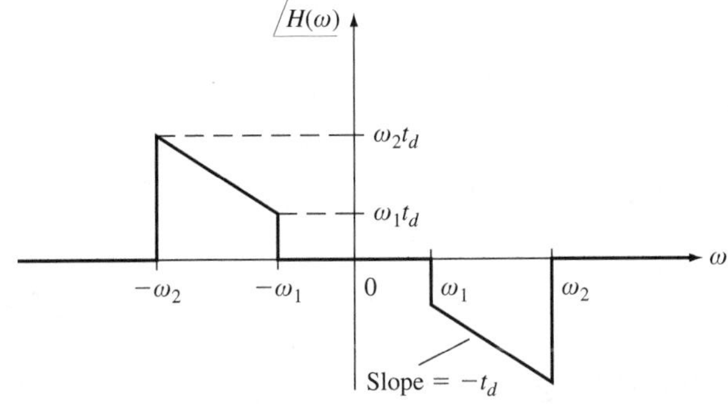

Figure 5.23 Phase function of ideal linear-phase bandpass filter.

the analysis of the sampling operation, the sampled version of $x(t)$ is often expressed in the form $x(t)p(t)$, where $p(t)$ is the impulse train given by

$$p(t) = \sum_{n=-\infty}^{\infty} \delta(t - nT). \tag{5.53}$$

Hence, the sampled waveform $x(t)p(t)$ is given by

$$x(t)p(t) = \sum_{n=-\infty}^{\infty} x(t)\delta(t - nT) = \sum_{n=-\infty}^{\infty} x(nT)\delta(t - nT) \tag{5.54}$$

Thus, the sampled waveform $x(t)p(t)$ is an impulse train whose weights (areas) are the sample values $x(nT)$ of the signal $x(t)$. The sampling process given by (5.54) is referred to as "idealized sampling."

To determine the Fourier transform of $x(t)p(t)$, first observe that since the impulse train $p(t)$ is a periodic signal with fundamental period T, $p(t)$ has the complex exponential Fourier series

$$p(t) = \sum_{k=-\infty}^{\infty} c_k e^{jk\omega_s t} \tag{5.55}$$

where $\omega_s = 2\pi/T$ is the *sampling frequency* in rad/sec. The coefficients c_k of the Fourier series are computed as follows:

$$\begin{aligned} c_k &= \frac{1}{T}\int_{-T/2}^{T/2} p(t)e^{-jk\omega_s t}\,dt, \qquad k = 0, \pm 1, \pm 2, \ldots \\ &= \frac{1}{T}\int_{-T/2}^{T/2} \delta(t)e^{-jk\omega_s t}\,dt \\ &= \frac{1}{T}[e^{-jk\omega_s t}]_{t=o} \\ &= \frac{1}{T} \end{aligned}$$

Inserting $c_k = 1/T$ into (5.55) yields

$$p(t) = \sum_{k=-\infty}^{\infty} \frac{1}{T} e^{jk\omega_s t}$$

and thus

$$x(t)p(t) = \sum_{k=-\infty}^{\infty} \frac{1}{T} x(t)e^{jk\omega_s t} \tag{5.56}$$

The Fourier transform of $x(t)p(t)$ can then be computed by transforming the right-hand side of (5.56) using the property of the Fourier transform involving multiplication by a complex exponential [see (4.55)]. With $X(\omega)$ equal to the Fourier transform of $x(t)$, the result is

$$X_s(\omega) = \sum_{k=-\infty}^{\infty} \frac{1}{T} X(\omega - k\omega_s) \tag{5.57}$$

where $X_s(\omega)$ is the Fourier transform of the sampled waveform $x_s(t) = x(t)p(t)$.

From (5.57) it is seen that the Fourier transform $X_s(\omega)$ consists of a sum of magnitude-scaled replicas of $X(\omega)$ sitting at integer multiples $k\omega_s$ of ω_s for $k = 0$, $\pm 1, \pm 2, \ldots$. For example, suppose that $x(t)$ has the bandlimited Fourier transform $X(\omega)$ shown in Figure 5.24a. If $\omega_s - B > B$ or $\omega_s > 2B$, the Fourier transform $X_s(\omega)$ of the sampled signal $x_s(t) = x(t)p(t)$ is as shown in Figure 5.24b. Note that in this case, the replicas of $X(\omega)$ in $X_s(\omega)$ do not overlap in frequency. As a result, it turns out that it is possible to reconstruct $x(t)$ from the sampled signal by using lowpass filtering. The reconstruction process is studied next.

Signal Reconstruction

Given a signal $x(t)$, the reconstruction of $x(t)$ from the sampled waveform $x(t)p(t)$ can be carried out as follows. First, suppose that $x(t)$ has bandwidth B; that is,

$$|X(\omega)| = 0 \qquad \text{for } \omega > B$$

Then if $\omega_s \geq 2B$, in the expression (5.57) for $X_s(\omega)$ the replicas of $X(\omega)$ do not overlap in frequency. For example, if $X(\omega)$ has the form shown in Figure 5.24a, then $X_s(\omega)$ has the form shown in Figure 5.24b in the case when $\omega_s \geq 2B$. Thus if the sampled signal $x_s(t)$ is applied to an ideal lowpass filter with the frequency function shown in Figure 5.25, the only component of $X_s(\omega)$ that is passed is $X(\omega)$. Hence the output of the filter is equal to $x(t)$, which shows that the original signal $x(t)$ can be completely and exactly reconstructed from the sampled waveform $x_s(t)$. So the reconstruction of $x(t)$ from the sampled signal $x_s(t) = x(t)p(t)$ can be accomplished by simply lowpass filtering the sampled signal. The process is illustrated in Figure 5.26. The filter in this figure is sometimes called an *interpolation filter,* since it reproduces $x(t)$ from the values of $x(t)$ at the time points $t = nT$.

By this result, which is called the *sampling theorem,* a signal with bandwidth B can be reconstructed completely and exactly from the sampled signal $x_s(t) = x(t)p(t)$ by lowpass filtering with cutoff frequency B if the sampling frequency ω_s is chosen to be greater than or equal to $2B$. The minimum sampling frequency $\omega_s = 2B$ is called the *Nyquist sampling frequency.*

Example 5.7 *Nyquist Sampling Frequency for Speech*

The spectrum of a speech signal is essentially zero for all frequencies above 10 kHz, so the bandwidth of a speech signal can be taken to be $2\pi \times 10^4$ rad/sec. Then the Nyquist sampling frequency for speech is

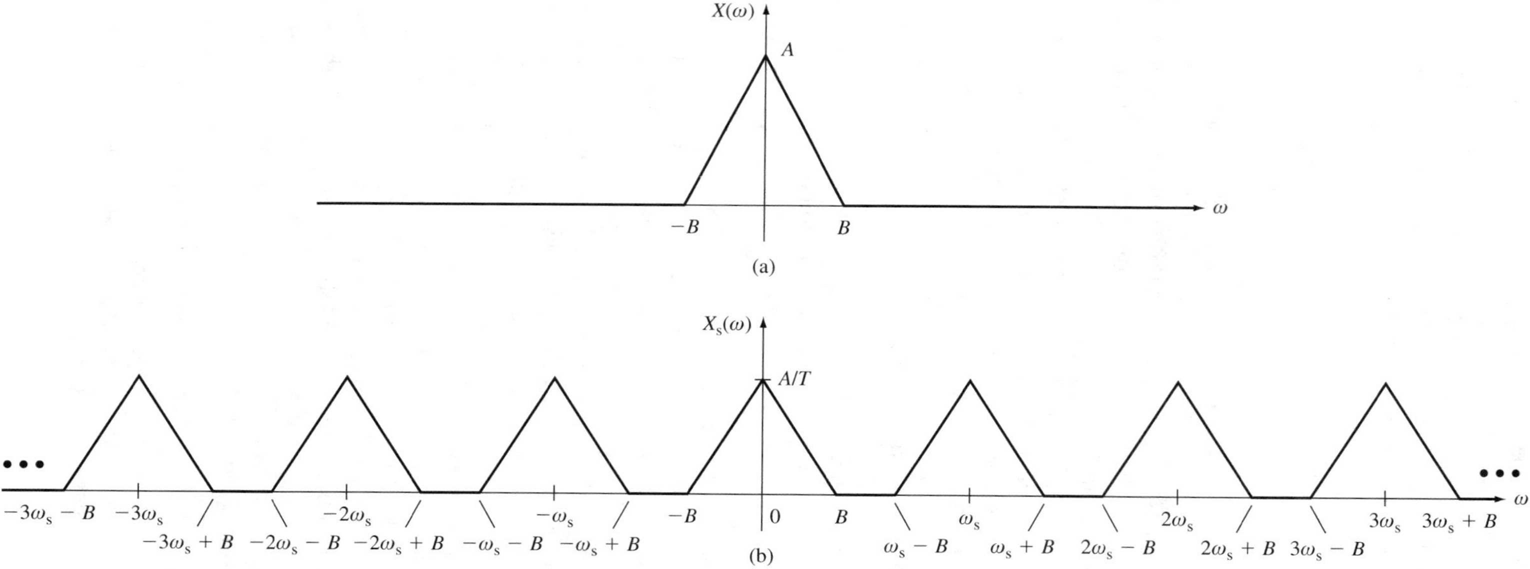

Figure 5.24 Fourier transform of (a) $x(t)$ and (b) $x_s(t) = x(t)p(t)$.

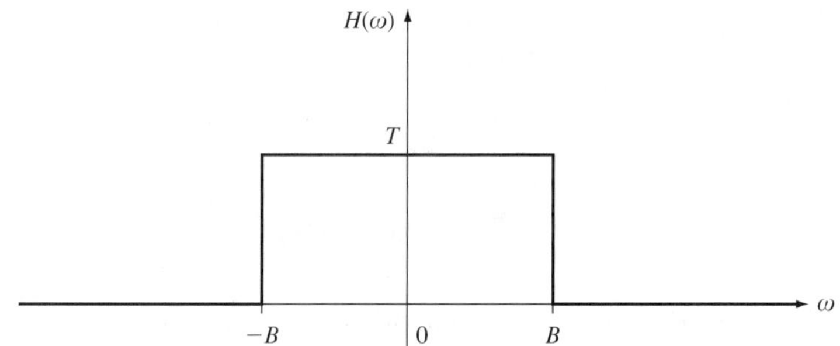

Figure 5.25 Frequency response function of ideal lowpass filter with bandwidth B.

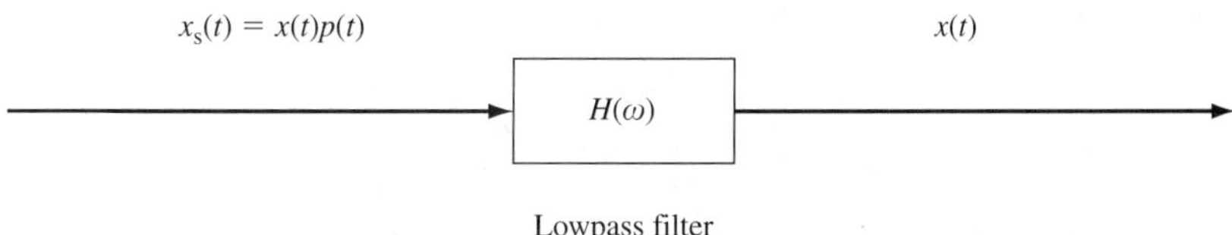

Figure 5.26 Reconstruction of $x(t)$ from $x_s(t) = x(t)p(t)$.

$$\omega_s = 2B = 4\pi \times 10^4 \text{ rad/sec}$$

Since $\omega_s = 2\pi/T$, the sampling interval T is equal to $2\pi/\omega_s = 50\ \mu\text{s}$. So the sampling interval corresponding to the Nyquist sampling rate is very small.

Interpolation Formula

From Figure 5.25 it is clear that the frequency response function $H(\omega)$ of the interpolating filter is given by

$$H(\omega) = \begin{cases} T, & -B \leq \omega \leq B \\ 0, & \text{all other } \omega \end{cases}$$

From the results in Section 5.4, the impulse $h(t)$ of this filter is given by

$$h(t) = \frac{BT}{\pi} \operatorname{sinc}\left(\frac{B}{\pi} t\right), \qquad -\infty < t < \infty \tag{5.58}$$

and the output $y(t)$ of the interpolating filter is given by

$$y(t) = h(t) * x_s(t) = \int_{-\infty}^{\infty} x_s(\tau)h(t - \tau)\, d\tau \tag{5.59}$$

But

$$x_s(\tau) = x(\tau)p(\tau) = \sum_{n=-\infty}^{\infty} x(nT)\delta(\tau - nT)$$

and inserting this into (5.59) gives

$$y(t) = \int_{-\infty}^{\infty} \sum_{n=-\infty}^{\infty} x(nT)\delta(\tau - nT)h(t - \tau)\, d\tau$$

$$= \sum_{n=-\infty}^{\infty} \int_{-\infty}^{\infty} x(nT)\delta(\tau - nT)h(t - \tau)\, d\tau \qquad (5.60)$$

From the sifting property of the impulse, (5.60) reduces to

$$y(t) = \sum_{n=-\infty}^{\infty} x(nT)h(t - nT) \qquad (5.61)$$

Finally, inserting (5.58) into (5.61) gives

$$y(t) = \frac{BT}{\pi} \sum_{n=-\infty}^{\infty} x(nT)\text{sinc}\left[\frac{B}{\pi}(t - nT)\right]$$

But $y(t) = x(t)$ and thus

$$x(t) = \frac{BT}{\pi} \sum_{n=-\infty}^{\infty} x(nT)\text{sinc}\left[\frac{B}{\pi}(t - nT)\right] \qquad (5.62)$$

The expression (5.60) is called the *interpolation formula* for the signal $x(t)$. In particular, it shows how the original signal $x(t)$ can be reconstructed from the sample values $x(nT), n = 0, \pm 1, \pm 2, \ldots$.

Aliasing

Sampling and Aliasing

In Chapter 4 it was noted that a time-limited signal cannot be bandlimited. Since all actual signals are time limited, they cannot be bandlimited. Therefore, if a time-limited signal is sampled with sampling interval T, no matter how small T is, the replicas of $X(\omega)$ in (5.57) will overlap. As a result of the overlap of frequency components, it is not possible to reconstruct $x(t)$ exactly by lowpass filtering the sampled signal $x_s(t) = x(t)p(t)$.

Although time-limited signals are not bandlimited, the amplitude spectrum $|X(\omega)|$ of a time-limited signal $x(t)$ will be small for suitably large values of ω. Thus for some finite B, all the significant components of $X(\omega)$ will be in the range $-B \leq \omega \leq B$. For instance, the signal $x(t)$ may have the amplitude spectrum shown in Figure 5.27. If B is chosen to have the value indicated, and if $x(t)$ is sampled with sampling frequency $\omega_s = 2B$, the amplitude spectrum of the resulting sampled signal $x_s(t)$ is shown in Figure 5.28.

Now if the sampled signal $x_s(t)$ is lowpass filtered with cutoff frequency B, the output spectrum of the filter will contain high-frequency components of $x(t)$ transposed to low-frequency components. This phenomenon is called *aliasing*.

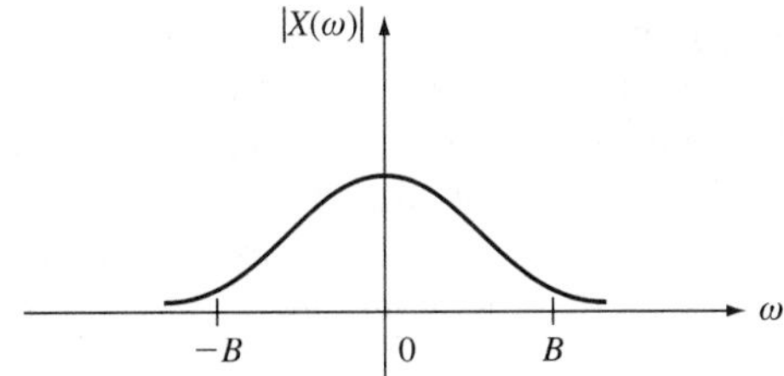

Figure 5.27 Amplitude spectrum of a time-limited signal.

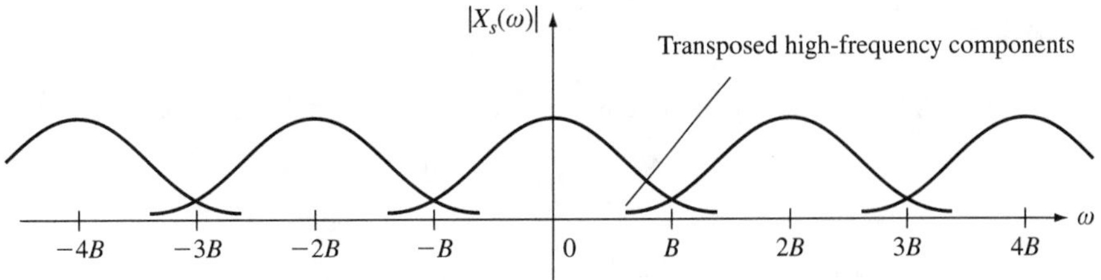

Figure 5.28 Amplitude spectrum of a sampled signal.

Aliasing will result in a distorted version of the original signal $x(t)$. It can be eliminated (theoretically) by first lowpass filtering $x(t)$ before $x(t)$ is sampled: If $x(t)$ is lowpass filtered so that all frequency components with values greater than B are removed, there will be no overlap of frequency components in the spectrum $X_s(\omega)$ of the sampled signal $x_s(t)$ assuming that $x(t)$ is sampled at the Nyquist rate $\omega_s = 2B$.

In practice, aliasing cannot be eliminated completely since a lowpass filter that cuts off all frequency components above a certain frequency cannot be synthesized (i.e., built). However, the magnitude of the aliased components can be reduced if the signal $x(t)$ is lowpass filtered before sampling. This approach is feasible as long as lowpass filtering $x(t)$ does not remove the "information content" of the signal $x(t)$.

Example 5.8 *Filtered Speech*

Again suppose that $x(t)$ is a speech waveform. Although a speech waveform may contain sizable frequency components above 4 kHz, voice recognition is possible for speech signals that have been filtered to a 4-kHz bandwidth. If B is chosen to be 4 kHz for filtered speech, the resulting Nyquist sampling frequency is

$$\omega_s = 2(2\pi)(4 \times 10^3) = 16\pi \times 10^3 \text{ rad/sec}$$

For this sampling frequency, the sampling interval T is

$$T = \frac{2\pi}{\omega_s} = 0.125 \text{ ms}$$

This is a much longer sampling interval than the 50-μs sampling interval required to transmit a 10-kHz bandwidth of speech. In general, the wider the bandwidth of a signal, the more expensive it is to transmit the signal. So it is much "cheaper" to send filtered speech over phone lines.

In many applications, the signal $x(t)$ cannot be lowpass filtered without removing information contained in $x(t)$. In such cases, the bandwidth B of the signal must be taken to be sufficiently large so that the aliased components do not seriously distort the reconstructed signal. Equivalently, for a given value of B, sampling can be performed at a rate higher than the Nyquist rate. For example, in applications to sampled-data control, the sampling frequency may be as large as 10 or 20 times B, where B is the bandwidth of the system being controlled.

PROBLEMS

5.1. A linear time-invariant continuous-time system has the frequency response function

$$H(\omega) = \begin{cases} 1, & 2 \le |\omega| \le 7 \\ 0, & \text{all other } \omega \end{cases}$$

Compute the output response $y(t)$ resulting from the input $x(t)$ given by

(a) $x(t) = 2 + 3\cos(3t) - 5\sin(6t - 30^\circ) + 4\cos(13t - 20^\circ), \quad -\infty < t < \infty$

(b) $x(t) = 1 + \sum_{k=1}^{\infty} \frac{1}{k}\cos(2kt), \quad -\infty < t < \infty$

(c) $x(t)$ as shown in Figure P5.1

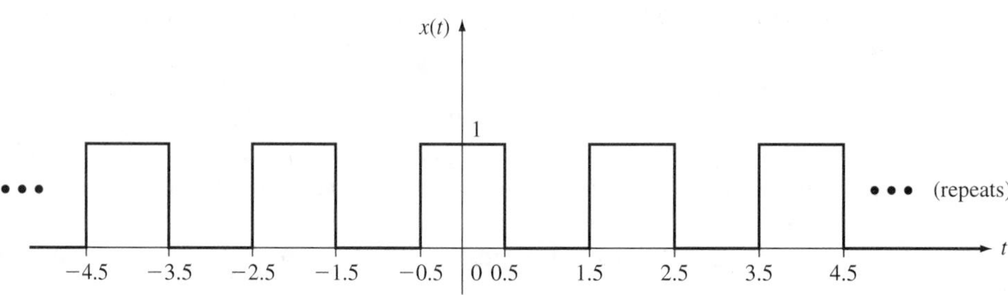

Figure P5.1

5.2. A linear time-invariant continuous-time system has the frequency response function

$$H(\omega) = \begin{cases} 2\exp(-|6 - |\omega||)\exp(-j3\omega), & 4 \le |\omega| \le 12 \\ 0, & \text{all other } \omega \end{cases}$$

(a) Plot the magnitude and phase functions for $H(\omega)$.

(b) Compute and plot the output response $y(t)$ resulting from the input $x(t)$ defined in Figure P5.1.

(c) Plot the amplitude and phase spectra of $x(t)$ and $y(t)$ for $k = 0, \pm1, \pm2, \pm3, \pm4, \pm5, \pm6$.

5.3. A linear time-invariant continuous-time system has the frequency response function

$$H(\omega) = \frac{1}{j\omega + 1}$$

Compute the output response $y(t)$ for $-\infty < t < \infty$ when the input $x(t)$ is

(a) $x(t) = \cos t,\ -\infty < t < \infty$

(b) $x(t) = \cos(t + 45°),\ -\infty < t < \infty$

5.4. The frequency response of a system is given below.

$$H(\omega) = \frac{10}{j\omega + 10}.$$

(a) Give the output to $x(t) = 2 + 2\cos(50t + \pi/2)$.

(b) Sketch $|H(\omega)|$. What is the bandwidth of the filter?

(c) Sketch the response of the filter to an input of $x(t) = 2e^{-2t}\cos(4t)u(t) + e^{-2t}\cos(20t)u(t)$. (See Figure P5.4.)

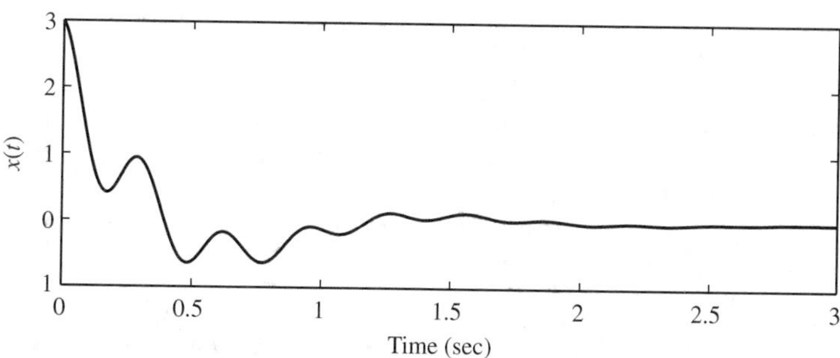

Figure P5.4

5.5. Repeat Problem 5.4, for the system given by

$$H(\omega) = \frac{40}{j\omega + 40}.$$

5.6. A linear time-invariant continuous-time system receives the periodic signal $x(t)$ as shown in Figure P5.6. The frequency response function is given by

$$H(\omega) = \frac{j\omega}{j\omega + 2}$$

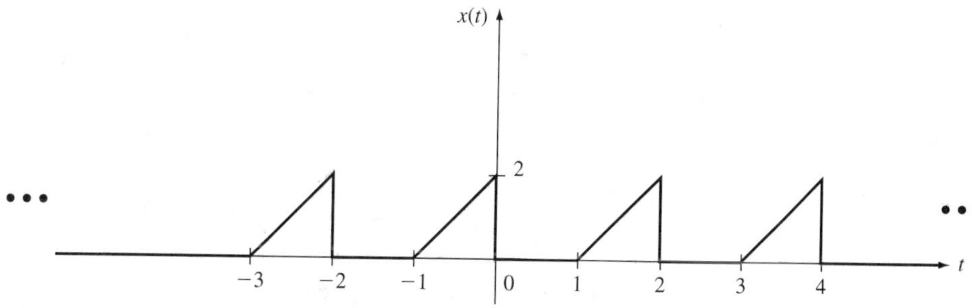

Figure P5.6

(a) Plot the amplitude and phase functions for $H(\omega)$.

(b) Compute the complex exponential Fourier series of the output response $y(t)$, and then sketch the amplitude and phase spectra for $k = 0, \pm1, \pm2, \pm3, \pm4, \pm5$ for both $x(t)$ and $y(t)$.

(c) Plot an approximation for $y(t)$ using the truncated complex exponential Fourier series from $k = -5$ to $k = 5$.

5.7. A periodic signal $x(t)$ with period T has the constant component $c_o^x = 2$. The signal $x(t)$ is applied to a linear time-invariant continuous-time system with frequency response function

$$H(\omega) = \begin{cases} 10e^{-j5\omega}, & \omega > \dfrac{\pi}{T}, \quad \omega < -\dfrac{\pi}{T} \\ 0, & \text{all other } \omega \end{cases}$$

(a) Show that the resulting output response $y(t)$ can be expressed in the form

$$y(t) = ax(t - b) + c$$

Compute the constants a, b, and c.

(b) Compute and plot the response of this system to the input $x(t)$ shown in Figure P5.1

5.8. The voltage $x(t)$ shown in Figure P5.8b is applied to the RL circuit shown in Figure P5.8a.

(a) Find the value of L so that the peak of the largest ac component (harmonic) in the output response $y(t)$ is 1/30 of the dc component of the output.

(b) Plot an approximation for $y(t)$ using the truncated complex exponential Fourier series from $k = -3$ to $k = 3$.

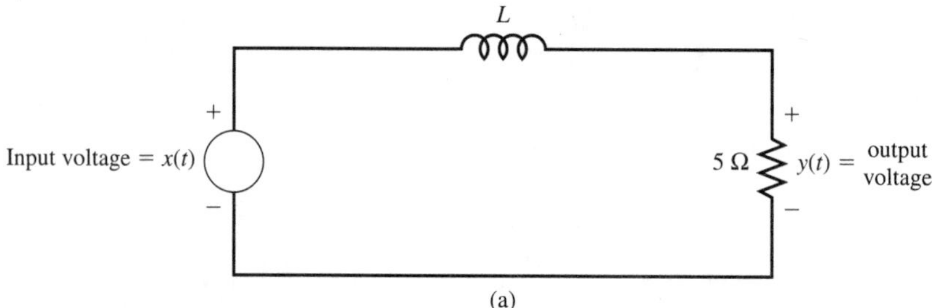

(a)

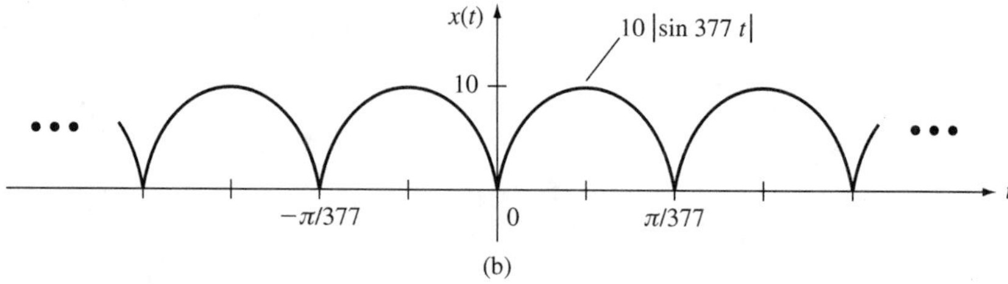

(b)

Figure P5.8

5.9. Consider the full-wave rectifier shown in Figure P5.9. The input voltage $v(t)$ is equal to $156\cos(120\pi t)$, $-\infty < t < \infty$. The voltage $x(t)$ is equal to $|v(t)|$.

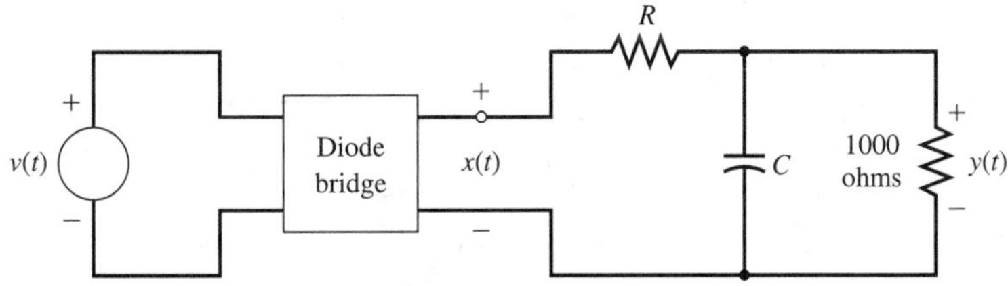

Figure P5.9

(a) Choose values for R and C such that the following two criteria are satisfied:
  1. The dc component of $y(t)$ is equal to 90% of the dc component of the input $x(t)$.
  2. The peak value of the largest harmonic in $y(t)$ is 1/30 of the dc component of $y(t)$.

(b) Plot an approximation for $y(t)$ using the truncated complex exponential Fourier series from $k = -3$ to $k = 3$.

5.10. The input

$$x(t) = 1.5 + \sum_{k=1}^{\infty}\left(\frac{1}{k\pi}\sin k\pi t + \frac{2}{k\pi}\cos k\pi t\right), \qquad -\infty < t < \infty$$

is applied to a linear time-invariant system with frequency function $H(\omega)$. This input produces the output response $y(t)$ shown in Figure P5.10. Compute $H(k\pi)$ for $k = 1,2,3,\ldots$.

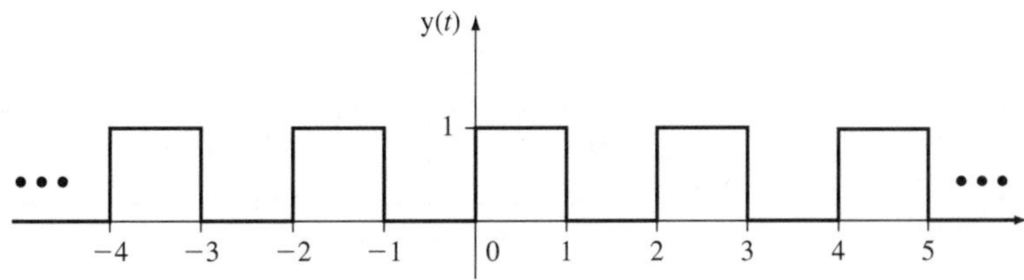

Figure P5.10

5.11. A linear time-invariant continuous-time system has frequency function $H(\omega)$ shown in Figure P5.11a. It is known that the system converts the sawtooth waveform in Figure P5.11b into the square waveform in Figure P5.11c; that is, the response to the sawtooth waveform is a square waveform. Compute the constants a and b in the plot of $H(\omega)$.

5.12. A linear time-invariant continuous-time system has the frequency function

$$H(\omega) = b - ae^{j\omega c}, \qquad -\infty < \omega < \infty$$

where a, b, and c are constants (real numbers). The input $x(t)$ shown in Figure 5.12a is applied to the system. Determine the constants a, b, and c so that the output response $y(t)$ resulting from $x(t)$ is given by the plot in Figure P5.12b.

5.13. A linear time-invariant continuous-time system has the frequency function $H(\omega)$. It is known that the input

$$x(t) = 1 + 4\cos 2\pi t + 8\sin(3\pi t - 90°)$$

produces the response

$$y(t) = 2 - 2\sin 2\pi t$$

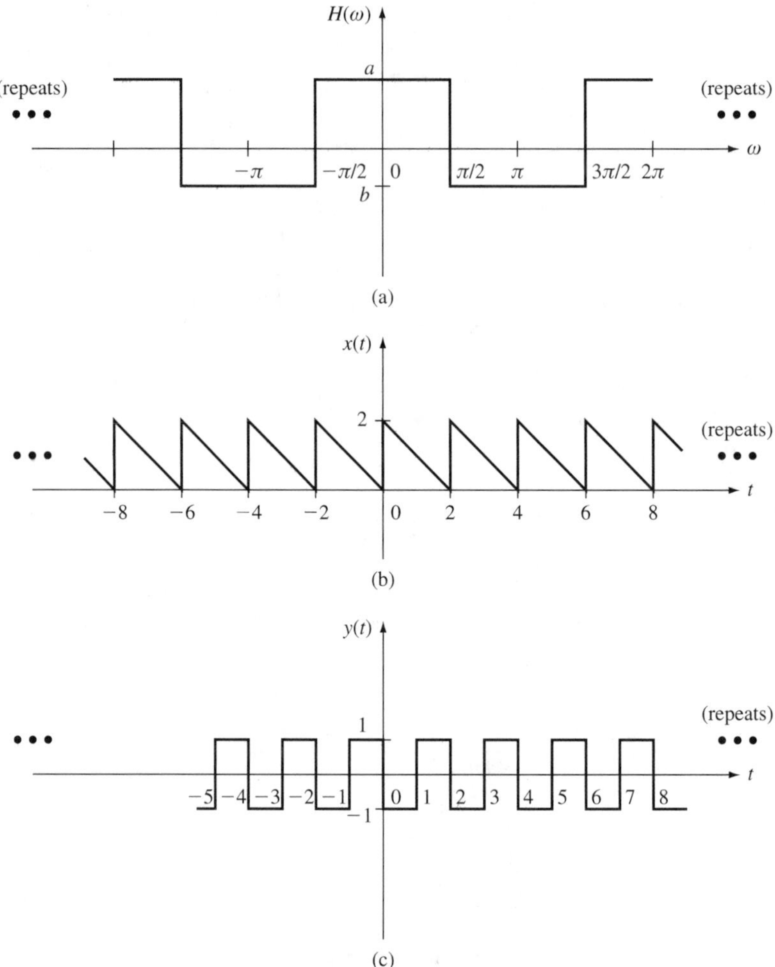

Figure P5.11

(a) For what values of ω is it possible to determine $H(\omega)$?

(b) Compute $H(\omega)$ for each of the values of ω determined in part (a).

5.14. An ideal linear-phase lowpass filter has the frequency response function

$$H(\omega) = \begin{cases} e^{-j\omega}, & -2 < \omega < 2 \\ 0, & \text{all other } \omega \end{cases}$$

Compute the filter's output response $y(t)$ for the different inputs $x(t)$ as given below. Plot each input $x(t)$ and the corresponding output $y(t)$. Also plot the magnitude and phase functions for $X(\omega)$, $H(\omega)$, and $Y(\omega)$.

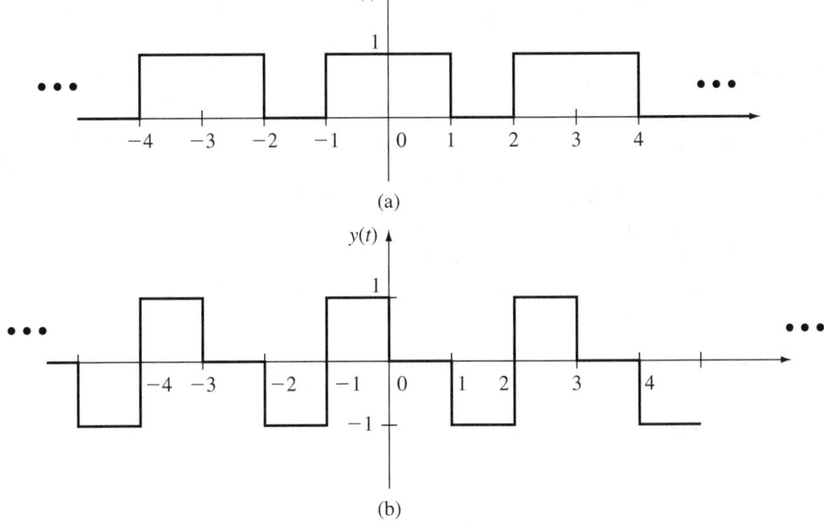

Figure P5.12

(a) $x(t) = 5\,\text{sinc}(3t/2\pi),\ -\infty < t < \infty$
(b) $x(t) = 5\,\text{sinc}(t/2\pi)\cos(2t),\ -\infty < t < \infty$
(c) $x(t) = \text{sinc}^2(t/2\pi),\ -\infty < t < \infty$
(d) $x(t) = \displaystyle\sum_{k=1}^{\infty} \frac{1}{k}\cos\left(\frac{k\pi}{2}t + 30^\circ\right), \qquad -\infty < t < \infty$

5.15. The triangular pulse shown in Figure P5.15 is applied to an ideal lowpass filter with frequency function $H(\omega) = p_{2B}(\omega)$. By using the Fourier transform approach and numerical integration, determine the filter output for the values of B given below. Express your results by plotting the output responses for $-1.5 \le t \le 1.5$. What do you conclude? You

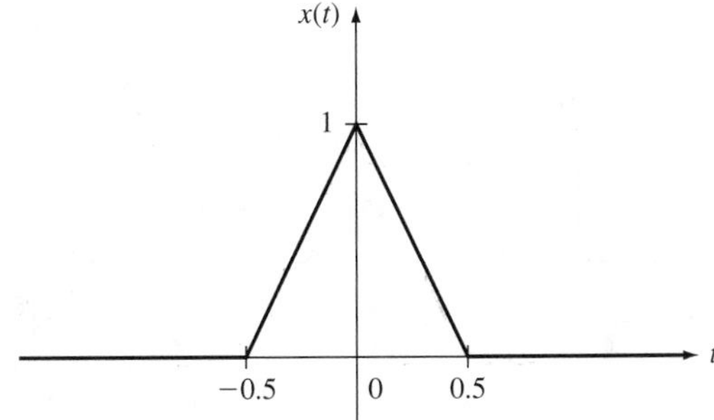

Figure P5.15

may wish to use the MATLAB M-file `quad` which can perform the integration for each value of t in the inverse Fourier transform.

(a) $B = 2\pi$
(b) $B = 4\pi$
(c) $B = 8\pi$

5.16. A lowpass filter has the frequency response function shown in Figure P5.16.
(a) Compute the impulse response $h(t)$ of the filter.
(b) Compute the response $y(t)$ when the input is $x(t) = \text{sinc}(t/2\pi)$, $-\infty < t < \infty$.
(c) Compute the response $y(t)$ when $x(t) = \text{sinc}(t/4\pi)$, $-\infty < t < \infty$.
(d) Compute the response $y(t)$ when $x(t) = \text{sinc}^2(t/2\pi)$, $-\infty < t < \infty$.
(e) For parts (b)–(d), plot $x(t)$ and the corresponding $y(t)$.

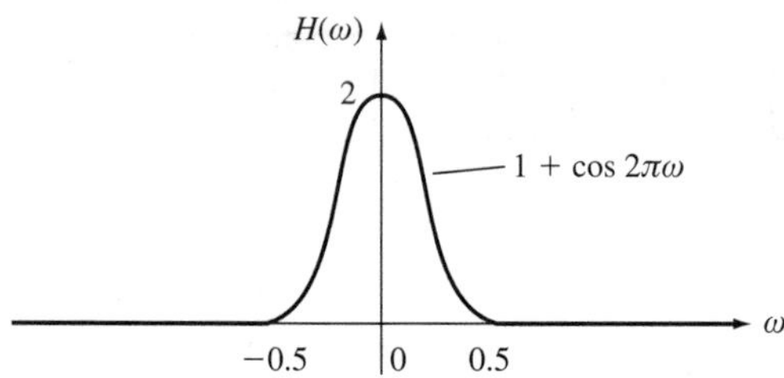

Figure P5.16

5.17. A lowpass filter has the frequency response curves shown in Figure P5.17.
(a) Compute the impulse response $h(t)$ of the filter.
(b) Compute the response $y(t)$ when $x(t) = 3\,\text{sinc}(t/\pi)\cos 4t$, $-\infty < t < \infty$.
(c) Plot $x(t)$ and $y(t)$.

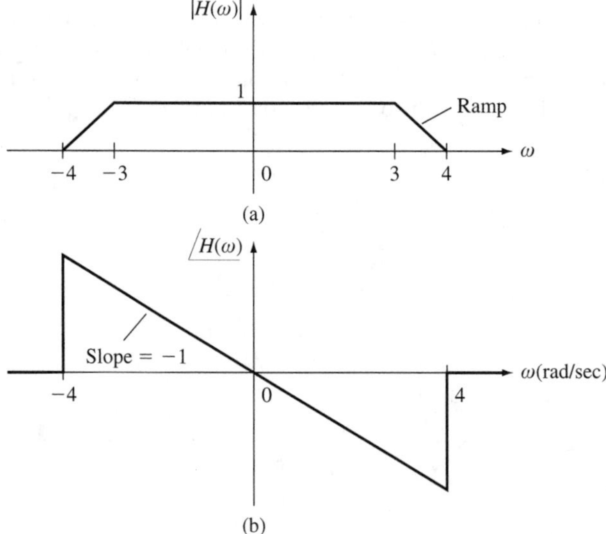

Figure P5.17

5.18. The input $x(t) = [\text{sinc}(t/\pi)](\cos 2t)$, $-\infty < t < \infty$, is applied to an ideal lowpass filter with frequency function $H(\omega) = 1$, $-a < \omega < a$, $H(\omega) = 0$ for all other ω. Determine the smallest possible value of a for which the resulting output response $y(t)$ is equal to the input $x(t) = [\text{sinc}(t/\pi)](\cos 2t)$.

5.19. An ideal linear-phase highpass filter has frequency response function

$$H(\omega) = \begin{cases} 6e^{-j2\omega}, & \omega > 3, \quad \omega < -3 \\ 0, & \text{all other } \omega \end{cases}$$

(a) Compute the impulse response $h(t)$ of the filter.
(b) Compute the output response $y(t)$ when the input $x(t)$ is given by $x(t) = \text{sinc}(5t/\pi)$, $-\infty < t < \infty$. Plot $x(t)$ and $y(t)$.
(c) Compute the output response $y(t)$ when the input $x(t)$ is the periodic signal shown in Figure P5.19. Plot $y(t)$.

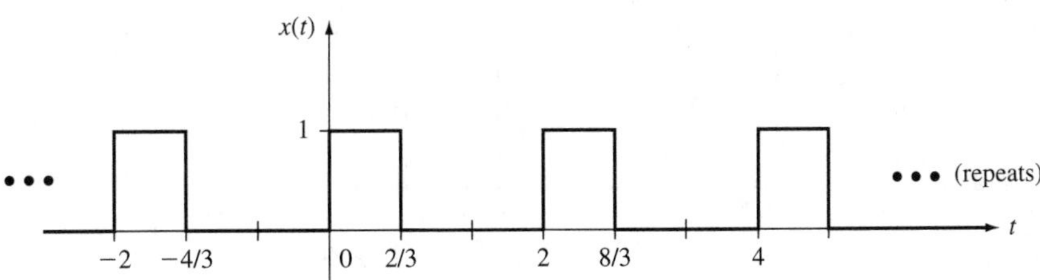

Figure P5.19

5.20. Given an input of

$$x(t) = 4 + 2\cos(10t + \pi/4) + 3\cos(30t - \pi/2),$$

find the output $y(t)$ to each of the following filters.

(a)

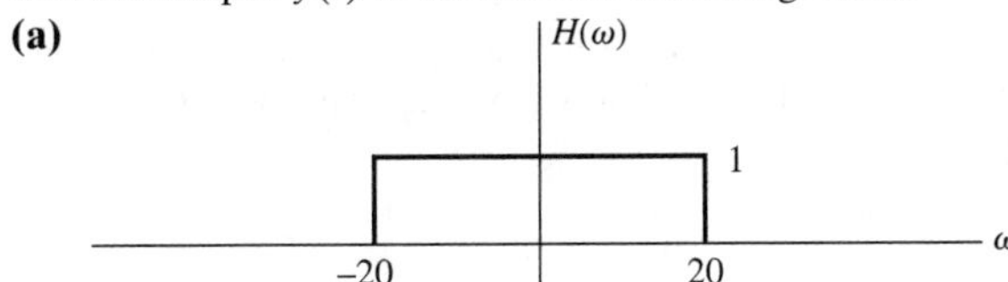

(b)

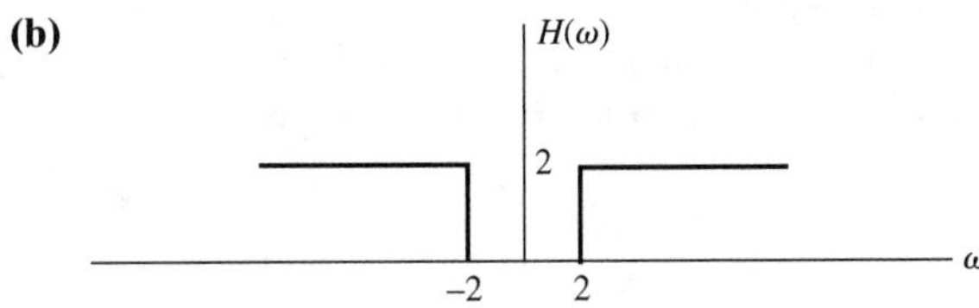

(c) $H(\omega) = \text{sin c}(\omega/20)$
(d) $H(\omega)$ as given in Figure P5.20.

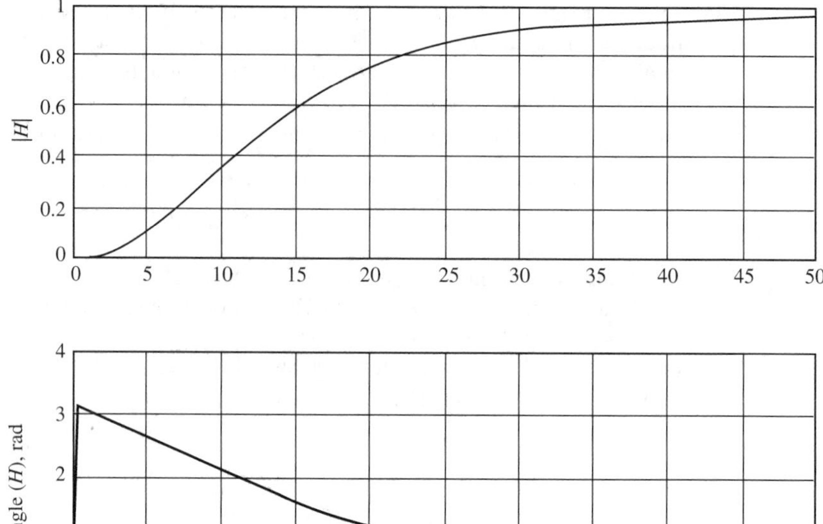

Figure P5.20

5.21. Design a filter to give a response of $y(t) = 6\cos(30t)$ for the input given in Problem 5.20.

5.22. The input

$$x(t) = \text{sinc}\left(\frac{t}{2\pi}\right)(\cos 3t)^2 + \text{sinc}\left(\frac{t}{2\pi}\right)\cos t, \qquad -\infty < t < \infty$$

is applied to a linear time-invariant continuous-time system with the frequency response function $H(\omega)$. Determine $H(\omega)$ so that the output response $y(t)$ resulting from this input is given by

$$y(t) = \text{sinc}\left(\frac{t}{2\pi}\right)$$

(a) Express your answer by giving $H(\omega)$ in analytical form.

(b) Plot $x(t)$ and $y(t)$ for $-30 < t < 30$ to see the filtering effect of $H(\omega)$ in the time domain. [To get sufficient resolution on $x(t)$, use a time increment of 0.1 second.]

5.23. An ideal linear-phase bandpass filter has frequency response

$$H(\omega) = \begin{cases} 10e^{-j4\omega}, & -4 < \omega < -2, \quad 2 < \omega < 4 \\ 0, & \text{all other } \omega \end{cases}$$

Compute the output response $y(t)$ of the filter when the input $x(t)$ is

(a) $x(t) = \text{sinc}(2t/\pi)$, $-\infty < t < \infty$

(b) $x(t) = \text{sinc}(3t/\pi)$, $-\infty < t < \infty$

(c) $x(t) = \text{sinc}(4t/\pi), -\infty < t < \infty$
(d) $x(t) = \text{sinc}(2t/\pi)\cos t, -\infty < t < \infty$
(e) $x(t) = \text{sinc}(2t/\pi)\cos 3t, -\infty < t < \infty$
(f) $x(t) = \text{sinc}(2t/\pi)\cos 6t, -\infty < t < \infty$
(g) $x(t) = \text{sinc}^2(t/\pi)\cos 2t, -\infty < t < \infty$

Plot $x(t)$ and the corresponding output $y(t)$ for each of the cases computed above. Use a small enough time increment on the plot to capture the high-frequency content of the signal.

5.24. A linear time-invariant continuous-time system has the frequency response function $H(\omega) = p_2(\omega + 4) + p_2(\omega - 4)$. Compute the output response for the inputs given below.
(a) $x(t) = \delta(t)$
(b) $x(t) = \cos t \sin \pi t, -\infty < t < \infty$
(c) $x(t) = \text{sinc}(4t/\pi), -\infty < t < \infty$
(d) $x(t) = \text{sinc}(4t/\pi)\cos 3t, -\infty < t < \infty$

Plot $x(t)$ and the corresponding output $y(t)$ for each of the cases computed above. Use a small enough time increment on the plot to capture the high-frequency content of the signal.

5.25. A periodic signal $x(t)$ with period $T = 2$ has the Fourier coefficients

$$c_k = \begin{cases} 0, & k = 0 \\ 0 & \text{if } k \text{ is even} \\ 1 & \text{if } k \text{ is odd} \end{cases}$$

The signal $x(t)$ is applied to a linear time-invariant continuous-time system with the magnitude and phase curves shown in Figure P5.25. Determine the system output.

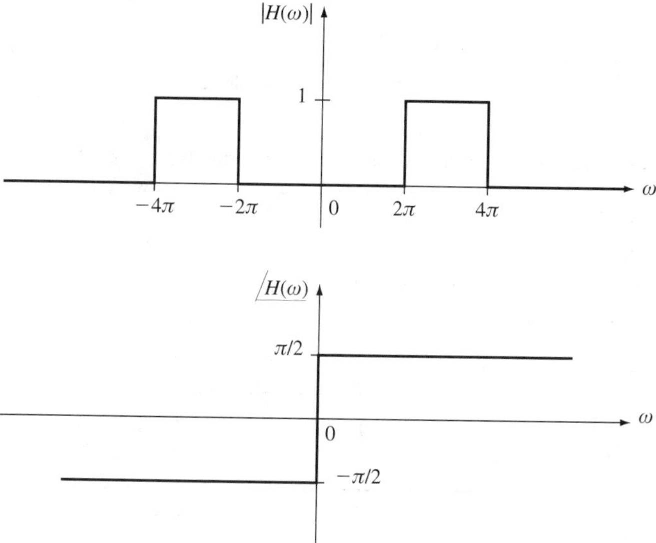

Figure P5.25

5.26. A linear time-invariant continuous-time system has frequency function $H(\omega) = 5\cos 2\omega$, $-\infty < \omega < \infty$.

(a) Sketch the system's magnitude function $|H(\omega)|$ and phase function $\angle H(\omega)$.

(b) Compute the system's impulse response $h(t)$.

(c) Derive an expression for the output response $y(t)$ resulting from an arbitrary input $x(t)$ with the system at rest prior to the application of $x(t)$.

5.27. A *Hilbert transformer* is a linear time-invariant continuous-time system with impulse response $h(t) = 1/t$, $-\infty < t < \infty$. Using the Fourier transform approach, determine the output response resulting from the input $x(t) = A\cos\omega_0 t$, $-\infty < t < \infty$, where ω_0 is an arbitrary strictly positive real number.

5.28. A linear time-invariant continuous-time system has frequency response function $H(\omega) = j\omega e^{-j\omega}$. The input $x(t) = \cos(\pi t/2)p_2(t)$ is applied to the system for $-\infty < t < \infty$.

(a) Determine the input spectrum $X(\omega)$ and the corresponding output spectrum $Y(\omega)$.

(b) Compute the output $y(t)$.

5.29. Consider the system in Figure P5.29, where $p(t)$ is an impulse train with period T and $H(\omega) = Tp_2(\omega)$. Compute $y(t)$ when,

(a) $x(t) = \text{sinc}^2(t/2\pi)$ for $-\infty < t < \infty$, $T = \pi$

(b) $x(t) = \text{sinc}^2(t/2\pi)$ for $-\infty < t < \infty$, $T = 2\pi$

(c) For each case above, compare the plots of $x(t)$ and the corresponding $y(t)$.

(d) Repeat part (a) using the interpolation formula to solve for $y(t)$ and plot your results with n ranging from $n = -5$ to $n = 5$.

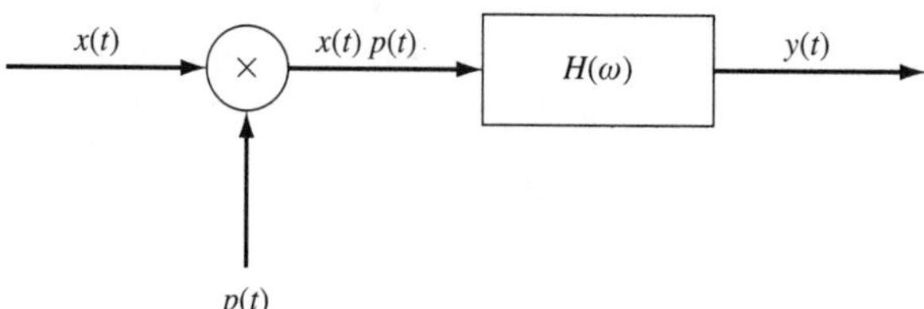

Figure P5.29

5.30. Consider the signal whose Fourier transform is shown in Figure P5.30. Let $x_s(t) = x(t)p(t)$ represent the sampled signal. Draw $|X_s(\omega)|$ for the following cases.

(a) $T = \pi/15$

(b) $T = 2\pi/15$

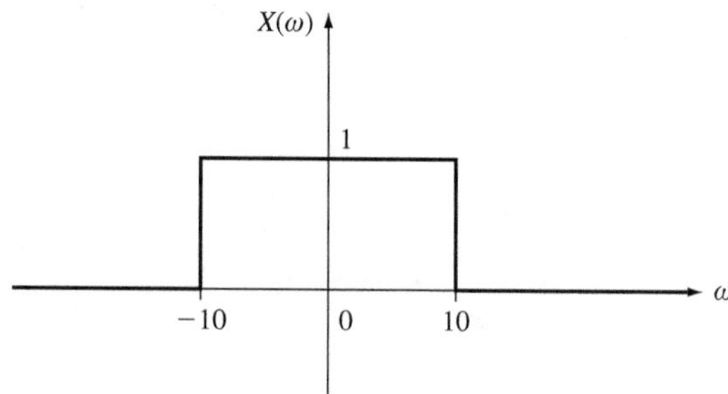

Figure P5.30

5.31. Repeat Problem 5.30 for the signal whose transform is shown in Figure P5.31.

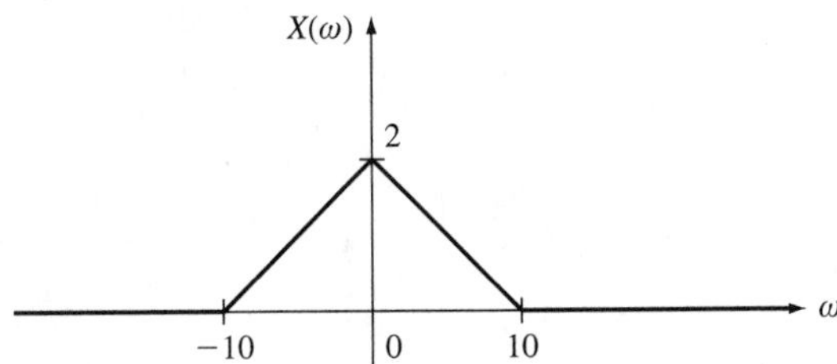

Figure P5.31

5.32. Consider the signal with the amplitude spectrum shown in Figure P5.32. Let $x_s(t) = x(t)p(t)$ represent the sampled signal. Draw $|X_s(\omega)|$ for the following cases.

(a) $T = \pi/4$ sec
(b) $T = \pi/2$ sec
(c) $T = 2\pi/3$ sec

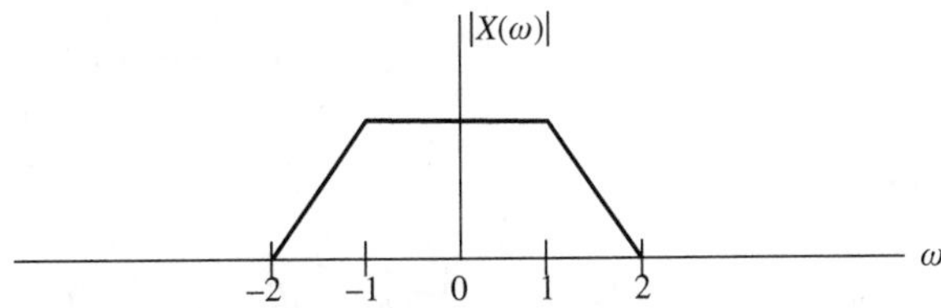

Figure P5.32

5.33. Repeat Problem 5.32 where now the signal is $x(t) = e^{-t/4}\cos(t)u(t)$. You can either sketch the plots by hand or use MATLAB for a more accurate plot. In order to examine the effects of aliasing in the time domain, plot $x(t)$ for each of the sampling times for $t = 0$ through 15 sec. In MATLAB, this is done by defining your time vector with the time increment set to the desired sampling period. MATLAB then "reconstructs" the signal by connecting the sampled points with straight lines. (This procedure is known as a linear interpolation.) Compare your sampled/reconstructed signals with a signal that is more accurate, one that is created by using a very small sampling period (such as $T = 0.05$ sec) by plotting them on the same graph.

5.34. Consider the following sampling and reconstruction configuration:

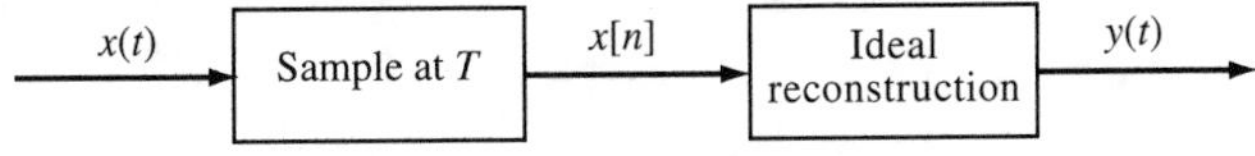

The output $y(t)$ of the ideal reconstruction can be found by sending the sampled signal $x_s(t) = x(t)p(t)$ through an ideal lowpass filter with the frequency response function

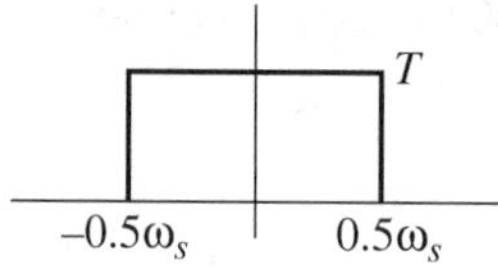

Let $x(t) = 2 + \cos(50\pi t)$ and $T = 0.01$ sec

(a) Draw $|X_s(\omega)|$, where $x_s(t) = x(t)p(t)$. Determine if aliasing occurs.

(b) Determine the expression for $y(t)$.

(c) Determine an expression for $x[n]$.

5.35. Repeat Problem 5.34 for $x(t) = 2 + \cos(50\pi t)$ and $T = 0.025$ sec.

5.36. Repeat Problem 5.34 for $x(t) = 1 + \cos(20\pi t) + \cos(60\pi t)$ and $T = 0.01$ sec

5.37. Consider the following sampling and reconstruction configuration:

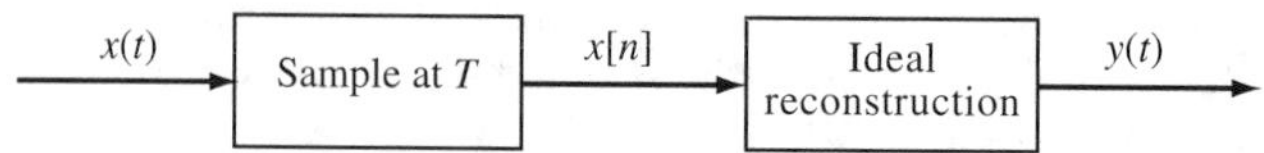

The output $y(t)$ of the ideal reconstruction can be found by sending the sampled signal $x_s(t) = x(t)p(t)$ through an ideal lowpass filter with the frequency response function

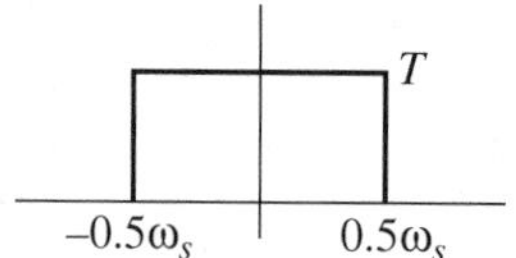

(a) Let $x(t) = 1 + \cos(15\pi t)$ and $T = 0.1$ sec. Draw $|X_s(\omega)|$, where $x_s(t) = x(t)p(t)$. Determine the expression for $y(t)$.

(b) Let $X(\omega) = 1/(j\omega + 1)$ and $T = 1$ sec. Draw $|X_s(\omega)|$, where $x_s(t) = x(t)p(t)$. Does aliasing occur? (Justify your answer.)

◆

CHAPTER 6

Application to Communications

To further illustrate the use of the Fourier transform, in this chapter we consider the transmission of information (in the form of a signal generated by a source) over a channel and the reception of the information by a user. The channel may consist of free space, twisted wire, coaxial cable, fiber optic cable, etc. A key component of the transmission process is the use of modulation to convert the signal source into an appropriate form for transmission over the channel. In the modulation process, some parameter of a carrier signal is varied based on the signal that is being transmitted. There are two basic types of modulation: analog and digital. In analog modulation, the parameter being varied can take on a continuous range of values, whereas in digital modulation, the parameter takes on only a finite number of different possible values. After transmission over a channel, the transmitted signal is reconstructed by a receiver that uses a demodulation process to extract the original signal. Analog modulation and demodulation are considered first in Sections 6.1 and 6.2. Then in Section 6.3, two techniques (FDM and TDM) are presented for the simultaneous transmission of several signals. An introduction to digital modulation is given in Section 6.4, and then in Sections 6.5 and 6.6, the focus is on popular types of digital communication processes employing baseband or passband pulse amplitude modulation (PAM). Digital PAM is commonly used in a variety of applications such as compact discs, digital cellular phones, digital phone lines, and computer modems. In the last section of the chapter, a simulation of the digital communication process is given using MATLAB.

6.1 ANALOG MODULATION

Let $x(t)$ be a continuous-time signal, such as an audio signal that is to be transmitted over a channel consisting of free space or a cable. As noted above, the signal is transmitted by modulating a carrier. The most common type of carrier is a sinusoid given by $A\cos\omega_c t$, where A is the amplitude and ω_c is the frequency in rad/sec. In *amplitude modulation* (AM), the amplitude of the sinusoidal carrier is modulated by the signal $x(t)$. In one form of AM transmission, the signal $x(t)$ and carrier $A\cos\omega_c t$ are simply multiplied together to produce the modulated carrier $s(t) = Ax(t)\cos\omega_c t$. The process is illustrated in Figure 6.1. The *local oscillator* in Figure 6.1 is a device that produces the sinusoidal signal $A\cos\omega_c t$. The signal multiplier may be realized by using a nonlinear element, such as a diode.

Example 6.1 *Amplitude Modulation*

Suppose that $x(t)$ is the signal shown in Figure 6.2a and that the carrier is equal to $\cos 5\pi t$. The modulated carrier $s(t) = x(t)\cos 5\pi t$ is plotted in Figure 6.2b.

The frequency spectrum of the modulated carrier $s(t) = Ax(t)\cos\omega_c t$ can be determined using the modulation property of the Fourier transform. First, we assume that the signal $x(t)$ is bandlimited with bandwidth B, that is,

$$|X(\omega)| = 0, \text{for all } \omega > B$$

where $X(\omega)$ is the Fourier transform of $x(t)$. It is also assumed that $\omega_c > B$; that is, the frequency ω_c of the carrier is greater than the bandwidth B of the signal. If $x(t)$ is an audio signal, such as a music waveform, the bandwidth B can be taken to be 20 kHz, since an audio signal is not likely to contain any significant frequency components above 20 kHz.

Now by the modulation property, the Fourier transform $S(\omega)$ of the modulated carrier $s(t) = Ax(t)\cos\omega_c t$ is given by

$$S(\omega) = \frac{A}{2}[X(\omega + \omega_c) + X(\omega - \omega_c)]$$

This result shows that the modulation process translates the Fourier transform $X(\omega)$ of $x(t)$ up to the frequency range from $\omega_c - B$ to $\omega_c + B$ (and to the negative

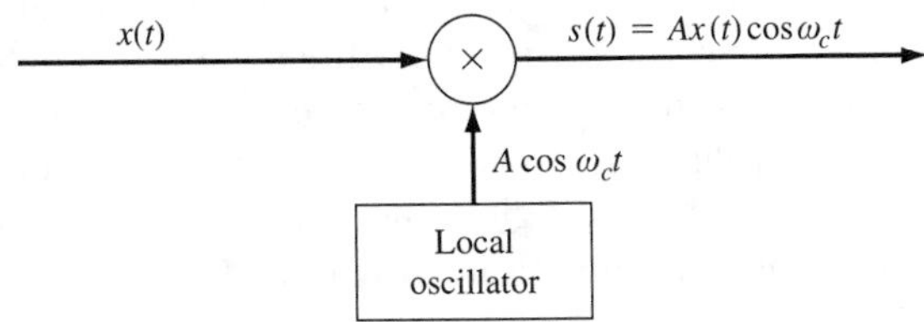

Figure 6.1 Amplitude modulation.

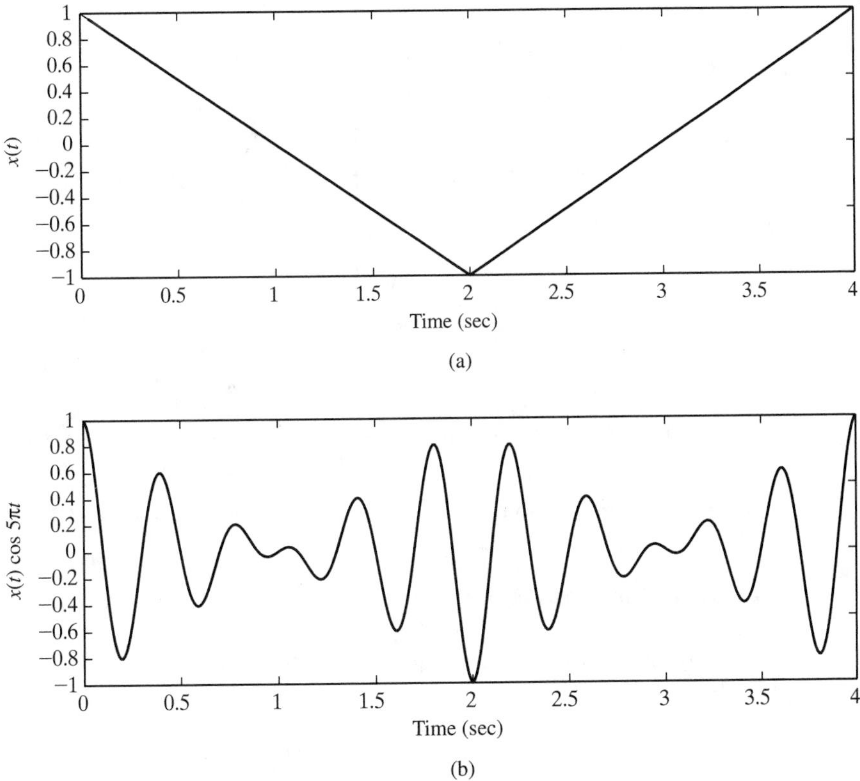

Figure 6.2 (a) Signal $x(t)$ and (b) modulated carrier $Ax(t)\cos\omega_c t$ in Example 6.1.

frequency range from $-\omega_c - B$ to $-\omega_c + B$). For example, if the transform $X(\omega)$ has the shape shown in Figure 6.3a, then the transform of the modulated carrier has the form shown in Figure 6.3b. As illustrated, the portion of $X(\omega - \omega_c)$ from $\omega_c - B$ to ω_c is called the lower *sideband*, and the portion of $X(\omega - \omega_c)$ from ω_c to $\omega_c + B$ is called the upper *sideband*. Each sideband contains all the spectral components of the signal $x(t)$. As a result, the signal $x(t)$ can be reconstructed (as shown in the next section) from either the upper or lower sideband.

A key property of amplitude modulation in the transmission of a signal $x(t)$ is the up conversion of the spectrum of $x(t)$. The higher frequency range of the modulated carrier makes it possible to achieve good propagation properties in transmission through cable or free space. For example, in optical communications, a beam of light is modulated with the result that the spectrum of the signal $x(t)$ is up converted to an optical frequency range. The up-converted signal is often referred to as the *passband signal*, since it consists of the frequencies in the up-converted spectrum. The source signal $x(t)$ is referred to as the *baseband signal*.

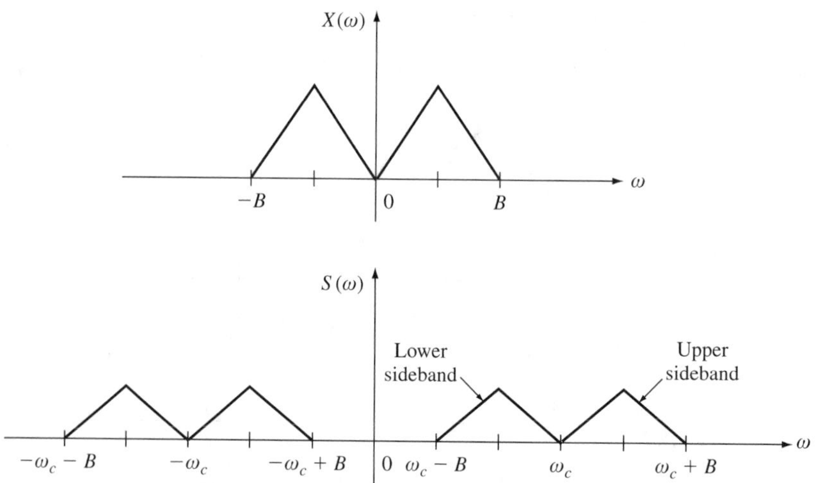

Figure 6.3 Fourier transform of (a) signal $x(t)$ and (b) modulated carrier $s(t) = Ax(t)\cos\omega_c t$.

Alternate Form of AM

In some types of AM transmission, such as AM radio, the modulated carrier $s(t)$ is given by

$$s(t) = A[1 + kx(t)]\cos\omega_c t \tag{6.1}$$

where k is a positive constant called the *amplitude sensitivity*, which is chosen so that $1 + kx(t) > 0$ for all t. This condition ensures that the envelope of the modulated carrier $s(t)$ is a replica of the signal $x(t)$. In this form of AM transmission, it is also assumed that the carrier frequency ω_c is much larger than the bandwidth B of $x(t)$.

Example 6.2 *Alternative Form of AM*

Again consider the signal in Figure 6.2a, and let the carrier be $\cos 5\pi t$. Then, with $k = 0.8$, the modulated signal $s(t) = [1 + kx(t)]\cos 5\pi t$ is shown in Figure 6.4. Note that the envelope is a replica of the signal $x(t)$, whereas such is not the case for the modulated signal in Figure 6.2b.

The frequency spectrum $S(\omega)$ of the modulated carrier $s(t) = A[1 + kx(t)]\cos\omega_c t$ can also be determined by taking the Fourier transform. This gives

$$S(\omega) = \pi A[\delta(\omega + \omega_c) + \delta(\omega - \omega_c)] + \frac{Ak}{2}[X(\omega + \omega_c) + X(\omega - \omega_c)]$$

Note the frequency components at $\omega = \pm\omega_c$, which are due to the presence of the carrier $A\cos\omega_c t$ in the modulated signal $A[1 + kx(t)]\cos\omega_c t$. Thus, in this alternative form of AM, the spectrum of the transmitted signal $s(t)$ contains the

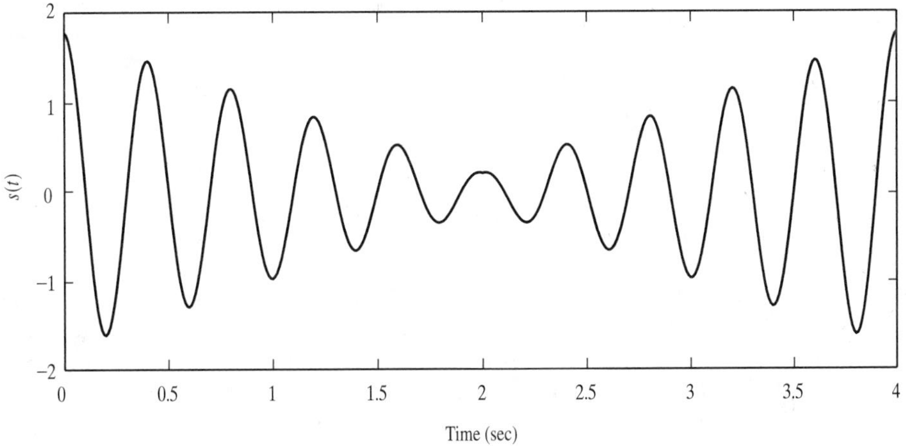

Figure 6.4 Modulated carrier in Example 6.2.

carrier and the upper and lower sidebands. In contrast, the spectrum of the modulated carrier $Ax(t)\cos\omega_c t$ contains only the upper and lower sidebands; the carrier is suppressed. Hence, when the modulated carrier is of the form $Ax(t)\cos\omega_c t$, it is referred to as *double-sideband–suppressed carrier* (DSB-SC) *transmission*. When the modulated carrier is of the form $s(t) = A[1 + kx(t)]\cos\omega_c t$, it is referred to as *double-sideband* (DSB) *transmission*. A major advantage of DSB-SC over DSB is that since DSB-SC does not require that the carrier be transmitted, it uses much less power than DSB to transmit the source signal $x(t)$. On the other hand, as will be seen in the next section, DSB signals can be demodulated using a simple envelope detector, whereas demodulation of DSB-SC signals requires synchronization between the transmitter and the receiver.

Single Sideband Transmission

Since both the upper and lower sidebands contain all the spectral components of $x(t)$, it is only necessary to transmit one of the sidebands. The resulting process is called *single-sideband* (SSB) *transmission*. SSB transmission can be realized by bandpass filtering the modulated carrier $Ax(t)\cos\omega_c t$. In particular, the upper sideband associated with $Ax(t)\cos\omega_c t$ can be generated by applying $Ax(t)\cos\omega_c t$ to a bandpass filter with passband from ω_c to $\omega_c + B$. The process is illustrated in Figure 6.5. The frequency function $H(\omega)$ of the bandpass filter in Figure 6.5 is given by

$$H(\omega) = \begin{cases} 1, & \omega_c \le |\omega| \le \omega_c + B \\ 0, & \text{all other } \omega \end{cases} \tag{6.1}$$

Note that the phase of the filter is equal to zero. A linear phase would result in a time delay that can be ignored since it does not cause signal distortion.

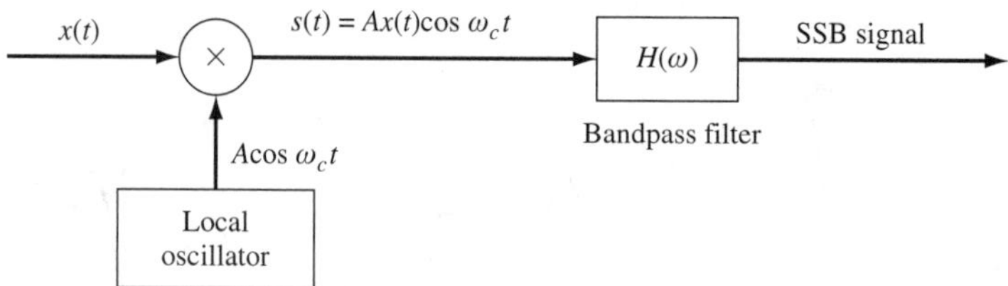

Figure 6.5 SSB Transmission.

In practice, the ideal bandpass filter characteristic given by (6.1) cannot be achieved, since there are no transition regions between the passband and the stopbands. An actual filter has a gradual transition between passbands and stopbands. (See Section 9.6.) However, if the baseband signal $x(t)$ has no spectral components below some suitably large value of frequency, the bandpass filter given by (6.1) can be modified to include a gradual transition from the stopband that blocks the lower sideband of $Ax(t)\cos\omega_c t$, and the passband that passes the upper sideband of $Ax(t)\cos\omega_c t$. To show this, suppose that $x(t)$ has the Fourier transform shown in Figure 6.6a and that the transform of the modulated carrier $s(t) = Ax(t)\cos\omega_c t$ is as shown in Figure 6.6. To generate the SSB signal, from Figure 6.6b it's clear that the passband of the bandpasss filter needs to extend from $\omega_c + A$ to $\omega_c + B$ and that the lower stopband of the filter needs to go up to $\omega_c - A$. Thus, if A is suitably large, a gradual transition from the lower stopband to the passband can be fitted into the frequency range from $\omega_c - B$ to $\omega_c + A$. The transition region from the passband to the upper stopband can be as wide as desired. A sketch of the filter's frequency response function $H(\omega)$ for positive ω is given in Figure 6.7.

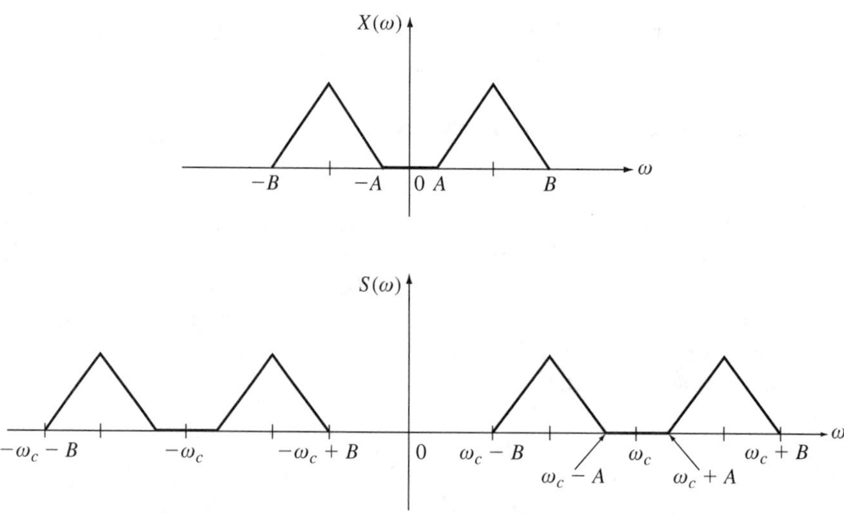

Figure 6.6 Fourier transform of (a) signal $x(t)$ and (b) modulated carrier $s(t) = Ax(t)\cos\omega_c t$.

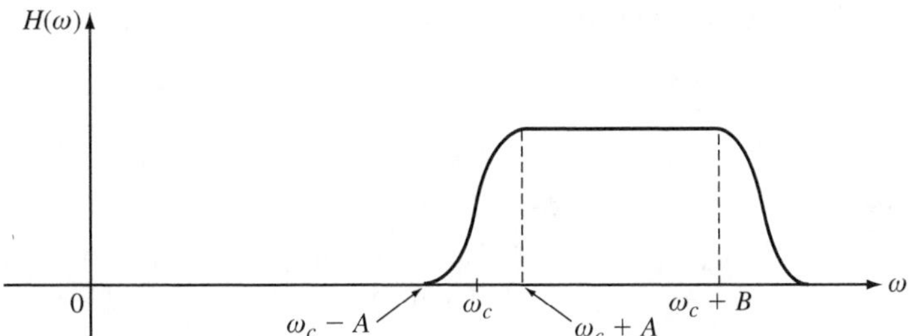

Figure 6.7 Frequency response function of bandpass filter.

SSB transmission with the bandpass filter characteristic shown in Figure 6.7 is feasible for voice signals where A can be taken to be around 300 Hz; that is, speech does not contain significant spectral components with frequencies below 300 Hz. Hence, for voice transmission, the transition region from the lower stopband to the passband of the bandpass filter in SSB transmission is around 600 Hz. However, there exist signals, such as TV signals, that have significant low-frequency content. In such cases, bandpass filtering the upper sideband will result in a small remnant of the lower sideband (called a *vestigal sideband*). This type of transmission is referred to as *vestigal sideband* (VSB) *modulation*. The frequency response function $H(\omega)$ of the bandpass filter in VSB transmission must satisfy certain conditions to insure that the baseband signal can be reconstructed from the modulated carrier. We leave this to a more advanced treatment of communication systems.

Angle Modulation

In addition to amplitude modulation, a signal $x(t)$ can be "put on" a sinusoidal carrier by modulating the angle of the carrier. In this form of transmission, called *angle modulation*, the modulated carrier is given by $s(t) = A\cos[\theta(t)]$, where the angle $\theta(t)$ is a function of the baseband signal $x(t)$. There are two basic types of angle modulation: *phase modulation* and *frequency modulation*. In phase modulation (PM), the angle is given by

$$\theta(t) = \omega_c t + k_p x(t)$$

where k_p is the *phase sensitivity* of the modulator. In frequency modulation (FM), the angle is given by

$$\theta(t) = \omega_c t + 2\pi k_f \int_0^t x(\tau)\, d\tau$$

where k_f is the *frequency sensitivity* of the modulator. Thus, the modulated carrier in PM transmission is equal to

$$s(t) = A\cos[\omega_c t + k_p x(t)]$$

and the modulated carrier in FM transmission is equal to

$$s(t) = A\cos\left[\omega_c t + 2\pi k_f \int_0^t x(\tau)\, d\tau\right] \tag{6.2}$$

Note that if $x(t)$ is the sinusoid $x(t) = \alpha \cos \omega_x t$, the FM signal (6.2) becomes

$$s(t) = A \cos\left[\omega_c t + \frac{2\pi k_f \alpha}{\omega_x} \sin\omega_x t\right]$$

Example 6.3 *PM and FM Modulation*

Suppose that $x(t) = \cos \pi t$, which is plotted in Figure 6.8a. Then, with $\omega_c = 10\pi$, $A = 1$, $k_p = 5$, and $k_f = 5/2$, the PM and FM signals are plotted in Figure 6.8b and 6.8c.

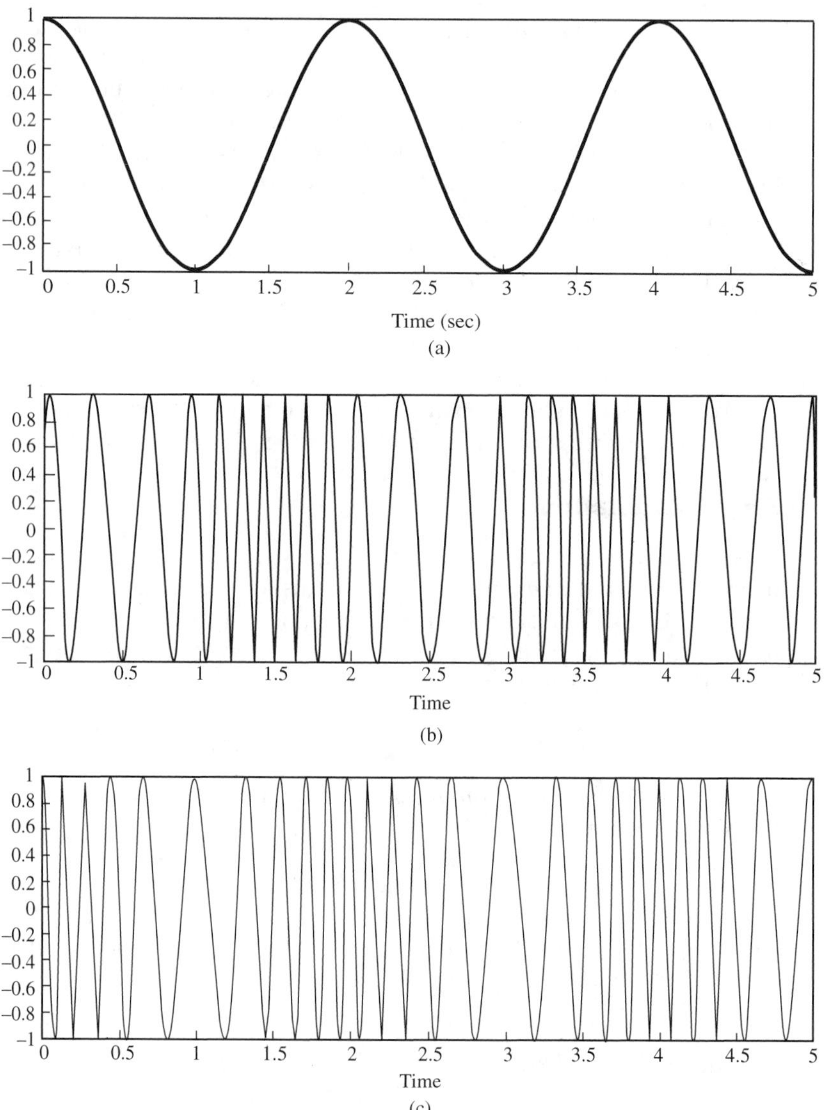

Figure 6.8 (a) Signal (a), (b) PM signal (b), and (c) FM signal (c) in Example 6.3.

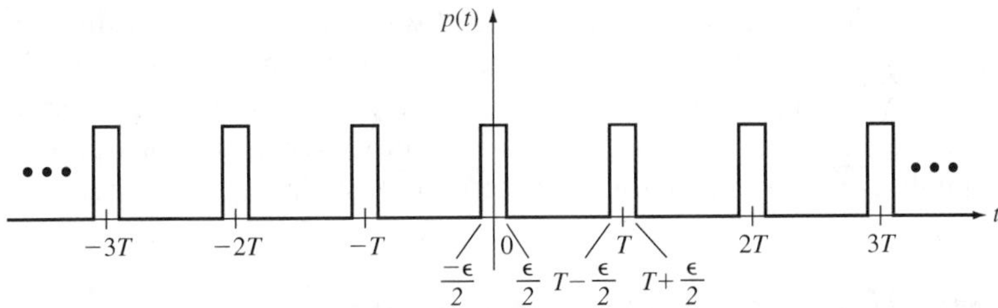

Figure 6.9 Pulse train with period T.

Pulse-Amplitude Modulation

Instead of modulating a sinusoid, information in the form of a signal $x(t)$ can be transmitted by modulating other types of waveforms, such as the pulse train $p(t)$ shown in Figure 6.9. The amplitude of $p(t)$ can be modulated by multiplying $x(t)$ and $p(t)$ together as illustrated in Figure 6.10. This process is called *pulse-amplitude modulation* (PAM).

Example 6.4 *PAM*

Consider the signal displayed in Figure 6.11a. With $T = 0.2$ and $\varepsilon \ll 0.2$, the PAM signal is shown in Figure 6.11b.

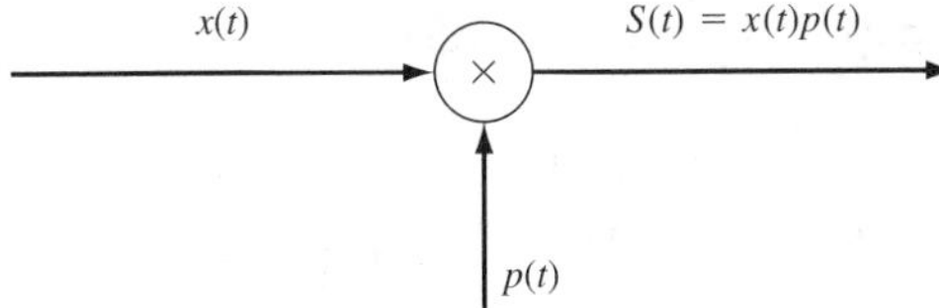

Figure 6.10 Pulse-amplitude Modulation.

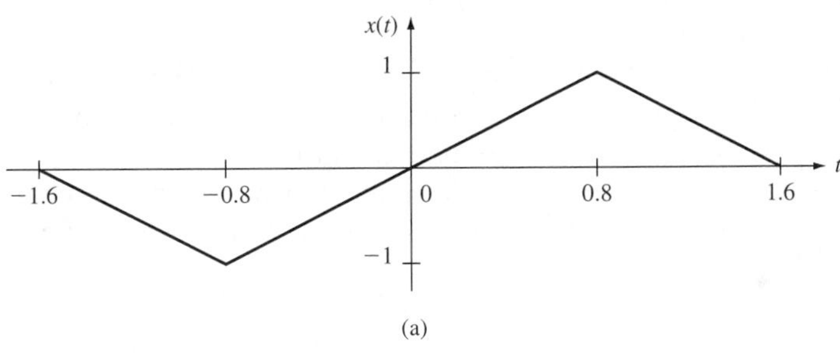

(a)

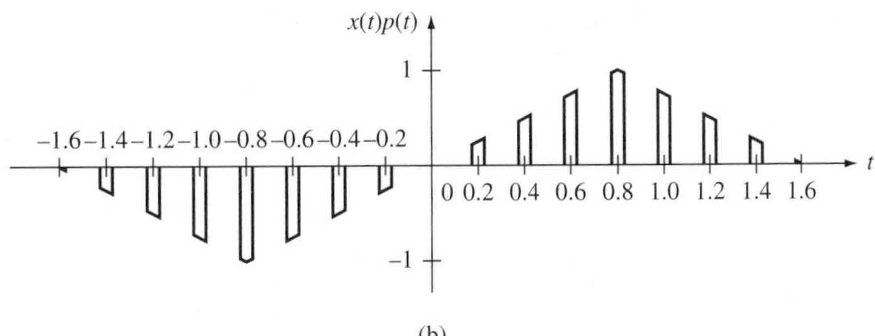

(b)

Figure 6.11 (a) Signal and (b) PAM signal.

A PAM signal can be generated by applying $x(t)$ to a switch that is closed ε seconds every T seconds. In the limit as $\varepsilon \to 0$, the modulated signal $s(t) = x(t)p(t)$ is actually a sampled version of $x(t)$, where T is the sampling interval. Thus, sampling is closely related to PAM. In fact, the Fourier transform of a PAM signal is approximately equal to that of the idealized sampled signal considered in Section 5.5.

6.2 DEMODULATION OF ANALOG SIGNALS

Demodulation is the process of reconstructing the baseband signal $x(t)$ from the modulated carrier (the passband signal). In the case of DSB-SC transmission, $x(t)$ can be reconstructed from the modulated carrier $s(t) = Ax(t)\cos\omega_c t$ by first applying $s(t)$ to the signal multiplier shown in Figure 6.12. Note that the local oscillator in Figure 6.12 is synchronized with the carrier signal $\cos\omega_c t$; that is, there is no phase shift between the carrier and the signal generated by the local oscillator. The type of demodulator under consideration here requires synchronization. (See Problem 6.4.)

Now the Fourier transform of the output $Ax(t)\cos^2\omega_c t$ of the multiplier in Figure 6.12 can be computed using the trigonometric identity

$$x(t)\cos^2\omega_c t = \frac{1}{2}(1 + \cos 2\omega_c t)x(t)$$

and the modulation property of the Fourier transform. This yields the result that the Fourier transform of $Ax(t)\cos^2\omega_c t$ is given by

$$\frac{1}{2}X(\omega) + \frac{1}{4}[X(\omega + 2\omega_c) + X(\omega - 2\omega_c)]$$

where $X(\omega)$ is the Fourier transform of $x(t)$. For example, if $X(\omega)$ has the shape shown in Figure 6.13a, the Fourier transform of $Ax(t)\cos^2\omega_c t$ has the form shown in Figure 6.13b. From Figure 6.13 it can be seen that $x(t)$ can be extracted from by applying $Ax(t)\cos^2\omega_c t$ to a lowpass filter with gain 2 and bandwidth B. In particular, if the frequency response function $G(\omega)$ of the lowpass filter is taken to be

$$G(\omega) = \begin{cases} 2, & -B \le \omega \le B \\ 0, & \text{all other } \omega \end{cases}$$

the output of the filter will be equal to $x(t)$.

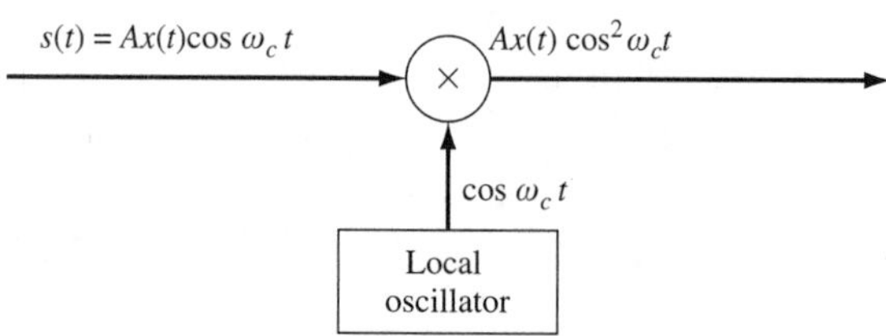

Figure 6.12 First stage of demodulation.

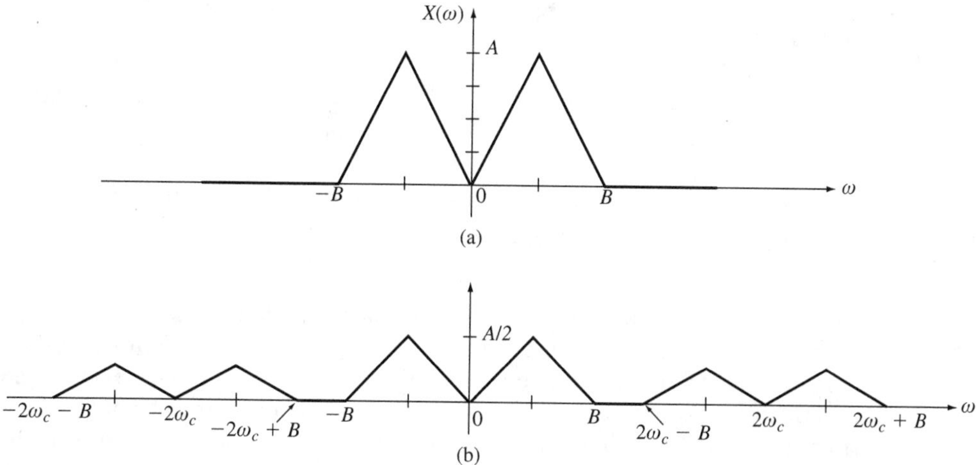

Figure 6.13 Fourier transform of (a) signal $x(t)$ and (b) $Ax(t)\cos^2\omega_c t$.

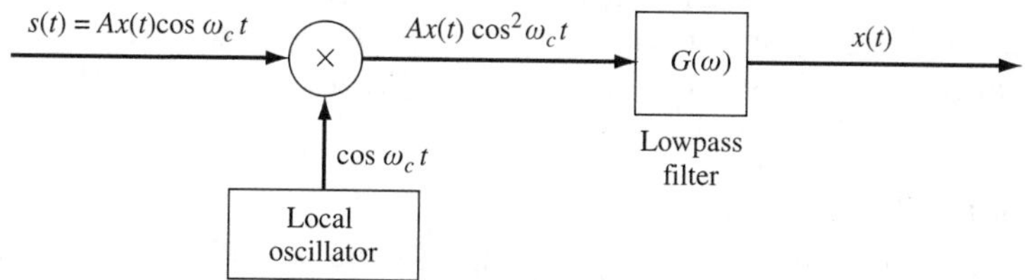

Figure 6.14 Synchronous demodulator.

Combining the multiplication process in Figure 6.12 and the lowpass filter results in the synchronous demodulator shown in Figure 6.14.

Demodulation of DSB Signals

In the case of DSB transmission where the carrier is not suppressed, $x(t)$ can be reconstructed from the modulated carrier $s(t) = A[1 + kx(t)]\cos\omega_c t$ by applying $s(t)$ to an *envelope detector* given by the circuit in Figure 6.15. As seen from the figure, the circuit consists of a source resistance R_s, diode, capacitor with capacitance C, and load resistance R_L. When there is no voltage on the capacitor and the modulated carrier $s(t)$ increases from 0 to some peak value, current flows through the diode, and the capacitor charges to a voltage equal to the peak value of $s(t)$. When $s(t)$ decreases in value from the peak value, the diode becomes an open circuit, and the voltage on the capacitor slowly discharges through the load resistance R_L. The discharging continues until $s(t)$ reaches a value that exceeds the value of the voltage across the

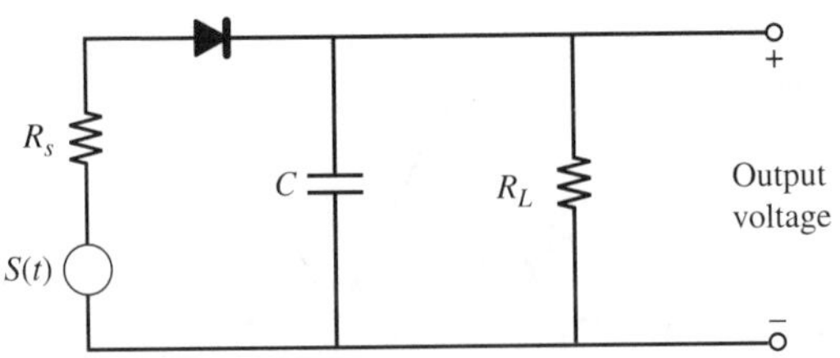

Figure 6.15 Envelope detector.

capacitor, at which time the capacitor again charges up to the peak value of $s(t)$, and then the process repeats. To insure that the charging of the capacitor is sufficiently fast so that the capacitor voltage reaches the peak value of $s(t)$ on every cycle, the charging time constant must be very small in comparison to the period $2\pi/\omega_c$ of the carrier $A\cos\omega_c t$. Assuming that the diode has zero resistance in the forward-biased region, the charging time constant of the envelope detector is equal to R_sC, and thus it is required that $R_sC \ll 2\pi/\omega_c$. In addition, the discharging time constant of the envelope detector must be large enough to insure that the capacitor discharge between positive peaks of $s(t)$ is sufficiently slow. The discharging time constant is equal to R_LC, and thus it must be true that $R_LC \gg 2\pi/\omega_c$. It also must be true that the discharging time constant is small in comparison with the maximum rate of change of $x(t)$. If $x(t)$ has bandwidth B, the maximum rate of change of $x(t)$ can be taken to be $2\pi/B$, and thus it is also required that $R_LC \ll 2\pi/B$.

Example 6.5

For the case when $x(t) = \cos\pi t$, $\omega_c = 20\pi$, $k = 0.5$, $R_s = 100$ ohms, $C = 10$ microfarads, and $R_L = 40{,}000$ ohms, the modulated carrier and the output of the envelop-detector are shown in Figure 6.16.

Demodulation of SSB Signals

Let $v(t)$ be a SSB signal that is generated by passing the modulated carrier $s(t) = Ax(t)\cos\omega_c t$ through a bandpass filter with frequency response function $H(\omega)$ given by equation (6.1). (See Figure 6.5.) The baseband signal $x(t)$ can be recovered from $v(t)$ by applying $v(t)$ to the synchronous demodulator shown in Figure 6.17, where it is again assumed that the bandwidth of $x(t)$ is equal to B. The frequency response function $G(\omega)$ of the lowpass filter in Figure 6.17 is given by

$$G(\omega) = \begin{cases} 4/A, & -B \le \omega \le B \\ 0, & \text{all other } \omega \end{cases} \tag{6.3}$$

To show that the output of the demodulator in Figure 6.17 is equal to $x(t)$, first note that the Fourier transform of $v(t)$ is given by

$$V(\omega) = \frac{A}{2}[X(\omega + \omega_c) + X(\omega - \omega_c)]H(\omega)$$

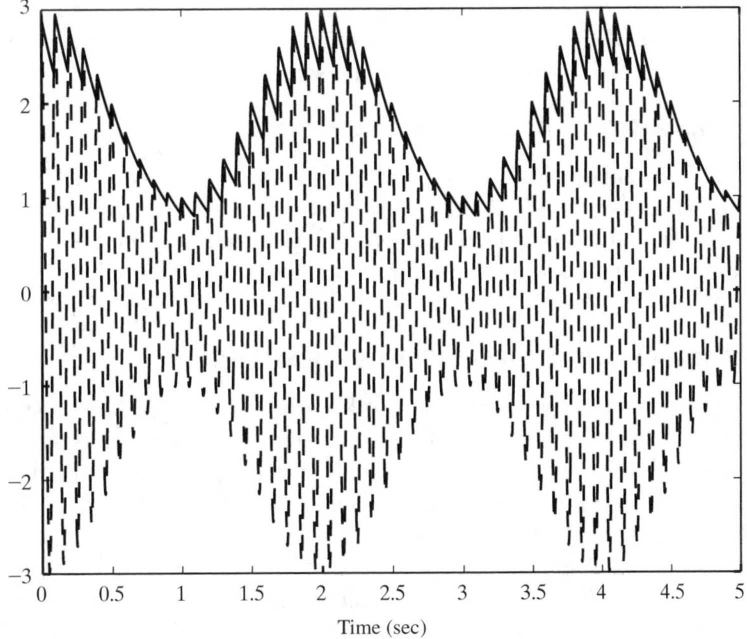

Figure 6.16 Modulated carrier and output of envelope detector in Example 6.5.

where $X(\omega)$ is the Fourier transform of $x(t)$. Thus, the Fourier transform of the output of the multiplier in Figure 6.17 is equal to

$$\frac{A}{4}\{X(\omega+2\omega_c)H(\omega+\omega_c)+X(\omega)[H(\omega+\omega_c)+H(\omega-\omega_c)]+X(\omega-2\omega_c)H(\omega-\omega_c)\}$$

The components $X(\omega + 2\omega_c)H(\omega + \omega_c)$ and $X(\omega - 2\omega_c)H(\omega - \omega_c)$ in the above expression are not in the passband of $G(\omega)$ given by (6.3), and therefore the Fourier transform of the output of the demodulator is equal to

$$\frac{A}{4}X(\omega)[H(\omega+\omega_c)+H(\omega-\omega_c)]G(\omega)$$

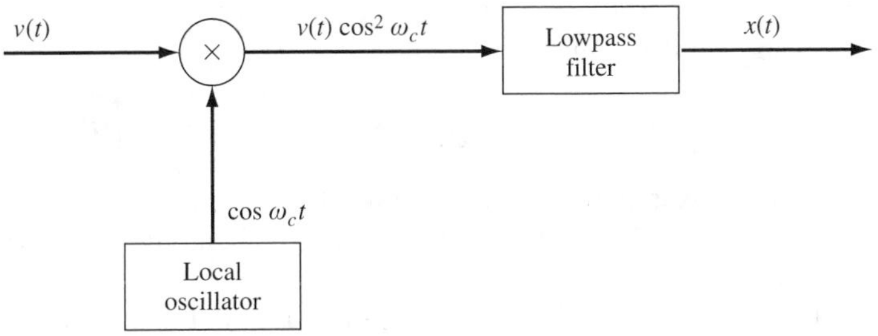

$v(t)$ | × | $v(t)\cos^2\omega_c t$ | Lowpass filter | $x(t)$

$\cos\omega_c t$

Local oscillator

Figure 6.17 Synchronous demodulator for SSB signals.

Now by (6.1), $H(\omega + \omega_c) + H(\omega - \omega_c) = 1$ for $-B \leq \omega \leq B$, and thus the Fourier transform of the demodulator output is equal to $X(\omega)$. Hence, the output of the demodulator is equal to the baseband signal $x(t)$.

Demodulation of Other Signal Types

The demodulation of VSB and FM signals is left to a more advanced treatment of communication systems. The demodulation of the PAM signal discussed in Section 6.1 can be carried out by lowpass filtering the PAM signal. The analysis is very similar to the reconstruction of a signal from samples of the signal, and is thus omitted. (See Section 5.5.)

6.3 SIMULTANEOUS TRANSMISSION OF SIGNALS

Suppose that the objective is to send N signals $x_1(t), x_2(t), \ldots, x_N(t)$ from one point to another. To keep the cost of the communication process as small as possible, it is desirable to be able to send the signals through the same channel (e.g., wire cable) at the same time. For example, if the signals are telephone calls, it would be desirable to be able to send them over the same wire cable or optical-fiber cable at the same time. The process of transmitting several signals over the same channel simultaneously is called *multiplexing*. Multiplexing can be achieved by using frequency-division multiplexing (FDM) or time-division multiplexing (TDM). We consider FDM first.

FDM

It is assumed that the spectra of the signals $x_1(t), x_2(t), \ldots, x_N(t)$ are contained in the frequency range from 0 to B (rad/sec); that is,

$$|X_i(\omega)| = 0, \qquad \omega > B, \quad \text{all } i$$

where $X_i(\omega)$ is the Fourier transform of $x_i(t)$. The idea of FDM is to up convert the spectra of the signals into adjacent (nonoverlapping) frequency "slots."

The up conversion of the spectra of the given signals is accomplished by putting the signals on carriers whose frequencies differ by an amount that is greater than or equal to $2B$. More precisely, given $\omega_c > B$, the frequencies of the carrier signals may be chosen to be $\omega_c, \omega_c + 2B, \omega_c + 4B, \ldots, \omega_c + 2(N-1)B$. This modulation process is illustrated in the first part of the block diagram in Figure 6.18.

As shown in Figure 6.18, the outputs of the modulators are added together, which results in the signal

$$v(t) = \sum_{i=1}^{N} x_i(t)\cos[\omega_c + 2(i-1)B]t$$

The signal $v(t)$ is then applied to a channel which may be a cable, a waveguide, or free space. It is assumed that the channel is an ideal channel; in other words, the output of the channel is equal to the input.

The spectrum $V(\omega)$ of the output $v(t)$ of the channel consists of the up-converted spectra of the signals $x_1(t), x_2(t), \ldots, x_N(t)$. Since the up-converted spectra

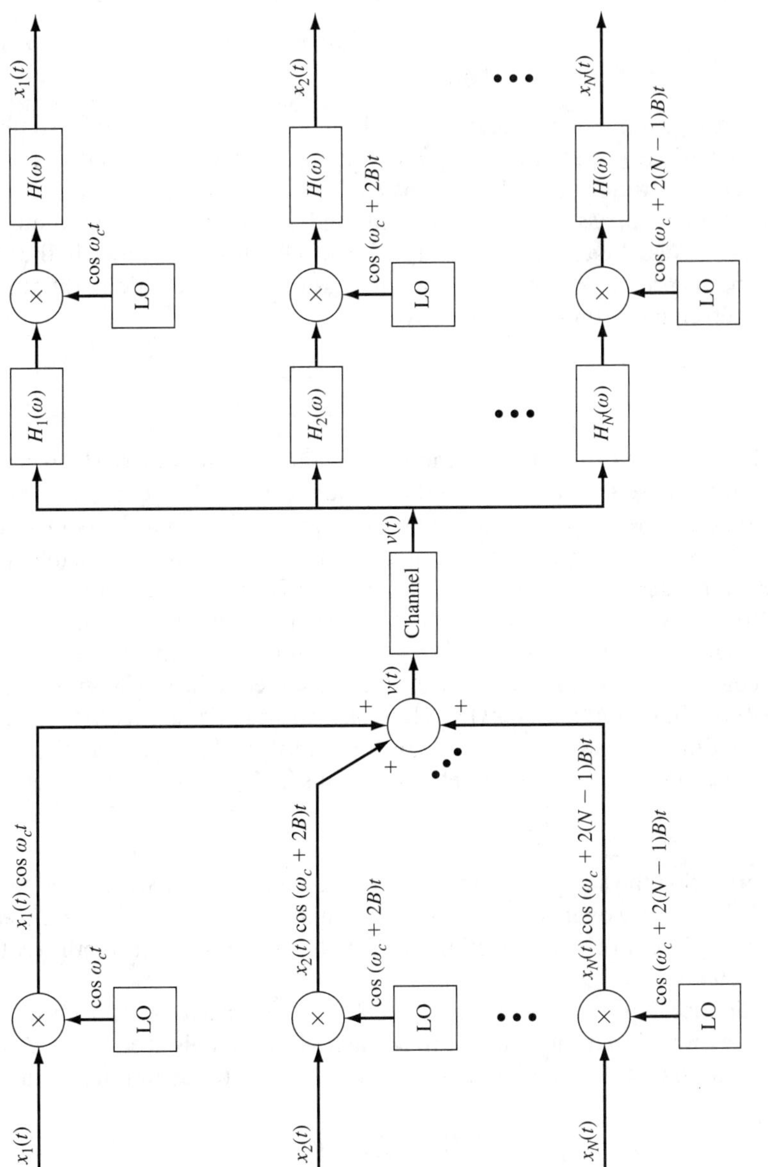

Figure 6.18 FDM process.

comprising $V(\omega)$ do not overlap, they can be separated by bandpass filtering. So as shown in Figure 6.18 the output of the channel is applied to a collection of N bandpass filters, with the frequency function $H_i(\omega)$ of the ith filter given by

$$H_i(\omega) = \begin{cases} 1, & \omega_c + 2(i - 1.5)B \le |\omega| \le \omega_c + 2(i - 0.5)B \\ 0, & \text{all other } \omega \end{cases}$$

To simplify the analysis, it is assumed that the filters are ideal with zero phase.

Now the spectrum of the output of the ith bandpass filter contains only the upconverted spectrum of the ith signal $x_i(t)$. Thus $x_i(t)$ can be extracted from the output of the ith bandpass filter by demodulation. The synchronous demodulator shown in Figure 6.14 can be used. The individual demodulators are shown in the last part of the block diagram in Figure 6.18. The lowpass filters in these demodulators all have the same frequency function, given by

$$H(\omega) = \begin{cases} 2, & -B \le \omega \le B \\ 0, & \text{all other } \omega \end{cases}$$

The number of signals that can be transmitted simultaneously using FDM depends on the base frequency ω_c of the carrier $\cos(\omega_c + 2iB)t$ and the bandwidth of the channel. For example, suppose that the channel is an optical-fiber cable and that telephone calls are transmitted. In this case, the carrier signals are visible light, which ranges in frequency from 4.0×10^{14} Hz (red) to 7.5×10^{14} Hz (violet).

To have voice recognition in the transmission of phone calls, as noted in Chapter 5, it is necessary to transmit only a 4-kHz bandwidth of the voice; that is, a person can recognize a filtered version of the voice of someone he or she knows if the bandwidth of the filter is at least 4 kHz. Telephone calls can be restricted to a 4-kHz bandwidth by first lowpass filtering the voice waveform. Hence B can be chosen to be 4 kHz for FDM transmission of telephone calls. This is done in practice.

TDM

By interleaving the samples from several time signals, many signals can be transmitted simultaneously over the same channel. This process is called *time-division multiplexing* (TDM). A brief description of TDM follows. For an in-depth treatment, see Couch [1996].

Suppose that the N signals $x_1(t), x_2(t), \ldots, x_N(t)$ are to be transmitted over a cable. It is assumed that the bandwidth of each of the signals is equal to B, so that the Nyquist sampling frequency is $\omega_s = 2B$. Let $p(t)$ denote the impulse train

$$v(t) = \sum_{i=-\infty}^{\infty} \delta(t - iT)$$

with $T = 2\pi/\omega_s = \pi/B$.

In the first part of the TDM process, the signals $x_1(t), x_2(t), \ldots, x_N(t)$ are sampled by multiplying the ith signal $x_i(t)$ by $p[t - (i - 1)(T/N)]$. As shown in Figure 6.19, the outputs of the multipliers are added together, which results in the signal $v(t)$ given by

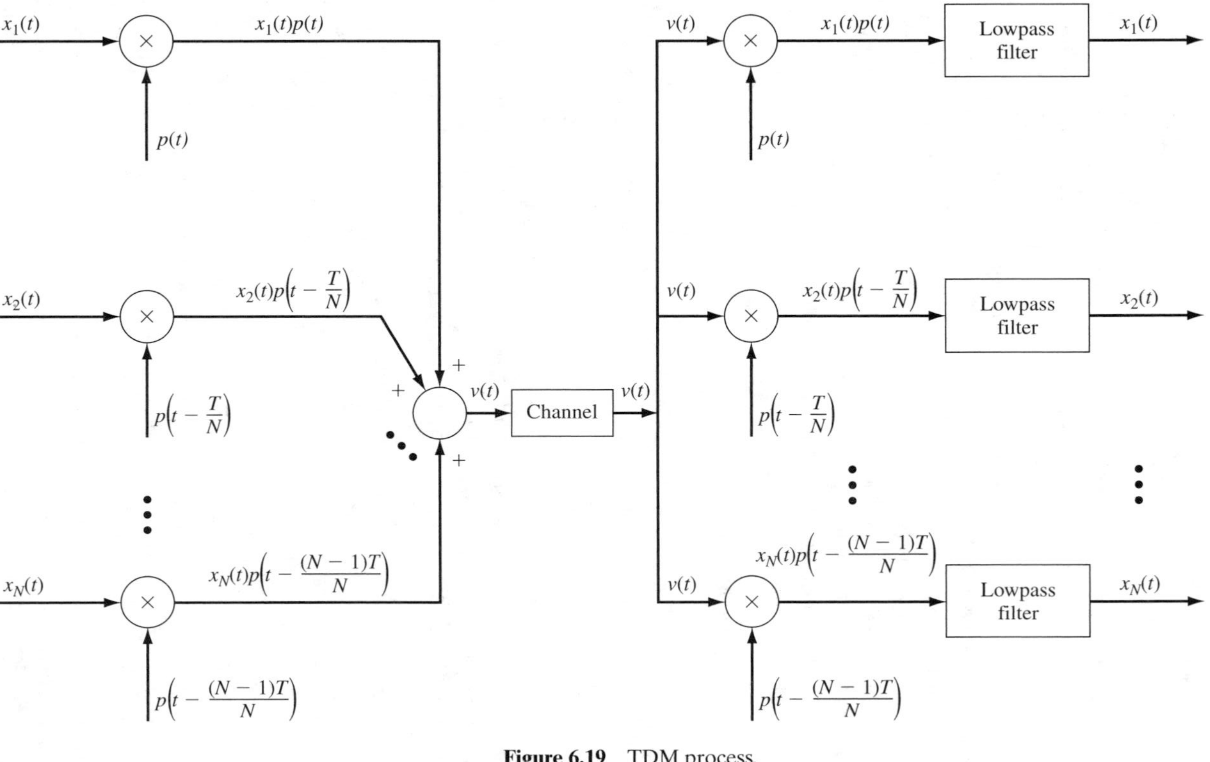

Figure 6.19 TDM process.

$$v(t) = \sum_{i=1}^{N} x_i(t)p\left[t - (i-1)\frac{T}{N}\right]$$

The signal $v(t)$ consists of the interleaved samples of the signals $x_1(t), x_2(t), \ldots, x_N(t)$.

The signal $v(t)$ is then applied to a channel whose output is assumed to be equal to $v(t)$. As shown in Figure 6.19, the original signals are recovered from $v(t)$ by first multiplying by $p[t - (i - 1)(T/N)]$. These multipliers (samplers) must be synchronized with the samplers in the first part of the process. If they are not synchronized, the sample values of one signal will be mixed with the sample values of the other signals, resulting in distortion in the reconstruction (demodulation) process. In the final stage of the TDM process, the original signals are recovered from the sampled signals $x_i(t)p[t - (i - 1)(T/N)]$ by lowpass filtering. The cutoff frequency of each lowpass filter is equal to B.

It should be noted that in digital communication systems employing TDM, the sampled signals are quantized in amplitude and then encoded before the sample values are interleaved. The process of quantization and encoding is omitted.

6.4 DIGITAL MODULATION

Digital modulation is rapidly becoming more popular than analog modulation for transmitting of information. Digital modulation is currently used in such common applications as compact discs, digital cellular service, digital phone lines, and computer modems. The signal to be modulated might originate from a digital source, such as data from a computer, in which case, digital modulation would seem natural. The signal might also originate from an analog source, such as a telephone conversation, where it has to be sampled first in order to be converted to a digital form. The advantages of digital modulation over analog modulation include:

1. Digital transmission can be made more accurate, since there are more accurate methods of removing signal degradation during transmission. This is especially important for long-distance transmissions, such as occurs over long-distance phone lines.
2. Multiple media, such as video, speech and data, can all be coded and modulated using the same type of processes resulting in signals of similar type that can be multiplexed together, so that digital modulation has a *multimedia capability*.
3. Digital transmission can be encrypted (or coded) easily for confidential or secure data transfer.
4. Electronic routing of data (such as done with email) is easier to perform digitally.
5. Algorithms that compress information (such as images) so that transmission requires less bandwidth are easier to implement digitally rather than in analog.
6. Digital storage devices allow for faster storing and retrieval of data than analog storage devices.
7. Optical fiber, an economical media, is well suited to digital communications.

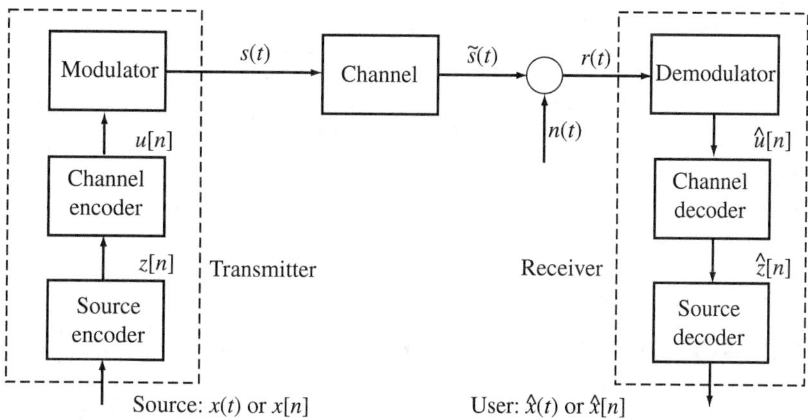

Figure 6.20 Digital modulation components.

On the other hand, digital transmission tends to require signal synchronization, which can be avoided in analog transmission.

The main components of a digital modulation system are shown in Figure 6.20. The left side of the diagram is the transmitter, and the right side is the receiver, with the channel in the middle. The original signal to be transmitted can be continuous time $x(t)$ or discrete time $x[n]$. If the signal is continuous time, it must be sampled or converted to a discrete time form. This step takes place in the *source encoder*. In addition as part of the source encoder block, data can be compressed to reduce bandwidth and/or encrypted for secure data transmission. The output of the source encoder block is a discrete-time signal $z[n]$. A *channel encoder* block is often inserted to reduce errors incurred in the channel. This is usually done by creating a signal $u[n]$ that contains the information from the input signal $z[n]$ plus some redundancy that is a function of $z[n]$. The redundancy (for example, parity checks) allows one to detect or correct errors introduced by channel impairments. The *modulator* block in Figure 6.20 converts the discrete-time signal $u[n]$ to a continuous-time signal $s(t)$ for transmission. A major difference between analog modulation and digital modulation is that the continuous-time signal $s(t)$ transmitted in the digital case is generated from a finite set of possible signals. In particular, the digital modulator is a mapping between a finite set $\{u_i, i = 0, 1, \dots, M - 1\}$ of possible values for $u[n]$, and a finite set $\{s_i(t), i = 0, 1, \dots, M - 1\}$ of corresponding signals for $s(t)$. The elements u_i are usually called *symbols*, and the set of symbols $\{u_i, i = 0, 1, \dots, M - 1\}$ is called an *alphabet*. In this case, the size of the alphabet is M, and thus this is called *M-ary modulation*. Generally, M is chosen to be a factor of 2; that is, $M = 2^N$, where $N = \log_2 M$ is the number of bits needed to represent each of the M signals $s_i(t)$ using binary code. If $M = 2$, then the modulation is called *binary*. The waveform for $s(t)$ is obtained by summing time-shifted versions of the signals $s_i(t)$, and thus $s(t)$ is given by

$$s(t) = \sum_n s_i(t - nT) \tag{6.4}$$

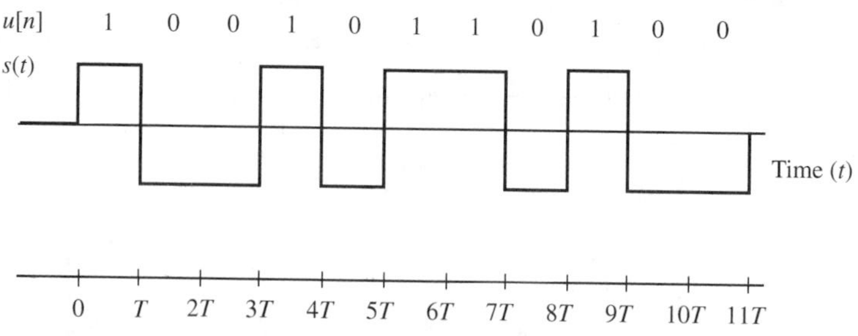

Figure 6.21 Binary modulation scheme.

where T is the signal interval and $s_i(t - nT)$ is the signal sent during the nth signal interval.

Example 6.6 *Binary Modulation*

A simple example is when the alphabet for $u[n]$ is binary; that is, it can only take on the values of 0 or 1 in any signal interval. A corresponding set of signals for $s_i(t)$ might be chosen as a set of square pulses:

$$s_0(t) = \begin{cases} -1, & \text{if } 0 \leq t \leq T \\ 0, & \text{otherwise} \end{cases} \qquad s_1(t) = \begin{cases} 1, & \text{if } 0 \leq t \leq T \\ 0, & \text{otherwise} \end{cases}$$

Correspondingly, $s_0(t - nT)$ is transmitted if $u[n] = 0$, and $s_1(t - nT)$ is transmitted if $u[n] = 1$. A timing diagram for a sample sequence of $u[n]$ is shown in Figure 6.21.

The *channel* is the medium over which the information is transferred. Examples include the atmosphere, twisted wire, coaxial cable, fiber optic cable, and optical or magnetic disks. These media are analog in nature, which explains why the signal $s(t)$ to be transmitted over the channel must be continuous time. The channel itself is a system with its own frequency response. If $H(\omega)$ represents the channel's frequency response, and if the Fourier transform of the input signal is $S(\omega)$, then the transform of the channel output due to the channel characteristics is

$$\tilde{S}(\omega) = H(\omega)S(\omega)$$

An ideal channel would have a flat frequency response over the frequency range in which $s(t)$ has significant components, so that $\tilde{s}(t) \approx s(t)$. However, most channels distort the signal because the frequency response is not flat. For example, a square pulse that is processed through a channel with a low-pass filter frequency characteristic is shown in Figure 6.22.

Returning to Figure 6.20, $n(t)$ is channel noise that adds additional error in the signal $r(t)$ received by the receiver. The goal of the *demodulator* block in Figure 6.20 is to extract from $r(t)$ the individual signals $s_i(t)$ that comprise $s(t)$ and then to recreate an estimate $\hat{u}[n]$ of the sequence of numbers $u[n]$ that generated them. Recall that the channel distorts the signal $s(t)$ because it acts as a filter and that it adds noise. The demodulator usually first filters the received signal $r(t)$ to remove noise and/or correct the signal distortion incurred by the channel. There needs to be syn-

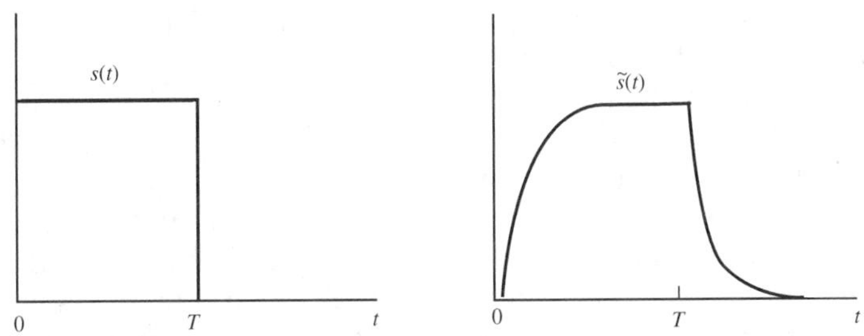

Figure 6.22 Distortion introduced by a channel with a low-pass frequency characteristic.

chronization between the timing of the transmitter and the receiver. This information can be transmitted in a standard code at the beginning of the digital transmission (and possibly periodically during the transmission to accommodate changing channel characteristics). It may also be detected from the signal with a special timing circuit.

In the remainder of the receiver flow diagram in Figure 6.20, the signal $\hat{u}[n]$ may be sent into a *channel decoder*, where additional error correction is accomplished. Encryption or data compression provided by the source encoder is reversed in the *source decoder* to create an estimate $\hat{x}[n]$ of the original signal $x[n]$, if the signal is discrete-time. If the signal is to be converted to analog, then it is sent through a digital to analog converter to create an estimate $\hat{x}(t)$ of the original signal $x(t)$.

Transmission and reception of the analog signals $s(t)$ and $r(t)$ are accomplished using analog electronic devices. For the case of metallic media (coaxial cable or twisted wire), the electronics create electrical signals for transmission. For fiber-optic media, the electrical signal controls a laser with varying intensity that transmits an optical signal, and the receiver is an optical detector, which converts light intensity into an electrical signal for further processing. There are many devices that both transmit and receive digital communications; the most common devices are called *modems* (short for *mo*dulator and *dem*odulator).

An advantage of digital modulation over analog modulation is that *regenerative repeaters* [that regenerate $s(t)$] can be installed periodically along the channel to make the end-to-end communication system as reliable as possible. A repeater acts as a transmitter and receiver combined, where the signal is detected and converted into estimates $\hat{u}[n]$ that can then be used to regenerate an analog signal. The repeaters are located close enough to one another so that the signal is not degraded significantly. (Generally, the longer the distance between repeaters, the greater the error is.) For example, repeaters are located about 40 km to 100 km apart in fiber optic channels. Regeneration is achievable with much greater accuracy in digital modulation than in analog modulation, because in the former case signals are generated from a finite set. In analog modulation, the only regeneration possible is through filtering, that is, sending the signal through a filter with a frequency response that is an approximate of the inverse of the frequency response of the channel. Having only a finite set of signals to match allows the regenerators in digital modulation schemes to "lock into" the desired signal before the actual signal has degraded beyond a certain threshold of error.

When the channel has a low-pass frequency characteristic, it is possible to transmit signals $s(t)$ relatively accurately if the signals have low-frequency content. An example is voice transmission, which spans from about 300 Hz to about 3300 Hz. In these cases, the signal can be transmitted directly using the low-frequency (or baseband) part of the spectrum. In other cases, the channel may have poor low-frequency characteristics, so the signal is first modulated to a higher frequency range by multiplying it with a sinusoidal carrier wave (as discussed in Section 6.1). The resulting transmission utilizes a mid-range band of frequencies (or passband). The vast majority of digital communications uses some form of *baseband PAM* or *passband PAM* to modulate the signal. These methods are described in the next two sections.

6.5 BASEBAND PAM

The digital modulator must have a predetermined set of signals $s_i(t)$ to generate the transmission signal $s(t)$ as defined by (6.4). There are many choices for the shape of the signals $s_i(t)$. One option is to define all the $s_i(t)$ signals to have the same pulse shape $g(t)$, but different amplitudes that are based on the corresponding values of u_i. In this case, the $s_i(t)$ are given by

$$s_i(t) = a_i g(t) \tag{6.5}$$

Substituting (6.5) into (6.4) gives the form of the transmission signal:

$$s(t) = \sum_n a_i g(t - nT) \tag{6.6}$$

Using a pulse with a varying amplitude to create the modulated signal $s(t)$ is known as *pulse amplitude modulation*, or PAM. It differs from the analog PAM method described in Section 6.1 in that the amplitude of the pulse can only take on a finite set of values. When a signal $s(t)$ is generated using (6.6) and transmitted directly, it generally has relatively low frequency content if $g(t)$ has low-frequency content; hence, the baseband part of the spectrum is utilized in the transmission. The resulting digital modulation scheme is commonly referred to as *baseband* PAM. The binary modulation described in Example 6.6 is a simple form of baseband PAM. Recall that if there are M different signals possible for $s_i(t)$, then the modulation is called M-ary.

Example 6.7 *4-ary PAM*

Suppose that the alphabet for $u[n]$ consists of four numbers, as shown in the table below, along with the corresponding square pulse levels for $s_i(t)$ where A is a constant.

The timing diagram for a sample input sequence is shown in Figure 6.23.

u_i	level of pulse for $s_i(t)$
−3	$-3A$
−1	$-A$
1	A
3	$3A$

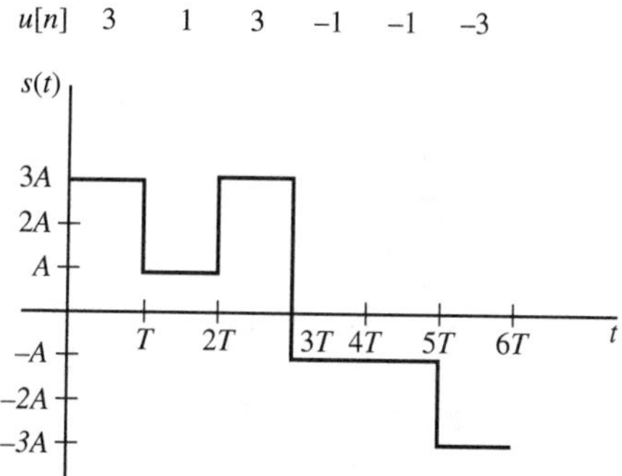

Figure 6.23 Timing diagram for a 4-ary PAM signal.

Demodulation in a PAM scheme requires that estimates of the signals $s_i(t - nT)$ be recovered from the received signal $r(t)$. As mentioned previously, there is often some filtering performed first to reduce the noise level in the signal and to alleviate some of the distortion generated in the channel. The amplitude levels of $s_i(t - nT)$ can be determined by sampling the signal $r(t)$ (or the filtered signal if filtering is used) where it is least distorted. After the signal is sampled, it must be converted to a string $\hat{u}[n]$ of discrete values that are estimates of the symbols contained in the string $u[n]$. In PAM, the conversion process is usually termed a *slicer*.

Example 6.8 *4-ary Demodulation*

Suppose that the 4-ary modulation scheme generated with square pulses from Example 6.7 is sent through a channel with a low-pass characteristic. The received signal $r(t)$ is shown in Figure 6.24. The signal is sampled at the peaks and dips of the received pulse string to obtain the best match for $s(t)$. (Usually there is some automatic gain adjustment that boosts the signal $r(t)$ to an appropriate level before sampling.) The sampled signals are then sent through an appropriate slicer, as defined in Figure 6.25, to create the sequence $u[n]$.

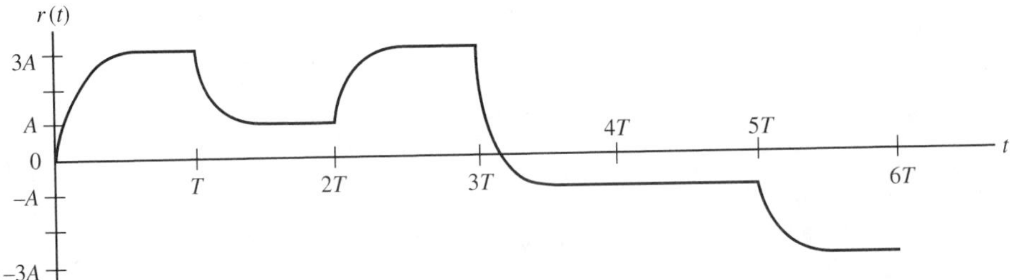

Figure 6.24 The signal $r(t)$ at the receiver, including channel effects, from the 4-ary signal transmitted in Figure 6.23.

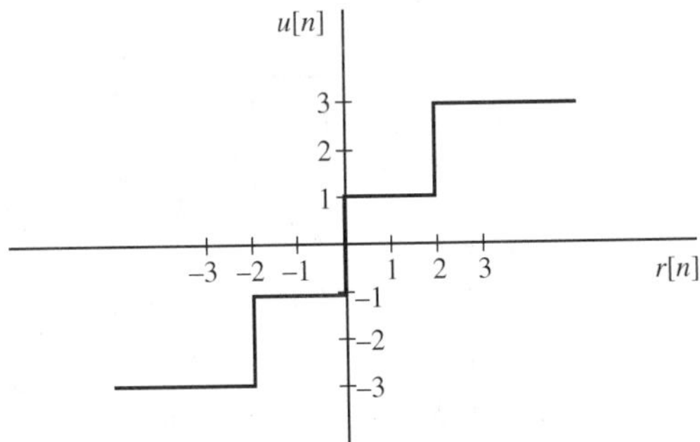

Figure 6.25 4-ary slicer with input as the sampled elements of $r(t)$ and output as an estimate of the elements of the sequence $u[n]$.

The goal of the communications designer is to transmit information at the highest possible rate within acceptable error tolerance levels. The *baud rate* is the inverse $1/T$ of the signal interval T, and it represents the number of digital symbols transmitted per second. The amount of information transmitted within each signal interval is determined by the number of bits (or binary digits) needed to represent the symbols. As noted previously, if the system is M-ary, there are M symbols in the alphabet for $u[n]$. If $M = 2^N$, then there are N bits needed to represent the symbol using binary code. Equivalently, the number of bits conveyed in a signal interval is $N = \log_2 M$.

The information transfer rate is computed as

$$R_b = \frac{\textit{number of bits per signal interval}}{\textit{signal interval}} = N \times \textit{baud rate} = \frac{N}{T} \textit{ bits per second (bps)}$$

Example 6.9 *Bit Rates*

Consider a standard modem connected to a phone line. The common alphabet size is $M = 512$, and the baud rate is $1/T = 3200$ baud. The bit rate is

$$R_b = \frac{\log_2 M}{T} = \frac{9}{T} = 28{,}800\ b/s = 28.8\ kb/s$$

The bit rate is limited by the number of bits per signal interval and by the baud rate. The number of bits is limited by the noise of the channel. A large number of bits per signal interval means that there is less distinction between the signals $s_i(t)$, making demodulation more susceptible to noise in the channel. The baud rate is often limited by the bandwidth of the media. For example, a channel with a low-pass characteristic cannot transmit a pulse train with a small signal interval, because the received pulses will be "smeared" together so much so that the receiver cannot achieve an accurate estimate of the pulses. This "smearing" can be seen in the signals shown in Figure 6.24. Though the transmitted signal has length T, the received signal has a length that extends beyond T. This "smearing" is known as *intersymbol interferance* (ISI).

In the above examples, square wave pulses were used for $g(t)$ in (6.6). However, there are better choices of pulse shapes that produce lower levels of ISI there by resulting in fewer errors at the same baud rate.

Example 6.10 *Sinc Pulse Shapes*

Consider the sinc pulse (plotted in Figure 6.26a) defined by

$$g(t) = \sin c\left(\frac{t}{T}\right) = \frac{\sin(\pi\tau/T)}{\pi\tau/T}$$

Note that the zero crossings of this signal occur at multiples of T for $t \neq 0$. Similarly, the delayed pulse, shown in Figure 6.26b, has zero crossings at multiples of T for $t \neq T$. Since $g(t)$ is not a time-limited signal, it is clear that neighboring signals $s_i(t - nT) = a_i g(t - nT)$ for $n = k$ and $n = k + 1$ do overlap when added together. However, the overlap is zero at the multiples of T. Therefore, the correct amplitudes a_i can be recovered exactly by sampling $s(t)$ at multiples of T. As a result, the sinc pulse is ideal for eliminating ISI. However, it is impossible to generate a sinc function, since it is not a time-limited signal.

Recall from Chapter 4 that the Fourier transform of a square pulse is a sinc function. A sinc function in the frequency domain is not bandlimited, which gives rise to the poor performance of square pulses when sent through a channel with bandwidth limitations. Also, duality shows that the Fourier transform of a sinc function is a square pulse in the frequency domain. Thus, the sinc pulse is bandlimited, which

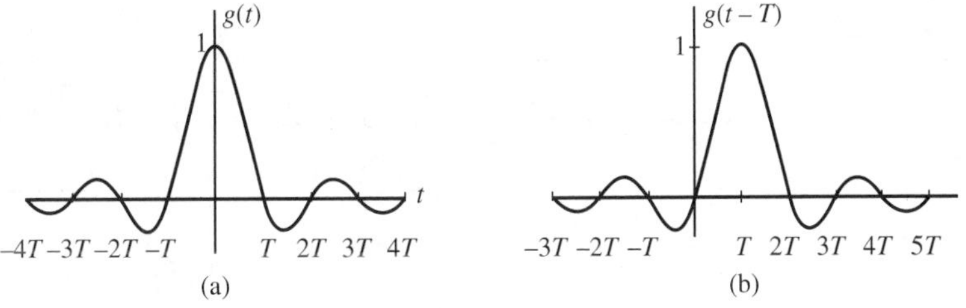

Figure 6.26 Sinc pulse functions.

makes it well suited for occasions when the channel has a low-pass frequency characteristic; however, it is impractical to implement. It is desirable to select a pulse shape that is easy to create with hardware (as is the case for a square wave), but has a sharp attenuation in spectral components above some value in frequency (so that it is nearly bandlimited such as a sinc pulse). One possible choice is a raised cosine pulse, which has some of the desired attributes. See [Lee and Messerschmitt, 1998] for further discussion on the selection criteria for pulse shapes.

6.6 PASSBAND PAM

Some media exhibit poor transmission characteristics at low frequencies, but do provide good transmission at higher frequency ranges. In other words, they have the characteristics of a bandpass (or passband) filter. An example of this type of media is the atmosphere, which is used in commercial radio and television broadcasts, satellite communication systems, mobile radio and cellular phone service. In these applications, the higher frequencies of the passband format also result in a reduction in the size of the antennas required to transmit and receive signals. (Antenna size is proportional to wavelength and inversely proportional to frequency.) As discussed in Section 6.1, AM radio takes advantage of the passband characteristics of the media by modulating the signal into a higher frequency range before transmission; that is, the signal is multiplied by a sinusoidal carrier wave. This same procedure can be used with digital modulation. A simple and straightforward approach is to modify the baseband PAM scheme shown in Figure 6.27 by first multiplying the signal $s(t)$ by a sinusoidal carrier wave before transmitting it, which results in the signal

$$s_m(t) = \sqrt{2}\, s(t) \cos(\omega_c t) \tag{6.7}$$

where the $\sqrt{2}$ factor is used to ensure that the energy of $s_m(t)$ is equal to that of $s(t)$. As discussed in Section 6.2, synchronous demodulation is performed by multiplying the received signal by the same carrier wave. A block diagram of this procedure is shown in Figure 6.27. The addition of carrier wave modulation to baseband PAM is termed *passband* PAM.

An alternative way of representing the modulation is to use complex notation:

$$s_m(t) = \sqrt{2}\, \mathrm{Re}\{s(t)e^{j\omega_c t}\} \tag{6.8}$$

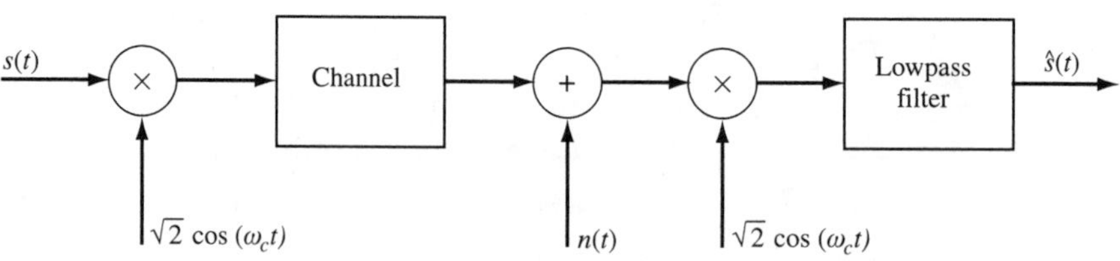

Figure 6.27 Sinusoidal carrier wave modulation and demodulation in passband PAM.

If $s(t)$ is real-valued, then (6.8) reduces to (6.7). However, there is no restriction requiring that $s(t)$ be real-valued. In fact, more information can be transmitted over the same bandwidth if $s(t)$ is allowed to be complex. In particular, information can be coded onto both the real part and the imaginary part of $s(t)$. Consider the standard expression for $s(t)$ in a PAM configuration:

$$s(t) = \sum_n s_i(t - nT) = \sum_n a_i g(t - nT) \tag{6.9}$$

Generally, the pulse $g(t)$ is selected to be real-valued, but the amplitudes a_i can be complex. With the a_i real or complex, substitution of (6.9) into (6.8) yields an expression for the transmitted signal:

$$s_m(t) = \sqrt{2}\mathrm{Re}\left\{e^{j\omega_c t}\sum_n a_i\, g(t - nT)\right\} \tag{6.10}$$

Phase-Shift Keying (PSK)

A common modulation scheme, *Phase-Shift Keying* (PSK), can be represented as a type of passband PAM when the amplitudes a_i in (6.10) are allowed to be complex. The transmitted signal generated with PSK has the form

$$s_m(t) = \sqrt{2}\, A \sum_n \cos(\omega_c t + \phi_i) g(t - nT) \tag{6.11}$$

Note that the information in PSK is actually coded on the phase component; that is, the alphabet for $u[n]$ is mapped to the set of phases $\{\phi_i, i = 0, \dots, M - 1\}$. To write this expression in the form (6.10) for passband PAM, consider the complex amplitudes written in polar notation:

$$a_i = Ae^{j\phi_i}$$

Substituting a_i into the expression (6.10) gives

$$s_m(t) = \sqrt{2}\mathrm{Re}\left\{e^{j\omega_c t}\sum_n Ae^{j\phi_i} g(t - nT)\right\}$$

$$= \sqrt{2}\, A \sum_n \mathrm{Re}\{e^{j(\omega_c t + \phi_i)}\} g(t - nT)$$

This is equivalent to the standard expression for a PSK signal given by (6.11). The factor A is used to normalize the signal and is generally chosen as $A = (E/T)^{1/2}$, where E is the averaged energy of the signals, calculated by

$$E = \frac{1}{M}\sum_{i=0}^{M-1}\int_0^T s_i^2(t)\, dt$$

PSK is a common choice for passband PAM, because the receivers are relatively inexpensive, especially for binary PSK.

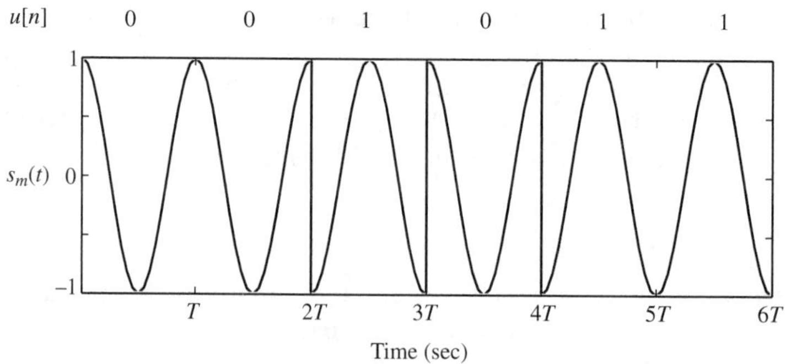

Figure 6.28 Binary PSK signal.

Example 6.11 *PSK*

Binary PSK is formed when the alphabet for $u[n]$ has two elements, say, $u_0 = 0$ and $u_1 = 1$. The phases of the modulated signal are chosen to be $\phi_0 = 0$ and $\phi_1 = \pi$, corresponding to u_0 and u_1, respectively. A sample sequence for square wave pulses is shown in Figure 6.28. In 4-ary PSK, also known as Quadrature PSK (QPSK), the phases are chosen to be $\phi_0 = 0, \phi_1 = \pi/2, \phi_2 = \pi$, and $\phi_3 = 3\pi/2$.

Frequency-Shift Keying (FSK)

In contrast to PSK, *Frequency Shift Keying* (FSK) codes the information on the frequency of the carrier wave as opposed to the phase. The transmitted signal for FSK has the form

$$s_m(t) = \sqrt{2}\,A\sum_n \cos(\omega_i t)g(t - nT) \tag{6.12}$$

where A is again chosen as $A = (E/T)^{1/2}$ to normalize the signal. In this case, the alphabet for $u[n]$ is mapped to the set of frequencies $\{\omega_i, i = 0, \dots, M - 1\}$. FSK can also be considered as a special type of passband PAM by defining the amplitudes a_i used to create the $s_i(t)$ signals as

$$a_i = Ae^{j\gamma_i t}$$

These amplitudes can be substituted into the expression (6.10) for $s_m(t)$. Defining the new frequencies to be $\omega_0 = \omega_c + \gamma_0, \omega_1 = \omega_c + \gamma_1, \dots$ results in the following expression for $s_m(t)$:

$$s_m(t) = \sqrt{2}\,\mathrm{Re}\left\{\sum_n Ae^{j\omega_i t}g(t - nT)\right\}$$

$$= \sqrt{2}\,A\sum_n \mathrm{Re}\{e^{j\omega_i t}\}g(t - nT)$$

which is equivalent to (6.12).

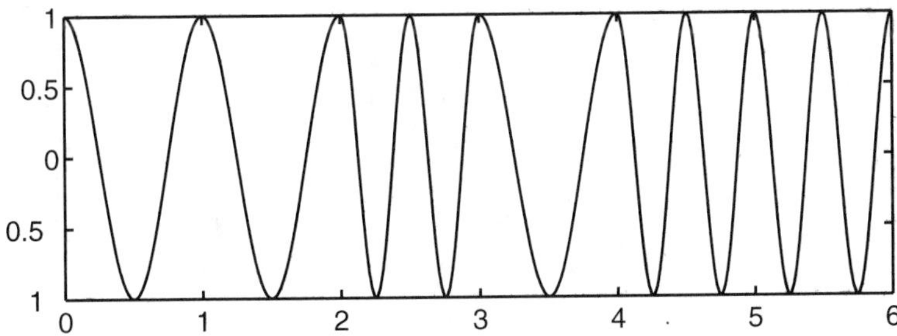

Figure 6.29 Binary FSK for a sample sequence.

Example 6.12 *Binary FSK*

In binary FSK, there are two different frequencies, ω_0 and ω_1, which correspond to the two-element alphabet, $u_0 = 0$ and $u_1 = 1$, respectively. The modulated signal corresponding to a sample sequence for $u[n]$ is shown in Figure 6.29.

Quadrature Amplitude Modulation (QAM)

In the general case when the amplitudes a_i in (6.10) are complex numbers given by

$$a_i = b_i + jc_i$$

Euler's formula can be used to write (6.10) in the form

$$s_m(t) = \sqrt{2}\cos(\omega_c t)\sum_n \mathrm{Re}\{a_i\}g(t - nT) - \sqrt{2}\sin(\omega_c t)\sum_n \mathrm{Im}\{a_i\}g(t - nT) \tag{6.13}$$

The signal $s_m(t)$ contains two modulated signals, one modulated by $\cos(\omega_c t)$ and one modulated by $-\sin(\omega_c t)$. Note that $s_m(t)$ can be viewed as a sum of a passband PAM signal and a PSK signal with a phase of 90°.

To see how $s_m(t)$ can be generated, consider two baseband PAM signals, both generated using the same mapping scheme from the alphabet $\{u_i\}$ to the set of signals $s_i(t)$:

$$s_{u1}(t) = \sum_n b_i g(t - nT) \quad \text{and} \quad s_{u2}(t) = \sum_n c_i g(t - nT)$$

The amplitudes represented by b_i correspond to a discrete sequence $u_1[n]$, while the amplitudes represented by c_i correspond to a discrete sequence $u_2[n]$. The sequences $u_1[n]$ and $u_2[n]$ can themselves be traced back to two original signals $x_1[n]$ and $x_2[n]$ (or $x_1(t)$ and $x_2(t)$ if continuous-time signals). Then, since $a_i = b_i + jc_i$, (6.13) is simply written as

$$s_m(t) = \sqrt{2}\cos(\omega_c t)s_{u1}(t) - \sqrt{2}\sin(\omega_c t)s_{u2}(t) \tag{6.14}$$

The signal $s_{u1}(t)$, which is modulated by $\sqrt{2}\cos(\omega_c t)$, is called the in-phase component. The signal $s_{u2}(t)$, which is modulated by $-\sqrt{2}\sin(\omega_c t)$, is called the quadrature component.

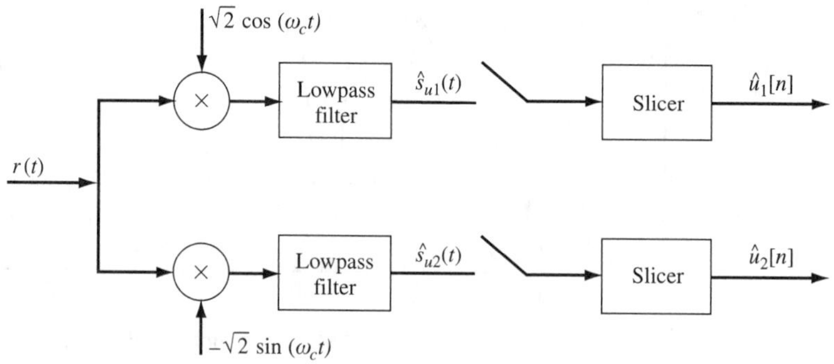

Figure 6.30 Demodulation method for a QAM signal.

The resulting technique, known as *Quadrature Amplitude Modulation* (QAM), is summarized by first using baseband PAM to modulate two separate signals $x_1[n]$ and $x_2[n]$ (or $x_1(t)$ and $x_2(t)$). One of the resulting baseband PAM signals, $s_{u1}(t)$, is modulated by $\sqrt{2}\cos(\omega_c t)$, and the other, $s_{u2}(t)$, is modulated by $-\sqrt{2}\sin(\omega_c t)$. The sum of the two modulated signals $s_m(t)$, given by (6.13) or (6.14), is then transmitted over the channel. Demodulation of a QAM signal is performed using the schematic given in Figure 6.30. The proof that the demodulation scheme does indeed recover the two baseband PAM signals is left to the reader. (See Problem 6.13.)

Example 6.13 *Binary QAM*

Suppose that a channel has the frequency response characteristic shown in Figure 6.31. A carrier wave should be selected to fall within the passband range, which extends from about 1G rad/sec to about 2G rad/sec. Consider a single square wave pulse PAM signal $s(t)$ shown in Figure 6.32a along with its Fourier transform $S(\omega)$. It is clear that $S(\omega)$ would be attenuated greatly if it were sent through the channel without a sinusoidal modulation. The Fourier transform of the signal modulated by $\sqrt{2}\cos(\omega_c t)$ is shown in Figure 6.32b, while the Fourier transform of the signal modulated by $-\sqrt{2}\sin(\omega_c t)$ is shown in Figure 6.32c, where $\omega_c = 1000$ rad/sec.

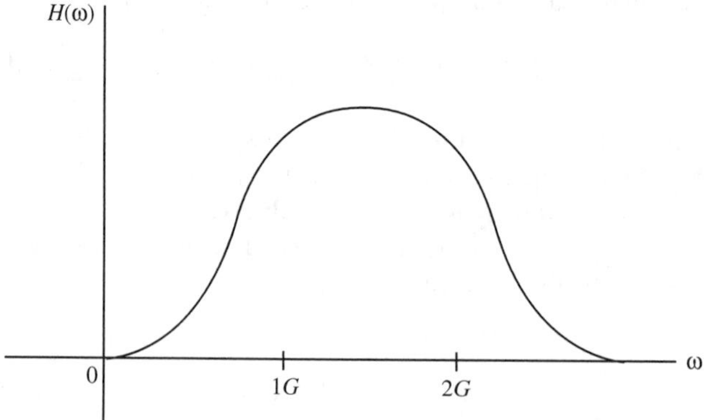

Figure 6.31 Frequency response of channel for Example 6.13.

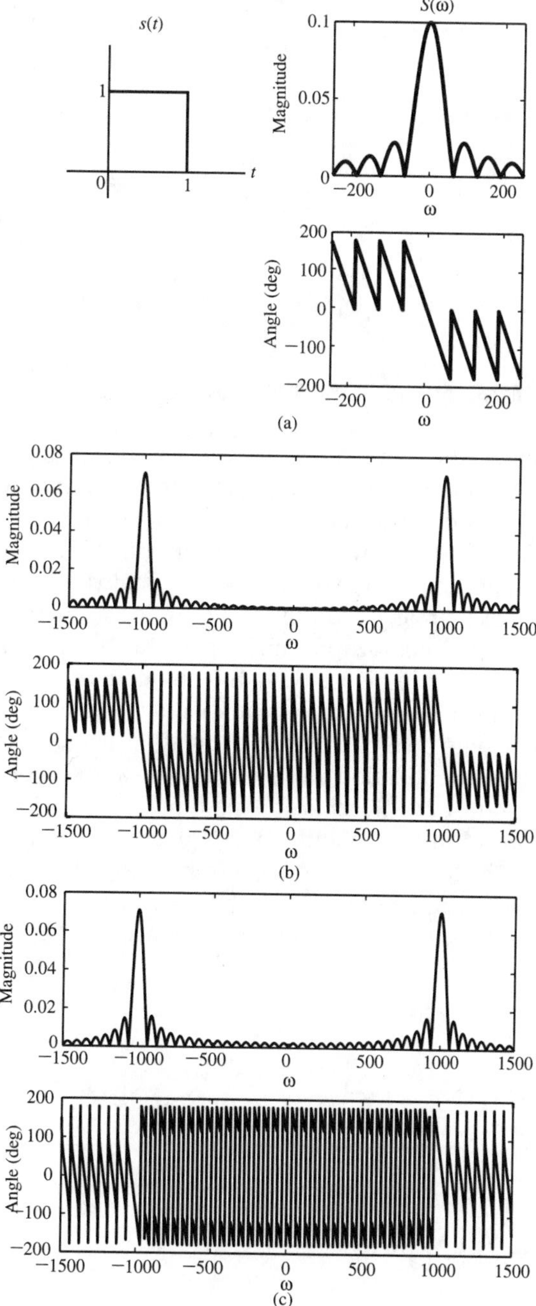

Figure 6.32 QAM process, (a) $s(t)$ and $S(\omega)$, (b) modulation of $s(t)$ by $\sqrt{2}\cos(\omega_c t)$, and (c) modulation of $s(t)$ by $-\sqrt{2}\sin(\omega_c t)$, where $\omega_c = 1000$ rad/sec.

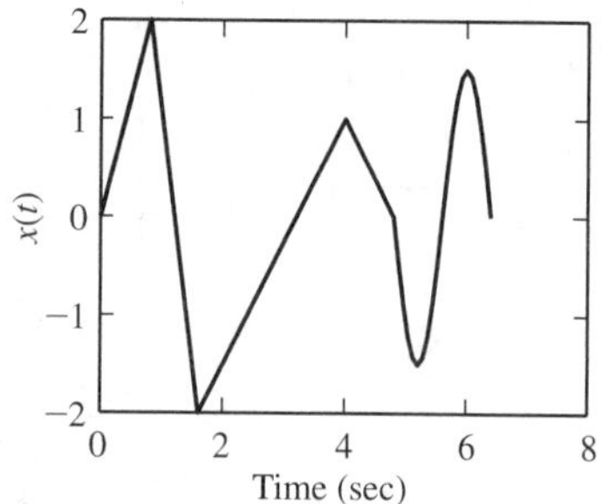

Figure 6.33 Sample signal used in the simulations.

6.7 DIGITAL COMMUNICATION SIMULATIONS

Digital Communications

In this section, the communication process shown in Figure 6.20 is simulated in MATLAB to demonstrate transmission of the sample waveform shown in Figure 6.33 using the modulation techniques: PAM with square pulses, PAM with sinc pulses, On–Off Keying (OOK), which is a special form of PAM, and FSK. In each case, the source encoder samples and quantizes the signal $x(t)$. Quantization is a means to digitize a signal by mapping a continuously varying signal to one that is discrete in value. Examples of two quantization mappings are shown in Figure 6.34. Note that the quantization shown in Figure 6.34a has an odd number of discrete levels, while that shown in Figure 6.34b has an even number of levels. For PAM transmission, these levels are coded onto an alphabet for modulation, and, therefore, it is desirable to have the number of quantization levels be equal to a power of 2, so that the alphabet size is also a power of 2. At each sampling time, the source encoder samples and quantizes the waveform $x(t)$, leading to a sequence of discrete values that correspond to the quantization levels. For the PAM transmission below, this sequence is multiplied by a scalar to form the sequence $z[n]$. In the case of OOK and FSK, this sequence is coded further resulting in a different sequence for $z[n]$.

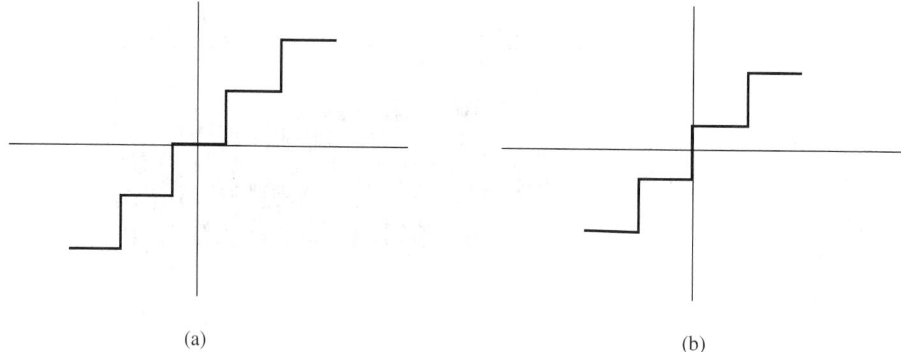

Figure 6.34 Quantization mappings.

For simplicity, no additional coding is performed in the channel encoder, that is, $u[n] = z[n]$. The sequence $u[n]$ is used to generate a transmission signal $s(t)$ that depends on the type of modulation method. The channel remains the same for each case (except for the FSK case) and has lowpass filter characteristics with bandwidth $\omega_b = 200\pi$ rad/sec. This channel limitation is not used for the FSK modulation case, since FSK produces a wide frequency spectrum and would not be used in a situation where there is a severe bandwidth limitation. The response of the channel to the signal $s(t)$ is found from the MATLAB command `lsim`. A random noise signal is added to the response to form the received signal $r(t)$. A lowpass receiver filter with bandwidth $\omega_b = 200\pi$ rad/sec is used to filter out high-frequency noise. The demodulation and decoding are performed differently, depending on the type of modulation scheme used. The final product is a sequence of numbers that is used to reconstruct the estimate $\hat{x}(t)$ of $x(t)$. For details, the reader is directed to the MATLAB files that were used to generate these results. The files can be obtained from the Web site that accompanies this text.

M-ary PAM with Square Pulses

The procedure outlined above is used to transmit the signal. For M-ary PAM, the sequence $u[n]$ is generated by quantizing the samples of $x(t)$ into M different levels and then normalizing the values. In particular, the alphabet for $u[n]$ is chosen to be $\{-1, -1/3, 1/3, 1\}$ for 4-ary PAM and is $\{-1, -3/5, -1/5, 1/5, 3/5, 1\}$ for 8-ary PAM. Equation 6.6 is used to generate the transmitted signal $s(t)$ from square pulses for $g(t)$. The resulting transmitted signal for $M = 8$ is shown in Figure 6.35 along with the received signal (including channel distortion and noise). The received signal $r(t)$ is first filtered using a lowpass filter with bandwidth $\omega_b = 200\pi$ rad/sec in order to remove some of the random noise. The filtered received signal is then sampled near the end of each signal interval in order to minimize the error due to the channel characteristics. These samples are sent through a slicer to obtain a sequence $\hat{u}[n]$ with discrete values taken from the alphabet for $u[n]$. Finally, $\hat{u}[n]$ is decoded to obtain a sequence of values $\hat{x}[n]$ that are reconstructed to form $\hat{x}(t)$.

With $M = 8$, there is no error between $u[n]$ and $\hat{u}[n]$. The error evident in $\hat{x}(t)$, shown in Figure 6.36, is due solely to the quantization error. Increasing the number

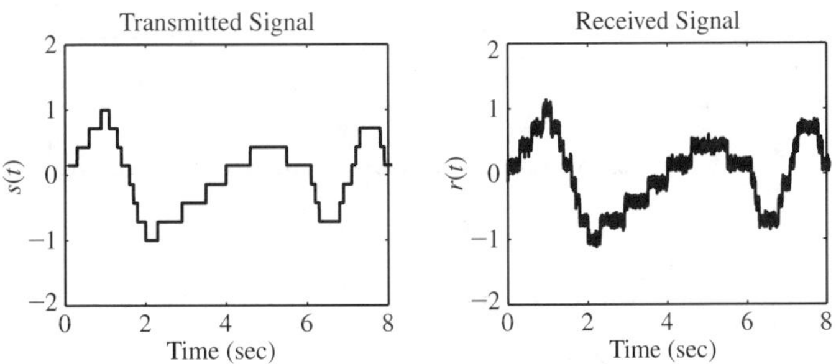

Figure 6.35 Transmitted and received signals for 8-ary PAM with square pulses.

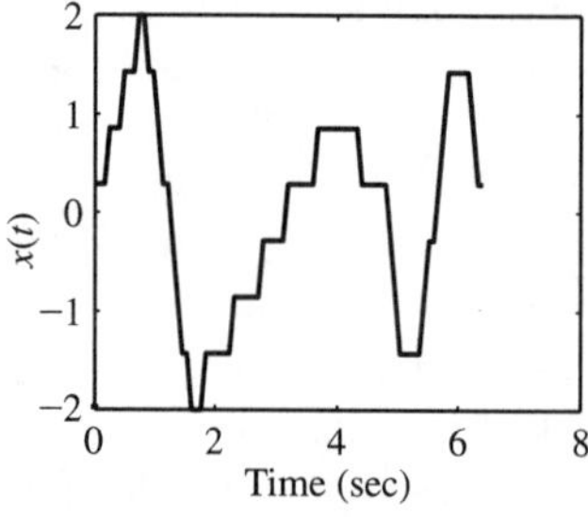

Figure 6.36 Reconstructed signal for 8-ary PAM with square pulses.

of quantization levels—that is, increasing M—should decrease the error between $x(t)$ and $\hat{x}(t)$. However, to maintain the same probability of symbol error due to noise, increasing the size of M necessitates an increase in the ratio of the signal energy to noise energy (or signal-to-noise ratio, SNR). This means that the average energy of the transmitted signal must be boosted if M is increased. As an example, consider the case when M is doubled to 16, but the average signal energy is maintained (requiring that the distance between the different pulse levels in the transmission is halved). The transmitted and received signals are shown in Figure 6.37. In comparison to Figure 6.35 with $M = 8$, the channel noise obscures the height of the individual pulses within in each signal interval making the decoding more difficult. In fact, with $M = 16$, there are some errors in $\hat{u}[n]$, as shown in Figure 6.38. These decoding errors are passed along to the reconstruction error $x(t) - \hat{x}(t)$.

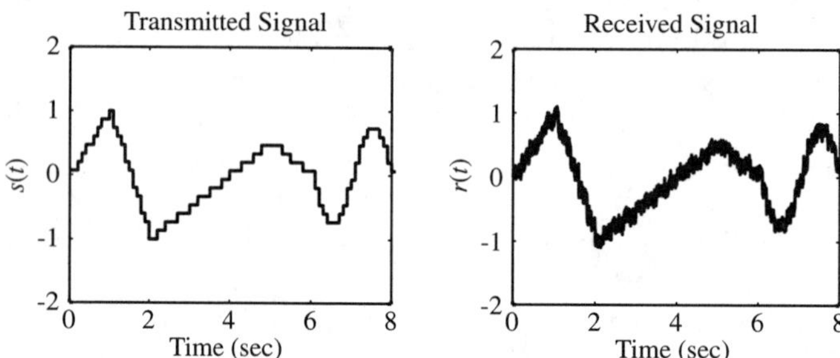

Figure 6.37 Transmitted and received signals for 16-ary PAM with square pulses.

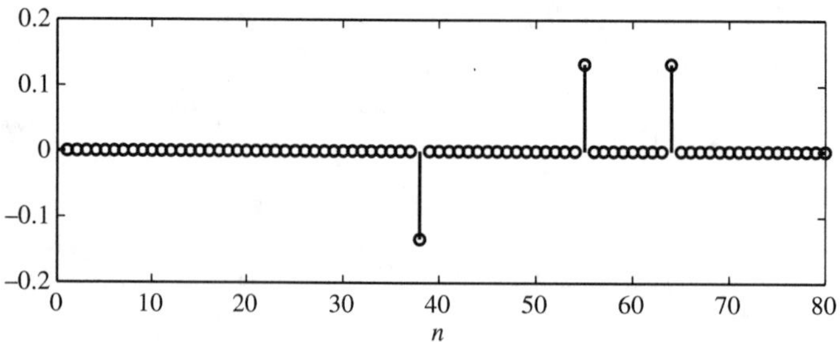

Figure 6.38 Error in $u[n]$ for 16-ary PAM with square pulse.

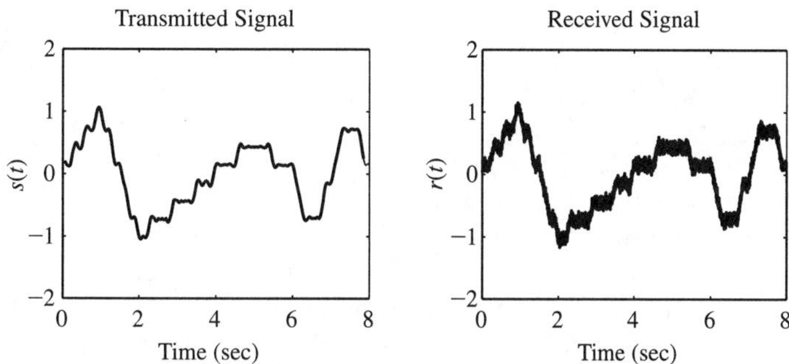

Figure 6.39 Transmitted and received signals for 8-ary PAM with sinc pulses.

M-ary PAM with sinc Pulses

As mentioned in Section 6.5, square pulses have a higher frequency content than sinc pulses. This can be a problem if the channel has a low bandwidth. Consider the PAM transmission of the signal displayed in Figure 6.33 and discussed in the previous section. In this section, the same signal is transmitted with using baseband PAM, except that the pulse $g(t)$ is selected to be a sinc function. The transmitted and the received signals are shown in Figure 6.39. Notice that the transmitted signal with the sinc pulses is smoother than the transmitted signal with the square pulses shown in Figure 6.35, indicating lower frequency content. This can be verified by examining the Fourier transforms of the transmitted signals $s(t)$ for both cases. Numerical estimates of the Fourier transforms of these signals are shown in Figure 6.40. (These estimates are obtained using a MATLAB command `contfft` that is discussed in detail in Chapter 7.) The plots clearly show that the sinc pulse transmission results in a lower frequency content than the transmission with square pulses.

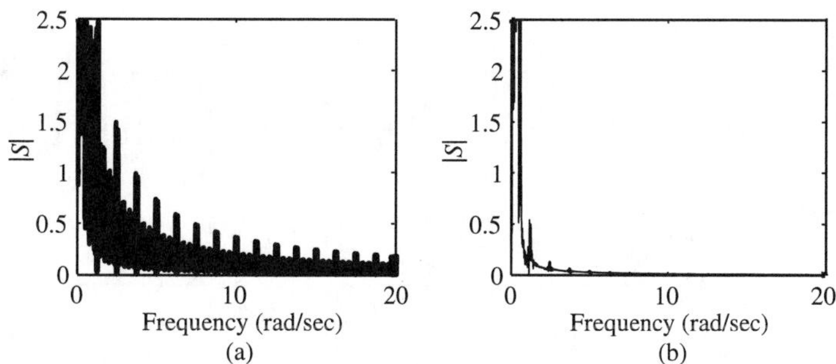

Figure 6.40 Fourier Transforms of $s(t)$ for (a) square pulses and (b) sinc pulses.

OOK Transmission

Binary PAM modulation requires having an alphabet consisting of two numbers. On–Off Keying (OOK) is the special case where the alphabet is {0, 1}, so that the transmitted signal is a sequence of zeros and ones. Previous results for PAM transmission show that a value of $M = 2$ should give relatively low sensitivity to noise; however, choosing two quantization levels would increase the quantization error considerably. Instead, a common method for binary coding called pulse code modulation (PCM) is used. The analog signal is first quantized into M levels, where M is a power of 2, as was done in the PAM transmission simulations. Then a corresponding binary representation of each quantization level is found. (See, for example, the table shown below when $M = 8$.) At each sample time, $x(t)$ is sampled and quantized. The corresponding quantization level of $x(nT)$ is then converted into its binary representation. The bits in the binary representation are then transmitted sequentially in signal intervals that are equal to the sample time divided by the number of bits needed to represent the quantization level ($\log_2 M$).

Consider the signal in Figure 6.33. Suppose that the number of quantization levels chosen is 8, so that the number of bits needed to represent each sample in $x(t)$ is $3 = \log_2 8$. If the sampling period is $T_s = 0.1$ sec, then the signal interval is $T = T_s/3$. The analog signal is coded in this manner with $M = 8$ (or 3 bits per sample); the transmitted signal and the received signals are shown in Figure 6.41. Note the time axis is expanded, so that details in the transmission can be viewed. The slicer has an easy task of decoding the received signal, since it must simply distinguish between 0 and 1. For this example, there is no error between $u[n]$ and $\hat{u}[n]$. The entire error in $\hat{x}(t)$ is due to quantization errors. If more bits are used, then the quantization error decreases without the signal becoming more sensitive to channel noise (because the modulation still only uses 0's and 1's, which are not very sensitive to noise). In fact, the OOK transmission in this example can transmit with no errors in $u[n]$ for quantization levels up to 65,536 (16 bits). Recall that there are errors in $u[n]$ with 16 quantization levels used in the 16-ary PAM simulation shown above. However, increasing the number of bits per channel in OOK transmission does decrease the sig-

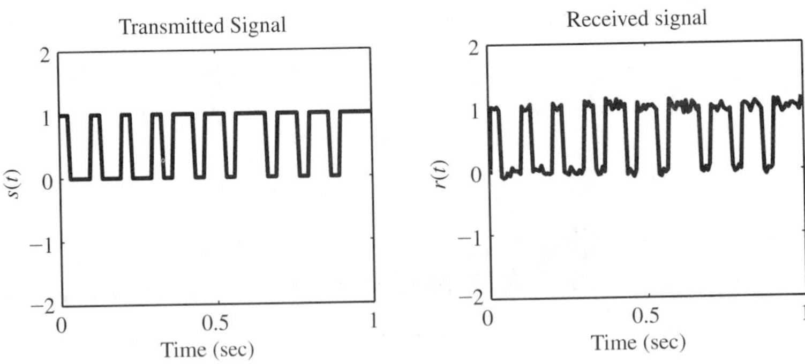

Figure 6.41 Transmitted and received signals for OOK.

nal interval, thereby increasing the frequency content of the transmitted signal $s(t)$. This may be a problem if the channel has severe bandwidth requirements.

Two common applications of PCM are for transmitting data in fiber-optic cables and for recording analog signals on compact dics. Fiber-optic cable has very large bandwidth capability. As a result, OOK (using PCM for coding data into a binary form) is the most common method of transmitting data in this medium. A laser simply flashes on to transmit a 1 or remains off to transmit a 0, and a photo detector picks up the flashes on the receiving end. Another application of PCM is in the recording (also known as *mastering*) of analog signals, such as speech and music on compact discs. The quantization and encoding remains the same as demonstrated in this example. Instead of transmitting a signal $s(t)$ over a channel, the signal is transferred onto the surface of a CD by using a laser to cut pits into the surface. Following a certain track, a laser has two choices: 1) to cut a pit into the disc; or 2) leave that part of the disc intact. This decision corresponds to the binary information that is being stored. The mastering of CD's actually uses a form of PCM coding that is different from that described in this example. Following a bit stream of numbers in $u[n]$, this code commands the laser to cut a pit whenever the bit stream changes from a 0 to a 1 or from a 1 to a 0. It does not cut a pit when the bit stream does not change.

Quantization Level	Relative Level	Binary Representation
7	7	111
5	6	110
3	5	101
1	4	100
−1	3	011
−3	2	010
−5	1	001
−7	0	000

FSK Transmission

Consider the OOK transmission simulated in the last section, where square pulses with amplitudes of either one or zero were transmitted in each symbol interval. In comparison, binary FSK transmits sinusoids of two different frequencies to represent the binary code. The frequencies are chosen so that the corresponding sinusoids are orthogonal to one another, that is

$$\int_T \cos(\omega_1 t)\cos(\omega_2 t)\,dt = 0$$

(Orthogonality of functions is used to derive the Fourier series. In fact, the harmonic sinusoids and the complex exponentials used in Fourier series are all orthogonal to one another. More discussion on orthogonal functions can be found in Problem 4.10.) In addition, the frequencies are often chosen so that there is no

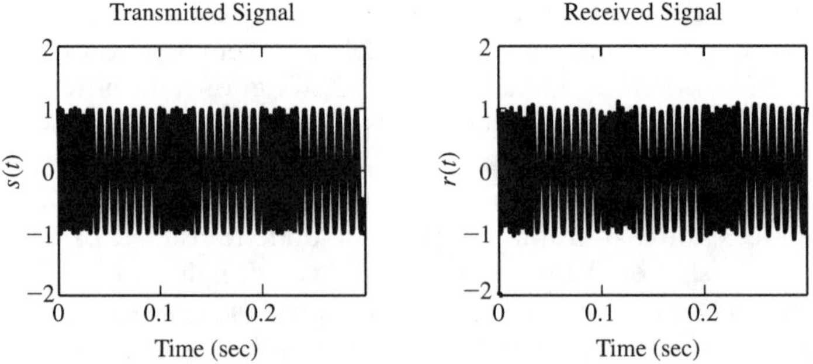

Figure 6.42 Transmitted and received signals for FSK.

discontinuity of the signal $s(t)$. Both criteria can be met by choosing frequencies of the form

$$\omega_1 = \frac{2\pi}{T}p \quad \text{and} \quad \omega_2 = \omega_1 + \frac{\pi}{T}r$$

where r and p are integers. For this example, values of $\omega_1 = 8\pi/T$ and $\omega_2 = 16\pi/T$ were chosen. The transmitted signal and the received signal are shown in Figure 6.42. Coherent demodulation can be performed by multiplying the signal $r(t)$ separately with each of the sinusoids and then integrating over one signal interval, as shown in Figure 6.43. Whichever has the higher value is determined to be the binary number that generated the signal. For the MATLAB simulation, the decision logic was simplified by multiplying the stored version of $r(t)$ with the stored version of $\cos(\omega_1 t)$ and with the stored version of $\cos(\omega_2 t)$ over one signal interval T to form the sums

$$\sum_{t_i} r(t_i)\cos(\omega_1 t_i) \quad \text{and} \quad \sum_{t_i} r(t_i)\cos(\omega_2 t_i)$$

where t_i represents the discrete values of time during the signal interval T for which the quantity $r(t)$ is defined. If these summations are scaled by $1/T$, they form approximations to the integrals shown in Figure 6.43. FSK transmission uses a large bandwidth, but very little power. Therefore, it is suitable to applications, such

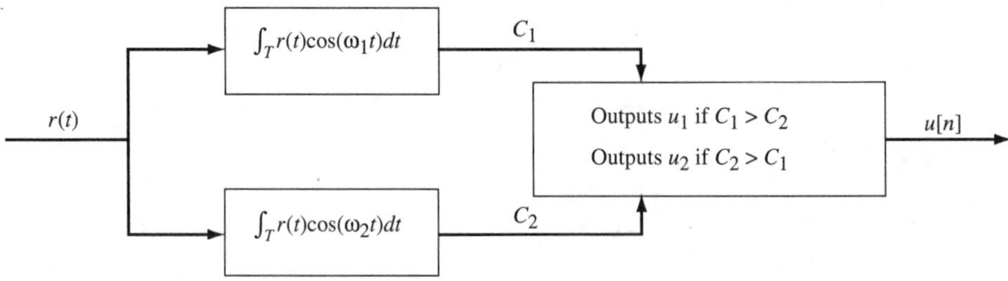

Figure 6.43 Demodulation for binary FSK.

as deep-space satellite communications, where bandwidth is not regulated, but power is at a premium.

In summary, there are several choices for the type of digital modulation used in a given application. Baseband PAM or passband PAM can be used, depending on the frequency characteristics of the channel. In actuality, most applications use some form of passband PAM; baseband PAM is still important, since it forms a stepping stone to generating a passband PAM signal. There are choices on the type of pulses $g(t)$ used to generate the transmitted signal; this choice depends on the channel characteristics, the noise level, and the allowable intersymbol interference. The size of the alphabet is selected based on the size of the noise in the channel, while the baud rate is determined by the bandwidth of the channel. The choice to use PSK, FSK, OOK, or QAM can be made based on the channel restrictions and/or government restrictions on bandwidth (such as those imposed on commercial radio broadcasts). Some channels are bandwidth limited (such as telephone lines using twisted-pair wires), and some are power limited (such as a satellite link). For example, QAM is the most common choice for data modems because it makes efficient use of the available bandwidth by coding two separate baseband PAM signals on the same transmission (one in phase and one quadrature). FSK is often used in situations when bandwidth is not a major concern, such as in satellite communications (e.g., deep space, geosynchronous satellites, etc.). PSK is used when bandwidth (allowable spectrum) is more precious (e.g., in cellular communications, low earth orbiting satellites, etc.). OOK is a common choice for transmissions using fiber-optic channels, since the on–off transmission is easy to achieve with optics, and fiber-optic cables have a very high bandwidth that can accommodate the needs of OOK. In communication systems, there are also numerous choices for the type of receiver, the compression or encrypting algorithm, the error coding (done in the channel encoder), and the filtering performed in the receiver to reduce ISI and noise. More information on digital communications can be found in [Lee and Messerschmitt, 1994], [Haykin, 1994], and [Wilson, 1996].

PROBLEMS

6.1. Consider the system in Figure P6.1a. The frequency response function $H(\omega)$ of the filter in Figure 6.1a is given by

$$H(\omega) = \begin{cases} 2e^{-j3\omega}, & -2 \le \omega \le 2 \\ 0, & \text{all other } \omega \end{cases}$$

(a) Compute the output response $y(t)$ when $x(t) = \cos t, -\infty < t < \infty$.
(b) Compute $y(t)$ when $x(t) = \cos 2t, -\infty < t < \infty$.
(c) Compute $y(t)$ when $x(t) = \text{sinc}(t/\pi)\cos 3t, -\infty < t < \infty$. Plot $x(t)$ and $y(t)$.
(d) Compute $y(t)$ when $x(t) = \text{sinc}^2(t/\pi), -\infty < t < \infty$. Plot $x(t)$ and $y(t)$.
(e) Compute $y(t)$ when $x(t)$ is the periodic signal shown in Figure P6.1b.
[*Hint:* Sketch $\cos 2tx(t)$.]

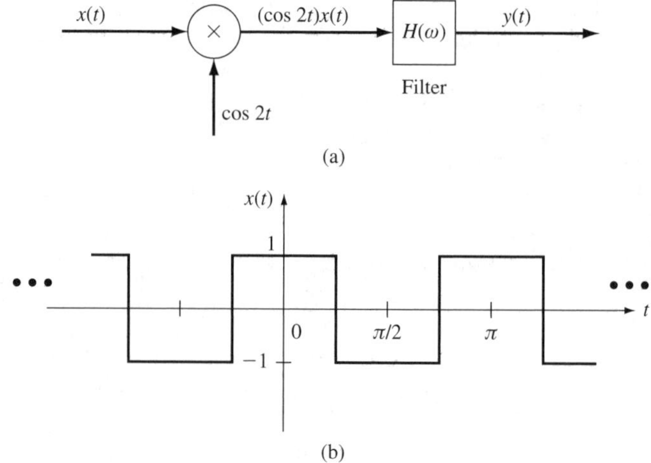

Figure P6.1

6.2. Consider the system in Figure P6.2a. The frequency functions of the filters in Figure P6.2a are given by

$$H_1(\omega) = \begin{cases} 3, & -2 \le \omega \le 2 \\ 0, & \text{all other } \omega \end{cases}, \qquad H_2(\omega) = \begin{cases} e^{-j\omega}, & \omega < -2, \omega > 2 \\ 0, & -2 \le \omega \le 2 \end{cases}$$

(a) Compute the output $y(t)$ when $x(t) = \text{sinc}(t/\pi)$, $-\infty < t < \infty$.

(b) Compute $y(t)$ when $x(t) = [\sin(t/\pi)](\cos 2t)$, $-\infty < t < \infty$.

(c) Compute the output response $y(t)$ when $x(t)$ is the periodic signal shown in Figure P6.2b.

(d) Plot the inputs $x(t)$ defined in parts (a) to (c) and the corresponding outputs $y(t)$.

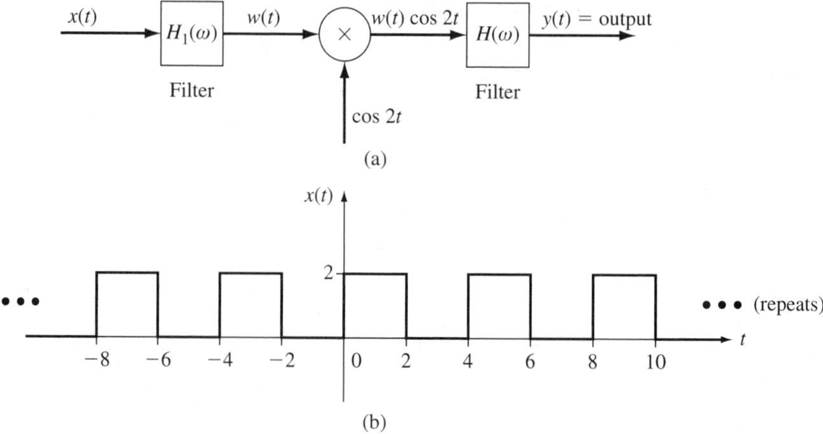

Figure P6.2

6.3. Consider the system in Figure P6.3a where the frequency functions of the filters are given by

$$H_1(\omega) = \begin{cases} 2, & 3 \le |\omega| \le 5 \\ 0, & \text{all other } \omega \end{cases}, \qquad H_2(\omega) = \begin{cases} 2, & |\omega| \le 3 \\ 0, & \text{all other } \omega \end{cases}$$

The input $x(t)$ has spectrum $X(\omega)$ shown in Figure P6.3b. Sketch $V(\omega)$, $W(\omega)$, $R(\omega)$, and $Y(\omega)$.

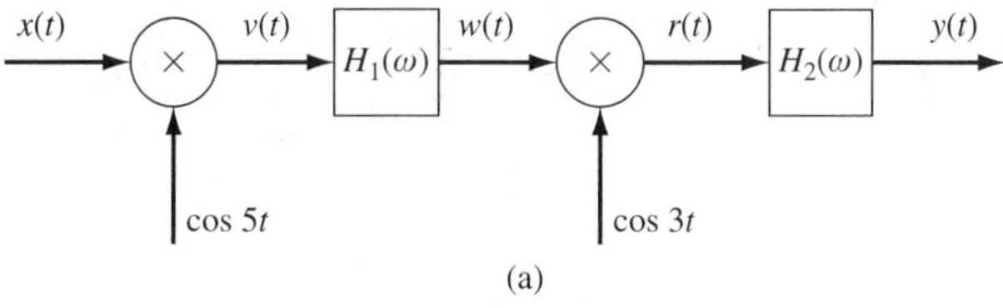

(a)

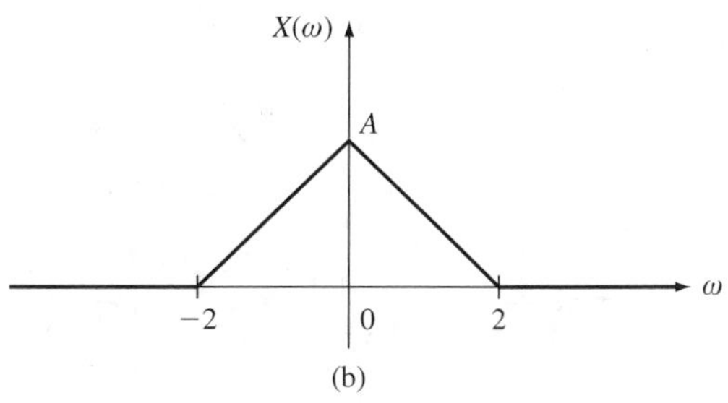

(b)

Figure P6.3

6.4. A signal $x(t)$ with bandwidth B is put on a carrier $\cos \omega_c t$ with $\omega_c >> B$. The modulated signal $x(t) \cos \omega_c t$ is then applied to the system shown in Figure P6.4. The frequency function $H(\omega)$ of the filter in the system is given by

$$H(\omega) = \begin{cases} 2, & -B < \omega < B \\ 0, & \text{all other } \omega \end{cases}$$

Compute $y(t)$ when:

(a) $\theta = 0°$

(b) $\theta = 90°$

(c) $\theta = 180°$

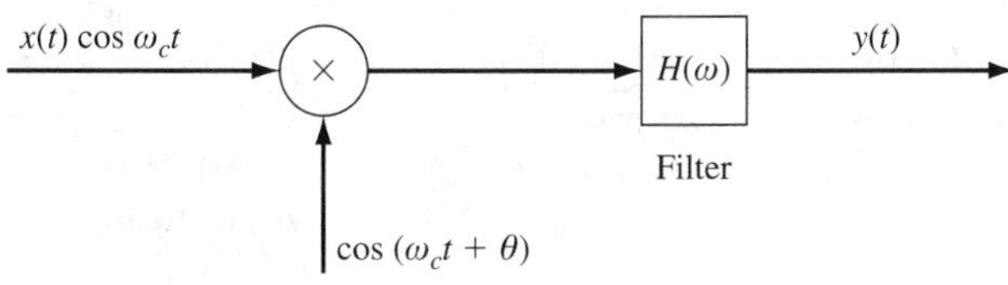

Figure P6.4

6.5. The input $x(t)$ to the demodulator in Figure P6.5a is an amplitude-modulated signal $s(t) \cos \omega_c t$ plus an interfering signal $\cos Bt \cos At$; that is,

$$x(t) = s(t) \cos \omega_c t + (\cos Bt)(\cos At), \qquad B = \text{bandwidth of } s(t)$$

The amplitude spectrum $|S(\omega)|$ of $s(t)$ is shown in Figure P6.5b.

(a) Sketch the amplitude spectrum $|V(\omega)|$ of $v(t)$, where $v(t)$ is the signal applied to the lowpass filter in Figure P6.5a.

(b) For what range of values of A will the interfering signal cause the output $y(t)$ to be not equal to $s(t)$?

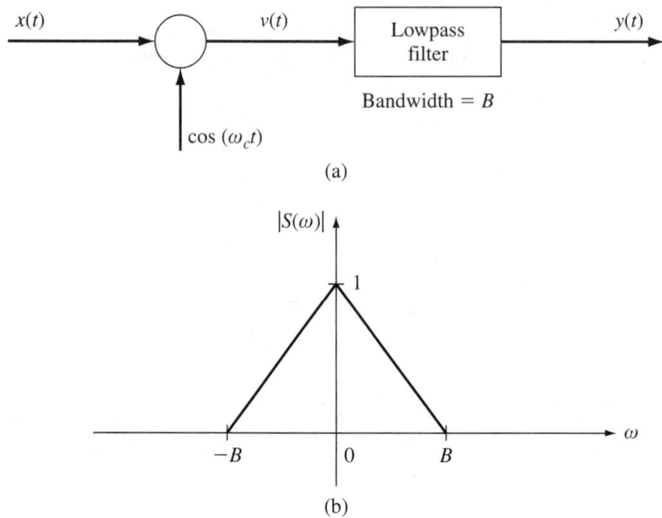

Figure P6.5

6.6. The objective of this problem is to show that the balanced modulator illustrated in Figure P6.6a can be used to recover the signal $x(t)$ from the amplitude-modulated waveform $x(t) \cos \omega_c t$. The signal $x(t)$ has the amplitude spectrum shown in Figure P6.6b.

(a) Determine the signal $e(t)$ in the output of the subtracter.

(b) Determine $E(\omega)$ and sketch the amplitude spectrum $|E(\omega)|$.

(c) What conditions on ω_c are necessary to ensure that $x(t)$ can be recovered from $x(t) \cos \omega_c t$ using the system in Figure P6.6a?

(d) Assuming that the conditions in part (c) are met, design the filter so that $y(t) = x(t)$. Give the filter type, bandwidth, gain, and so on.

6.7. A signal $x(t)$ with bandwidth B is applied to the system shown in Figure P6.7. Design a continuous-time system that reproduces $x(t)$ from the output of the system in Figure P6.7; that is, the response of the system to input $4x(t) + x(t) \cos Bt$ is equal to $x(t)$.

6.8. A speech signal $x(t)$ with bandwidth B is applied to the speech scrambler shown in Figure P6.8. Filter 1 is an ideal zero-phase highpass filter that stops all frequencies below ω_c (rad/sec), where $\omega_c >> B$. The second filter is an ideal zero-phase lowpass filter with bandwidth ω_c. Design a descrambler that reproduces $x(t)$ from the output $y(t)$ of the scrambler.

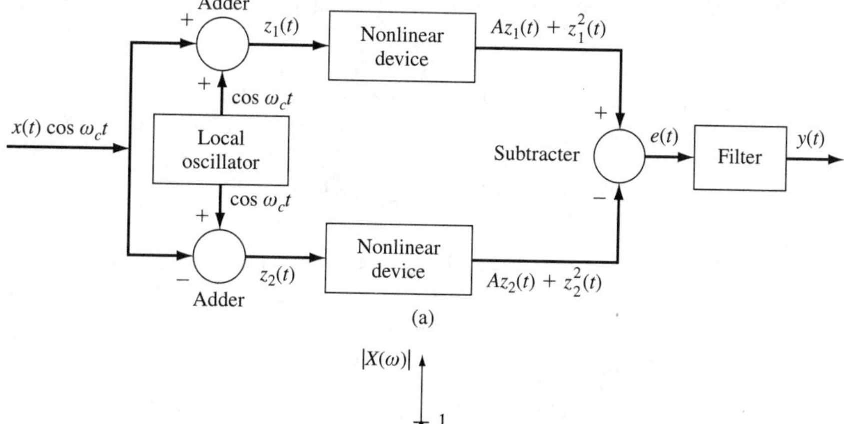

(a)

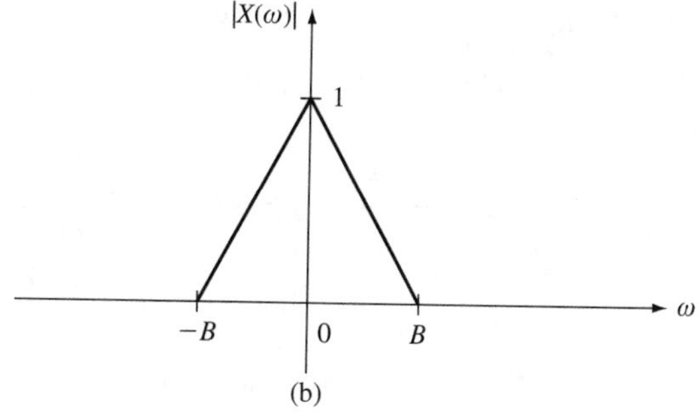

(b)

Figure P6.6

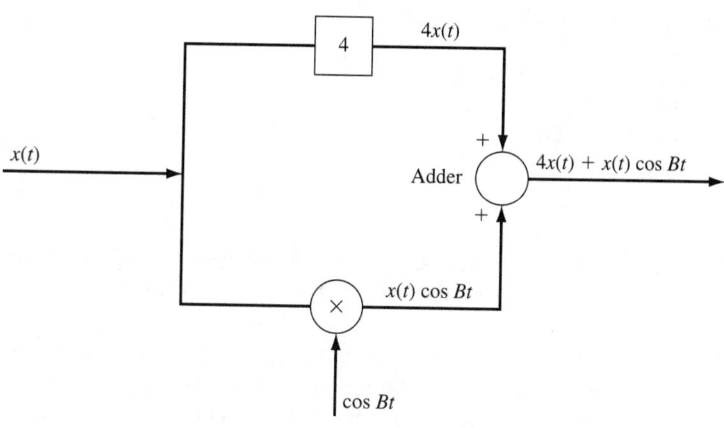

Figure P6.7

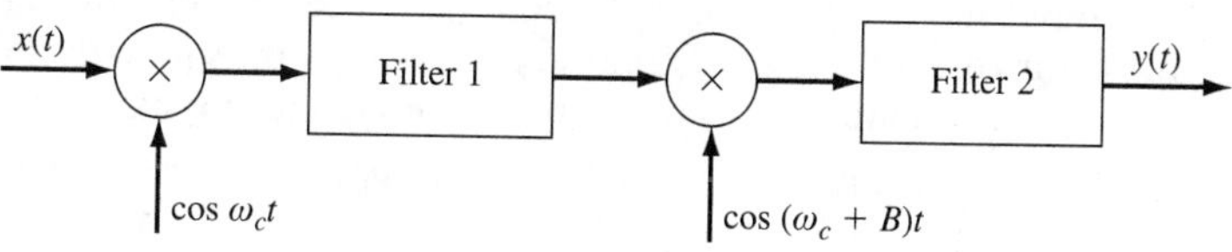

Figure P6.8

6.9. A special multiplexing technique for two signals $x_1(t)$ and $x_2(t)$ produces the signal

$$x(t) = x_1(t) \cos \omega_c t + x_2(t) \sin \omega_c t$$

where ω_c is much greater than the bandwidth B_1 of $x_1(t)$ and the bandwidth B_2 of $x_2(t)$. The multiplexed signal $x(t)$ is applied to the system shown in Figure P6.9. Design the filters in this system so that the first output $y_1(t)$ is equal to $x_1(t)$ and the second output $y_2(t)$ is equal to $x_2(t)$. Give the filter types, bandwidths, and so on.

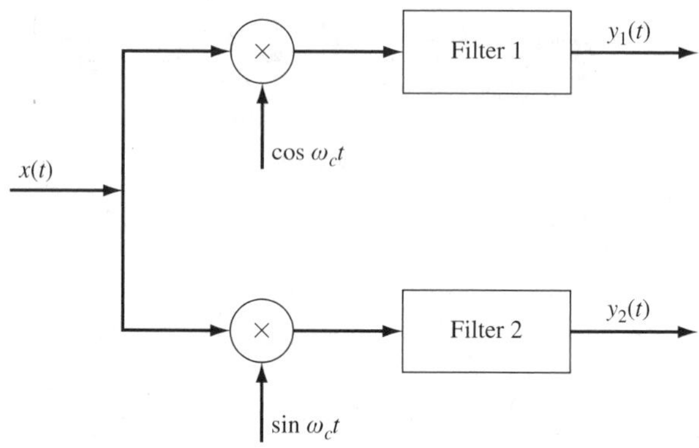

Figure P6.9

6.10. Sketch the Fourier transforms of the following signals. Identify important values on the axes.

(a) $x(t) = e^{-10t} \cos(100t)u(t)$
(b) $x(t) = (1 - |t|)p_2(t) \cos(10t)$
(c) $x(t) = p_2(t) \cos(10t)$
(d) $x(t) = 2 \operatorname{sinc}^2\left(\dfrac{t}{2p}\right) \cos(10t)$
(e) $x(t) = 4 \cos(10t) \cos(100t)$

6.11. Suppose that a channel has a bandwidth of 5000 Hz. Select a baud rate of 80% of the bandwidth. Determine the size of M in M-ary PAM needed to transmit at a rate of at least 9600 bps.

6.12. Suppose that a channel has a bandwidth of 10 kHz. Determine the smallest signal interval to achieve a baud rate of no more than 90% of the bandwidth. What happens when you choose a signal interval that gives a baud rate much larger than the channel bandwidth?

6.13. Prove that the demodulation scheme for QAM shown in Figure 6.30 does recover the two Baseband PAM signals, $s_{u1}(t)$ and $s_{u2}(t)$. In particular, first compute the Fourier transforms of the signals after the multiplication by the sinusoids (in terms of $S_{u1}(\omega)$ and $S_{u2}(\omega)$). Then compute the Fourier transforms of the outputs of the lowpass filters.

6.14. Prove the following sets of signals are orthogonal over the time interval T.

(a) $\sin\left(\dfrac{2\pi t}{T}\right), \cos\left(\dfrac{2\pi t}{T}\right)$

(b) $\cos\left(\frac{2\pi t}{T}\right), \cos\left(\frac{4\pi t}{T}\right)$

(c) $\cos\left(\frac{2n\pi t}{T}\right), \cos\left(\frac{4n\pi t}{T}\right), \cos\left(\frac{6n\pi t}{T}\right)$

6.15. Consider a sequence of $u[n] = [-3\ 1\ 3\ -1\ -3]$.

(a) Draw the corresponding waveform for $s(t)$ if using 4-ary Baseband PAM with square pulses for $g(t)$.

(b) Draw the corresponding waveform for $s_m(t)$ using 4-ary PSK.

(c) Draw a possible waveform for $s_m(t)$ using 4-ary FSK.

6.16. Consider a sequence $u[n] = [5\ 7\ 1\ 3\ -1\ -5\ -3]$.

(a) Draw the corresponding waveform for $s(t)$ if using 8-ary Baseband PAM with square pulses for $g(t)$.

(b) Use MATLAB to draw the corresponding waveform for $s_m(t)$ using 8-ary PSK.

(c) Use MATLAB to draw a possible waveform for $s_m(t)$ using 8-ary FSK.

6.17. Consider the signal $x(t) = e^{-2t}\cos(10t)u(t)$ for $t = 0$ sec to $t = 1$ sec. The goal is to transmit this signal to a remote location using digital modulation.

(a) Sketch the sampled signal $x[n]$ if the sampling period is $T_s = 0.1$ sec.

(b) Create a mapping between the signal $x[n]$ and an 8-ary alphabet consisting of the numbers $\{-7, -5, -3, -1, 1, 3, 5, 7\}$. Use uniform quantization for the mapping. The resulting sequence is $u[n]$.

(c) Create a signal to be transmitted $s(t)$ using square pulses for $g(t)$, and sketch $s(t)$.

(d) Suppose that the channel distortion and noise are negligible, so that $r(t) \approx s(t)$. Now draw the mapping in the slicer that gives the relationship between the sampled $r[n]$ and the reconstructed sequence $\hat{u}[n]$.

(e) Pass the signal through an appropriate source decoder to create a signal $\hat{x}[n]$. Sketch the reconstructed signal $\hat{x}(t)$.

(Quantization need not be uniform; that is, the difference between the quantization levels may be smaller near the origin and larger near the maximum and minimum values of $x(t)$. Nonuniform quantization is often used when the signal $x(t)$ spends far less time near its maximum and minimum than near the origin. You may improve your overall results in this problem by experimenting with nonuniform quantization.)

6.18. Consider the signal $x(t) = e^{-2t}\cos(10t)u(t)$ for $t = 0$ sec to $t = 2$ sec. Repeat Problem 6.17 parts (a)–(e) using MATLAB to simulate the transmission and to plot the signals. Use a sampling period of $T_s = 0.05$ sec.

6.19. Repeat Problem 6.18, except use 16-ary transmission. Compare the results generated here with those generated in Problem 6.18. Considering the improved results obtained using 16-ary, describe any drawbacks between using 16-ary and 8-ary.

6.20. Repeat Problem 6.18 with MATLAB, except use OOK (with PCM) to code the sequence for $u[n]$. Compare the results obtained with 8 levels of quantization to results obtained with 32 levels of quantization.

6.21. Repeat Problem 6.18 with MATLAB, except use binary FSK (with PCM) to code the sequence for $u[n]$. Compare the results obtained with 8 levels of quantization to results obtained with 32 levels of quantization.

6.22. Consider two signals $x_1(t) = \sin(800\pi t)$ and $x_2(t) = \sin(400\pi(80t^2 + 2t))$. The objective in this problem is to perform a simulation of QAM transmission of these signals.

(a) Plot the signals for $t = 0$ sec to 0.01 sec.

(b) Sample the signals at $T_s = 0.05$ msec to obtain the sequences $x_1[n]$ and $x_2[n]$. Use uniform quantization with 16 levels to obtain sequences $u_1[n]$ and $u_2[n]$ corresponding to $x_1(t)$ and $x_2(t)$, respectively.

(c) Compute the corresponding signals $s_{u1}(t)$ and $s_{u2}(t)$.

(d) Compute the quadrature and in-phase components to form the signal to be transmitted, $s_m(t)$. Use a carrier signal $\omega_c = 20000\pi$ rad/sec. Plot $s_m(t)$ for $t = 0$ sec to 0.01 sec.

(e) Assume that the received signal is the same as the transmitted signal. (That is, assume that the channel distortion and the noise are negligible.) Perform the demodulation indicated in Figure 6.20 to obtain the estimates $\hat{u}_1[n]$ and $\hat{u}_2[n]$. Use `lsim` to simulate a lowpass filter. For example, the following commands can be used to implement a second order lowpass Butterworth filter that recovers $\hat{s}_{u1}(t)$ from $r_1(t)$:

```
wb = 7500*pi; % bandwidth of lowpass filter
num = wb^2;
den = [1 2*.707*wb wb^2];
su1hat = lsim(num,den,r1,t);
```

(f) Compare the estimates $\hat{u}_1[n]$ and $\hat{u}_2[n]$ to $u_1[n]$ and $u_2[n]$, and plot the differences to determine if there were any errors in the transmission.

(g) Reconstruct the estimates $\hat{x}_1(t)$ and $\hat{x}_2(t)$ and plot them.

◆

CHAPTER 7

Fourier Analysis of Discrete-Time Signals and Systems

The discrete-time counterpart to the theory developed in Chapters 4 and 5 is presented in this chapter. The development begins in Section 7.1 with the study of the discrete-time Fourier transform (DTFT), which is the discrete-time counterpart to the Fourier transform. As is the case for the Fourier transform of a continuous-time signal, the DTFT of a discrete-time signal is a function of a continuum of frequencies, but unlike the continuous-time case, the DTFT is always a periodic function with period 2π.

In Section 7.2 a transform of a discrete-time signal is defined that is a function of a finite number of frequencies. This transform is called the discrete Fourier transform (DFT). For time-limited discrete-time signals, it is shown that the DFT is equal to the DTFT with the frequency variable evaluated at a finite number of points. Hence, the DFT can be viewed as a "discretization in frequency" of the DTFT. Since the DFT is a function of a finite number of frequencies, it is the transform that is often used in practice. In particular, the DFT is used extensively in digital signal processing and digital communications.

In Section 7.3 the properties of the DFT are studied, and then in Section 7.4 the DTFT and DFT are applied to the study of linear time-invariant discrete-time systems. Section 7.5 deals with a fast method for computing the DFT, called the fast Fourier transform (FFT) algorithm. In Section 7.6 the FFT algorithm is utilized to compute the Fourier transform of a continuous-time signal and to compute the output response of a linear time-invariant discrete-time system.

7.1 DISCRETE-TIME FOURIER TRANSFORM

In Section 4.3 the Fourier transform $X(\omega)$ of a continuous-time signal $x(t)$ was defined by

$$X(\omega) = \int_{-\infty}^{\infty} x(t)e^{-j\omega t}\, dt \tag{7.1}$$

Given a discrete-time signal $x[n]$, the discrete-time Fourier transform (DTFT) of $x[n]$ is defined by

$$X(\Omega) = \sum_{n=-\infty}^{\infty} x[n]e^{-j\Omega n} \tag{7.2}$$

The DTFT $X(\Omega)$ defined by (7.2) is in general a complex-valued function of the real variable Ω (the frequency variable). Note that (7.2) is a natural discrete-time counterpart of (7.1) in that the integral is replaced by a summation. The uppercase omega (Ω) is utilized for the frequency variable to distinguish between the continuous- and discrete-time cases.

A discrete-time signal $x[n]$ is said to have a DTFT in the *ordinary sense* if the bi-infinite sum in (7.2) converges (i.e., is finite) for all real values of Ω. A sufficient condition for $x[n]$ to have a DTFT in the ordinary sense is that $x[n]$ be absolutely summable; that is,

$$\sum_{n=-\infty}^{\infty} |x[n]| < \infty \tag{7.3}$$

If $x[n]$ is a time-limited discrete-time signal (i.e., there is a positive integer N such that $x[n] = 0$ for all $n \leq -N$ and $n \geq N$), then obviously the sum in (7.3) is finite, and thus any such signal has a DTFT in the ordinary sense.

Example 7.1 *Computation of DTFT*

Consider the discrete-time signal $x[n]$ defined by

$$x[n] = \begin{cases} 0, & n < 0 \\ a^n, & 0 \leq n \leq q \\ 0, & n > q \end{cases}$$

where a is a nonzero real constant and q is a positive integer. This signal is clearly time limited, and thus it has a DTFT in the ordinary sense. To compute the DTFT, insert $x[n]$ into (7.2), which yields

$$X(\Omega) = \sum_{n=0}^{q} a^n e^{-j\Omega n}$$

$$= \sum_{n=0}^{q} (ae^{-j\Omega})^n \tag{7.4}$$

The summation above can be written in "closed form" by using the relationship

$$\sum_{n=q_1}^{q_2} r^n = \frac{r^{q_1} - r^{q_2+1}}{1 - r} \tag{7.5}$$

where q_1 and q_2 are integers with $q_2 > q_1$ and r is a real or complex number. The reader is asked to prove (7.5) in Problem 7.2. Then using (7.5) with $q_1 = 0, q_2 = q$ and $r = ae^{-j\Omega}$, (7.4) can be written in the form

$$X(\Omega) = \frac{1 - (ae^{-j\Omega})^{q+1}}{1 - ae^{-j\Omega}} \tag{7.6}$$

For any discrete-time signal $x[n]$, the DTFT $X(\Omega)$ is a periodic function of Ω with period 2π; that is,

$$X(\Omega + 2\pi) = X(\Omega) \qquad \text{for all } \Omega, -\infty < \Omega < \infty$$

To prove the periodicity property, note that

$$X(\Omega + 2\pi) = \sum_{n=-\infty}^{\infty} x[n]e^{-jn(\Omega+2\pi)}$$

$$= \sum_{n=-\infty}^{\infty} x[n]e^{-jn\Omega}e^{-jn2\pi}$$

But

$$e^{-jn2\pi} = 1 \qquad \text{for all integers } n$$

and thus

$$X(\Omega + 2\pi) = X(\Omega) \qquad \text{for all } \Omega$$

An important consequence of periodicity of $X(\Omega)$ is that $X(\Omega)$ is completely determined by computing $X(\Omega)$ over any 2π interval such as $0 \le \Omega \le 2\pi$ or $-\pi \le \Omega \le \pi$.

Given the discrete-time signal $x[n]$ with DTFT $X(\Omega)$, since $X(\Omega)$ is complex valued in general, $X(\Omega)$ can be expressed in either rectangular or polar form. Using Euler's formula yields the following rectangular form of $X(\Omega)$:

$$X(\Omega) = R(\Omega) + jI(\Omega) \tag{7.7}$$

where $R(\Omega)$ and $I(\Omega)$ are real-valued functions of Ω given by

$$R(\Omega) = \sum_{n=-\infty}^{\infty} x[n] \cos n\Omega$$

$$I(\Omega) = -\sum_{n=-\infty}^{\infty} x[n] \sin n\Omega$$

The polar form of $X(\Omega)$ is

$$X(\Omega) = |X(\Omega)| \exp[j\angle X(\Omega)] \tag{7.8}$$

where $|X(\Omega)|$ is the magnitude of $X(\Omega)$ and $\angle X(\Omega)$ is the angle of $X(\Omega)$. Note that since $X(\Omega)$ is periodic with period 2π, both $|X(\Omega)|$ and $\angle X(\Omega)$ are periodic with period 2π. Thus both $|X(\Omega)|$ and $\angle X(\Omega)$ need to be specified only over some interval of length 2π such as $0 \le \Omega \le 2\pi$ or $-\pi \le \Omega \le \pi$.

Assuming that $x[n]$ is real valued, the magnitude function $|X(\Omega)|$ is an even function of Ω and the angle function $X(\Omega)$ is an odd function of Ω; that is,

$$|X(-\Omega)| = |X(\Omega)| \qquad \text{for all } \Omega \tag{7.9}$$

$$\angle X(-\Omega) = -\angle X(\Omega) \qquad \text{for all } \Omega \tag{7.10}$$

To verify (7.9) and (7.10), first replace Ω by $-\Omega$ in (7.2), which gives

$$\begin{aligned} X(-\Omega) &= \sum_{n=-\infty}^{\infty} x[n]e^{j\Omega n} \\ &= \overline{X(\Omega)} \end{aligned} \tag{7.11}$$

where $\overline{X(\Omega)}$ is the complex conjugate of $X(\Omega)$. Now replacing Ω by $-\Omega$ in the polar form (7.8) gives

$$X(-\Omega) = |X(-\Omega)| \exp[j\angle X(-\Omega)] \tag{7.12}$$

and taking the complex conjugate of both sides of (7.8) gives

$$\overline{X(\Omega)} = |X(\Omega)| \exp[-j\angle X(\Omega)] \tag{7.13}$$

Finally, combining (7.11)–(7.13) yields

$$|X(-\Omega)| \exp[j\angle X(-\Omega) = |X(\Omega)| \exp[-j\angle X(\Omega)]$$

Hence it must be true that

$$|X(-\Omega)| = |X(\Omega)|$$

and

$$\angle X(-\Omega) = -\angle X(\Omega)$$

which verifies (7.9) and (7.10).

If the DTFT is given in the rectangular form (7.7), it is possible to generate the polar form (7.8) by using the relationships

$$|X(\Omega)| = \sqrt{R^2(\Omega) + I^2(\Omega)} \tag{7.14}$$

$$\angle X(\Omega) = \tan^{-1} \frac{I(\Omega)}{R(\Omega)} \tag{7.15}$$

Example 7.2 *Rectangular and Polar Forms*

Consider the discrete-time signal $x[n] = a^n u[n]$ where a is a nonzero real constant and $u[n]$ is the discrete-time unit-step function. For the case $a = 0.5$, the signal is displayed in Figure 7.1. The signal $x[n] = a^n u[n]$ is equal to the signal in Example 7.1 in the limit

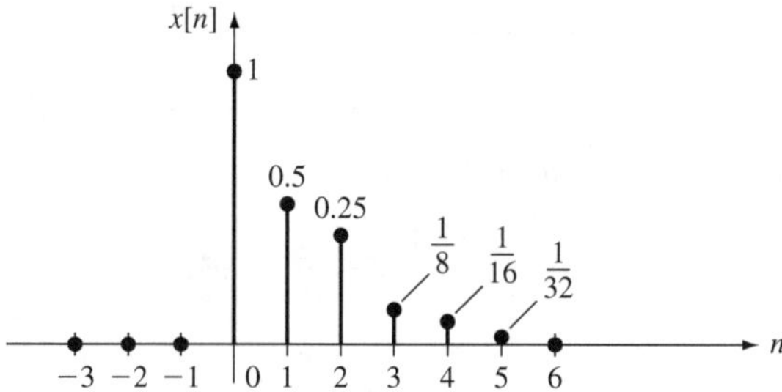

Figure 7.1 The signal $x[n] = (0.5)^n u[n]$.

as $q \to \infty$. Hence the DTFT $X(\Omega)$of $x[n]$ is equal to the limit as $q \to \infty$ of the DTFT of the signal in Example 7.1. That is, using (7.5), the DTFT is

$$X(\Omega) = \lim_{q\to\infty} \frac{1 - (ae^{-j\Omega})^{q+1}}{1 - ae^{-j\Omega}}$$

Now the limit above exists if and only if $|a| < 1$, in which case

$$\lim_{q\to\infty} (ae^{-j\Omega})^{q+1} = 0$$

Thus, when $|a| < 1$, the signal $x[n] = a^n u[n]$ has a DTFT in the ordinary sense given by

$$X(\Omega) = \frac{1}{1 - ae^{-j\Omega}} \tag{7.16}$$

When $|a| \geq 1, x[n] = a^n u[n]$ does not have a DTFT in the ordinary sense.

To express the DTFT given by (7.16) in rectangular form, first multiply the right-hand side of (7.16) by $[1 - a\exp(j\Omega)]/[1 - a\exp(j\Omega)]$, which gives

$$\begin{aligned} X(\Omega) &= \frac{1 - ae^{j\Omega}}{(1 - ae^{-j\Omega})(1 - ae^{j\Omega})} \\ &= \frac{1 - ae^{j\Omega}}{1 - a(e^{-j\Omega} + e^{j\Omega}) + a^2} \end{aligned}$$

Using Euler's formula, $X(\Omega)$ becomes

$$X(\Omega) = \frac{1 - a\cos\Omega - ja\sin\Omega}{1 - 2a\cos\Omega + a^2}$$

and thus the rectangular form of $X(\Omega)$ is

$$X(\Omega) = \frac{1 - a\cos\Omega}{1 - 2a\cos\Omega + a^2} + j\frac{-a\sin\Omega}{1 - 2a\cos\Omega + a^2} \tag{7.17}$$

To compute the polar form of $X(\Omega)$, first take the magnitude of both sides of (7.16), which gives

$$|X(\Omega)| = \frac{1}{|1 - ae^{-j\Omega}|}$$

$$= \frac{1}{|1 - a\cos\Omega + ja\sin\Omega|}$$

$$= \frac{1}{\sqrt{(1 - a\cos\Omega)^2 + a^2\sin^2\Omega}}$$

$$= \frac{1}{\sqrt{1 - 2a\cos\Omega + a^2}}$$

Finally, taking the angle of the right-hand side of (7.16) yields

$$\angle X(\Omega) = -\angle 1 - ae^{-j\Omega}$$

$$= -\angle(1 - a\cos\Omega + ja\sin\Omega)$$

$$= -\tan^{-1}\frac{a\sin\Omega}{1 - a\cos\Omega}$$

Therefore, the polar form of $X(\Omega)$ is

$$X(\Omega) = \frac{1}{\sqrt{1 - 2a\cos\Omega + a^2}}\exp\left(-j\tan^{-1}\frac{a\sin\Omega}{1 - a\cos\Omega}\right) \tag{7.18}$$

Note that the polar form (7.18) could also have been determined directly from the rectangular form (7.17) by using the relationships (7.14) and (7.15). The reader is invited to check that this results in the same answer as (7.18).

For the case $a = 0.5$, the magnitude function $|X(\Omega)|$ and the angle function $\angle X(\Omega)$ of the DTFT are plotted in Figure 7.2.

Signals with Even or Odd Symmetry

A real-valued discrete-time signal $x[n]$ is an even function of n if $x[-n] = x[n]$ for all integers $n \geq 1$. If $x[n]$ is an even signal, it follows from Euler's formula that the DTFT $X(\Omega)$ given by (7.2) can be expressed in the form

$$X(\Omega) = x[0] + \sum_{n=1}^{\infty} 2x[n]\cos\Omega n \tag{7.19}$$

From (7.19) it is seen that $X(\Omega)$ is a real-valued function of Ω, and thus the DTFT of an even signal is always real valued.

If $x[n]$ is an odd signal; that is, $x[-n] = -x[n]$ for all integers $n \geq 1$, the DTFT $X(\Omega)$ can be written in the form

$$X(\Omega) = x[0] - \sum_{n=1}^{\infty} j2x[n]\sin\Omega n \tag{7.20}$$

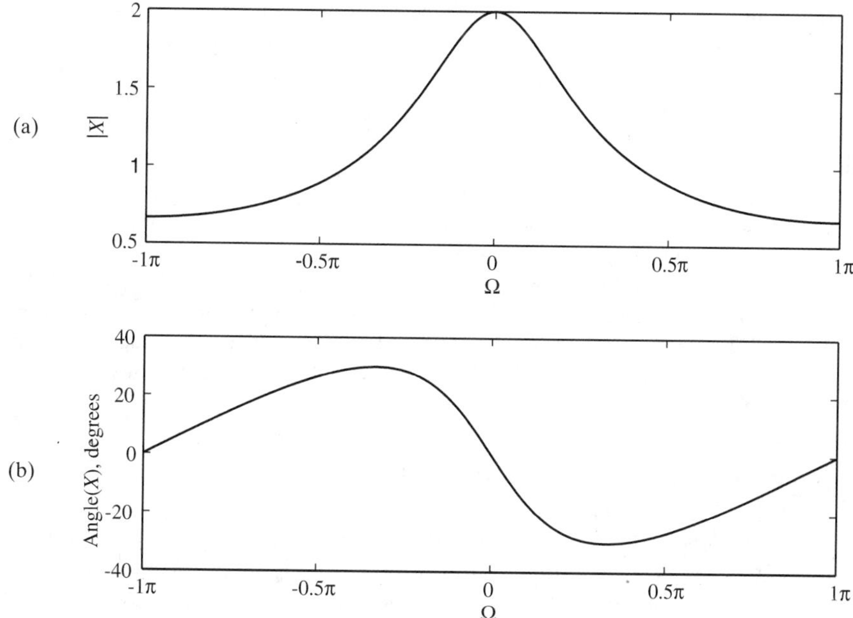

Figure 7.2 (a) Magnitude and (b) angle of the DTFT of the signal $x[n] = (0.5)^n u[n]$.

The reader is asked to verify (7.20) [and (7.19)] in Problem 7.3. From (7.20) it is seen that the DTFT of an odd signal is equal to the constant $x[0]$ plus a purely imaginary-valued function.

Example 7.3 *DTFT of Rectangular Pulse*

Given a positive integer q, let $p[n]$ denote the discrete-time rectangular pulse function defined by

$$p[n] = \begin{cases} 1, & n = -q, -q + 1, \ldots, -1, 0, 1, \ldots, q \\ 0, & \text{all other } n \end{cases}$$

This signal is even, and thus the DTFT is a real-valued function of Ω. To compute the DTFT $P(\Omega)$ of the pulse $p[n]$, it turns out to be easier to use (7.2) instead of (7.19): Inserting $p[n]$ into (7.2) gives

$$P(\Omega) = \sum_{n=-q}^{q} e^{-j\Omega n} \tag{7.21}$$

Then using (7.5) with $q_1 = -q, q_2 = q$, and $r = e^{-j\Omega}$, (7.21) becomes

$$P(\Omega) = \frac{e^{j\Omega q} - e^{-j\Omega(q+1)}}{1 - e^{-j\Omega}} \tag{7.22}$$

Multiplying the top and bottom of the right-hand side of (7.22) by $e^{j\Omega/2}$ gives

$$P(\Omega) = \frac{e^{j\Omega(q+1/2)} - e^{-j\Omega(q+1/2)}}{e^{j(\Omega/2)} - e^{-j(\Omega/2)}} \tag{7.23}$$

Finally, using Euler's formula, (7.23) reduces to

$$P(\Omega) = \frac{\sin[(q + 1/2)\Omega]}{\sin(\Omega/2)} \tag{7.24}$$

Hence the DTFT of the rectangular pulse $p[n]$ is given by (7.24). It is interesting to note that as the value of the integer q is increased, the plot of the DTFT $P(\Omega)$ looks more and more like a sinc function of the variable Ω. For example, in the case $q = 10$, $P(\Omega)$ is plotted in Figure 7.3 for $-\pi \le \Omega \le \pi$. The transform (7.24) is the discrete-time counterpart to the transform of the rectangular pulse in the continuous-time case (see Example 4.7).

Spectrum of a Discrete-Time Signal

Fourier analysis can be used to determine the frequency components of a discrete-time signal just as it was used for continuous-time signals. The decomposition of a periodic discrete-time signal $x[n]$ into sinusoidal components can be viewed as a generalization of the Fourier series representation of a periodic discrete-time signal. To keep the presentation as simple as possible, the discrete-time version of the Fourier series is not considered in this book.

For a discrete-time signal $x[n]$ that is not equal to a sum of sinusoids, the frequency spectrum consists of a continuum of frequency components that make up the signal. As in the continuous-time case, the DTFT $X(\Omega)$ displays the various sinusoidal components (with frequency Ω) that comprise $x[n]$, and thus $X(\Omega)$ is called the *frequency spectrum* of $x[n]$. The magnitude function $|X(\Omega)|$ is called the *amplitude spectrum* of the signal, and the angle function $\angle X(\Omega)$ is called the *phase spectrum* of the signal. In this book the plots of $|X(\Omega)|$ and $\angle X(\Omega)$ will be over the interval $-\pi \le \Omega \le \pi$. [Recall that $X(\Omega)$ is a periodic function of Ω with period 2π.] The

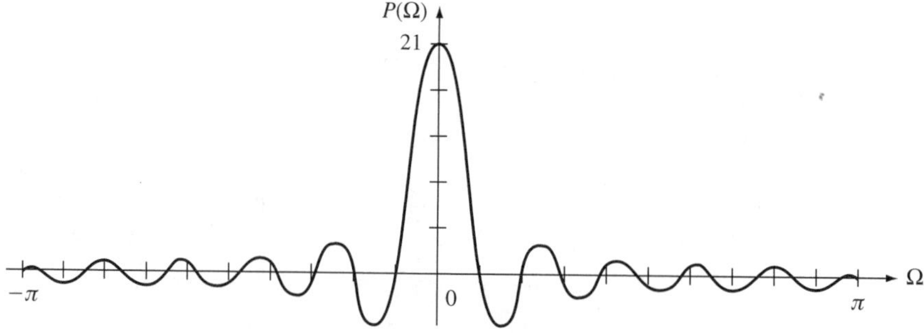

Figure 7.3 DTFT of the rectangular pulse $p[n]$ with $q = 10$.

sinusoidal components comprising $x[n]$ have positive frequencies ranging from 0 to π. Thus the highest possible frequency that may be in the spectrum of $x[n]$ is $\Omega = \pi$.

Example 7.4 *Signal with Low-Frequency Components*

Consider the discrete-time signal $x[n] = (0.5)^n u[n]$, which is plotted in Figure 7.1. The amplitude and phase spectrum of the signal were determined in Example 7.2, with the results plotted in Figure 7.2. Note that over the frequency range from 0 to π, most of the spectral content of the signal is concentrated near the zero frequency $\Omega = 0$. Thus the signal has a preponderance of low-frequency components.

Example 7.5 *Signal with High-Frequency Components*

Now consider the signal $x[n] = (-0.5)^n u[n]$, which is plotted in Figure 7.4. Note that due to the sign changes, the time variations of this signal are much greater than those of the signal in Example 7.4. Hence it is expected that the spectrum of this signal should contain a much larger portion of high-frequency components in comparison to the spectrum of the signal in Example 7.4.

From the results in Example 7.2, the DTFT of $x[n] = (-0.5)^n u[n]$ is

$$X(\Omega) = \frac{1}{1 + 0.5e^{-j\Omega}}$$

and the amplitude and phase spectra are given by

$$|X(\Omega)| = \frac{1}{\sqrt{1.25 + \cos \Omega}} \tag{7.25}$$

$$\angle X(\Omega) = -\tan^{-1} \frac{-0.5 \sin \Omega}{1 + 0.5 \cos \Omega} \tag{7.26}$$

Plots of $|X(\Omega)|$ and $\angle X(\Omega)$ are shown in Figure 7.5. From the figure, note that over the frequency range from 0 to π, the spectral content of the signal is concentrated near the highest possible frequency $\Omega = \pi$, and therefore this signal has a preponderance of high-frequency components.

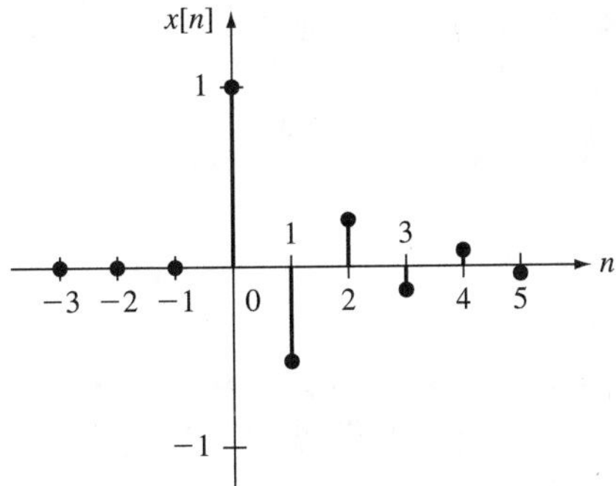

Figure 7.4 The signal $x[n] = (-0.5)^n u[n]$.

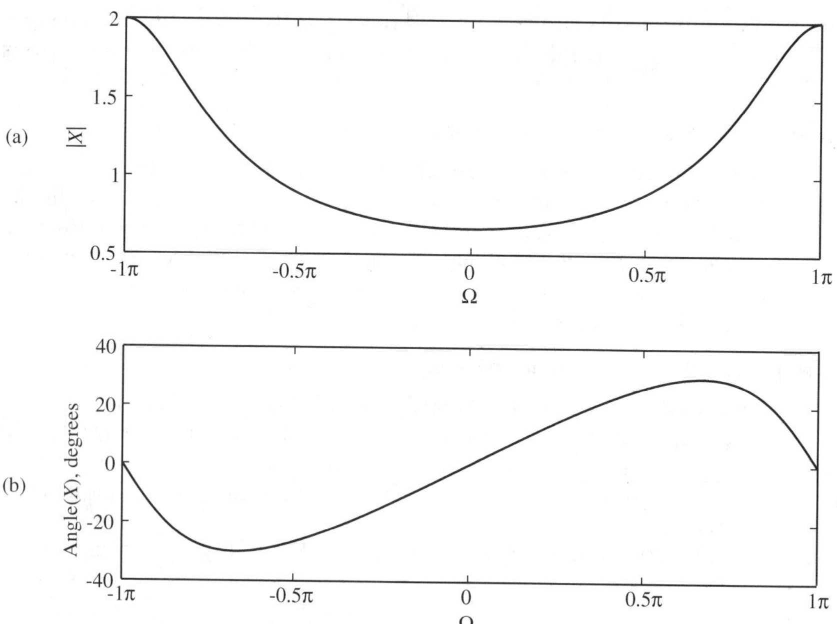

Figure 7.5 (a) Amplitude and (b) phase spectra of the signal $x[n] = (-0.5)^n u[n]$.

Inverse DTFT

Given a signal $x[n]$ with DTFT $X(\Omega)$, $x[n]$ can be recomputed from $X(\Omega)$ by applying the inverse DTFT to $X(\Omega)$. The inverse DTFT is defined by

$$x[n] = \frac{1}{2\pi}\int_0^{2\pi} X(\Omega)e^{jn\Omega}\,d\Omega \tag{7.27}$$

Since $X(\Omega)$ and $e^{jn\Omega}$ are both periodic functions of Ω with period 2π, the product $X(\Omega)e^{jn\Omega}$ is also a periodic function of Ω with period 2π. As a result, the integral in (7.27) can be evaluated over any interval of length 2π. For example,

$$x[n] = \frac{1}{2\pi}\int_{-\pi}^{\pi} X(\Omega)e^{jn\Omega}\,d\Omega \tag{7.28}$$

Generalized DTFT

As in the continuous-time Fourier transform, there are discrete-time signals that do not have a DTFT in the ordinary sense, but do have a generalized DTFT. One such signal is given in the following example.

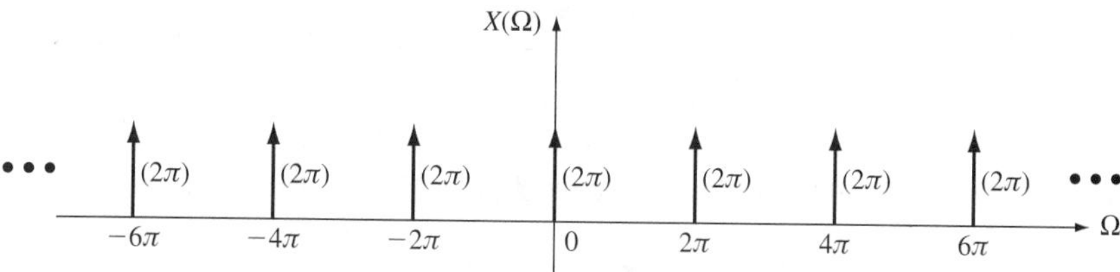

Figure 7.6 DTFT of the discrete-time constant signal.

Example 7.6 *DTFT of Constant Signal*

Consider the constant signal $x[n] = 1$ for all integers n. Since

$$\sum_{n=-\infty}^{\infty} x[n] = \infty$$

this signal does not have a DTFT in the ordinary sense. The constant signal does have a generalized DTFT that is defined to be the impulse train

$$X(\Omega) = \sum_{k=-\infty}^{\infty} 2\pi\delta(\Omega - 2\pi k)$$

This transform is displayed in Figure 7.6.

The justification for taking the transform in Figure 7.6 to be the generalized DTFT of the constant signal follows from the property that the inverse DTFT of $X(\Omega)$ is equal to the constant signal. To see this, by (7.28),

$$\begin{aligned} x[n] &= \frac{1}{2\pi}\int_{-\pi}^{\pi} X(\Omega)e^{jn\Omega}\,d\Omega \\ &= \frac{1}{2\pi}\int_{-\pi}^{\pi} 2\pi\delta(\Omega)e^{0}\,d\Omega = \int_{-\pi}^{\pi}\delta(\Omega)\,d\Omega \\ &= 1 \quad \text{for all } n \end{aligned}$$

Transform Pairs

As in the Fourier transform theory of continuous-time signals, the transform pair notation

$$x[n] \leftrightarrow X(\Omega)$$

will be used to denote the fact that $X(\Omega)$ is the DTFT of $x[n]$, and conversely, that $x[n]$ is the inverse DTFT of $X(\Omega)$. For the convenience of the reader, a list of common DTFT pairs is given in Table 7.1.

TABLE 7.1 COMMON DTFT PAIRS

$$1,\ \text{all } n \leftrightarrow \sum_{k=-\infty}^{\infty} 2\pi\delta(\Omega - 2\pi k)$$

$$\text{sgn}[n] \leftrightarrow \frac{2}{1 - e^{-j\Omega}}, \quad \text{where } \text{sgn}[n] = \begin{cases} 1, & n = 0, 1, 2, \ldots \\ -1, & n = -1, -2, \ldots \end{cases}$$

$$u[n] \leftrightarrow \frac{1}{1 - e^{-j\Omega}} + \sum_{k=-\infty}^{\infty} \pi\delta(\Omega - 2\pi k)$$

$$\delta[n] \leftrightarrow 1$$

$$\delta[n - N] \leftrightarrow e^{-jN\Omega}, \quad N = \pm 1, \pm 2, \ldots$$

$$a^n u[n] \leftrightarrow \frac{1}{1 - ae^{-j\Omega}}, \quad |a| < 1$$

$$e^{j\Omega_0 n} \leftrightarrow \sum_{k=-\infty}^{\infty} 2\pi\delta(\Omega - \Omega_0 - 2\pi k)$$

$$p[n] \leftrightarrow \frac{\sin[(q + \frac{1}{2})\Omega]}{\sin(\Omega/2)}$$

$$\frac{B}{\pi}\text{sinc}\left(\frac{B}{\pi}n\right) \leftrightarrow \sum_{k=-\infty}^{\infty} p_{2B}(\Omega + 2\pi k)$$

$$\cos \Omega_0 n \leftrightarrow \sum_{k=-\infty}^{\infty} \pi[\delta(\Omega + \Omega_0 - 2\pi k) + \delta(\Omega - \Omega_0 - 2\pi k)]$$

$$\sin \Omega_0 n \leftrightarrow \sum_{k=-\infty}^{\infty} j\pi[\delta(\Omega + \Omega_0 - 2\pi k) - \delta(\Omega - \Omega_0 - 2\pi k)]$$

$$\cos\left(\Omega_0 n + \theta\right) \leftrightarrow \sum_{k=-\infty}^{\infty} \pi[e^{-j\theta}\delta(\Omega + \Omega_0 - 2\pi k) + e^{j\theta}\delta(\Omega - \Omega_0 - 2\pi k)]$$

Properties of the DTFT

The DTFT has several properties, most of which are discrete-time versions of the properties of the continuous-time Fourier transform (CTFT). The properties of the DTFT are listed in Table 7.2. Except for the last property in Table 7.2, the proofs of these properties closely resemble the proofs of the corresponding properties of the CTFT. The details are omitted.

It should be noted that in contrast to the CTFT, there is no duality property for the DTFT. However, there is a relationship between the inverse DTFT and the inverse CTFT. This is the last property listed in Table 7.2. This property is stated and proved below.

Given a discrete-time signal $x[n]$ with DTFT $X(\Omega)$, let $X(\omega)$ denote $X(\Omega)$ with Ω replaced by ω and let $p_{2\pi}(\omega)$ denote the rectangular frequency function with width equal to 2π. Then the product $X(\omega)p_{2\pi}(\omega)$ is equal to $X(\omega)$ for $-\pi \le \omega < \pi$ and is equal to zero for all other values of ω. Let $\gamma(t)$ denote the inverse CTFT of $X(\omega)p_{2\pi}(\omega)$. Then the last property in Table 7.2 states that $x[n] = \gamma(n)$. To prove this, first observe that by definition of the inverse CTFT,

TABLE 7.2 PROPERTIES OF THE DTFT

Property	Transform Pair/Property
Linearity	$ax[n] + bv[n] \leftrightarrow aX(\Omega) + bV(\Omega)$
Right or left shift in time	$x[n-q] \leftrightarrow X(\Omega)e^{-jq\Omega}$, q any integer
Time reversal	$x[-n] \leftrightarrow X(-\Omega) = \overline{X(\Omega)}$
Multiplication by n	$nx[n] \leftrightarrow j\dfrac{d}{d\Omega}X(\Omega)$
Multiplication by a complex exponential	$x[n]e^{jn\Omega_0} \leftrightarrow X(\Omega - \Omega_0)$, Ω_0 real
Multiplication by $\sin \Omega_0 n$	$x[n]\sin \Omega_0 n \leftrightarrow \dfrac{j}{2}[X(\Omega+\Omega_0) - X(\Omega-\Omega_0)]$
Multiplication by $\cos \Omega_0 n$	$x[n]\cos \Omega_0 n \leftrightarrow \dfrac{1}{2}[X(\Omega+\Omega_0) + X(\Omega-\Omega_0)]$
Convolution in the time domain	$x[n] * v[n] \leftrightarrow X(\Omega)V(\Omega)$
Summation	$\sum_{i=0}^{n} x[i] \leftrightarrow \dfrac{1}{1-e^{-j\Omega}}X(\Omega) + \sum_{n=-\infty}^{\infty} \pi X(2\pi n)\delta(\Omega - 2\pi n)$
Multiplication in the time domain	$x[n]v[n] \leftrightarrow \dfrac{1}{2\pi}\int_{-\pi}^{\pi} X(\Omega - \lambda)V(\lambda)\,d\lambda$
Parseval's theorem	$\sum_{n=-\infty}^{\infty} x[n]v[n] = \dfrac{1}{2\pi}\int_{-\pi}^{\pi} \overline{X(\Omega)}V(\Omega)\,d\Omega$
Special case of Parseval's theorem	$\sum_{n=-\infty}^{\infty} x^2[n] = \dfrac{1}{2\pi}\int_{-\pi}^{\pi} \lvert X(\Omega)\rvert^2\,d\Omega$
Relationship to inverse CTFT	If $x[n] \leftrightarrow X(\Omega)$ and $\gamma(t) \leftrightarrow X(\omega)p_{2\pi}(\omega)$, then $x[n] = \gamma(t)\rvert_{t=n} = \gamma(n)$

$$\gamma(t) = \frac{1}{2\pi}\int_{-\infty}^{\infty} X(\omega)p_{2\pi}(\omega)e^{j\omega t}\,d\omega \tag{7.29}$$

By the definition of $X(\omega)p_{2\pi}(\omega)$, (7.29) reduces to

$$\gamma(t) = \frac{1}{2\pi}\int_{-\pi}^{\pi} X(\omega)e^{j\omega t}\,d\omega \tag{7.30}$$

Setting $t = n$ in (7.30) gives

$$\gamma(t)\rvert_{t=n} = \gamma(n) = \frac{1}{2\pi}\int_{-\pi}^{\pi} X(\omega)e^{j\omega n}\,d\omega \tag{7.31}$$

and replacing ω by Ω in (7.31) gives

$$\gamma(n) = \frac{1}{2\pi}\int_{-\pi}^{\pi} X(\Omega)e^{j\Omega n}\,d\Omega \tag{7.32}$$

The right-hand side of (7.32) is equal to the inverse DTFT of $X(\Omega)$, and thus $\gamma(n)$ is equal to $x[n]$.

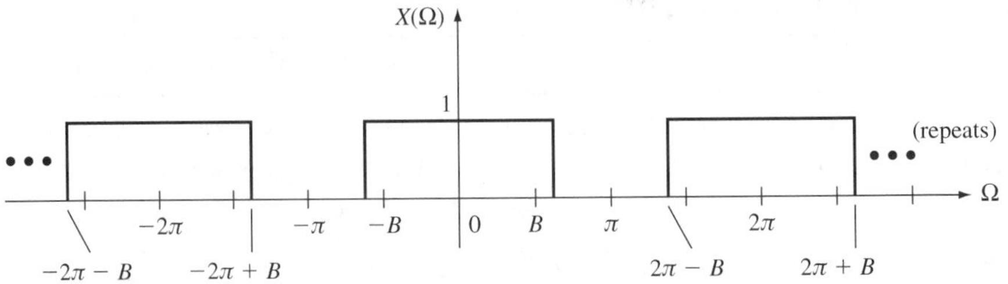

Figure 7.7 Transform in Example 7.7.

The relationship between the inverse CTFT and inverse DTFT can be used to generate DTFT pairs from CTFT pairs, as illustrated in the following example.

Example 7.7 *DTFT Pair from CTFT*

Suppose that

$$X(\Omega) = \sum_{k=-\infty}^{\infty} p_{2B}(\Omega + 2\pi k)$$

where $B < \pi$. The transform $X(\Omega)$ is plotted in Figure 7.7. It is seen that

$$X(\omega)p_{2\pi}(\omega) = p_{2B}(\omega)$$

From Table 4.2 the inverse CTFT of $p_{2B}(\omega)$ is equal to

$$\frac{B}{\pi}\operatorname{sinc}\left(\frac{B}{\pi}t\right), \qquad -\infty < t < \infty$$

Thus

$$x[n] = \gamma[n] = \frac{B}{\pi}\operatorname{sinc}\left(\frac{B}{\pi}n\right), \qquad n = 0, \pm 1, \pm 2, \ldots$$

which yields the DTFT pair

$$\frac{B}{\pi}\operatorname{sinc}\left(\frac{B}{\pi}n\right) \leftrightarrow \sum_{k=-\infty}^{\infty} p_{2B}(\Omega + 2\pi k)$$

7.2 DISCRETE FOURIER TRANSFORM

Let $x[n]$ be a discrete-time signal with DTFT $X(\Omega)$. Since $X(\Omega)$ is a function of the continuous variable Ω, it cannot be stored in the memory of a digital computer unless $X(\Omega)$ can be expressed in a closed form. To implement DTFT techniques on a digital computer, it is necessary to discretize in frequency. This leads to the concept of the discrete Fourier transform, which is defined below.

Suppose that the discrete-time signal $x[n]$ is zero for all integers $n < 0$ and all integers $n \geq N$, where N is a fixed positive integer. The integer N could be very large, for example $N = 2^{10} = 1024$. The N-point discrete Fourier transform (DFT) X_k of $x[n]$ is defined by

$$X_k = \sum_{n=0}^{N-1} x[n]e^{-j2\pi kn/N}, \qquad k = 0, 1, \ldots, N-1 \tag{7.33}$$

From (7.33) it is seen that the DFT X_k is a function of the discrete (integer) variable k. Also note that in contrast to the DTFT, the DFT X_k is completely specified by the N values $X_0, X_1, X_2, \ldots, X_{N-1}$. In general, these values are complex and thus X_k can be expressed in either polar or rectangular form. The polar form is

$$X_k = |X_k| \exp[j\angle X_k], \qquad k = 0, 1, \ldots, N-1 \tag{7.34}$$

where $|X_k|$ is the magnitude of X_k and $\angle X_k$ is the angle of X_k. The rectangular form is

$$X_k = R_k + jI_k, \qquad k = 0, 1, \ldots, N-1 \tag{7.35}$$

where R_k is the real part of X_k given by

$$R_k = x(0) + \sum_{n=1}^{N-1} x[n] \cos \frac{2\pi kn}{N}$$

and I_k is the imaginary part of X_k given by

$$I_k = -\sum_{n=1}^{N-1} x[n] \sin \frac{2\pi kn}{N}$$

Since the summation in (7.33) is finite, the DFT X_k always exists. Further, X_k can be computed by simply evaluating the finite summation in (7.33). A MATLAB program for computing the DFT is given in Figure 7.8.

Example 7.8 *Computation of DFT*

Suppose that $x[0] = 1, x[1] = 2, x[2] = 2, x[3] = 1$, and $x[n] = 0$ for all other integers n. Then $N = 4$ and from (7.33), the DFT is

```
%
%  Discrete Fourier Transform
%
function Xk = dft(x)
[N,M] = size(x);
if M ~=1,      % makes sure that x is a column vector
  x = x';
  N = M;
end
Xk=zeros(N,1);
n = 0:N-1
for k=0:N-1
  Xk(k+1) = exp(-j*2*pi*k*n/N)*x;
end
```

Figure 7.8 MATLAB program for evaluating the DFT.

$$X_k = \sum_{n=0}^{3} x[n]e^{-j\pi kn/2}, \qquad k = 0, 1, 2, 3$$
$$= x[0] + x[1]e^{-j\pi k/2} + x[2]e^{-j\pi k} + x[3]e^{-j\pi 3k/2}, \qquad k = 0, 1, 2, 3$$
$$= 1 + 2e^{-j\pi k/2} + 2e^{-j\pi k} + e^{-j\pi 3k/2}, \qquad k = 0, 1, 2, 3$$

The real part R_k of X_k is

$$R_k = 1 + 2\cos\frac{-\pi k}{2} + 2\cos -\pi k + \cos\frac{-\pi 3k}{2}, \qquad k = 0, 1, 2, 3$$

Thus

$$R_k = \begin{cases} 6, & k = 0 \\ -1, & k = 1 \\ 0, & k = 2 \\ -1, & k = 3 \end{cases}$$

The imaginary part I_k of X_k is

$$I_k = -2\sin\frac{\pi k}{2} - 2\sin \pi k - \sin\frac{\pi 3k}{2}, \qquad k = 0, 1, 2, 3$$

Hence

$$I_k = \begin{cases} 0, & k = 0 \\ -1, & k = 1 \\ 0, & k = 2 \\ 1, & k = 3 \end{cases}$$

and the rectangular form of X_k is

$$X_k = \begin{cases} 6, & k = 0 \\ -1 - j, & k = 1 \\ 0, & k = 2 \\ -1 + j, & k = 3 \end{cases}$$

As a check, these values for X_k were also obtained by using the MATLAB program in Figure 7.8 with the commands

```
x = [1 2 2 1];
Xk = dft(x)
```

The polar form of X_k is

$$X_k = \begin{cases} 6e^{j0}, & k = 0 \\ \sqrt{2}\, e^{j5\pi/4}, & k = 1 \\ 0e^{j0}, & k = 2 \\ \sqrt{2}\, e^{j3\pi/4}, & k = 3 \end{cases}$$

If X_k is the N-point DFT of $x[n]$, then $x[n]$ can be determined from X_k by applying the inverse DFT given by

$$x[n] = \frac{1}{N}\sum_{k=0}^{N-1} X_k e^{j2\pi kn/N}, \qquad n = 0, 1, \ldots, N-1 \tag{7.36}$$

Since the sum in (7.36) is finite, the inverse DFT can be computed by simply evaluating the summation in (7.36). A MATLAB program for computing the inverse DFT is given in Figure 7.9.

Example 7.9 *Computation of Inverse DFT*

Again consider the signal in Example 7.8 with the rectangular form of the DFT given by

$$X_k = \begin{cases} 6, & k = 0 \\ -1 - j, & k = 1 \\ 0, & k = 2 \\ -1 + j, & k = 3 \end{cases}$$

Evaluating (7.36) with $N = 4$ gives

$$x[n] = \frac{1}{4}[X_0 + X_1 e^{j\pi n/2} + X_2 e^{j\pi n} + X_3 e^{j3\pi n/2}], \qquad n = 0, 1, 2, 3$$

Thus

$$x[0] = \frac{1}{4}[X_0 + X_1 + X_2 + X_3] = 1$$

$$x[1] = \frac{1}{4}[X_0 + jX_1 - X_2 - jX_3] = \frac{1}{4}[8] = 2$$

$$x[2] = \frac{1}{4}[X_0 - X_1 + X_2 - X_3] = 2$$

$$x[3] = \frac{1}{4}[X_0 - jX_1 - X_2 + jX_3] = \frac{1}{4}[4] = 1$$

```
%
%  Inverse Discrete Fourier Transform
%
function x = idft(Xk)
[N,M] = size(Xk);
  if M ~= 1,     % makes sure that Xk is a column vector
     Xk = Xk.'
     N = M;
  end
x=zeros(N,1);
k = 0:N-1;
for n=0:N-1
  x(n+1) = exp(j*2*pi*k*n/N)*Xk;
end
x = x/N;
```

Figure 7.9 MATLAB program for computing the inverse DFT.

These values are equal to the values of $x[n]$ specified in Example 7.8. Also, these values for $x[n]$ result when the program in Figure 7.9 is run with the commands

```
Xk = [6 -1-j 0 -1+j].';
x = idft(Xk)
```

Relationship to DTFT

Given a discrete-time signal $x[n]$ with $x[n] = 0$ for $n < 0$ and $n \geq N$, let X_k denote the N-point DFT defined by (7.33) and let $X(\Omega)$ denote the DTFT of $x[n]$ defined by

$$X(\Omega) = \sum_{n=-\infty}^{\infty} x[n]e^{-j\Omega n} \tag{7.37}$$

Since $x[n] = 0$ for $n < 0$ and $n \geq N$, (7.37) reduces to

$$X(\Omega) = \sum_{n=0}^{N-1} x[n]e^{-j\Omega n} \tag{7.38}$$

Comparing (7.33) and (7.38) reveals that

$$X_k = X(\Omega)\big|_{\Omega=2\pi k/N} = X\left(\frac{2\pi k}{N}\right) \tag{7.39}$$

Thus the DFT X_k can be viewed as a frequency-sampled version of the DTFT $X(\Omega)$; more precisely, X_k is equal to $X(\Omega)$ with Ω evaluated at the frequency points $\Omega = 2\pi k/N$ for $k = 0, 1, \ldots, N-1$.

Example 7.10 *DTFT and DFT of a Pulse*

DTFT and DFT of Pulse

With $p[n]$ equal to the rectangular pulse defined in Example 7.3, let $x[n] = p[n-q]$. Then by definition of $p[n]$,

$$x[n] = \begin{cases} 1, & n = 0, 1, 2, \ldots, 2q \\ 0, & \text{all other } n \end{cases}$$

From the result in Example 7.3,

$$p[n] \leftrightarrow \frac{\sin[(q+\frac{1}{2})\Omega]}{\sin(\Omega/2)}$$

and by the time-shift property of the DTFT (see Table 7.2), the DTFT of $x[n] = p[n-q]$ is given by

$$X[\Omega] = \frac{\sin[(q+\frac{1}{2})\Omega]}{\sin(\Omega/2)} e^{-jq\Omega}$$

Thus the amplitude spectrum of $x[n]$ is

$$|X[\Omega]| = \frac{|\sin[(q+\frac{1}{2})\Omega]|}{|\sin(\Omega/2)|}$$

In the case $q = 5$, $|X(\Omega)|$ is plotted in Figure 7.10. Note that $|X(\Omega)|$ is plotted for Ω ranging from 0 to 2π (as opposed to $-\pi$ to π). In this section the amplitude spectrum $|X(\Omega)|$ is displayed from 0 to 2π since X_k for $k = 0, 1, \ldots, N-1$ corresponds to the values of $X(\Omega)$ for values of Ω over the interval from 0 to 2π.

Now letting X_k denote the N-point DFT of $x[n]$ with $N = 2q + 1$, from (7.39)

$$|X_k| = \left|X\left(\frac{2\pi k}{N}\right)\right| = \frac{|\sin[(q+\frac{1}{2})(2\pi k/N)]|}{|\sin(\pi k/N)|}, \qquad k = 0, 1, \ldots, 2q$$

Replacing N by $2q + 1$ gives

$$|X_k| = \frac{\left|\sin\left[\left(\frac{2q+1}{2}\right)\frac{2\pi k}{2q+1}\right]\right|}{\left|\sin\left(\frac{\pi k}{2q+1}\right)\right|}, \qquad k = 0, 1, \ldots, 2q$$

$$|X_k| = \frac{|\sin \pi k|}{\left|\sin\left(\frac{\pi k}{2q+1}\right)\right|}, \qquad k = 0, 1, \ldots, 2q$$

$$|X_k| = \begin{cases} 2q+1, & k = 0 \\ 0, & k = 1, 2, \ldots, 2q \end{cases}$$

The value of $|X_k|$ for $k = 0$ was computed using l'Hôpital's rule.

Note that since $|X_k| = 0$ for $k = 1, 2, \ldots, 2q$, the sample values $X(2\pi k/N)$ of $X(\Omega)$ are all equal to zero for these values of k. This is a consequence of sampling $X(\Omega)$ at the zero points located between the sidelobes of $X(\Omega)$ (see Figure 7.10). Since X_k is nonzero for only $k = 0$, X_k bears very little resemblance to the spectrum $X(\Omega)$ of the rectangular pulse $x[n] = p[n - q]$. However, by making N larger, so that the sampling frequencies $2\pi k/N$ are closer together, it is expected that the DFT X_k should be a better representation of the spectrum $X(\Omega)$. For example, when $N = 2(2q + 1)$, the resulting N-point DFT X_k is equal to the values of $X(\Omega)$ at the frequency points corresponding to the peaks and zero points of the sidelobes in $X(\Omega)$. A plot of the amplitude $|X_k|$ for $q = 5$ and $N = 2(2q + 1) = 22$ is given in Figure 7.11. Clearly, $|X_k|$ now bears some resemblance to

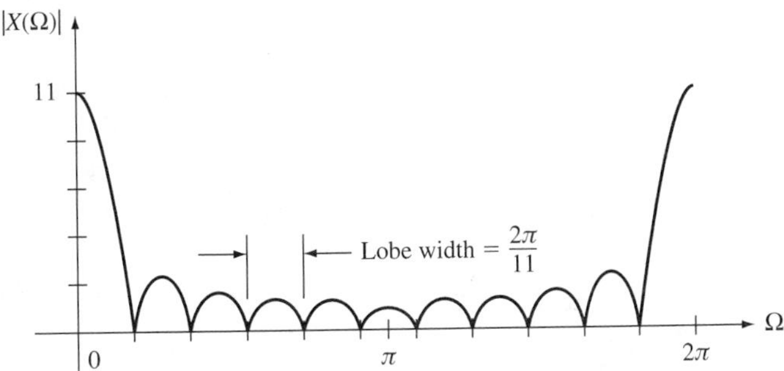

Figure 7.10 Amplitude spectrum in the case $q = 5$.

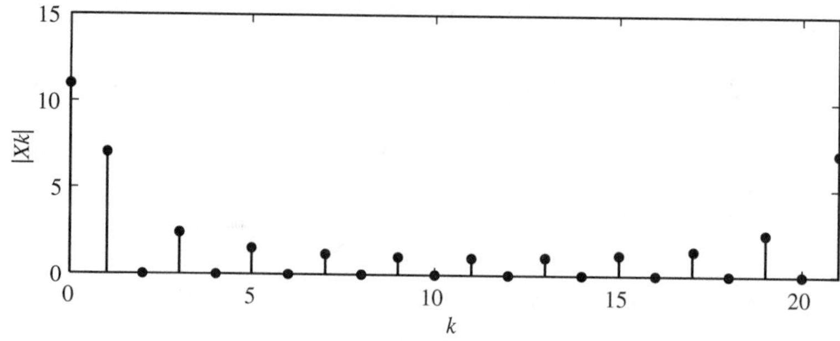

Figure 7.11 Amplitude of DFT in the case $q = 5$ and $N = 22$.

$|X(\Omega)|$ displayed in Figure 7.10. To obtain an even closer correspondence, N can be increased again. For instance, $|X_k|$ is plotted in Figure 7.12 for the case when $q = 5$ and $N = 88$. Here the DFT was computed by running the program in Figure 7.8 with

```
q = 5; N = 88;
x = [ones(1,2*q+1) zeros(1,N-2*q-1)];
Xk = dft(x);
k = 0:N-1;
stem(k,abs(Xk))    % plots the magnitude
```

Example 7.11 *DTFT and DFT of Finite-Duration Sinusoid*

Now consider the finite-duration sinusoid $x[n] = (\cos \Omega_0 n)p[n - q]$, where $0 \leq \Omega_0 \leq \pi$. By definition of the rectangular pulse $p[n]$,

$$x[n] = \begin{cases} \cos \Omega_0 n, & n = 0, 1, \ldots, 2q \\ 0, & \text{all other } n \end{cases}$$

From Table 7.1 the DTFT of the infinite-duration sinusoid $\cos \Omega_0 n$, $-\infty < n < \infty$, is the impulse train

$$\sum_{i=-\infty}^{\infty} \pi[\delta(\Omega + \Omega_0 - 2\pi i) + \delta(\Omega - \Omega_0 - 2\pi i)]$$

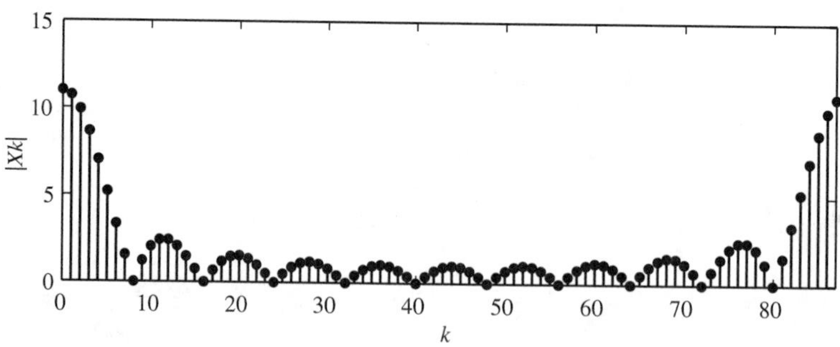

Figure 7.12 Amplitude of DFT in the case $q = 5$ and $N = 88$.

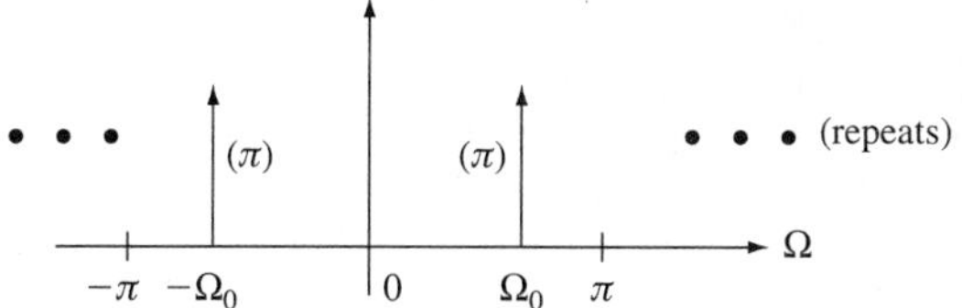

Figure 7.13 DTFT of $\cos \Omega_0 n$ with $-\pi \le \Omega_0 \le \pi$.

A plot of the DTFT of $\cos \Omega_0 n$ for $-\pi \le \Omega \le \pi$ is shown in Figure 7.13. From the figure it is seen that over the frequency range $-\pi \le \Omega \le \pi$ all the spectral content of the signal $\cos \Omega_0 n$ is concentrated at $\Omega = \Omega_0$ and $\Omega = -\Omega_0$. Now let $P(\Omega)$ denote the DTFT of the rectangular pulse $p[n - q]$. From the result in Example 7.10,

$$P(\Omega) = \frac{\sin\left[\left(q + \frac{1}{2}\right)\Omega\right]}{\sin(\Omega/2)} e^{-jq\Omega}$$

Then by the DTFT property involving multiplication of signals (see Table 7.2), the DTFT $X(\Omega)$ of $x[n]$ is given by

$$X(\Omega) = \frac{1}{2\pi}\int_{-\pi}^{\pi} P(\Omega - \lambda)\pi[\delta(\lambda + \Omega_0) + \delta(\lambda - \Omega_0)]\, d\lambda$$

Using the shifting property of the impulse (see Section 1.2) yields

$$X(\Omega) = \frac{1}{2}[P(\Omega + \Omega_0) + P(\Omega - \Omega_0)]$$

Now setting $N = 2q + 1$, by (7.39) the N-point DFT of $x[n]$ is given by

$$X_k = X\left(\frac{2\pi k}{2q+1}\right) = \frac{1}{2}\left[P\left(\frac{2\pi k}{2q+1} + \Omega_0\right) + P\left(\frac{2\pi k}{2q+1} - \Omega_0\right)\right], \qquad k = 0, 1, \ldots, 2q$$

where

$$P\left(\frac{2\pi k}{2q+1} \pm \Omega_0\right) = \frac{\sin\left[\left(q + \frac{1}{2}\right)\left(\frac{2\pi k}{2q+1} \pm \Omega_0\right)\right]}{\sin\left[\left(\frac{2\pi k}{2q+1} \pm \Omega_0\right)\Big/2\right]} \exp\left[-jq\left(\frac{2\pi k}{2q+1} \pm \Omega_0\right)\right]$$

Suppose that $\Omega_0 = (2\pi r)/(2q + 1)$ for some integer r where $0 \le r \le q$. This is equivalent to assuming that $\cos \Omega_0 n$ goes through r complete periods as n is varied from $n = 0$ to $n = 2q$. Then

$$P\left(\frac{2\pi k}{2q+1} \pm \Omega_0\right) = \frac{\sin\left[\left(q + \frac{1}{2}\right)\left(\frac{2\pi k}{2q+1} \pm \Omega_0\right)\right]}{\sin\left[\left(\frac{2\pi k}{2q+1} \pm \Omega_0\right)\Big/2\right]} \exp\left[-jq\left(\frac{2\pi k}{2q+1} \pm \Omega_0\right)\right], \qquad k = 0, 1, \ldots, 2q$$

$$P\left(\frac{2\pi k}{2q+1} \pm \Omega_0\right) = \frac{\sin\left[\left(\frac{2q+1}{2}\right)\left(\frac{2\pi k \pm 2\pi r}{2q+1}\right)\right]}{\sin\left[\left(\frac{2\pi k \pm 2\pi r}{2q+1}\right)\Big/2\right]} \exp\left[-jq\frac{2\pi k \pm 2\pi r}{2q+1}\right], \quad k = 0, 1, \ldots, 2q$$

$$= \frac{\sin(\pi k \pm \pi r)}{\sin\left(\dfrac{\pi k \pm \pi r}{2q+1}\right)} \exp\left[-jq\frac{2\pi(k \pm r)}{2q+1}\right], \quad k = 0, 1, \ldots, 2q$$

and thus

$$P\left(\frac{2\pi k}{2q+1} - \Omega_0\right) = \begin{cases} 2q+1, & k = r \\ 0, & k = 0, 1, \ldots, r-1, r+1, \ldots, 2q \end{cases}$$

$$P\left(\frac{2\pi k}{2q+1} + \Omega_0\right) = \begin{cases} 2q+1, & k = 2q+1-r \\ 0, & k = 0, 1, \ldots, 2q-r, 2q+2-r, \ldots, 2q \end{cases}$$

Finally, the DFT X_k is given by

$$X_k = \begin{cases} q + \frac{1}{2}, & k = r \\ q + \frac{1}{2}, & k = 2q+1-r \\ 0, & \text{all other } k \text{ for } 0 \le k \le 2q \end{cases}$$

Since $k = r$ corresponds to the frequency point $\Omega_0 = (2\pi r)/(2q+1)$, this result shows that the portion of the DFT corresponding to the frequency range from 0 to π is concentrated at the expected point. Hence the DFT is a "faithful" representation of the DTFT of the sinusoid $\cos \Omega_0 n$. The DFT X_k is plotted in Figure 7.14 for the case when $q = 10$, $N = 2q + 1 = 21$, and $\Omega_0 = 10\pi/21$ (which implies that $r = 5$). This plot was generated by applying the MATLAB command `dft` directly to $x[n]$. The MATLAB program is

```
q=10; N = 2*q+1;
omo=10*pi/21;
n=0:N-1;
x=cos(omo*n);
Xk=dft(x);
k=0:N-1;
stem(k,abs(Xk))
```

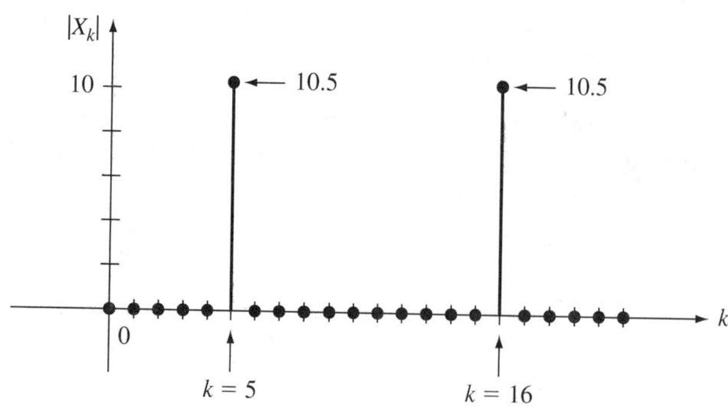

Figure 7.14 DFT of $\cos \Omega_0 n$ in the case when $q = 10$, $N = 21$, and $\Omega_0 = 10\pi/21$.

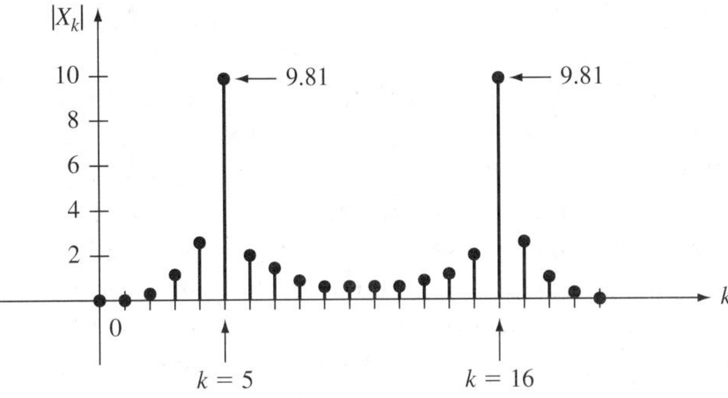

Figure 7.15 Amplitude of DFT of $\cos \Omega_0 n$ in the case when $q = 10, N = 21$, and $\Omega_0 = 9.5\pi/21$.

Now suppose that Ω_0 is not equal to $2\pi r/(2q + 1)$ for any integer r. Let β denote the integer for which

$$\left|\Omega_0 - \frac{2\pi\beta}{2q + 1}\right|$$

has the smallest possible value. Then the DFT X_k will have nonzero values distributed in a neighborhood of the point $k = \beta$. This characteristic is referred to as *leakage,* meaning that the spectral component concentrated at Ω_0 is spread (or "leaks") into the frequency components in a neighborhood of $2\pi\beta/(2q + 1)$. For the case $q = 10, N = 21$, and $\Omega_0 = 9.5\pi/21$, the amplitude $|X_k|$ of the DFT is plotted in Figure 7.15. Here $\beta = 5$ and thus the values of $|X_k|$ are distributed about the point $k = 5$ (and the corresponding point $k = 2q + 1 - r$). The plot given in Figure 7.15 was generated by modifying the MATLAB commands above with $N = 21$ and `omo = 9.5*pi/21`.

DFT of Truncated Signals

In most applications, a discrete-time signal $x[n]$ would be known over only a finite interval of time; that is, only the values $x[0], x[1], \ldots, x[N - 1]$ would be known, where N (the record length) is some positive integer. When the N-point DFT is computed based on the values $x[0], x[1], \ldots, x[N - 1]$, the result is the N-point DFT of the truncated signal

$$\tilde{x}[n] = \begin{cases} x[n], & n = 0, 1, \ldots, N - 1 \\ 0, & n \geq N \end{cases} \tag{7.40}$$

As noted in Example 7.10, in order for the DFT to be a good representation of the spectrum $X(\Omega)$ of the signal $x[n]$, it may be necessary to consider the L-point DFT $\tilde{X}_k$ of the truncated signal $\tilde{x}[n]$, where $L > N$.

Letting $\tilde{X}(\Omega)$ denote the DTFT of the truncated signal $\tilde{x}[n]$, from (7.39)

$$\tilde{X}_k = \tilde{X}\left(\frac{2\pi k}{L}\right), \qquad k = 0, 1, \ldots, L - 1 \tag{7.41}$$

where again $\tilde{X}_k$ is the L-point DFT of the truncated signal. However, if the values of $x[n]$ are not small for $n \geq N$, the DTFT $\tilde{X}(\Omega)$ of the truncated signal may differ significantly from the DTFT $X(\Omega)$ of the actual signal $x[n]$. Thus the values of the L-point DFT $\tilde{X}_k$ may not bear much of a correspondence to the DTFT $X(\Omega)$, in which case $\tilde{X}_k$ may not yield an accurate representation of the spectrum $X(\Omega)$ of the signal $x[n]$. An example of this occurred when the truncated sinusoid $x[n] = (\cos \Omega_0 n)p[n - q]$ was considered in Example 7.11. In particular, if the sinusoid does not go through an integer number of complete periods over the record length, it was observed that the spectral component (at $\Omega = \Omega_0$) of the sinusoid $\cos \Omega_0 n$ is spread (leaks) into other frequency locations in the DFT of the truncated signal. The analysis given in Example 7.11 can be extended to arbitrary signals as follows.

Again consider the truncated signal given by (7.40), where now it is assumed that N is an odd integer. By definition of the rectangular pulse $p[n]$, setting $q = (N - 1)/2$ gives

$$p\left[n - \frac{N-1}{2}\right] = \begin{cases} 1, & n = 0, 1, \ldots, N-1 \\ 0, & \text{all other } n \end{cases}$$

Hence the truncated signal can be expressed in the form

$$\tilde{x}[n] = x[n]p\left[n - \frac{N-1}{2}\right] \tag{7.42}$$

Now letting $P(\Omega)$ denote the DTFT of $p\left[n - \frac{N-1}{2}\right]$, from the results in Example 7.11,

$$P(\Omega) = \frac{\sin(N/2)\Omega}{\sin(\Omega/2)} e^{-j(N-1)\Omega/2}$$

Taking the DTFT of both sides of (7.42) yields

$$\tilde{X}(\Omega) = X(\Omega) * P(\Omega) = \frac{1}{2\pi}\int_{-\pi}^{\pi} X(\Omega - \lambda)P(\lambda)\, d\lambda$$

Thus the L-point DFT $\tilde{X}_k$ of the truncated signal $\tilde{x}[n]$ [defined by (7.40)] is given by

$$\tilde{X}_k = [X(\Omega) * P(\Omega)]_{\Omega = 2\pi k/L}, \qquad k = 0, 1, \ldots, L-1 \tag{7.43}$$

By (7.43) it is seen that the distortion in $\tilde{X}_k$ from the desired values $X(2\pi k/L)$ can be characterized in terms of the effect of convolving $P(\Omega)$ with the spectrum $X(\Omega)$ of the signal. If $x[n]$ is not suitably small for $n \geq N$, in general the sidelobes that exist in the amplitude spectrum $|P(\Omega)|$ will result in sidelobes in the amplitude spectrum $|X(\Omega) * P(\Omega)|$. This produces the leakage phenomenon first observed in Example 7.11.

Example 7.12 *N-Point DFT*

Consider the discrete-time signal $x[n] = (0.9)^n$, $n \geq 0$, which is plotted in Figure 7.16. From the results in Example 7.2, the DTFT is

$$X(\Omega) = \frac{1}{1 - 0.9e^{-j\Omega}}$$

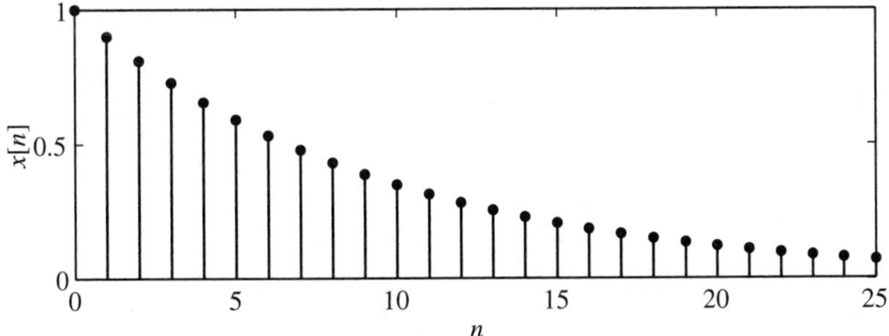

Figure 7.16 Signal in Example 7.12.

The amplitude spectrum $|X(\Omega)|$ is plotted in Figure 7.17. Note that since the signal varies rather slowly, most of the spectral content over the frequency range from 0 to π is concentrated near the zero point $\Omega = 0$. For $N = 21$ the amplitude of the N-point DFT of the signal is shown in Figure 7.18. This plot was obtained using the commands

```
N = 21; n = 0:N-1;
x = 0.9.^n;
Xk=dft(x);
k=n;
stem(k,abs(Xk))
```

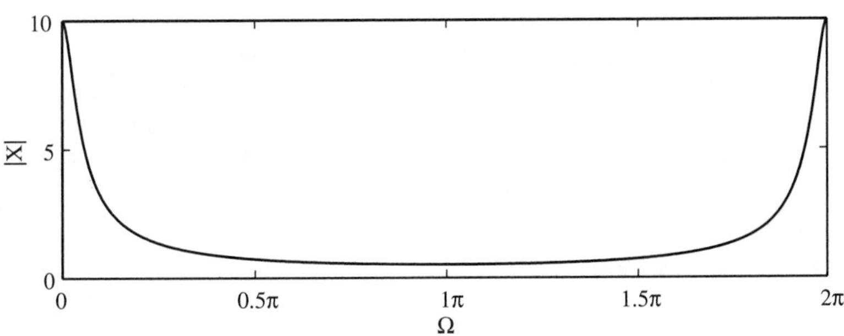

Figure 7.17 Amplitude spectrum of signal in Example 7.12.

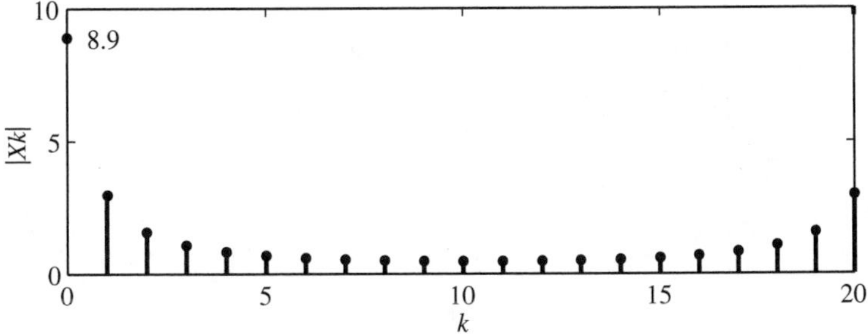

Figure 7.18 Amplitude of 21-point DFT.

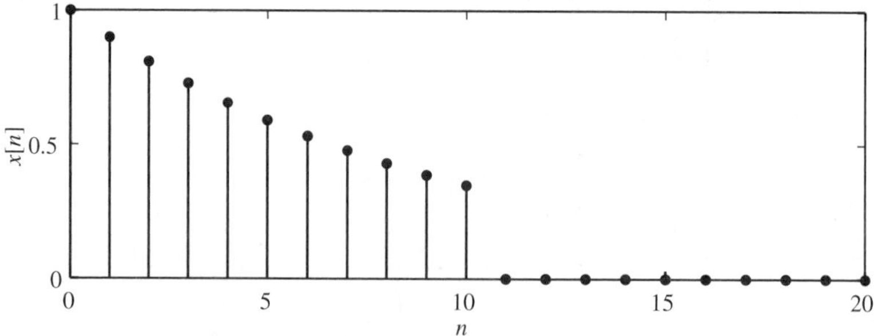

Figure 7.19 Truncated signal in Example 7.12.

Comparing Figures 7.17 and 7.18, it is seen that the amplitude of the 21-point DFT is a close approximation to the amplitude spectrum $|X(\Omega)|$. This turns out to be the case since $x[n]$ is small for $n \geq 21$. Now consider the truncated signal $x[n]$ shown in Figure 7.19. The amplitude of the 21-point DFT of the truncated signal is plotted in Figure 7.20. This plot was generated using the following commands

```
N = 21; n = 0:N-1;
x = 0.9.^n;
x(12:21) = zeros(1,10);
Xk = dft(x);
k=n;
stem(k,abs(Xk))
```

Comparing Figures 7.20 and 7.18 reveals that the spectral content of the truncated signal has higher frequency components than those of the signal displayed in Figure 7.16. The reason for this is that the truncation causes an abrupt change in the signal magnitude which introduces high-frequency components in the signal spectrum (as displayed by the DFT).

7.3 PROPERTIES OF THE DFT

Since the DFT is applied to discrete-time signals $x[n]$ that are zero for $n < 0$ and $n \geq N$, some of the DFT properties (such as those involving time shifts and

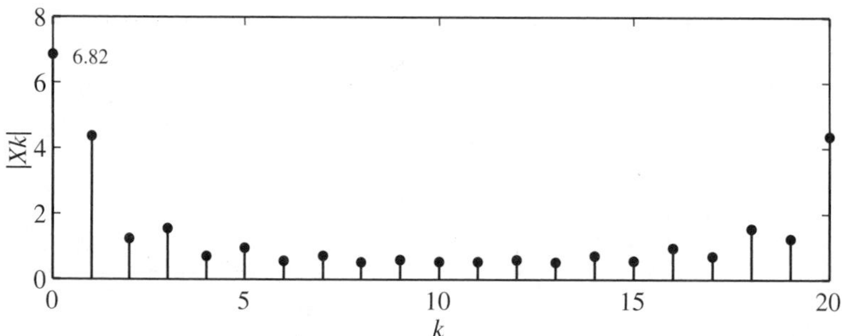

Figure 7.20 Amplitude of 21-point DFT of the truncated signal.

convolution) are based on "modulo N" or "mod N" operations that differ from the ordinary operations on discrete-time signals. This section deals with the mod N operations and the DFT properties that are based on these operations. First, the property of linearity is considered.

Linearity

The DFT is a linear operation, as is the DTFT. Hence if $x[n]$ and $v[n]$ have N-point DFTs X_k and V_k, then for any real or complex scalars a and b, the N-point DFT of $ax[n] + bv[n]$ is $aX_k + bV_k$.

Circular Time Shift

Given a discrete-time signal $x[n]$ with $x[n] = 0$ for $n < 0$ and $n \geq N$, for any positive or negative integer q, the time shift $x[n - q]$ will have nonzero values in general for n outside the range $n = 0, 1, \ldots, N - 1$. Since the N-point DFT involves only those values of a signal for $n = 0, 1, \ldots, N - 1$, the shifted signal must be defined so that it is also zero outside the range $n = 0, 1, \ldots$.

In the definition of a q-step right shift (with $0 < q \leq N$), the values of the signal $x[n]$ for $N - q \leq n \leq N - 1$ are "wrapped around" to form the first part of the shifted signal. The resulting operation is referred to as a time shift (mod N) and is denoted by $x[n - q, \text{mod } N]$. The precise definition of the shift is

$$x[n - q, \text{mod } N] = \begin{cases} x[N - q], & n = 0 \\ x[N - q + 1], & n = 1 \\ \vdots & \\ x[N - 1], & n = q - 1 \\ x[0], & n = q \\ x[1], & n = q + 1 \\ \vdots & \\ x[N - q - 1], & n = N - 1 \end{cases}$$

Note that $x[n - N, \text{mod } N] = x[n]$.

To obtain some additional insight on the definition of the time shift (mod N), suppose that the values of $x[n]$ are displayed on a circle as shown in Figure 7.21a. Then for $0 < q \leq N$, the shift $x[n - q, \text{mod } N]$ is a q-step counterclockwise rotation of the circle as shown in Figure 7.21b. This is why the shift $x[n - q, \text{mod } N]$ is often referred to a *circular time shift.*

Example 7.13 *Circular Time Shift*

For the case $N = 4$, $x[n]$ and the circular time shifts of $x[n]$ are illustrated in Figure 7.22.

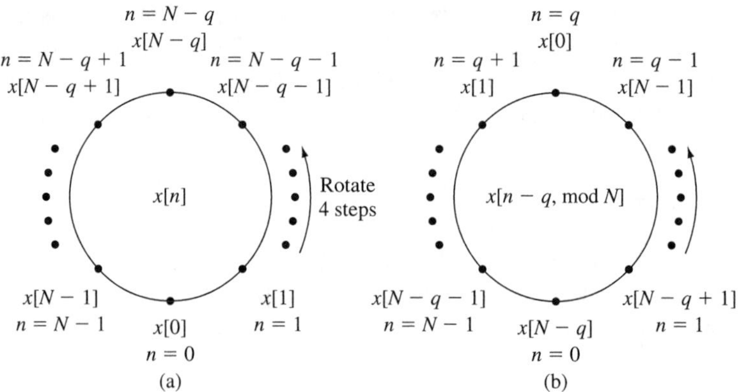

Figure 7.21 Circle representations of (a) $x[n]$ and (b) $x[n - q, \text{mod } N]$.

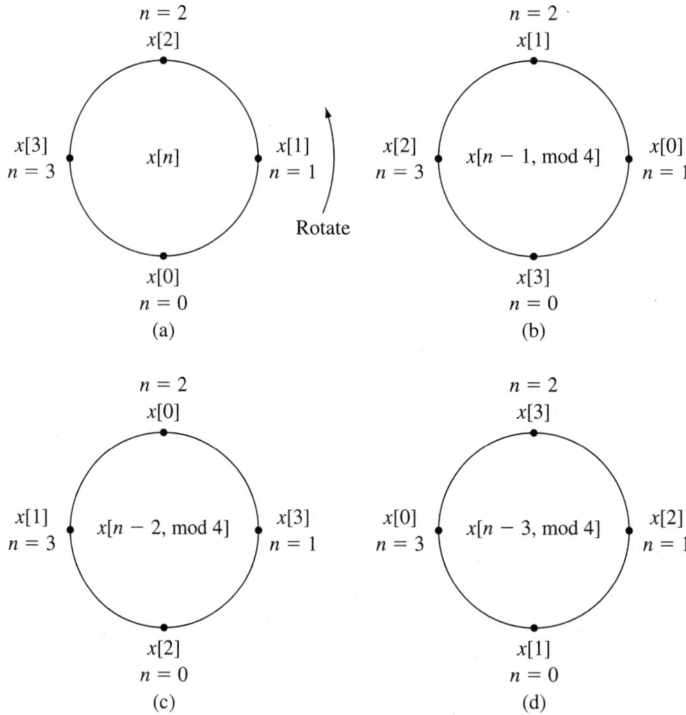

Figure 7.22 Circle representations of (a) $x[n]$, (b) $x[n - 1, \text{mod } 4]$, (c) $x[n - 2, \text{mod } 4]$, and (d) $x[n - 3, \text{mod } 4]$.

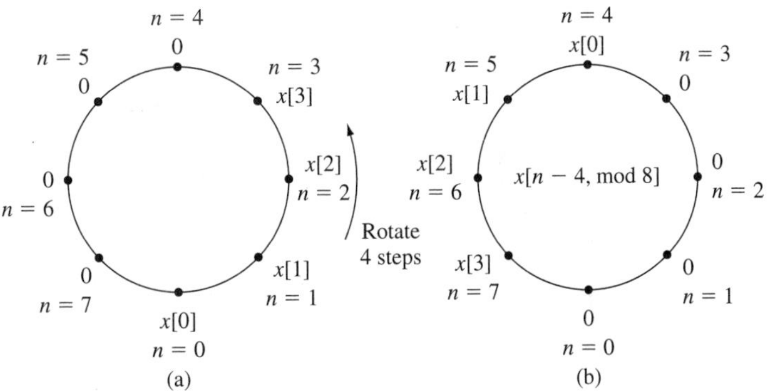

Figure 7.23 Circle representations of (a) padded $x[n]$ and (b) $x[n-4, \text{mod } 8]$.

Now suppose that $q > N$ and let r be the largest integer which is less than or equal to q/N. Then $q \geq Nr$ and

$$x[n-q, \text{mod } N] = x[n-q+Nr, \text{mod } N]$$

and thus $x[n-q, \text{mod } N]$ can be computed by rotating the circle representation of $x[n]$ in the counterclockwise direction for $q - Nr$ steps.

Suppose that $x[n]$ is padded with zeros by setting $x[n] = 0$ for $n = N, N+1, \ldots, L-1$, where $L > N$. Then when $0 < q \leq L - N$, the shift $x[n-q, \text{mod } L]$ is equal to the ordinary shift $x[n-q]$. For example, when $N = 4$ and $L = 8$, the circle representations of the padded signal $x[n]$ and the shift $x[n-4, \text{mod } 8]$ are shown in Figure 7.23. Clearly, the circle representation of $x[n-4, \text{mod } 8]$ is equivalent to the circle representation of the ordinary shift $x[n-4]$ (with n restricted to the range $0 \leq n \leq 7$).

If q is negative, the shift $x[n-q, \text{mod } N]$ is computed by rotating the circle representation of $x[n]$ in the clockwise direction for $|q|$ steps. For example, when $N = 4$, $x[n]$ and the shifts $x[n-q, \text{mod } 4]$ for $q = -1, -2, -3$ are illustrated in Figure 7.24.

Now let q be a positive or negative integer and let X_k be the N-point DFT of the signal $x[n]$. Then the N-point DFT of $x[n-q, \text{mod } N]$ is equal to $X_k e^{-j2\pi kq/N}$. Hence a circular time shift corresponds to multiplication by a complex exponential in the DFT domain. This result is analogous to the property that the DTFT of the ordinary shift $x[n-q]$ is equal to the complex exponential $e^{-jq\Omega}$ times the DTFT $X(\Omega)$ of $x[n]$.

Time Reversal

Given $x[n]$ with $x[n] = 0$ for $n < 0$ and $n \geq N$, the time-reversed signal is denoted by $x[-n, \text{mod } N]$ and is defined by

$$x[-n, \text{mod } N] = \begin{cases} x[0], & n = 0 \\ x[N-n], & 0 < n \leq N-1 \end{cases}$$

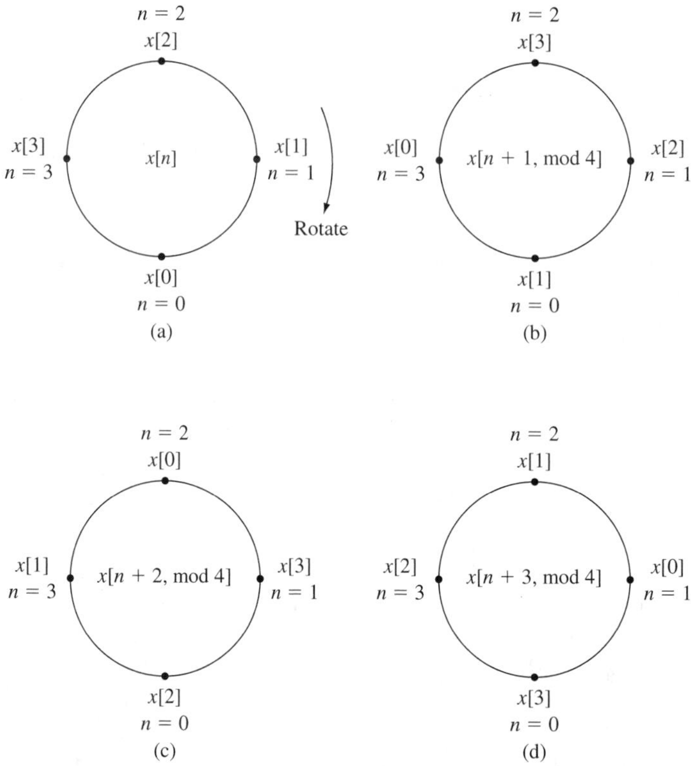

Figure 7.24 Circle representations of (a) $x[n]$, (b) $x[n+1, \text{mod } 4]$, (c) $x[n+2, \text{mod } 4]$, and (d) $x[n+3, \text{mod } 4]$.

The circle representation of the time-reversed signal $x[-n, \text{mod } N]$ is constructed by reversing the values about the zero point on the circle representation for $x[n]$. For example, for $N = 8$ the circle representations of $x[n]$ and $x[-n, \text{mod } N]$ are shown in Figure 7.25.

If X_k is the N-point DFT of $x[n]$, the N-point DFT of $x[-n, \text{mod } N]$ is equal to X_0 for $k = 0$ and equal to X_{N-k} for $0 < k \leq N - 1$. Now suppose that the values of X_k are displayed on a circle, which results in the circle representation of the DFT X_k illustrated in Figure 7.26. Then the circle representation of the DFT of $x[-n, \text{mod } N]$ is constructed by reversing the values about the zero point on the circle representation of X_k. For $N = 8$, the circle representations of X_k and the DFT of $x[-n, \text{mod } N]$ are shown in Figure 7.27.

Multiplication by a Complex Exponential

Given $x[n]$ with N-point DFT X_k, consider the circle representation of X_k shown in Figure 7.26. For any positive integer q, let $X_{k-q, \text{mod } N}$ denote the transform

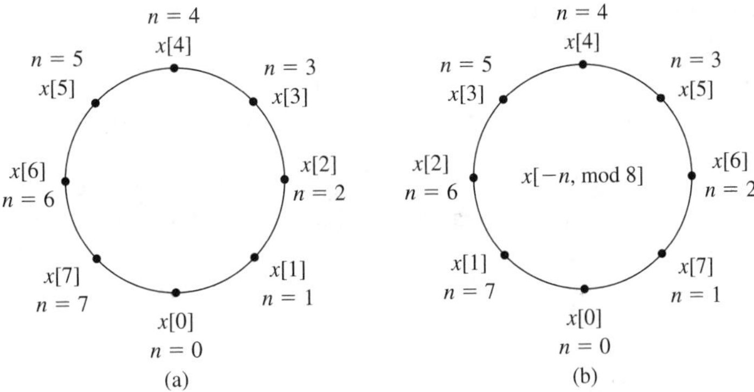

Figure 7.25 Circle representation of (a) $x[n]$ and (b) $x[-n, \text{mod } 8]$.

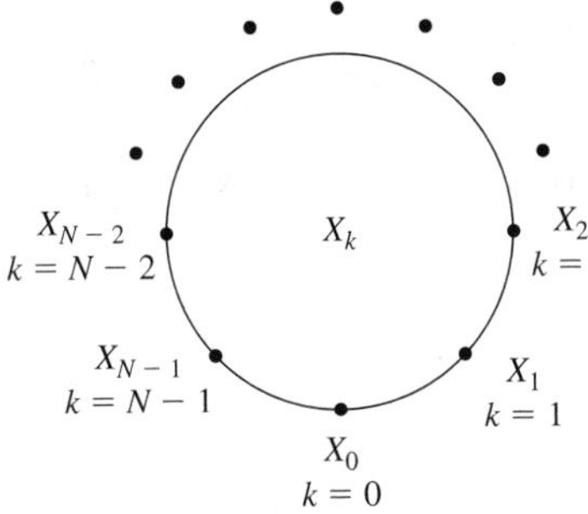

Figure 7.26 Circle representation of X_k.

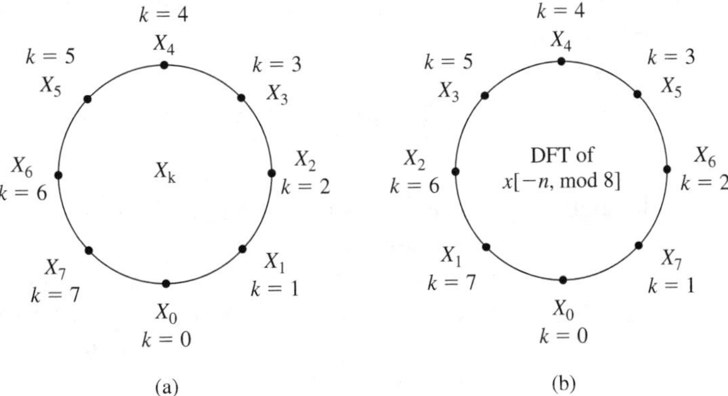

Figure 7.27 Circle representations of (a) X_k and (b) the DFT of $x[-n, \text{mod } 8]$.

constructed by rotating the circle representation of X_k in the counterclockwise direction for q steps. For a negative integer q, let $X_{k-q,\,\mathrm{mod}\,N}$ denote the transform constructed by rotating the circle representation of X_k in the clockwise direction for $|q|$ steps. The transform $X_{k-q,\,\mathrm{mod}\,N}$ is referred to as a *circular frequency shift* of the DFT X_k.

Now the inverse DFT of $X_{k-q,\,\mathrm{mod}\,N}$ turns out to be equal to $x[n]e^{j2\pi qn/N}$. Thus multiplication by a complex exponential in the time domain corresponds to a circular frequency shift. This result is analogous to the property that the DTFT of a signal $x[n]$ multiplied by $e^{jn\Omega_0}$ is equal to the frequency shift $X(\Omega - \Omega_0)$ of the DTFT $X(\Omega)$ of $x[n]$ (see Table 7.2).

Circular Convolution

Given discrete-time signals $x[n]$ and $v[n]$, in Chapter 3 the convolution of $x[n]$ and $v[n]$ was defined by

$$x[n] * v[n] = \sum_{i=-\infty}^{\infty} x[i]v[n-i] \tag{7.44}$$

If both $x[n]$ and $v[n]$ are zero for $n < 0$ and $n \geq N$, (6.44) reduces to

$$x[n] * v[n] = \sum_{i=0}^{N-1} x[i]v[n-i] \tag{7.45}$$

When i ranges over the values $i = 0, 1, \ldots, N-1$, the folded and shifted signal $v[n-i]$ is zero for $n < 0$ and $n \geq 2N-1$, and thus the convolution $x[n] * v[n]$ is zero for $n < 0$ and $n \geq 2N-1$. However, in general $x[n] * v[n]$ is not zero for all $n \geq N$, and therefore the N-point DFT of $x[n] * v[n]$ will not incorporate these values. It is necessary to define a convolution operation so that the convolved signal $x[n] * v[n]$ is zero outside the range $n = 0, 1, \ldots, N-1$. This leads to the notion of *circular convolution* defined by

$$x[n] \circledN v[n] = \sum_{i=0}^{N-1} x[i]v[n-i, \mathrm{mod}\, N] \tag{7.46}$$

To distinguish the ordinary convolution (7.45) from the circular convolution (7.46), the operation (7.45) is sometimes called *linear convolution.* Note that the only difference between linear convolution and circular convolution is that the folded and shifted signal $v[n-i]$ in (7.45) is replaced by the folded and shifted (mod N) signal $v[n-i, \mathrm{mod}\, N]$. For $n = 0$, $v[n-i, \mathrm{mod}\, N] = v[-i, \mathrm{mod}\, N]$ is the time reversal (mod N) of $v[i]$, and $v[n-i, \mathrm{mod}\, N]$ is the n-step counterclockwise circular time shift of $v[-i, \mathrm{mod}\, N]$. For $N = 4$, the circle representations of $v[i]$, $v[-i, \mathrm{mod}\, N]$, and $v[n-i, \mathrm{mod}\, N]$ are given in Figure 7.28.

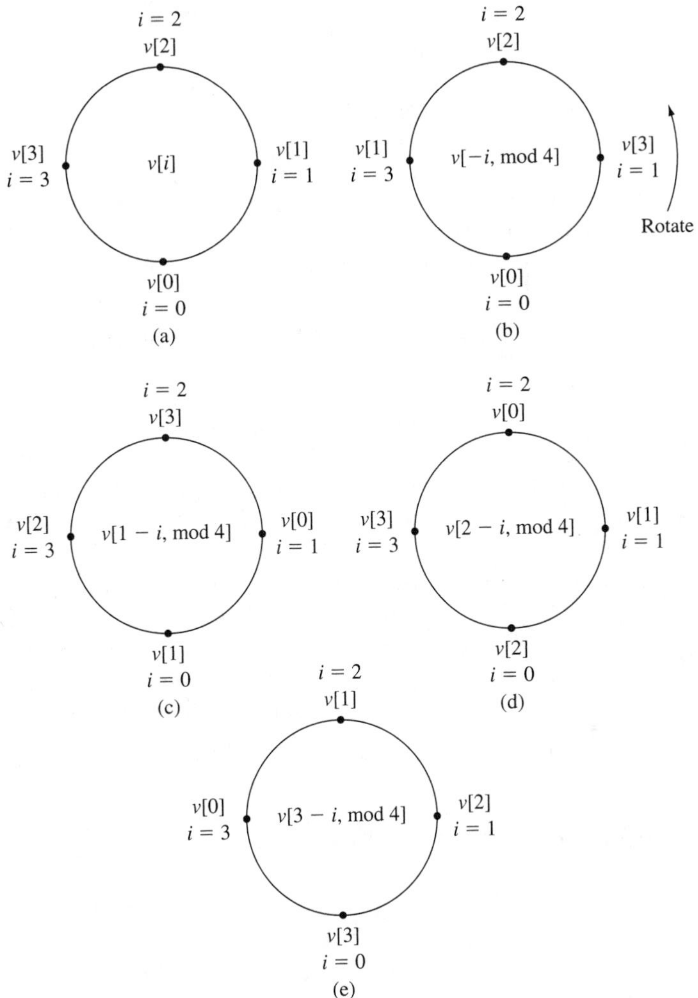

Figure 7.28 Circle representations of (a) $v[i]$, (b) $v[-i, \text{mod } 4]$, (c) $v[1 - i, \text{mod } 4]$, (d) $v[2 - i, \text{mod } 4]$, and (e) $v[3 - i, \text{mod } 4]$.

From (7.46) and the circle representations in Figure 7.28, the circular convolution for $N = 4$ is given by

$$x[n] \circledast v[n] = \begin{cases} x[0]v[0] + x[1]v[3] + x[2]v[2] + x[3]v[1], & n = 0 \\ x[0]v[1] + x[1]v[0] + x[2]v[3] + x[3]v[2], & n = 1 \\ x[0]v[2] + x[1]v[1] + x[2]v[0] + x[3]v[3], & n = 2 \\ x[0]v[3] + x[1]v[2] + x[2]v[1] + x[3]v[0], & n = 3 \end{cases}$$

Given discrete-time signals $x[n]$ and $v[n]$ defined for $n = 0, 1, \ldots, N - 1$, suppose that the signals are padded with zeros for $n = N, N + 1, \ldots, 2N - 2$. Then since $x[n]$ and $v[n]$ are zero for $n = N, N + 1, \ldots, 2N - 2$, setting $L = 2N - 1$ gives

$$x[n] * v[n] = x[n] \, \textcircled{L} \, v[n], \qquad n = 1, 2, \ldots, L - 1 \tag{7.47}$$

In other words, over an L-step interval (where $L = 2N - 1$) linear convolution and circular convolution give the same result.

Again let $x[n]$ and $v[n]$ be discrete-time signals that are zero outside the interval $n = 0, 1, \ldots, N - 1$, and let X_k and V_k denote the N-point DFTs of $x[n]$ and $v[n]$. Then the N-point DFT of the circular convolution $x[n] \, \textcircled{N} \, v[n]$ is equal to $X_k V_k$. This results in the transform pair

$$x[n] \, \textcircled{N} \, v[n] \leftrightarrow X_k V_k$$

This transform pair is the DFT counterpart of the result that the DTFT of the linear convolution $x[n] * v[n]$ is equal to $X(\Omega)V(\Omega)$, where $X(\Omega)$ and $V(\Omega)$ are the DTFTs of $x[n]$ and $v[n]$.

In general the N-point DFT of the linear convolution $x[n] * v[n]$ is not equal to the N-point DFT of the circular convolution $x[n] \, \textcircled{N} \, v[n]$. However, if $x[n]$ and $v[n]$ are padded with zeros for $n = N, N + 1, \ldots, L - 1$, where $L = 2N - 1$, then since $x[n] * v[n]$ and $x[n] \, \textcircled{L} \, v[n]$ are the same [see (7.47)], the L-point DFT of the linear convolution $x[n] * v[n]$ is identical to the L-point DFT of the circular convolution $x[n] \, \textcircled{L} \, v[n]$.

Multiplication in the Time Domain

Again consider the discrete-time signals $x[n]$ and $v[n]$ with N-point DFTs X_k and V_k. Let $V_{-i, \bmod N}$ denote the DFT-domain reversal (mod N) of V_i. The circle representation of $V_{-i, \bmod N}$ is constructed by reversing the values about the zero point on the circle representation of V_i. For $k = 0, 1, \ldots, N - 1$, let $V_{k-i, \bmod N}$ denote the k-step counterclockwise circular frequency shift of $V_{-i, \bmod N}$. The circular convolution $X_k \, \textcircled{N} \, V_k$ in the DFT domain can then be defined by

$$X_k \, \textcircled{N} \, V_k = \sum_{i=0}^{N-1} X_i V_{k-i, \bmod N}$$

It can be shown that the DFT of the product $x[n]v[n]$ is equal to the circular convolution $(1/N)X_k \, \textcircled{N} \, V_k$. Thus multiplication in the time domain corresponds to circular convolution in the DFT domain.

The properties of the DFT are summarized in Table 7.3. Included in the table is the DFT version of Parseval's theorem.

TABLE 7.3 PROPERTIES OF THE DFT

Property	Transform Pair
Linearity	$ax[n] + bv[n] \leftrightarrow aX_k + bV_k$
Circular time shift	$x[n - q, \text{mod } N] \leftrightarrow X_k e^{-j2\pi kq/N}$
Time reversal	$x[-n, \text{mod } N] \leftrightarrow X_{-k, \text{ mod } N}$
Multiplication by a complex exponential	$x[n]e^{j2\pi qn/N} \leftrightarrow X_{k-q, \text{ mod } N}$
Circular convolution	$x[n] \circledN v[n] \leftrightarrow X_k V_k$
Multiplication in the time domain	$x[n]v[n] \leftrightarrow (1/N)X_k \circledN V_k$
Parseval's theorem	$\sum_{n=0}^{N-1} x[n]v[n] \leftrightarrow (1/N) \sum_{i=0}^{N-1} X_i \overline{Y}_i$, where $\overline{Y}_i$ = complex conjugate of Y_i

7.4 SYSTEM ANALYSIS VIA THE DTFT AND DFT

In this section, the DTFT domain and DFT domain representations are generated for a linear time-invariant discrete-time system. The development begins with the DTFT domain representation, which is the discrete-time counterpart of the Fourier transform representation of a linear time-invariant continuous-time system.

Consider a linear time-invariant discrete-time system with unit-pulse response $h[n]$. By the results in Chapter 3, the output response $y[n]$ resulting from the application of input $x[n]$ with no initial energy is given by

$$y[n] = h[n] * x[n] = \sum_{i=-\infty}^{\infty} h[i]x[n - i] \tag{7.48}$$

In this section it is not assumed that the system is necessarily causal, and thus $h[n]$ may be nonzero for values of $n < 0$. It is assumed that the unit-pulse response h$[n]$ satisfies the absolute summability condition

$$\sum_{n=-\infty}^{\infty} |h[n]| < \infty \tag{7.49}$$

As a result of the summability condition (7.49), the ordinary DTFT H(Ω) of the unit-pulse response $h[n]$ exists and is given by

$$H(\Omega) = \sum_{n=-\infty}^{\infty} h[n]e^{-j\Omega n}$$

Now as given in Table 7.2, the DTFT of a convolution of two signals is equal to the product of the DTFTs of the two signals. Hence taking the DTFT of both sides of the input/output relationship (7.48) gives

$$Y(\Omega) = H(\Omega)X(\Omega) \tag{7.50}$$

where $Y(\Omega)$ is the DTFT of the output $y[n]$ and $X(\Omega)$ is the DTFT of the input $x[n]$. Equation (7.50) is the DTFT domain (or Ω domain) representation of the given discrete-time system.

The function $H(\Omega)$ in (7.50) is called the *frequency response function* of the system defined by (7.48). Thus the DTFT of the unit-pulse response $h[n]$ is equal to the frequency response function of the system. The frequency function $H(\Omega)$ is the discrete-time counterpart of the frequency response function $H(\omega)$ of a linear time-invariant continuous-time system (as defined in Section 5.1).

Given a discrete-time system with frequency function $H(\Omega)$, the magnitude $|H(\Omega)|$ is the magnitude function of the system and $\angle H(\Omega)$ is the phase function of the system. Taking the magnitude and angle of both sides of (7.50) yields

$$|Y(\Omega)| = |H(\Omega)|\,|X(\Omega)| \tag{7.51}$$

$$\angle Y(\Omega) = \angle H(\Omega) + \angle X(\Omega) \tag{7.52}$$

By (7.51) the amplitude spectrum $|Y(\Omega)|$ of the output is the product of the amplitude spectrum $|X(\Omega)|$ of the input and the system's magnitude function $|H(\Omega)|$. By (7.52) the phase spectrum $\angle Y(\Omega)$ of the output is the sum of the phase spectrum $\angle X(\Omega)$ of the input and the system's phase function $\angle H(\Omega)$.

Response to a Sinusoidal Input

Suppose that the input $x[n]$ to the system defined by (7.48) is the sinusoid

$$x[n] = A\cos(\Omega_0 n + \theta), n = 0, \pm 1, \pm 2, \ldots$$

where $\Omega_0 \geq 0$. To find the output response $y[n]$ resulting from $x[n]$, first note that from Table 7.1, the DTFT $X(\Omega)$ of $x[n]$ is given by

$$X(\Omega) = \sum_{k=-\infty}^{\infty} A\pi[e^{-j\theta}\delta(\Omega + \Omega_0 - 2\pi k) + e^{j\theta}\delta(\Omega - \Omega_0 - 2\pi k)]$$

From equation (7.50), we see that the DTFT $Y(\Omega)$ of $y[n]$ is equal to the product of $H(\Omega)$ and $X(\Omega)$, and thus

$$Y(\Omega) = \sum_{k=-\infty}^{\infty} A\pi H(\Omega)[e^{-j\theta}\delta(\Omega + \Omega_0 - 2\pi k) + e^{j\theta}\delta(\Omega - \Omega_0 - 2\pi k)]$$

Now, $H(\Omega)\delta(\Omega + c) = H(-c)\delta(\Omega + c)$ for any constant c, and thus

$$Y(\Omega) = \sum_{k=-\infty}^{\infty} A\pi[H(-\Omega_0 + 2\pi k)e^{-j\theta}\delta(\Omega + \Omega_0 - 2\pi k) + H(\Omega_0 + 2\pi k)e^{j\theta}\delta(\Omega - \Omega_0 - 2\pi k)] \tag{7.53}$$

Since $H(\Omega)$ is periodic with period 2π, $H(-\Omega_0 + 2\pi k) = H(-\Omega_0)$ and $H(\Omega_0 + 2\pi k) = H(\Omega_0)$. In addition, assuming that $|H(-\Omega)| = |H(\Omega)|$ and $\angle H(-\Omega) = -\angle H(\Omega)$ yields

$$H(-\Omega_0) = |H(\Omega_0)|e^{-j\angle H(\Omega_0)} \text{ and } H(\Omega_0) = |H(\Omega_0)|e^{j\angle H(\Omega_0)}$$

Hence (7.53) becomes

$$Y(\Omega) = \sum_{k=-\infty}^{\infty} A\pi|H(\Omega_0)|[e^{-j(\angle H(\Omega_0)+\theta)}\delta(\Omega + \Omega_0 - 2\pi k) + e^{j(\angle H(\Omega_0)+\theta)}\delta(\Omega - \Omega_0 - 2\pi k)] \quad (7.54)$$

Taking the inverse DTFT of (7.54) gives

$$y[n] = A|H(\Omega_0)|\cos(\Omega_0 n + \theta + \angle H(\Omega_0)) \quad (7.55)$$

Equation (7.55) is the discrete-time counterpart of the output response of a continuous-time system to a sinusoidal input as derived in Chapter 5. [See (5.15).]

Example 7.14 *Response to a Sinusoidal Input*

Suppose that $H(\Omega) = 1 + e^{-j\Omega}$ and that the objective is the find the output $y[n]$ resulting from the sinusoidal input $x[n] = 2 + 2\sin\left(\frac{\pi}{2}n\right)$. By linearity, $y[n]$ is equal to the sum of the responses to $x_1[n] = 2$ and $x_2[n] = 2\sin\left(\frac{\pi}{2}n\right)$. The response to $x_1[n] = 2 = 2\cos(0n)$ is

$$y_1[n] = 2|H(0)|\cos(0n + \angle H(0)) = 4$$

The response to $x_2[n] = 2\sin\left(\frac{\pi}{2}n\right)$ is

$$y_2[n] = 2\left|H\left(\frac{\pi}{2}\right)\right|\cos\left(\frac{\pi}{2}n + \angle H\left(\frac{\pi}{2}\right)\right)$$

where $H\left(\frac{\pi}{2}\right) = 1 + e^{-j\pi/2} = \sqrt{2}\,e^{-j\pi/4}$. Hence,

$$y_2[n] = 2\sqrt{2}\cos\left(\frac{\pi}{2}n - \frac{\pi}{4}\right)$$

Combining $y_1[n]$ and $y_2[n]$ yields

$$y[n] = 4 + 2\sqrt{2}\cos\left(\frac{\pi}{2}n - \frac{\pi}{4}\right)$$

Analysis of an Ideal Lowpass Filter

As an illustration of the use of the DTFT representation, it will be applied to the study of an ideal lowpass filter. Consider the discrete-time system with the frequency function

$$H(\Omega) = \sum_{k=-\infty}^{\infty} p_{2B}(\Omega + 2\pi k) \quad (7.56)$$

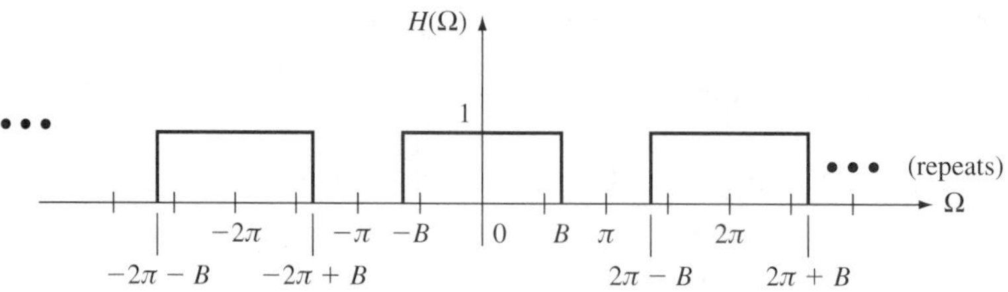

Figure 7.29 Frequency function $H(\Omega)$ given by (7.56).

where $B < \pi$. The function $H(\Omega)$ is plotted in Figure 7.29. Note that in this example, the magnitude function $|H(\Omega)|$ is equal to $H(\Omega)$, and the phase function $\angle H(\Omega)$ is identically zero.

The Ω domain representation (7.50) will be utilized to compute the output response $y[n]$ resulting from the sinusoidal input

$$x[n] = A\cos(\Omega_0 n), \qquad n = 0, \pm 1, \pm 2, \ldots$$

where $\Omega_0 \geq 0$. From Table 7.1, the DTFT $X(\Omega)$ of $x[n]$ is given by

$$X(\Omega) = \sum_{k=-\infty}^{\infty} A\pi[\delta(\Omega + \Omega_0 - 2\pi k) + \delta(\Omega - \Omega_0 - 2\pi k)]$$

For the case $0 \leq \Omega_0 < \pi$, $X(\Omega)$ is plotted in Figure 7.30.

Now by (7.50), the DTFT $Y(\Omega)$ of the output response $y[n]$ is equal to the product of $H(\Omega)$ and $X(\Omega)$. From the plots of $H(\Omega)$ and $X(\Omega)$ given in Figures 7.29 and 7.30, it is clear that $Y(\Omega)$ is equal to $X(\Omega)$ when $\Omega_0 \leq B$ and $Y(\Omega)$ is identically zero if $B < \Omega_0 < \pi$. Thus, in mathematical terms,

$$Y(\Omega) = \begin{cases} X(\Omega), & \Omega_0 \leq B \\ 0, & B < \Omega_0 < \pi \end{cases}$$

Taking the inverse DTFT then gives

$$y[n] = \begin{cases} A\cos(\Omega_0 n), & \Omega_0 \leq B \\ 0, & B < \Omega_0 \end{cases} \tag{7.57}$$

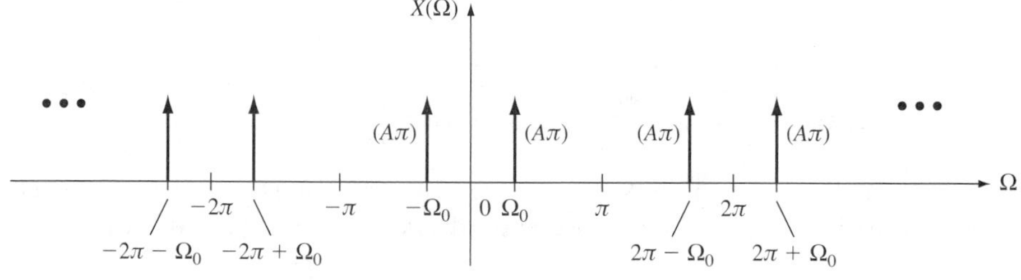

Figure 7.30 DTFT of $x[n]$ with $0 \leq \Omega_0 < \pi$.

As a result of the periodicity of $X(\Omega)$ and $H(\Omega)$, the output response $y[n]$ is equal to $A \cos \Omega_0 n$ when

$$2\pi k - BT \leq \Omega_0 \leq 2\pi k + BT, \qquad k = 0, 1, 2, \ldots \tag{7.58}$$

For all other positive values of Ω_0, the response $y[n]$ is zero.

From (7.57) it is seen that the system passes all input sinusoids $A \cos \Omega_0 n$ with $\Omega_0 \leq B$, while it completely stops all such inputs with $B \leq \Omega_0 < \pi$. However, as a result of periodicity of the frequency function, this discrete-time filter is not a "true" lowpass filter, since it passes input sinusoids $A \cos \Omega_0 n$ with Ω_0 belonging to the intervals given by (7.58). If Ω_0 is restricted to lie in the range $0 \leq \Omega_0 < \pi$, the filter can be viewed as an ideal lowpass filter with bandwidth B.

Digital-Filter Realization of an Ideal Analog Lowpass Filter

The filter with frequency function (7.56) can be used as a digital-filter realization (or discrete-time realization) of an ideal lowpass zero-phase analog filter with bandwidth B. To see this, suppose that the input

$$x(t) = A \cos \omega_0 t, \qquad -\infty < t < \infty$$

is applied to an analog filter with the frequency function $p_{2B}(\omega)$. From the results in Section 5.4, the output of the filter is equal to $x(t)$ when $\omega_0 \leq B$ and is equal to zero when $\omega_0 > B$.

Now suppose that the sampled version of the input

$$x[n] = x(t)\big|_{t=nT} = A \cos \Omega_0 n, \text{ where } \Omega_0 = \omega_0 T$$

is applied to the discrete-time filter with frequency function (7.56). Then by (7.57), the output is equal to $x[n]$ when $\Omega_0 \leq B$ and is equal to zero when $B < \Omega_0 < \pi$. Thus as long as $\Omega_0 < \pi$ or $\omega_0 < \pi/T$, the output $y[n]$ of the discrete-time filter will be equal to a sampled version of the output of the analog filter. An analog signal can then be generated from the sampled output using a hold circuit as discussed in Chapter 12. Hence this results in a digital-filter realization of the given analog filter.

The requirement that the frequency ω_0 of the input sinusoid $A \cos \omega_0 t$ be less than π/T is not an insurmountable constraint, since π/T can be increased by decreasing the sampling interval T, which is equivalent to increasing the sampling frequency $\omega_s = 2\pi/T$ (see Section 5.5). Therefore, as long as a suitably fast sample rate can be achieved, the upper bound π/T on the input frequency ω_0 is not a problem.

Unit-Pulse Response of Ideal Lowpass Filter

From the transform pairs in Table 7.1, the unit-pulse response $h[n]$ of the filter with the frequency function (7.56) is given by

$$h[n] = \frac{B}{\pi} \operatorname{sinc}\left(\frac{B}{\pi} n\right), \qquad n = 0, \pm 1, \pm 2, \ldots$$

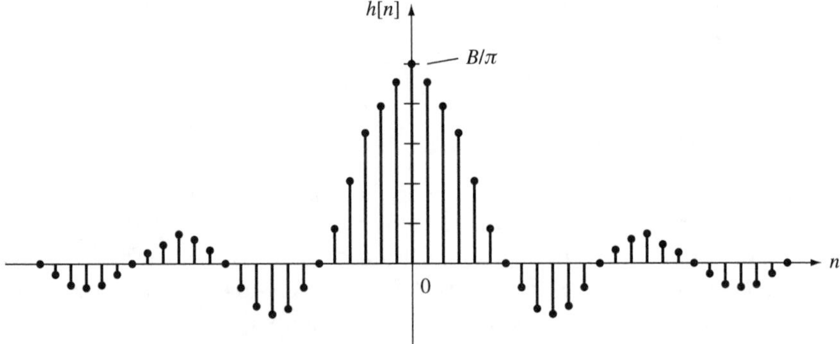

Figure 7.31 Unit-pulse response of ideal lowpass discrete-time filter.

The unit-pulse response is displayed in Figure 7.31. Note that the sinc function form of the unit-pulse response is very similar to the form of the impulse response of an ideal analog lowpass filter (see Section 5.4).

From Figure 7.31 it is seen that $h[n]$ is not zero for $n < 0$, and thus the filter is noncausal. Therefore, the filter cannot be implemented on-line (in real time); but it can be implemented off-line. In an off-line implementation the filtering process is applied to the values of signals that have been stored in the memory of a digital computer or stored by some other means. So in the discrete-time case, ideal filters can be used in practice, as long as the filtering process is carried out off-line.

Causal Lowpass Filter

As noted above, an ideal lowpass filter cannot be implemented in real time, since the filter is noncausal. For "real-time filtering" it is necessary to consider a causal lowpass filter. One very simple example is the averager, which is defined by the input/output difference equation

$$y[n] = \tfrac{1}{2}(x[n] + x[n-1]) \tag{7.59}$$

or

$$y[n+1] = \tfrac{1}{2}(x[n+1] + x[n])$$

Since the system (7.59) averages the past two values of the input, the time variation of the input should be smoothed out to some extent in the output; in other words, the system should behave like a lowpass filter. To verify this, the magnitude function $|H(\Omega)|$ of the system will be determined. First, from (7.59) the unit-pulse response $h[n]$ is given by

$$h[n] = \tfrac{1}{2}(\delta[n] + \delta[n-1])$$

Then taking the DTFT of $h[n]$ yields

$$H(\Omega) = \tfrac{1}{2}[1 + e^{-j\Omega}] = \tfrac{1}{2}[(1 + \cos\Omega) - j\sin\Omega]$$

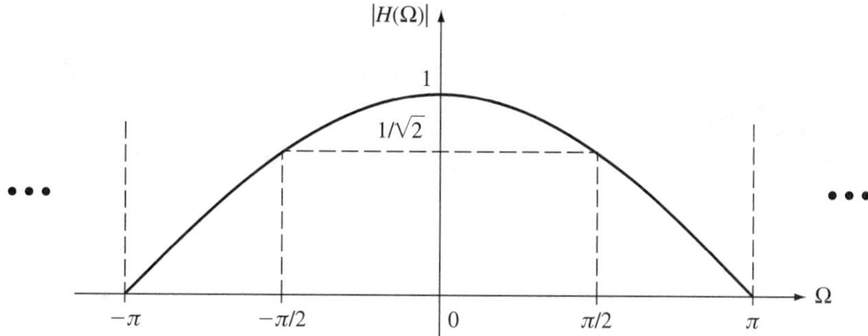

Figure 7.32 Magnitude function of the averager.

Taking the magnitude of $H(\Omega)$ gives

$$|H(\Omega)| = \frac{1}{\sqrt{2}}\sqrt{1 + \cos\Omega}$$

The magnitude function $|H(\Omega)|$ is plotted in Figure 7.32. From this it is seen that the averager is a (nonideal) lowpass filter with $|H(\Omega)|$ down by $1/\sqrt{2}$ when $\Omega = \pi/2$. If the input to the averager is the sampled sinusoid

$$x[n] = A\cos\Omega_0 n, \qquad n = 0, \pm 1, \pm 2, \ldots$$

where $\Omega_0 = \omega_0 T$, the averager will pass the sinusoid if $\Omega_0 < \pi/2$ or $\omega_0 < \pi/2T$; otherwise, the averager will attenuate the sinusoid. Thus the effective bandwidth of the filter can be set by choosing the sampling interval T.

In this example the separation between the passband and the stopband is not very sharp. A filter with a sharper frequency cutoff is the filter with the input/output difference equation

$$y[n] = \frac{1}{N}\sum_{i=0}^{N-1} x[n-i] \tag{7.60}$$

where $N \geq 3$. The filter (7.60) is called the *mean filter*. The mean filter is an example of a causal lowpass discrete-time filter.

System Analysis via the DFT

Now suppose that the signal $x[n]$ is applied to a linear time-invariant discrete-time system with unit-pulse response $h[n]$. It is assumed that $x[n] = 0$ for $n < 0$ and $n \geq N$, and that $h[n] = 0$ for $n < 0$ and $n > Q$, where Q is some positive integer. The resulting output response $y[n]$ with the system at rest at time $n = 0$ is given by

$$y[n] = h[n] * x[n] = \sum_{i=0}^{\infty} h[i]x[n-i], \qquad n \geq 0$$

Since $h[i] = 0$ for $i > Q$,

$$y[n] = \sum_{i=0}^{Q} h[i]x[n-i], \qquad n \geq 0 \tag{7.61}$$

Since $x[n - i] = 0$ when $n - i \geq N$, from (7.61)

$$y[n] = 0 \qquad \text{for all integers } n \geq N + Q$$

Now the DFT domain representation can be generated by taking the $(N + Q)$-point DFT of both sides of the convolution representation (7.61). First it is necessary to pad $x[n]$ and $h[n]$ with zeros so that

$$x[n] = 0, \qquad n = N, N + 1, \ldots, N + Q - 1$$

$$h[n] = 0, \qquad n = Q + 1, Q + 2, \ldots, Q + N - 1$$

The $(N + Q)$-point DFTs of $x[n]$ and $h[n]$ are then given by

$$X_k = \sum_{n=0}^{N-1} x[n]e^{-j2\pi kn/(N+Q)}, \qquad k = 0, 1, \ldots, N + Q - 1$$

$$H_k = \sum_{n=0}^{Q} h[n]e^{-j2\pi kn/(N+Q)}, \qquad k = 0, 1, \ldots, N + Q - 1$$

Since $y[n] = 0$ for $n \geq N + Q$, over the interval $n = 0, 1, \ldots, N + Q - 1$, the linear convolution $x[n] * h[n]$ is equal to the circular convolution $x[n]$ Ⓛ $h[n]$ (where $L = N + Q$), and thus by the convolution property of the DFT (see Table 7.3),

$$Y_k = H_k X_k \qquad k = 0, 1, \ldots, N + Q - 1 \tag{7.62}$$

where Y_k is the $(N + Q)$-point DFT of $y[n]$ given by

$$Y_k = \sum_{n=0}^{N+Q-1} y[n]e^{-j2\pi kn/(N+Q)}, \qquad k = 0, 1, \ldots, N + Q - 1$$

Equation (7.62) is the DFT domain representation of the given discrete-time system. The representation (7.62) is a frequency-sampled version of the DTFT domain representation (7.50). More precisely, setting $\Omega = 2\pi k/(N + Q)$ in (7.50) yields

$$Y\left(\frac{2\pi k}{N + Q}\right) = H\left(\frac{2\pi k}{N + Q}\right)X\left(\frac{2\pi k}{N + Q}\right), \qquad k = 0, 1, \ldots, N + Q - 1 \tag{7.63}$$

But the $(N + Q)$-point DFTs X_k, H_k, and Y_k are equal to the sampled DTFTs $X(2\pi k/N + Q)$, $H(2\pi k/N + Q)$, and $Y(2\pi k/N + Q)$, respectively; and thus (7.63) is identical to the DFT-domain representation (7.62).

The output response $y[n]$ can be computed by taking the $(N + Q)$-point inverse DFT of the right-hand side of (7.62). Therefore,

$$y[n] = \frac{1}{N + Q} \sum_{k=0}^{N+Q-1} H_k X_k e^{j2\pi kn/(N+Q)}, \qquad n = 0, 1, \ldots, N + Q - 1 \tag{7.64}$$

It is important to note that the right-hand side of (7.64) is only an approximation of $y[n]$ if $x[n]$ is not zero for $n \geq N$ and/or $h[n]$ is not zero for $n > Q$. The expression (7.64) for $y[n]$ will be a close approximation to the true values of $y[n]$ if $|x[n]|$ is small for $n \geq N$ and if $|h[n]|$ is small for $n > Q$.

7.5 FFT ALGORITHM

Again let $x[n]$ be a discrete-time signal with $x[n] = 0$ for $n < 0$ and $n \geq N$. In Section 7.2 the N-point DFT and inverse DFT were defined by

$$X_k = \sum_{n=0}^{N-1} x[n]e^{-j2\pi kn/N}, \qquad k = 0, 1, \ldots, N-1 \tag{7.65}$$

$$x[n] = \frac{1}{N}\sum_{k=0}^{N-1} X_k e^{j2\pi kn/N}, \qquad n = 0, 1, \ldots, N-1 \tag{7.66}$$

From (7.65) it is seen that for each value of k, the computation of X_k from $x[n]$ requires N multiplications. Thus the computation of X_k for $k = 0, 1, \ldots, N-1$ requires N^2 multiplications. Similarly, from (7.66) it follows that the computation of $x[n]$ from X_k also requires N^2 multiplications.

It should be mentioned that the multiplications in (7.65) and (7.66) are complex multiplications in general; that is, the numbers being multiplied are complex numbers. The multiplication of two complex numbers requires four real multiplications. In the following analysis, the number of complex multiplications is counted. The number of additions required to compute the DFT or inverse DFT will not be considered.

Since the direct evaluation of (7.65) or (7.66) requires N^2 multiplications, this can result in a great deal of computation if N is large. It turns out that (7.65) or (7.66) can be computed using a fast Fourier transform (FFT) algorithm, which requires on the order of $(N \log_2 N)/2$ multiplications. This is a significant decrease in the N^2 multiplications required in the direct evaluation of (7.65) or (7.66). For instance, if $N = 1024$, the direct evaluation requires $N^2 = 1{,}048{,}576$ multiplications. In contrast, the FFT algorithm requires

$$\frac{1024(\log_2 1024)}{2} = 5120 \text{ multiplications}$$

There are different versions of the FFT algorithm. Here the development is limited to one particular approach based on decimation in time. For an in-depth treatment of the FFT algorithm, the reader is referred to Brigham [1988] or Rabiner and Gold [1975].

The basic idea of the decimation-in-time approach is to subdivide the time interval into intervals having a smaller number of points. This is illustrated by first showing that the computation of X_k can be broken up into two parts. First, to simplify the notation, let W_N equal $\exp(-j2\pi/N)$. The complex number W_N is an Nth root of unity; that is,

$$W_N^N = e^{-j2\pi} = 1$$

It is assumed that $N > 1$, and thus $W_N \neq 1$. In terms of W_N, the N-point DFT and inverse DFT are given by

$$X_k = \sum_{n=0}^{N-1} x[n]W_N^{kn}, \qquad k = 0, 1, \ldots, N-1 \tag{7.67}$$

$$x[n] = \frac{1}{N}\sum_{k=0}^{N-1} X_k W_N^{-kn}, \qquad n = 0, 1, \ldots, N-1 \tag{7.68}$$

Now let N be an even integer, so that $N/2$ is an integer. Given the signal $x[n]$ with $x[n] = 0$ for $n < 0$ and $n \geq N$, define the signals

$$a[n] = x[2n], \qquad n = 0, 1, 2, \ldots, \frac{N}{2} - 1$$

$$b[n] = x[2n + 1], \qquad n = 0, 1, 2, \ldots, \frac{N}{2} - 1$$

Note that the signal $a[n]$ consists of the values of $x[n]$ at the even values of the time index n, while $b[n]$ consists of the values at the odd time points.

Let A_k and B_k denote the $(N/2)$-point DFTs of $a[n]$ and $b[n]$; that is,

$$A_k = \sum_{n=0}^{(N/2)-1} a[n]W_{N/2}^{kn}, \qquad k = 0, 1, \ldots, \frac{N}{2} - 1 \tag{7.69}$$

$$B_k = \sum_{n=0}^{(N/2)-1} b[n]W_{N/2}^{kn}, \qquad k = 0, 1, \ldots, \frac{N}{2} - 1 \tag{7.70}$$

Let X_k denote the N-point DFT of $x[n]$. Then it is claimed that

$$X_k = A_k + W_N^k B_k, \qquad k = 0, 1, \ldots, \frac{N}{2} - 1 \tag{7.71}$$

$$X_{(N/2)+k} = A_k - W_N^k B_k, \qquad k = 0, 1, \ldots, \frac{N}{2} - 1 \tag{7.72}$$

To verify (7.71), insert the expressions (7.69) and (7.70) for A_k and B_k into the right-hand side of (7.71). This gives

$$A_k + W_N^k B_k = \sum_{n=0}^{(N/2)-1} a[n]W_{N/2}^{kn} + \sum_{n=0}^{(N/2)-1} b[n]W_N^k W_{N/2}^{kn}$$

Now $a[n] = x[2n]$ and $b[n] = x[2n + 1]$, and thus

$$A_k + W_N^k B_k = \sum_{n=0}^{(N/2)-1} x[2n]W_{N/2}^{kn} + \sum_{n=0}^{(N/2)-1} x[2n + 1]W_N^k W_{N/2}^{kn}$$

Using the properties

$$W_{N/2}^{kn} = W_N^{2kn}, \qquad W_N^k W_{N/2}^{kn} = W_N^{(1+2n)k}$$

yields the result

$$A_k + W_N^k B_k = \sum_{n=0}^{(N/2)-1} x[2n]W_N^{2kn} + \sum_{n=0}^{(N/2)-1} x[2n + 1]W_N^{(1+2n)k} \tag{7.73}$$

Defining the change of index $\bar{n} = 2n$ in the first sum of the right-hand side of (7.73) and the change of index $\bar{n} = 2n + 1$ in the second sum yields

$$A_k + W_N^k B_k = \sum_{\substack{\bar{n}=0 \\ \bar{n}\text{ even}}}^{N-2} x[\bar{n}]W_N^{\bar{n}k} + \sum_{\substack{\bar{n}=0 \\ \bar{n}\text{ odd}}}^{N-1} x[\bar{n}]W_N^{\bar{n}k}$$

$$A_k + W_N^k B_k = \sum_{\bar{n}=0}^{N-1} x[\bar{n}]W_N^{\bar{n}k}$$

$$A_k + W_N^k B_k = X_k$$

Hence (7.71) is verified. The proof of (7.72) is similar and is therefore omitted.

The computation of X_k using (7.71) and (7.72) requires $N^2/2 + N/2$ multiplications. To see this, first note that the computation of A_k requires $(N/2)^2 = N^2/4$ multiplications, as does the computation of B_k. The computation of the products $W_N^k B_n$ in (7.71) and (7.72) requires $N/2$ multiplications. So the total number of multiplications is equal to $N^2/2 + N/2$. This is $N^2/2 - N/2$ multiplications less than N^2 multiplications. Therefore, when N is large, the computation of X_k using (7.71) and (7.72) requires significantly fewer multiplications than the computation of X_k using (7.67).

If $N/2$ is even, each of the signals $a[n]$ and $b[n]$ can be expressed in two parts, and then the process described above can be repeated. If $N = 2^q$ for some positive integer q, the subdivision process can be continued until signals with only one nonzero value (with each value equal to one of the values of the given signal $x[n]$) are obtained.

In the case $N = 8$, a block diagram of the FFT algorithm is given in Figure 7.33. On the far left-hand side of the diagram, the values of the given signal $x[n]$ are inputed. Note the order (in terms of row position) in which the signal values $x[n]$ are applied to the process. The order can be determined by a process called *bit reversing*. Suppose that $N = 2^q$. Given an integer n ranging from 0 to $N - 1$, the time index n can be represented by the q-bit binary word for the integer n. Reversing the bits comprising this word results in the integer corresponding to the reversed-bit word; which is the row at which the signal value $x[n]$ is applied to the FFT algorithm. For example, when $N = 8$ the binary words and bit-reversed words corresponding to the time index n are shown in Table 7.4. The last column in Table 7.4 gives the order for which the signal values are applied to the FFT algorithm shown in Figure 7.33.

The MATLAB software package contains commands for computing the FFT and the inverse FFT, denoted by `fft` and `ifft`. The commands `fft` and `ifft` are interchangeable with the commands `dft` and `idft` used in the examples given in Section 7.2. Examples in the next section further demonstrate the use of these commands.

7.6 APPLICATIONS OF THE FFT ALGORITHM

The FFT algorithm is very useful in a wide range of applications involving digital signal processing and digital communications. In this section it is first shown that the FFT algorithm can be used to compute the Fourier transform of a continuous-time

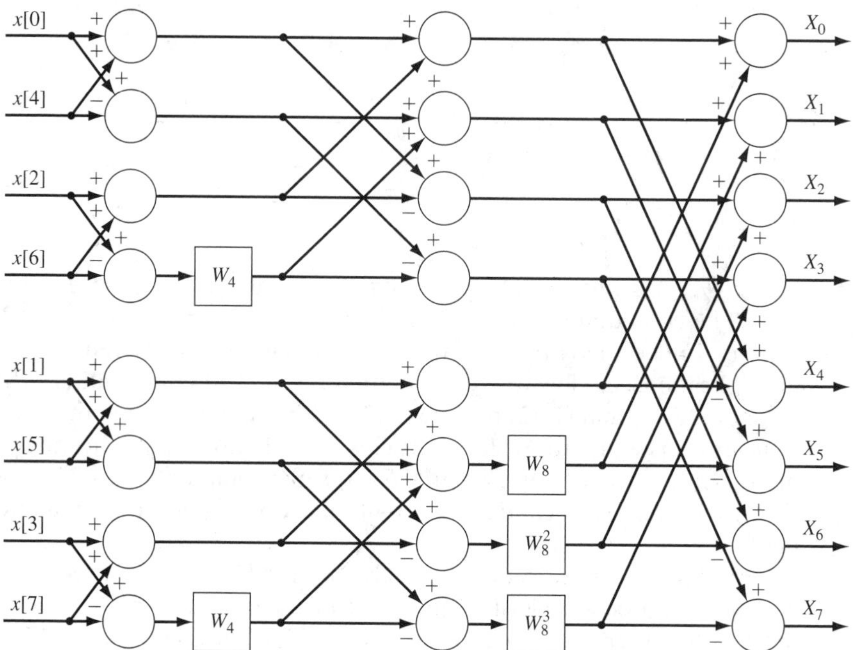

Figure 7.33 Block diagram of FFT algorithm when $N = 8$.

TABLE 7.4 BIT REVERSING IN THE CASE $N = 8$

Time Point (n)	Binary Word	Reversed-Bit Word	Order
0	000	000	$x[0]$
1	001	100	$x[4]$
2	010	010	$x[2]$
3	011	110	$x[6]$
4	100	001	$x[1]$
5	101	101	$x[5]$
6	110	011	$x[3]$
7	111	111	$x[7]$

signal. Then the FFT algorithm is applied to the problem of computing the output response of a linear time-invariant discrete-time system.

Computation of the Fourier Transform via the FFT

Let $x(t)$ be a continuous-time signal with Fourier transform $X(\omega)$. It is assumed that $x(t) = 0$ for all $t < 0$ so that the Fourier transform $X(\omega)$ of $x(t)$ is given by

$$X(\omega) = \int_0^\infty x(t)e^{-j\omega t}\,dt \tag{7.74}$$

Let Γ be a fixed positive real number and let N be a fixed positive integer. It will be shown that by using the FFT algorithm, $X(\omega)$ can be computed for $\omega = k\Gamma, k = 0, 1, 2, \ldots, N-1$.

Given a fixed positive number T, the integral in (7.74) can be written in the form

$$X(\omega) = \sum_{n=0}^{\infty} \int_{nT}^{nT+T} x(t)e^{-j\omega t}\,dt \tag{7.75}$$

Suppose that T is chosen small enough so that the variation in $x(t)$ is small over each T-second interval $nT \le t < nT + T$. Then the sum in (7.75) can be approximated by

$$\begin{aligned} X(\omega) &= \sum_{n=0}^{\infty} \left(\int_{nT}^{nT+T} e^{-j\omega t}\,dt \right) x(nT) \\ &= \sum_{n=0}^{\infty} \left[\left(\frac{-1}{j\omega} e^{-j\omega t} \right) \Bigg|_{t=nT}^{t=nT+T} \right] x(nT) \\ &= \frac{1 - e^{-j\omega T}}{j\omega} \sum_{n=0}^{\infty} e^{-j\omega nT} x(nT) \end{aligned} \tag{7.76}$$

Now suppose that for some sufficiently large positive integer N, the magnitude $|x(nT)|$ is small for all integers $n \ge N$. Then (7.76) becomes

$$X(\omega) = \frac{1 - e^{-j\omega T}}{j\omega} \sum_{n=0}^{N-1} e^{-j\omega nT} x(nT) \tag{7.77}$$

Evaluating both sides of (7.77) at $\omega = 2\pi k/NT$ gives

$$X\left(\frac{2\pi k}{NT}\right) = \frac{1 - e^{-j2\pi k/N}}{j2\pi k/NT} \sum_{n=0}^{N-1} e^{-j2\pi nk/N} x(nT) \tag{7.78}$$

Now let X_k denote the N-point DFT of the sampled signal $x[n] = x(nT)$. By definition of the DFT

$$X_k = \sum_{n=0}^{N-1} x[n]e^{-j2\pi kn/N}, \qquad k = 0, 1, \ldots, N-1 \tag{7.79}$$

Comparing (7.78) and (7.79) reveals that

$$X\left(\frac{2\pi k}{NT}\right) = \frac{1 - e^{-j2\pi k/N}}{j2\pi k/NT} X_k \tag{7.80}$$

Finally, letting $\Gamma = 2\pi/NT$, (7.77) can be rewritten in the form

$$X(k\Gamma) = \frac{1 - e^{-jk\Gamma T}}{jk\Gamma} X_k, \qquad k = 0, 1, 2, \ldots, N - 1 \tag{7.81}$$

By first calculating X_k via the FFT algorithm and then using (6.78), $X(k\Gamma)$ can be computed for $k = 0, 1, \ldots, N - 1$.

It should be stressed that the relationship (7.81) is an approximation, so the values of $X(\omega)$ computed using (7.81) are only approximate values. Better accuracy can be obtained by taking a smaller value for the sampling interval T and/or by taking a larger value for N. If the amplitude spectrum $|X(\omega)|$ is small for $\omega > B$, a good choice for T is the sampling interval π/B corresponding to the Nyquist sampling frequency $\omega_s = 2B$. If the given signal $x(t)$ is known only for the time interval $0 \le t \le t_1$, N can still be selected to be as large as desired by padding the sampled signal $x[n] = x(nT)$ with zeros for those values of n for which $nT > t_1$ (or $n > t_1/T$).

Example 7.15 *Computation of Fourier Transform Via the FFT*

Consider the continuous-time signal $x(t)$ shown in Figure 7.34. The FFT program in MATLAB can be used to compute $X(\omega)$ via the following procedure. First, a sampled version of $x(t)$ is obtained and is denoted by $x(nT)$, where T is a small sampling interval and $n = 0, 1, \ldots, N - 1$. Then the FFT X_k of $x[n] = x(nT)$ is determined. Finally, X_k is rescaled using (7.81) to obtain the approximation $X(k\Gamma)$ of the actual Fourier transform $X(\omega)$. The MATLAB commands for obtaining the approximation are given below. For comparison's sake, the program also plots the actual Fourier transform $X(\omega)$, which can be computed as follows: Let $x_1(t) = tp_2(t)$ where $p_2(t)$ is the two-second rectangular pulse centered at the origin. Then $x(t) = x_1(t - 1)$ and from Example 4.12, the Fourier transform of $x_1(t)$ is

$$X_1(\omega) = j2\frac{\omega \cos \omega - \sin \omega}{\omega^2}$$

Using the shift in time property,

$$X(\omega) = j2\frac{\omega \cos \omega - \sin \omega}{\omega^2} e^{-j\omega}$$

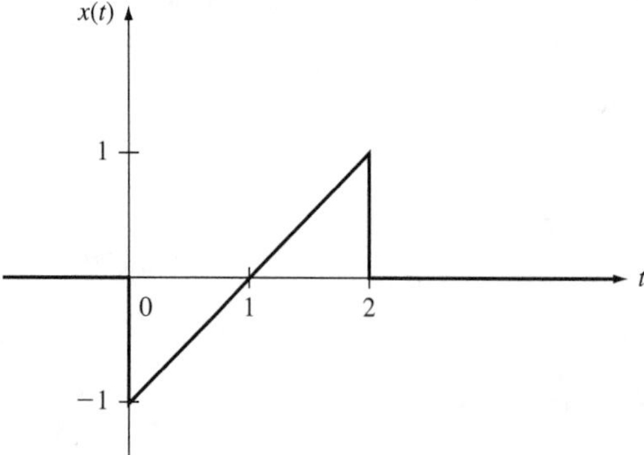

Figure 7.34 Continuous-time signal in Example 7.15.

Now the evaluation of the approximate and exact Fourier transforms of $x(t)$ can be carried out using the following MATLAB commands:

```
N = input('Input N: ');
T = input('Input T: ');
%
% compute the approximation of X(w)
t=0:T:2;
x = [t-1 zeros(1,N-length(t))];
Xk=fft(x);
gam = 2*pi/N/T;
k = 0:10/gam;     % for plotting purposes
Xapp = (1-exp(-j*k*gam*T))/j/k/gam*Xk;
%
%  compute the actual X(w)
w = 0.05:.05:10;
Xact = j*2*exp(-j*w).*(w.*cos(w)-sin(w))./(w.*w);
plot(k*gam,abs(Xapp(1:length(k))),'o',w,abs(Xact))
```

To run this program, the user first inputs the desired values for N and T, and then the program plots the approximate Fourier transform, denoted by `Xapp`, and the actual Fourier transform, denoted by `Xact`. The program was run with $N = 2^7 = 128$ and $T = 0.1$, in which case $\Gamma = 2\pi/NT = 0.4909$. The resulting amplitude spectra of the actual and the approximate Fourier transforms are plotted in Figure 7.35. Note that the approximation is reasonably accurate. More detail in the plot is achieved by increasing NT, and more accuracy is achieved by decreasing T. The program was rerun with $N = 2^9 = 512$ and $T = 0.05$ so that $\Gamma = 0.2454$. The resulting amplitude spectrum is displayed in Figure 7.36.

The computations required to perform the approximation of the Fourier transform for a generic signal are contained in an M-file named `contfft.m` that is given on the Web site that accompanies this text. This M-file is used by first defining the signal x and the time interval T. The Fourier transform is computed via the command

```
[X,w] = contfft(x,T)
```

where `X` $= X(\omega)$ and `w` $= 2\pi k/NT$.

Fast Convolution

The FFT algorithm can be used to perform a fast version of convolution. Since the output of a linear time-invariant discrete-time system is the convolution of the input and the unit-pulse response, this can be used to compute the output response as follows.

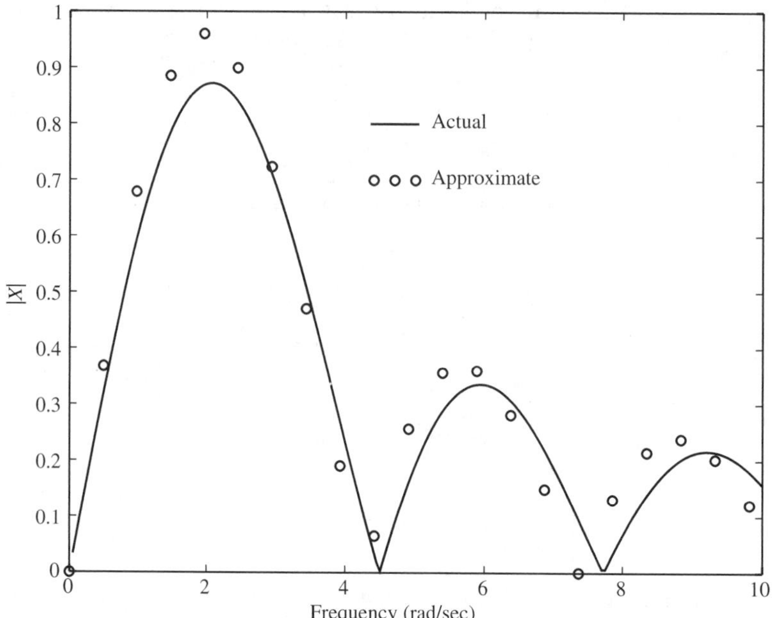

Figure 7.35 Amplitude spectrum in the case $N = 128$ and $T = 0.1$.

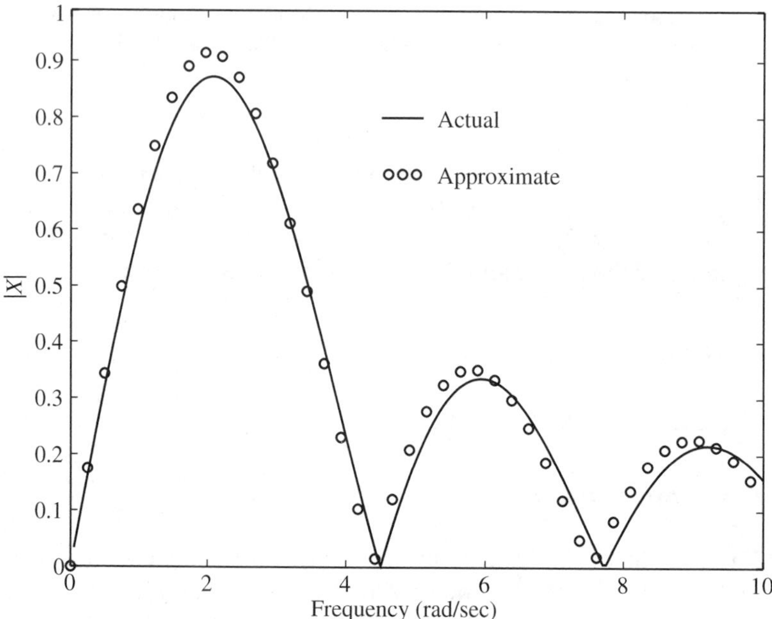

Figure 7.36 Amplitude spectrum in the case $N = 512$ and $T = 0.05$.

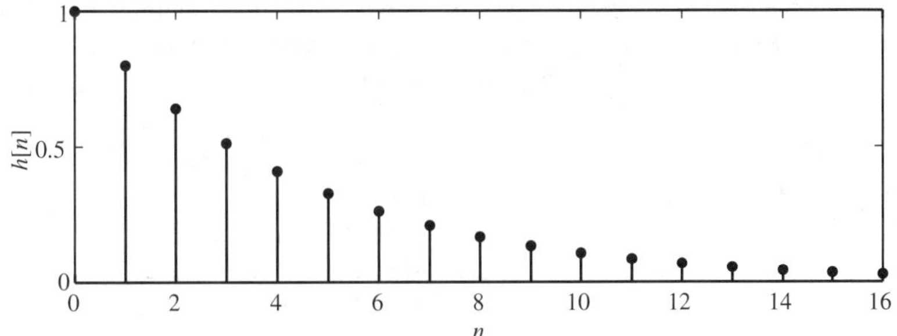

Figure 7.37 Unit-pulse response of system in Example 6.15.

Given signals $x[n]$ and $h[n]$, with $x[n] = 0$ for $n \geq N$ and $h[n] = 0$ for $n > Q$, $x[n]$ and $h[n]$ are first padded with zeros as discussed in Section 6.4. With r equal to the smallest positive integer such that $N + Q \leq 2^r$ and with $L = 2^r$, the L-point DFTs of $x[n]$ and $h[n]$ can be computed using the FFT algorithm. With X_k and H_k equal to the DFTs, the output response $y[n] = h[n] * x[n]$ is equal to the inverse L-point DFT of the product $H_k X_k$, which also can be computed using the FFT algorithm. This approach requires on the order of $(1.5L)\log_2 L + L$ multiplications. In contrast, the computation of $y[n]$ using the convolution sum requires on the order of $0.5L^2 + 1.5L$ multiplications.

Example 7.16 *Computation of Output Response*

Consider the discrete-time system with the unit-phase response

$$h[n] = (0.8)^n u[n]$$

which is plotted in Figure 7.37. The objective is to compute the output response $y[n]$ of the system resulting from the rectangular input shown in Figure 7.38 with no initial energy in the system at time $n = 0$. The output $y[n]$ could be calculated by evaluating the convolution $h[n] * x[n]$; however, here $y[n]$ will be computed using the FFT approach.

In this example there is no finite integer Q for which $h[n] = 0$ for all $n > Q$. However, from Figure 7.37 it is seen that $h[n]$ is very small for $n > 16$, and thus Q can be taken to be equal to 16. Since the input $x[n]$ is zero for all $n \geq 10$, the integer N defined above is equal to 10. Thus with $Q = 16$, $N + Q = 26$, and the smallest integer value of r for

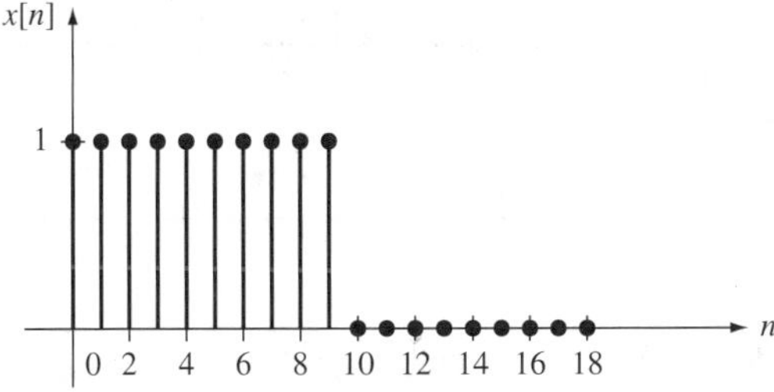

Figure 7.38 System input $x[n]$ in Example 7.16.

which $N + Q \leq 2^r$ is $r = 5$. With $L = 2^5 = 32$, the L-point DFT of the input $x[n]$ and unit-pulse response $h[n]$ can be computed using the MATLAB `fft` file. The MATLAB commands for generating the L-point DFTs are

```
n=0:16; L = 32;
h = (.8).^n;
Hk = fft(h,L);
x = [ones(1,10)];
Xk = fft(x,L);
```

MATLAB plots of the magnitude and phase functions of the system and the amplitude and phase spectra of the input are displayed in Figures 7.39 and 7.40. Note that from the amplitude plot in Figure 7.39a, it is seen that the system is a lowpass filter.

Now the L-point DFT Y_k of the output response $y[n]$ and the inverse DFT of Y_k (using the inverse FFT algorithm) are computed by using the MATLAB commands

```
Yk = Hk.*Xk;
y = ifft(Yk,L);
```

A MATLAB plot of the output response is given in Figure 7.41.

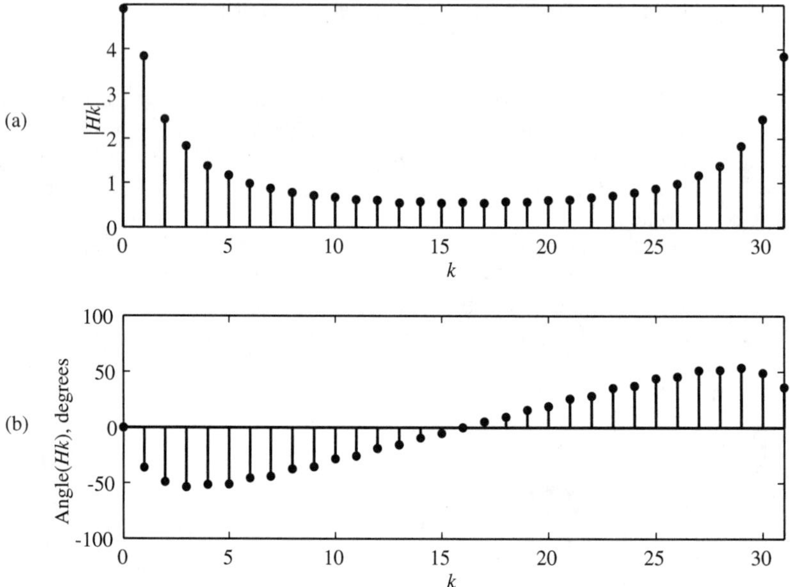

Figure 7.39 (a) Magnitude and (b) phase functions of the system in Example 7.16.

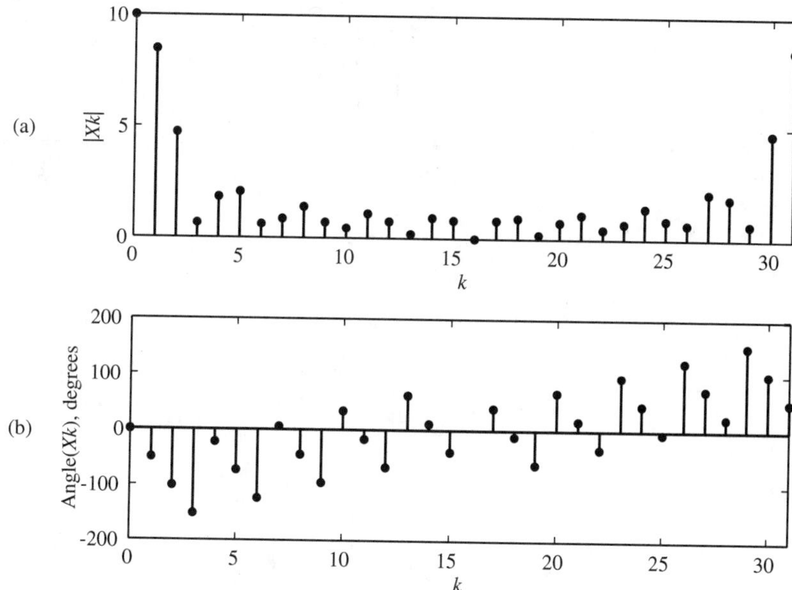

Figure 7.40 (a) Amplitude and (b) phase spectra of the input in Example 7.15.

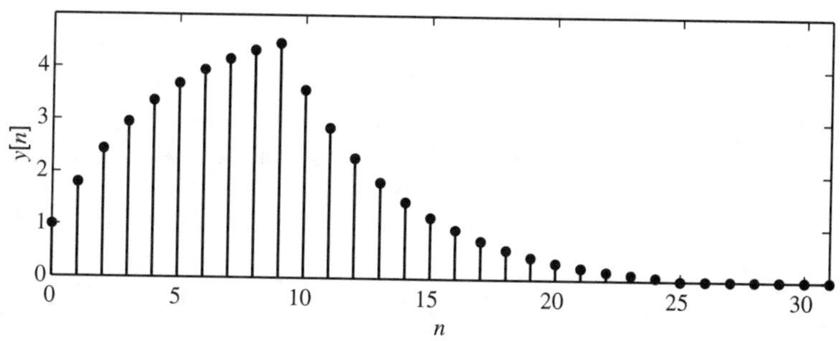

Figure 7.41 Output response in Example 7.16.

PROBLEMS

7.1. Compute the DTFTs of the discrete-time signals shown in Figure P7.1. Express the DTFTs in the simplest possible form. Plot the amplitude and phase spectrum for each signal.

7.2. Prove the following relationship:

$$\sum_{n=q_1}^{q_2} r^n = \frac{r^{q_1} - r^{q_2+1}}{1 - r}$$

Hint: Multiply both sides of the equation by $(1 - r)$.

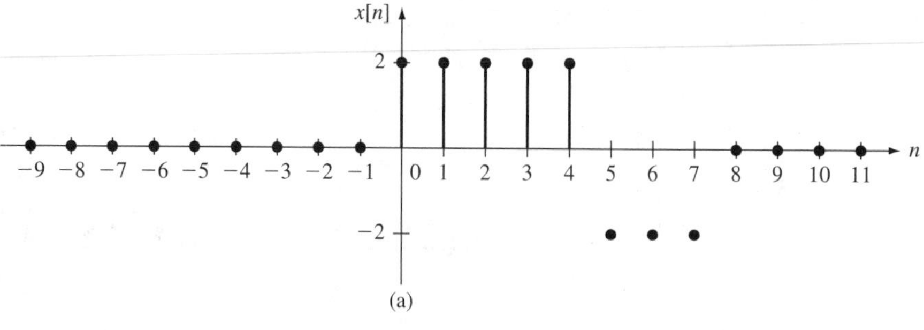
(a)

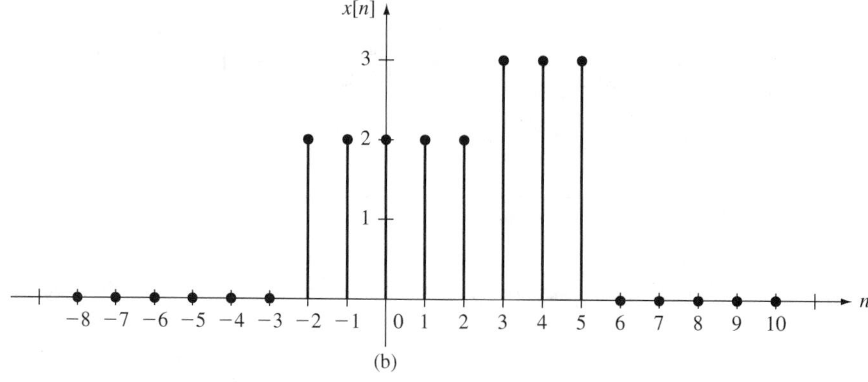
(b)

Figure P7.1

7.3. By breaking the DTFT $X(\Omega)$ into three summations, (from $n = -\infty$ to $n = -1$; $n = 0$; and $n = 1$ to $n = \infty$) and applying Euler's formula, prove the following:

(a) If $x[n]$ is an even function of n, then $X(\Omega) = x[0] + \sum_{n=1}^{\infty} 2x[n] \cos \Omega n$.

(b) If $x[n]$ is an odd function of n, then $X(\Omega) = x[0] - \sum_{n=1}^{\infty} j2x[n] \sin \Omega n$.

7.4. Compute the DTFT of the following discrete-time signals. Plot the amplitude and the phase spectrum for each signal.

(a) $x[n] = (0.8)^n u[n]$
(b) $x[n] = (0.5)^n \cos 4n\, u[n]$
(c) $x[n] = n(0.5)^n u[n]$
(d) $x[n] = n(0.5)^n \cos 4n\, u[n]$
(e) $x[n] = (0.5)^n \cos^2 4n\, u[n]$
(f) $x[n] = (0.5)^{|n|}, -\infty < n < \infty$
(g) $x[n] = (0.5)^{|n|} \cos 4n, -\infty < n < \infty$

7.5. A discrete-time signal $x[n]$ has DTFT

$$X(\Omega) = \frac{1}{e^{j\Omega} + b}$$

where b is an arbitrary constant. Determine the DTFT $V(\Omega)$ of the following:

(a) $v[n] = x[n-5]$
(b) $v[n] = x[-n]$
(c) $v[n] = nx[n]$
(d) $v[n] = x[n] - x[n-1]$
(e) $v[n] = x[n] * x[n]$
(f) $v[n] = x[n] \cos 3n$
(g) $v[n] = x^2[n]$
(h) $v[n] = x[n]e^{j2n}$

7.6. Use (7.28) to compute the inverse DTFT of the frequency functions $X(\Omega)$ shown in Figure P7.6.

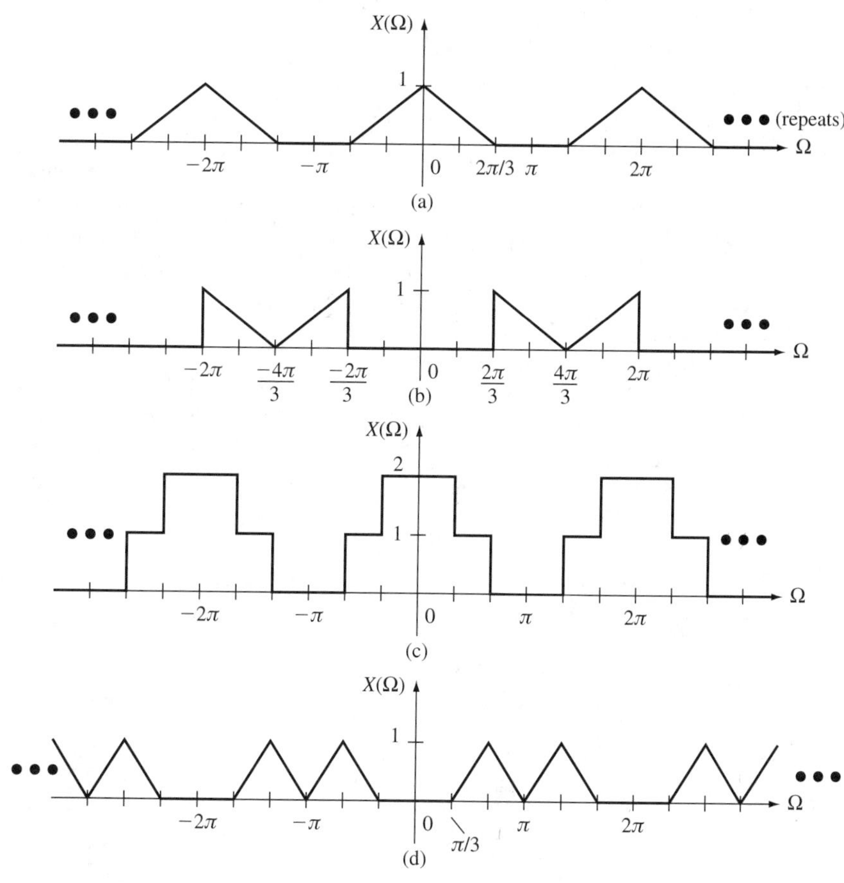

Figure P7.6

7.7. Determine the inverse DTFT of the following frequency functions.
(a) $X(\Omega) = \sin \Omega$
(b) $X(\Omega) = \cos \Omega$
(c) $X(\Omega) = \cos^2 \Omega$
(d) $X(\Omega) = \sin \Omega \cos \Omega$

7.8. The autocorrelation of a discrete-time signal $x[n]$ is defined by

$$R_x[n] = \sum_{i=-\infty}^{\infty} x[i]x[n+i]$$

Let $P_x(\Omega)$ denote the DTFT of $R_x[n]$.

(a) Derive an expression for $P_x(\Omega)$ in terms of the DTFT $X(\Omega)$ of $x[n]$.
(b) Derive an expression for $R_x[-n]$ in terms of $R_x[n]$.
(c) Express $P_x(0)$ in terms of $x[n]$.

7.9. Compute the rectangular form of the four-point DFT of the following signals, all of which are zero for $n < 0$ and $n \geq 4$.

(a) $x[0] = 1, x[1] = 0, x[2] = 1, x[3] = 0$
(b) $x[0] = 1, x[1]\, 0, x[2] = -1, x[3] = 0$
(c) $x[0] = 1, x[1] = 1, x[2] = -1, x[3] = -1$
(d) $x[0] = -1, x[1] = 1, x[2] = 1, x[3] = 1$
(e) $x[0] = -1, x[1] = 0, x[2] = 1, x[3] = 2$
(f) $x[0] = 1, x[1] = -1, x[2] = 1, x[3] = -1$
(g) Compute the DFT for each of the signals above using the MATLAB M-file `dft`. Compare these results to the results obtained analytically in parts (a) to (f).

7.10. Using the MATLAB M-file `dft`, compute the 32-point DFT of the following signals. Express your answer by plotting the amplitude $|X_k|$ and phase $\angle X_k$ of the DFTs.

(a) $x[n] = 1, 0 \leq n \leq 10, x[n] = 0$ for all other n
(b) $x[n] = 1, 0 \leq n \leq 10, x[n] = -1, 11 \leq n \leq 20, x[n] = 0$ for all other n
(c) $x[n] = n, 0 \leq n \leq 20, x[n] = 0$ for all other n
(d) $x[n] = n, 0 \leq n \leq 10, x[n] = 20 - n, 11 \leq n \leq 20, x[n] = 0$ for all other n
(e) $x[n] = \cos(10\pi n/11), 0 \leq n \leq 10, x[n] = 0$ for all other n
(f) $x[n] = \cos(9\pi n/11), 0 \leq n \leq 10, x[n] = 0$ for all other n

7.11. Using the MATLAB M-file `dft`, compute the magnitude of the 32-point DFT X_k of the following signals.

(a) $x[n] = \begin{cases} 1, & n = 0 \\ \dfrac{1}{n}, & n = 1, 2, 3, \ldots, 31 \\ 0, & n = 32, 33, \ldots \end{cases}$

(b) $x[n] = \begin{cases} 1, & n = 0 \\ \dfrac{1}{n^2}, & n = 1, 2, 3, \ldots, 31 \\ 0, & n = 32, 33, \ldots \end{cases}$

(c) $x[n] = \begin{cases} 1, & n = 0 \\ \dfrac{1}{n!}, & n = 1, 2, 3, \ldots, 31 \\ 0, & n = 32, 33, \ldots \end{cases}$

(d) Compare the results obtained for parts (a) to (c). Explain the differences in the results.

7.12. Consider the discrete-time signal

$$x[n] = \begin{cases} r[n] - 0.5, & n = 0, 1, 2, \ldots, 31 = N - 1 \\ 0, & \text{all other } n \end{cases}$$

where r is a sequence of random numbers uniformly distributed between 0 and 1. This sequence can be generated in MATLAB using `rand(N,1)`. The signal $x[n]$ can be interpreted as random noise. Using the `dft` M-file, compute the magnitude of the 32-point DFT of $x[n]$. What frequencies would you expect to see in the amplitude spectrum of $x[n]$? Explain.

7.13. For each of the signals $x[n]$ in Problem 7.9, compute the four-point DFT of the circular time shift $x[n-2, \text{mod } 4]$. Express the DFTs in rectangular form.

7.14. For each of the signals $x[n]$ in Problem 7.9, compute the four-point DFT of $x[n]e^{j\pi n}$. Express the DFTs in rectangular form.

7.15. Compute the following circular convolutions.

(a) $x[n] \circledast_4 v[n]$, where $x[0] = 1$, $x[1] = 0$, $x[2] = 1$, $x[3] = 0$, and $v[0] = 1$, $v[1] = 0$, $v[2] = -1$, $v[3] = 0$

(b) $x[n] \circledast_4 v[n]$, where $x[0] = 1$, $x[1] = 0$, $x[2] = 1$, $x[3] = 0$, and $v[0] = -1$, $v[1] = 1$, $v[2] - 1$, $v[3] = 1$

(c) $x[n] \circledast_4 v[n]$, where $x[0] = -1$, $x[1] = 0$, $x[2] = 1$, $x[3] = 2$, and $v[0] = -1$, $v[1] = 0$, $v[2] = 1$, $v[3] = 2$

(d) $x[n] \circledast_4 v[n]$, where $x[0] = 1$, $x[1] = 1$, $x[2] = -1$, $x[3] = -1$, and $v[0] = -1$, $v[1] = 0$, $v[2] = 1$, $v[3] = 2$

7.16. Using the property that the DFT of $x[n] \circledast_N v[n]$ is equal to $X_k V_k$, compute the four-point DFTs of the circular convolutions in Problem 7.15.

7.17. Compute the four-point DFTs of the products $x[n]v[n]$, where $x[n]$ and $v[n]$ are the signals defined in each part of Problem 7.15.

7.18. Use the MATLAB M-file `dft` with $N = 10$ to approximate the DTFT of the signal plotted in Figure P7.1a. Plot the amplitude and phase spectrum for X_k versus $\Omega = 2\pi k/N$. Compare this result to the DTFT obtained in Problem 7.1 over the frequency range $\Omega = 0$ to $\Omega = 2\pi$. Repeat for $N = 20$.

7.19. Repeat Problem 7.18 for the signal plotted in Figure P7.1b.

7.20. To determine the effect of truncation in computing the approximation of a DTFT by a DFT, consider the signal defined by $x[n] = n(0.5)^n u[n]$.

(a) Determine the minimum value of N so that the signal has magnitude $|x[n]| \leq 20\%$ of its maximum value for all $n \geq N$.

(b) Use MATLAB to compute the 50-point DFT of the truncated signal $\tilde{x}[n]$ defined by

$$\tilde{x}[n] = \begin{cases} x[n], & 0 \leq n \leq N - 1 \\ 0, & \text{all other } n \end{cases}$$

where N was determined in part (a). Plot the amplitude and phase spectrum of X_k versus $\Omega = 2\pi k/50$.

(c) Compare the result obtained in part (b) to the DTFT of $x[n]$ found in Problem 6.4(c).

(d) Repeat parts (a) to (c) for the value of N such that the signal has magnitude $|x[n]| \leq 5\%$ of its maximum value for all $n \geq N$.

7.21. An ideal lowpass discrete-time filter has the frequency function $H(\Omega)$ given by

$$H(\Omega) = \begin{cases} 1, & 0 \le |\Omega| \le \dfrac{\pi}{4} \\ 0, & \dfrac{\pi}{4} < |\Omega| \le \pi \end{cases}$$

(a) Determine the unit-pulse response $h[n]$ of the filter.

(b) Compute the output response $y[n]$ of the filter when the input $x[n]$ is given by

(i) $x[n] = \cos(\pi n/8), n = 0, \pm 1, \pm 2, \ldots$

(ii) $x[n] = \cos(3\pi n/4) + \cos(\pi n/16), n = 0, \pm 1, \pm 2, \ldots$

(iii) $x[n] = \text{sinc}(n/2), n = 0, \pm 1, \pm 2, \ldots$

(iv) $x[n] = \text{sinc}(n/4), n = 0, \pm 1, \pm 2, \ldots$

(v) $x[n] = \text{sinc}(n/8)\cos(\pi n/8), n = 0, \pm 1, \pm 2, \ldots$

(vi) $x[n] = \text{sinc}(n/8)\cos(\pi n/4), n = 0, \pm 1, \pm 2, \ldots$

(c) For each signal defined in part (b), plot the input $x[n]$ and the corresponding output $y[n]$ to determine the effect of the filter.

7.22. An ideal linear-phase highpass discrete-time filter has frequency function $H(\Omega)$, where for one period, $H(\Omega)$ is given by

$$H(\Omega) = \begin{cases} e^{-j3\Omega}, & \dfrac{\pi}{2} \le |\Omega| \le \pi \\ 0, & 0 \le |\Omega| < \dfrac{\pi}{2} \end{cases}$$

(a) Determine the unit-pulse response $h[n]$ of the filter.

(b) Compute the output response $y[n]$ of the filter when the input $x[n]$ is given by

(i) $x[n] = \cos(\pi n/4), n = 0, \pm 1, \pm 2, \ldots$

(ii) $x[n] = \cos(3\pi n/4), n = 0, \pm 1, \pm 2, \ldots$

(iii) $x[n] = \text{sinc}(n/2), n = 0, \pm 1, \pm 2, \ldots$

(iv) $x[n] = \text{sinc}(n/4), n = 0, \pm 1, \pm 2, \ldots$

(v) $x[n] = \text{sinc}(n/4)\cos(\pi n/8), n = 0, \pm 1, \pm 2, \ldots$

(vi) $x[n] = \text{sinc}(n/2)\cos(\pi n/8), n = 0, \pm 1, \pm 2, \ldots$

(c) For each signal defined in part (b), plot the input $x[n]$ and the corresponding output $y[n]$ to determine the effect of the filter.

7.23. A linear time-invariant discrete-time system has the frequency response function $H(\Omega)$ shown in Figure P7.23.

(a) Determine the unit-pulse response $h[n]$ of the system.

(b) Compute the output response $y[n]$ when the input $x[n]$ is equal to $\delta[n] - \delta[n-1]$.

(c) Compute the output response $y[n]$ when the input is $x[n] = 2 + \sin(\pi n/4) + 2\sin(\pi n/2)$.

(d) Compute the output response $y[n]$ when $x[n] = \text{sinc}(n/4), n = 0, \pm 1, \pm 2$.

(e) For the signals defined in parts (b)–(c), plot the input $x[n]$ and the corresponding output $y[n]$ to determine the effect of the filter.

7.24. Consider the mean filter given by the input/output difference equation

$$y[n] = \frac{1}{N}\sum_{i=0}^{N-1} x[n-i]$$

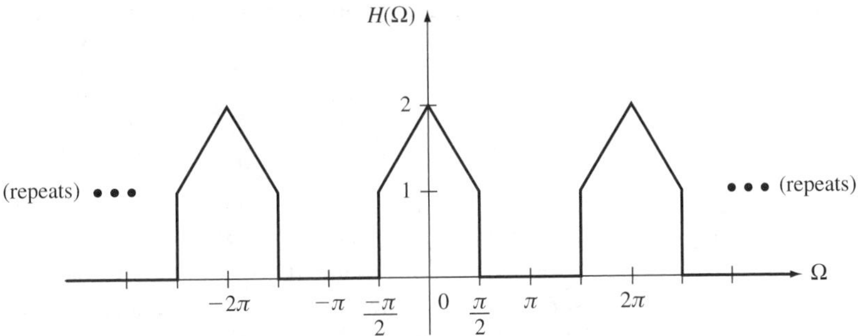

Figure P7.23

(a) Determine the unit-pulse response $h[n]$ of the filter.

(b) Show that the frequency response function $H(\Omega)$ of the filter can be expressed in the form

$$H(\Omega) = \begin{cases} 1, & \Omega = 0 \\ \dfrac{1 - \cos(N\Omega) + j\sin(N\Omega)}{N(1 - \cos(\Omega) + j\sin(\Omega))}, & 0 < |\Omega| < \pi \end{cases}$$

(c) Sketch $|H(\Omega)|$ for $-\pi \le \Omega \le \pi$ in the case when $N = 3$ and $N = 4$.

(d) Sketch $\angle H(\Omega)$ for $-\pi \le \Omega \le \pi$ in the case when $N = 3$ and $N = 4$.

(e) Compute the output response $y[n]$ when $N = 3$ and

(i) $x[n] = 1, n = 0, \pm 1, \pm 2, \ldots$

(ii) $x[n] = \cos(\pi n/4), n = 0, \pm 1, \pm 2, \ldots$

(iii) $x[n] = \cos(\pi n/2), n = 0, \pm 1, \pm 2, \ldots$

(iv) $x[n] = \cos(\pi n/2)\sin(\pi n/4), n = 0, \pm 1, \pm 2, \ldots$

(f) For each signal defined in part (e), plot the input $x[n]$ and the corresponding output $y[n]$ to determine the effect of the filter.

7.25. As shown in Figure P7.25, a sampled version $x[n]$ of an analog signal $x(t)$ is applied to a linear time-invariant discrete-time system with frequency response function $H(\Omega)$. Choose the sampling interval T and determine the frequency response function $H(\Omega)$ so that

$$y[n] = \begin{cases} x[n], & \text{when } x(t) = A\cos\omega_0 t, \quad 100 < \omega_0 < 1000 \\ 0, & \text{when } x(t) = A\cos\omega_0 t, \quad 0 \le \omega_0 \le 100 \end{cases}$$

Express $H(\Omega)$ in analytical form.

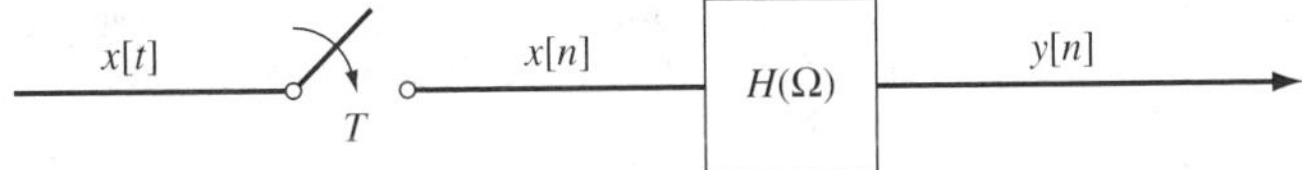

Figure P7.25

7.26. Consider the ideal lowpass discrete-time filter with frequency response function $H(\Omega)$ defined by

$$H(\Omega) = \begin{cases} e^{-j10\Omega}, & 0 \le |\Omega| \le \dfrac{\pi}{4} \\ 0, & \dfrac{\pi}{4} < |\Omega| < \pi \end{cases}$$

(a) The input $x[n] = u[n] - u[n - 10]$ is applied to the filter.

(i) Using `fft` in MATLAB, compute the 32-point DFT of the resulting output response. *Note:* To calculate the DFT of the output, Y_k, write an M-file that carries out the multiplication of the DFT of the input, X_k, with $H(2\pi k/N)$, where $H(\Omega)$ is the frequency response function of the filter. Take $N = 32$.

(ii) Using `ifft` in MATLAB, compute the output response $y[n]$ for $n = 0, 1, 2, \ldots, 31$.

(b) Repeat part (a) for the input $x[n] = u[n] - u[n - 5]$.

(c) Compare the output response obtained in parts (a) and (b). In what respects do the responses differ? Explain.

(d) Repeat part (a) for the input

$$x[n] = \begin{cases} r[n] - 0.5, & n = 0, 1, 2, \ldots, 10 \\ 0, & \text{all other } n \end{cases}$$

where $r[n]$ is a sequence of random numbers uniformly distributed between 0 and 1 (see Problem 7.12). How does the magnitude of the response compare with the magnitude of the input? Explain.

7.27. A mean filter is given by the input/output difference equation

$$y[n] = \frac{x[n] + x[k - 1] + x[k - 2]}{3}$$

(a) Show that the frequency response function is given by

$$H(\Omega) = \frac{1}{3}(1 + e^{-j\Omega} + e^{-j2\Omega})$$

(b) Repeat Problem 7.26 for this filter.

7.28. Repeat Problem 7.26 for the linear time-invariant discrete-time system with frequency response function

$$H(\Omega) = \frac{0.04}{e^{j2\Omega} - 1.6e^{j\Omega} + 0.64}$$

7.29. Consider the discrete-time system given by the input/output difference equation

$$y[n + 1] + 0.9y[n] = 1.9x[n + 1]$$

(a) Show that the impulse response is given by $h[n] = 1.9(-0.9)^n u[n]$.

(b) Compute the frequency response function and sketch the magnitude function $|H(\Omega)|$ for $-\pi \le \Omega \le \pi$.

(c) Compute the output response $y[n]$ to an input of $x[n] = 1 + \sin(\pi n/4) + \sin(\pi n/2)$.

(d) Compute the output response $y[n]$ resulting from the input $x[n] = u[n] - u[n - 3]$.

(e) Use the `fft` command to compute and plot the response of the system to the input in given in part (d) for $n = 0$ through 30. Compare your numerical answer with the answer found in part (d). Does the response match what you might expect from the plot of the frequency response function determined in part (b)? Explain.

7.30. Consider the discrete-time system given by the input/output difference equation

$$y[n + 1] - 0.9y[n] = 0.1x[n + 1]$$

(a) Show that the impulse response is given by $h[n] = 0.1(0.9)^n u[n]$.
(b)–(e) Repeat Problem 7.29, parts (b)–(e) for this system.

7.31. This problem explores the use of the FFT in approximating the Fourier Transform of continuous-time signals.

(a) Compute the Fourier transform of $x(t) = 4e^{-4t}u(t)$.

(b) Create a sampled version of the signal $x(t)$ in MATLAB for the cases (i)–(iv) below, where T is the sampling time and N is the total number of points. Use the M-file `contfft.m` from the textbook to compute the approximation to $X(\omega)$. Plot $|X(\omega)|$ versus ω for the exact Fourier transform obtained in part (a) and for the approximated Fourier transform obtained from `contfft`, both on the same graph. (Use the range $0 \leq \omega \leq 50$ rad/sec.)
(i) $T = 0.5, N = 10$
(ii) $T = 0.1, N = 50$
(iii) $T = 0.05, N = 100$
(iv) $T = 0.05, N = 400$

(c) Identify the trends in accuracy and resolution in the plots, as T is decreased and as NT is increased.

7.32. This problem shows how you can use the `contfft` program to identify the frequency response of a system from its inputs and outputs.

(a) Generate a random input signal $x(t)$ in MATLAB by using the command `randn`. In particular, let `x = randn(1000,1)`. This will create a 1000 point vector of random numbers that represents the samples of $x(t)$. Assume that the sampling period was $T = 0.1$, and create a corresponding time vector.

(b) Use MATLAB to determine the corresponding output $y(t)$ of a particular system. In particular, type the command `y = lsim(1,[1 1],x,t);` to generate $y(t)$. Plot x versus t and y versus t, and compare the two signals in terms of frequency content. (In the time domain, we characterize high-frequency content qualitatively by sharp transitions or quick motion of the signal.)

(c) Suppose that you have a spectrum analyzer (an instrument that computes the FFT of sampled signals from voltage inputs) available to identify the system using only the signals $x(t)$ and $y(t)$. Use `contfft` to compute an approximation X to $X(\omega)$ and an approximation Y to $Y(\omega)$. Plot $|X(\omega)|$ and $|Y(\omega)|$ versus ω, and compare the difference in frequency content of these two signals.

(d) Recall that $Y(\omega) = X(\omega)H(\omega)$. Compute the approximation to $H(\omega)$ by dividing the elements of Y by the elements of X. Plot the frequency response $|H(\omega)|$. The approximation errors are large at high frequencies, so only consider the plot in the range $0 \leq \omega \leq 40$ rad/sec. What sort of filter does this represent? What is its bandwidth?

◆

CHAPTER 8

The Laplace Transform and the Transfer Function Representation

In this chapter the Laplace transform of a continuous-time signal is introduced, and then this operator is used to generate the transfer function representation of a causal linear time-invariant continuous-time system. It will be seen that the transfer function representation gives an algebraic relationship between the Laplace transforms of the input and output of a system, and in terms of this setup, the output response resulting from a large class of input signals can be computed using a purely algebraic procedure.

The Laplace transform is named after Pierre Simon Laplace (1749–1827), a French mathematician and astronomer. The chapter begins in Section 8.1 with the definition of the Laplace transform of a continuous-time signal. It is shown that the Laplace transform can be viewed as a generalization of the Fourier transform; more precisely, the addition of an exponential factor to the integrand in the definition of the Fourier transform results in the two-sided Laplace transform. The one-sided Laplace transform is then defined, which is the form of the transform that is studied in this book. In Section 8.2 the basic properties of the (one-sided) Laplace transform are given. By using these properties it is shown that many new transforms can be generated from a small set of transforms. Then in Section 8.3 the computation of the inverse Laplace transform is developed in terms of partial fraction expansions.

In Sections 8.4 through 8.6 the Laplace transform is applied to the study of causal linear time-invariant continuous-time systems. The development begins in Section 8.4 with systems defined by an input/output differential equation. For any such system the transfer function representation can be generated by taking the Laplace transform of the input/output differential equation. This results in an s-domain framework that can be used to solve the input/output differential equation via an algebraic procedure. In Section 8.5 the transfer function representation is generated by applying the Laplace transform to the input/output convolution relationship of the system. Techniques for generating the transfer function model are then given for RLC circuits and interconnections of integrators. The last section of the chapter deals with the computation of transfer functions for systems specified by block diagrams.

8.1 LAPLACE TRANSFORM OF A SIGNAL

Given a continuous-time signal $x(t)$, in Chapter 4 the Fourier transform $X(\omega)$ of $x(t)$ was defined by

$$X(\omega) = \int_{-\infty}^{\infty} x(t)e^{-j\omega t}\,dt \tag{8.1}$$

As discussed in Section 4.3, the Fourier transform $X(\omega)$ displays the frequency components comprising the signal $x(t)$.

It was also observed in Chapter 4 that for some common signals, the integral in (8.1) does not exist, and thus there is no Fourier transform (in the ordinary sense). An example is the unit-step function $u(t)$, for which (8.1) becomes

$$X(\omega) = \int_{0}^{\infty} e^{-j\omega t}\,dt \tag{8.2}$$

Although the integral in (8.2) does not exist, it is possible to circumvent this problem by adding an exponential convergence factor $e^{-\sigma t}$ to the integrand, where σ is a real number. Then (8.2) becomes

$$X(\omega) = \int_{0}^{\infty} e^{-\sigma t}e^{-j\omega t}\,dt$$

which can be written as

$$X(\omega) = \int_{0}^{\infty} e^{-(\sigma+j\omega)t}\,dt \tag{8.3}$$

Now $X(\omega)$ given by (8.3) is actually a function of the complex number $\sigma + j\omega$, so X should be expressed as a function of $\sigma + j\omega$ rather than ω. Then rewriting (8.3) gives

$$X(\sigma + j\omega) = \int_{0}^{\infty} e^{-(\sigma+j\omega)t}\,dt \tag{8.4}$$

Evaluating the right-hand side of (8.4) gives

$$X(\sigma + j\omega) = -\frac{1}{\sigma + j\omega}\left[e^{-(\sigma+j\omega)t}\right]_{t=0}^{t=\infty} \tag{8.5}$$

Now

$$\lim_{t\to\infty} e^{-(\sigma+j\omega)t}$$

exists if and only if $\sigma > 0$, in which case

$$\lim_{t\to\infty} e^{-(\sigma+j\omega)t} = 0$$

and (8.5) reduces to

$$\begin{aligned} X(\sigma + j\omega) &= -\frac{1}{\sigma + j\omega}\left[0 - e^{-(\sigma+j\omega)(0)}\right] \\ &= \frac{1}{\sigma + j\omega} \end{aligned} \tag{8.6}$$

Finally, by letting s denote the complex number $\sigma + j\omega$, (8.6) can be rewritten as

$$X(s) = \frac{1}{s} \tag{8.7}$$

The function $X(s)$ given by (8.7) is the Laplace transform of the unit step function $u(t)$. Note that $X(s)$ is a complex-valued function of the complex number s. Here complex valued means that if a particular complex number s is inserted into $X(s)$, the resulting value of $X(s)$ is in general a complex number. For example, inserting $s = 1 + j$ into (8.7) gives

$$X(1 + j) = \frac{1}{1 + j} = \frac{1 - j}{(1 + j)(1 - j)} = \frac{1}{2} - j\frac{1}{2} = \frac{1}{\sqrt{2}}e^{-j(\pi/4)}$$

so the value of $X(s)$ at $s = 1 + j$ is the complex number $\frac{1}{2} - j\frac{1}{2}$.

It is important to note that the function $X(s)$ given by (8.7) is defined for only those complex numbers s for which the real part of s (which is equal to σ) is strictly positive. In this particular example, the function $X(s)$ is not defined for $\sigma = 0$ or $\sigma < 0$ since the integral in (8.4) does not exist for such values of σ. The set of all complex numbers $s = \sigma + j\omega$ for which

$$\text{Re}\, s = \sigma > 0$$

is called the *region of convergence* of the Laplace transform of the unit-step function.

The construction above can be generalized to a very large class of signals $x(t)$ as follows: Given a signal $x(t)$, the exponential factor $e^{-\sigma t}$ is again added to the integrand in the definition (8.1) of the Fourier transform, which yields

$$X(\sigma + j\omega) = \int_{-\infty}^{\infty} x(t)e^{-(\sigma+j\omega)t}\, dt \tag{8.8}$$

With s equal to the complex number $\sigma + j\omega$, (8.8) becomes

$$X(s) = \int_{-\infty}^{\infty} x(t)e^{-st}\,dt \tag{8.9}$$

The function $X(s)$ given by (8.9) is the *two-sided* (or *bilateral*) *Laplace transform* of $x(t)$. As is the case for the unit-step function, the Laplace transform $X(s)$ is in general a complex-valued function of the complex number s.

Clearly, the two-sided Laplace transform $X(s)$ of a signal $x(t)$ can be viewed as a generalization of the Fourier transform of $x(t)$. More precisely, as was done above, $X(s)$ can be generated directly from the definition of the Fourier transform by adding the exponential factor $e^{-\sigma t}$ to the integrand in the definition of the Fourier transform.

The *one-sided* (or *unilateral*) *Laplace transform* of $x(t)$, also denoted by $X(s)$, is defined by

$$X(s) = \int_{0}^{\infty} x(t)e^{-st}\,dt \tag{8.10}$$

From (8.10) it is clear that the one-sided transform depends only on the values of the signal $x(t)$ for $t \geq 0$. For this reason the definition (8.10) of the transform is called the *one-sided Laplace transform.* The one-sided transform can be applied to signals $x(t)$ that are nonzero for $t < 0$; however, any nonzero values of $x(t)$ for $t < 0$ do not have any effect on the one-sided transform of $x(t)$. If $x(t)$ is zero for all $t < 0$, the expression (8.9) reduces to (8.10), and thus in this case the one- and two-sided Laplace transforms are the same.

As will be seen in the following section, the initial values of a signal and its derivatives can be explicitly incorporated into the s-domain framework by using the one-sided Laplace transform (as opposed to the two-sided Laplace transform). This is particularly useful for some problems, such as solving a differential equation with initial conditions. Hence in this book the development is limited to the one-sided transform, which will be referred to as the *Laplace transform.* In addition, as was the case for the Fourier transform, in this book the Laplace transform of a signal will always be denoted by an uppercase letter, with signals denoted by lowercase letters.

Given a signal $x(t)$, the set of all complex numbers s for which the integral in (8.10) exists is called the *region of convergence* of the Laplace transform $X(s)$ of $x(t)$. Hence the Laplace transform $X(s)$ of $x(t)$ is well defined (i.e., exists) for all values of s belonging to the region of convergence. It should be stressed that the region of convergence depends on the given function $x(t)$. For example, when $x(t)$ is the unit-step function $u(t)$, as noted above the region of convergence is the set of all complex numbers s such that Re $s > 0$. This is also verified in the example given below.

Example 8.1 *Laplace Transform of Exponential Function*

Let $x(t) = e^{-bt}u(t)$, where b is an arbitrary real number. The Laplace transform is

$$X(s) = \int_0^\infty e^{-bt}e^{-st}\,dt$$

$$= \int_0^\infty e^{-(s+b)t}\,dt$$

$$= -\frac{1}{s+b}[e^{-(s+b)t}]_{t=0}^{t=\infty} \tag{8.11}$$

To evaluate the right-hand side of (8.11), it is necessary to determine

$$\lim_{t\to\infty} e^{-(s+b)t} \tag{8.12}$$

Setting $s = \sigma + j\omega$ in (8.12) gives

$$\lim_{t\to\infty} e^{-(\sigma+j\omega+b)t} \tag{8.13}$$

The limit in (8.13) exists if and only if $\sigma + b > 0$, in which case the limit is zero, and from (8.11) the Laplace transform is

$$X(s) = \frac{1}{s+b} \tag{8.14}$$

The region of convergence of the transform $X(s)$ given by (8.14) is the set of all complex numbers s such that $\text{Re}\, s > -b$. Note that if $b = 0$, so that $x(t)$ is the unit-step function $u(t)$, then $X(s) = 1/s$ and the region of convergence is $\text{Re}\, s > 0$. This corresponds to the result that was obtained above.

Relationship Between the Fourier and Laplace Transforms

As shown above, the two-sided Laplace transform of a signal $x(t)$ can be viewed as a generalization of the definition of the Fourier transform of $x(t)$; that is, the two-sided Laplace transform is the Fourier transform with the addition of an exponential factor. Given a signal $x(t)$ with $x(t) = 0$ for all $t < 0$, from the constructions given above it may appear that the Fourier transform $X(\omega)$ can be computed directly from the (one-sided) Laplace transform $X(s)$ by setting $s = j\omega$. However, this is often not the case. To see this, let $x(t)$ be a signal that is zero for all $t < 0$ and suppose that $x(t)$ has the Laplace transform $X(s)$ given by (8.10). Since $x(t)$ is zero for $t < 0$, the Fourier transform $X(\omega)$ of $x(t)$ is given by

$$X(\omega) = \int_0^\infty x(t)e^{-j\omega t}\,dt \tag{8.15}$$

By comparing (8.10) and (8.15), it is tempting to conclude that the Fourier transform $X(\omega)$ is equal to the Laplace transform $X(s)$ with $s = j\omega$, or in mathematical terms,

$$X(\omega) = X(s)\big|_{s=j\omega} \tag{8.16}$$

However, (8.16) is valid if and only if the region of convergence for $X(s)$ includes $\sigma = 0$. For example, if $x(t)$ is the unit-step function u(t), then (8.16) is *not* valid since the region of convergence is $\text{Re}\, s > 0$, which does not include the point $\text{Re}\, s = \sigma = 0$.

This is simply a restatement of the fact that the unit-step function has a Laplace transform, but does not have a Fourier transform (in the ordinary sense).

Example 8.2 *Fourier Transform from Laplace Transform*

Let $x(t) = e^{-bt}u(t)$, where b is an arbitrary real number. From the result in Example 8.1, the Laplace transform $X(s)$ of $x(t)$ is equal to the $1/(s + b)$ and the region of convergence is Re $s > -b$. Thus, if $b > 0$, the region of convergence includes $\sigma = 0$, and the Fourier transform $X(\omega)$ of $x(t)$ is given by

$$X(\omega) = X(s)\big|_{s=j\omega} = \frac{1}{j\omega + b} \tag{8.17}$$

As was first observed in Example 4.6, when $b \le 0$, $x(t)$ does not have a Fourier transform in the ordinary sense, but it does have the Laplace transform $X(s) = 1/(s + b)$.

As was done in the case of the Fourier transform, the transform pair notation

$$x(t) \leftrightarrow X(s)$$

will sometimes be used to denote the fact that $X(s)$ is the Laplace transform of $x(t)$, and conversely, that $x(t)$ is the inverse Laplace transform of $X(s)$. Some authors prefer to use the operator notation

$$X(s) = \mathcal{L}[x(t)]$$
$$x(t) = \mathcal{L}^{-1}[X(s)]$$

where $\mathcal{L}$ denotes the Laplace transform operator and $\mathcal{L}^{-1}$ denotes the inverse Laplace transform operator. In this book the transform pair notation will be used.

In giving a transform pair, the region of convergence will not be specified, since in most applications it is not necessary to consider the region of convergence (as long as the transform does have a region of convergence). An example of a transform pair arising from the result of Example 8.1 is

$$e^{-bt}u(t) \leftrightarrow \frac{1}{s + b} \tag{8.18}$$

It should be noted that in the transform pair (8.18), the scalar b may be real or complex. The verification of this transform pair in the case when b is complex is an easy modification of the derivation given in Example 8.1. The details are omitted.

The Laplace transform of many signals of interest can be determined by table lookup. Hence it is often not necessary to have to evaluate the integral in (8.10) in order to compute a transform. By using the properties of the Laplace transform given in the next section, it will be shown that transform pairs for many common signals can be generated. These results will then be displayed in a table of transforms.

8.2 PROPERTIES OF THE LAPLACE TRANSFORM

The Laplace transform satisfies a number of properties that are useful in a wide range of applications, such as the derivation of new transform pairs from a given transform pair. In this section various fundamental properties of the Laplace transform are presented. Most of these properties correspond directly to the properties of the Fourier transform that were studied in Section 4.4. The properties of the Laplace transform that correspond to properties of the Fourier transform can be proved simply by replacing $j\omega$ by s in the proof of the Fourier transform property. Thus the proofs of these properties follow easily from the constructions given in Section 4.4 and will not be considered here.

The Fourier transform does enjoy some properties for which there is no version in the Laplace transform theory. Two examples are the duality property and Parseval's theorem. Hence the reader will notice that there are no versions of these properties stated below.

Linearity

The Laplace transform is a linear operation, as is the Fourier transform. Hence, if $x(t) \leftrightarrow X(s)$ and $v(t) \leftrightarrow V(s)$, then for any real or complex scalars a, b,

$$ax(t) + bv(t) \leftrightarrow aX(s) + bV(s) \tag{8.19}$$

Example 8.3 *Linearity*

Consider the signal $u(t) + e^{-t}u(t)$. Using the transform pair (8.18) and the property of linearity results in the transform pair

$$u(t) + e^{-t}u(t) \leftrightarrow \frac{1}{s} + \frac{1}{s+1} = \frac{2s+1}{s(s+1)} \tag{8.20}$$

Right Shift in Time

If $x(t) \leftrightarrow X(s)$, then for any positive real number c,

$$x(t-c)u(t-c) \leftrightarrow e^{-cs}X(s) \tag{8.21}$$

In (8.21), note that $x(t-c)u(t-c)$ is equal to the c-second right shift of $x(t)u(t)$. Here multiplication of $x(t)$ by $u(t)$ is necessary to eliminate any nonzero values of $x(t)$ for $t < 0$.

From (8.21) it is seen that a c-second right shift in the time domain corresponds to multiplication by e^{-cs} in the Laplace transform domain (or *s-domain*). The proof of the right-shift property is analogous to the one given for the Fourier transform and is thus omitted.

Example 8.4 *Laplace Transform of a Pulse*

Let $x(t)$ denote the c-second rectangular pulse function defined by

$$x(t) = \begin{cases} 1, & 0 \le t < c \\ 0, & \text{all other } t \end{cases}$$

where c is an arbitrary positive real number. Expressing $x(t)$ in terms of the unit-step function $u(t)$ gives

$$x(t) = u(t) - u(t - c)$$

By linearity, the Laplace transform $X(s)$ of $x(t)$ is the sum of the transform of $u(t)$ and the transform of $u(t - c)$. Now $u(t - c)$ is the c-second right shift of $u(t)$, and thus by the right-shift property (8.21), the Laplace transform of $u(t - c)$ is equal to e^{-cs}/s. Hence

$$u(t) - u(t - c) \leftrightarrow \frac{1}{s} - \frac{e^{-cs}}{s} = \frac{1 - e^{-cs}}{s} \tag{8.22}$$

It should be noted that there is no comparable result for a left shift in time. To see this, let c be an arbitrary positive real number and consider the time-shifted signal $x(t + c)$. Since $c > 0$, $x(t + c)$ is a c-second left shift of the signal $x(t)$. The Laplace transform of $x(t + c)$ is equal to

$$\int_0^\infty x(t + c)e^{-st}\, dt \tag{8.23}$$

However, (8.23) cannot be expressed in terms of the Laplace transform $X(s)$ of $x(t)$. (Try it!) In particular, (8.23) is not equal to $e^{cs}X(s)$.

Time Scaling

If $x(t) \leftrightarrow X(s)$, for any positive real number a,

$$x(at) \leftrightarrow \frac{1}{a} X\left(\frac{s}{a}\right) \tag{8.24}$$

As discussed in Section 4.4, the signal $x(at)$ is a time-scaled version of $x(t)$. By (8.24) it is seen that time scaling corresponds to scaling the complex variable s by the factor of $1/a$ in the Laplace transform domain (plus multiplication of the transform by $1/a$). The transform pair (8.24) can be proved in the same way that the corresponding transform pair in the Fourier theory was proved (see Section 4.4).

Example 8.5 *Time Scaling*

Consider the time-scaled unit-step function $u(at)$, where a is an arbitrary positive real number. By (8.24)

$$u(at) \leftrightarrow \frac{1}{a}\left(\frac{1}{s/a}\right) = \frac{1}{s}$$

This result is not unexpected, since $u(at) = u(t)$ for any real number $a > 0$.

Multiplication by a Power of t

If $x(t) \leftrightarrow X(s)$, then for any positive integer N,

$$t^N x(t) \leftrightarrow (-1)^N \frac{d^N}{ds^N} X(s) \tag{8.25}$$

In particular, for $N = 1$,

$$tx(t) \leftrightarrow -\frac{d}{ds}X(s) \tag{8.26}$$

and for $N = 2$,

$$t^2x(t) \leftrightarrow \frac{d^2}{ds^2}X(s) \tag{8.27}$$

The proof of (8.26) is very similar to the proof of the multiplication-by-t property given in the Fourier theory, so the details are again omitted.

Example 8.6 *Unit-Ramp Function*

Consider the unit-ramp function $r(t) = tu(t)$. From (8.26), the Laplace transform $R(s)$ of $r(t)$ is given by

$$R(s) = -\frac{d}{ds}\frac{1}{s} = \frac{1}{s^2}$$

Generalizing Example 8.6 to the case $t^N x(t), N = 1, 2, \ldots,$ yields the transform pair

$$t^N u(t) \leftrightarrow \frac{N!}{s^{N+1}} \tag{8.28}$$

where N! is N factorial.

Example 8.7

Let $v(t) = te^{-bt}u(t)$, where b is any real number. Using the transform pairs (8.18) and (8.26) yields

$$V(s) = -\frac{d}{ds}\frac{1}{s+b} = \frac{1}{(s+b)^2}$$

Generalizing Example 8.7 to the case $t^N e^{-bt}u(t)$ results in the transform pair

$$t^N e^{-bt}u(t) \leftrightarrow \frac{N!}{(s+b)^{N+1}} \tag{8.29}$$

Multiplication by an Exponential

If $x(t) \leftrightarrow X(s)$, then for any real or complex number a,

$$e^{at}x(t) \leftrightarrow X(s-a) \tag{8.30}$$

By the property (8.30), multiplication by an exponential function in the time domain corresponds to a shift of the s variable in the Laplace transform domain. The proof of (8.30) follows directly from the definition of the Laplace transform. The details are left to the reader.

Example 8.8 *Multiplication by an Exponential*

Let $v(t) = [u(t) - u(t - c)]e^{at}$, where c is a positive real number and a is any real number. The function $v(t)$ is the product of the c-second pulse $u(t) - u(t - c)$ and exponential function e^{at}. The function $v(t)$ is displayed in Figure 8.1 for the case $a < 0$. Now from the result in Example 8.4,

$$u(t) - u(t - c) \leftrightarrow \frac{1 - e^{-cs}}{s}$$

Then using (8.30) yields

$$V(s) = \frac{1 - e^{-c(s-a)}}{s - a}$$

Multiplication by a Sinusoid

If $x(t) \leftrightarrow X(s)$, then for any real number ω,

$$x(t) \sin \omega t \leftrightarrow \frac{j}{2}[X(s + j\omega) - X(s - j\omega)] \tag{8.31}$$

$$x(t) \cos \omega t \leftrightarrow \frac{1}{2}[X(s + j\omega) + X(s - j\omega)] \tag{8.32}$$

The transform pairs (8.31) and (8.32) can be proved by first writing $x(t) \sin \omega t$ and $x(t) \cos \omega t$ in the form

$$x(t) \sin \omega t = \frac{j}{2}x(t)[e^{-j\omega t} - e^{j\omega t}] \tag{8.33}$$

$$x(t) \cos \omega t = \frac{1}{2}x(t)[e^{-j\omega t} \pm e^{j\omega t}] \tag{8.34}$$

By (8.30),

$$e^{\pm j\omega t}x(t) \leftrightarrow X(s \pm j\omega)$$

Combining this with (8.33) and (8.34) yields (8.31) and (8.32).

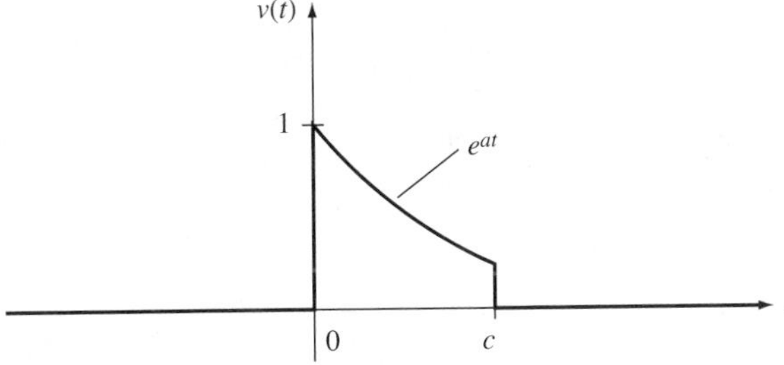

Figure 8.1 The function $v(t)$ in Example 8.8.

Example 8.9 *Multiplication by Cosine*

Let $v(t) = (\cos \omega t)u(t)$. Now $u(t) \leftrightarrow 1/s$, and using (8.32) with $x(t) = u(t)$ gives

$$V(s) = \frac{1}{2}\left(\frac{1}{s + j\omega} + \frac{1}{s - j\omega}\right)$$

$$= \frac{1}{2}\frac{s - j\omega + s + j\omega}{s^2 + \omega^2}$$

$$= \frac{s}{s^2 + \omega^2}$$

Example 8.9 yields the transform pair

$$(\cos \omega t)u(t) \leftrightarrow \frac{s}{s^2 + \omega^2} \tag{8.35}$$

Similarly, it is possible to verify the transform pair

$$(\sin \omega t)u(t) \leftrightarrow \frac{\omega}{s^2 + \omega^2} \tag{8.36}$$

Example 8.10 *Multiplication by Cosine and Sine*

Now let $v(t) = (e^{-bt}\cos \omega t)u(t)$. The Laplace transform of $v(t)$ can be computed by setting $x(t) = e^{-bt}u(t)$ and then using the multiplication by $\cos \omega t$ property. It is also possible to set $x(t) = (\cos \omega t)u(t)$ and use the multiplication by an exponential property. The latter is simpler to carry out, so it will be done that way. Replacing s by $s + b$ on the right-hand side of (8.35) results in the transform pair

$$(e^{-bt} \cos \omega t)u(t) \leftrightarrow \frac{s + b}{(s + b)^2 + \omega^2} \tag{8.37}$$

Similarly, it is possible to verify the transform pair

$$(e^{-bt} \sin \omega t)u(t) \leftrightarrow \frac{\omega}{(s + b)^2 + \omega^2} \tag{8.38}$$

Example 8.11 *Multiplication by Sine*

Let $v(t) = (\sin^2 \omega t)u(t)$. Setting $x(t) = (\sin \omega t)u(t)$ and using the multiplication by $\sin \omega t$ property yields

$$V(s) = \frac{j}{2}\left[\frac{\omega}{(s + j\omega)^2 + \omega^2} - \frac{\omega}{(s - j\omega)^2 + \omega^2}\right]$$

$$= \frac{j}{2}\frac{\omega(s - j\omega)^2 + \omega^3 - \omega(s + j\omega)^2 - \omega^3}{(s + j\omega)^2(s - j\omega)^2 + \omega^2(s - j\omega)^2 + \omega^2(s + j\omega)^2 + \omega^4}$$

$$= \frac{j}{2}\frac{-j4\omega^2 s}{s^4 + 4\omega^2 s^2}$$

$$= \frac{2\omega^2}{s(s^2 + 4\omega^2)} \tag{8.39}$$

Differentiation in the Time Domain

If $x(t) \leftrightarrow X(s)$, then

$$\dot{x}(t) \leftrightarrow sX(s) - x(0) \tag{8.40}$$

where $\dot{x}(t) = dx(t)/dt$. Thus differentiation in the time domain corresponds to multiplication by s in the Laplace transform domain [plus subtraction of the initial value $x(0)$]. The property (8.40) will be proved by computing the transform of the derivative of $x(t)$. The transform of $\dot{x}(t)$ is given by

$$\int_0^\infty \dot{x}(t)e^{-st}\,dt \tag{8.41}$$

The integral in (8.41) will be evaluated by parts: Let $v = e^{-st}$ so that $dv = -se^{-st}$, and let $w = x(t)$ so that $dw = \dot{x}(t)dt$. Then

$$\begin{aligned}\int_0^\infty \dot{x}(t)e^{-st}\,dt &= vw\Big|_{t=0}^{t=\infty} - \int_0^\infty w\,dv \\ &= e^{-st}x(t)\Big|_{t=o}^{t=\infty} - \int_0^\infty x(t)(-s)e^{-st}\,dt \\ &= \lim_{t\to\infty}[e^{-st}x(t)] - x(0) + sX(s) \end{aligned} \tag{8.42}$$

When $|x(t)| < ce^{at}$, $t > 0$, for some constants a and c, it follows that for any s such that Re $s > a$,

$$\lim_{t\to\infty} e^{-st}x(t) = 0$$

Thus, from (8.42),

$$\int_0^\infty \dot{x}(t)e^{-st}\,dt = -x(0) + sX(s)$$

which verifies (8.40).

It should be pointed out that if $x(t)$ is discontinuous at $t = 0$ or if $x(t)$ contains an impulse or derivative of an impulse located at $t = 0$, it is necessary to take the initial time in (8.40) to be at 0^-. In other words, the transform pair (8.40) becomes

$$\dot{x}(t) \leftrightarrow sX(s) - x(0^-) \tag{8.43}$$

Note that if $x(t) = 0$ for $t < 0$, then $x(0^-) = 0$ and

$$\dot{x}(t) \leftrightarrow sX(s) \tag{8.44}$$

Example 8.12 *Differentiation*

Let $x(t) = u(t)$. Then $\dot{x}(t) = \delta(t)$. Since $\dot{x}(t)$ is the unit impulse located at $t = 0$, it is necessary to use (8.43) to compute the Laplace transform of $\dot{x}(t)$. This gives

$$\dot{x}(t) \leftrightarrow s\frac{1}{s} - u(0^-) = 1 - 0 = 1$$

Hence the Laplace transform of the unit impulse $\delta(t)$ is equal to the constant function 1. This result could have been obtained by directly applying the definition (8.10) of the transform. Putting this result in the form of a transform pair yields

$$\delta(t) \leftrightarrow 1 \tag{8.45}$$

The Laplace transform of the second- and higher-order derivatives of a signal $x(t)$ can also be expressed in terms of $X(s)$ and initial conditions. For example, the transform pair in the second-order case is

$$\frac{d^2x(t)}{dt^2} \leftrightarrow s^2X(s) - sx(0) - \dot{x}(0) \tag{8.46}$$

The transform pair (8.46) can be proved by using integration by parts twice on the integral expression for the transform of the second derivative of $x(t)$. The details are omitted. It should be noted that if the second derivative of $x(t)$ is discontinuous or contains an impulse or a derivative of an impulse located at $t = 0$, it is necessary to take the initial conditions in (8.46) at time $t = 0^-$.

Now let N be an arbitrary positive integer and let $x^{(N)}(t)$ denote the Nth derivative of a given signal $x(t)$. Then the transform of $x^{(N)}(t)$ is given by the transform pair

$$x^{(N)}(t) \leftrightarrow s^NX(s) - s^{N-1}x(0) - s^{N-2}\dot{x}(0) - \cdots - sx^{(N-2)}(0) - x^{(N-1)}(0) \tag{8.47}$$

Integration

If $x(t) \leftrightarrow X(s)$, then

$$\int_0^t x(\lambda)\, d\lambda \leftrightarrow \frac{1}{s}X(s) \tag{8.48}$$

By (8.48), the Laplace transform of the integral of $x(t)$ is equal to $X(s)$ divided by s. The transform pair (8.48) follows directly from the derivative property given above. To see this, let $v(t)$ denote the integral of $x(t)$ given by

$$v(t) = \begin{cases} \int_0^t x(\lambda)\, d\lambda, & t \geq 0 \\ 0, & t < 0 \end{cases}$$

Then $x(t) = \dot{v}(t)$ for $t > 0$, and since $v(t) = 0$ for $t < 0$, by (8.44), $X(s) = sV(s)$. Therefore, $V(s) = (1/s)X(s)$, which verifies (8.48).

Example 8.13 *Integration*

Let $x(t) = u(t)$. Then the integral of $x(t)$ is the unit-ramp function $r(t) = tu(t)$. By (8.48), the Laplace transform of $r(t)$ is equal to $1/s$ times the transform of $u(t)$. The result is the transform pair

$$r(t) \leftrightarrow \frac{1}{s^2}$$

Recall that this transform pair was derived previously by using the multiplication-by-t property.

Convolution

Given two signals $x(t)$ and $v(t)$ with both $x(t)$ and $v(t)$ equal to zero for all $t < 0$, consider the convolution $x(t) * v(t)$ given by

$$x(t) * v(t) = \int_0^t x(\lambda)v(t - \lambda)\, d\lambda$$

Now with $X(s)$ equal to the Laplace transform of $x(t)$ and $V(s)$ equal to the Laplace transform of $v(t)$, it turns out that the transform of the convolution $x(t) * v(t)$ is equal to the product $X(s)V(s)$; that is, the following transform pair is valid:

$$x(t) * v(t) \leftrightarrow X(s)V(s) \tag{8.49}$$

By (8.49), convolution in the time domain corresponds to a product in the Laplace transform domain. It will be seen in Section 8.5 that this property results in an algebraic relationship between the transforms of the inputs and outputs of a linear time-invariant continuous-time system. The proof of (8.49) is very similar to the proof of the corresponding property in the Fourier theory, and thus will not be given.

The transform pair (8.49) yields a procedure for computing the convolution $x(t) * v(t)$ of two signals $x(t)$ and $v(t)$ [with $x(t) = v(t) = 0$ for all $t < 0$]: First compute the Laplace transforms $X(s)$, $V(s)$ of $x(t)$, $v(t)$, and then compute the inverse Laplace transform of the product $X(s)V(s)$. The result is the convolution $x(t) * v(t)$. The process is illustrated by the following example.

Example 8.14 *Convolution*

Let $x(t)$ denote the 1-second pulse given by $x(t) = u(t) - u(t - 1)$. The objective is to determine the convolution $x(t) * x(t)$ of this signal with itself. From Example 8.4 the transform $X(s)$ of $x(t)$ is $X(s) = (1 - e^{-s})/s$. Thus, from (8.49) the transform of the convolution $x(t) * x(t)$ is equal to $X^2(s)$, where

$$X^2(s) = \left(\frac{1 - e^{-s}}{s}\right)^2 = \frac{1 - 2e^{-s} + e^{-2s}}{s^2}$$

Now the convolution $x(t) * x(t)$ is equal to the inverse Laplace transform of $X^2(s)$, which can be computed in this case by using linearity, the right-shift property, and the transform pairs $u(t) \leftrightarrow 1/s$, $tu(t) \leftrightarrow 1/s^2$. The result is

$$x(t) * x(t) = tu(t) - 2(t - 1)u(t - 1) + (t - 2)u(t - 2)$$

The convolution $x(t) * x(t)$ is displayed in Figure 8.2. The plot shows that the convolution of a rectangular pulse with itself results in a triangular pulse. Note that this result was first observed in Chapter 3 by direct evaluation of the convolution integral.

Initial-Value Theorem

Given a signal $x(t)$ with transform $X(s)$, the initial values $x(0)$ and $\dot{x}(0)$ can be computed using the expressions

$$x(0) = \lim_{s \to \infty} sX(s) \tag{8.50}$$

$$\dot{x}(0) = \lim_{s \to \infty} [s^2 X(s) - sx(0)] \tag{8.51}$$

In the general case, for an arbitrary positive integer N,

$$x^{(N)}(0) = \lim_{s \to \infty} [s^{N+1} X(s) - s^N x(0) - s^{N-1}\dot{x}(0) - \cdots - sx^{(N-1)}(0)] \tag{8.52}$$

It should be noted that the relationship (8.52) is not valid if the Nth derivative $x^{(N)}(t)$ contains an impulse or a derivative of an impulse at time $t = 0$.

The relationship (8.52) for $N = 0, 1, 2, \ldots$ is called the *initial-value theorem.* It will be proved for the case $N = 0$, assuming that $x(t)$ has the *Taylor series* expansion:

$$x(t) = \sum_{i=0}^{\infty} x^{(i)}(0) \frac{t^i}{i!} \tag{8.53}$$

where $i!$ is i factorial and $x^{(i)}(0)$ is the ith derivative of $x(t)$ evaluated at $t = 0$. Now

$$sX(s) = \int_0^{\infty} sx(t)e^{-st}\,dt \tag{8.54}$$

and using (8.53) in (8.54) gives

$$sX(s) = \int_0^{\infty} \sum_{i=0}^{\infty} sx^{(i)}(0) \frac{t^i}{i!} e^{-st}\,dt \tag{8.55}$$

Interchanging the integral and summation in (8.55) and using the transform pair (8.28) yields

$$sX(s) = \sum_{i=0}^{\infty} x^{(i)}(0) \frac{s}{s^{i+1}} = \sum_{i=0}^{\infty} x^{(i)}(0) \frac{1}{s^i}$$

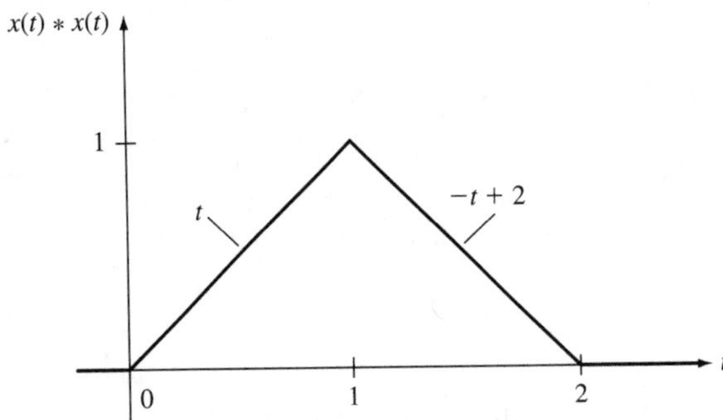

Figure 8.2 Plot of the convolution $x(t) * x(t)$.

Then taking the limit as $s \to \infty$ gives

$$\lim_{s\to\infty} sX(s) = x(0)$$

which proves (8.50).

The initial-value theorem is useful since it allows for computation of the initial values of a function $x(t)$ and its derivatives directly from the Laplace transform $X(s)$ of $x(t)$. If $X(s)$ is known but $x(t)$ is not, it is possible to compute these initial values without having to compute the inverse Laplace transform of $x(t)$. It should also be noted that these initial values are at $t = 0$ or $t = 0^+$, not at $t = 0^-$. The initial values at $t = 0^-$ cannot be determined from the one-sided Laplace transform (unless the signal is continuous at t = 0) since the transform is based on the signal $x(t)$ for $t \geq 0$ only.

Example 8.15 *Initial Value*

Suppose that the signal $x(t)$ has the Laplace transform

$$X(s) = \frac{-3s^2 + 2}{s^3 + s^2 + 3s + 2}$$

Then

$$\lim_{s\to\infty} sX(s) = \lim_{s\to\infty} \frac{-3s^3 + 2s}{s^3 + s^2 + 3s + 2} = \frac{-3}{1}$$

Thus $x(0) = -3$.

Final-Value Theorem

Given the signal $x(t)$ with transform $X(s)$, suppose that $x(t)$ has a limit $x(\infty)$ as $t \to \infty$; that is,

$$x(\infty) = \lim_{t\to\infty} x(t)$$

The existence of the limit $x(\infty)$ turns out to be equivalent to requiring that the region of convergence of $sX(s)$ includes the value $s = 0$.

If $x(\infty)$ exists, the *final-value theorem* states that

$$\lim_{t\to\infty} x(t) = \lim_{s\to 0} sX(s) \qquad (8.56)$$

To prove (8.56), first note that by the derivative property,

$$\int_0^\infty \dot{x}(t)e^{-st}\,dt = sX(s) - x(0) \qquad (8.57)$$

Taking the limit as $s \to 0$ of both sides of (8.57) gives

$$\lim_{s\to 0} \int_0^\infty \dot{x}(t)e^{-st}\,dt = \int_0^\infty \dot{x}(t)\,dt = \lim_{s\to 0} [sX(s) - x(0)] \qquad (8.58)$$

Now if $x(t)$ has a limit $x(\infty)$ as $t \to \infty$, using integration by parts yields

$$\int_0^{\infty} \dot{x}(t)\,dt = x(\infty) - x(0) \tag{8.59}$$

Combining (8.58) and (8.59) gives (8.56).

The final-value theorem is a very useful property since the limit as $t \to \infty$ of a time signal $x(t)$ can be computed directly from the Laplace transform $X(s)$. However, care must be used in applying the final-value theorem since the limit of $sX(s)$ as $s \to 0$ may exist even though $x(t)$ does not have a limit as $t \to \infty$. For example, suppose that

$$X(s) = \frac{1}{s^2 + 1}$$

Then

$$\lim_{s\to 0} sX(s) = \lim_{s \to 0} \frac{s}{s^2 + 1} = 0$$

But $x(t) = \sin t$, and $\sin t$ does not have a limit as $t \to \infty$. In the next section it is shown that in many cases of interest, whether or not a signal $x(t)$ has a limit as $t \to \infty$ can be determined by checking the transform $X(s)$.

For the convenience of the reader, the properties of the Laplace transform are summarized in Table 8.1. Table 8.2 contains a collection of common transform pairs, which includes the transform pairs that were derived in this section using the properties of the Laplace transform.

8.3 COMPUTATION OF THE INVERSE LAPLACE TRANSFORM

Given a signal $x(t)$ with Laplace transform $X(s)$, $x(t)$ can be computed from $X(s)$ by taking the inverse Laplace transform of $X(s)$. The inverse transform operation is given by

$$x(t) = \frac{1}{2\pi j} \int_{c-j\infty}^{c+j\infty} X(s)e^{st}\,ds \tag{8.60}$$

The integral in (8.60) is evaluated along the path $s = c + j\omega$ in the complex plane from $c - j\infty$ to $c + j\infty$, where c is any real number for which the path $s = c + j\omega$ lies in the region of convergence of $X(s)$. For a detailed treatment of complex integration, see Churchill et al. [1976].

The integral in (8.60) is usually difficult to evaluate, and thus it is desirable to avoid having to use (8.60) to compute the inverse transform. In this section an algebraic procedure is given for computing the inverse transform in the case when $X(s)$ is a rational function of s. The development begins below with the definition of a rational Laplace transform.

TABLE 8.1 PROPERTIES OF THE LAPLACE TRANSFORM

Property	Transform Pair/Property
Linearity	$ax(t) + bv(t) \leftrightarrow aX(s) + bV(s)$
Right shift in time	$x(t-c)u(t-c) \leftrightarrow e^{-cs}X(s), \quad c > 0$
Time scaling	$x(at) \leftrightarrow \frac{1}{a}X\left(\frac{s}{a}\right), \quad a > 0$
Multiplication by a power of t	$t^N x(t) \leftrightarrow (-1)^N \frac{d^N}{ds^N}X(s), \quad N = 1, 2, \ldots$
Multiplication by an exponential	$e^{at}x(t) \leftrightarrow X(s-a), \quad a$ real or complex
Multiplication by $\sin \omega t$	$x(t)\sin \omega t \leftrightarrow \frac{j}{2}[X(s+j\omega) - X(s-j\omega)]$
Multiplication by $\cos \omega t$	$x(t)\cos \omega t \leftrightarrow \frac{1}{2}[X(s+j\omega) + X(s-j\omega)]$
Differentiation in the time domain	$\dot{x}(t) \leftrightarrow sX(s) - x(0)$
Second derivative	$\ddot{x}(t) \leftrightarrow s^2X(s) - sx(0) - \dot{x}(0)$
nth derivative	$x^{(N)}(t) \leftrightarrow s^NX(s) - s^{N-1}x(0) - s^{N-2}\dot{x}(0) - \cdots - sx^{(N-2)}(0) - x^{(N-1)}(0)$
Integration	$\int_0^t x(\lambda)\,d\lambda \leftrightarrow \frac{1}{s}X(s)$
Convolution	$x(t) * v(t) \leftrightarrow X(s)V(s)$
Initial-value theorem	$x(0) = \lim_{s\to\infty} sX(s)$
	$\dot{x}(0) = \lim_{s\to\infty} [s^2X(s) - sx(0)]$
	$x^{(N)}(0) = \lim_{s\to\infty} [s^{N+1}X(s) - s^Nx(0) - s^{N-1}\dot{x}(0) - \cdots - sx^{(N-1)}(0)]$
Final-value theorem	If $\lim_{t\to\infty} x(t)$ exists, then $\lim_{t\to\infty} x(t) = \lim_{s\to 0} sX(s)$

Rational Laplace Transforms

Suppose that $x(t)$ has Laplace transform $X(s)$ with

$$X(s) = \frac{B(s)}{A(s)} \tag{8.61}$$

where $B(s)$ and $A(s)$ are polynomials in the complex variable s given by

$$B(s) = b_M s^M + b_{M-1}s^{M-1} + \cdots + b_1 s + b_0 \tag{8.62}$$

$$A(s) = a_N s^N + a_{N-1}s^{N-1} + \cdots + a_1 s + a_0 \tag{8.63}$$

In (8.62) and (8.63), M and N are positive integers and the coefficients $b_M, b_{M-1}, \ldots, b_1, b_0$ and $a_N, a_{N-1}, \ldots, a_1, a_0$ are real numbers. Assuming that $b_M \neq 0$ and $a_N \neq 0$, the degree of the polynomial $B(s)$ is equal to M and the degree of the polynomial $A(s)$ is equal to N. The polynomial $B(s)$ is the "numerator polynomial" of $X(s)$ and $A(s)$ is the "denominator polynomial" of $X(s)$. It is always assumed that $B(s)$ and $A(s)$ do not have any common factors. If there are common factors, they should be divided out.

TABLE 8.2 COMMON LAPLACE TRANSFORM PAIRS

$$u(t) \leftrightarrow \frac{1}{s}$$

$$u(t) - u(t-c) \leftrightarrow \frac{1-e^{-cs}}{s}, \quad c > 0$$

$$t^N u(t) \leftrightarrow \frac{N!}{s^{N+1}}, \quad N = 1, 2, 3, \ldots$$

$$\delta(t) \leftrightarrow 1$$

$$\delta(t-c) \leftrightarrow e^{-cs}, \quad c > 0$$

$$e^{-bt}u(t) \leftrightarrow \frac{1}{s+b}, \quad b \text{ real or complex}$$

$$t^N e^{-bt}u(t) \leftrightarrow \frac{N!}{(s+b)^{N+1}}, \quad N = 1, 2, 3, \ldots$$

$$(\cos \omega t)u(t) \leftrightarrow \frac{s}{s^2+\omega^2}$$

$$(\sin \omega t)u(t) \leftrightarrow \frac{\omega}{s^2+\omega^2}$$

$$(\cos^2 \omega t)u(t) \leftrightarrow \frac{s^2+2\omega^2}{s(s^2+4\omega^2)}$$

$$(\sin^2 \omega t)u(t) \leftrightarrow \frac{2\omega^2}{s(s^2+4\omega^2)}$$

$$(e^{-bt}\cos \omega t)u(t) \leftrightarrow \frac{s+b}{(s+b)^2+\omega^2}$$

$$(e^{-bt}\sin \omega t)u(t) \leftrightarrow \frac{\omega}{(s+b)^2+\omega^2}$$

$$(t\cos \omega t)u(t) \leftrightarrow \frac{s^2-\omega^2}{(s^2+\omega^2)^2}$$

$$(t\sin \omega t)u(t) \leftrightarrow \frac{2\omega s}{(s^2+\omega^2)^2}$$

$$(te^{-bt}\cos \omega t)u(t) \leftrightarrow \frac{(s+b)^2-\omega^2}{[(s+b)^2+\omega^2]^2}$$

$$(te^{-bt}\sin \omega t)u(t) \leftrightarrow \frac{2\omega(s+b)}{[(s+b)^2+\omega^2]^2}$$

The transform $X(s) = B(s)/A(s)$ with $B(s)$ and $A(s)$ given by (8.62) and (8.63) is said to be a *rational function of s* since it is a ratio of polynomials in s. The degree N of the denominator polynomial $A(s)$ is called the *order* of the rational function. For a large class of signals $x(t)$, the Laplace transform $X(s)$ is rational. For example, most of the signals in Table 8.2 have a rational Laplace transform. An exception is the c-second rectangular pulse $u(t) - u(t-c)$, whose Laplace transform is

$$\frac{1-e^{-cs}}{s}$$

Due to the presence of the complex exponential e^{-cs}, this transform cannot be expressed as a ratio of polynomials in s, and thus the transform of the rectangular pulse is not a rational function of s.

Given a rational transform $X(s) = B(s)/A(s)$, let $p_1, p_2, \ldots, p_N$ denote the roots of the equation

$$A(s) = 0$$

Then $A(s)$ can be written in the factored form

$$A(s) = a_N(s - p_1)(s - p_2)\cdots(s - p_N) \tag{8.64}$$

The roots $p_1, p_2, \ldots, p_N$, which may be real or complex, are also said to be the *zeros* of the polynomial $A(s)$ since $A(s)$ is equal to zero when s is set equal to p_i for any value of i ranging from 1 to N. Note that if any one of the zeros (say p_1) is complex, there must be another zero that is equal to the complex conjugate of p_1. In other words, complex zeros always appear in complex-conjugate pairs.

MATLAB can be used to find the zeros of a polynomial $A(s)$ using the command `roots`. For example, to find the zeros of

$$A(s) = s^3 + 4s^2 + 6s + 4 = 0$$

use the commands

```
den = [1 4 6 4];        % store the coefficients of A(s)
p = roots(den)
```

MATLAB returns the zeros:

```
p =

   -2.0000
   -1.0000 + 1.0000i
   -1.0000 - 1.0000i
```

Hence $A(s)$ has the factored form

$$A(s) = (s + 2)(s + 1 - j)(s + 1 + j)$$

Now given the rational transform $X(s) = B(s)/A(s)$, if $A(s)$ is specified by the factored form (8.64), the result is

$$X(s) = \frac{B(s)}{a_N(s - p_1)(s - p_2)\cdots(s - p_N)} \tag{8.65}$$

The p_i for $i = 1, 2, \ldots, N$ are called the *poles* of the rational function $X(s)$ since if the value $s = p_i$ is inserted into $X(s)$, the result is ∞. So the poles of the rational function $X(s)$ are equal to the zeros (or roots) of the denominator polynomial $A(s)$.

The inverse Laplace transform of $X(s)$ can be computed by first carrying out a partial fraction expansion of (8.65). The procedure is described below. In the following development it is assumed that $M < N$, that is, the degree of $B(s)$ is strictly less than the degree of $A(s)$. Such a rational function is said to be *strictly proper in s*. The case when $X(s)$ is not strictly proper is considered later.

Distinct Poles

The poles $p_1, p_2, \ldots, p_N$ of $X(s)$ are now assumed to be distinct (or nonrepeated); that is, $p_i \neq p_j$ when $i \neq j$. Then $X(s)$ has the partial fraction expansion

$$X(s) = \frac{c_1}{s - p_1} + \frac{c_2}{s - p_2} + \cdots + \frac{c_N}{s - p_N} \tag{8.66}$$

where

$$c_i = [(s - p_i)X(s)]_{s=p_i}, \qquad i = 1, 2, \ldots, N \tag{8.67}$$

The expression (8.67) for the c_i can be verified by first multiplying both sides of (8.66) by $s - p_i$. This gives

$$(s - p_i)X(s) = c_i + \sum_{\substack{r=1 \\ r \neq i}}^{N} c_r \frac{s - p_i}{s - p_r} \tag{8.68}$$

Evaluating both sides of (8.68) at $s = p_i$ eliminates all terms inside the summation, which yields (8.67).

The constants c_i in (8.66) are called the *residues* and the computation of the c_i using (8.67) is called the *residue method.* The constant c_i is real if the corresponding pole p_i is real. In addition, since the poles $p_1, p_2, \ldots, p_N$ appear in complex-conjugate pairs, the c_i must also appear in complex-conjugate pairs. Hence, if c_i is complex, one of the other constants must be equal to the complex conjugate of c_i.

It is worth noting that to compute the partial fraction expansion (8.66), it is not necessary to factor the numerator polynomial $B(s)$. However, it is necessary to compute the poles of $X(s)$, since the expansion is given directly in terms of the poles.

The inverse Laplace transform $x(t)$ of $X(s)$ can then be determined by taking the inverse transform of each term in (8.66) and using linearity of the inverse transform operation. The result is

$$x(t) = c_1 e^{p_1 t} + c_2 e^{p_2 t} + \cdots + c_N e^{p_N t}, \qquad t \geq 0 \tag{8.69}$$

It is very important to note that the form of the time variation of the function $x(t)$ given by (8.69) is determined by the poles of the rational function $X(s)$; more precisely, $x(t)$ is a sum of exponentials in time whose exponents are completely specified in terms of the poles of $X(s)$. As a consequence, it is the poles of $X(s)$ that determine the characteristics of the time variation of $x(t)$. This fundamental result will be utilized extensively in Chapter 9 in the study of system behavior.

It should also be noted that if all the p_i are real, the terms comprising the function $x(t)$ defined by (8.69) are all real. However, if two or more of the p_i are complex, the corresponding terms in (8.69) will be complex, and thus in this case, the complex terms must be combined to obtain a real form. This will be considered after the example given below.

Given the rational function $X(s) = B(s)/A(s)$ with the polynomials $B(s)$ and $A(s)$ defined as above, the MATLAB software can be used to compute the residues and the poles of $X(s)$. The commands are as follows:

```
num = [bM bM-1 ... b1 b0];
den = [aN aN-1 ... a1 a0];
[r,p] = residue(num,den);
```

The MATLAB program will then produce a vector r consisting of the residues and a vector p consisting of the corresponding poles. The process is illustrated in the example below.

Example 8.16 *Distinct Pole Case*

Suppose that

$$X(s) = \frac{s+2}{s^3 + 4s^2 + 3s}$$

Here

$$A(s) = s^3 + 4s^2 + 3s = s(s+1)(s+3)$$

The roots of $A(s) = 0$ are $0, -1, -3$, and thus the poles of $X(s)$ are $p_1 = 0$, $p_2 = -1$, $p_3 = -3$. Therefore,

$$X(s) = \frac{c_1}{s-0} + \frac{c_2}{s-(-1)} + \frac{c_3}{s-(-3)}$$

$$X(s) = \frac{c_1}{s} + \frac{c_2}{s+1} + \frac{c_3}{s+3}$$

where

$$c_1 = [sX(s)]_{s=0} = \left.\frac{s+2}{(s+1)(s+3)}\right|_{s=0} = \frac{2}{3}$$

$$c_2 = [(s+1)X(s)]_{s=-1} = \left.\frac{s+2}{s(s+3)}\right|_{s=-1} = \frac{1}{-2}$$

$$c_3 = [(s+3)X(s)]_{s=-3} = \left.\frac{s+2}{s(s+1)}\right|_{s=-3} = \frac{-1}{6}$$

Hence the inverse Laplace transform $x(t)$ of $X(s)$ is given by

$$x(t) = \frac{2}{3} - \frac{1}{2}e^{-t} - \frac{1}{6}e^{-3t}, \qquad t \geq 0$$

The computation of the residues and poles can be checked using the MATLAB commands

```
num = [1 2];
den = [1 4 3 0];
[r,p] = residue(num,den);
```

The MATLAB program produces the vectors

```
r =                                  p =

   -0.1667                              -3
   -0.5000                              -1
    0.6667                               0
```

which checks with the results obtained above.

Distinct Poles with Two or More Poles Complex

It is still assumed that the poles of $X(s)$ are distinct, but now two or more of the poles of $X(s)$ are complex so that the corresponding exponentials in (8.69) are complex. As shown below, it is possible to combine the complex terms in order to express $x(t)$ in real form.

Suppose that $p_1 = \sigma + j\omega$ is complex, so that $\omega \neq 0$. Then the complex conjugate $\bar{p}_1 = \sigma - j\omega$ is another pole of $X(s)$. Let p_2 denote this pole. Then the residue c_2 corresponding to the pole p_2 is equal to the conjugate $\bar{c}_1$ of the residue corresponding to the pole p_1 and $X(s)$ has the partial fraction expansion

$$X(s) = \frac{c_1}{s - p_1} + \frac{\bar{c}_1}{s - \bar{p}_1} + \frac{c_3}{s - p_3} + \cdots + \frac{c_N}{s - p_N}$$

where $c_1, c_3, \ldots, c_N$ are again given by (8.67). Hence the inverse transform is

$$x(t) = c_1 e^{p_1 t} + \bar{c}_1 e^{\bar{p}_1 t} + c_3 e^{p_3 t} + \cdots + c_N e^{p_N t} \tag{8.70}$$

Now the first two terms on the right-hand side of (8.70) can be expressed in real form as follows:

$$c_1 e^{p_1 t} + \bar{c}_1 e^{\bar{p}_1 t} = 2\,|c_1|\, e^{\sigma t} \cos(\omega t + \angle c_1) \tag{8.71}$$

where $|c_1|$ is the magnitude of the complex number c_1 and $\angle c_1$ is the angle of c_1. The verification of the relationship (8.71) is considered in the homework problems (see Problem 8.9).

Using (8.71), the inverse transform of $X(s)$ is given by

$$x(t) = 2\,|c_1|\, e^{\sigma t} \cos(\omega t + \angle c_1) + c_3 e^{p_3 t} + \cdots + c_N e^{p_N t} \tag{8.72}$$

The expression (8.72) for $x(t)$ shows that if $X(s)$ has a pair of complex poles $p_1, p_2 = \sigma \pm j\omega$, the signal $x(t)$ contains a term of the form

$$c e^{\sigma t} \cos(\omega t + \theta)$$

Note that the coefficient σ of t in the exponential function is the real part of the pole $p_1 = \sigma + j\omega$ and the frequency ω of the cosine is equal to the imaginary part of the pole p_1.

The computation of the inverse Laplace transform using (8.71) is illustrated by the following example.

Example 8.17 *Complex Pole Case*

Suppose that

$$X(s) = \frac{s^2 - 2s + 1}{s^3 + 3s^2 + 4s + 2}$$

Here

$$A(s) = s^3 + 3s^2 + 4s + 2 = (s + 1 - j)(s + 1 + j)(s + 1)$$

The roots of $A(s) = 0$ are

$$p_1 = -1 + j, \qquad p_2 = -1 - j, \qquad p_3 = -1$$

Thus $\sigma = -1$ and $\omega = 1$, and

$$X(s) = \frac{c_1}{s - (-1 + j)} + \frac{\bar{c}_1}{s - (-1 - j)} + \frac{c_3}{s - (-1)}$$

$$X(s) = \frac{c_1}{s + 1 - j} + \frac{\bar{c}_1}{s + 1 + j} + \frac{c_3}{s + 1}$$

where

$$c_1 = [(s + 1 - j)X(s)]_{s=-1+j} = \left.\frac{s^2 - 2s + 1}{(s + 1 + j)(s + 1)}\right|_{s=-1+j}$$

$$= \frac{-3}{2} + j2$$

$$c_3 = [(s + 1)X(s)]_{s=-1} = \left.\frac{s^2 - 2s + 1}{s^2 + 2s + 2}\right|_{s=-1} = 4$$

Now

$$|c_1| = \sqrt{\frac{9}{4} + 4} = \frac{5}{2}$$

and since c_1 lies in the second quadrant:

$$\angle c_1 = 180^\circ + \tan^{-1}\frac{-4}{3} = 126.87^\circ$$

Then using (8.71) and (8.72) gives

$$x(t) = 5e^{-t}\cos(t + 126.87^\circ) + 4e^{-t}, \qquad t \geq 0$$

MATLAB will generate the residues and poles in the case when $X(s)$ has complex (and real) poles. In this example the commands are

```
num = [1 -2 1];
den = [1 3 4 2];
[r,p] = residue(num,den);
```

which yields

```
r =                                  p =
   -1.5000 + 2.0000i                    -1.0000 + 1.0000i
   -1.5000 - 2.0000i                    -1.0000 - 1.0000i
    4.0000                              -1.0000
```

This matches the poles and residues calculated above.

When $X(s)$ has complex poles, it is possible to avoid having to work with complex numbers by not factoring quadratic terms whose zeros are complex. For example, suppose that $X(s)$ is the second-order rational function given by

$$X(s) = \frac{b_1 s + b_0}{s^2 + a_1 s + a_0}$$

"Completing the square" in the denominator of $X(s)$ gives

$$X(s) = \frac{b_1 s + b_0}{(s + a_1/2)^2 + a_0 - a_1^2/4}$$

It follows from the quadratic formula that the poles of $X(s)$ are complex if and only if

$$a_0 - \frac{a_1^2}{4} > 0$$

in which case the poles of $X(s)$ are

$$p_1, p_2 = -\frac{a_1}{2} \pm j\omega$$

where

$$\omega = \sqrt{a_0 - \frac{a_1^2}{4}}$$

With $X(s)$ expressed in the form

$$X(s) = \frac{b_1 s + b_0}{(s + a_1/2)^2 + \omega^2} = \frac{b_1(s + a_1/2) + (b_0 - b_1 a_1/2)}{(s + a_1/2)^2 + \omega^2}$$

the inverse Laplace transform can be computed by table lookup. This is illustrated by the following example.

Example 8.18 *Completing the Square*

Suppose that

$$X(s) = \frac{3s + 2}{s^2 + 2s + 10}$$

Completing the square in the denominator of $X(s)$ gives

$$X(s) = \frac{3s + 2}{(s + 1)^2 + 9}$$

Then since $9 > 0$, the poles of $X(s)$ are complex and are equal to $-1 \pm j3$ (here $\omega = 3$). Now $X(s)$ can be expressed in the form

$$X(s) = \frac{3(s+1) - 1}{(s+1)^2 + 9} = \frac{3(s+1)}{(s+1)^2 + 9} - \frac{1}{3}\frac{3}{(s+1)^2 + 9}$$

and using the transform pairs in Table 8.2, the inverse transform is

$$x(t) = 3e^{-t}\cos 3t - \frac{1}{3}e^{-t}\sin 3t, \qquad t \geq 0$$

Finally, using the trigonometric identity

$$C\cos\omega t - D\sin\omega t = \sqrt{C^2 + D^2}\cos(\omega t + \theta), \qquad \text{where } \theta = \tan^{-1}\frac{D}{C} \tag{8.73}$$

$x(t)$ can be written in the form

$$x(t) = ce^{-t}\cos(3t + \theta), \qquad t \geq 0$$

where

$$c = \sqrt{(3)^2 + \left(\frac{1}{3}\right)^2} = 3.018$$

and

$$\theta = \tan^{-1}\frac{1/3}{3} = 83.7°$$

Now suppose that $X(s)$ has a pair of complex poles $p_1, p_2 = \sigma \pm j\omega$ and real distinct poles $p_3, p_4, \ldots, p_N$. Then

$$X(s) = \frac{B(s)}{[(s-\sigma)^2 + \omega^2](s - p_3)(s - p_4)\cdots(s - p_N)}$$

which can be expanded into the form

$$X(s) = \frac{b_1 s + b_0}{(s-\sigma)^2 + \omega^2} + \frac{c_3}{s - p_3} + \frac{c_4}{s - p_4} + \cdots + \frac{c_N}{s - p_N} \tag{8.74}$$

where the coefficients b_0 and b_1 of the second-order term are real numbers. The residues $c_3, c_4, \ldots, c_N$ are real numbers and are computed from (8.67) as before; however, b_0 and b_1 cannot be calculated from this formula. The constants b_0 and b_1 can be computed by putting the right-hand side of (8.74) over a common denominator and then equating the coefficients of the resulting numerator with the numerator of $X(s)$. The inverse Laplace transform can then be computed from (8.74). The process is illustrated by the following example.

Example 8.19 *Equating Coefficients*

Again consider the rational function $X(s)$ in Example 8.17 given by

$$X(s) = \frac{s^2 - 2s + 1}{s^3 + 3s^2 + 4s + 2}$$

In this example

$$A(s) = (s^2 + 2s + 2)(s + 1) = [(s + 1)^2 + 1](s + 1)$$

and thus $X(s)$ has the expansion

$$X(s) = \frac{b_1 s + b_0}{(s + 1)^2 + 1} + \frac{c_3}{s + 1} \tag{8.75}$$

where

$$c_3 = [(s + 1)X(s)]_{s=-1} = 4$$

The right-hand side of (8.75) can be put over a common denominator and then the resulting numerator can be equated to the numerator of $X(s)$. This yields

$$X(s) = \frac{(b_1 s + b_0)(s + 1) + 4[(s + 1)^2 + 1]}{[(s + 1)^2 + 1](s + 1)} = \frac{s^2 - 2s + 1}{s^3 + 3s^2 + 4s + 2}$$

and equating numerators yields

$$s^2 - 2s + 1 = (b_1 s + b_0)(s + 1) + 4(s^2 + 2s + 2)$$

$$s^2 - 2s + 1 = (b_1 + 4)s^2 + (b_1 + b_0 + 8)s + b_0 + 8$$

Hence

$$b_1 + 4 = 1$$

and

$$b_0 + 8 = 1$$

which implies that $b_1 = -3$ and $b_0 = -8$. Thus, from (8.75),

$$X(s) = \frac{-3s - 7}{(s + 1)^2 + 1} + \frac{4}{s + 1}$$

Writing $X(s)$ in the form

$$X(s) = \frac{-3(s + 1)}{(s + 1)^2 + 1} - 4\frac{1}{(s + 1)^2 + 1} + \frac{4}{s + 1}$$

and using Table 8.2 results in the inverse transform:

$$x(t) = -3e^{-t}\cos t - 4e^{-t}\sin t + 4e^{-t}, \qquad t \geq 0$$

Finally, using the trigonometric identity (8.73) yields

$$x(t) = 5e^{-t}\cos(t + 126.87^\circ) + 4e^{-t}, \qquad t \geq 0$$

which agrees with the result obtained in Example 8.17.

Repeated Poles

Again consider the general case where

$$X(s) = \frac{B(s)}{A(s)}$$

It is still assumed that $X(s)$ is strictly proper; that is, the degree M of $B(s)$ is strictly less than the degree N of $A(s)$. Now suppose that pole p_1 of $X(s)$ is repeated r times and the other $N - r$ poles (denoted by $p_{r+1}, p_{r+2}, \ldots, p_N$) are distinct. Then $X(s)$ has the partial fraction expansion

$$X(s) = \frac{c_1}{s - p_1} + \frac{c_2}{(s - p_1)^2} + \cdots + \frac{c_r}{(s - p_1)^r} + \frac{c_{r+1}}{s - p_{r+1}} + \cdots + \frac{c_N}{s - p_N} \tag{8.76}$$

In (8.76), the residues $c_{r+1}, c_{r+2}, \ldots, c_N$ are calculated as in the distinct-pole case; that is,

$$c_i = [(s - p_i)X(s)]_{s=p_i}, \qquad i = r + 1, r + 2, \ldots, N$$

The constant c_r is given by

$$c_r = [(s - p_1)^r X(s)]_{s=p_1}$$

and the constants $c_1, c_2, \ldots, c_{r-1}$ are given by

$$c_{r-i} = \frac{1}{i!}\left[\frac{d^i}{ds^i}[(s - p_1)^r X(s)]\right]_{s=p_1}, \qquad i = 1, 2, \ldots, r - 1 \tag{8.77}$$

In particular, setting the index i equal to 1,2 in (8.77) gives

$$c_{r-1} = \left[\frac{d}{ds}[(s - p_1)^r X(s)]\right]_{s=p_1}$$

$$c_{r-2} = \frac{1}{2}\left[\frac{d^2}{ds^2}[(s - p_1)^r X(s)]\right]_{s=p_1}$$

The constants $c_1, c_2, \ldots, c_{r-1}$ in (8.76) can also be computed by putting the right-hand side of (8.76) over a common denominator and then equating coefficients of the resulting numerator with the numerator of $X(s)$.

If the poles of $X(s)$ are all real numbers, the inverse Laplace transform can then be determined by using the transform pairs

$$\frac{t^{N-1}}{(N - 1)!}e^{-at} \leftrightarrow \frac{1}{(s + a)^N}, \qquad N = 1, 2, 3, \ldots$$

The process is illustrated by the following example.

Example 8.20 *Repeated Poles*

Consider the rational function

$$X(s) = \frac{5s - 1}{s^3 - 3s - 2}$$

The roots of $A(s) = 0$ are $-1, -1, 2$, so $r = 2$ and therefore the partial fraction expansion has the form

$$X(s) = \frac{c_1}{s + 1} + \frac{c_2}{(s + 1)^2} + \frac{c_3}{s - 2}$$

where

$$c_1 = \left[\frac{d}{ds}[(s+1)^2X(s)]\right]_{s=-1} = \left[\frac{d}{ds}\frac{5s-1}{s-2}\right]_{s=-1}$$

$$= \left.\frac{-9}{(s-2)^2}\right|_{s=-1} = -1$$

$$c_2 = [(s+1)^2X(s)]_{s=-1} = \left.\frac{5s-1}{s-2}\right|_{s=-1} = 2$$

$$c_3 = [(s-2)X(s)]_{s=2} = \left.\frac{5s-1}{(s+1)^2}\right|_{s=2} = 1$$

Thus

$$x(t) = -e^{-t} + 2te^{-t} + e^{2t}, \qquad t \geq 0$$

Instead of having to differentiate with respect to s, the constant c_1 can be computed by putting the partial fraction expansion

$$X(s) = \frac{c_1}{s+1} + \frac{2}{(s+1)^2} + \frac{1}{s-2}$$

over a common denominator and then equating numerators. This yields

$$5s - 1 = c_1(s+1)(s-2) + 2(s-2) + (s+1)^2$$

$$5s - 1 = (c_1 + 1)s^2 + (-c_1 + 4)s + (-2c_1 - 3)$$

Hence $c_1 = -1$, which is consistent with the value above.

MATLAB can also handle the case of repeated roots. In this example, the MATLAB commands are

```
num = [5 -1];
den = [1 0 -3 -2];
[r,p] = residue(num,den);
```

which yields

```
r =                                 p =

     1.0000                              2.0000
    -1.0000                             -1.0000
     2.0000                             -1.0000
```

where the second residue, -1, corresponds to the $1/(s+1)$ term and the third residue, 2, corresponds to the $1/(s+1)^2$ term.

If $X(s)$ has repeated complex poles, it is possible to avoid having to use complex arithmetic by expressing the complex part of $A(s)$ in terms of powers of quadratic terms. This solution technique is illustrated via the following example.

Example 8.21 *Powers of Quadratic Terms*

Suppose that

$$X(s) = \frac{s^3 + 3s^2 - s + 1}{s^5 + s^4 + 2s^3 + 2s^2 + s + 1}$$

Using the MATLAB command roots reveals that the poles of $X(s)$ are equal to $-1, -j, -j, j, j$. Thus there are a pair of repeated complex poles corresponding to the factor

$$[(s + j)(s - j)]^2 = (s^2 + 1)^2$$

Therefore,

$$X(s) = \frac{s^3 + 3s^2 - s + 1}{(s^2 + 1)^2(s + 1)}$$

and the expansion of $X(s)$ has the form

$$X(s) = \frac{cs + d}{s^2 + 1} + \frac{w(s)}{(s^2 + 1)^2} + \frac{c_5}{s + 1}$$

where $w(s)$ is a polynomial in s. To be able to compute the inverse Laplace transform of $w(s)/(s^2 + 1)^2$ from the transform pairs in Table 8.2 it is necessary to write this term in the form

$$\frac{w(s)}{(s^2 + 1)^2} = \frac{g(s^2 - 1) + hs}{(s^2 + 1)^2}$$

for some real constants g and h. Table 8.2 yields the transform pair

$$\left(gt \cos t + \frac{h}{2} t \sin t\right)u(t) \leftrightarrow \frac{g(s^2 - 1) + hs}{(s^2 + 1)^2}$$

Now

$$c_5 = [(s + 1)X(s)]_{s=-1} = \left.\frac{s^3 + 3s^2 - s + 1}{(s^2 + 1)^2}\right|_{s=-1} = 1$$

so

$$X(s) = \frac{cs + d}{s^2 + 1} + \frac{g(s^2 - 1) + hs}{(s^2 + 1)^2} + \frac{1}{s + 1}$$

Putting the right-hand side over a common denominator and equating numerators gives

$$s^3 + 3s^2 - s + 1 = (cs + d)(s^2 + 1)(s + 1) + [g(s^2 - 1) + hs](s + 1) + (s^2 + 1)^2$$

$$s^3 + 3s^2 - s + 1 = (c + 1)s^4 + (c + d + g)s^3 + (c + d + h + g + 2)s^2$$
$$+ (c + d - g + h)s + d - g + 1$$

Equating coefficients of the polynomials gives

$$c = -1$$
$$c + d + g = 1$$
$$c + d + h + g + 2 = 3$$
$$c + d - g + h = -1$$
$$d - g + 1 = 1$$

Solving these equations gives $d = 1, g = 1$, and $h = 0$, and thus the inverse transform is

$$x(t) = -\cos t + \sin t + t\cos t + e^{-t}, \qquad t \geq 0$$

Pole Locations and the Form of a Signal

Given a signal $x(t)$ with rational Laplace transform $X(s) = B(s)/A(s)$, again suppose that $M < N$, where N is the degree of $A(s)$ and M is the degree of $B(s)$. As seen from the development above, there is a direct relationship between the poles of $X(s)$ and the form of the signal $x(t)$. In particular, if $X(s)$ has a nonrepeated pole p which is real, then $x(t)$ contains a term of the form ce^{pt} for some constant c, and if the pole p is repeated twice, $x(t)$ contains the term $c_1e^{pt} + c_2te^{pt}$ for some constants c_1 and c_2. If $X(s)$ has a nonrepeated pair $\sigma \pm j\omega$ of complex poles, then $x(t)$ contains a term of the form $ce^{\sigma t}\cos(\omega t + \theta)$ for some constants c and θ. If the complex pair $\sigma \pm j\omega$ is repeated twice, $x(t)$ contains the term $c_1e^{\sigma t}\cos(\omega t + \theta_1) + c_2te^{\sigma t}\cos(\omega t + \theta_2)$ for some constants c_1, c_2, θ_1, and θ_2.

As a result of these relationships, the form of a signal $x(t)$ can be determined directly from the poles of the transform $X(s)$. This is illustrated by the following example.

Example 8.22 *General Form of a Signal*

Consider the signal $x(t)$ with transform

$$X(s) = \frac{1}{s(s+1)(s-4)^2((s+2)^2+3^2)}$$

The poles of $X(s)$ are $0, -1, 4, 4, -2 \pm 3j$, and thus the general form of the time signal $x(t)$ is

$$x(t) = c_1 + c_2e^{-t} + c_3e^{4t} + c_4te^{4t} + c_5e^{-2t}\cos(3t + \theta)$$

where the c_i's and θ are constants. Note that the pole at the origin corresponds to the constant term (given by c_1e^{0t}), the pole at -1 corresponds to the term c_2e^{-t}, the repeated poles at -4 corresponds to the c_3e^{4t} and c_4te^{4t} terms, and the complex poles $-2 \pm 3j$ correspond to the term $c_5e^{-2t}\cos(3t + \theta)$.

It is important to note that modifying the numerator of $X(s)$ does not change the general form of $x(t)$, but it will change the values of the constants (the coefficients of terms). For instance, in Example 8.22, if the numerator were changed from $B(s) = 1$ to any polynomial $B(s)$ with degree less than the denominator $A(s)$, the form of $x(t)$ remains the same.

A very important consequence of the relationship between the poles of $X(s)$ and the form of the signal $x(t)$ is that the behavior of the signal in the limit as $t \to \infty$ can be determined directly from the poles. In particular, it follows from the results given above that $x(t)$ converges to 0 as $t \to \infty$ if and only if the poles $p_1, p_2, \ldots, p_N$ all have real parts that are strictly less than zero, or in mathematical terms,

$$\text{Re}(p_i) < 0 \qquad \text{for } i = 1, 2, \ldots, N \tag{8.78}$$

where $\text{Re}(p_i)$ denotes the real part of the pole p_i.

It also follows from the relationship between poles and the form of the signal that $x(t)$ has a limit as $t \to \infty$ if and only if (8.78) is satisfied, except that one of the poles of $X(s)$ may be at the origin ($p = 0$). If $X(s)$ has a single pole at $s = 0$ [and all other poles satisfy (8.78)], the limiting value of $x(t)$ is equal to the value of the residue corresponding to the pole at 0; that is,

$$\lim_{t\to\infty} x(t) = [sX(s)]_{s=0}$$

This result is consistent with the final-value theorem [see (8.56)] since for any rational function $X(s)$,

$$\lim_{s\to 0} sX(s) = [sX(s)]_{s=0}$$

Example 8.23 *Limiting Value*

Suppose that

$$X(s) = \frac{2s^2 - 3s + 4}{s^3 + 3s^2 + 2s}$$

The poles of $X(s)$ are $0, -1, -2$, and thus (8.78) is satisfied for all poles of $X(s)$ except for one pole which is equal to zero. Thus $x(t)$ has a limit at $t \to \infty$ and

$$\lim_{t\to\infty} x(t) = [sX(s)]_{s=0} = \left[\frac{2s^2 - 3s + 4}{s^2 + 3s + 2}\right]_{s=0} = \frac{4}{2} = 2$$

Numerical Solution for Inverse Laplace Transform

Given a rational transform $X(s) = B(s)/A(s)$ with the degree of $B(s)$ less than the degree of $A(s)$, MATLAB can be used to compute (and plot) the inverse Laplace transform $x(t)$ at discrete points in time. This is accomplished by using the command `impulse`. The meaning of this command will become clear later when transfer functions are introduced.

Suppose that $X(s)$ has the form

$$X(s) = \frac{B(s)}{A(s)} = \frac{b_M s^M + b_{M-1} s^{M-1} + \cdots + b_0}{a_N s^N + a_{N-1} s^{N-1} + \cdots + a_0}$$

In using MATLAB, the numerator and denominator coefficients are stored in vectors, and the commands to compute and plot $x(t)$ are

```
num = [b_M b_M-1 . . . b_0];
den = [a_N a_N-1 . . . a_0];
impulse(num,den);
```

Example 8.24 *Use of MATLAB*

Consider the Laplace transform given in Example 8.16:

$$X(s) = \frac{s + 2}{s^3 + 4s^2 + 3s}$$

To compute $x(t)$, use the commands

```
num = [1 2];
den = [1 4 3 0];
impulse(num,den);
```

Running MATLAB with the commands above yields the plot shown in Figure 8.3. Note that the plot is consistent with the analytical expression of $x(t)$ found in Example 8.16:

$$x(t) = \tfrac{2}{3} - \tfrac{1}{2}e^{-t} - \tfrac{1}{6}e^{-3t}, \qquad t \geq 0$$

In particular, from both the plot in Figure 8.3 and the expression above for $x(t)$, it is clear that $x(t)$ converges to the value $\frac{2}{3}$ as $t \to \infty$. Convergence of $x(t)$ to the value $\frac{2}{3}$ can also be verified by using the final theorem, which is applicable here since the poles of $X(s)$ have negative real parts (except for the pole at $p = 0$). Applying the final-value theorem gives

$$\begin{aligned}\lim_{t\to\infty} x(t) &= \lim_{s\to 0} sX(s)\\ &= [sX(s)]_{s=0}\\ &= \left[\frac{s+2}{s^2+4s+3}\right]_{s=0}\\ &= \frac{2}{3}\end{aligned}$$

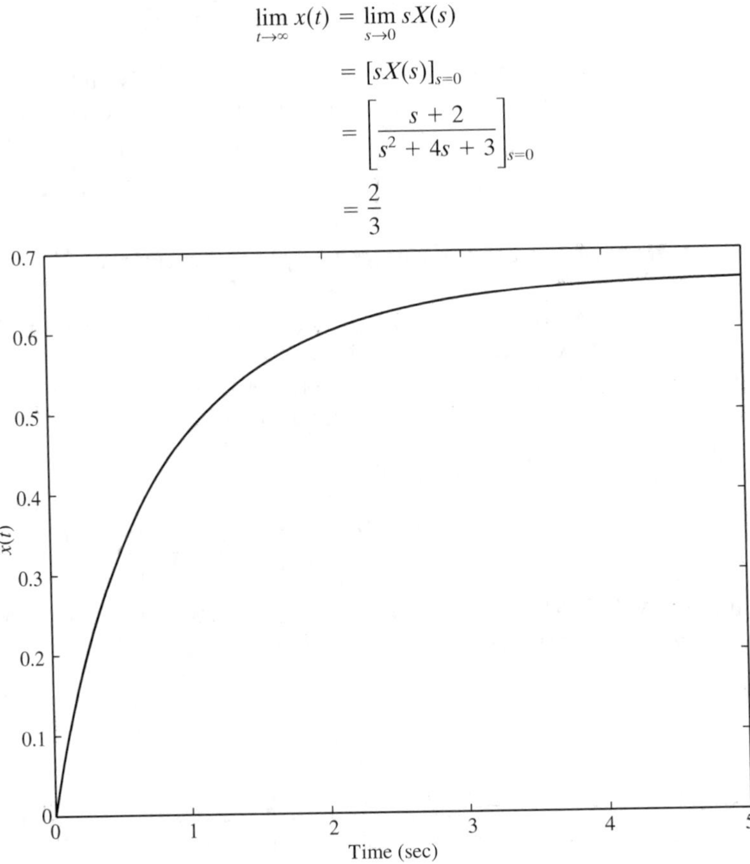

Figure 8.3 Plot of the inverse transform $x(t)$ in Example 8.24.

If $X(s)$ is given in factored form, the MATLAB command conv can be used to multiply the polynomials together to obtain the coefficients of the numerator $B(s)$ and the denominator polynomial $A(s)$ comprising $X(s)$. Hence it is possible to generate the vectors num and den [containing the coefficients of $B(s)$ and $A(s)$] directly from the factored forms. For example, suppose that

$$A(s) = (s^2 + 3s)(s + 4)$$

Then the command

```
den = conv([1 3 0],[1 4)];
```

yields den = [1 7 12 0] and thus $A(s) = s^3 + 7s^2 + 12s$.

The command conv can also be "concatenated" to multiply three or more factors. For example, to expand

$$A(s) = (s + 1)(s + 3)(s + 4)$$

use the command

```
A = conv([1 1],conv([1 3],[1 4]));
```

which first computes the product $(s + 3)(s + 4)$ and then multiplies $(s + 1)$ by this product. The result is den = [1 8 19 12].

Case When $M \geq N$

Again consider the rational function $X(s) = B(s)/A(s)$, with the degree of $B(s)$ equal to M and the degree of $A(s)$ equal to N. If $M \geq N$, by long division $X(s)$ can be written in the form

$$X(s) = Q(s) + \frac{R(s)}{A(s)}$$

where the quotient $Q(s)$ is a polynomial in s with degree $M - N$, and the remainder $R(s)$ is a polynomial in s with degree strictly less than N. The computation of the quotient $Q(s)$ and remainder $R(s)$ can be accomplished using MATLAB with the commands

```
num = [bM  bM-1  ...  b1 b0];
den = [aN  aN-1  ...  a1 a0];
[Q,R] = deconv(num,den)
```

Once $Q(s)$ and $R(s)$ are determined, the inverse Laplace transform of $X(s)$ can then be computed by determining the inverse Laplace transform of $Q(s)$ and the inverse Laplace transform of $R(s)/A(s)$. Since $R(s)/A(s)$ is strictly proper [i.e., degree $R(s) < N$], the inverse transform of $R(s)/A(s)$ can be computed by first expanding into partial fractions as given above. The `residue` command can be used to perform partial fraction expansion on $R(s)/A(s)$ as was done in the examples above. The inverse transform of the quotient $Q(s)$ can be computed by using the transform pair

$$\frac{d^N}{dt^N}\delta(t) \leftrightarrow s^N, \qquad N = 1, 2, 3, \ldots$$

The process is illustrated by the following example.

Example 8.25 $M = 3, N = 2$

Suppose that

$$X(s) = \frac{s^3 + 2s - 4}{s^2 + 4s - 2}$$

Using the MATLAB commands

```
num = [1 0 2 -4];
den = [1 4 -2];
[Q,R] = deconv(num,den)
```

gives `Q = [1 -4]` and `R = [20 -12]`. Thus the quotient is $Q(s) = s - 4$ and the remainder is $R(s) = 20s - 12$. Then

$$X(s) = s - 4 + \frac{20s - 12}{s^2 + 4s - 2}$$

and thus

$$x(t) = \frac{d}{dt}\delta(t) - 4\delta(t) + v(t)$$

where $v(t)$ is the inverse Laplace transform of

$$V(s) = \frac{20s - 12}{s^2 + 4s - 2}$$

Using the MATLAB commands

```
num = [20 -12];
den = [1 4 -2];
[r,p] = residue(num,den);
```

results in the following residues and poles for $V(s)$:

```
r =

    20.6145
    -0.6145

p =

    -4.4495
     0.4495
```

Hence the partial fraction expansion of $V(s)$ is

$$V(s) = \frac{20.6145}{s + 4.4495} - \frac{0.6145}{s - 0.4495}$$

and the inverse Laplace transform of $V(s)$ is

$$v(t) = 20.6145e^{-4.4495t} - 0.6145e^{0.4495t}, \qquad t \geq 0$$

Transforms Containing Exponentials

In many cases of interest, a function $x(t)$ will have a transform $X(s)$ of the form

$$X(s) = \frac{B_0(s)}{A_0(s)} + \frac{B_1(s)}{A_1(s)}\exp(-h_1 s) + \cdots + \frac{B_q(s)}{A_q(s)}\exp(-h_q s) \tag{8.79}$$

In (8.79) the h_i are distinct positive real numbers, the $A_i(s)$ are polynomials in s with real coefficients, and the $B_i(s)$ are polynomials in s with real coefficients. Here it is assumed that $B_i(s) \neq 0$ for at least one value of $i \geq 1$.

The function $X(s)$ given by (8.79) is not rational in s. In other words, it is not possible to express $X(s)$ as a ratio of polynomials in s with real coefficients. This is a result of the presence of the exponential terms $\exp(-h_i s)$, which cannot be written as ratios of polynomials in s. Functions of the form (8.79) are examples of *irrational functions* of s. They are also called *transcendental functions* of s.

Functions $X(s)$ of the form (8.79) arise when the Laplace transform is applied to a piecewise-continuous function $x(t)$. For instance, as shown in Example 8.4 the transform of the c-second pulse $u(t) - u(t - c)$ is equal to

$$\frac{1}{s} - \frac{1}{s}e^{-cs}$$

Clearly, this transform is in the form (8.79). Take

$$B_0(s) = 1, \qquad B_1(s) = -1, \qquad A_0(s) = A_1(s) = s, \qquad h_1 = c$$

Since $X(s)$ given by (8.79) is not rational in s, it is not possible to apply the partial fraction expansion directly to (8.79). However, partial fraction expansions can still be used to compute the inverse transform of $X(s)$. The procedure is as follows.

First, $X(s)$ can be written in the form

$$X(s) = \frac{B_0(s)}{A_0(s)} + \sum_{i=1}^{q} \frac{B_i(s)}{A_i(s)}\exp(-h_i s) \tag{8.80}$$

Now each $B_i(s)/A_i(s)$ in (8.80) is a rational function of s. If $\deg B_i(s) < \deg A_i(s)$ for $i = 0, 1, 2, \ldots, q$, each rational function $B_i(s)/A_i(s)$ can be expanded by partial fractions. In this way, the inverse Laplace transform of $B_i(s)/A_i(s)$ can be computed for $i = 0, 1, 2, \ldots, q$. Let $x_i(t)$ denote the inverse transform of $B_i(s)/A_i(s)$.

Then by linearity and the right-shift property, the inverse Laplace transform $x(t)$ is given by

$$x(t) = x_0(t) + \sum_{i=1}^{q} x_i(t - h_i)u(t - h_i), \qquad t \geq 0$$

Example 8.26 *Transform Containing an Exponential*

Suppose that

$$X(s) = \frac{s+1}{s^2+1} - \frac{1}{s+1}e^{-s} + \frac{s+2}{s^2+1}\exp(-1.5s)$$

Using linearity and the transform pairs in Table 8.2 gives

$$(\cos t + \sin t)u(t) \leftrightarrow \frac{s+1}{s^2+1}$$

$$(\cos t + 2\sin t)u(t) \leftrightarrow \frac{s+2}{s^2+1}$$

Thus

$$x(t) = \cos t + \sin t - \exp[-(t-1)]u(t-1) + [\cos(t-1.5) + 2\sin(t-1.5)]u(t-1.5), \qquad t \geq 0$$

8.4 TRANSFORM OF THE INPUT/OUTPUT DIFFERENTIAL EQUATION

Application of the Laplace transform to the study of linear time-invariant continuous-time systems is initiated in this section. The development begins with systems defined by an input/output differential equation. An "s-domain description" of any such system can be generated by taking the Laplace transform of the input/output differential equation. It will be shown that this yields an algebraic procedure for solving the input/output differential equation. Systems given by a first-order differential equation are considered first.

First-Order Case

Consider the linear time-invariant continuous-time system given by the first-order input/output differential equation

$$\frac{dy(t)}{dt} + ay(t) = bx(t) \tag{8.81}$$

where a and b are real numbers, $y(t)$ is the output, and $x(t)$ in the input. Taking the Laplace transform of both sides of (8.81) and using linearity and the differentiation-in-time property (8.43) gives

$$sY(s) - y(0^-) + aY(s) = bX(s) \tag{8.82}$$

where $Y(s)$ is the Laplace transform of the output $y(t)$ and $X(s)$ is the Laplace transform of the input $x(t)$. Note that the initial condition $y(0^-)$ is at time $t = 0^-$.

Rearranging terms in (8.82) yields

$$(s + a)Y(s) = y(0^-) + bX(s)$$

and solving for $Y(s)$ gives

$$Y(s) = \frac{y(0^-)}{s + a} + \frac{b}{s + a}X(s) \tag{8.83}$$

Equation (8.83) is the *s-domain representation* of the system given by the input/output differential equation (8.81). The first term on the right-hand side of (8.81) is the Laplace transform of the part of the output response due to the initial condition $y(0^-)$, and the second term on the right-hand side of (8.81) is the Laplace transform of the part of the output response resulting from the input $x(t)$ applied for $t \geq 0$.

The given system has no initial energy at time $t = 0^-$ if and only if $y(0^-) = 0$. Hence, if the system has no initial energy at time $t = 0^-$, the transform of the output is given by

$$Y(s) = \frac{b}{s + a}X(s) \tag{8.84}$$

Defining

$$H(s) = \frac{b}{s + a}$$

(8.84) becomes

$$Y(s) = H(s)X(s) \tag{8.85}$$

The function $H(s)$ is called the *transfer function* of the system since it specifies the transfer from the input to the output in the s-domain (assuming no initial energy), and (8.85) is referred to as the *transfer function representation* of the system.

For any initial condition $y(0^-)$ and any input $x(t)$ with Laplace transform $X(s)$, the output $y(t)$ can be computed by taking the inverse Laplace transform of $Y(s)$ given by (8.83). The process is illustrated by the following example.

Example 8.27 *RC Circuit*

Consider the RC circuit in Figure 8.4, where the input $x(t)$ is the voltage applied to the circuit and the output $y(t)$ is the voltage across the capacitor. The input/output differential equation of the circuit is

$$\frac{dy(t)}{dt} + \frac{1}{RC}y(t) = \frac{1}{RC}x(t) \tag{8.86}$$

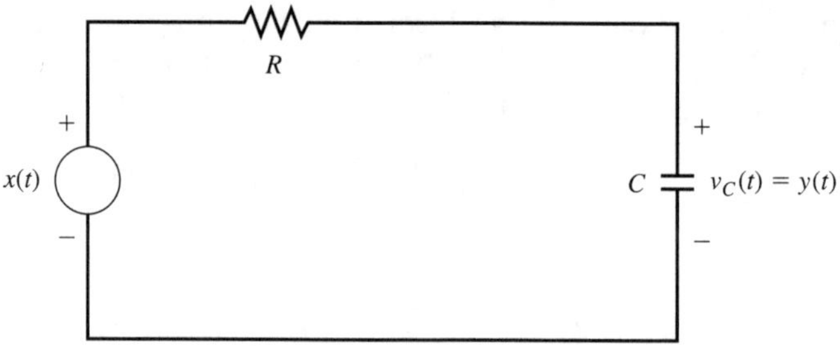

Figure 8.4 *RC* circuit in Example 8.27.

Clearly, (8.86) has the form (8.81) with a $= 1/RC$ and $b = 1/RC$, and thus the s-domain representation (8.83) for the RC circuit is given by

$$Y(s) = \frac{y(0^-)}{s + 1/RC} + \frac{1/RC}{s + 1/RC} X(s) \tag{8.87}$$

Now for any input $x(t)$, the response $y(t)$ can be determined by first computing $Y(s)$ given by (8.87) and then taking the inverse Laplace transform. To illustrate this, suppose that the input $x(t)$ is the unit step $u(t)$ so the $X(s) = 1/s$. Then (8.87) becomes

$$Y(s) = \frac{y(0^-)}{s + 1/RC} + \frac{1/RC}{(s + 1/RC)s} \tag{8.88}$$

Expanding the second term on the right-hand side of (8.88) gives

$$Y(s) = \frac{y(0^-)}{s + 1/RC} + \frac{1}{s} - \frac{1}{s + 1/RC}$$

Then taking the inverse Laplace transform of $Y(s)$ yields the output response:

$$y(t) = y(0^-)e^{-(1/RC)t} + 1 - e^{-(1/RC)t}, \qquad t \geq 0 \tag{8.89}$$

Note that if the initial condition $y(0^-)$ is zero, (8.89) reduces to

$$y(t) = 1 - e^{-(1/RC)t}, \qquad t \geq 0 \tag{8.90}$$

As defined in Section 3.4, the response $y(t)$ given by (8.90) is the step response, since it is the output when $x(t)$ is the unit step $u(t)$ with no initial energy.

Second-Order Case

Now consider the linear time-invariant continuous-time system given by the second-order input/output differential equation

$$\frac{d^2y(t)}{dt^2} + a_1 \frac{dy(t)}{dt} + a_0 y(t) = b_1 \frac{dx(t)}{dt} + b_0 x(t) \tag{8.91}$$

where a_1, a_0, b_1, and b_0 are real numbers. Assuming that $x(0^-) = 0$, taking the Laplace transform of both sides of (8.91) gives

$$s^2Y(s) - y(0^-)s - \dot{y}(0^-) + a_1[sY(s) - y(0^-)] + a_0Y(s) = b_1sX(s) + b_0X(s) \quad (8.92)$$

Solving (8.92) for Y(s) gives

$$Y(s) = \frac{y(0^-)s + \dot{y}(0^-) + a_1y(0^-)}{s^2 + a_1s + a_0} + \frac{b_1s + b_0}{s^2 + a_1s + a_0}X(s) \quad (8.93)$$

Equation (8.93) is the s-domain representation of the system with the input/output differential equation (8.91). The first term on the right-hand side of (8.93) is the Laplace transform of the part of the output response resulting from initial conditions, and the second term is the transform of the part of the response resulting from the application of the input $x(t)$ for $t \geq 0$. There is no initial energy in the system at time $t = 0$ if and only if $y(0^-) = 0$ and $\dot{y}(0^-) = 0$, in which case (8.93) reduces to the transfer function representation:

$$Y(s) = \frac{b_1s + b_0}{s^2 + a_1s + a_0}X(s) \quad (8.94)$$

In this case, the transfer function $H(s)$ is a second-order rational function of s given by

$$H(s) = \frac{b_1s + b_0}{s^2 + a_1s + a_0} \quad (8.95)$$

Example 8.28 *Second-Order Case*

Consider the system given by the input/output differential equation

$$\frac{d^2y(t)}{dt^2} + 6\frac{dy(t)}{dt} + 8y(t) = 2x(t)$$

Hence, the input/output differential equation is in the form (8.91) with $a_1 = 6, a_0 = 8$, $b_1 = 0$, and $b_0 = 2$, and from (8.95), the transfer function is

$$H(s) = \frac{2}{s^2 + 6s + 8}$$

Now to compute the step response of the system, set $x(t) = u(t)$ so that $X(s) = 1/s$. Then since there is no initial energy in the system (by definition of the step response), the transform of the response is

$$Y(s) = H(s)X(s) = \frac{2}{s^2 + 6s + 8}\frac{1}{s}$$

Expanding $Y(s)$ gives

$$Y(s) = \frac{0.25}{s} - \frac{0.5}{s + 2} + \frac{0.25}{s + 4}$$

and thus the step response is

$$y(t) = 0.25 - 0.5e^{-2t} + 0.25e^{-4t}, \qquad t \geq 0$$

Now suppose that $x(t) = u(t)$ with the initial conditions $y(0^-) = 1$ and $\dot{y}(0^-) = 2$. In this case, the system does have initial energy at time $t = 0^-$, and thus it is *not* true that $Y(s) = H(s)X(s)$. To compute $Y(s)$, it is necessary to use the s-domain representation (8.93). This gives

$$Y(s) = \frac{s+8}{s^2+6s+8} + \frac{2}{s^2+6s+8}\frac{1}{s}$$

$$= \frac{s^2+8s+2}{s(s^2+6s+8)}$$

Expanding yields

$$Y(s) = \frac{0.25}{s} + \frac{2.5}{s+2} - \frac{1.75}{s+4}$$

Thus

$$y(t) = 0.25 + 2.5e^{-2t} - 1.75e^{-4t}, \qquad t \geq 0$$

*N*th-Order Case

Now consider the general case where the system is given by the Nth-order input/output differential equation

$$\frac{d^N y(t)}{dt^N} + \sum_{i=0}^{N-1} a_i \frac{d^i y(t)}{dt^i} = \sum_{i=0}^{M} b_i \frac{d^i x(t)}{dt^i} \tag{8.96}$$

where $M \leq N$. It is assumed that $x^{(i)}(0^-) = 0$ for $i = 0, 1, 2, \ldots, M-1$. By taking the Laplace transform of both sides of (8.96) with initial conditions at time $t = 0^-$, it is possible to express the Laplace transform $Y(s)$ of the output $y(t)$ in the form

$$Y(s) = \frac{C(s)}{A(s)} + \frac{B(s)}{A(s)}X(s) \tag{8.97}$$

where $B(s)$ and $A(s)$ are polynomials in s given by

$$B(s) = b_M s^M + b_{M-1}s^{M-1} + \cdots + b_1 s + b_0$$

$$A(s) = s^N + a_{N-1}s^{N-1} + \cdots + a_1 s + a_0$$

The numerator $C(s)$ of the first term on the right-hand side of (8.97) is also a polynomial in s whose coefficients are determined by the initial conditions $y(0^-)$, $y^{(1)}(0^-), \ldots, y^{(N-1)}(0^-)$. For example, if $N = 2$, then by the results above,

$$C(s) = y(0^-)s + \dot{y}(0^-) + a_1 y(0^-)$$

Equation (8.97) is the s-domain representation of the system with the Nth-order input/output differential equation (8.96).

Since it is assumed that $x^{(i)}(0^-) = 0$ for $i = 0, 1, 2, \ldots, M - 1$, the system is at rest at time $t = 0^-$ if and only if the initial conditions $y(0^-), y^{(1)}(0^-), \ldots, y^{(N-1)}(0^-)$ are zero, which is equivalent to the condition that $C(s) = 0$. If the system is at rest at time $t = 0^-$, the transform $Y(s)$ of the output response is given by

$$Y(s) = \frac{B(s)}{A(s)} X(s) = \frac{b_M s^M + \cdots + b_1 s + b_0}{s^N + a_{N-1}s^{N-1} + \cdots + a_1 s + a_0} X(s) \tag{8.98}$$

From (8.98) it is seen that the transfer function $H(s)$ is an Nth-order rational function in s given by

$$H(s) = \frac{b_M s^M + \cdots + b_1 s + b_0}{s^N + a_{N-1}s^{N-1} + \cdots + a_1 s + a_0} \tag{8.99}$$

Example 8.29 *Automobile Suspension System*

Consider the automobile suspension system defined in Example 2.2. The system is described by the equations

$$M_1 \frac{d^2q(t)}{dt^2} + k_t[q(t) - x(t)] = k_s[y(t) - q(t)] + k_d\left[\frac{dy(t)}{dt} - \frac{dq(t)}{dt}\right] \tag{8.100}$$

$$M_2 \frac{d^2y(t)}{dt^2} + k_s[y(t) - q(t)] + k_d\left[\frac{dy(t)}{dt} - \frac{dq(t)}{dt}\right] = 0 \tag{8.101}$$

where $x(t)$ is the input and $y(t)$ is the output. Assuming that the initial conditions are all zero, taking the transform of (8.100) and (8.101) yields

$$M_1 s^2 Q(s) + k_t[Q(s) - X(s)] = (k_s + k_d s)[Y(s) - Q(s)] \tag{8.102}$$

$$M_2 s^2 Y(s) + (k_s + k_d s)[Y(s) - Q(s)] = 0 \tag{8.103}$$

where $Q(s)$ is the Laplace transform of $q(t)$. Now it is necessary to combine (8.102) and (8.103) so that $Q(s)$ is eliminated: Inserting (8.103) into (8.102) gives

$$M_1 s^2 Q(s) + k_t(Q(s) - X(s)) = -M_2 s^2 Y(s)$$

Thus

$$(M_1 s^2 + k_t)Q(s) = k_t X(s) - M_2 s^2 Y(s)$$

which implies that

$$Q(s) = \frac{k_t X(s) - M_2 s^2 Y(s)}{M_1 s^2 + k_t} \tag{8.104}$$

Inserting (8.104) into (8.103) gives

$$M_2 s^2 Y(s) + (k_s + k_d s)\left[Y(s) - \frac{k_t X(s) - M_2 s^2 Y(s)}{M_1 s^2 + k_t}\right] = 0 \tag{8.105}$$

Multiplying both sides of (8.105) by $M_1s^2 + k_t$ and combining terms yields

$$[(M_1s^2 + k_t)(M_2s^2 + k_s + k_ds) + (k_s + k_ds)M_2s^2]Y(s) - (k_s + k_ds)k_tX(s) = 0$$

Hence the transfer function is

$$H(s) = \frac{(k_s + k_ds)k_t}{(M_1s^2 + k_t)(M_2s^2 + k_s + k_ds) + (k_s + k_ds)M_2s^2}$$

Computation of output response. Again consider the transfer function representation (8.98). If the transform $X(s)$ of the input $x(t)$ is a rational function of s, the product $H(s)X(s)$ is a rational function of s. In this case the output $y(t)$ can be computed by first expanding $H(s)X(s)$ by partial fractions. The process is illustrated by the following example.

Example 8.30 *Computation of Output Response*

Consider the system with transfer function

$$H(s) = \frac{s^2 + 2s + 16}{s^3 + 4s^2 + 8s}$$

The output response $y(t)$ resulting from input $x(t) = e^{-2t}u(t)$ will be computed assuming there is no initial energy in the system at time $t = 0$. The transform of $x(t)$ is

$$X(s) = \frac{1}{s + 2}$$

and thus

$$Y(s) = H(s)X(s) = \frac{s^2 + 2s + 16}{(s^3 + 4s^2 + 8s)(s + 2)} = \frac{s^2 + 2s + 16}{[(s + 2)^2 + 4]s(s + 2)}$$

Expanding by partial fractions yields

$$Y(s) = \frac{cs + d}{(s + 2)^2 + 4} + \frac{c_3}{s} + \frac{c_4}{s + 2} \qquad (8.106)$$

where

$$c_3 = [sY(s)]_{s=0} = \frac{16}{2(8)} = 1$$

$$c_4 = [(s + 2)Y(s)]_{s=-2} = \frac{(-2)^2 - (2)(2) + 16}{(-2)(4)} = -2$$

Putting the right-hand side of (8.106) over a common denominator and equating numerators gives

$$s^2 + 2s + 16 = (cs + d)s(s + 2) + c_3[(s + 2)^2 + 4](s + 2) + c_4[(s + 2)^2 + 4]s$$

Collecting terms with like powers of s yields

$$s^3 - 2s^3 + cs^3 = 0$$

$$6s^2 - 8s^2 + (d + 2c)s^2 = s^2$$

which implies that $c = 1$ and $d = 1$. Hence

$$Y(s) = \frac{s+1}{(s+2)^2+4} + \frac{1}{s} + \frac{-2}{s+2}$$

$$Y(s) = \frac{s+2}{(s+2)^2+4} + \frac{-1}{(s+2)^2+4} + \frac{1}{s} + \frac{-2}{s+2}$$

Using Table 8.1 gives

$$y(t) = e^{-2t}\cos 2t - \tfrac{1}{2}e^{-2t}\sin 2t + 1 - 2e^{-2t}, \qquad t \geq 0$$

and by using the trigonometric identity (8.73), $y(t)$ can be expressed in the form

$$y(t) = \frac{\sqrt{5}}{2}e^{-2t}\cos(2t + 26.565°) + 1 - 2e^{-2t}, \qquad t \geq 0$$

8.5 TRANSFER FUNCTION REPRESENTATION

If a linear time-invariant system is given by an input/output differential equation, it was seen in Section 8.4 that an s-domain representation (or transfer function representation) of the system can be generated by taking the Laplace transform of the input/output differential equation. As shown below, the transfer function representation can be generated for any causal linear time-invariant system by taking the Laplace transform of the input/output convolution expression given by

$$y(t) = h(t) * x(t) = \int_0^{\infty} h(\lambda)x(t-\lambda)\,d\lambda \tag{8.107}$$

where $h(t)$ is the impulse response of the system. From the results in Chapter 3, recall that $y(t)$ given by (8.107) is the output response resulting from the input $x(t)$ with no initial energy in the system prior to the application of the input. Also recall that causality implies that $h(t)$ is zero for $t < 0$ and thus the lower limit of the integral in (8.107) is taken at $\lambda = 0$. Finally, it is important to note that in contrast to the Fourier transform approach developed in Chapter 5, here there is no requirement that the impulse response $h(t)$ be absolutely integrable [see (5.2)].

Now if the input $x(t)$ is zero for all $t < 0$, the (one-sided) Laplace transform can be applied to both sides of (8.107), which results in the transfer function representation

$$Y(s) = H(s)X(s) \tag{8.108}$$

In (8.108), $H(s)$ is the transfer function of the system. Note that if the input $x(t)$ is equal to the impulse $\delta(t)$, then $X(s) = 1$ and (8.108) reduces to $Y(s) = H(s)$. Hence the transfer function $H(s)$ is equal to the Laplace transform of the impulse response $h(t)$. This relationship between the impulse response $h(t)$ and the transfer function $H(s)$ can be expressed in terms of the transform pair notation

$$h(t) \leftrightarrow H(s) \tag{8.109}$$

The transform pair (8.109) is of fundamental importance. In particular, it provides a bridge between the time domain representation given by the convolution relationship (8.107) and the s-domain representation (8.108) given in terms of the transfer function $H(s)$.

It is important to note that if the input $x(t)$ is not the zero function, so that $X(s)$ is not zero, both sides of (8.108) can be divided by $X(s)$, which yields

$$H(s) = \frac{Y(s)}{X(s)} \tag{8.110}$$

From (8.110) it is seen that the transfer function $H(s)$ is equal to the ratio of the transform $Y(s)$ of the output and the transform $X(s)$ of the input.

Since $H(s)$ is the transform of the impulse response $h(t)$, and a system has only one $h(t)$, each system has a unique transfer function. Therefore, although $Y(s)$ will change as the input $x(t)$ ranges over some collection of signals, by (8.110) the ratio $Y(s)/X(s)$ cannot change [assuming that there is no initial energy in the system before $x(t)$ is applied].

It also follows from (8.110) that the transfer function $H(s)$ can be determined from knowledge of the response $y(t)$ to any nonzero input signal $x(t)$. It should be stressed that this result is valid only if it is known that the given system is both linear and time invariant. If the system is time varying or is nonlinear, there is no transfer function, and thus (8.110) has no meaning in such cases.

Example 8.31 *Determining the Transfer Function*

Suppose that the input $x(t) = e^{-t}u(t)$ is applied to a causal linear time-invariant continuous-time system, and that the resulting output response with the system at rest at time $t = 0$ is

$$y(t) = 2 - 3e^{-t} + e^{-2t}\cos 2t, \qquad t \geq 0$$

Then

$$Y(s) = \frac{2}{s} - \frac{3}{s+1} + \frac{s+2}{(s+2)^2 + 4}$$

and

$$X(s) = \frac{1}{s+1}$$

Inserting these expressions for $Y(s)$ and $X(s)$ into (8.110) gives

$$\begin{aligned} H(s) &= \frac{\dfrac{2}{s} - \dfrac{3}{s+1} + \dfrac{s+2}{(s+2)^2 + 4}}{\dfrac{1}{s+1}} \\ &= \frac{2(s+1)}{s} - 3 + \frac{(s+1)(s+2)}{(s+2)^2 + 4} \\ &= \frac{[2(s+1) - 3s][(s+2)^2 + 4] + s(s+1)(s+2)}{s[(s+2)^2 + 4]} \\ &= \frac{s^2 + 2s + 16}{s^3 + 4s^2 + 8s} \end{aligned}$$

Example 8.32 *Infinite-Dimensional System*

A causal linear time-invariant continuous-time system has the impulse response

$$h(t) = u(t) - u(t - c)$$

where c is a strictly positive real number. In Example 3.9 it was shown that the system processes an input $x(t)$ by integrating it over the past c-second interval. That is, the response $y(t)$ due to $x(t)$ is equal to the integral of $x(\tau)$ over the interval $t - c \leq \tau \leq t$. Taking the Laplace transform of $h(t)$ yields the transfer function

$$H(s) = \frac{1 - e^{-cs}}{s}$$

As noted before, this function cannot be written as a ratio of two polynomials in s, and thus $H(s)$ is not a rational function of s. It follows from the development given below that this system is infinite dimensional.

Finite-Dimensional Systems

Recall from the definition of basic system properties in Section 1.5 that a linear time-invariant continuous-system is finite dimensional if and only if it has an input/output differential equation. As shown in Section 8.4, for any system given by an input/output differential equation, the transfer function is a rational function of s. Hence any finite-dimensional system has a rational transfer function. Conversely, if the transfer function $H(s)$ in the s-domain representation (8.108) is rational, the system is finite dimensional. To see this, suppose that $H(s)$ has the rational form

$$H(s) = \frac{b_M s^M + b_{M-1} s^{M-1} + \cdots + b_1 s + b_0}{s^N + a_{N-1} s^{N-1} + \cdots + a_1 s + a_0} \tag{8.111}$$

Note that the leading coefficient of the denominator polynomial of $H(s)$ is equal to 1. Usually, a rational transfer function $H(s)$ is written in this form, which is the convention followed in this book. Clearly, if the leading coefficient of the denominator were equal to some constant a_N, $H(s)$ can be expressed in the form (8.111) by dividing the numerator and denominator polynomials by a_N.

Now multiplying both sides of (8.111) by the denominator polynomial gives

$$(s^N + a_{N-1} s^{N-1} + \cdots + a_1 s + a_0)Y(s) = (b_M s^M + \cdots + b_1 s + b_0)X(s) \tag{8.112}$$

Inverse transforming both sides of (8.112) yields

$$\frac{d^N y(t)}{dt^N} + \sum_{i=0}^{N-1} a_i \frac{d^i y(t)}{dt^i} = \sum_{i=0}^{M} b_i \frac{d^i x(t)}{dt^i} \tag{8.113}$$

Thus the system can be described by an input/output differential equation, which proves that the system is finite dimensional.

The observations above yield the fundamental result that a causal linear time-invariant continuous-time system is finite dimensional if and only if the transfer function is a rational function of s.

When the system is finite dimensional, it is important to emphasize that the transfer function $H(s)$ given by (8.111) can be determined directly from the coefficients of the system's input/output differential (8.113). Hence it is possible to go directly from the time-domain representation (8.113) to the transfer function representation (8.108). Conversely, if a system has transfer function $H(s)$ given by (8.111), the input/output differential equation is given by (8.113). So it is also possible to go from the transfer function representation to the input/output differential equation. This produces another fundamental link between the time domain and the s-domain. (*Exercise:* Find the input/output differential equation of the automobile suspension system in Example 8.29.)

Poles and zeros of a system. Given a finite-dimensional system with the transfer function

$$H(s) = \frac{b_M s^M + b_{M-1}s^{M-1} + \cdots + b_1 s + b_0}{s^N + a_{N-1}s^{N-1} + \cdots + a_1 s + a_0} \tag{8.114}$$

$H(s)$ can be expressed in the factored form

$$H(s) = \frac{b_M(s - z_1)(s - z_2)\cdots(s - z_M)}{(s - p_1)(s - p_2)\cdots(s - p_N)} \tag{8.115}$$

where $z_1, z_2, \ldots, z_M$ are the zeros of $H(s)$ and $p_1, p_2, \ldots, p_N$ are the poles of $H(s)$. The zeros of $H(s)$ are said to be the *zeros of the system* and the poles of $H(s)$ are said to be the *poles of the system.* Note that the number of poles of the system is equal to the order N of the system (as defined in Section 1.5). From (8.115) it is seen that except for the constant b_M, the transfer function is determined completely by the values of the poles and zeros of the system. The poles and zeros of a finite-dimensional system are often displayed in a *pole–zero diagram,* which is a plot in the complex plane showing the location of all the poles (marked by $\times$) and all zeros (marked by C). As will be seen in Chapter 9, the location of the poles and zeros is of fundamental importance in determining the behavior of the system.

Example 8.33 *Mass–Spring–Damper System*

Mass–Spring–Damper System

For the mass–spring–damper system defined in Chapter 1 (Figure 1.28), the input/output differential equation of the system is given by

$$M\frac{d^2y(t)}{dt^2} + D\frac{dy(t)}{dt} + Ky(t) = x(t)$$

where M is the mass, D is the damping constant, K is the stiffness constant, $x(t)$ is the force applied to the mass, and $y(t)$ is the displacement of the mass relative to the equilibrium position. Then using the relationship between (8.111) and (8.113) reveals that the transfer function $H(s)$ of the mass–spring–damper system is given by

$$H(s) = \frac{1}{Ms^2 + Ds + K} = \frac{1/M}{s^2 + (D/M)s + (K/M)}$$

The system is a second-order system; that is, the system has two poles, p_1 and p_2. From the quadratic formula, p_1 and p_2 are given by

$$p_1 = -\frac{D}{M} + \frac{1}{2}\sqrt{\frac{D^2}{M^2} - 4\frac{K}{M}}$$

$$p_2 = -\frac{D}{M} - \frac{1}{2}\sqrt{\frac{D^2}{M^2} - 4\frac{K}{M}}$$

It turns out that the poles p_1 and p_2 may be real numbers, or they may be complex numbers with one pole equal to the complex conjugate of the other. The poles are real if and only if

$$\frac{D^2}{M^2} - 4\frac{K}{M} \geq 0$$

which is equivalent to the condition $D^2 \geq 4KM$. The poles are real and distinct (nonrepeated) if and only if $D^2 > 4KM$. Assuming that both the poles are real and distinct, the Laplace transform $Y(s)$ of the output response $y(t)$ resulting from the constant input $x(t) = A, t \geq 0$, is given by

$$Y(s) = \frac{1/M}{(s - p_1)(s - p_2)}\frac{A}{s} = \frac{A/(p_1p_2M)}{s} + \frac{A/[M(p_1 - p_2)]}{s - p_1} + \frac{A/[M(p_2 - p_1)]}{s - p_2}$$

Taking the inverse Laplace transform results in the output response

$$y(t) = \frac{A}{p_1p_2M} + \frac{A}{(p_1 - p_2)M}[e^{p_1t} - e^{p_2t}], t \geq 0$$

If M, D, and K are strictly positive (greater than 0) and $D^2 > 4KM$, it turns out that p_1 and p_2 are strictly-negative real numbers (less than 0), and thus the response $y(t)$ converges to the constant $A/(p_1p_2M)$ as $t \to \infty$. The reader can see an animation of this behavior by running the on-line demo on the Web site.

Example 8.34 *Third-Order System*

Consider the system with the transfer function

$$H(s) = \frac{2s^2 + 12s + 20}{s^3 + 6s^2 + 10s + 8}$$

Factoring $H(s)$ gives

$$H(s) = \frac{2(s + 3 - j)(s + 3 + j)}{(s + 4)(s + 1 - j)(s + 1 + j)}$$

Thus the zeros of the system are

$$z_1 = -3 + j \quad \text{and} \quad z_2 = -3 - j$$

and the poles of the system are

$$p_1 = -4, \quad p_2 = -1 + j, \quad p_3 = -1 - j$$

The pole–zero diagram is shown in Figure 8.5.

Computation of output response using MATLAB. Given any finite-dimensional system, the system's response to an arbitrary input can be calculated using MATLAB as follows. First, any transfer function $H(s)$ given by (8.114) is represented in MATLAB by two vectors, one containing the coefficients of the numerator polynomial and one containing the coefficients of the denominator polynomial, with both polynomials given in descending powers of s. Hence the transfer function (8.114) is stored in MATLAB via the commands

```
num = [bM  bM-1  ...  b0];
den = [1  aN-1  ...  a0];
```

Then to compute the system output resulting from an input $x(t)$ with no initial energy, a time vector is defined that contains the values of t for which $y(t)$ will be computed. The step response is found using the command `step` while the responses to other types of inputs can be obtained using the commands `impulse` and `lsim`. The use of these commands is illustrated in the following example.

Example 8.35 *Output Response Using MATLAB*

Consider the system with the transfer function given in Example 8.30:

$$H(s) = \frac{s^2 + 2s + 16}{s^3 + 4s^2 + 8s}$$

The step response of this system is obtained from the following commands:

```
num = [1 2 16];
den = [1 4 8 0];
t = 0:10/300:10;
y = step(num,den,t);
```

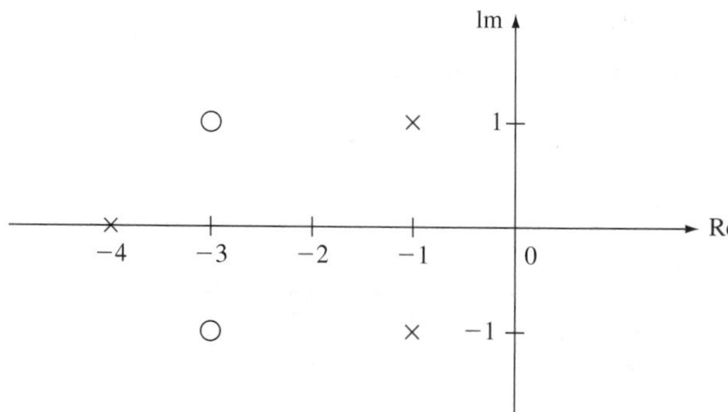

Figure 8.5 Pole–zero diagram in Example 8.34.

Running the MATLAB program results in the plot of the step response $y(t)$ shown in Figure 8.6. Note from the plot that the step response contains a ramp. This can be predicted from the form of $Y(s)$ obtained by substituting $X(s) = 1/s$ and $H(s)$ into (8.108), which gives

$$Y(s) = H(s)X(s) = \frac{s^2 + 2s + 16}{s^3 + 4s^2 + 8s}\frac{1}{s}$$

Thus $Y(s)$ has two poles at the origin, and the inverse transform of $Y(s)$ will therefore contain a ramp.

The impulse response of the system with the transfer function $H(s)$ above is obtained by replacing the `step` command used above with

```
y = impulse(num,den,t);
```

The output response resulting from an arbitrary input with no initial energy can be found using the command `lsim`. For example, if $x(t) = e^{-t}$ for $t \geq 0$, the commands are

```
num = [1 2 16];
den = [1 4 8 0];
t = 0:10/300:10;
x = exp(-t);
y = lsim(num,den,x,t);
```

The reader is invited to plot the response $y(t)$ and check the result with the analytical expression for $y(t)$ computed in Example 8.30.

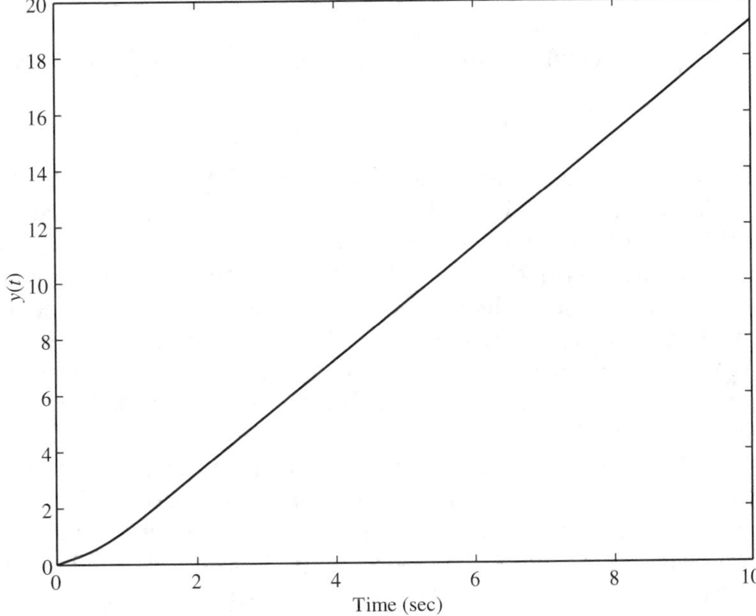

Figure 8.6 Step response of system in Example 8.35.

Direct Construction of the Transfer Function

The transfer function of a system is often determined directly from a wiring diagram of the system, so it is not always necessary first to determine the impulse response or the input/output differential equation. This can be done for *RLC* circuits and systems consisting of interconnections of integrators. The following development begins with *RLC* circuits.

RLC circuits. In Section 2.2 the resistor, capacitor, and inductor were defined in terms of the voltage–current relationships

$$v(t) = Ri(t) \tag{8.116}$$

$$\frac{dv(t)}{dt} = \frac{1}{C}i(t) \tag{8.117}$$

$$v(t) = L\frac{di(t)}{dt} \tag{8.118}$$

In (8.116)–(8.118), $i(t)$ is the current into the circuit element and $v(t)$ is the voltage across the element (see Figure 2.1). The voltage–current relationships can be expressed in the s-domain by taking the Laplace transform of both sides of (8.116)–(8.118). Using the differentiation property of the Laplace transform yields

$$V(s) = RI(s) \tag{8.119}$$

$$sV(s) - v(0) = \frac{1}{C}I(s) \quad \text{or} \quad V(s) = \frac{1}{Cs}I(s) + \frac{1}{s}v(0) \tag{8.120}$$

$$V(s) = LsI(s) - Li(0) \tag{8.121}$$

where $V(s)$ is the Laplace transform of the voltage and $I(s)$ is the Laplace transform of the current. In (8.120), $v(0)$ is the initial voltage on the capacitor at time $t = 0$ and in (8.121), $i(0)$ is the initial current in the inductor at time $t = 0$.

Using (8.119)–(8.121) results in the s-domain representations of the resistor, capacitor, and inductor shown in Figure 8.7. Here the circuit elements are represented in terms of their impedances; that is, the resistor has impedance R, the capacitor has (complex) impedance $1/Cs$, and the inductor has (complex) impedance Ls. Note that the initial voltage on the capacitor and the initial current in the inductor are treated as voltage sources in the s-domain representations.

Now given an interconnection of *RLC*s, an s-domain representation of the circuit can be constructed by taking the Laplace transform of the voltages and currents in the circuit and by expressing resistors, capacitors, and inductors in terms of their s-domain representations. The resulting s-domain representation satisfies the same circuit laws as a purely resistive circuit with voltage and current sources. In particular, the voltage and current division rules for resistive circuits can be applied. For example, consider two impedances $Z_1(s)$ and $Z_2(s)$ connected in series as shown in

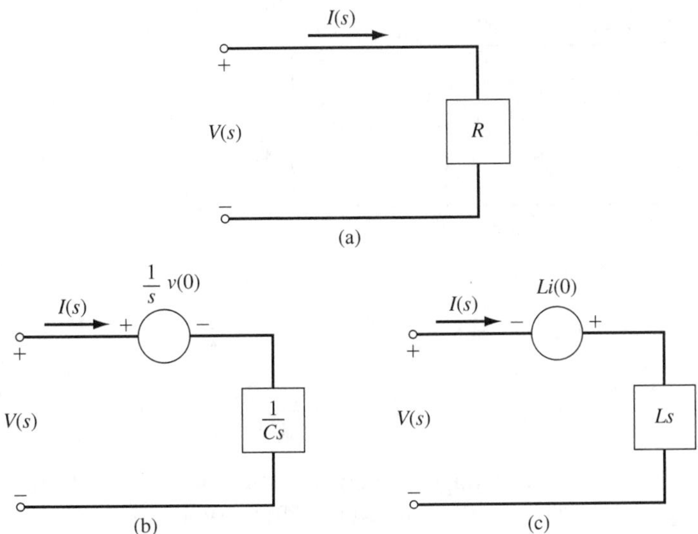

Figure 8.7 *s*-Domain representations: (a) resistor; (b) capacitor; (c) inductor.

Figure 8.8. With $V(s)$ equal to the Laplace transform of the applied voltage, and with $V_1(s)$ and $V_2(s)$ equal to the Laplace transforms of the voltages across the impedances $Z_1(s)$ and $Z_2(s)$, by the voltage division rule

$$V_1(s) = \frac{Z_1(s)}{Z_1(s) + Z_2(s)} V(s)$$

$$V_2(s) = \frac{Z_2(s)}{Z_1(s) + Z_2(s)} V(s)$$

Now suppose that the two impedances $Z_1(s)$ and $Z_2(s)$ are connected in parallel as shown in Figure 8.9. With $I(s)$ equal to the transform of the current into the connection and with $I_1(s)$ and $I_2(s)$ equal to the transforms of the currents in the impedances, by the current division rule

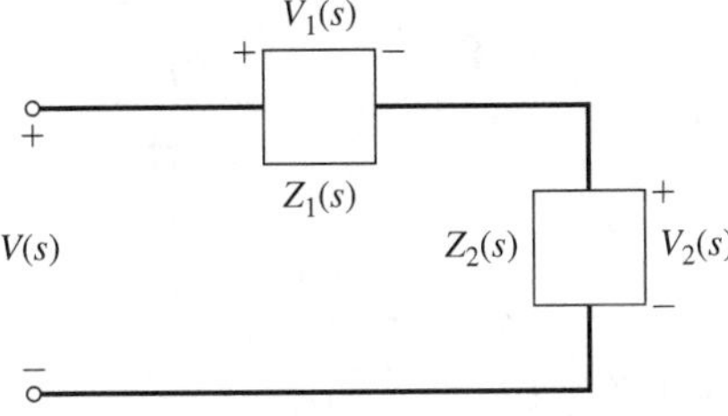

Figure 8.8 Series connection of two impedances.

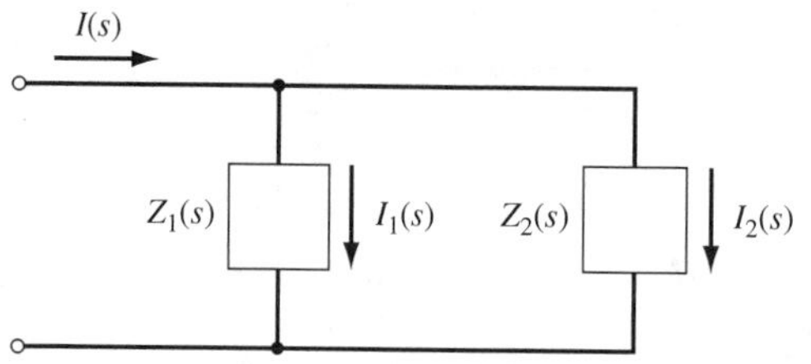

Figure 8.9 Two impedances in parallel.

$$I_1(s) = \frac{Z_2(s)}{Z_1(s) + Z_2(s)} I(s)$$

$$I_2(s) = \frac{Z_1(s)}{Z_1(s) + Z_2(s)} I(s)$$

By using the voltage and current division rules and other basic circuit laws, it is possible to determine the transfer function of an RLC circuit directly from the s-domain representation of the circuit. In computing the transfer function it is assumed that there is no initial energy in the circuit at time $t = 0$, and thus the initial voltages on the capacitors and the initial currents in the inductors are all assumed to be zero. The construction of the transfer function is illustrated by the following two examples.

Example 8.36 *Series RLC Circuit*

Consider the series RLC shown in Figure 8.10. As shown, the input is the voltage $x(t)$ applied to the series connection and the output is the voltage $v_c(t)$ across the capacitor. Assuming that the initial voltage on the capacitor and the initial current in the inductor are both zero, the s-domain representation of the circuit is shown in Figure 8.11. Working with the s-domain representation and using voltage division gives

$$V_c(s) = \frac{1/Cs}{Ls + R + (1/Cs)} X(s) = \frac{1/LC}{s^2 + (R/L)s + (1/LC)} X(s) \qquad (8.122)$$

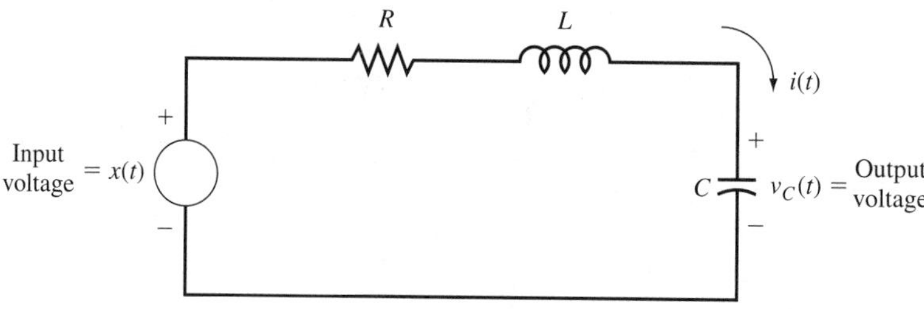

Figure 8.10 Series RLC circuit.

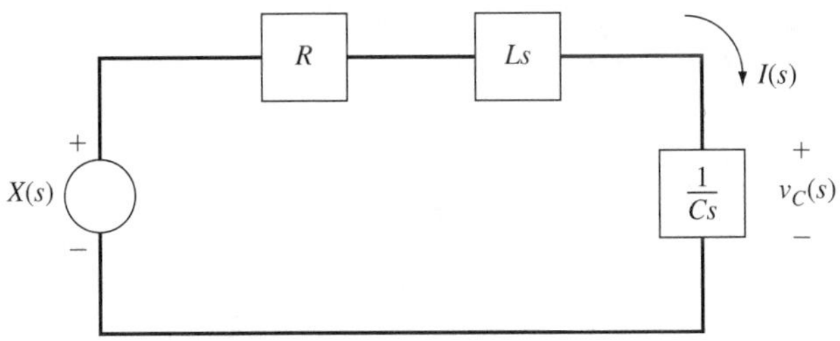

Figure 8.11 Representation of series *RLC* circuit in *s*-domain.

Comparing (8.122) with the general form (8.108) of the transfer function representation reveals that the transfer function $H(s)$ of the circuit is

$$H(s) = \frac{1/LC}{s^2 + (R/L)s + (1/LC)} \tag{8.123}$$

It is interesting to note that if a different choice for the input and output of the circuit had been taken, the transfer function would not equal the result given in (8.123). For instance, if the definition of the input is kept the same, but the output is taken to be the voltage $v_R(t)$ across the resistor, by voltage division

$$V_R(s) = \frac{R}{Ls + R + (1/Cs)}X(s) = \frac{(R/L)s}{s^2 + (R/L)s + (1/LC)}X(s)$$

The resulting transfer function is

$$H(s) = \frac{(R/L)s}{s^2 + (R/L)s + (1/LC)}$$

which differs from (8.123).

Example 8.37 *Computation of Transfer Function*

In the circuit shown in Figure 8.12, the input $x(t)$ is the applied voltage and the output $y(t)$ is the current in the capacitor with capacitance C_1. If it is assumed that the initial voltages on the capacitors and the initial currents in the inductors are all zero, then the s-domain representation of the circuit is as shown in Figure 8.13. The impedance of the parallel connection consisting of the capacitance C_1 and the inductance L_2 in series with the capacitance C_2 is equal to

$$\frac{(1/C_1s)[L_2s + (1/C_2s)]}{(1/C_1s) + L_2s + (1/C_2s)} = \frac{C_2L_2s^2 + 1}{C_1C_2L_2s^3 + (C_1 + C_2)s}$$

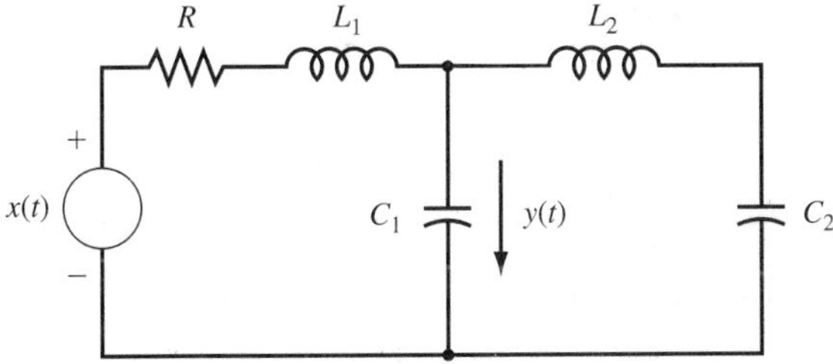

Figure 8.12 Circuit in Example 8.37.

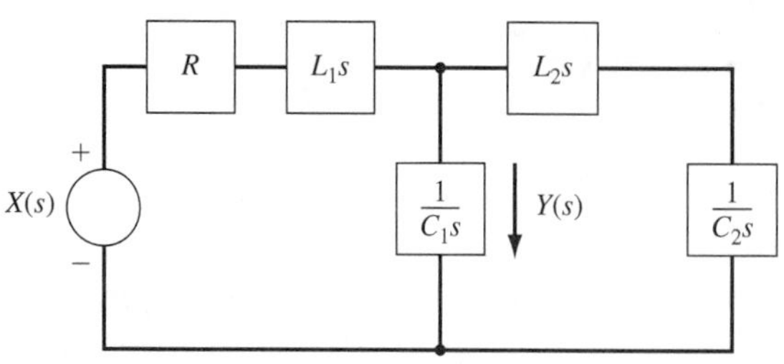

Figure 8.13 *s*-Domain representation of circuit in Example 8.37.

Letting $V_1(s)$ denote the transform of the voltage across the capacitance C_1[with $V_1(s)$ defined so that $Y(s) = C_1sV_1(s)$], by voltage division

$$V_1(s) = \frac{\dfrac{C_2L_2s^2 + 1}{C_1C_2L_2s^3 + (C_1 + C_2)s}}{R + L_1s + \dfrac{C_2L_2s^2 + 1}{C_1C_2L_2s^3 + (C_1 + C_2)s}} X(s)$$

$$V_1(s) = \frac{C_2L_2s^2 + 1}{(R + L_1s)(C_1C_2L_2s^3 + (C_1 + C_2)s) + C_2L_2s^2 + 1} X(s)$$

$$V_1(s) = \frac{C_2L_2s^2 + 1}{C_1C_2L_1L_2s^4 + RC_1C_2L_2s^3 + [L_1(C_1 + C_2) + L_2C_2]s^2 + R(C_1 + C_2)s + 1} X(s)$$

Finally, since $Y(s) = C_1sV_1(s)$,

$$Y(s) = \frac{C_1C_2L_2s^3 + C_1s}{C_1C_2L_1L_2s^4 + RC_1C_2L_2s^3 + [L_1(C_1 + C_2) + L_2C_2]s^2 + R(C_1 + C_2)s + 1} X(s)$$

and thus the transfer function is

$$H(s) = \frac{C_1C_2L_2s^3 + C_1s}{C_1C_2L_1L_2s^4 + RC_1C_2L_2s^3 + [L_1(C_1 + C_2) + L_2C_2]s^2 + R(C_1 + C_2)s + 1}$$

Interconnections of integrators. Continuous-time systems are sometimes given in terms of an interconnection of integrators, adders, subtracters, and scalar multipliers. These basic system components are illustrated in Figure 8.14. As shown in Figure 8.14a, the output $y(t)$ of the integrator is equal to the initial value of $y(t)$ plus the integral of the input (hence the term integrator). In mathematical terms

$$y(t) = y(0) + \int_0^t x(\lambda)\, d\lambda \tag{8.124}$$

Differentiating both sides of (8.124) yields the input/output differential equation

$$\frac{dy(t)}{dt} = x(t) \tag{8.125}$$

From (8.125) it is seen that if the input to the integrator is the derivative of a signal $v(t)$, the resulting output is $v(t)$. This makes sense since integration "undoes" differentiation.

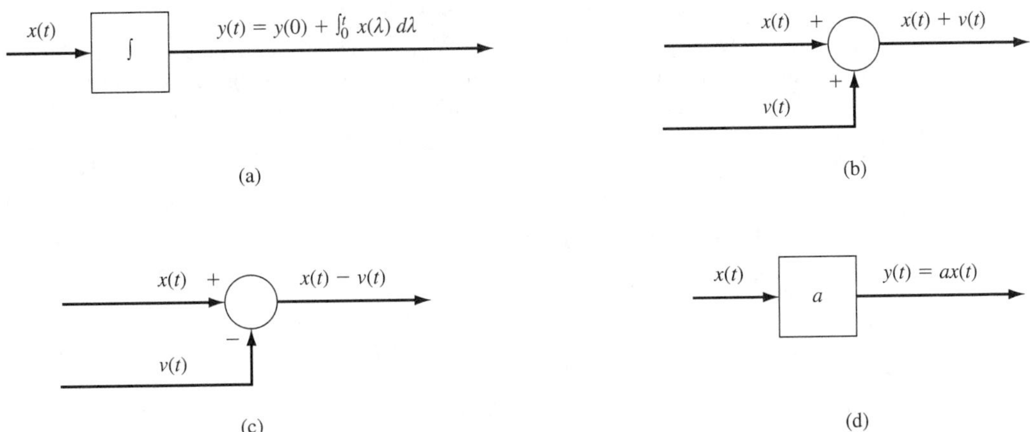

Figure 8.14 Basic system components: (a) integrator; (b) adder; (c) subtracter; (d) scaler multiplier.

As shown in Figure 8.14b and c, the adder simply adds the inputs and the subtracter subtracts the inputs. As the name implies, the scalar multiplier scales the input by the factor a, where a is any real number.

Now taking the Laplace transform of both sides of (8.125) results in the s-domain representation of the integrator:

$$sY(s) - y(0) = X(s)$$

or

$$Y(s) = \frac{1}{s}y(0) + \frac{1}{s}X(s)$$

There is no initial energy in the integrator at time $t = 0$ if and only if $y(0) = 0$, in which case the s-domain representation reduces to

$$Y(s) = \frac{1}{s}X(s) \tag{8.126}$$

From (8.126) it is seen that the integrator has transfer function $1/s$.

Now consider an interconnection of integrators, adders, subtracters, and scalar multipliers. To compute the transfer function, the interconnection can first be redrawn in the s-domain by taking transforms of all signals in the interconnection and by representing integrators by $1/s$. An equation for the Laplace transform of the output of each integrator in the interconnection can then be written. An equation for the transform of the output can also be written in terms of the transforms of the outputs of the integrators. These equations can then be combined algebraically to derive the transfer function relationship. The procedure is illustrated by the following example.

Example 8.38 *Integrator Interconnection*

Consider the system shown in Figure 8.15. The output of the first integrator is denoted by $q_1(t)$ and the output of the second integrator is donated by $q_2(t)$. Assuming that $q_1(0) = q_2(0) = 0$, the s-domain representation of the system is shown in Figure 8.16. Then

$$sQ_1(s) = -4Q_1(s) + X(s) \tag{8.127}$$

$$sQ_2(s) = Q_1(s) - 3Q_2(s) + X(s) \tag{8.128}$$

$$Y(s) = Q_2(s) + X(s) \tag{8.129}$$

Solving (8.127) for $Q_1(s)$ gives

$$Q_1(s) = \frac{1}{s+4}X(s) \tag{8.130}$$

Solving (8.128) for $Q_2(s)$ and using (8.130) yields

$$\begin{aligned} Q_2(s) &= \frac{1}{s+3}[Q_1(s) + X(s)] = \frac{1}{s+3}\left(\frac{1}{s+4} + 1\right)X(s) \\ &= \frac{s+5}{(s+3)(s+4)}X(s) \end{aligned} \tag{8.131}$$

Inserting the expression (8.131) for $Q_2(s)$ into (8.129) gives

$$Y(s) = \left[\frac{s+5}{(s+3)(s+4)} + 1\right]X(s) = \frac{s^2+8s+17}{(s+3)(s+4)}X(s)$$

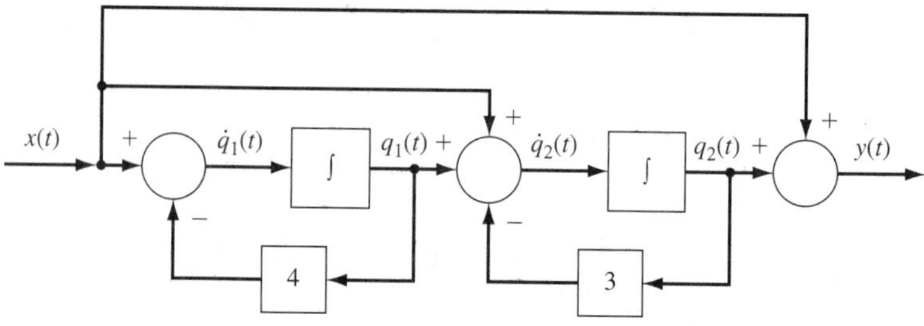

Figure 8.15 System with two integrators.

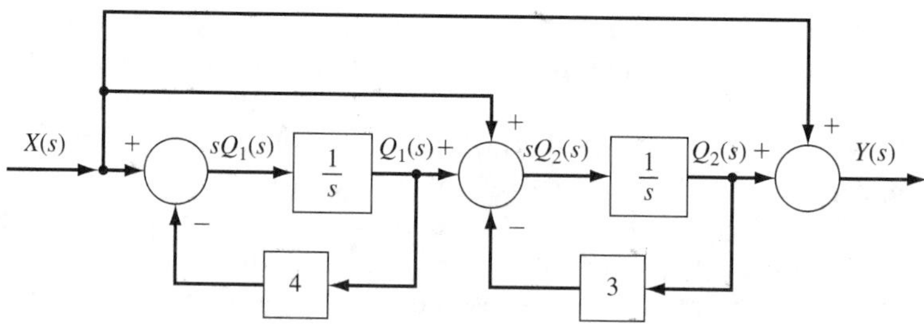

Figure 8.16 Representation of system in s-domain.

Thus the transfer function $H(s)$ is

$$H(s) = \frac{s^2 + 8s + 17}{(s + 3)(s + 4)} = \frac{s^2 + 8s + 17}{s^2 + 7s + 12}$$

8.6 TRANSFER FUNCTION OF BLOCK DIAGRAMS

A linear time-invariant continuous-time system is sometimes specified by a block diagram consisting of an interconnection of "blocks," with each block represented by a transfer function. The blocks can be thought of as subsystems comprising the given system. The transfer function of a system given by a block diagram can be determined by combining blocks in the diagram. The process is called *block-diagram reduction.* In this section the transfer functions for three basic types of interconnections are first determined and then the process of block-diagram reduction is considered.

Parallel Interconnection

Consider a parallel interconnection of two linear time-invariant continuous-time systems with transfer functions $H_1(s)$ and $H_2(s)$. The interconnection is shown in Figure 8.17. The Laplace transform $Y(s)$ of the output of the parallel connection is given by

$$Y(s) = Y_1(s) + Y_2(s) \tag{8.132}$$

If each system in the connection has no initial energy, then

$$Y_1(s) = H_1(s)X(s) \qquad \text{and} \qquad Y_2(s) = H_2(s)X(s)$$

Inserting these expressions into (8.132) gives

$$\begin{aligned} Y(s) &= H_1(s)X(s) + H_2(s)X(s) \\ &= (H_1(s) + H_2(s))X(s) \end{aligned} \tag{8.133}$$

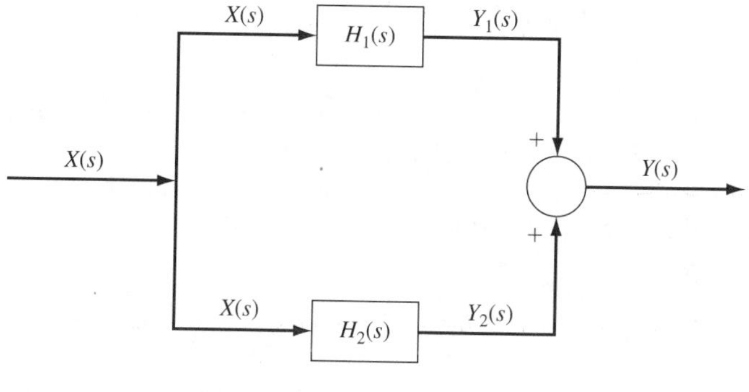

Figure 8.17 Parallel interconnection of two systems.

From (8.133) it is seen that the transfer function $H(s)$ of the parallel interconnection is equal to the sum of the transfer functions of the systems in the connection; that is,

$$H(s) = H_1(s) + H_2(s) \tag{8.134}$$

Series Connection

Now consider the series connection shown in Figure 8.18. It is assumed that each system in the series connection has no initial energy and that the second system does not load the first system. No loading means that

$$Y_1(s) = H_1(s)X(s) \tag{8.135}$$

If $y_1(t)$ is a voltage waveform, it may be assumed that there is no loading if the output impedance of the first system is much less than the input impedance of the second system.

Now since

$$Y_2(s) = H_2(s)Y_1(s)$$

using (8.135) gives

$$Y(s) = Y_2(s) = H_2(s)H_1(s)X(s) \tag{8.136}$$

From (8.136) it follows that the transfer function $H(s)$ of the series connection is equal to the product of the transfer functions of the systems in the connection; that is,

$$H(s) = H_2(s)H_1(s)$$

Since $H_1(s)$ and $H_2(s)$ are scalar-valued functions of s,

$$H_2(s)H_1(s) = H_1(s)H_2(s)$$

and thus $H(s)$ can also be expressed in the form

$$H(s) = H_1(s)H_2(s) \tag{8.137}$$

Feedback Connection

Now consider the interconnection shown in Figure 8.19. In this connection the output of the first system is fed back to the input through the second system, and thus the connection is called a *feedback connection.* Note that if the feedback loop is disconnected, the transfer function from $X(s)$ to $Y(s)$ is $H_1(s)$. The system with transfer function $H_1(s)$ is called the *open-loop system* since the transfer function from $X(s)$ to $Y(s)$

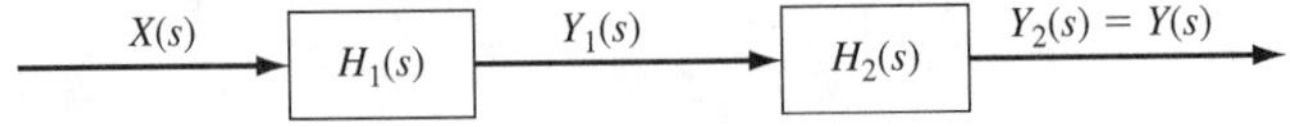

Figure 8.18 Series connection.

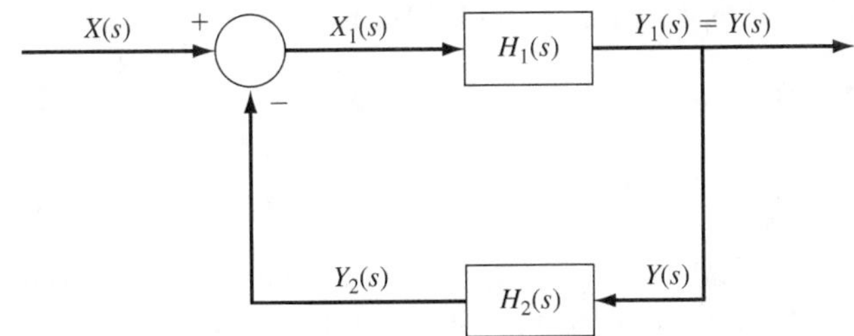

Figure 8.19 Feedback connection.

is equal to $H_1(s)$ if the feedback is disconnected. [Some authors refer to $H_1(s)H_2(s)$ as the *open-loop transfer function.*] The system with transfer function $H_2(s)$ is called the *feedback system,* and the feedback connection is called the *closed-loop system.* The objective here is to compute the transfer function of the closed-loop system.

It is assumed that there is no initial energy in either system and that the feedback system does not load the open-loop system. Then from Figure 8.19,

$$Y(s) = H_1(s)X_1(s) \tag{8.138}$$

$$X_1(s) = X(s) - Y_2(s) = X(s) - H_2(s)Y(s) \tag{8.139}$$

Inserting the expression (8.139) for $X_1(s)$ into (8.138) yields

$$Y(s) = H_1(s)[X(s) - H_2(s)Y(s)] \tag{8.140}$$

Rearranging terms in (8.140) gives

$$[1 + H_1(s)H_2(s)]Y(s) = H_1(s)X(s) \tag{8.141}$$

Solving (8.141) for $Y(s)$ gives

$$Y(s) = \frac{H_1(s)}{1 + H_1(s)H_2(s)}X(s) \tag{8.142}$$

From (8.142) it is seen that the transfer function $H(s)$ of the feedback connection is given by

$$H(s) = \frac{H_1(s)}{1 + H_1(s)H_2(s)} \tag{8.143}$$

It follows from (8.143) that the closed-loop transfer function $H(s)$ is equal to the open-loop transfer function $H_1(s)$ divided by 1 plus the product $H_1(s)H_2(s)$ of the transfer functions of the open-loop system and feedback system. Note that if the subtracter in Figure 8.19 were changed to an adder, the transfer function $H(s)$ of the closed-loop system would change to

$$H(s) = \frac{H_1(s)}{1 - H_1(s)H_2(s)} \tag{8.144}$$

It is worth noting that MATLAB can be used to compute the transfer function for feedback, series, or parallel connections. Some details on this can be found in the tutorial that is available from The MathWorks.

Block-Diagram Reduction

Block-diagram reduction is greatly facilitated by using the following rules for moving pick-off points and adders or subtracters.

Moving a pick-off point. As illustrated in Figure 8.20, a pick-off point may be moved from one side of a system to another. The symbol "≡" in the figure means that the two diagrams are equivalent. This will be verified to be the case for the diagrams in Figure 8.20b. From the diagram on the left-hand side of Figure 8.20b

$$Y_1(s) = H(s)X(s)$$

and

$$Y_2(s) = X(s)$$

From the diagram on the right-hand side of Figure 8.20b

$$Y_1(s) = H(s)X(s)$$

and

$$Y_2(s) = H^{-1}(s)H(s)X(s) = X(s)$$

The equations for the two diagrams in Figure 8.20b are the same, and thus the two systems in Figure 8.20b are equivalent. It is obvious that the two systems in Figure 8.20a are equivalent.

Moving an adder or subtracter. An adder or subtracter may be moved from one side of a system to another as illustrated in Figure 8.21. The verification that

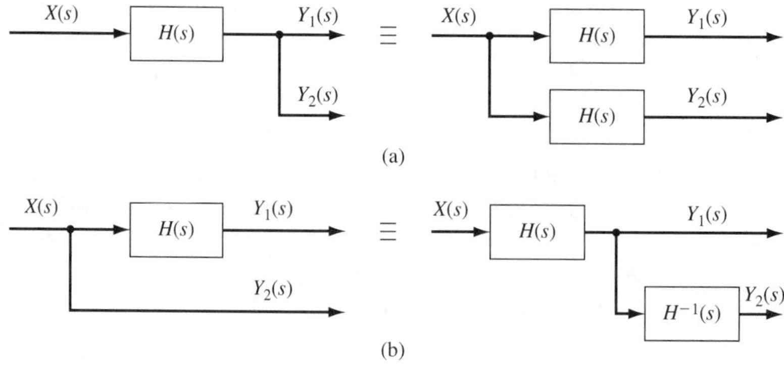

Figure 8.20 Moving a pick-off point.

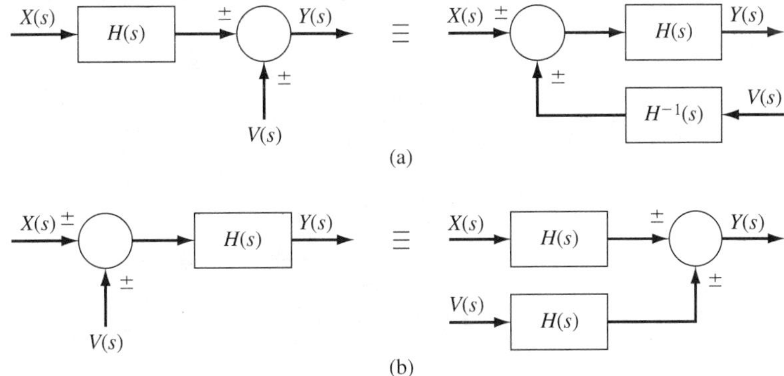

Figure 8.21 Moving an adder or subtracter.

the diagrams in Figure 8.21a and b are equivalent is an easy exercise that is left to the reader.

Block-diagram reduction. The transfer function of a system specified by a block diagram can be determined by using the equivalences in Figures 8.20 and 8.21 and the expressions for the transfer functions of parallel, series, and feedback connections. The process is illustrated by the following example.

Example 8.39 *Block-Diagram Reduction*

The transfer function of the system given by the block diagram shown in Figure 8.22 will be computed. The block-diagram reduction can be started by moving the pick-off point before the integrator to the other side of the integrator. The result is displayed in Figure 8.23. In this diagram the blocks with transfer functions $2/(s + 2)$ and s are in series, and thus they can be combined. Also, there is a feedback connection to the left of the last adder in the diagram. From the expression for the transfer function of a feedback connection, the feedback connection has transfer function

$$\frac{1/s}{1 + 3/s} = \frac{1}{s + 3}$$

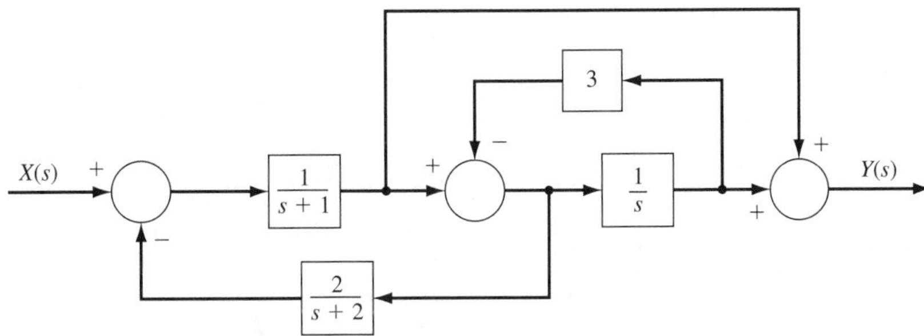

Figure 8.22 System in Example 8.39.

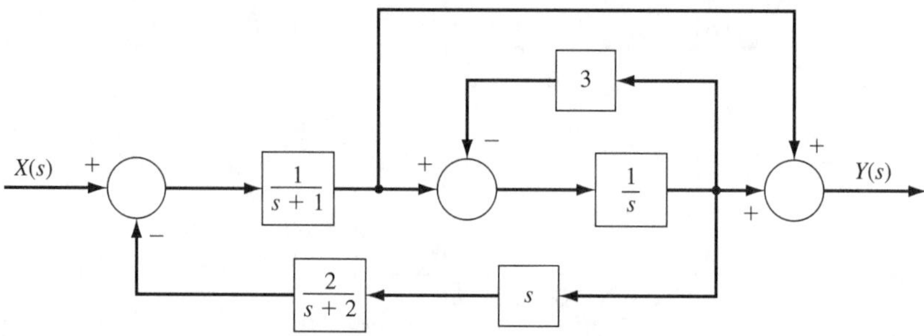

Figure 8.23 First stage of reduction.

The resulting block diagram is shown in Figure 8.24. In this diagram the pick-off point before the $1/(s + 3)$ block can be moved to the other side of the block. This results in the diagram shown in Figure 8.25. The blocks $1/(s + 1)$ and $1/(s + 3)$, which are in series, can be combined. Then computing the transfer function of the parallel connection after the $1/(s + 3)$ block results in the diagram in Figure 8.26. Now the transfer function of the feedback connection is

$$\frac{\dfrac{1}{(s+1)(s+3)}}{1 + \dfrac{2s}{(s+1)(s+2)(s+3)}} = \frac{s+2}{(s+1)(s+2)(s+3) + 2s}$$

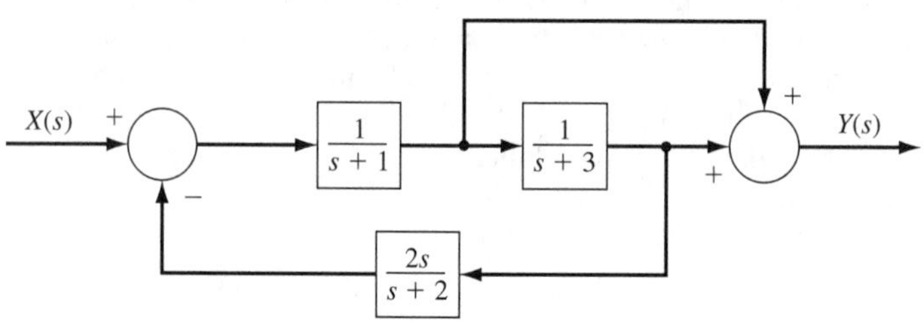

Figure 8.24 Next stage of reduction.

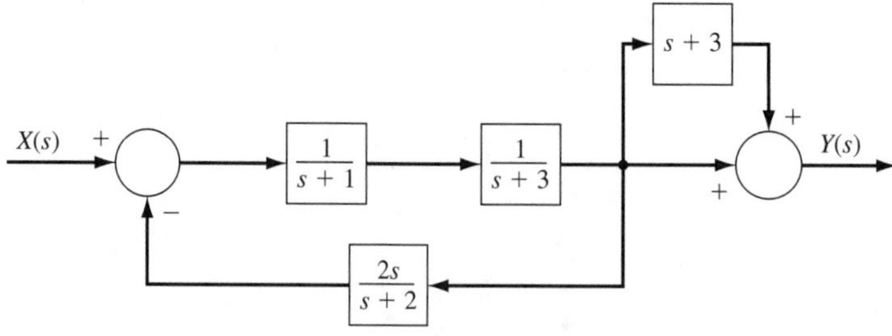

Figure 8.25 Another stage of the reduction.

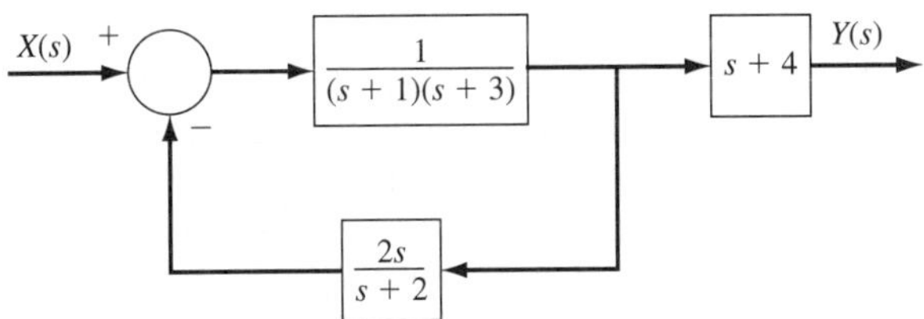

Figure 8.26 Next-to-last step in reduction.

Thus the diagram reduces to the result shown in Figure 8.27 and the transfer function of the system is

$$H(s) = \frac{(s+2)(s+4)}{(s+1)(s+2)(s+3)+2s} = \frac{s^2+6s+8}{s^3+6s^2+13s+6}$$

Mason's Theorem

Another technique for determining the transfer function of a block diagram is based on Mason's theorem. This result is specified in terms of the transfer functions of the paths from the input to the output and the transfer functions of the feedback loops comprising the block diagram. In particular, suppose that there are N paths from the input to the output, and that there are Q feedback loops. Let $P_i(s)$ denote the transfer function of the ith path from the input to the output, and let $L_i(s)$ denote the transfer function of the ith feedback loop. Let $\Delta(s)$ denote the system determinant defined by

$$\Delta(s) = 1 - \sum_{i=1}^{Q} L_i(s) + \sum L_q(s)L_r(s) - \sum L_m(s)L_n(s)L_p(s) + \cdots$$

where the second sum consists of the products $L_q(s)L_r(s)$ of the transfer functions of all nontouching feedback loops. (Two loops are nontouching if the signals in one loop do not appear in the other loop, and conversely.) The third sum consists of all the products $L_m(s)L_n(s)L_p(s)$, where each loop in a product does not touch the other two loops in the product.

Mason's theorem states that the transfer function $H(s)$ of the system is given by

$$H(s) = \frac{1}{\Delta(s)} \sum_{i=1}^{N} P_i(s)\,\Delta_i(s)$$

where $\Delta_i(s)$ is the system determinant after excluding all feedback loops that intersect the ith path from the input to the output.

Mason's formula will be applied to the system with the block diagram shown in Figure 8.22. The system has two paths from the input to the output with the transfer functions of the paths given by

$$P_1(s) = \frac{1}{s+1}, \qquad P_2(s) = \frac{1}{s(s+1)}$$

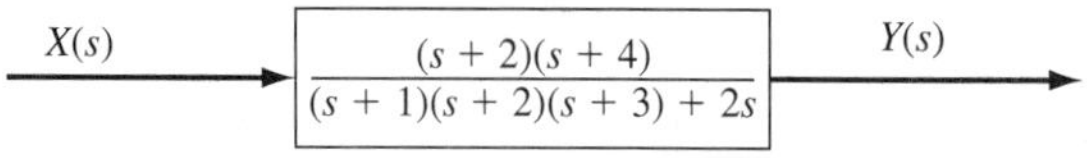

$$X(s) \longrightarrow \boxed{\frac{(s+2)(s+4)}{(s+1)(s+2)(s+3)+2s}} \longrightarrow Y(s)$$

Figure 8.27 Reduced diagram.

There are two feedback loops in the block diagram with transfer functions

$$L_1(s) = \frac{-2}{(s+1)(s+2)}, \qquad L_2(s) = \frac{-3}{s}$$

Note the minus signs that result from the subtracters in the feedback loops. In this example, the two feedback loops are touching, and thus the system determinant is

$$\Delta(s) = 1 - L_1(s) - L_2(s)$$

Inserting the expressions for $L_1(s)$ and $L_2(s)$ gives

$$\Delta(s) = 1 + \frac{2}{(s+1)(s+2)} + \frac{3}{s}$$

Now the path with transfer function $1/(s+1)$ intersects the feedback loop with transfer function $-2/[(s+1)(s+2)]$, but does not intersect the feedback loop with transfer function $-3/s$. Thus $\Delta_1(s)$ is the system determinant with the first feedback loop excluded. This gives

$$\Delta_1(s) = 1 - L_2(s) = 1 + \frac{3}{s}$$

The other path from the input to the output intersects both of the feedback loops, so $\Delta_2(s)$ is the system determinant with both loops excluded. Hence $\Delta_2(s) = 1$. Finally, from Mason's formula

$$H(s) = \frac{\dfrac{1}{s+1}\left(1 + \dfrac{3}{s}\right) + \dfrac{1}{s(s+1)}(1)}{1 + \dfrac{2}{(s+1)(s+2)} + \dfrac{3}{s}}$$

$$= \frac{s(s+2)(1+3/s) + s + 2}{s(s+1)(s+2) + 2s + 3(s+1)(s+2)}$$

$$= \frac{s^2 + 6s + 8}{s^3 + 6s^2 + 13s + 6}$$

This result agrees with the result derived above using block-diagram reduction.

Signal-Flow Graphs

Linear time-invariant systems consisting of an interconnection of blocks or subsystems are sometimes represented by a signal-flow graph. A *signal-flow graph* consists of a collection of nodes interconnected by branches. The nodes represent signal

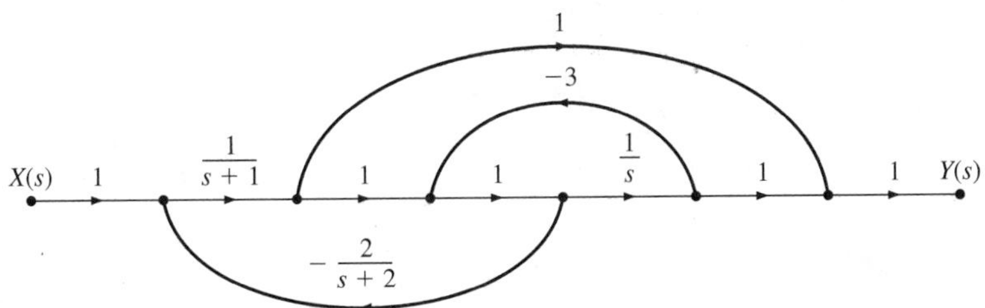

Figure 8.28 Signal-flow graph of system in Example 8.39.

points in the system, while the branches represent the transfer functions between the nodes. Each branch has an arrow showing the "flow" of signals from node to node. As an example, the signal-flow graph of the system in Example 8.39 is displayed in Figure 8.28.

PROBLEMS

8.1. Determine the Laplace transform of the following signals:

(a) $\cos(3t)u(t)$

(b) $e^{-10t}u(t)$

(c) $e^{-10t}\cos(3t)u(t)$

(d) $e^{-10t}\cos(3t - 1)u(t)$

(e) $(2 - 2e^{-4t})u(t)$

(f) $(t - 1 + e^{-10t}\cos(4t - \pi/3))u(t)$

8.2. A continuous-time signal $x(t)$ has the Laplace transform

$$X(s) = \frac{s+1}{s^2 + 5s + 7}$$

Determine the Laplace transform $V(s)$ of the following signals.

(a) $v(t) = x(3t - 4)u(3t - 4)$

(b) $v(t) = tx(t)$

(c) $v(t) = \dfrac{d^2x(t)}{dt^2}$

(d) $v(t) = \displaystyle\int_0^t x(\tau)\, d\tau$

(e) $v(t) = x(t)\sin 2t$

(f) $v(t) = e^{-3t}x(t)$

(g) $v(t) = x(t)*x(t)$

8.3 Compute the Laplace transform of each of the signals displayed in Figure P8.3.

8.4. Using the transform pairs in Table 8.2 and the properties of the Laplace transform in Table 8.1, determine the Laplace transform of the following signals.

(a) $x(t) = (e^{-bt}\cos^2\omega t)u(t)$

(b) $x(t) = (e^{-bt}\sin^2\omega t)u(t)$

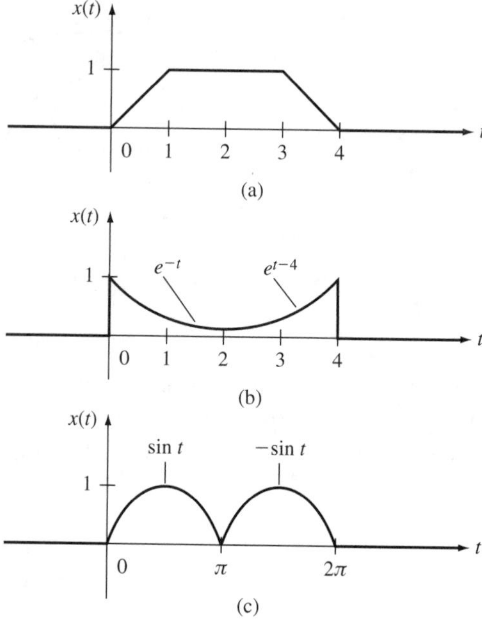

Figure P8.3

(c) $x(t) = (t \cos^2 \omega t)u(t)$
(d) $x(t) = (t \sin^2 \omega t)u(t)$
(e) $x(t) = (\cos^3 \omega t)u(t)$
(f) $x(t) = (\sin^3 \omega t)u(t)$
(g) $x(t) = (t^2 \cos \omega t)u(t)$
(h) $x(t) = (t^2 \sin \omega t)u(t)$

8.5. Determine the final values $[\lim_{t\to\infty} x(t)]$ of each of the signals whose Laplace transforms are given below. If there is no final value, state why not. Do not attempt to compute the inverse Laplace transforms.

(a) $X(s) = \dfrac{4}{s^2 + s}$

(b) $X(s) = \dfrac{3s + 4}{s^2 + s}$

(c) $X(s) = \dfrac{4}{s^2 - s}$

(d) $X(s) = \dfrac{3s^2 + 4s + 1}{s^3 + 2s^2 + s + 2}$

(e) $X(s) = \dfrac{3s^2 + 4s + 1}{s^3 + 3s^2 + 3s + 2}$

(f) $X(s) = \dfrac{3s^2 + 4s + 1}{s^4 + 3s^3 + 3s^2 + 2s}$

8.6. Determine the initial values $x(0)$ for each of the signals whose transforms are given in Problem 8.5.

8.7. A signal $x(t)$ which is zero for all $t < 0$ repeats itself every T seconds for $t \geq 0$; that is, $x(t + T) = x(t)$ for all $t \geq 0$. Let $x_0(t) = x(t)[u(t) - u(t - T)]$ and suppose that the Laplace transform of $x_0(t)$ is $X_0(s)$. Derive a closed-form expression for the Laplace transform $X(s)$ of $x(t)$ in terms of $X_0(s)$.

8.8. By using the Laplace transform, compute the convolution $x(t) * v(t)$, where:

(a) $x(t) = e^{-t}u(t), v(t) = (\sin t)u(t)$

(b) $x(t) = (\cos t)u(t), v(t) = (\sin t)u(t)$

(c) $x(t) = (\sin t)u(t), v(t) = (t \sin t)u(t)$

(d) $x(t) = (\sin^2 t)u(t), v(t) = tu(t)$

8.9. Let p be a complex number given by $p = \sigma + j\omega$. Use Euler's formula and trigonometric identities to verify the following expression:

$$ce^{pt} + \bar{c}e^{\bar{p}t} = 2|c|e^{\sigma t}\cos(\omega t + \angle c)$$

8.10. Determine the inverse Laplace transform of each of the following functions. Compute the partial fraction expansion analytically for each case. You may use `residue` to check your answers for parts (a) to (e).

(a) $X(s) = \dfrac{s + 2}{s^2 + 7s + 12}$

(b) $X(s) = \dfrac{s + 1}{s^3 + 5s^2 + 7s}$

(c) $X(s) = \dfrac{2s^2 - 9s - 35}{s^2 + 4s + 2}$

(d) $X(s) = \dfrac{3s^2 + 2s + 1}{s^3 + 5s^2 + 8s + 4}$

(e) $X(s) = \dfrac{s^2 + 1}{s^5 + 18s^3 + 81s}$

(f) $X(s) = \dfrac{s + e^{-s}}{s^2 + s + 1}$

(g) $X(s) = \dfrac{s}{s + 1} + \dfrac{se^{-s} + e^{-2s}}{s^2 + 2s + 1}$

8.11. Compute the inverse Laplace transform of the signals defined in Problem 8.10 (a) to (e) using the numerical solution method. Plot the results and compare them to those found analytically in Problem 8.10.

8.12. Determine the inverse Laplace transform of each of the following functions. Compute the partial fraction expansion analytically for each case. You may use `residue` to check your answers for parts (a) to (h).

(a) $X(s) = \dfrac{s^2 - 2s + 1}{s(s^2 + 4)}$

(b) $X(s) = \dfrac{s^2 - 2s + 1}{s(s^2 + 4)^2}$

(c) $X(s) = \dfrac{s^2 - 2s + 1}{s^2(s^2 + 4)}$

(d) $X(s) = \dfrac{s^2 - 2s + 1}{s^2(s^2 + 4)^2}$

(e) $X(s) = \dfrac{s^2 - 2s + 1}{(s + 2)^2 + 4}$

(f) $X(s) = \dfrac{s^2 - 2s + 1}{s[(s+2)^2 + 4]}$

(g) $X(s) = \dfrac{s^2 - 2s + 1}{[(s+2)^2 + 4]^2}$

(h) $X(s) = \dfrac{s^2 - 2s + 1}{s[(s+2)^2 + 4]^2}$

(i) $X(s) = \dfrac{1}{s + se^{-s}}$

(j) $X(s) = \dfrac{1}{(s+1)(1+e^{-s})}$

8.13. Compute the inverse Laplace transform of the signals defined in Problem 8.12 (a) to (h) using the numerical solution method. Plot the results and compare them to those found analytically in Problem 8.12.

8.14. Compute the solution to the following differential equations.

(a) $\dfrac{dy}{dt} + 2y = u(t), \quad y(0) = 0.$

(b) $\dfrac{dy}{dt} - 2y = u(t), \quad y(0) = 1.$

(c) $\dfrac{dy}{dt} + 10y = 4\sin(2t)u(t), \quad y(0) = 1.$

(d) $\dfrac{dy}{dt} + 10y = 8e^{-10t}u(t), \quad y(0) = 0.$

(e) $\dfrac{d^2y}{dt^2} + 6\dfrac{dy}{dt} + 8y = u(t), \quad y(0) = 0, \dot{y}(0) = 1.$

(f) $\dfrac{d^2y}{dt^2} + 6\dfrac{dy}{dt} + 9y = \sin(2t)u(t), \quad y(0) = 0, \dot{y}(0) = 0.$

(g) $\dfrac{d^2y}{dt^2} + 6\dfrac{dy}{dt} + 13y = u(t), \quad y(0) = 1, \dot{y}(0) = 1.$

8.15. A continuous-time system is given by the input/output differential equation

$$\frac{d^2y(t)}{dt^2} + 2\frac{dy(t-1)}{dt} - y(t) + 3y(t-2) = \frac{dx(t)}{dt} + x(t-2)$$

(a) Compute the transfer function $H(s)$ of the system.
(b) Compute the impulse response $h(t)$.

8.16. A continuous-time system is given by the input/output differential equation

$$\frac{d^2y(t)}{dt^2} + 4\frac{dy(t)}{dt} + 3y(t) = 2\frac{d^2x(t)}{dt^2} - 4\frac{dx(t)}{dt} - x(t)$$

In each of the following parts, compute the response $y(t)$ for all $t \geq 0$.

(a) $y(0^-) = -2, \dot{y}(0^-) = 1, x(t) = 0$ for all $t \geq 0^-$.
(b) $y(0^-) = 0, \dot{y}(0^-) = 0, x(t) = \delta(t), \delta(t) =$ unit impulse
(c) $y(0^-) = 0, \dot{y}(0^-) = 0, x(t) = u(t)$
(d) $y(0^-) = -2, \dot{y}(0^-) = 1, x(t) = u(t)$
(e) $y(0^-) = -2, \dot{y}(0^-) = 1, x(t) = u(t+1)$

8.17. Consider the field-controlled dc motor shown in Figure P8.17. The input/output differential equation of the motor is

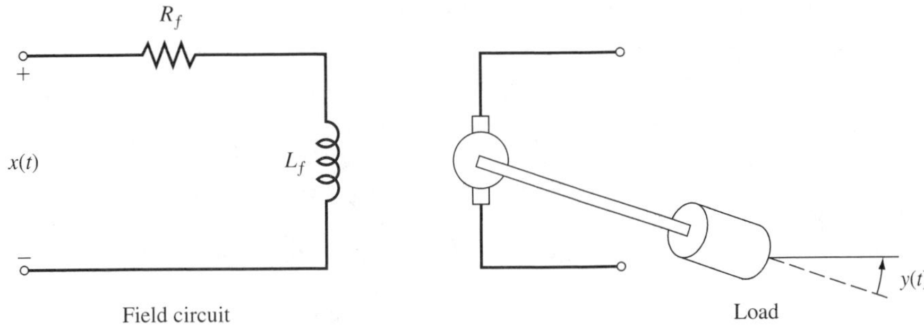

Figure P8.17

$$IL_f\frac{d^3y(t)}{dt^3} + (k_dL_f + R_fI)\frac{d^2y(t)}{dt^2} + R_fk_d\frac{dy(t)}{dt} = kx(t)$$

where the input $x(t)$ is the voltage applied to the field winding and the output $y(t)$ is the angle of the motor shaft and load, I is the moment of inertia of the motor and load, k_d is the viscous friction coefficient of the motor and load, and k is a constant.

(a) Determine the transfer function of the system.

(b) Find the impulse response $h(t)$ of the system.

8.18. Consider the simple pendulum given by the small-signal model

$$I\frac{d^2\theta(t)}{dt^2} + MgL\,\theta(t) = Lx(t)$$

Recall that the output $y(t)$ of the system is equal to the angle $\theta(t)$ of the pendulum from the vertical reference.

(a) Determine the transfer function $H(s)$ of the system.

(b) Suppose that $\theta(0) = 0, \dot{\theta}(0) = 1$, and the input force $x(t) = A\delta(t)$, where A is a constant and $\delta(t)$ is the unit impulse. Determine the value of A so that the resulting response $\theta(t)$ is zero for *all* $t > 0$. In other words, we want the impulsive input force $A\delta(t)$ to "cancel" the nonzero initial velocity $\dot{\theta}(0)$.

(c) Now suppose that $\theta(0) = 30°$ ($\pi/6$ radians) and $\dot{\theta}(0) = 1$. Find an input force $x(t)$ so that $\theta(t) = 30°$ for all $t > 0$; in other words, the input force holds the pendulum at its initial position $\theta(0)$.

8.19. Consider the automobile on a level surface given by the input/output differential equation

$$\frac{d^2y(t)}{dt^2} + \frac{k_f}{M}\frac{dy(t)}{dt} = \frac{1}{M}x(t)$$

(a) Determine the transfer function of the system.

(b) Suppose that $y(0) = 0$ and $\dot{y}(0) = 0$. By using the Laplace transform, compute the position $y(t)$ of the car for *all* $t \geq 0$ when the input force is $x(t) = t$, $0 \leq t \leq 10$, $x(t) = 20 - t$, $10 \leq t \leq 20$, $x(t) = 0$ for all other t.

(c) Suppose that $k_f = 0$. Compute a force of the form $x(t) = au(t) - bu(t - 10) + (b - a)u(t - 20)$ so that when $y(0) = 0, \dot{y}(0) = 5$, the car stops at time $t = 20$ at position $y(20) = 50$.

8.20. Consider the single-eye system studied in Problem 3.32. The model for rapid eye movement is given by the equations

$$T_e \frac{d\theta_E(t)}{dt} + \theta_E(t) = R(t)$$

$$R(t) = b\theta_T(t-d) - b\theta_T(t-d-c) + \theta_T(t-d)$$

where the input is the angular position $\theta_T(t)$ of the target and the output is the angular position $\theta_E(t)$ of the eye (see Figure P3.32).

(a) Determine the transfer function $H(s)$ of the system.

(b) Find the impulse response $h(t)$ of the system.

(c) Using the transfer function representation, compute $\theta_E(t)$ for all $t > 0$ when $\theta_T(t) = Au(t)$ and $\theta_E(0) = 0$.

(d) Repeat part (c) when $\theta_T(t) = Atu(t)$ and $\theta_E(0) = 0$.

(e) Sketch the output $\theta_E(t)$ found in parts (c) and (d) assuming that $T_e = c = 0.1$, $d = 0.2, A = 1$, and $b = 0.58$. Does the eye lock onto the target? Discuss your results.

8.21. In Problems 1.20 and 2.3, the ingestion and metabolism of a drug in a human was modeled by the equations

$$\frac{dq(t)}{dt} = -k_1 q(t) + x(t)$$

$$\frac{dy(t)}{dt} = k_1 q(t) - k_2 y(t)$$

where the input $x(t)$ is the ingestion rate of the drug, the output $y(t)$ is the mass of the drug in the bloodstream, and $q(t)$ is the mass of the drug in the gastrointestinal tract.

(a) Determine the transfer function $H(s)$.

(b) Determine the impulse response $h(t)$. Assume that $k_1 \neq k_2$.

(c) By using the Laplace transform, compute $y(t)$ for $t > 0$ when $x(t) = 0$ for $t \geq 0$, $q(0) = M_1$ and $y(0) = M_2$. Assume that $k_1 \neq k_2$.

(d) Sketch your answer in part (c) assuming that $M_1 = 100$ mg, $M_2 = 10$ mg, $k_1 = 0.05$, and $k_2 = 0.02$. Does your result "make sense"? Explain.

(e) By using the Laplace transform, compute $y(t)$ for $t > 0$ when $x(t) = e^{-at}u(t)$, $q(0) = y(0) = 0$. Assume that $a \neq k_1 \neq k_2$.

(f) Sketch your answer in part (e) assuming that $a = 0.1, k_1 = 0.05$, and $k_2 = 0.02$. When is the mass of the drug in the bloodstream equal to its maximum value? When is the mass of the drug in the gastrointestinal tract equal to its maximum value?

8.22. Determine the transfer function of the mass/spring system in Problem 2.15.

8.23. Determine the transfer function for the small-signal model of the two connected pendulums in Problem 2.16.

8.24. Again consider the automobile suspension system defined in Example 2.2. Assume that $M_1 = 100, M_2 = 1500, k_t = 1000, k_s = 200$, and $k_d = 10$.

(a) By using the transfer function representation, compute the step response.

(b) Use MATLAB to compute the step response numerically and compare your answer to that found analytically in part (a) by plotting both answers.

8.25. For each of the continuous-time systems defined below, determine the system's transfer function $H(s)$ if the system has a transfer function. If there is no transfer function, state why not.

(a) $\dfrac{dy(t)}{dt} + e^{-t}y(t) = x(t)$

(b) $\dfrac{dy(t)}{dt} + v(t) * y(t) = x(t)$, where $v(t) = (\sin t)u(t)$

(c) $\dfrac{d^2y(t)}{dt^2} + \displaystyle\int_0^t y(\lambda)\, d\lambda = \dfrac{dx(t)}{dt} - x(t)$

(d) $\dfrac{dy(t)}{dt} = y(t) * x(t)$

(e) $\dfrac{dy(t)}{dt} - 2y(t) = tx(t)$

8.26. A linear time-invariant continuous-time system has transfer function $H(s) = (s + 7)/(s^2 + 4)$. Derive an expression for the output response $y(t)$ in terms of $y(0^-)$, $\dot{y}(0^-)$ and the input $x(t)$. Assume that $x(0^-) = 0$.

8.27. The transfer function of a linear time-invariant continuous-time system is given by

$$H(s) = \frac{2s - 1}{s^2 + 3s + 2}$$

The following parts are independent.

(a) An input $x(t)$ with $x(0^-) = 1$ produces the output response

$$y(t) = -\tfrac{1}{2} + 3e^{-t} - \tfrac{5}{2}e^{-2t}, \qquad t \geq 0^-$$

Was the system at rest at time $t = 0^-$? Justify your answer.

(b) An input $x(t)$ with $x(0^-) = 1$ produces the output response

$$y(t) = -\tfrac{1}{2} + 5e^{-t} - \tfrac{9}{2}e^{-2t}, \qquad t \geq 0^-$$

Was the system at rest at time $t = 0^-$? Justify your answer.

8.28. The input $x_1(t) = e^{-t}u(t)$ is applied to a linear time-invariant continuous-time system with nonzero initial conditions $y(0)$, $\dot{y}(0)$. The resulting response is $y_1(t) = 3t + 2 - e^{-t}$, $t \geq 0$. A second input $x_2(t) = e^{-2t}u(t)$ is applied to the system with the *same* initial conditions $y(0)$, $\dot{y}(0)$. The resulting response is $y_2(t) = 2t + 2 - e^{-2t}$, $t \geq 0$. Compute $y(0)$, $\dot{y}(0)$, and the impulse response $h(t)$ of the system.

8.29. The input

$$x(t) = \begin{cases} \sin t, & 0 \leq t \leq \pi \\ -\sin t, & \pi \leq t \leq 2\pi \\ 0, & \text{all other } t \end{cases}$$

is applied to a linear time-invariant continuous-time system with no initial energy in the system at time $t = 0$. The resulting response is displayed in Figure P8.29. Determine the transfer function $H(s)$ of the system.

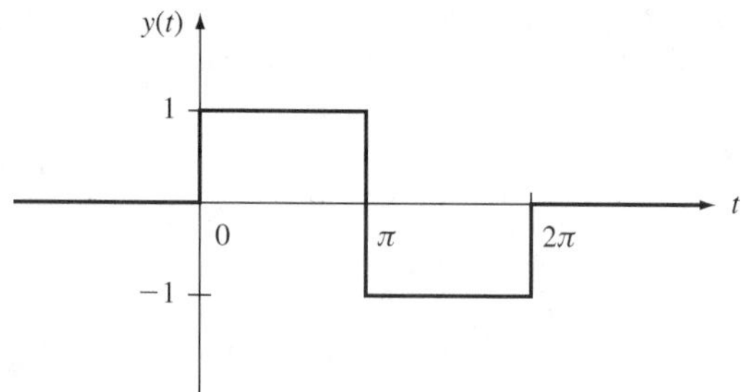

Figure P8.29

8.30. Using the s-domain representation, compute the transfer functions of the RC circuits shown in Figure P8.30.

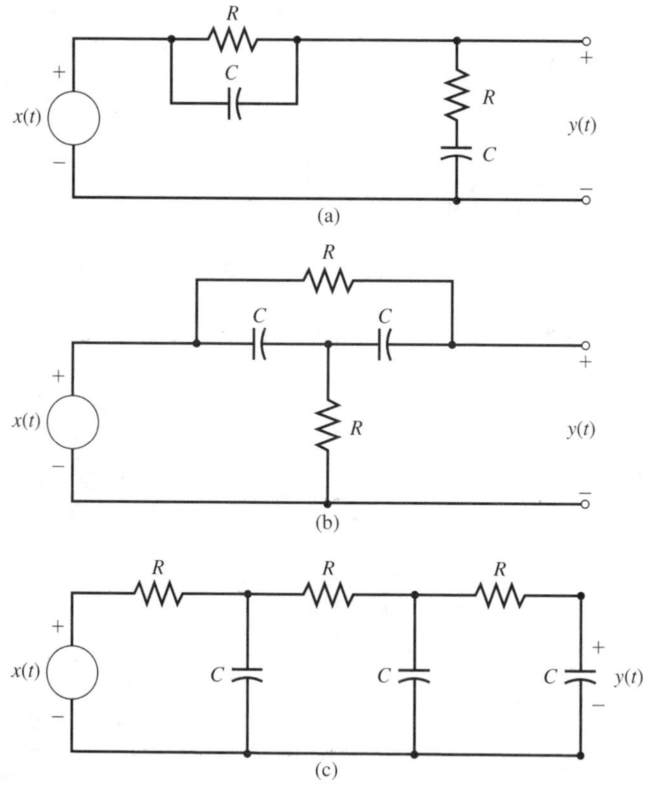

Figure P8.30

8.31. Using the s-domain representation, compute the transfer functions of the circuits shown in Figure P8.31.

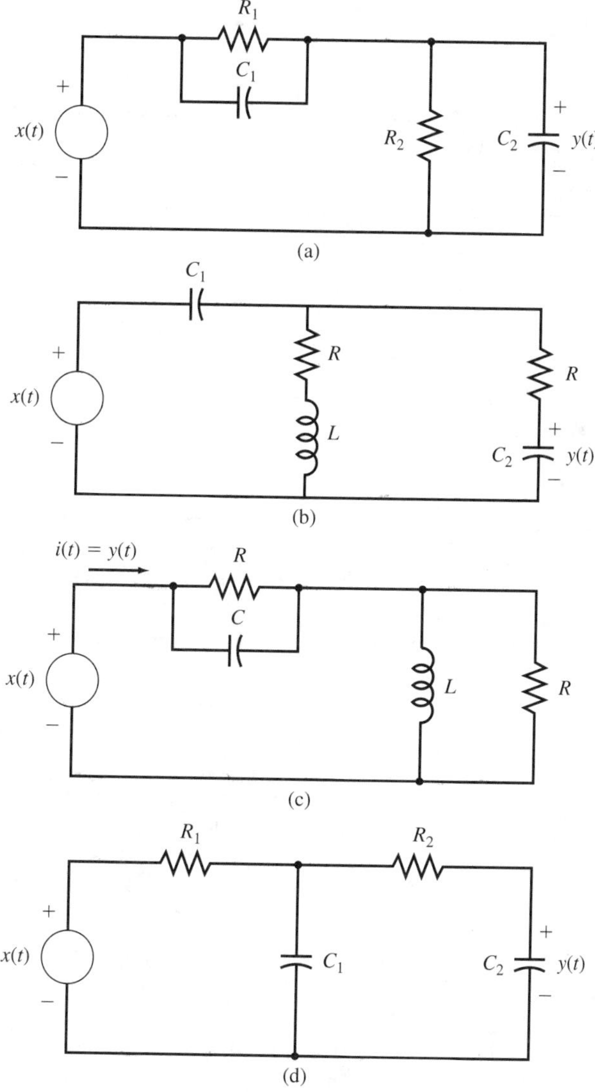

Figure P8.31

8.32. For the circuit in Figure P8.31c, determine all values of R, L, and C such that $H(s) = K$, where K is a constant.

8.33. Using the s-domain representation, compute the transfer function for the systems displayed in Figure P8.33.

8.34. A linear time-invariant continuous-time system has the impulse response $h(t) = \cos 2t + 4 \sin 2t$.

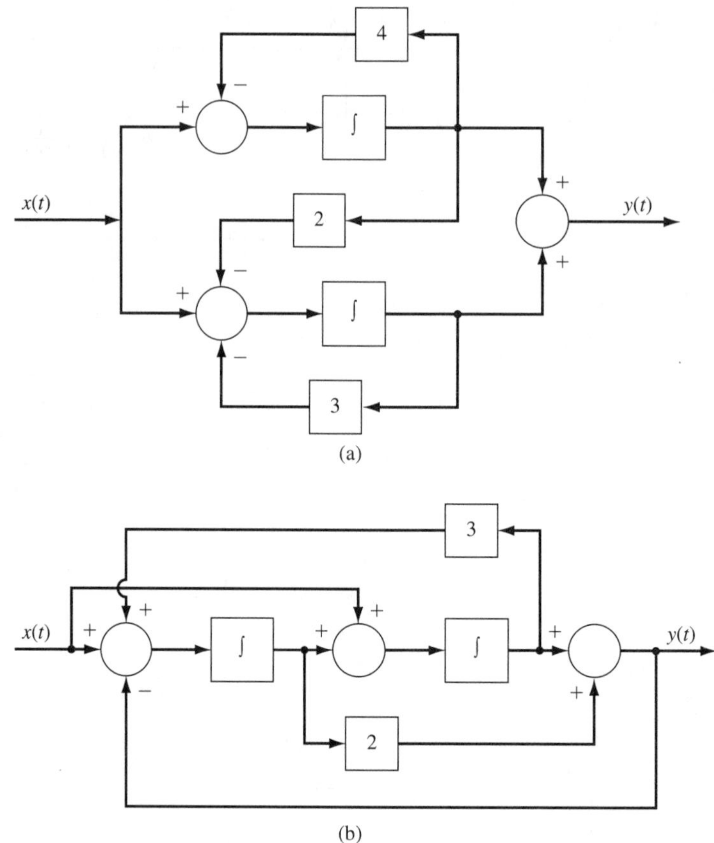

Figure P8.33

(a) Determine the transfer function $H(s)$ of the system.

(b) By using the Laplace transform, compute the output response $y(t)$ when the input $x(t)$ is equal to $\frac{5}{7}e^{-t} - \frac{12}{7}e^{-8t}$ for $t \geq 0$ with no initial energy in the system at time $t = 0$.

(c) Use MATLAB to find $y(t)$ numerically and compare this answer to that obtained analytically in part (b) by plotting both answers.

8.35. A linear time-invariant continuous-time system has impulse response $h(t)$ given by

$$h(t) = \begin{cases} e^{-t}, & 0 \leq t \leq 2 \\ e^{t-4}, & 2 \leq t \leq 4 \\ 0, & \text{all other } t \end{cases}$$

By using the Laplace transform, compute the output response $y(t)$ resulting from the input $x(t) = (\sin t)u(t)$ with no initial energy.

8.36. In this problem the objective is to design the oscillator illustrated in Figure P8.36. Using two integrators and subtracters, adders, and scalar multipliers, design the oscillator so that when the switch is closed at time $t = 0$, the output voltage $v(t)$ is $\sin 200t$, $t \geq 0$. We are assuming that the oscillator is at rest at time $t = 0$.

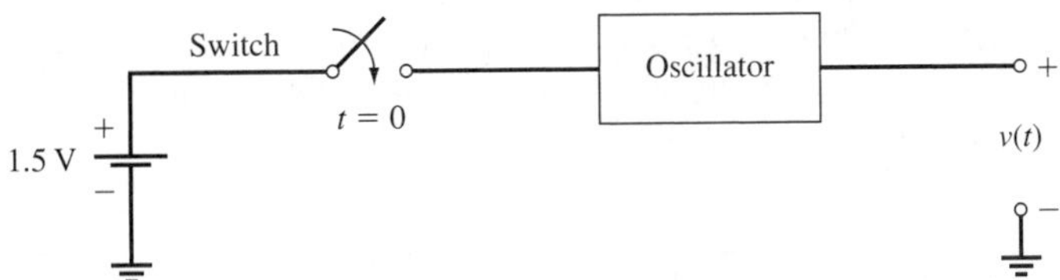

Figure P8.36

8.37. A linear time-invariant continuous-time system has impulse response $h(t) = [e^{-t}\cos(2t - 45°)]u(t) - tu(t)$. Determine the input/output differential equation of the system.

8.38. Use block diagram reduction to obtain the transfer functions of the systems shown in Figure P8.38.

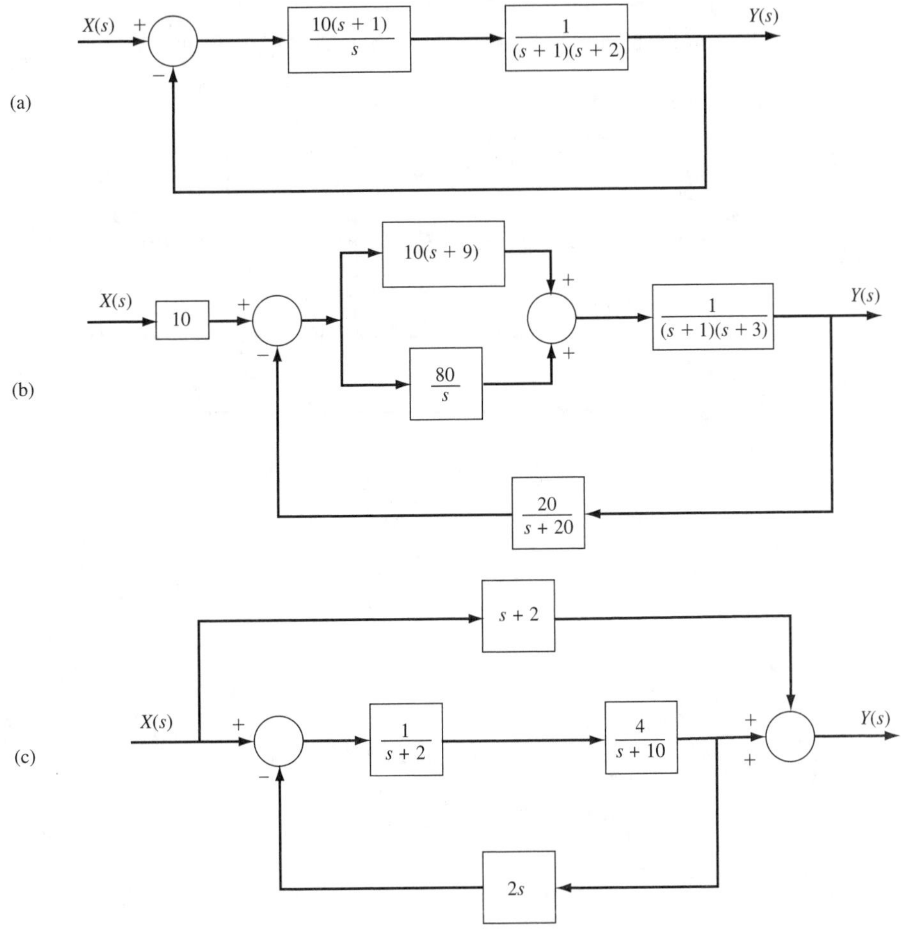

Figure P8.38

8.39. Consider the system shown in Figure P8.39.

(a) Using block-diagram reduction, determine the transfer function of the system.

(b) Using the Laplace transform, compute $y(t)$ for $t \geq 0$ when $x(t) = u(t)$ with $q_1(0^-) = 1$ and $q_2(0^-) = -3$.

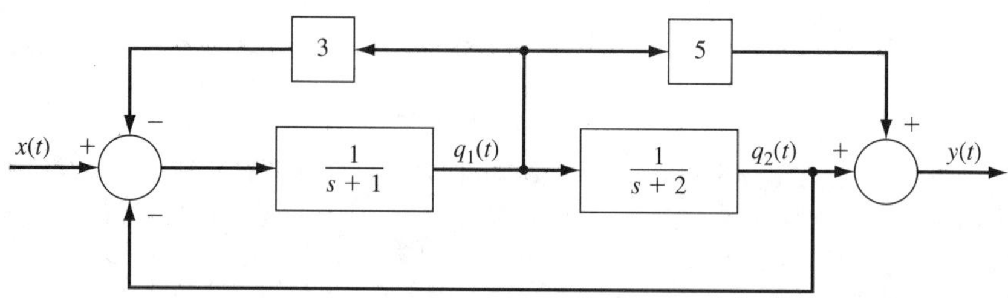

Figure P8.39

8.40. Repeat Problem 8.39 for the system shown in Figure P8.40.

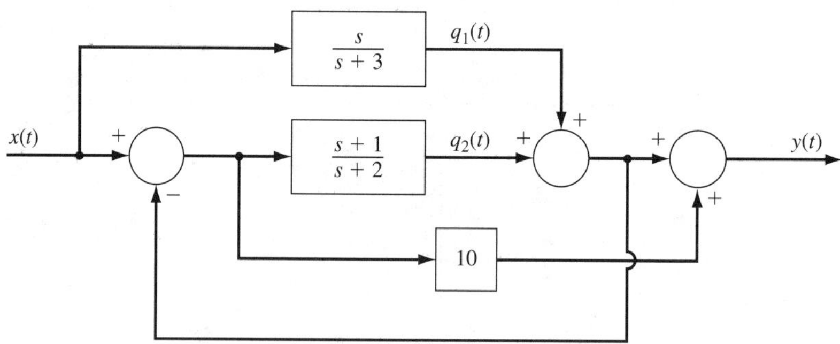

Figure P8.40

◆

CHAPTER 9

System Analysis Using the Transfer Function Representation

In addition to giving an algebraic procedure for computing the output response resulting from an input, the transfer function representation can also be used to study the basic properties and behavior of a system. Various fundamental aspects of transfer function analysis are explored in this chapter for the class of causal linear time-invariant finite-dimensional continuous-time systems. The presentation begins in the next section with the study of stability of a system, which is a time-domain property of a system since it involves the behavior of output responses. It is shown that stability or instability of a given system can be characterized in terms of the location of the poles of the system. Since the poles of a system are a feature of the transfer function representation, the characterization of stability in terms of pole locations is an excellent illustration of how the transfer function model can be utilized to study time-domain properties.

In Section 9.3 the transfer function representation is used to study the basic characteristics of the output response resulting from an input, with the focus on the case of a step input. Here it is shown that the poles of a system determine the basic features of the transient response resulting from a step input. Then in Section 9.4 the transfer function representation is utilized to study the steady-state response resulting from a sinusoidal input. This leads to the concept of a system's frequency response function that was first considered in Chapter 5. A detailed development of the frequency response function is given in Section 9.5, including the description of frequency response data in terms of Bode diagrams. In the last section of the chapter, the frequency function analysis is applied to the study of causal filters.

9.1 STABILITY AND THE IMPULSE RESPONSE

Consider a causal linear time-invariant continuous-time system with input $x(t)$ and output $y(t)$. Throughout this chapter it is assumed that the system is finite dimensional, and thus the system's transfer function $H(s)$ is rational in s; that is,

$$H(s) = \frac{b_M s^M + b_{M-1}s^{M-1} + \cdots + b_1 s + b_0}{s^N + a_{N-1}s^{N-1} + \cdots + a_1 s + a_0} \tag{9.1}$$

In the following development it is assumed that the order N of the system is less than or equal to M and that the transfer function $H(s)$ does not have any common poles and zeros. If there are common poles and zeros, they should be canceled.

As first observed in Section 8.5, the transfer function $H(s)$ is the Laplace transform of the system's impulse response $h(t)$. That is, if the input $x(t)$ applied to the system is the impulse $\delta(t)$, the transform of the resulting output response is $H(s)$ (assuming no initial energy in the system at time 0^-). Since $H(s)$ is the Laplace transform of the impulse response $h(t)$, it follows directly from the results in Section 8.3 that the form of the impulse response is determined directly by the poles of the system [i.e., the poles of $H(s)$]. In particular, if $H(s)$ has a real pole p, then $h(t)$ contains a term of the form ce^{pt}, and if $H(s)$ has a complex pair of $\sigma \pm j\omega$ of poles, then $h(t)$ contains a term of the form $ce^{\sigma t}\cos(\omega t + \theta)$. If $H(s)$ has repeated poles, $h(t)$ will contain terms of the form $ct^i e^{pt}$ and/or $ct^i e^{\sigma t}\cos(\omega t + \theta)$.

It follows from the relationship between the form of $h(t)$ and the poles of $H(s)$ that the impulse response $h(t)$ converges to zero as $t \to \infty$ if and only if

$$\text{Re}(p_i) < 0 \qquad \text{for } i = 1, 2, \ldots, N \tag{9.2}$$

where $p_1, p_2, \ldots, p_N$ are the poles of $H(s)$. The condition (9.2) is equivalent to requiring that all the poles of the system are located in the open left-half plane, where the open left-half plane is the region of the complex plane consisting of all points to the left of the $j\omega$-axis but not including the $j\omega$-axis. The open left-half plane is indicated by the hatched region shown in Figure 9.1.

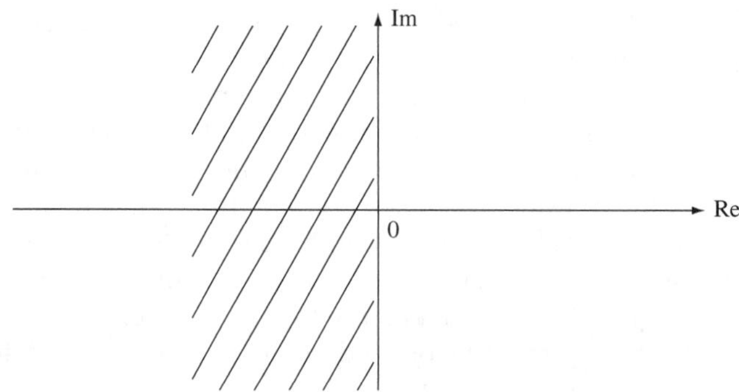

Figure 9.1 Open left-half plane.

A system with transfer function $H(s)$ given by (9.1) is said to be *stable* if its impulse response $h(t)$ converges to zero as $t \to \infty$. Hence a system is stable if and only if all the poles are located in the open left-half plane.

A system with transfer function $H(s)$ is said to be *marginally stable* if its impulse response $h(t)$ is bounded; that is, there exists a finite positive constant c such that

$$|h(t)| \leq c \qquad \text{for all } t \tag{9.3}$$

Again from the relationship between the form of $h(t)$ and the poles of $H(s)$, it follows that a system is marginally stable if and only if $\text{Re}(p_i) \leq 0$ for all nonrepeated poles of $H(s)$ and $\text{Re}(p_i) < 0$ for all repeated poles. Thus a system is marginally stable if and only if all the poles are in the open left-half plane, except that there can be nonrepeated poles on the $j\omega$-axis.

Finally, a system is *unstable* if the impulse response $h(t)$ grows without bound as $t \to \infty$. In mathematical terms, the system is unstable if and only if

$$|h(t)| \to \infty \qquad \text{as } t \to \infty \tag{9.4}$$

The relationship between the form of $h(t)$ and the poles reveals that a system is unstable if and only if there is at least one pole p_i with $\text{Re}(p_i) > 0$ or if there is at least one repeated pole p_i with $\text{Re}(p_i) = 0$. Hence a system is unstable if and only if there are one or more poles in the open right-half plane (the region to the right of the $j\omega$-axis) or if there are repeated poles on the $j\omega$-axis.

Example 9.1 *Automobile on a Level Surface*

Consider the automobile on a level surface that was studied in Section 1.4. The car is given by the input/output differential equation

$$\frac{d^2y(t)}{dt^2} + \frac{k_f}{M}\frac{dy(t)}{dt} = \frac{1}{M}x(t)$$

where the input $x(t)$ is the drive or braking force applied to the car at time t and the output $y(t)$ is the position of the car at time t. The transfer function of the system is

$$H(s) = \frac{1/M}{s^2 + (k_f/M)s} = \frac{1/M}{s(s + k_f/M)}$$

Thus the poles of the system are $p_1 = 0$ and $p_2 = -k_f/M$. Since the coefficient k_f corresponding to viscous friction must be strictly positive and the mass M must be strictly positive, $p_2 < 0$, and thus p_2 is in the open left-half plane. However, due to the pole $p_1 = 0$, the system is not stable, and thus the impulse response $h(t)$ does not converge to zero as $t \to \infty$ [but $h(t)$ is bounded]. This is an expected result since if the car is "hit" with an impulsive force at time $t = 0$ with $y(0^-) = 0$ and $\dot{y}(0^-) = 0$, the car will start to move, but it will not return to the zero position ($y = 0$). The car will move to some finite position $y(\infty)$ in the limit as $t \to \infty$. The stopping position $y(\infty)$ can be determined by applying the final-value theorem

$$y(\infty) = \lim_{t\to\infty} y(t) = \lim_{s\to 0} sH(s) = \lim_{s\to 0}\frac{1/M}{s + k_f/M} = \frac{1}{k_f}$$

Example 9.2 *Series RLC Circuit*

Consider the series RLC circuit that was studied in Example 8.36. The circuit is redrawn in Figure 9.2. In the following analysis it is assumed that $R > 0$, $L > 0$, and $C > 0$. As computed in Example 8.36, the transfer function $H(s)$ of the circuit is

$$H(s) = \frac{1/LC}{s^2 + (R/L)s + 1/LC}$$

From the quadratic formula, the poles of the system are

$$p_1, p_2 = -\frac{R}{2L} \pm \sqrt{b}$$

where

$$b = \left(\frac{R}{2L}\right)^2 - \frac{1}{LC}$$

Now if $b < 0$, both poles are complex with real part equal to $-R/2L$, and thus in this case the circuit is stable. If $b \geq 0$, both poles are real. In this case,

$$-\frac{R}{2L} - \sqrt{b} < 0$$

In addition, $b \geq 0$ implies that

$$b < \left(\frac{R}{2L}\right)^2$$

and thus

$$\sqrt{b} < \frac{R}{2L}$$

Therefore,

$$-\frac{R}{2L} + \sqrt{b} < 0$$

and thus the circuit is still stable. So the circuit is stable for any values of $R, L, C > 0$. This means that if an impulsive input voltage $x(t)$ is applied to the circuit with zero initial conditions, the voltage $v_c(t)$ across the capacitor decays to zero as $t \to \infty$.

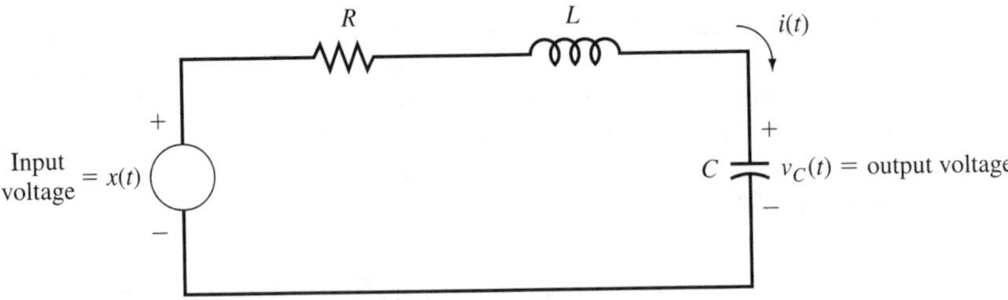

Figure 9.2 Series RLC circuit in Example 9.2.

Alternative Characterizations of Stability

Again consider the system with the rational transfer function $H(s)$ given by (9.1). It turns out that the system's impulse response $h(t)$ will converge to zero as $t \to \infty$ if and only if $h(t)$ is absolutely integrable; that is,

$$\int_0^\infty |h(t)|\, dt < \infty \tag{9.5}$$

Hence stability of the system is equivalent to absolute integrability of the system's impulse response.

Now let $y(t)$ denote the output response of the system resulting from input $x(t)$ applied for $t \geq 0$ with no initial energy in the system at time $t = 0$ (or 0^-). The system is said to be *bounded-input bounded-output (BIBO) stable* if $y(t)$ is bounded whenever the input $x(t)$ is bounded. In mathematical terms, this means that if $|x(t)| \leq c_1$ for all t where c_1 is a finite positive constant, then the resulting output response (with zero initial energy) satisfies the condition

$$|y(t)| \leq c_2 \qquad \text{for all } t$$

where c_2 is a finite positive constant [depending in general on $x(t)$]. It turns out that BIBO stability is equivalent to the integrability condition (9.5). Then since (9.5) is equivalent to the condition that $h(t) \to 0$ as $t \to \infty$, it is seen that BIBO stability is equivalent to stability as defined above. It should be stressed that the validity of this result is based on the assumption that the transfer function $H(s)$ does not have any common poles and zeros.

9.2 ROUTH–HURWITZ STABILITY TEST

By the results in Section 9.1, stability of a system with rational transfer function $H(s) = B(s)/A(s)$ can be checked by first determining the poles of $H(s)$, which are the roots of $A(s) = 0$. The poles of $H(s)$ can be computed by using the MATLAB command `roots`. The use of this command was illustrated in Chapter 8.

It turns out that there are procedures for testing for stability that do not require the computation of the poles of the system. One such procedure is the Routh–Hurwitz stability test, which is based on simple computations involving the coefficients of the polynomial $A(s)$. The details are as follows.

Suppose that

$$A(s) = a_N s^N + a_{N-1} s^{N-1} + \cdots + a_1 s + a_0, \qquad a_N > 0 \tag{9.6}$$

Note that the leading coefficient a_N of $A(s)$ may be any nonzero positive number. By the results in Section 9.1, the system is stable if and only if all the zeros of $A(s)$ are in the open left-half plane (OLHP). A necessary (but in general not sufficient) condition for this to be the case is that all the coefficients of $A(s)$ must be strictly positive; that is,

$$a_i > 0 \qquad \text{for } i = 0, 1, 2, \ldots, N-1 \tag{9.7}$$

Thus, if $A(s)$ has one or more coefficients that are zero or negative, there is at least one pole not in the OLHP and the system is not stable. Here the expression "pole not in the OLHP" means a pole located on the $j\omega$-axis or located in the open right-half plane.

It should be stressed that the condition (9.7) is not a sufficient condition for stability in general. In other words, there are unstable systems for which (9.7) is satisfied.

Now the Routh–Hurwitz stability test will be stated which gives necessary and sufficient conditions for stability. Given the polynomial $A(s)$ defined by (9.6) the first step is to construct the Routh array shown in Table 9.1.

As seen from Table 9.1, the Routh array has $N + 1$ rows, with the rows indexed by the powers of s. The number of columns of the array is $(N/2) + 1$ if N is even or $(N + 1)/2$ if N is odd. The first two rows of the Routh array are filled by the coefficients of $A(s)$, starting with the leading coefficient a_N. The elements in the third row are given by

$$b_{N-2} = \frac{a_{N-1}a_{N-2} - a_N a_{N-3}}{a_{N-1}} = a_{N-2} - \frac{a_N a_{N-3}}{a_{N-1}}$$

$$b_{N-4} = \frac{a_{N-1}a_{N-4} - a_N a_{N-5}}{a_{N-1}} = a_{N-4} - \frac{a_N a_{N-5}}{a_{N-1}}$$

$$\vdots$$

The elements in the fourth row are given by

$$c_{N-3} = \frac{b_{N-2}a_{N-3} - a_{N-1}b_{N-4}}{b_{N-2}} = a_{N-3} - \frac{a_{N-1}b_{N-4}}{b_{N-2}}$$

$$c_{N-5} = \frac{b_{N-2}a_{N-5} - a_{N-1}b_{N-6}}{b_{N-2}} = a_{N-5} - \frac{a_{N-1}b_{N-6}}{b_{N-2}}$$

$$\vdots$$

The other rows (if there are any) are computed in a similar fashion. As a check on the computations, it should turn out that the last nonzero element in each column of the array is equal to the coefficient a_0 of $A(s)$.

TABLE 9.1 ROUTH ARRAY

s^N	a_N	a_{N-2}	a_{N-4}	$\cdots$
s^{N-1}	a_{N-1}	a_{N-3}	a_{N-5}	$\cdots$
s^{N-2}	b_{N-2}	b_{N-4}	b_{N-6}	$\cdots$
s^{N-3}	c_{N-3}	c_{N-5}	c_{N-7}	$\cdots$
$\vdots$	$\vdots$	$\vdots$	$\vdots$	
s^2	d_2	d_0	0	$\cdots$
s^1	e_1	0	0	$\cdots$
s^0	f_0	0	0	$\cdots$

In calculating the Routh array, it may happen that $b_{N-2} = 0$, in which case it is not possible to perform the division in computing the elements in the fourth row. If $b_{N-2} = 0$, set $b_{N-2} = \varepsilon$ (a very small positive number), and then continue. Similarly, if $c_{N-3} = 0$, set $c_{N-3} = \varepsilon$, and continue. If any zero elements are set equal to small positive numbers, the last nonzero element of the columns of the Routh array will, in general, not be equal to a_0.

The Routh–Hurwitz stability test states that the system is stable (all poles in OLHP) if and only if all the elements in the first column of the Routh array are strictly positive (> 0). In addition, the number of poles not in the OLHP is equal to the number of sign changes in the first column. The application of the Routh–Hurwitz stability test is illustrated in the examples given below. The proof of the Routh–Hurwitz stability test is well beyond the scope of this book.

As shown in the following examples, when the degree N of $A(s)$ is less than or equal to 3, the Routh–Hurwitz test can be used to derive simple conditions for stability given directly in terms of the coefficients of $A(s)$.

Example 9.3 *Second-Order Case*

Let $N = 2$ and $a_2 = 1$, so that

$$A(s) = s^2 + a_1 s + a_0$$

The Routh array for this case is given in Table 9.2. The elements in the first column of the Routh array are $1, a_1, a_0$, and thus the poles are in the OLHP if and only if the coefficients a_1 and a_0 are both positive. So in this case, the positive-coefficient condition (9.7) is necessary and sufficient for stability. Now suppose that $a_1 > 0$ and $a_0 < 0$. Then there is one sign change in the first column of the Routh array, which means that there is one pole not in the OLHP. If $a_1 < 0$ and $a_0 < 0$, there still is one sign change and thus there still is one pole not in the OLHP. If $a_1 < 0$ and $a_0 > 0$, there are two sign changes in the first column, and therefore both poles are not in the OLHP.

Example 9.4 *Third-Order Case*

Consider the third-order case

$$A(s) = s^3 + a_2 s^2 + a_1 s + a_0$$

The Routh array is displayed in Table 9.3. Since

$$a_1 - \frac{a_0}{a_2} > 0$$

TABLE 9.2 ROUTH ARRAY IN THE $N = 2$ CASE

s^2	1	a_0
s^1	a_1	0
s^0	$\dfrac{a_1 a_0 - (1)(0)}{a_1} = a_0$	0

TABLE 9.3 THE $N = 3$ CASE

s^3	1	a_1
s^2	a_2	a_0
s^1	$\dfrac{a_2a_1 - (1)a_0}{a_2} = a_1 - \dfrac{a_0}{a_2}$	0
s^0	a_0	0

if and only if

$$a_1 > \frac{a_0}{a_2}$$

all three poles are in the OLHP if and only if

$$a_2 > 0, \qquad a_1 > \frac{a_0}{a_2}, \qquad a_0 > 0$$

This result shows that when $N = 3$, it is not true in general that positivity of a_2, a_1 and a_0 implies that the system is stable. Note that if $a_2 < 0$ and $a_0 > 0$, there are two sign changes in the first column of the Routh array, and thus there are two poles not in the OLHP. If $a_2 < 0$, $a_1 > a_0/a_2$, and $a_0 < 0$, there are three sign changes, and therefore all three poles are not in the OLHP. If $a_2 < 0, a_1 < a_0/a_2$, and $a_0 < 0$, there is one sign change, which means that there is one pole not in the OLHP.

As N is increased above the value $N = 3$, the conditions for stability in terms of the coefficients of $A(s)$ get rather complicated. For $N \geq 4$, the Routh–Hurwitz test can still be applied on a case-by-case basis.

Example 9.5 *Higher-Order Case*

Suppose that $H(s) = B(s)/A(s)$, where

$$A(s) = 6s^5 + 5s^4 + 4s^3 + 3s^2 + 2s + 1$$

Then $N = 5$ and

$$a_0 = 1, \quad a_1 = 2, \quad a_2 = 3, \quad a_3 = 4, \quad a_4 = 5, \quad a_5 = 6$$

The Routh array for this example is shown in Table 9.4. There are two sign changes in the first column of the Routh array, and thus two of the five poles are not located in the open left-half plane. The system is therefore not stable.

The Routh–Hurwitz test can also be used to determine if there is a pair of complex poles located on the $j\omega$-axis: Given a rational transfer function $H(s) = B(s)/A(s)$, there is a pair of poles on the $j\omega$-axis with all other poles in the open left-half plane if and only if all the entries in the first column of the Routh array are strictly positive, except for the entry in the row indexed by s^1, which is zero. If this is the case, there is a pair of poles at $\pm j\sqrt{a_0/\gamma_2}$, where γ_2 is the entry in the first row of the Routh array indexed by s^2 and a_0 is the constant term of $A(s)$.

TABLE 9.4 ROUTH ARRAY FOR EXAMPLE 9.5

s^5	$a_5 = 6$	$a_3 = 4$	$a_1 = 2$
s^4	$a_4 = 5$	$a_2 = 3$	$a_0 = 1$
s^3	$\dfrac{(5)(4) - (6)(3)}{5} = 0.4$	$\dfrac{(5)(2) - (6)(1)}{5} = 0.8$	0
s^2	$\dfrac{(0.4)(3) - (5)(0.8)}{0.4} = -7$	$a_0 = 1$	0
s^1	$\dfrac{(-7)(0.8) - (0.4)(1)}{-7} = 6/7$	0	0
s^0	$a_0 = 1$	0	0

Example 9.6 *Fourth-Order Case*

Suppose that

$$A(s) = s^4 + s^3 + 3s^2 + 2s + 2$$

The Routh array for this example is shown in Table 9.5. Note that all the entries in the first column of the array are strictly positive except for the entry in the row indexed by s^1, which is zero. As a result, two of the poles are in the open left-half plane, and the other two poles are on the $j\omega$-axis, located at $\pm j\sqrt{2/1} = \pm j\sqrt{2}$.

9.3 ANALYSIS OF THE STEP RESPONSE

Again consider the system with the rational transfer function $H(s) = B(s)/A(s)$, where the degree of $B(s)$ is less than or equal to the degree of $A(s)$. If an input $x(t)$ is applied to the system for $t \geq 0$ with no initial energy in the system, then from the results in Section 8.5 the transform $Y(s)$ of the resulting output response is given by

$$Y(s) = \frac{B(s)}{A(s)} X(s) \tag{9.8}$$

Now suppose that $x(t)$ is the unit-step function $u(t)$, so that $X(s) = 1/s$. Then inserting $X(s) = 1/s$ into (9.8) results in the transform of the step response

$$Y(s) = \frac{B(s)}{A(s)s} \tag{9.9}$$

TABLE 9.5 ROUTH ARRAY FOR EXAMPLE 9.6

s^4	1	3	2
s^3	1	2	0
s^2	$\dfrac{3-2}{1} = 1$	2	0
s^1	$\dfrac{2-2}{1} \approx \varepsilon$	0	0
s^0	$\dfrac{2\varepsilon - 0}{\varepsilon} = 2$	0	0

Then if $A(0) \neq 0$, "pulling out" the $1/s$ term from $Y(s)$ and using the residue formula (8.67) yields

$$Y(s) = \frac{E(s)}{A(s)} + \frac{c}{s} \tag{9.10}$$

where $E(s)$ is a polynomial in s and c is the constant given by

$$c = [sY(s)]_{s=0} = H(0)$$

Then taking the inverse Laplace transform of $Y(s)$ results in the step response:

$$y(t) = y_1(t) + H(0), \qquad t \geq 0 \tag{9.11}$$

where $y_1(t)$ is the inverse Laplace transform of $E(s)/A(s)$.

Note that if the system is stable so that all the roots of $A(s) = 0$ are in the open left-half plane, the term $y_1(t)$ in (9.11) converges to zero as $t \to \infty$, in which case $y_1(t)$ is the *transient part of the response.* So if the system is stable, the step response contains a transient that decays to zero and it contains a constant with value $H(0)$. The constant $H(0)$ is the *steady-state* value of the step response.

It is very important to note that since the transform of the term $y_1(t)$ in (9.11) is equal to $E(s)/A(s)$, the form of $y_1(t)$ depends on the system's poles [the poles of $H(s)$]. This is examined in detail below, beginning with first-order systems.

First-Order Systems

Consider the system given by the first-order transfer function

$$H(s) = \frac{k}{s - p} \tag{9.12}$$

where k is a real constant and p is the pole (which is real). Then with $x(t) = u(t)$, the transform of $y(t)$ is equal to $H(s)/s$ and the partial fraction expansion for $Y(s)$ is

$$Y(s) = \frac{-k/p}{s} + \frac{k/p}{s - p}$$

Taking the inverse Laplace transform of $Y(s)$ yields the step response:

$$y(t) = -\frac{k}{p}(1 - e^{pt}), \qquad t \geq 0 \tag{9.13}$$

In this case, the step response $y(t)$ can be expressed in the form (9.11) with

$$y_1(t) = \frac{k}{p}e^{pt}, \qquad t \geq 0$$

$$H(0) = -\frac{k}{p}$$

Note that the time behavior of the term $y_1(t) = (k/p)\exp(pt)$ depends directly on the location of the pole p in the complex plane. In particular, if p is in the right-half plane, the system is unstable and $y_1(t)$ grows without bound. Furthermore, the farther the pole is to the right in the right-half plane, the faster the rate of growth of $y_1(t)$. On the other hand, if the system is stable, so that p lies in the open left-half plane, then $y_1(t)$ decays to zero and thus $y_1(t)$ is the transient part of the response. Note that the rate at which the transient decays to zero depends on how far over to the left the pole is in the open left-half plane. Also, since the step response is equal to $y_1(t) - k/p$, the rate at which the step response converges to the constant $-k/p$ is equal to the rate at which the transient decays to zero. In the example below, these properties of the step response are verified by using the `step` command introduced in Chapter 8.

Example 9.7 *First-Order System*

Pole Positions and Step Response

Consider the first-order system given by the transfer function (9.12) with $k = 1$. Given any specific value for p, the MATLAB commands to generate the step response are

```
num = 1; den = [1 -p];
t = 0:0.05:10;
y = step(num,den,t);
```

The step responses for $p = 1, 2,$ and 3 are displayed in Figure 9.3. Note that in all three cases the step response grows without bound, which shows unstable behavior. Note also

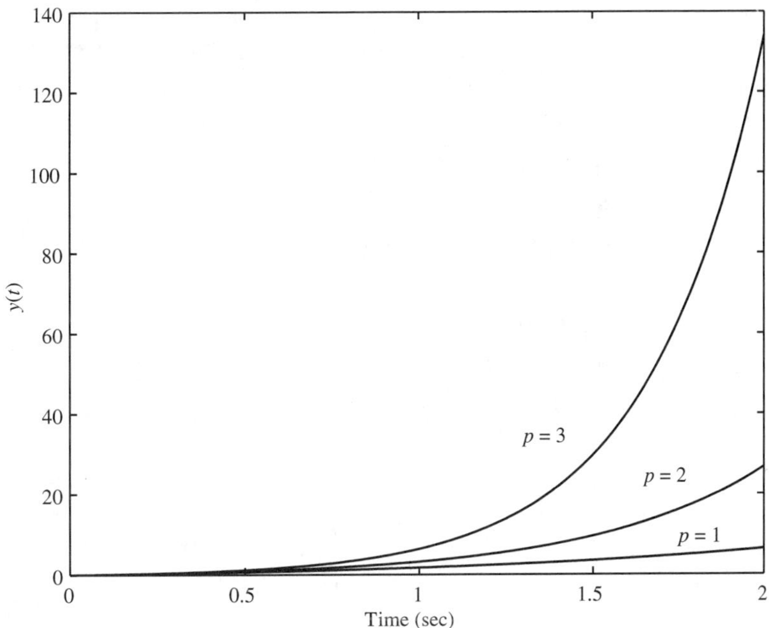

Figure 9.3 Step response when $p = 1, 2, 3$.

that the response for $p = 3$ grows with the fastest rate, since the pole $p = 3$ is farther to the right in the right-half plane (in comparison to $p = 1$ or $p = 2$).

If the pole p is negative, then the system is stable and $y_1(t) = (k/p)\exp(pt)$ is the transient part of the step response. In this case the step response will converge to the constant value $H(0) = -k/p$. For $k = -p$ [which yields $H(0) = 1$] and $p = -1, -2, -3$, the step response is plotted in Figure 9.4. As seen in figure, the step response approaches the steady-state value of 1 at a faster rate as p becomes more negative, that is, as the pole moves farther to the left in the open left-half plane. This also corresponds to the fact that the rate of decay to zero of the transient is fastest for the case $p = -5$.

An important quantity that characterizes the rate of decay to zero of the transient part of a response is the *time constant* τ, which is defined to be the amount of time that it takes for the transient to decay to $1/e$ ($\approx 37\%$) of its initial value. Since the transient for the first-order system (9.12) is equal to $(k/p)\exp(pt)$, the time constant τ is equal to $-1/p$ assuming that $p < 0$. To verify that τ is equal to $-1/p$, first let $y_{tr}(t)$ denote the transient so that $y_{tr}(t) = (k/p)\exp(pt)$. Then setting $t = \tau = -1/p$ in $y_{tr}(t)$ gives

$$y_{tr}(\tau) = \frac{k}{p}e^{p(-1/p)} = \frac{k}{p}e^{-1} = \frac{1}{e}y_{tr}(0)$$

In Example 9.7 the time constants for $p = -1, -2$, and -5 are $\tau = 1, 0.5$, and 0.2 second, respectively. Note that the smaller the value of τ, the faster the rate of decay of the transient.

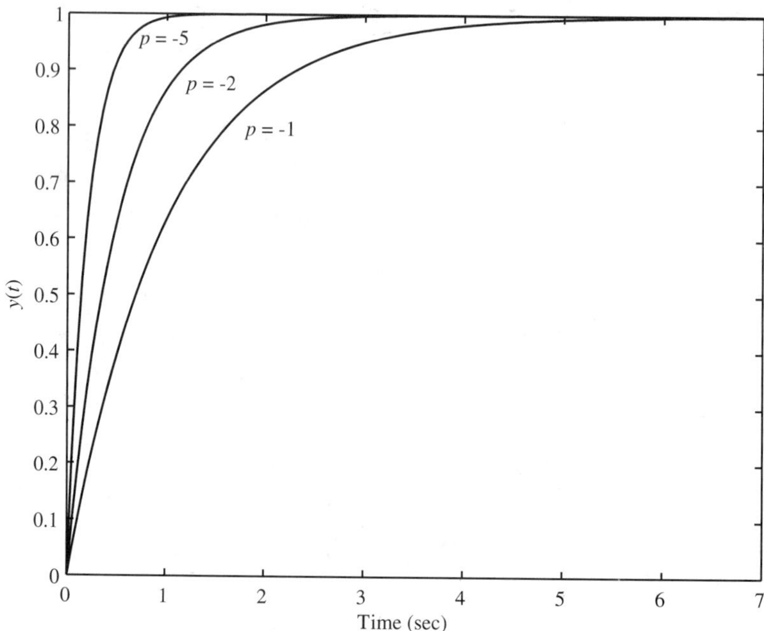

Figure 9.4 Step response when $p = -1, -2, -5$.

Example 9.8 *Determining the Pole Location from the Step Response*

Consider a first-order system $H(s) = k/(s - p)$ with the step response shown in Figure 9.5. From the plot it is possible to determine both k and the pole position (i.e., the value of p). First, since the step response displayed in Figure 9.5 is bounded, the system must be stable, and thus p must be negative. From the plot it is seen that the steady-state value of the step response is equal to 2. Hence $H(0) = -k/p = 2$, and from (9.13) the step response is

$$y(t) = 2(1 - e^{pt}) \tag{9.14}$$

Now from the plot in Figure 9.5, $y(0.1) = 1.73$, and thus evaluating both sides of (9.14) at $t = 0.1$ gives

$$y(0.1) = 1.73 = 2[1 - e^{p(0.1)}] \tag{9.15}$$

Solving (9.15) for p gives $p = -20$.

Second-Order Systems

Now consider the second-order system given by the transfer function

$$H(s) = \frac{k}{s^2 + 2\zeta\omega_n s + \omega_n^2} \tag{9.16}$$

The real parameter ζ in (9.16) is called the *damping ratio* and the real parameter ω_n is called the *natural frequency*. The reason for this terminology will become clear

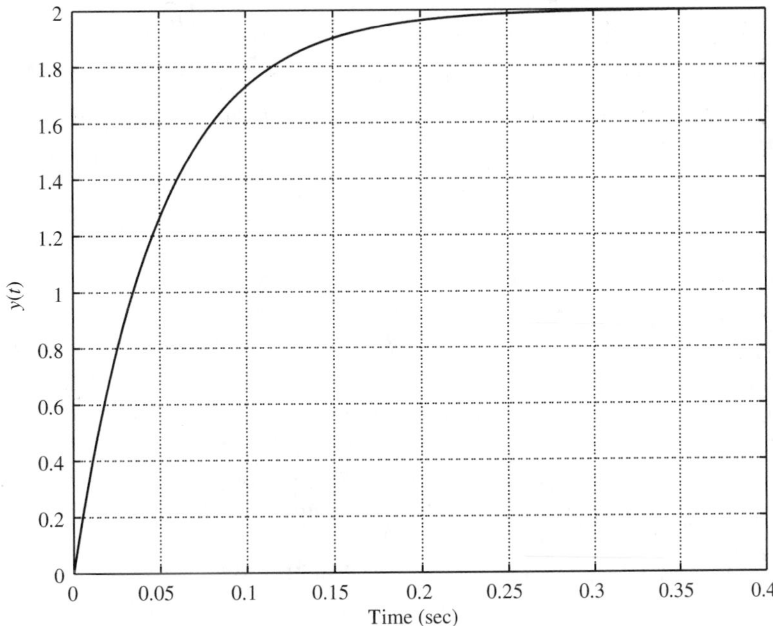

Figure 9.5 Step response in Example 9.8

from the results given below. In the following development it is assumed that $\zeta > 0$ and $\omega_n > 0$, and thus by the Routh–Hurwitz criterion, the system is stable.

Using the quadratic formula reveals that the poles of $H(s)$ are

$$p_1 = -\zeta\omega_n + \omega_n\sqrt{\zeta^2 - 1} \tag{9.17}$$

$$p_2 = -\zeta\omega_n - \omega_n\sqrt{\zeta^2 - 1} \tag{9.18}$$

From (9.17) and (9.18) it is seen that both the poles are real when $\zeta > 1$, the poles are real and repeated when $\zeta = 1$, and the poles are a complex-conjugate pair when $0 < \zeta < 1$. The step response for these three cases is considered below.

Case when both poles are real. When $\zeta > 1$, the poles p_1 and p_2 given by (9.17) and (9.18) are both real and nonrepeated, in which case $H(s)$ can be expressed in the factored form

$$H(s) = \frac{k}{(s - p_1)(s - p_2)} \tag{9.19}$$

The transform $Y(s)$ of the step response is then given by

$$Y(s) = \frac{k}{(s - p_1)(s - p_2)s}$$

Performing a partial fraction expansion on $Y(s)$ yields the step response:

$$y(t) = \frac{k}{p_1 p_2}(k_1 e^{p_1 t} + k_2 e^{p_2 t} + 1), \qquad t \geq 0 \tag{9.20}$$

where k_1 and k_2 are real constants whose values depend on the poles p_1 and p_2. Hence in this case the transient part $y_{tr}(t)$ of the step response is a sum of two exponentials given by

$$y_{tr}(t) = \frac{k}{p_1 p_2}(k_1 e^{p_1 t} + k_2 e^{p_2 t}), \qquad t \geq 0$$

and the steady-state value of the step response is

$$H(0) = \frac{k}{\omega_n^2} = \frac{k}{p_1 p_2}$$

One of the exponential terms in (9.20) usually dominates the other exponential term; that is, the magnitude of one of the exponential terms is often much larger than the other. In this case the pole corresponding to the dominant exponential term is called the *dominant pole.* The dominant pole is usually the one nearest the imaginary axis since it has the largest time constant (equal to $-1/p$, where p is the dominant pole). If one of the poles is dominant, the transient part of the step response (9.20) looks similar to the transient part of the step response in the first-order case considered above.

Example 9.9 *Case When Both Poles are Real*

Pole Positions and Step Response

In (9.19), let $k = 2$, $p_1 = -1$, and $p_2 = -2$. Then expanding $H(s)/s$ via partial fractions gives

$$Y(s) = \frac{-2}{s+1} + \frac{1}{s+2} + \frac{1}{s}$$

Thus the step response is

$$y(t) = -2e^{-t} + e^{-2t} + 1, \qquad t \geq 0$$

and the transient response is

$$y_{\text{tr}}(t) = -2e^{-t} + e^{-2t}, \qquad t \geq 0$$

The step response obtained from MATLAB is shown in Figure 9.6. In this example it turns out that the transient response is dominated by p_1 since the term in the transient due to p_2 decays faster. The response displayed in Figure 9.6 is similar to the response of the first-order system with $p = -1$ shown in Figure 9.4.

Case when poles are real and repeated. When $\zeta = 1$, the poles p_1 and p_2 given by (9.17) and (9.18) are real and are both equal to $-\omega_n$. In this case the transfer function $H(s)$ given by (9.16) has the factored form

$$H(s) = \frac{k}{(s+\omega_n)^2} \tag{9.21}$$

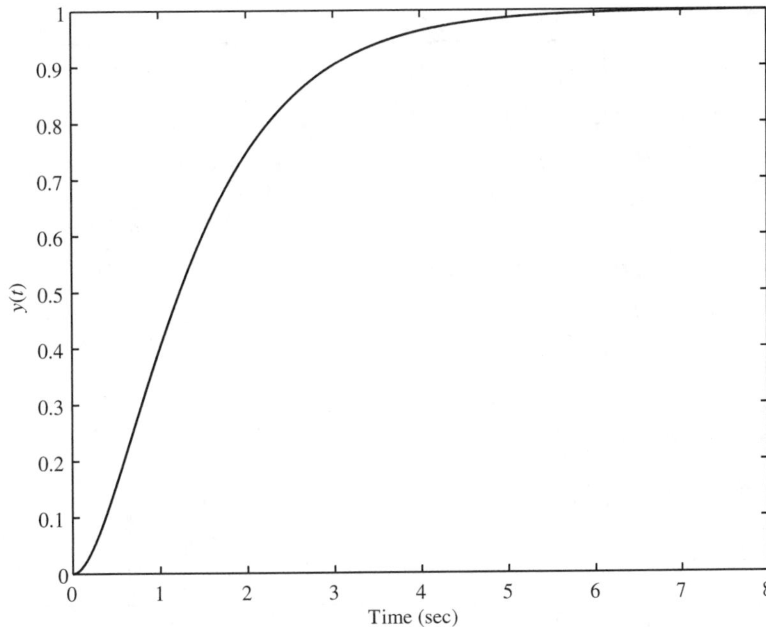

Figure 9.6 Step response of system in Example 9.9.

Then expanding $H(s)/s$ via partial fractions and taking the inverse transform yields the step response:

$$y(t) = \frac{k}{\omega_n^2}[1 - (1 + \omega_n t)e^{-\omega_n t}], \qquad t \geq 0 \tag{9.22}$$

Hence in this case the transient response is

$$y_{\text{tr}}(t) = -\frac{k}{\omega_n^2}(1 + \omega_n t)e^{-\omega_n t}, \qquad t \geq 0$$

Example 9.10 *Both Poles Real and Repeated*

Pole Positions and Step Response

In (9.21), let $k = 4$ and $\omega_n = 2$. Then both poles are equal to -2 and from (9.22) the step response is

$$y(t) = 1 - (1 + 2t)e^{-2t}, \qquad t \geq 0$$

A MATLAB plot of the step response is given in Figure 9.7.

Case when poles are a complex pair. Now suppose that $0 < \zeta < 1$, so that the poles p_1 and p_2 are a complex pair. With $\omega_d = \omega_n\sqrt{1 - \zeta^2}$, the poles are $p_1, p_2 = -\zeta\omega_n \pm j\omega_d$. Note that the real part of the poles is equal to $-\zeta\omega_n$ and the imaginary part of the poles is equal to $\pm\omega_d$.

Then given the transfer function (9.16), completing the square in the denominator of $H(s)$ yields

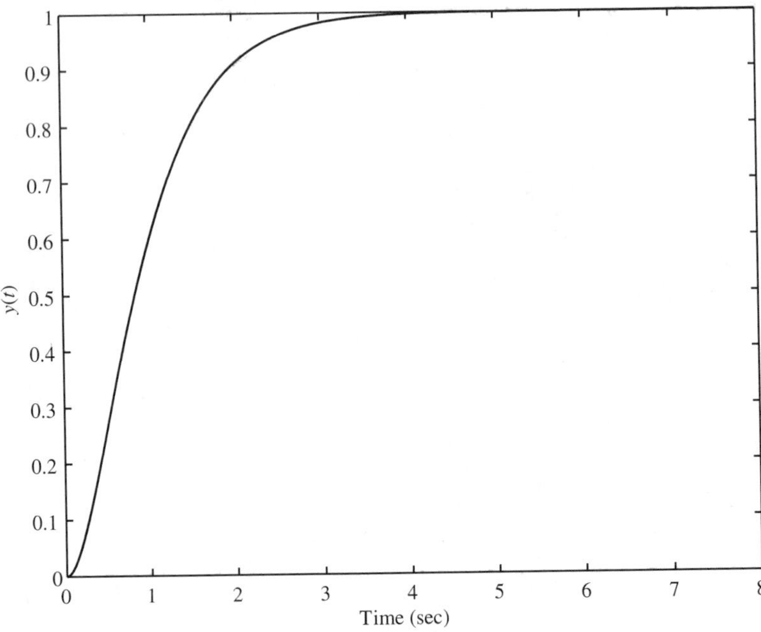

Figure 9.7 Step response in Example 9.10.

$$H(s) = \frac{k}{(s + \zeta\omega_n)^2 + \omega_d^2} \tag{9.23}$$

Expanding $Y(s) = H(s)/s$ gives

$$\begin{aligned} Y(s) &= \frac{-(k/\omega_n^2)s - 2k\zeta/\omega_n}{(s + \zeta\omega_n)^2 + \omega_d^2} + \frac{k/\omega_n^2}{s} \\ &= \frac{-(k/\omega_n^2)(s + \zeta\omega_n)}{(s + \zeta\omega_n)^2 + \omega_d^2} - \frac{(k\zeta/\omega_n)s}{(s + \zeta\omega_n)^2 + \omega_d^2} + \frac{k/\omega_n^2}{s} \end{aligned}$$

Thus, from Table 8.2 the step response is

$$y(t) = -\frac{k}{\omega_n^2} e^{-\zeta\omega_n t} \cos \omega_d t - \frac{k\zeta}{\omega_n \omega_d} e^{-\zeta\omega_n t} \sin \omega_d t + \frac{k}{\omega_n^2}, \qquad t \geq 0$$

Finally, using the trigonometric identity

$$C \cos \beta + D \sin \beta = \sqrt{C^2 + D^2} \sin(\beta + \theta), \qquad \text{where } \theta = \tan^{-1}(C/D)$$

results in the following form for the step response:

$$y(t) = \frac{k}{\omega_n^2}\left[1 - \frac{\omega_n}{\omega_d} e^{-\zeta\omega_n t} \sin(\omega_d t + \phi)\right], \qquad t \geq 0 \tag{9.24}$$

where $\phi = \tan^{-1}(\omega_d/\zeta\omega_n)$. Note that the steady-state value is equal to k/ω_n^2 and the transient response is an exponentially decaying sinusoid with frequency ω_d rad/sec. Thus second-order systems with complex poles have an oscillatory step response with the frequency of the oscillation equal to ω_d.

Example 9.11 *Poles are a Complex Pair*

Pole Positions and Step Response

Consider the second-order system given by the transfer function

$$H(s) = \frac{17}{s^2 + 2s + 17}$$

Writing $H(s)$ in the form (9.23) reveals that $k = 17$, $\zeta = 0.242$, and $\omega_n = \sqrt{17}$. Also, $\omega_d = 4$ and the poles of the system are $-1 \pm j4$. The step response is found from (9.24) to be given by

$$y(t) = 1 - \frac{\sqrt{17}}{4} e^{-t} \sin(4t + 1.326)$$

The step response can be obtained numerically from MATLAB using the commands

```
num = 17; den = [1 2 17];
t = 0:10/300:10;
y = step(num,den,t);
```

The step response is shown in Figure 9.8. Notice that the response oscillates with frequency $\omega_d = 4$ rad/sec and that the oscillations decay exponentially. An envelope corresponding to the decay of the transient part of the step response is shown as a dashed line in Figure 9.8. As seen from (9.24), the rate of decay of the transient is

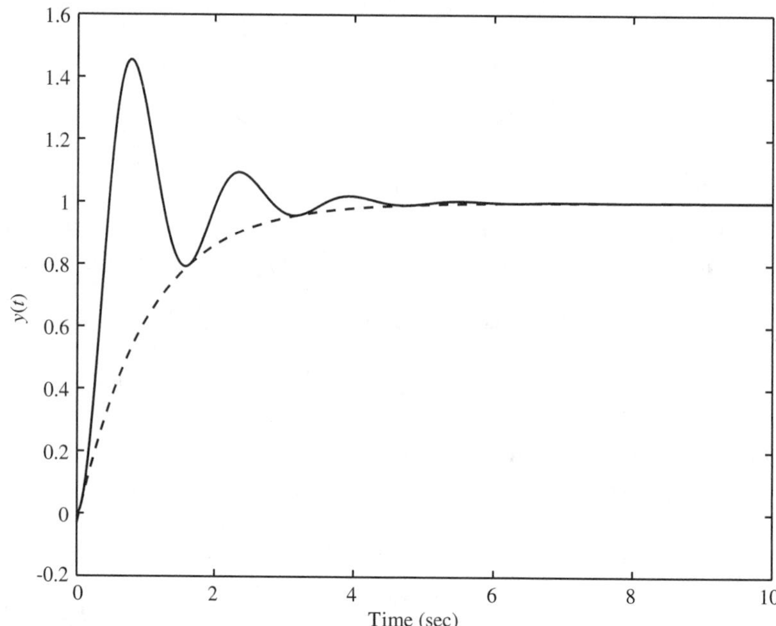

Figure 9.8 Step response in Example 9.11.

determined by the real part of the poles, $\zeta\omega_n = -1$. The time constant corresponding to the poles is equal to $1/\zeta\omega_n = 1$ sec.

As noted above, when $0 < \zeta < 1$, the poles are given by the complex pair $-\zeta\omega_n \pm j\omega_d$, where $\omega_d = \omega_n\sqrt{1 - \zeta^2}$. The location of the poles in the complex plane is illustrated in Figure 9.9. As shown in the figure, ω_n is equal to the distance from the origin to the poles and ζ is equal to the cosine of the angle formed from the negative real axis. If ω_n is held constant (at some strictly positive value) and ζ is varied from one to zero, the pole positions trace a circular arc in the left-half plane starting on the negative real axis (when $\zeta = 1$) and ending on the imaginary axis (when $\zeta = 0$). This is illustrated by the dashed line in Figure 9.9. As a general rule, the closer the poles are to the $j\omega$-axis, the more oscillatory the response is. Hence, as ζ is decreased from 1 to 0 (with ω_n held constant), the step response becomes more oscillatory. This is verified by the following example.

Example 9.12 *Effect of Damping Ratio on the Step Response*

Consider the transfer function (9.23) with $\omega_n = 1$ and $k = 1$. The step responses for $\zeta = 0.1$, $\zeta = 0.25$, and $\zeta = 0.7$ are shown in Figure 9.10. Note that the smaller the value of ζ, the more pronounced the oscillation is.

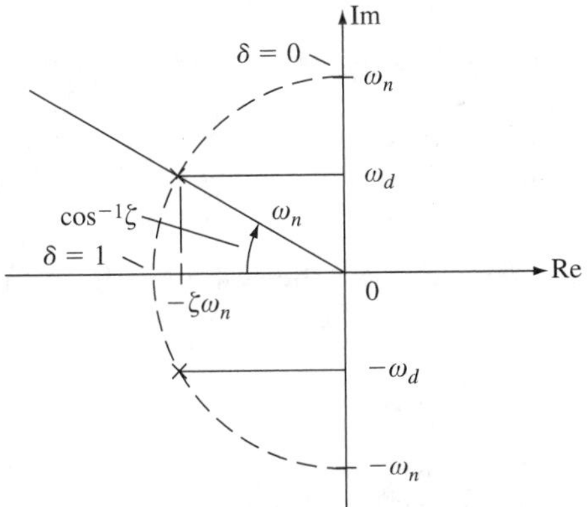

Figure 9.9 Location of poles in the complex plane.

Again consider the step response given by (9.24). In addition to the parameter ζ, the value of the natural frequency ω_n also has a substantial effect on the response. To see this, suppose that ζ is held constant and ω_n is varied. Since ζ determines the angle of the pole in polar coordinates (see Figure 9.9), keeping ζ constant will keep the angle constant, and thus in the plot of the poles, increasing the value of ω_n will generate a radial line starting from the origin and continuing outward into the left-half plane (see Figure 9.9). It follows that the transient response should decay faster and the frequency of the oscillation should increase as ω_n is increased. This is investigated in the next example.

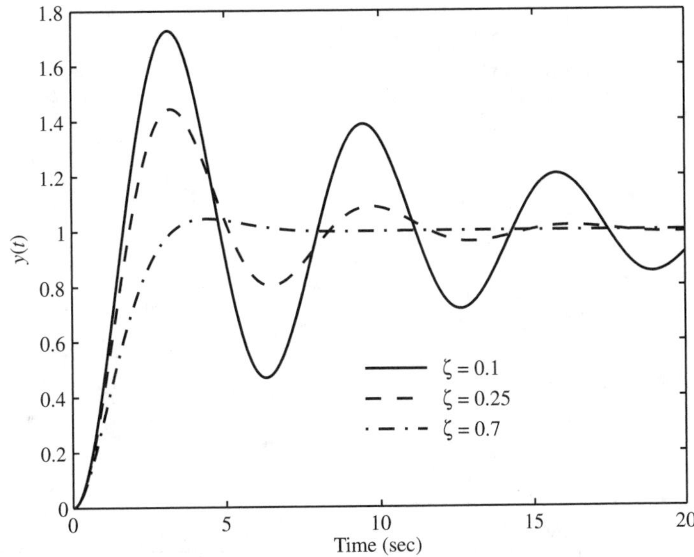

Figure 9.10 Step responses in Example 9.12.

Example 9.13 *Effect of ω_n on Step Response*

Consider the transfer function (9.23) with $\zeta = 0.4$ and $k = \omega_n^2$ (so that the steady-state value is equal to 1). The step responses for $\omega_n = 0.5, 1$, and 2 rad/sec are shown in Figure 9.11. Note that the larger the value of ω_n, the smaller the time constant and the higher the frequency of oscillations. Note also that since ζ is kept constant, the peak values of the oscillations are the same for each value of ω_n.

Comparison of cases. Again consider the system with transfer function

$$H(s) = \frac{k}{s^2 + 2\zeta\omega_n s + \omega_n^2} \tag{9.25}$$

From the results above it is seen that when $0 < \zeta < 1$, the transient part of the step response is oscillatory with "damped natural frequency" equal to $\omega_n\sqrt{1 - \zeta^2}$, and the oscillation is more pronounced as ζ is decreased to zero. For $\zeta \geq 1$ there is no oscillation in the transient. The existence of the oscillation implies a lack of "damping" in the system, and thus ζ does give a measure of the degree of damping in the system. When $0 < \zeta < 1$, the system is said to be *underdamped* since in this case the damping ratio ζ is not large enough to prevent an oscillation in the transient resulting from a step input. When $\zeta > 1$, the system is said to be *overdamped,* since in this case ζ is larger than necessary to prevent an oscillation in the transient. When $\zeta = 1$

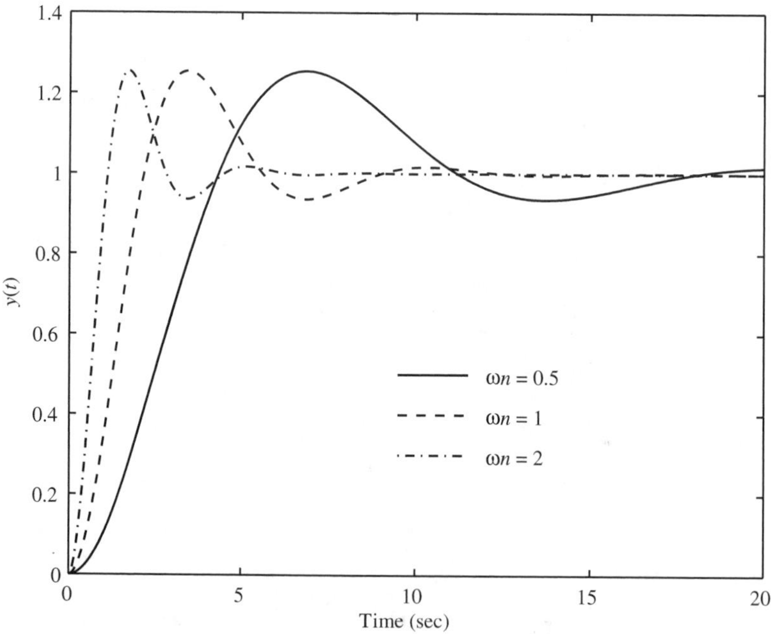

Figure 9.11 Step response in Example 9.13.

the system is said to be *critically damped* since this is the smallest value of ζ for which there is no oscillation in the transient response to a step input.

Example 9.14 *Comparison of Cases*

Consider the transfer function (9.25) with $k = 4$ and $\omega_n = 2$. To compare the underdamped, critically damped, and overdamped cases, the step response of the system will be computed for $\zeta = 0.5$, 1, and 1.5. The results are shown in Figure 9.12. Note that if the overshoot can be tolerated, the fastest response displayed in Figure 9.12 is the one for which $\zeta = 0.5$. Here "the fastest" refers to the step response that reaches the steady-state value (equal to 1 here) in the fastest time of all three of the responses shown in Figure 9.12.

Returning to the system with transfer function $H(s)$ given by (9.25), it is worth noting that if $\zeta < 0$ and $\omega_n > 0$, both of the poles are in the open right-half plane, and thus the system is unstable. In this case the "transient part" of the step response will grow without bound as $t \to \infty$. Hence the transient is not actually a transient since it does not decay to zero as $t \to \infty$. The transient will decay to zero if and only if the system is stable. This follows directly from the analysis of stability given in Section 9.1.

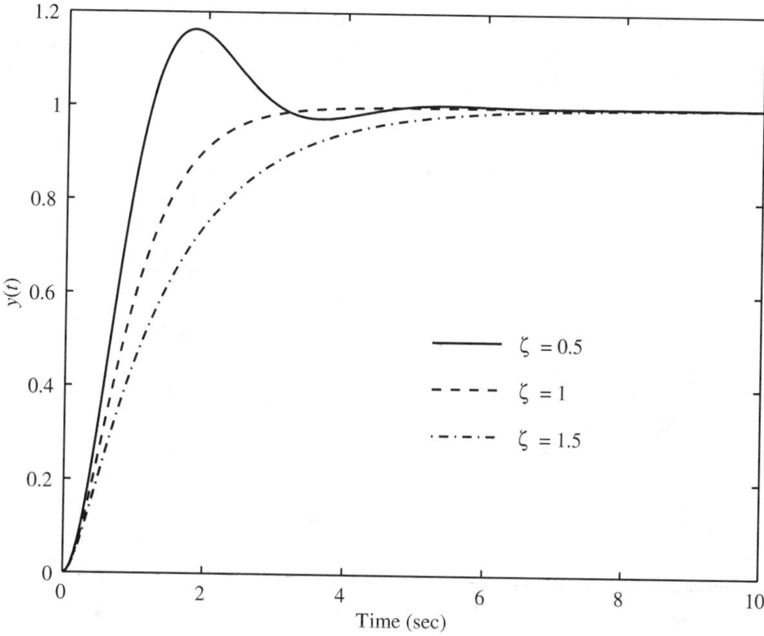

Figure 9.12 Step responses in Example 9.14.

Example 9.15 *Mass–Spring–Damper System*

Mass–Spring–Damper System

Consider the mass–spring–damper system (see Example 8.33) with the input/output differential equation

$$M\frac{d^2y(t)}{dt^2} + D\frac{dy(t)}{dt} + Ky(t) = x(t)$$

where M is the mass, D is the damping constant, K is the stiffness constant, $x(t)$ is the force applied to the mass, and $y(t)$ is the displacement of the mass relative to the equilibrium position. It is assumed that M, D, and K are strictly positive real numbers (greater than 0), which is the condition for stability of the system. The transfer function of the system is given by

$$H(s) = \frac{1}{Ms^2 + Ds + K} = \frac{1/M}{s^2 + (D/M)s + (K/M)} \tag{9.26}$$

Equating the coefficients of the polynomials in the denominators of (9.25) and (9.26) results in the relationships

$$2\zeta\omega_n = \frac{D}{M} \text{ and } \omega_n^2 = \frac{K}{M}$$

Solving for the damping ratio ζ and the natural frequency ω_n yields

$$\zeta = \frac{D}{2\sqrt{MK}},\ \omega_n = \sqrt{\frac{K}{M}}$$

Note that the damping ratio ζ is directly proportional to the damping constant D, and thus the damping in the system is a result of the term $Ddy(t)/dt$ in the input/output differential equation. In particular, there is no damping in the system if $D = 0$. The system is underdamped when

$$0 < \frac{D}{2\sqrt{MK}} < 1$$

which is equivalent to the following condition on the damping constant D:

$$0 < D < 2\sqrt{MK}$$

The system is critically damped when $D = 2\sqrt{MK}$, and the system is overdamped when $D > 2\sqrt{MK}$. The reader is invited to check out animations of the step response for these three cases by using the on-line demo on the Web site. To generate the three cases, values of M, D, and K need to be selected based on the ranges for D given above.

Higher-Order Systems

Higher-order systems can sometimes be approximated as first- or second-order systems since one or two of the poles are usually more dominant than the other poles, and thus these other poles can simply be neglected. An example where one pole dominates over another pole occurred in Example 9.9. In this example the two-pole system behaves similarly to the one-pole system with the dominant pole, and thus the system can be approximated by the dominant pole.

There are two situations when care must be exercised in carrying out system approximation based on the concept of dominant poles. First, if the dominant poles are not significantly different from the other poles, the approximation made by neglecting the faster poles (i.e., the poles with the smaller time constants) may not be very accurate. Second, a zero near a pole causes the residue of the pole to be small, making the magnitude of the corresponding term in the transient response small. Hence, although such a pole may appear to be dominant, it in fact is not. This is illustrated in the following example.

Example 9.16 *Third-Order System*

Consider the third-order system with the following transfer function:

$$H(s) = \frac{25}{(s^2 + 7s + 25)(s + 1)} \tag{9.27}$$

The plot of the poles is given in Figure 9.13. From the plot it is seen that the pole at $s = -1$ is the most dominant since it is closest to the imaginary axis. Thus it should be

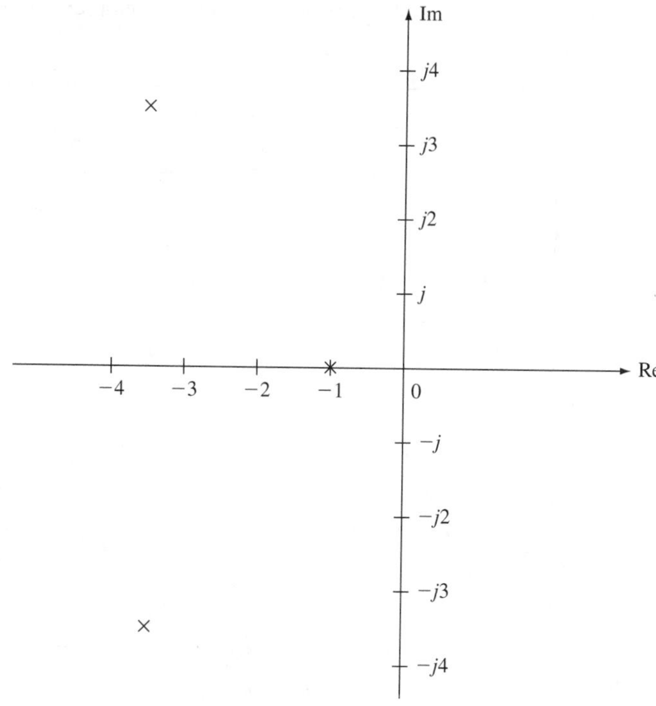

Figure 9.13 Location of poles in Example 9.16.

possible to neglect the other two complex poles. To verify that this can be done, the step response of the system will be calculated by first expanding $Y(s) = H(s)/s$ in the partial fraction expansion:

$$Y(s) = \frac{1}{s} - 1.316\frac{1}{s+1} + 0.05263\frac{6s+17}{s^2+7s+25}$$

Taking the inverse transform gives the step response:

$$y(t) = 1 - 1.316e^{-t} + 0.321e^{-3.5t}\sin(3.57t + 1.754), \qquad t \geq 0 \tag{9.28}$$

Note that the second term on the right-hand side of (9.28) is larger than the third term and will decay slower making it a dominant term. So this corresponds to the observation made above that the pole at $s = -1$ is dominant. To check this out further, the step response will be obtained from MATLAB by using the commands

```
num1 = 25; den1 = [1 7 25];
num2 = 1; den2 = [1 1];
num = conv(num1,num2);        % this multiplies the polynomials
den = conv(den1,den2);
t = 0:0.01:0.4;
y = step(num,den,t);
```

The resulting step response, shown in Figure 9.14, is very similar to a simple exponential first-order response with a pole at $s = -1$. Hence this again confirms the observation that the pole at $s = -1$ is dominant, and thus the other two poles can be neglected.

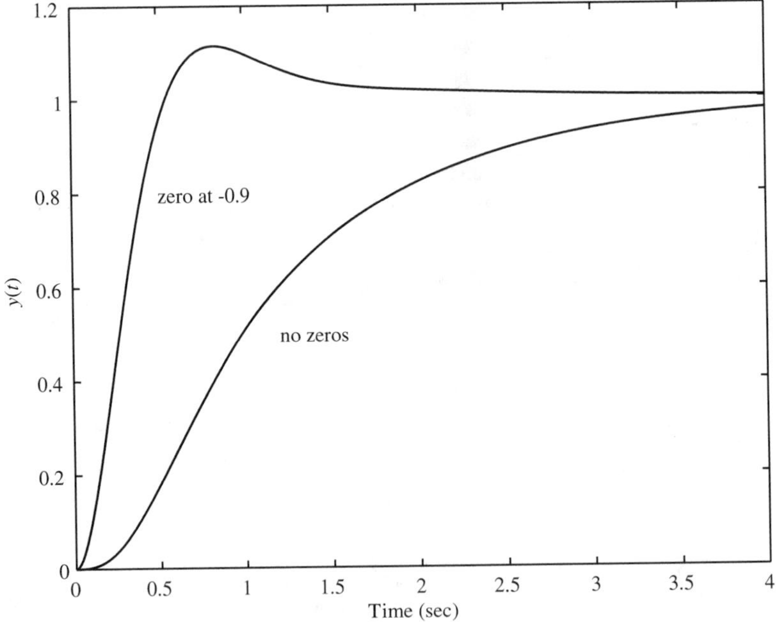

Figure 9.14 Step response in Example 9.16.

Now suppose that a zero at $s = -0.9$ is added to the transfer function (9.27) and the constant in the numerator is adjusted so that the steady-state value is still 1. The resulting transfer function is

$$H(s) = \frac{(25/0.9)s + 25}{(s^2 + 7s + 25)(s + 1)}$$

Note that the zero at $s = -0.9$ very nearly cancels the pole at $s = -1$. The partial fraction expansion of $H(s)/s$ is then given by

$$Y(s) = \frac{1}{s} + 0.1462\frac{1}{s + 1} + \frac{-1.146s - 7.87}{s^2 + 7s + 25}$$

Note that the residue corresponding to the pole -1 is now very small. This is a result of the zero being near the pole. Taking the inverse Laplace transform of $Y(s)$ then gives

$$y(t) = 1 + 0.1462e^{-t} + 1.575e^{-3.5t}\sin(3.57t - 2.327), \qquad t \geq 0$$

Since the coefficient multiplying e^{-t} is much smaller in this case, the pole at $s = -1$ is less significant than the other two poles, and thus the dominant poles are now the complex-conjugate pair (at $s = -3.5 \pm 3.57j$) even though they are farther from the imaginary axis than the pole -1. The resulting step response obtained from MATLAB is shown in Figure 9.14. The reader is invited to compare this response to a second-order system with poles at $s = -3.5 \pm 3.57j$.

9.4 RESPONSE TO SINUSOIDS AND ARBITRARY INPUTS

Again consider the system with the rational transfer function $H(s) = B(s)/A(s)$ with the degree of $B(s)$ less than the degree of $A(s)$. In the first part of this section the objective is to determine the output response of the system when $x(t)$ is the sinusoid

$$x(t) = C\cos\omega_0 t, \qquad t \geq 0$$

where the magnitude C and the frequency ω_0 (in rad/sec) are arbitrary constants. From Table 8.2, the Laplace transform of the input is

$$X(s) = \frac{Cs}{s^2 + \omega_0^2} = \frac{Cs}{(s + j\omega_0)(s - j\omega_0)}$$

Hence, in this case the transform $X(s)$ of the input has a zero at $s = 0$ and two poles at $s = \pm j\omega_0$.

If the system has no initial energy at time $t = 0$, the transform $Y(s)$ of the resulting output is given by

$$Y(s) = \frac{CsB(s)}{A(s)(s + j\omega_0)(s - j\omega_0)} \tag{9.29}$$

The computation of the output response $y(t)$ from (9.29) will be carried out in a manner similar to that done in Section 9.3 for the case of a step input. Here the terms

$s + j\omega_0$ and $s - j\omega_0$ in (9.29) are "pulled out" by using the partial fraction expansion assuming that $A(\pm j\omega_0) \neq 0$. This gives

$$Y(s) = \frac{\gamma(s)}{A(s)} + \frac{c}{s - j\omega_0} + \frac{\bar{c}}{s + j\omega_0} \tag{9.30}$$

where $\gamma(s)$ is a polynomial in s, c is a complex constant, and $\bar{c}$ is the complex conjugate of c. From the residue formula (8.67), c is given by

$$c = [(s - j\omega_0)Y(s)]_{s=j\omega_0} = \left[\frac{CsB(s)}{A(s)(s + j\omega_0)}\right]_{s=j\omega_0}$$

$$= \frac{jC\omega_0 B(j\omega_0)}{A(j\omega_0)(j2\omega_0)} = \frac{C}{2}H(j\omega_0)$$

Then inserting the values for c and $\bar{c}$ into (9.30) gives

$$Y(s) = \frac{\gamma(s)}{A(s)} + \frac{(C/2)H(j\omega_0)}{s - j\omega_0} + \frac{(C/2)\overline{H(j\omega_0)}}{s + j\omega_0} \tag{9.31}$$

where $\overline{H(j\omega_0)}$ is the complex conjugate of $H(j\omega_0)$.

Now let $y_1(t)$ denote the inverse Laplace transform of $\gamma(s)/A(s)$. Then taking the inverse Laplace transform of both sides of (9.31) yields

$$y(t) = y_1(t) + \frac{C}{2}[H(j\omega_0)e^{j\omega_0 t} + \overline{H(j\omega_0)}e^{-j\omega_0 t}] \tag{9.32}$$

Finally, using the identity [see (8.71)]

$$\beta e^{j\omega_0 t} + \bar{\beta}e^{-j\omega_0 t} = 2|\beta|\cos(\omega_0 t + \angle\beta)$$

expression (9.32) for $y(t)$ can be written in the form

$$y(t) = y_1(t) + C|H(j\omega_0)|\cos(\omega_0 t + \angle H(j\omega_0)), \qquad t \geq 0 \tag{9.33}$$

When the system is stable [all poles of $H(s)$ are in the open left-half plane], the term $y_1(t)$ in (9.33) decays to zero as $t \to \infty$, and thus $y_1(t)$ is the transient part of the response. The sinusoidal term in the right-hand side of (9.33) is the steady-state part of the response, which is denoted by $y_{ss}(t)$; that is,

$$y_{ss}(t) = C|H(j\omega_0)|\cos(\omega_0 t + \angle H(j\omega_0)), \qquad t \geq 0 \tag{9.34}$$

From (9.34) it is seen that the steady-state response $y_{ss}(t)$ to the sinusoidal input $x(t) = C\cos(\omega_0 t)$, $t \geq 0$, has the same frequency as the input, but it is scaled in magnitude by the amount $|H(j\omega_0)|$ and it is phase shifted by the amount $\angle H(j\omega_0)$. This result resembles the development given in Section 5.1, where it was shown that the response to the input

$$x(t) = C\cos\omega_0 t, \qquad -\infty < t < \infty$$

is given by

$$y(t) = C|H(\omega_0)| \cos(\omega_0 t + \angle H(\omega_0)), \qquad -\infty < t < \infty \tag{9.35}$$

where $H(\omega_0)$ is the Fourier transform $H(\omega)$ of the impulse response $h(t)$ with $H(\omega)$ evaluated at $\omega = \omega_0$. Note that in the expression (9.35) for the output, there is no transient since the input is applied at time $t = -\infty$. A transient response is generated only when the input is applied at some finite value of time (not $t = -\infty$).

If the given system is causal and stable, there is a direct correspondence between the derivation of $y_{ss}(t)$ given above and the result given by (9.35). To see this, first recall from Section 9.1, that stability implies the integrability condition

$$\int_0^{\infty} |h(t)|\, dt < \infty$$

It then follows from the discussion given in Section 8.1 that the Fourier transform $H(\omega)$ of $h(t)$ is equal to the Laplace transform $H(s)$ evaluated at $s = j\omega$; that is,

$$H(\omega) = H(j\omega) = H(s)|_{s=j\omega} \tag{9.36}$$

Note that $H(j\omega)$ is denoted by $H(\omega)$. This notation will be followed from here on.

As a consequence of (9.36), the expressions (9.34) and (9.35) are identical for $t \geq 0$, and thus there is a direct correspondence between the two results. In addition, by (9.36) the frequency response function of the system [which was first defined in Section 5.1 to be the Fourier transform $H(\omega)$ of $h(t)$] is equal to the transfer function $H(s)$ evaluated at $s = j\omega$. Hence the frequency response behavior of a stable system can be determined directly from the transfer function $H(s)$. In particular, the magnitude function $|H(\omega)|$ and the phase function $\angle H(\omega)$ can both be directly generated from the transfer function $H(s)$.

Example 9.17 *First-Order System*

Consider the first-order system with the transfer function

$$H(s) = \frac{k}{s - p} \tag{9.37}$$

It is assumed that $k > 0$ and $p < 0$, so that the system is stable. With $k = -p = 1/RC$, the system with transfer function (9.37) could be the RC circuit shown in Figure 9.15.

Now setting $s = j\omega$ in $H(s)$ gives

$$H(\omega) = \frac{k}{j\omega - p}$$

and taking the magnitude and angle of $H(\omega)$ yields

$$|H(\omega)| = \frac{|k|}{|j\omega - p|} = \frac{k}{\sqrt{\omega^2 + p^2}} \tag{9.38}$$

$$\angle H(\omega) = -\angle(j\omega - p) = -\tan^{-1}\frac{-\omega}{p} \tag{9.39}$$

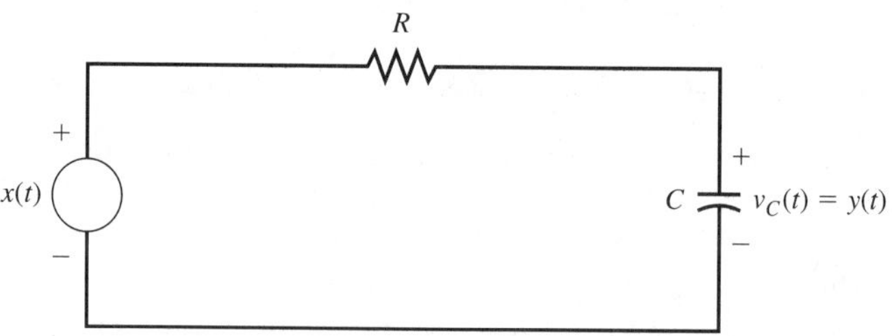

Figure 9.15 *RC* circuit with transfer function $H(s) = k/(s - p)$, where $k = -p = 1/RC$.

The output response resulting from the sinusoidal input $x(t) = C\cos\omega_0 t$, $t \geq 0$ (with no initial energy) can be computed using (9.38) and (9.39) as follows. First, the Laplace transform of the output response is given by

$$Y(s) = H(s)X(s) = \frac{kCs}{(s-p)(s^2+\omega_0^2)}$$

In this case, the partial fraction expression (9.30) for $Y(s)$ becomes

$$Y(s) = \frac{\gamma}{s-p} + \frac{c}{s-j\omega_0} + \frac{\bar{c}}{s+j\omega_0}$$

where

$$\gamma = [(s-p)Y(s)]|_{s=p} = \frac{kCp}{p^2+\omega_0^2}$$

Then the transient part of the output response is given by

$$y_{\text{tr}}(t) = \gamma e^{pt} = \frac{kCp}{p^2+\omega_0^2}e^{pt}, \qquad t \geq 0$$

and from (9.33), the complete output response is

$$y(t) = \frac{kCp}{p^2+\omega_0^2}e^{pt} + C|H(\omega_0)|\cos(\omega_0 t + \angle H(\omega_0)), \qquad t \geq 0 \qquad (9.40)$$

Finally, inserting (9.38) and (9.39) into (9.40) yields the output response

$$y(t) = \frac{kCp}{p^2+\omega_0^2}e^{pt} + \frac{Ck}{\sqrt{\omega_0^2+p^2}}\cos\left[\omega_0 t - \tan^{-1}\left(-\frac{\omega_0}{p}\right)\right], \qquad t \geq 0 \qquad (9.41)$$

Equation (9.41) is the complete response resulting from the input $x(t) = C\cos\omega_0 t$ applied for $t \geq 0$. Note that the transient part of the response is a decaying exponential since $p < 0$, with the rate of decay depending on the value of the pole p.

Now suppose that $k = 1$, $p = -1$, and the input is $x(t) = 10\cos(1.5t)$, $t \geq 0$, so that $C = 10$ and $\omega_0 = 1.5$ rad/sec. Then

$$\gamma = \frac{(1)(10)(-1)}{1 + (1.5)^2} = -3.08$$

$$|H(1.5)| = 5.55$$

$$\angle H(1.5) = -56.31°$$

and thus from (9.41), the output response is

$$y(t) = -3.08e^{-t} + 5.55\cos(1.5t - 56.31°), \qquad t \geq 0 \tag{9.42}$$

The response obtained using the following MATLAB commands is shown in Figure 9.16.

```
t = 0:0.05:20;
num = 1; den = [1 1];
x = 10*cos(1.5*t);
y = lsim(num,den,x,t);
```

From Figure 9.16, note that the transient $-3.08e^{-t}$ can be seen for small values of t, but then disappears. Since the time constant τ associated with the pole at -1 is $\tau = 1$, the transient decays to $1/e = 37\%$ of its initial value at $t = 1$ second. It can also be observed from the plot in Figure 9.16 that the amplitude and the phase of the steady-state part of the response matches with the values obtained above analytically [see (9.42)]. It should be noted that when the transient has died out, the resulting (steady-state) response obtained above is identical to the solution obtained using the Fourier theory in Example 5.2 for the RC circuit (with $RC = 1, A = C = 10$, and $\omega_0 = 1.5$). As remarked above, in the Fourier setup considered in Chapter 5 there is no transient response since the input is applied at time $t = -\infty$. The reader should check that the two solutions do in fact correspond.

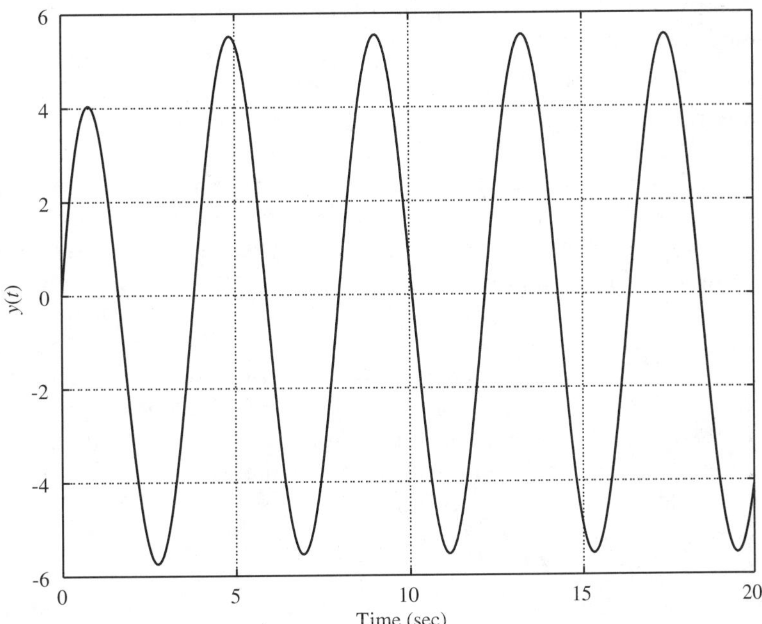

Figure 9.16 Output response in Example 9.17.

Example 9.18 *Mass–Spring–Damper System*

Mass–Spring–Damper System

For the mass–spring–damper system (see Example 9.15), an animation of the output response resulting from the input $x(t) = \cos \omega_0 t$, $t \geq 0$ can be generated by running the demo on the Web site. This provides a nice visualization of the transient response and the eventual convergence of the output response to the steady-state behavior. The reader is invited to run the demo for various values of M, D, and K and to compare the results with an analytical computation of the output response using the Laplace transform.

Response to Arbitrary Inputs

The analysis given above for a sinusoidal input, and the one given in Section 9.3 for a step input, generalize to arbitrary inputs as follows. Suppose that the transform $X(s)$ of the input $x(t)$ is a rational function; that is, $X(s) = C(s)/D(s)$, where $C(s)$ and $D(s)$ are polynomials in s with the degree of $C(s)$ less than the degree of $D(s)$. In terms of this notation, the poles of $X(s)$ are the roots of $D(s) = 0$.

Now if $x(t)$ is applied to a system with transfer function $H(s) = B(s)/A(s)$, the transform of the resulting response (with zero initial energy) is

$$Y(s) = \frac{B(s)C(s)}{A(s)D(s)}$$

If there are no common poles between $H(s)$ and $X(s)$, $Y(s)$ can be expressed in the form

$$Y(s) = \frac{E(s)}{A(s)} + \frac{F(s)}{D(s)} \tag{9.43}$$

where $E(s)$ and $F(s)$ are polynomials in s. Then taking the inverse transform of both sides of (9.43) gives

$$y(t) = y_1(t) + y_2(t) \tag{9.44}$$

where $y_1(t)$ is the inverse transform of $E(s)/A(s)$ and $y_2(t)$ is the inverse transform of $F(s)/D(s)$.

It is very important to note that the form of $y_1(t)$ depends directly on the poles of $H(s)$ and the form of $y_2(t)$ depends directly on the poles of $X(s)$. When the system is stable, $y_1(t)$ converges to zero as $t \to \infty$, in which case $y_1(t)$ is identified as the transient part of the response [although there may be terms in $y_2(t)$ that are also converging to zero]. A key point here is that the form of $y_1(t)$ (i.e., the transient) depends only on the poles on the system regardless of the particular form of the input signal $x(t)$.

If $X(s)$ has poles on the $j\omega$-axis, these poles appear in the transform of $y_2(t)$, and thus $y_2(t)$ will not converge to zero. Hence $y_2(t)$ is identified as the steady-state part of the response. It should be stressed that the form of $y_2(t)$ (i.e., the steady-state response) depends only on the poles of the input transform $X(s)$, regardless of what the system transfer function $H(s)$ is.

Example 9.19 *Form of Output Response*

Suppose that a system has transfer function $H(s)$ with two real poles a, b and a complex pair of poles $\sigma \pm jc$, where $a < 0$, $b < 0$, and $\sigma < 0$, so that the system is stable. Let $x(t)$ be any input whose transform $X(s)$ is rational in s and whose poles are different from those of $H(s)$. Then the form of the transient response is

$$y_{tr}(t) = k_1 e^{at} + k_2 e^{bt} + k_3 e^{\sigma t} \cos(ct + \theta), \qquad t \geq 0$$

where k_1, k_2, k_3, and θ are all constants that depend on the specific input and the zeros of the system.

If $x(t)$ is the step function, the form of the complete response is

$$y(t) = k_1 e^{at} + k_2 e^{bt} + k_3 e^{\sigma t}\cos(ct + \theta) + A, \qquad t \geq 0$$

where A is a constant. When $x(t)$ is the ramp $x(t) = tu(t)$, the form of the complete response is

$$y(t) = k_1 e^{at} + k_2 e^{bt} + k_3 e^{\sigma t}\cos(ct + \theta) + A + Bt, \qquad t \geq 0$$

where A and B are constants. When $x(t)$ is a sinusoid with frequency ω_0, the form of the complete response is

$$y(t) = k_1 e^{at} + k_2 e^{bt} + k_3 e^{\sigma t}\cos(ct + \theta) + B\cos(\omega_0 t + \phi), \qquad t \geq 0$$

for some constants B and ϕ. It should be noted that the values of the k_is in the above expressions are not the same.

9.5 FREQUENCY RESPONSE FUNCTION

Given a stable system with rational transfer function $H(s) = B(s)/A(s)$, in Section 9.4 it was shown that the steady-state response to the sinusoid $x(t) = C\cos\omega_0 t$, $t \geq 0$, with zero initial energy is given by

$$y_{ss}(t) = C|H(\omega_0)|\cos(\omega_0 t + \angle H(\omega_0)), \qquad t \geq 0 \tag{9.45}$$

where $H(\omega)$ is the frequency response function [which is equal to $H(s)$ with $s = j\omega$]. As a result of the fundamental relationship (9.45), the behavior of the system in terms of the response to sinusoidal inputs can be studied in terms of the frequency response curves given by the plots of the magnitude function $|H(\omega)|$ and the phase function $\angle H(\omega)$.

The magnitude function $|H(\omega)|$ is sometimes given in *decibels*, denoted by $|H(\omega)|_{\text{dB}}$ and defined by

$$|H(\omega)|_{\text{dB}} = 20\log_{10}|H(\omega)|$$

The term *decibel* (denoted by dB) was first defined as a unit of power gain in an electrical circuit. Specifically, the power gain through a circuit is defined to be 10 times the logarithm (to the base 10) of the output power divided by the input power. Since power in an electrical circuit is proportional to the square of the voltage or current and for any constant K,

$$10\log_{10}(K^2) = 20\log_{10}K$$

the definition of $|H(\omega)|_{\text{dB}}$ given above can be viewed as a generalization of the original meaning of the term *decibel.*

Note that

$$|H(\omega)|_{\text{dB}} < 0\text{ dB} \qquad \text{when } |H(\omega)| < 1$$

$$|H(\omega)|_{\text{dB}} = 0\text{ dB} \qquad \text{when } |H(\omega)| = 1$$

$$|H(\omega)|_{\text{dB}} > 0\text{ dB} \qquad \text{when } |H(\omega)| > 1$$

Thus it follows from (9.45) that when $|H(\omega)|_{\text{dB}} < 0$ dB, the system attenuates the sinusoidal input $x(t) = C\cos\omega_0 t$; when $|H(\omega)|_{\text{dB}} = 0$ dB, the system passes $x(t)$ with no attenuation; and when $|H(\omega)|_{\text{dB}} > 0$ dB, the system amplifies $x(t)$.

The plots of $|H(\omega)|$ (or $|H(\omega)|_{\text{dB}}$) versus ω and $\angle H(\omega)$ versus ω with ω on a logarithmic scale are called the *Bode diagrams* of the system. A technique for generating the Bode diagrams using asymptotes is given in the development below.

The frequency response curves (or the Bode diagrams) are often determined experimentally by measuring the steady-state response resulting from the sinusoidal input $x(t) = C\cos\omega_0 t$. By performing this experiment for various values of ω_0, it is possible to extrapolate the results to obtain the magnitude function $|H(\omega)|$ and phase function $\angle H(\omega)$ for all values of ω ($\omega \geq 0$). This then determines the frequency function $H(\omega)$ since

$$H(\omega) = |H(\omega)|\exp[j\angle H(\omega)]$$

The frequency response curves can be generated directly from the transfer function $H(s)$ by using the MATLAB command `bode`. The use of this command is illustrated in the examples given below. It will also be shown below that if the number of poles and zeros of the system is not large, the general shape of the frequency response curves can be determined from vector representations in the complex plane of the factors comprising $H(s)$. The development begins with the first-order case.

First-Order Case

Consider the first-order system given by the transfer function

$$H(s) = \frac{k}{s + B} \tag{9.46}$$

where $k > 0$ and $B > 0$. The frequency response function is $H(\omega) = k/(j\omega + B)$ and the magnitude and phase functions are given by

$$|H(\omega)| = \frac{k}{\sqrt{\omega^2 + B^2}} \tag{9.47}$$

$$\angle H(\omega) = -\tan^{-1}\frac{\omega}{B} \tag{9.48}$$

The frequency response curves can be generated by evaluating (9.47) and (9.48) for various values of ω. Instead of doing this, it will be shown that the shape of the frequency response curves can be determined from the vector representation of the factor $j\omega + B$ comprising $H(\omega)$. The vector representation of $j\omega + B$ is shown in Figure 9.17.

The magnitude $|j\omega + B|$ and the angle $\angle j\omega + B$ can be computed from the vector representation of $j\omega + B$ shown in Figure 9.17. Here the magnitude $|j\omega + B|$ is the length of the vector from the pole $s = -B$ to the point $s = j\omega$ on the imaginary axis, and the angle $\angle j\omega + B$ is the angle between this vector and the real axis of

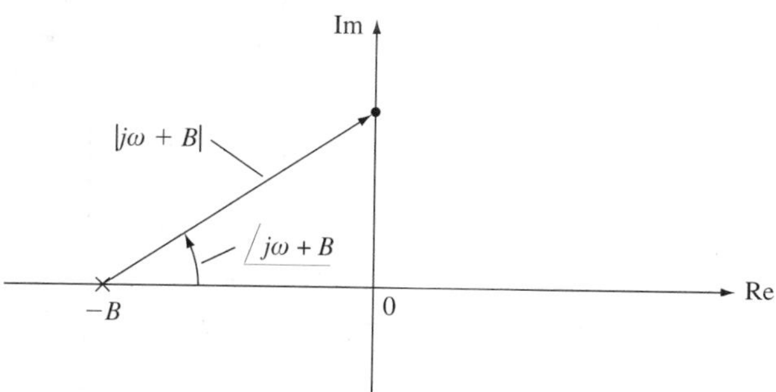

Figure 9.17 Vector representation of $j\omega + B$.

the complex plane. From Figure 9.17, it is clear that $|j\omega + B|$ becomes infinite as $\omega \to \infty$ and $\angle j\omega + B$ approaches 90° as $\omega \to \infty$. Then from (9.47) and (9.48), it is seen that the magnitude function $|H(\omega)|$ starts with value k/B when $\omega = 0$ and approaches zero as $\omega \to \infty$, while the phase $\angle H(\omega)$ starts with value 0° when $\omega = 0$ and approaches $-90°$ as $\omega \to \infty$. This provides a good indication as to the shape of the frequency response curves.

To generate accurate plots, the MATLAB command bode can be used. For example when $k = B = 2$, the curves can be generated using the commands

```
num = [2];
den = [1 2];
w = 0:.05:10;
[mag,phase] = bode(num,den,w)
subplot(211),plot(w,mag);
subplot(212),plot(w,phase);
```

The results are displayed in Figure 9.18. The magnitude plot in Figure 9.18a reveals that the system is a lowpass filter since it passes sinusoids whose frequency is less than 2 rad/sec, while it attenuates sinusoids whose frequency is above 2 rad/sec. Recall that the lowpass frequency response characteristic was first encountered in Example 5.2 which was given in terms of the Fourier analysis.

For an arbitrary value of $B > 0$, when $k = B$ the system with transfer function $H(s) = k/(s + B)$ is a lowpass filter since the magnitude function $|H(\omega)|$ starts with value $H(0) = k/B = 1$ and then rolls off to zero as $\omega \to \infty$. The point $\omega = B$ is called the 3-dB point of the filter since this is the value of ω for which $|H(\omega)|_{\text{dB}}$ is down by 3 dB from the peak value of $|H(0)|_{\text{dB}} = 0$ dB. This lowpass filter is said to have a *3-dB bandwidth* of B rad/sec since it passes (with less than 3 dB of attenuation) sinusoids whose frequency is less than B rad/sec. The *passband* of the filter is the frequency range from 0 rad/sec to B rad/sec. The *stopband* of the filter is the frequency range from B rad/sec to ∞. As seen from the magnitude plot in Figure 9.18a, for this filter the cutoff between the passband and the stopband is not very sharp. It will be shown in the next section that a much sharper cutoff can be obtained by increasing the number of poles of the system.

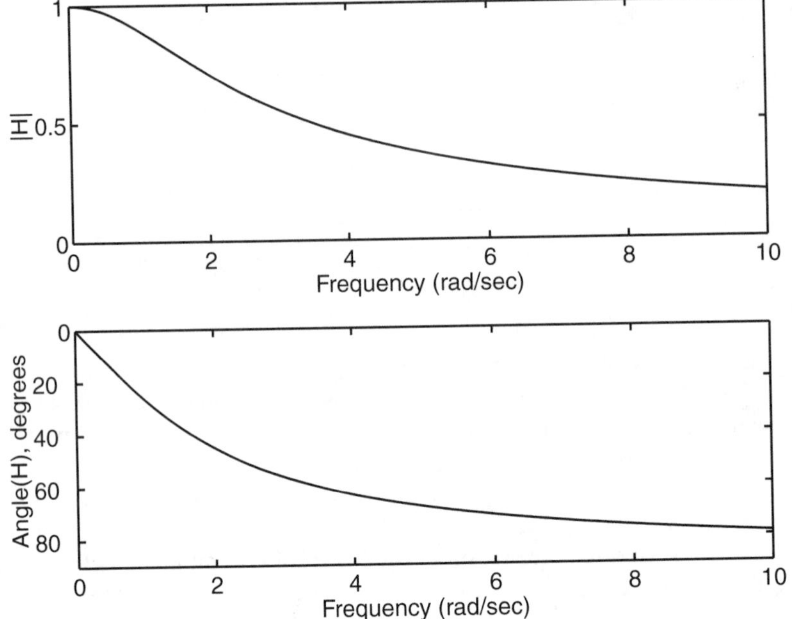

Figure 9.18 Frequency response curves for $H(s) = 2/(s + 2)$: (a) magnitude; (b) phase curve.

Single-pole systems with a zero. From the results derived above, it was discovered that a single-pole system with no zero is a lowpass filter. This frequency response characteristic can be changed by adding a zero to the system. In particular, consider the single-pole system with the transfer function

$$H(s) = \frac{s + C}{s + B}$$

It is assumed that $B > 0$ and $C > 0$. Setting $s = j\omega$ in $H(s)$ gives

$$H(\omega) = \frac{j\omega + C}{j\omega + B}$$

Then the magnitude and phase functions are given by

$$|H(\omega)| = \frac{|j\omega + C|}{|j\omega + B|} = \sqrt{\frac{\omega^2 + C^2}{\omega^2 + B^2}}$$

$$\begin{aligned} \angle H(\omega) &= \angle j\omega + C - \angle j\omega + B \\ &= \tan^{-1}\frac{\omega}{C} - \tan^{-1}\frac{\omega}{B} \end{aligned}$$

The frequency response curves will be determined in the case when $0 < C < B$. First consider the vector representations of $j\omega + B$ and $j\omega + C$ shown in Figure 9.19. As seen from the figure, both $|j\omega + B|$ and $|j\omega + C|$ increase as ω is increased from zero; however, the percent increase in $|j\omega + C|$ is larger. Thus $|H(\omega)|$ starts with value $|H(0)| = C/B$ and then $|H(\omega)|$ increases as ω is increased from zero. For large values of ω, the difference in $|j\omega + B|$ and $|j\omega + C|$ is very small, and thus $|H(\omega)| \to 1$ as $\omega \to \infty$.

From Figure 9.19, it is seen that the angles $\angle(j\omega + B)$ and $\angle(j\omega + C)$ both increase as ω is increased from zero; however, at first the increase in $\angle(j\omega + C)$ is larger. Hence $\angle H(\omega)$ starts with value $\angle H(0) = 0°$ and then $\angle H(\omega)$ increases as ω is increased from zero. For $\omega > B$, the percent increase in $\angle(j\omega + B)$ is greater than that of $\angle(j\omega + C)$, and thus $\angle H(\omega)$ decreases as ω is increased from the value $\omega = B$. Hence $\angle H(\omega)$ will have a maximum value at some point ω between $\omega = C$ and $\omega = B$. As $\omega \to \infty$, both angles $\angle(j\omega + B)$ and $\angle(j\omega + C)$ approach $90°$, and therefore $\angle H(\omega) \to 0°$ as $\omega \to \infty$.

In the case $C = 1$ and $B = 20$, the exact frequency response curves were computed using the MATLAB command `bode`. The results are shown in Figure 9.20. From 9.20a it is seen that the system is a *highpass filter* since it passes with little attenuation all frequencies above $B = 20$ rad/sec. Although highpass filters exist in theory, they do not exist in practice, since actual systems cannot pass sinusoids with arbitrarily large frequencies. In other words, no actual system can have an infinite bandwidth. Thus any implementation of the transfer function $H(s) = (s + C)/(s + B)$ would only be an approximation of a highpass filter (in the case $0 < C < B$).

Second-Order Systems

Now consider the system with the transfer function

$$H(s) = \frac{k}{s^2 + 2\zeta\omega_n s + \omega_n^2}$$

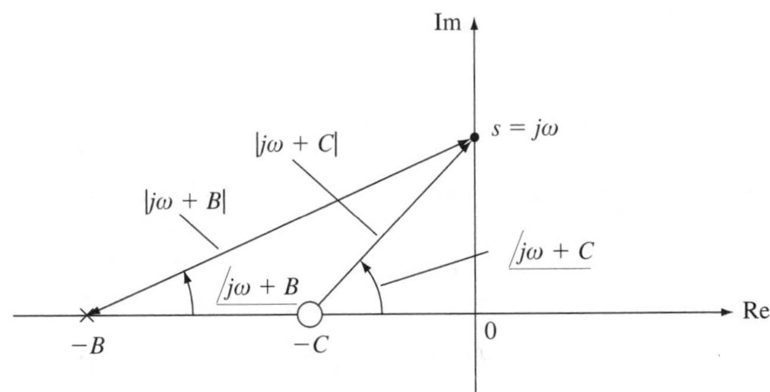

Figure 9.19 Vector representations of $j\omega + B$ and $j\omega + C$ when $0 < C < B$.

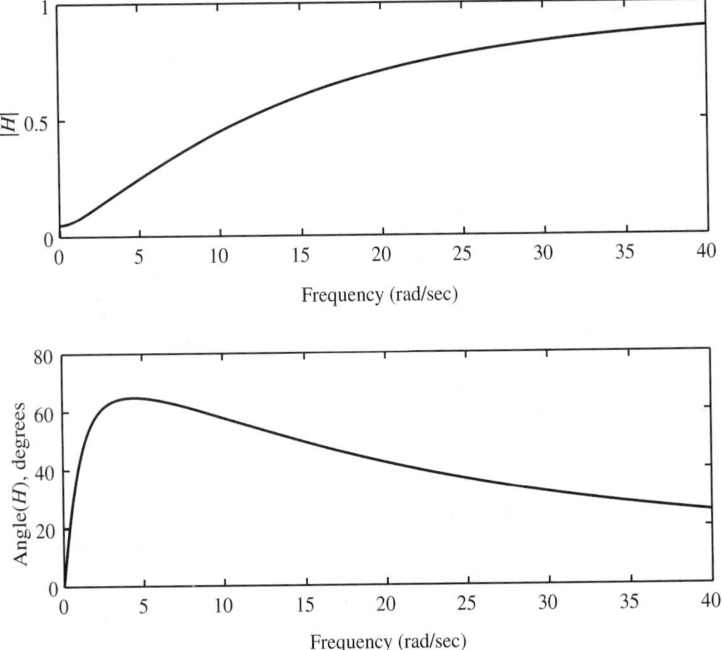

Figure 9.20 Frequency response curves for $H(s) = (s + 1)/(s + 20)$: (a) magnitude curve; (b) phase curve.

where $k > 0$, $\zeta > 0$, and $\omega_n > 0$, so that the system is stable. As discussed in Section 9.3, the poles of the system are

$$p_1 = -\zeta\omega_n + \omega_n\sqrt{\zeta^2 - 1}$$
$$p_2 = -\zeta\omega_n - \omega_n\sqrt{\zeta^2 - 1}$$

Expressing $H(s)$ in terms of p_1 and p_2 gives

$$H(s) = \frac{k}{(s - p_1)(s - p_2)}$$

and thus the magnitude and phase functions are given by

$$|H(\omega)| = \frac{k}{|j\omega - p_1|\,|j\omega - p_2|} \tag{9.49}$$

$$\angle H(\omega) = -\angle j\omega - p_1 - \angle j\omega - p_2 \tag{9.50}$$

As noted in Section 9.3, the poles p_1 and p_2 are real if and only if $\zeta \geq 1$. In this case the shape of the frequency response curves can be determined by considering the vector representations of $j\omega - p_1$ and $j\omega - p_2$ shown in Figure 9.21. Here the magnitudes $|j\omega - p_1|$ and $|j\omega - p_2|$ become infinite as $\omega \to \infty$, and the angles $\angle j\omega - p_1$ and $\angle j\omega - p_2$ approach 90° as $\omega \to \infty$. Then from (9.49) and (9.50), it is seen that the magnitude $|H(\omega)|$ starts with value $|k/p_1p_2| = k/\omega_n^2$ at $\omega = 0$ and approaches zero as $\omega \to \infty$. The phase $\angle H(\omega)$ starts with value 0° when $\omega = 0$ and approaches $-180°$ as $\omega \to \infty$. Thus when $k = \omega_n^2$, $H(0) = k/\omega_n^2 = 1$ and the system is a lowpass filter whose 3-dB bandwidth depends on the values of ζ and ω_n. When $\zeta = 1$, the 3-dB bandwidth is equal to $\left(\sqrt{\sqrt{2} - 1}\right)\omega_n^2$. This can be verified by considering the vector representations in Figure 9.21. The details are left to the reader.

With $\zeta = 1, \omega_n = 3.1$ rad/sec and $k = \omega_n^2$, the 3-dB bandwidth of the lowpass filter is approximately equal to 2 rad/sec. The frequency response curves for this case were computed using the MATLAB command `bode`. The results are shown in Figure 9.22. Also displayed in Figure 9.22 are the frequency response curves of the one-pole lowpass filter with transfer function $H(s) = 2/(s + 2)$. Note that the two-pole filter has a sharper cutoff than the one-pole filter.

Complex pole case. Now it is assumed that $0 < \zeta < 1$, so that the poles p_1 and p_2 are complex. With $\omega_d = \omega_n\sqrt{1 - \zeta^2}$ (as defined in Section 9.3), the poles are p_1, $p_2 = -\zeta\omega_n \pm j\omega_d$. Then $H(\omega)$ can be expressed in the form

$$H(\omega) = \frac{k}{(j\omega + \zeta\omega_n + j\omega_d)(j\omega + \zeta\omega_n - j\omega_d)}$$

The vector representations of $j\omega + \zeta\omega_n + j\omega_d$ and $j\omega + \zeta\omega_n - j\omega_d$ are shown in Figure 9.23. Note that as ω increases from $\omega = 0$, the magnitude $|j\omega + \zeta\omega_n - j\omega_d|$ decreases, while the magnitude $|j\omega + \zeta\omega_n + j\omega_d|$ increases. For $\omega > \omega_d$, both these magnitudes grow until they become infinite, and thus $|H(\omega)| \to 0$ as $\omega \to \infty$. However, it is not clear if $|H(\omega)|$ first increases or decreases as ω is increased from $\omega = 0$. It turns out that when $\zeta < 1/\sqrt{2}$, the magnitude $|H(\omega)|$ increases as ω is increased from 0, and when $\zeta \geq 1/\sqrt{2}$ the magnitude $|H(\omega)|$ decreases as ω is increased from 0. The proof of this follows by taking the derivative of $|H(\omega)|$ with respect to ω. The details are left to a homework problem (see Problem 9.26).

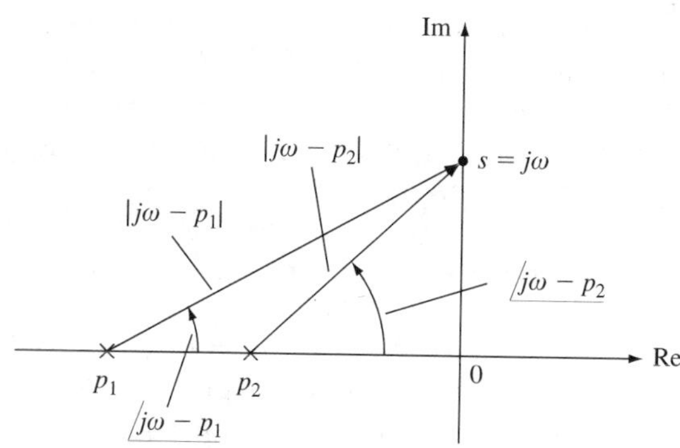

Figure 9.21 Vector representations of $j\omega - p_1$ and $j\omega - p_2$ when p_1 and p_2 are real.

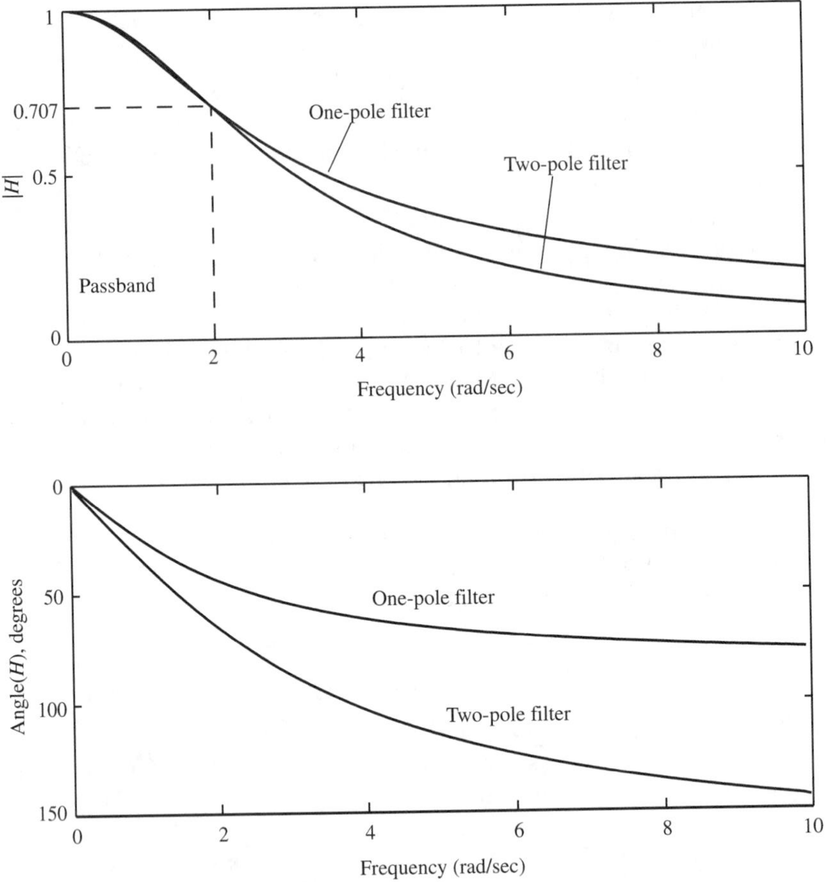

Figure 9.22 Frequency response curves of two-pole lowpass filter with $\zeta = 1$ and one-pole lowpass filter: (a) magnitude curve; (b) phase curve.

Since the magnitude function $|H(\omega)|$ has a peak when $\zeta < 1/\sqrt{2}$, the system is said to have a *resonance* when $\zeta < 1/\sqrt{2}$. In addition, it can be shown that the peak occurs when $\omega = \omega_r = \omega_n\sqrt{1 - 2\zeta^2}$, and thus ω_r is called the *resonant frequency* of the system. The magnitude of the resonance (i.e., the peak value of $|H(\omega)|$) increases as $\zeta \to 0$, which corresponds to the poles approaching the $j\omega$-axis (see Figure 9.23). When $\zeta \geq 1/\sqrt{2}$, the system does not have a resonance, and there is no resonant frequency.

When $\zeta < 1/\sqrt{2}$ and the peak value of $|H(\omega)|$ is equal to 1 (i.e., $|H(\omega_r)| = 1$) the system behaves like a *bandpass filter* since it will pass input sinusoids whose frequencies are in a neighborhood of the resonant frequency ω_r. The *center frequency* of the filter is equal to ω_r. The 3-dB bandwidth of the filter is defined to be all those frequencies ω for which the magnitude $|H(\omega)|$ is greater than or equal to $M_p/\sqrt{2}$, where $M_p = |H(\omega_r)|$ is the peak value of $|H(\omega)|$. It follows from the vector representations in Figure 9.23 that the 3-dB bandwidth is approximately equal to $2\zeta\omega_n$. This

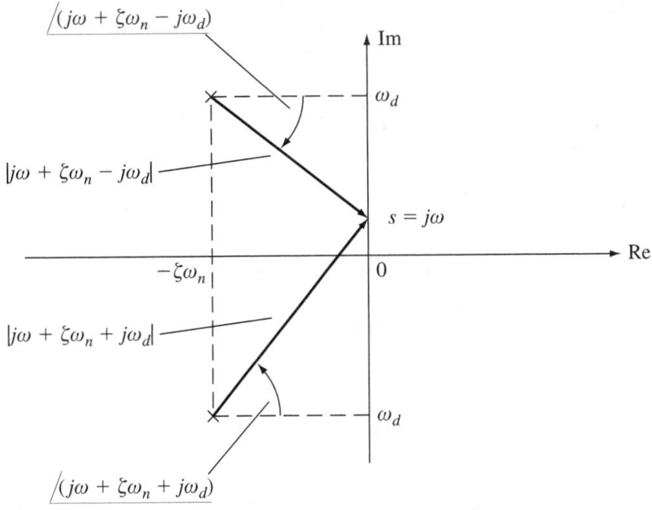

Figure 9.23 Vector representations in the complex-pole case.

bandpass filter characteristic is illustrated in the following example. An in-depth treatment of filtering is given in Section 9.6.

Example 9.20 *Two-Pole Bandpass Filter*

Suppose that the objective is to design a two-pole bandpass filter with center frequency $\omega_r = 10$ rad/sec, and with 3-dB bandwidth equal to 2 rad/sec. Then

$$10 = \omega_r = \omega_n\sqrt{1 - 2\zeta^2}$$

and

$$2 = 2\zeta\omega_n$$

Solving the second equation for ω_n and inserting the result into the first equation gives

$$10 = \frac{\sqrt{1 - 2\zeta^2}}{\zeta}$$

and thus

$$\frac{1 - 2\zeta^2}{\zeta^2} = 100$$

Solving for ζ gives

$$\zeta = \frac{1}{\sqrt{102}} \approx 0.099$$

Then

$$\omega_n = \frac{1}{\zeta} = 10.1$$

and the transfer function of the desired filter is

$$H(s) = \frac{k}{s^2 + 2s + 102}$$

Now the constant k should be chosen so that the peak value of $|H(\omega)|$ is equal to 1. Since the center frequency ω_r of the filter is equal to 10 rad/sec, the peak occurs at $\omega = 10$, and thus k must be chosen so that $|H(10)| = 1$. Then setting $s = j10$ in $H(s)$ and taking the magnitude gives

$$|H(10)| = \frac{k}{|-100 + j20 + 100|} = \frac{k}{20.1}$$

Hence $k = 20.1$. Using the MATLAB command bode results in the frequency response curves shown in Figure 9.24. From the plot it can be seen that the desired center frequency and 3-dB bandwidth have been obtained.

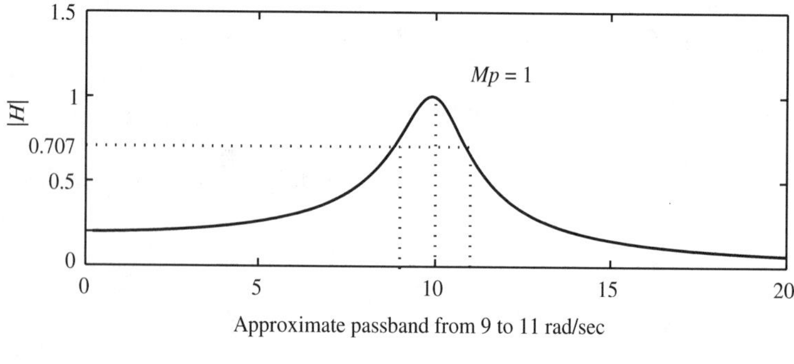

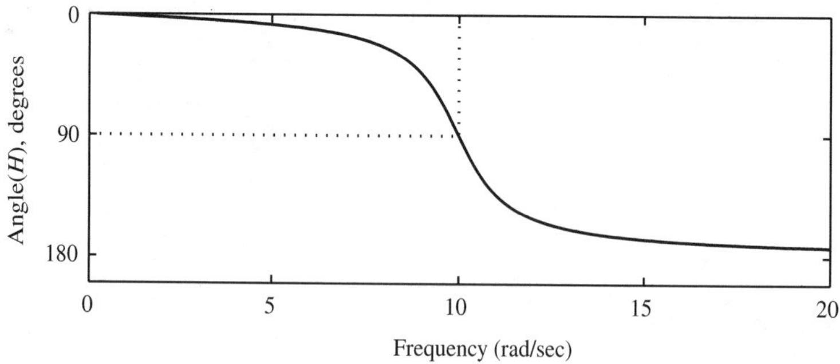

Figure 9.24 Frequency response curves for Example 9.20: (a) magnitude curve; (b) phase curve.

Example 9.21 *Mass–Spring–Damper System*

Mass–Spring–Damper System

For the mass–spring–damper system (see Example 9.15), recall that the damping ratio ζ and the natural frequency ω_n are given by

$$\zeta = \frac{D}{2\sqrt{MK}}, \omega_n = \sqrt{\frac{K}{M}}$$

When $\zeta < 1/\sqrt{2}$, which implies that $D < \sqrt{2MK}$, the system has a resonance with the resonance frequency ω_r given by

$$\omega_r = \omega_n\sqrt{1 - 2\zeta^2} = \sqrt{\frac{K}{M}}\sqrt{1 - \frac{D^2}{4MK}} = \frac{\sqrt{4MK - D^2}}{M} \tag{9.51}$$

For various positive values of M, D, and K satisfying the condition $D < \sqrt{2MK}$, the reader is invited to run the on-line demo with the input equal to the sine sweep. Verify that the resonance frequency observed in the demo is the same as the value computed using (9.51).

Construction of Bode Plots Via Asymptotes

Given a system with transfer function $H(s)$, recall that the Bode diagrams are the plots of the magnitude function $|H(\omega)|_{\text{dB}} = 20 \log|H(\omega)|$ and the phase function $\underline{/H(\omega)}$, where the scale for the frequency variable ω is logarithmic. The use of the log function in the definition of the magnitude $|H(\omega)|_{\text{dB}}$ and the log scale for ω enable the Bode plots to be approximated by straight lines, referred to as *asymptotes,* which can be drawn easily. To see this construction, first consider the system with the transfer function

$$H(s) = \frac{A(s + C_1)(s + C_2)\cdots(s + C_M)}{s(s + B_1)(s + B_2)\cdots(s + B_{N-1})} \tag{9.52}$$

In (9.52), A is a real constant, the zeros $-C_1, -C_2, \ldots, -C_M$ are real numbers, and the poles $-B_1, -B_2, \ldots, -B_{N-1}$ are real numbers. (The case of complex poles and/or zeros will be considered later.) Then setting $s = j\omega$ in (9.52) gives

$$H(\omega) = \frac{A(j\omega + C_1)(j\omega + C_2)\cdots(j\omega + C_M)}{j\omega(j\omega + B_1)(j\omega + B_2)\cdots(j\omega + B_{N-1})}$$

Dividing each factor $j\omega + C_i$ in the numerator by C_i and dividing each factor $j\omega + B_i$ in the denominator by B_i yields

$$H(\omega) = \frac{K\left(j\dfrac{\omega}{C_1} + 1\right)\left(j\dfrac{\omega}{C_2} + 1\right)\cdots\left(j\dfrac{\omega}{C_M} + 1\right)}{j\omega\left(j\dfrac{\omega}{B_1} + 1\right)\left(j\dfrac{\omega}{B_2} + 1\right)\cdots\left(j\dfrac{\omega}{B_{N-1}} + 1\right)} \tag{9.53}$$

where K is the real constant given by

$$K = \frac{AC_1C_2\cdots C_M}{B_1B_2\cdots B_{N-1}}$$

Now since $\log(AB) = \log(A) + \log(B)$ and $\log(A/B) = \log(A) - \log(B)$, from (9.53) the magnitude in dB of $H(\omega)$ is given by

$$|H(\omega)|_{\text{dB}} = 20\log|K| + 20\log\left|j\frac{\omega}{C_1} + 1\right| + \cdots + 20\log\left|j\frac{\omega}{C_M} + 1\right|$$

$$- 20\log|j\omega| - 20\log\left|j\frac{\omega}{B_1} + 1\right| - \cdots - 20\log\left|j\frac{\omega}{B_{N-1}} + 1\right|$$

The phase of $H(\omega)$ is given by

$$\angle H(\omega) = \angle K + \angle\left(j\frac{\omega}{C_1} + 1\right) + \cdots + \angle\left(j\frac{\omega}{C_M} + 1\right)$$

$$- \angle(j\omega) - \angle\left(j\frac{\omega}{B_1} + 1\right) - \cdots - \angle\left(j\frac{\omega}{B_{N-1}} + 1\right)$$

Thus the magnitude and phase functions can be decomposed into a sum of individual factors. The Bode diagrams can be computed for each factor and then can be added graphically to obtain the Bode diagrams for $H(\omega)$. To carry out this procedure, it is first necessary to determine the Bode plots for the three types of factors in $H(\omega)$: a constant K, the factor $j\omega$, and factors of the form $j\omega T + 1$, where T is a real number. Straight-line approximations (asymptotes) to the Bode plots are derived below for each factor, and from this the actual curves can be sketched. In this development it is assumed that $T > 0$ in the factor $j\omega T + 1$.

Constant factors. The magnitude plot for the constant factor K is a constant line versus ω given by

$$|K|_{\text{dB}} = 20\log|K|$$

Similarly, the phase of the factor K is a constant line versus ω:

$$\angle K = \begin{cases} 0^\circ & \text{for } K > 0 \\ -180^\circ & \text{for } K < 0 \end{cases}$$

($j\omega T + 1$) factors. The magnitude of $(j\omega T + 1)$ in dB is given by

$$|j\omega T + 1|_{\text{dB}} = 20\log\sqrt{\omega^2T^2 + 1}$$

Define the corner frequency ω_{cf} to be the value of ω for which $\omega T = 1$; that is, $\omega_{\text{cf}} = 1/T$. Then for $\omega < \omega_{\text{cf}}$, ωT is less than 1 and hence the magnitude can be approximated by

$$|j\omega T + 1|_{\text{dB}} \approx 20\log(1) = 0 \text{ dB}$$

For frequencies $\omega > \omega_{\text{cf}}$, ωT is greater than 1 and the magnitude can be approximated by

$$|j\omega T + 1|_{\text{dB}} \approx 20\log(\omega T)$$

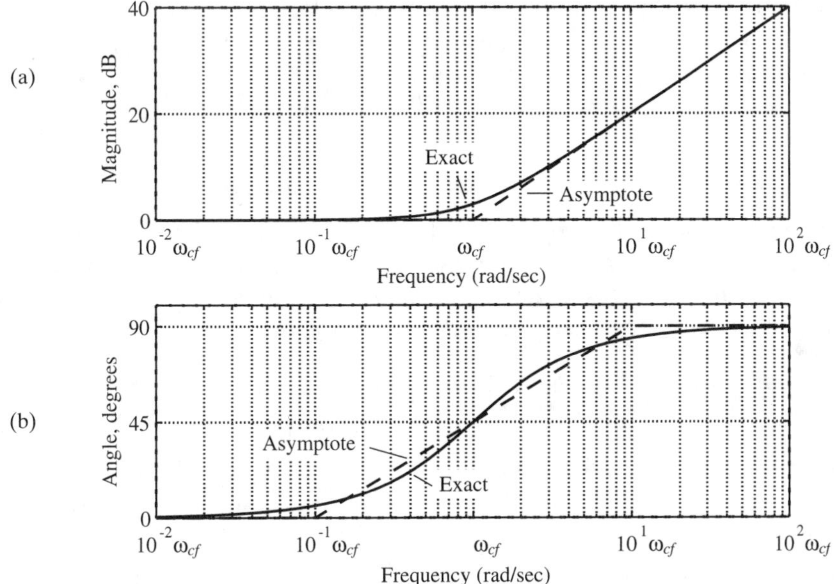

Figure 9.25 Magnitude (a) and phase (b) plots for the factor $j\omega T + 1$.

When plotted on a logarithmic scale for ω, the term 20 log(ωT) is a straight line with slope equal to -20 dB/decade, where a decade is a factor of 10 in frequency.

The plot of the constant 0 dB for $\omega < \omega_{cf}$ and the plot of the line 20 log(ωT) for $\omega > \omega_{cf}$ are the asymptotes for the magnitude term $|j\omega T + 1|_{dB}$. The asymptotes are plotted in Figure 9.25a along with the exact Bode magnitude function for the factor $j\omega T + 1$. As seen from the figure, the asymptotes provide a good approximation for frequencies away from the corner frequency. At the corner frequency, the asymptote approximation is off by exactly 3 dB.

The angle of the factor ($j\omega T + 1$) is given by

$$\underline{/j\omega T + 1} = \tan^{-1} \omega T$$

For very small frequencies, $\underline{/j\omega T + 1} \approx 0°$ and for very large frequencies, $\underline{/j\omega T + 1} \approx 90°$. A straight-line (asymptote) approximation of $\underline{/j\omega T + 1}$ for $\omega \leq \omega_{cf}/10$ is $\underline{/j\omega T + 1} = 0°$ and for $\omega \geq 10\omega_{cf}$, $\underline{/j\omega T + 1} = 90°$. The transition from 0° to 90° can be approximated by a straight line with slope 45°/decade drawn over a two decade range from $\omega_{cf}/10$ to $10\omega_{cf}$. A plot of the asymptote approximations as well as the exact angle plot are shown in Figure 9.25b. The approximations are fairly accurate with errors of about 5° at the corners.

When the factor ($Tj\omega + 1$) is in the numerator of $H(\omega)$, it represents a zero at $s = -1/T$ in the transfer function $H(s)$. From the analysis above it is seen that each (real) zero in the transfer function contributes a phase angle of approximately +90°

at high frequencies and a slope of +20 dB/decade in the magnitude at high frequencies. When the factor $j\omega T + 1$ is in denominator of $H(\omega)$, it corresponds to a pole of $H(s)$ at $-1/T$. Since

$$|(j\omega T + 1)^{-1}|_{\text{dB}} = -|j\omega T + 1|_{\text{dB}}$$

and

$$\angle(j\omega T + 1)^{-1} = -\angle j\omega T + 1$$

a pole factor $j\omega T + 1$ contributes a phase angle of approximately $-90°$ at high frequencies and a slope of -20 dB/decade in magnitude at high frequencies. The magnitude and phase curves for a pole factor $j\omega T + 1$ are the negative of the magnitude and phases curves for a zero factor $j\omega T + 1$ given in Figure 9.25.

***jω* factors.** The magnitude of $j\omega$ is given by

$$|j\omega|_{\text{dB}} = 20\log(\omega)$$

This is a straight line with slope 20 dB/decade when plotted on a logarithmic scale for ω. The line crosses the 0-dB line at $\omega = 1$. To see this, note that when $\omega = 1$,

$$|j\omega|_{\text{dB}} = 20\log(1) = 0 \text{ dB}$$

The plot of $|j\omega|_{\text{dB}}$ is given in Figure 9.26a. In this case, an approximation is not needed since the exact plot is already a straight line. The phase plot is also a straight line with a constant value of $\angle j\omega = 90°$, as shown in Figure 9.26b. Clearly, a $j\omega$ factor in the nu-

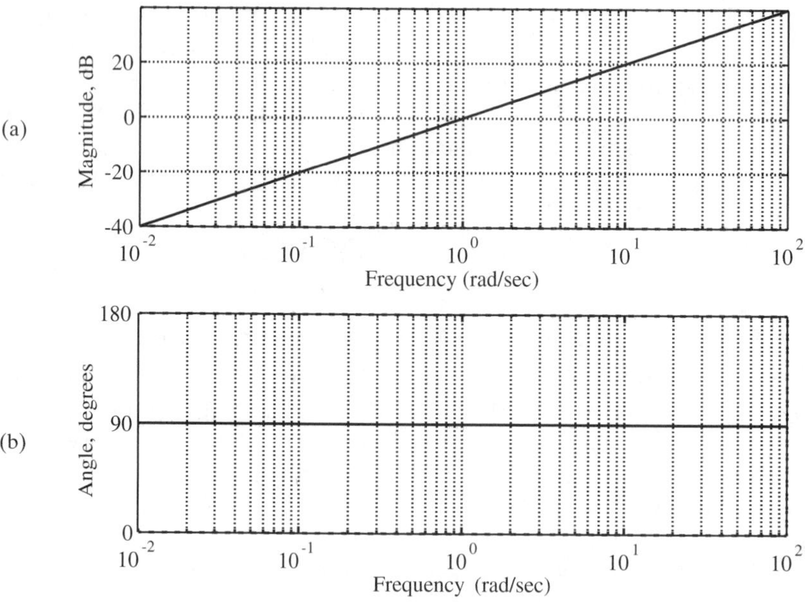

Figure 9.26 Magnitude (a) and phase (b) plots for the factor $j\omega$.

merator of $H(\omega)$ corresponds to a zero of $H(s)$ at $s = 0$. When the $j\omega$ factor is in the denominator of $H(\omega)$, the plots in Figure 9.26 are negated, resulting in a slope of -20 dB/decade in the magnitude plot and $-90°$ in the phase plot.

Plotting Bode diagrams. Now to compute the Bode diagrams for a system with $H(\omega)$ given by (9.53), the Bode diagrams for the various factors can simply be added together. The procedure is illustrated in the following example.

Example 9.22 *Bode Diagrams*

Consider the system with transfer function

$$H(s) = \frac{1000(s + 2)}{(s + 10)(s + 50)}$$

Writing $H(\omega)$ in the form (9.53) yields

$$H(\omega) = \frac{1000(j\omega + 2)}{(j\omega + 10)(j\omega + 50)} = \frac{4(j\omega(0.5) + 1)}{(j\omega(0.1) + 1)(j\omega(0.02) + 1)}$$

The factors comprising $H(\omega)$ are 4, $[j\omega(0.5) + 1]$, $[j\omega(0.1) + 1]^{-1}$, and $[j\omega(0.02) + 1]^{-1}$. The constant factor has a magnitude in decibels of $20\log(4) = 12.04$ dB and an angle of $0°$. The other factors have corner frequencies of $\omega_{cf} = 2, 10$, and 50, respectively. The asymptote approximations of the magnitude and phase for each factor (numbered 1 through 4) are shown in Figure 9.27 using dashed lines. The addition of all the asymptotes is also shown via a solid line in Figure 9.27.

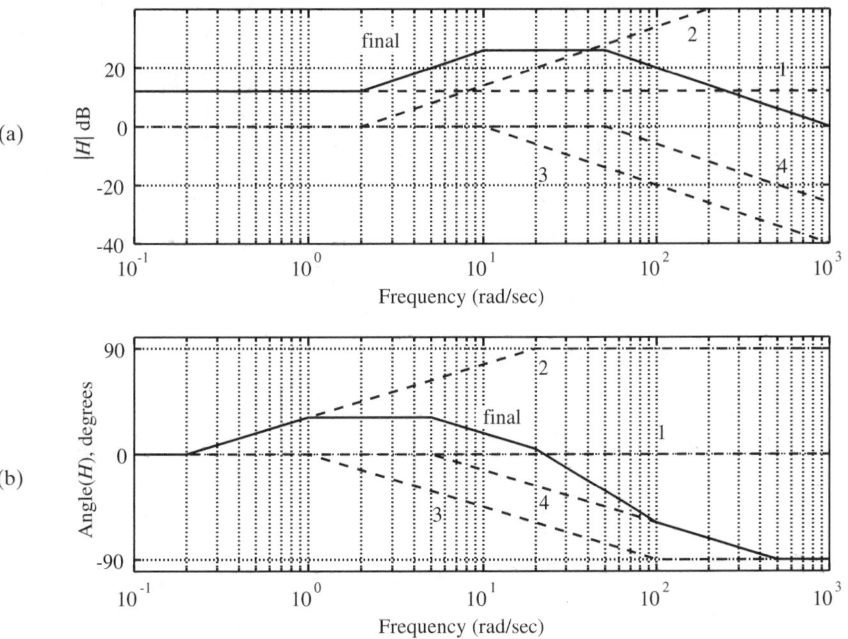

Figure 9.27 Magnitude (a) and phase (b) asymptotes for Example 9.22.

Note that when adding plots, it is easiest to add the slopes algebraically. For example, the slope of the magnitude curve for $\omega < 2$ is 0 dB/decade; the slope between $\omega = 2$ and $\omega = 10$ is 20 dB/decade; the slope between $\omega = 10$ and $\omega = 50$ is 0 dB/decade; and the slope for $\omega > 50$ is -20 dB/decade. The slopes of the angle plot add similarly. The exact plot obtained using MATLAB is shown in Figure 9.28 along with the summed asymptotes. The MATLAB commands to obtain the Bode diagrams are

```
num = 1000*[1 2];               % define the numerator
den = conv([1 10],[1 50]);      % define the denominator
bode(num,den);
```

This use of the command bode plots the Bode diagrams automatically. Other options can be used to customize the Bode diagrams.

The addition of slopes to compute the final curve as shown in the Example 9.22 is the foundation for a shorter way of constructing magnitude plots: First determine the lowest corner frequency. Below that frequency, the only nonzero factors are the constant and $j\omega$ factors. Each $j\omega$ factor in the numerator results in a slope of 20 dB/decade with a 0-dB intercept at $\omega = 1$ rad/sec, while each $j\omega$ factor in the denominator results in a slope of -20 dB/decade with a 0-dB intercept of $\omega =$ 1 rad/sec. Therefore, a factor of $(j\omega)^q$ in the numerator (where q is a positive or negative integer) corresponds to a line with slope $20q$ dB/decade and a 0-dB intercept at $\omega = 1$ rad/sec. This line is then offset by the magnitude in decibels of the

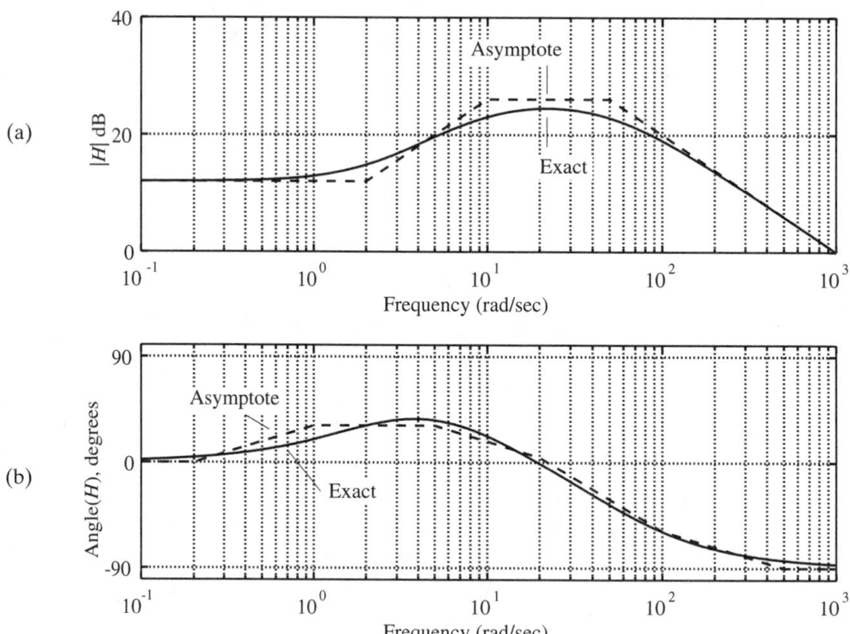

Figure 9.28 Summed asymptotes and exact Bode diagrams for Example 8.22.

constant factor to yield the low-frequency asymptote. This asymptote extends until the lowest corner frequency at which a change in slope will occur. A change in slope will occur also at each of the other corner frequencies. The slope change at a corner frequency ω_{cf} is -20 dB/decade if the corner frequency corresponds to a single pole, and $+20$ dB/decade if ω_{cf} corresponds to a single zero. The slope at high frequency should be $-20(N - M)$ dB/decade, where N is the number of poles and M is the number of zeros.

Similarly, the angle plot can be graphed using a similar shortcut since every pole adds $-90°$ at high frequencies with a transitional slope of $-45°$/decade and every zero adds $+90°$ at high frequencies with a transitional slope of $+45°$/decade. Since there are two slope changes with each pole or zero, it is easier to plot the individual terms and then add them graphically. A good way to check the final plot is to verify that the angle at high frequencies is equal to $-90(N - M)$, where N is the number of poles and M is the number of zeros.

Complex poles or zeros. Suppose that $H(s)$ contains a quadratic factor of the form $s^2 + 2\zeta\omega_n s + \omega_n^2$ with $0 < \zeta < 1$ and $\omega_n > 0$ (so that the zeros are complex). Setting $s = j\omega$ and dividing by ω_n^2 results in the factor $(j\omega/\omega_n)^2 + (2\zeta/\omega_n)j\omega + 1$. The magnitude in decibels for this quadratic term is

$$\left|\left(\frac{j\omega}{\omega_n}\right)^2 + \frac{2\zeta}{\omega_n}j\omega + 1\right|_{\text{dB}} = 20 \log \sqrt{\left(1 - \frac{\omega^2}{\omega_n^2}\right)^2 + \left(\frac{2\zeta\omega}{\omega_n}\right)^2}$$

Define the corner frequency to be the frequency for which $\omega/\omega_n = 1$; that is, $\omega_{cf} = \omega_n$. Then an asymptote construction can be carried out by making the following approximation for low frequencies $\omega < \omega_n$:

$$\left|\left(\frac{j\omega}{\omega_n}\right)^2 + \left(\frac{2\zeta}{\omega_n}\right)j\omega + 1\right|_{\text{dB}} \approx 20 \log(1) = 0 \text{ dB}$$

and for large frequencies $\omega > \omega_n$:

$$\left|\left(\frac{j\omega}{\omega_n}\right)^2 + \frac{2\zeta}{\omega_n}j\omega + 1\right|_{\text{dB}} \approx 20 \log\left(\frac{\omega^2}{\omega_n^2}\right) = 40 \log\left(\frac{\omega}{\omega_n}\right)$$

Thus the high-frequency asymptote is a straight line with slope of 40 dB/decade. The asymptote approximation of the magnitude of the quadratic term is shown in Figure 9.29a. In this case the difference between the asymptote approximation and the exact plot depends on the value of ζ, as will be shown later.

The angle for low frequencies $\omega < \omega_n$ is approximated by

$$\underline{/(j\omega/\omega_n)^2 + (2\zeta/\omega_n)j\omega + 1} = \tan^{-1}\frac{2\zeta\omega/\omega_n}{1 - (\omega/\omega_n)^2} \approx \tan^{-1}\frac{0}{1} = 0°$$

For high frequencies $\omega > \omega_n$, the angle is approximated by

$$\underline{/(j\omega/\omega_n)^2 + (2\zeta/\omega_n)j\omega + 1} = \tan^{-1}\frac{2\zeta\omega/\omega_n}{1 - (\omega/\omega_n)^2} \approx \tan^{-1}\frac{2\zeta\omega_n}{-\omega} \approx 180°$$

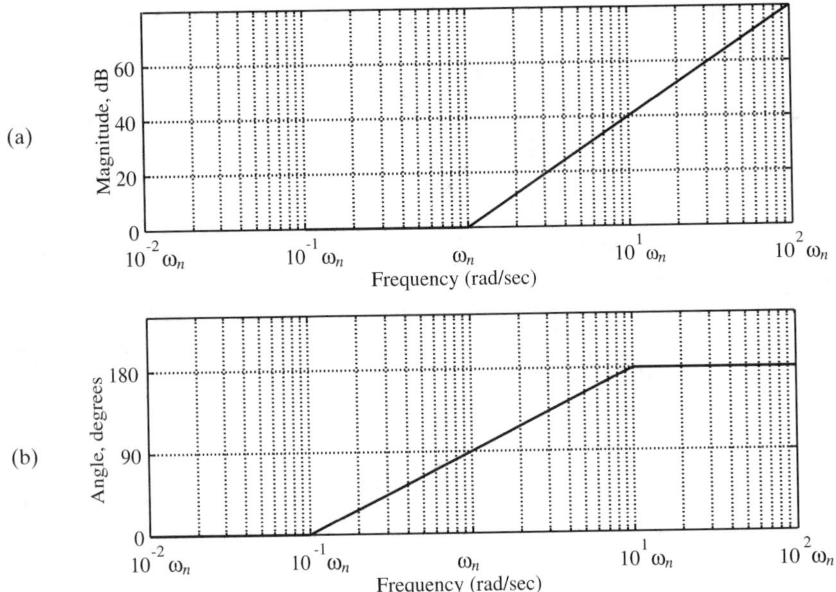

Figure 9.29 Magnitude (a) and phase (b) asymptote approximations for the quadratic term.

The transition between the low- and high-frequency asymptotes is a line spanning two decades from $(0.1)\omega_n$ to $10\omega_n$, with a slope of 90°/decade as shown in Figure 9.29b.

When the quadratic term is in the numerator of $H(\omega)$, it corresponds to a complex pair of zeros in $H(s)$. Thus, by the analysis above it is seen that a pair of complex zeros contribute a phase angle of approximately 180° and a slope of +40 dB/decade in magnitude at high frequencies. This corresponds to the comment made previously that each real zero of a transfer function contributes a phase angle of +90° and a slope of +20 dB/decade in magnitude to the Bode diagrams at high frequencies.

When the quadratic term is in the denominator, it corresponds to a complex pair of poles of $H(s)$. In this case the asymptote approximations shown in Figure 9.29 are negated to yield the Bode plots for $[(j\omega/\omega_n)^2 + (2\zeta/\omega_n)j\omega + 1]^{-1}$. Hence a pair of complex poles contributes to the Bode diagrams a phase angle of −180° and a slope in the magnitude plot of −40 dB/decade at high frequencies. As discussed above, every real pole contributes −90° and −20 dB/decade at high frequencies.

As mentioned before, the exact Bode plot in the quadratic case depends on the value of ζ. The exact Bode diagrams are plotted in Figure 9.30 for various values of ζ. Note from the figure that as ζ approaches 0, the exact magnitude curve has a peak at ω_n which grows in magnitude while the angle transition in the phase curve becomes sharper. Therefore, the asymptote approximation is not very accurate for small values of ζ. As ζ approaches 1, the quadratic term approaches a term containing two re-

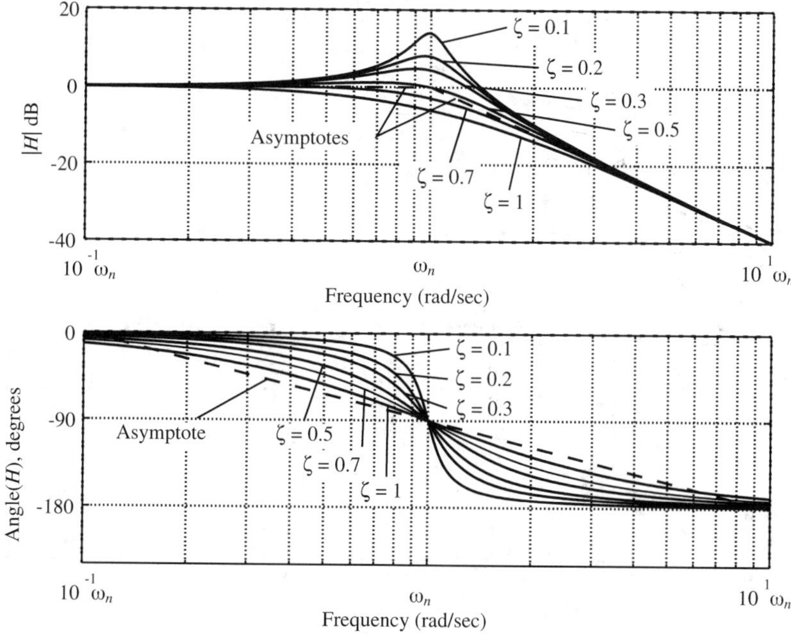

Figure 9.30 Exact Bode diagrams for a quadratic term.

peated real poles. The resulting Bode plot is equal to the sum of two single-pole plots. In this case the error in the magnitude approximation is 6 dB at the corners (3 dB at each corner frequency corresponding to the real poles).

Example 9.23 *Quadratic Term in Denominator*

Consider the system with transfer function

$$H(s) = \frac{63(s + 1)}{s(s^2 + 6s + 100)}$$

First, rewrite $H(\omega)$ in the standard form

$$H(\omega) = \frac{0.63(j\omega + 1)}{j\omega[(j\omega/10)^2 + 0.06j\omega + 1]}$$

The factors of $H(\omega)$ are 0.63, $j\omega + 1$, $(j\omega)^{-1}$, and $[(j\omega/10)^2 + 0.06j\omega + 1]^{-1}$. The corner frequencies are $\omega_{cf} = 1$ rad/sec for the zero and $\omega_n = 10$ rad/sec for the quadratic in the denominator. To obtain the magnitude plot, note that for $\omega < 1$ rad/sec, the only nonzero factors are the constant with magnitude $20 \log(0.63) = -4$ dB and $(j\omega)^{-1}$, which is a line with slope -20 dB/decade and an intercept of 0 dB at 1 rad/sec. Combining these factors for $\omega < 1$ simply offsets the -20 dB/decade line by -4 dB. This low-frequency asymptote is plotted in Figure 9.31a. The corner frequency at $\omega_{cf} = 1$ corresponds to a zero, which means that there is a change in slope of $+20$ dB/decade.

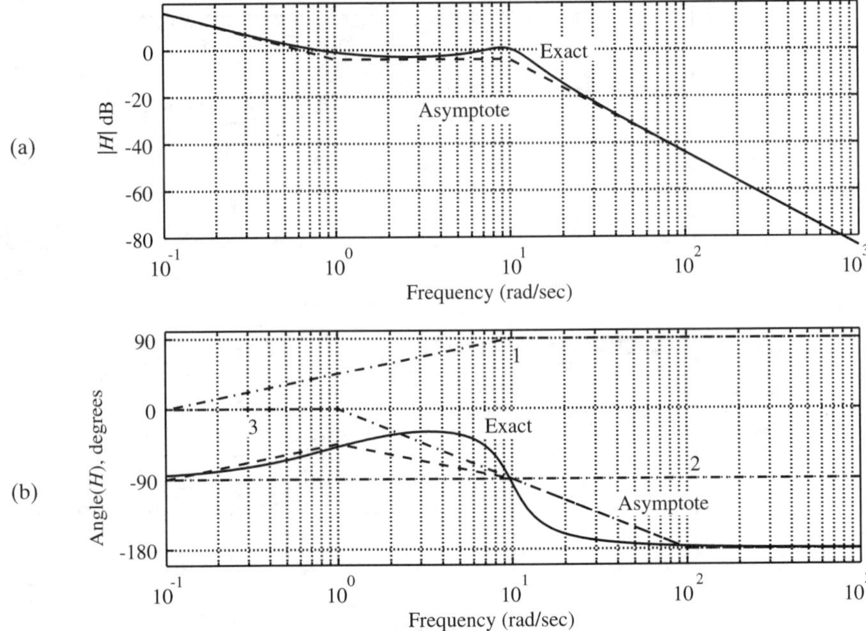

Figure 9.31 Asymptotes and exact Bode plots for Example 9.23: (a) magnitude; (b) phase.

Therefore, the slope for $1 < \omega < 10$ is 0 dB/decade. The corner frequency at $\omega = 10$ corresponds to a quadratic in the denominator, which means that the slope changes by -40 dB/decade for $\omega > 10$. The final magnitude plot is shown in Figure 9.31a along with the exact Bode plot obtained from MATLAB. The phase plots for the individual factors are shown in Figure 9.31b along with the final asymptote plot and the exact plot. The MATLAB commands used to generate the exact Bode plot are

```
num = 63*[1 1];
den = [1 6 100 0];
bode(num,den);
```

9.6 CAUSAL FILTERS

In real-time filtering applications, it is not possible to utilize ideal filters since they are noncausal (see Section 5.4). In such applications it is necessary to use causal filters which are nonideal; that is, the transition from the passband to the stopband (and vice versa) is gradual. In particular, the magnitude functions of causal versions of lowpass, highpass, bandpass, and bandstop filters have gradual transitions from the passband to the stopband. Examples of magnitude functions for these basic types of filters are shown in Figure 9.32.

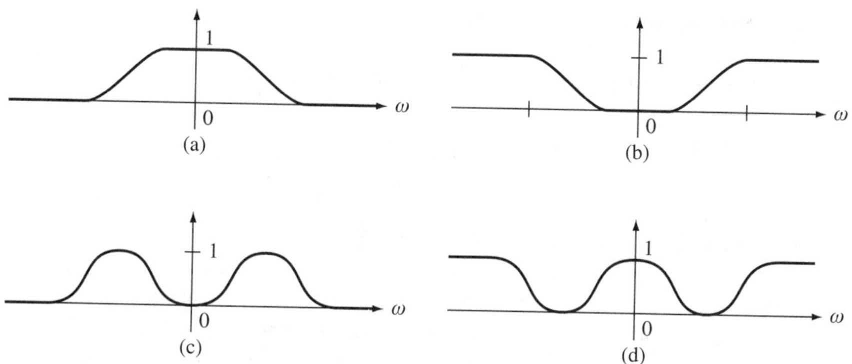

Figure 9.32 Causal filter magnitude functions: (a) lowpass; (b) highpass; (c) bandpass; (d) bandstop.

Consider a causal filter with frequency function $H(\omega)$ and with the peak value of $|H(\omega)|$ equal to 1. Then the passband is defined as the set of all frequencies ω for which

$$|H(\omega)| \geq \frac{1}{\sqrt{2}} = 0.707 \tag{9.54}$$

Note that (9.54) is equivalent to the condition that $|H(\omega)|_{\text{dB}}$ is less than 3 dB down from the peak value of 0 dB. For lowpass or bandpass filters, the width of the passband is called the 3-dB bandwidth.

A stopband in a causal filter is a set of frequencies ω for which $|H(\omega)|_{\text{dB}}$ is down some desired amount (e.g., 40 or 50 dB) from the peak value of 0 dB. The range of frequencies between a passband and a stopband is call a *transition region.* In causal filter design, a key objective is to have the transition regions be suitably small in extent. Later in this section it will be seen that the sharpest transitions can be achieved by allowing for ripple in the passband and/or stopband (as opposed to the monotone characteristics shown in Figure 9.32).

To be able to build a causal filter from circuit components, it is necessary that the filter transfer function $H(s)$ be rational in s. For ease of implementation, the order of $H(s)$ (i.e., the degree of the denominator) should be as small as possible. However, there is always a trade-off between the magnitude of the order and desired filter characteristics such as the amount of attenuation in the stopbands and the width of the transition regions.

As in the case of ideal filters, to avoid phase distortion in the output of a causal filter, the phase function should be linear over the passband of the filter. However, the phase function of a causal filter with rational transfer function cannot be exactly linear over the passband, and thus there will always be some phase distortion. The amount of phase distortion that can be tolerated is often included in the list of filter specifications in the design process.

Butterworth Filters

For the two-pole system with the transfer function

$$H(s) = \frac{\omega_n^2}{s^2 + 2\zeta\omega_n s + \omega_n^2}$$

it follows from the results in Section 9.5 that the system is a lowpass filter when $\zeta \geq 1/\sqrt{2}$. If $\zeta = 1/\sqrt{2}$, the resulting lowpass filter is said to be *maximally flat,* since the variation in the magnitude $|H(\omega)|$ is as small as possible across the passband of the filter. This filter is called the two-pole *Butterworth filter.*

The transfer function of the two-pole Butterworth filter is

$$H(s) = \frac{\omega_n^2}{s^2 + \sqrt{2}\omega_n s + \omega_n^2}$$

Factoring the denominator of $H(s)$ reveals that the poles are located at

$$s = -\frac{\omega_n}{\sqrt{2}} \pm j\frac{\omega_n}{\sqrt{2}}$$

Note that the magnitude of each of the poles is equal to ω_n.

Setting $s = j\omega$ in $H(s)$ yields the magnitude function of the two-pole Butterworth filter:

$$\begin{aligned} |H(\omega)| &= \frac{\omega_n^2}{\sqrt{(\omega_n^2 - \omega^2)^2 + 2\omega_n^2\omega^2}} \\ &= \frac{\omega_n^2}{\sqrt{\omega_n^4 - 2\omega_n^2\omega^2 + \omega^4 + 2\omega_n^2\omega^2}} \\ &= \frac{\omega_n^2}{\sqrt{\omega_n^4 + \omega^4}} \\ &= \frac{1}{\sqrt{1 + (\omega/\omega_n)^4}} \end{aligned} \tag{9.55}$$

From (9.55) it is seen that the 3-dB bandwidth of the Butterworth filter is equal to ω_n; that is, $|H(\omega_n)|_{\text{dB}} = -3$ dB. For a lowpass filter, the point where $|H(\omega)|_{\text{dB}}$ is down by 3 dB is often referred to as the *cutoff frequency.* Hence ω_n is the cutoff frequency of the lowpass filter with magnitude function given by (9.55).

For the case $\omega_n = 2$ rad/sec, the frequency response curves of the Butterworth filter are plotted in Figure 9.33. Also displayed are the frequency response curves for the one-pole lowpass filter with transfer function $H(s) = 2/(s + 2)$, and the two-pole lowpass filter with $\zeta = 1$ and with cutoff frequency equal to 2 rad/sec. Note that the Butterworth filter has the sharpest transition of all three filters.

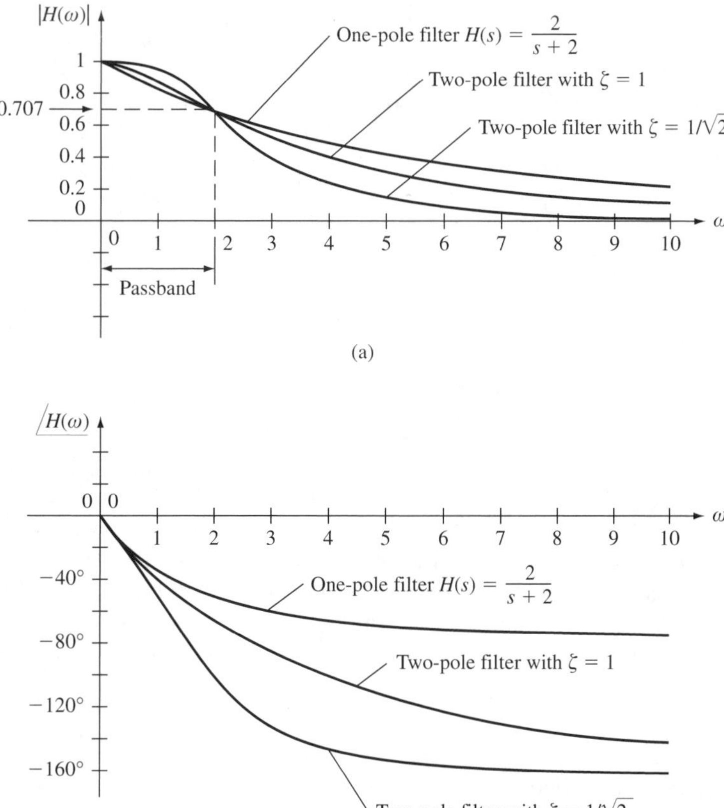

(b)

Figure 9.33 Frequency curves of one- and two-pole lowpass filters: (a) magnitude curves; (b) phase curves.

***N*-pole Butterworth filter.** For any positive integer N, the *N-pole Butterworth filter* is the lowpass filter of order N with a maximally flat frequency response across the passband. The distinguishing characteristic of the Butterworth filter is that the poles lie on a semicircle in the open left-half plane. The radius of the semicircle is equal to ω_c, where ω_c is the cutoff frequency of the filter. In the third-order case, the poles are as displayed in Figure 9.34.

The transfer function of the three-pole Butterworth filter is

$$H(s) = \frac{\omega_c^3}{(s + \omega_c)(s^2 + \omega_c s + \omega_c^2)} = \frac{\omega_c^3}{s^3 + 2\omega_c s^2 + s\omega_c^2 s + \omega_c^3}$$

Setting $s = j\omega$ in $H(s)$ and taking the magnitude results in the magnitude function of the three-pole filter:

$$|H(\omega)| = \frac{1}{\sqrt{1 + (\omega/\omega_c)^6}}$$

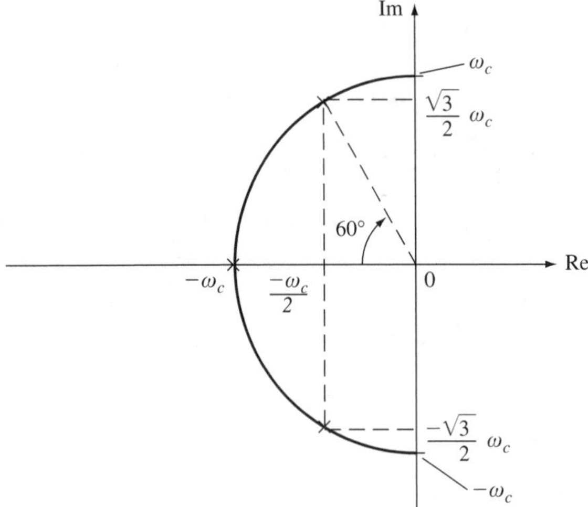

Figure 9.34 Pole locations for the three-pole Butterworth filter.

The magnitude function is plotted in Figure 9.35 for the case $\omega_c = 2$. Also plotted is the magnitude function of the two-pole Butterworth filter with cutoff frequency equal to 2. Clearly, the three-pole filter has a sharper transition than the two-pole filter.

In the general case, the magnitude function of the N-pole Butterworth filter is

$$|H(\omega)| = \frac{1}{\sqrt{1 + (\omega/\omega_c)^{2N}}}$$

The transfer function can be determined from a table for Butterworth polynomials. For example, when $N = 4$, the transfer function is

$$H(s) = \frac{\omega_c^4}{(s^2 + 0.765\omega_c s + \omega_c^2)(s^2 + 1.85\omega_c s + \omega_c^2)}$$

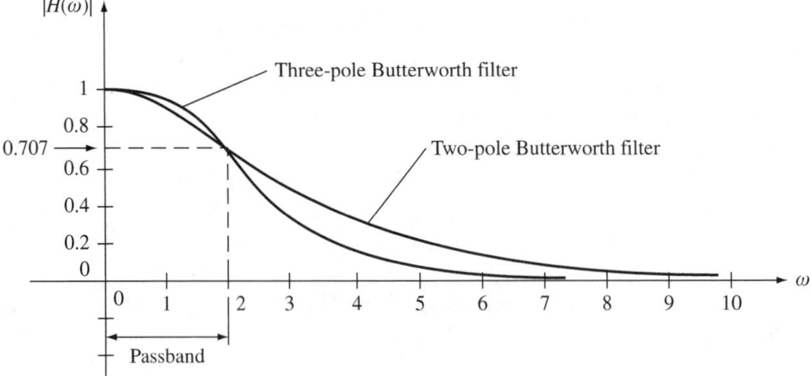

Figure 9.35 Magnitude curves of two- and three-pole Butterworth filters.

When $N = 5$, the transfer function is

$$H(s) = \frac{\omega_c^5}{(s + \omega_c)(s^2 + 0.618\omega_c s + \omega_c^2)(s^2 + 1.62\omega_c s + \omega_c^2)}$$

MATLAB contains a command for designing Butterworth filters with the cutoff frequency normalized to 1 rad/sec. For example, the following commands may be used to create the two-pole Butterworth filter and to obtain the magnitude and phase functions of the filter:

```
[z,p,k] = buttap(2);          % 2 pole filter
[b,a] = zp2tf(z,p,k);     % convert to polynomials
w = 0:0.01:4;
[mag,phase] = bode(b,a,w);
```

Executing the commands above yields

```
b = 1

a = [1 1.414 1]
```

which are the coefficients of the numerator and denominator polynomials of the filter transfer function. Running the MATLAB software for the two-, five-, and 10-pole Butterworth filters results in the frequency response curves shown in Figure 9.36. Note that the higher the order of the filter, the sharper the transition is from the passband to the stopband.

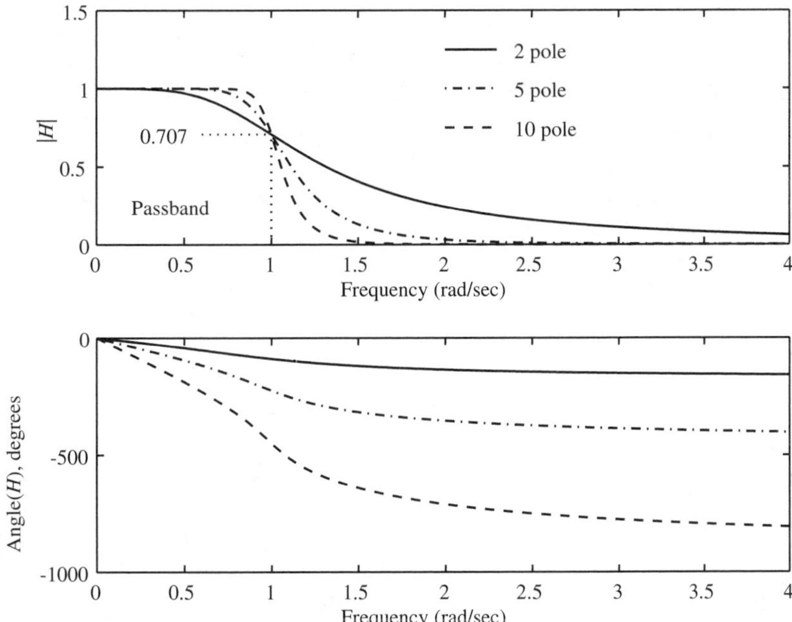

Figure 9.36 Frequency response curves for the two-, five-, and 10-pole Butterworth filters.

Chebyshev Filters

The magnitude function of the N-pole Butterworth filter has a monotone characteristic in both the passband and stopband of the filter. Here *monotone* means that the magnitude curve is gradually decreasing over the passband and stopband. In contrast to the Butterworth filter, the magnitude function of a type 1 Chebyshev filter has ripple in the passband and is monotone decreasing in the stopband (a type 2 Chebyshev filter has the opposite characteristic). By allowing ripple in the passband or stopband, it is possible to achieve a sharper transition between the passband and stopbands in comparison with the Butterworth filter.

The N-pole type 1 Chebyshev filter is given by the frequency function

$$|H(\omega)| = \frac{1}{\sqrt{1 + \varepsilon^2 T_N^2(\omega/\omega_1)}} \tag{9.56}$$

where $T_N(\omega/\omega_1)$ is the Nth-order Chebyshev polynomial and ε is a positive number. The Chebyshev polynomials can be generated from the recursion

$$T_N(x) = 2xT_{N-1}(x) - T_{N-2}(x)$$

where $T_0(x) = 1$ and $T_1(x) = x$. The polynomials for $N = 2, 3, 4, 5$ are

$$\begin{aligned} T_2(x) &= 2x(x) - 1 = 2x^2 - 1 \\ T_3(x) &= 2x(2x^2 - 1) - x = 4x^3 - 3x \\ T_4(x) &= 2x(4x^3 - 3x) - (2x^2 - 1) = 8x^4 - 8x^2 + 1 \\ T_5(x) &= 2x(8x^4 - 8x^2 + 1) - (4x^3 - 3x) = 16x^5 - 20x^3 + 5x \end{aligned} \tag{9.57}$$

Using (9.57) yields the two-pole type 1 Chebyshev filter with frequency function

$$|H(\omega)| = \frac{1}{\sqrt{1 + \varepsilon^2[2(\omega/\omega_1)^2 - 1]^2}}$$

For the N-pole filter defined by (9.56), it can be shown that the magnitude function $|H(\omega)|$ of the filter oscillates between the value 1 and the value $1/\sqrt{1 + \varepsilon^2}$ as ω is varied from 0 to ω_1, with

$$H(0) = \begin{cases} 1 & \text{if } N \text{ is odd} \\ \dfrac{1}{\sqrt{1 + \varepsilon^2}} & \text{if } N \text{ is even} \end{cases}$$

and

$$|H(\omega_1)| = \frac{1}{\sqrt{1 + \varepsilon^2}}$$

The magnitude function $|H(\omega)|$ is monotone decreasing for $\omega > \omega_1$, and thus the filter is a lowpass filter with ripple over the passband. In general, ω_1 is not equal to the cutoff frequency (the 3-dB point) of the filter; however, if

$$\frac{1}{\sqrt{1 + \varepsilon^2}} = \frac{1}{\sqrt{2}}$$

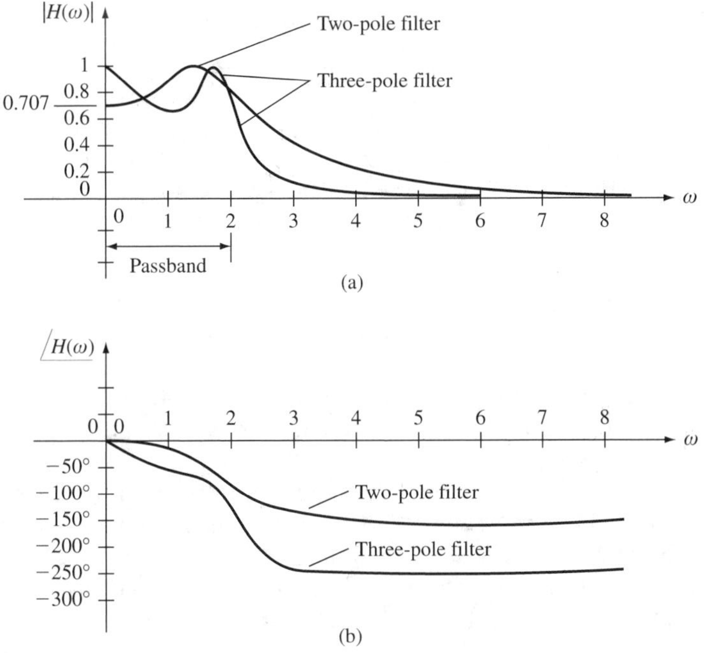

Figure 9.37 Frequency curves of two- and three-pole Chebyshev filters with $\omega_c = 2$ radians: (a) magnitude curves; (b) phase curves.

so that $\varepsilon = 1$, then $|H(\omega_1)| = 1/\sqrt{2}$, and in this case ω_1 is the cutoff frequency. When $\varepsilon = 1$, the ripple varies by 3 dB across the passband of the filter.

For the case of a 3-dB ripple ($\varepsilon = 1$), the transfer functions of the two- and three-pole type 1 Chebyshev filters are

$$H(s) = \frac{0.50\omega_c^2}{s^2 + 0.645\omega_c s + 0.708\omega_c^2}$$

$$H(s) = \frac{0.251\omega_c^3}{s^3 + 0.597\omega_c s^2 + 0.928\omega_c^2 s + 0.251\omega_c^3}$$

where ω_c is the cutoff frequency. The frequency response curves for these two filters are plotted in Figure 9.37 for the case $\omega_c = 2$ radians.

The magnitude response functions of the three-pole Butterworth filter and the three-pole type 1 Chebyshev filter are compared in Figure 9.38 with the cutoff frequency of both filters equal to 2 rad/sec. Note that the transition from passband to stopband is sharper in the Chebyshev filter; however, the Chebyshev filter does have the 3-dB ripple over the passband.

The transition from the passband to the stopband can be made sharper (in comparison with the Chebyshev filter) by allowing ripple in both the passband and

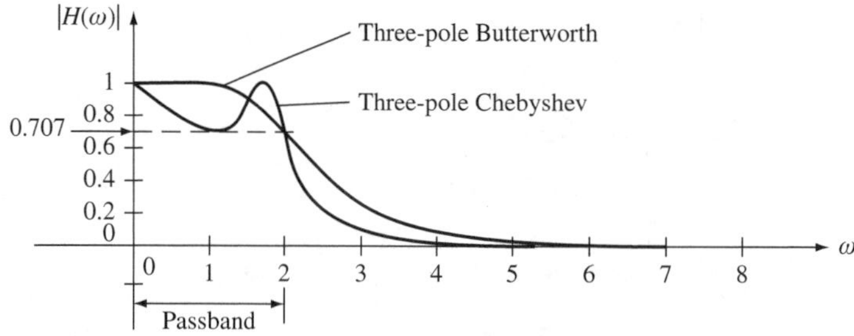

Figure 9.38 Magnitude curves of three-pole Butterworth and three-pole Chebyshev filters with cutoff frequency equal to 2 radians.

stopband. *Elliptic filters* are examples of a type of filter that yields sharp transitions by permitting ripple in the passband and stopband. This type of filter is not considered here (see Ludeman [1986]).

The MATLAB command `cheb1ap` can be used for designing type 1 Chebyshev filters. To run this command, the user must input the order of the filter and the amount of ripple in dB allowed in the passband. The resulting filter will have a normalized cutoff frequency of 1 rad/sec. For example, to design a two-pole Chebyshev filter that allows a 3-dB ripple in the passband, the commands are

```
[z,p,k] = cheb1ap(2,3);
[b,a] = zp2tf(z,p,k);        % convert to polynomials
w = 0:0.01:4;
[mag,phase] = bode(b,a,w);
```

The frequency response curves of the resulting filter are plotted in Figure 9.39. Also plotted in Figure 9.39 are the response curves of the Chebyshev three- and five-pole filters with a maximum 3 dB ripple in the passband. Note that the transition is sharper and the ripple is more pronounced as the number of poles is increased. Also note that in all three cases, the ripple remains within the 3-dB (0.707) limit.

Frequency Transformations

The Butterworth and Chebyshev filters discussed above are examples of lowpass filters. Starting with any lowpass filter having transfer function $H(s)$, it is possible to modify the cutoff frequency of the filter, or to construct highpass, bandpass, and bandstop filters, by transforming the frequency variable s. For example, if the cutoff frequency of a lowpass filter is $\omega_c = \omega_1$ and it is desired to have the cutoff frequency changed to ω_2, replace s in $H(s)$ by $s\omega_1/\omega_2$. To convert a lowpass filter with a cutoff frequency of ω_1 to a highpass filter with a 3-dB passband running from $\omega = \omega_2$ to $\omega = \infty$, replace s in $H(s)$ by $\omega_1\omega_2/s$. To obtain a bandpass filter with a 3-dB passband running from $\omega = \omega_1$ to $\omega = \omega_2$, replace s in $H(s)$ by

$$\omega_c \frac{s^2 + \omega_1\omega_2}{s(\omega_2 - \omega_1)}$$

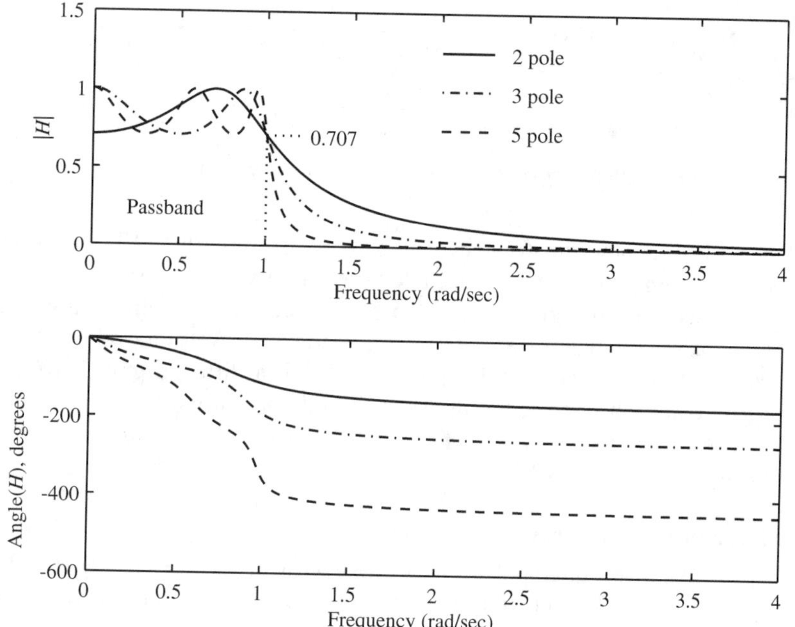

Figure 9.39 Frequency response curves of the two-, three-, and five-pole Chebyshev filters.

Finally, to obtain a bandstop filter with a 3-dB passband running from $\omega = 0$ to $\omega = \omega_1$ and from $\omega = \omega_2$ to $\omega = \infty$, replace s in $H(s)$ by

$$\omega_c \frac{s(\omega_2 - \omega_1)}{s^2 + \omega_1\omega_2}$$

Example 9.24 *Three-Pole Butterworth Filter*

Consider the three-pole Butterworth filter with transfer function

$$H(s) = \frac{\omega_c^3}{s^3 + 2\omega_c s^2 + 2\omega_c^2 s + \omega_c^3}$$

where ω_c is the cutoff frequency. Suppose that the objective is to design a bandpass filter with passband running from $\omega_1 = 3$ to $\omega_2 = 5$. With s replaced by

$$\omega_c \frac{s^2 + 15}{2s}$$

the transfer function of the resulting bandpass filter is

$$H(s) = \frac{\omega_c^3}{\omega_c^3\left(\frac{s^2 + 15}{2s}\right)^3 + 2\omega_c^3\left(\frac{s^2 + 15}{2s}\right)^2 + 2\omega_c^3\left(\frac{s^2 + 15}{2s}\right) + \omega_c^3}$$

$$H(s) = \frac{8s^3}{(s^2 + 15)^3 + 4s(s^2 + 15)^2 + 8s^2(s^2 + 15) + 8s^3}$$
$$= \frac{8s^3}{s^6 + 4s^5 + 53s^4 + 128s^3 + 795s^2 + 900s + 3375}$$

The frequency curves for this filter are displayed in Figure 9.40.

Frequency transformations are very useful with MATLAB since the standard filter design programs produce a lowpass filter with a normalized cutoff frequency of 1 rad/sec. The design process can be started by first generating a Butterworth or Chebyshev lowpass filter with normalized cutoff frequency. Then the resulting filter can be transformed to a lowpass filter with a different cutoff frequency, or transformed to a highpass, bandpass, or bandstop filter using the commands: `lp2lp`, `lp2hp`, `lp2bp`, and `lp2bs`. The following examples show how to use MATLAB to design various types of filters.

Example 9.25 *Lowpass Filter Design*

To design a three-pole Butterworth lowpass filter with a bandwidth of 5 Hz, first design a three-pole filter with cutoff frequency of 1 rad/sec using the `buttap` command. Then

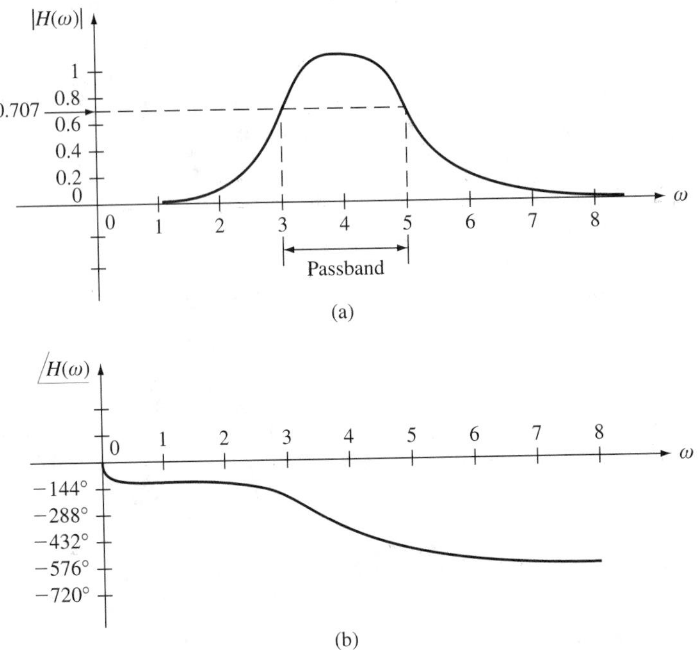

Figure 9.40 Frequency response curves of six-pole bandpass filter: (a) magnitude; (b) phase.

transform the frequency using the command lp2lp. The MATLAB commands are as follows:

```
[z,p,k] = buttap(3); % 3 pole filter
[b,a] = zp2tf(z,p,k);% convert to polynomials
wb = 5*2*pi;            % new bandwidth in rad/sec
[b,a] = lp2lp(b,a,wb);     % transforms to the new bandwidth
f = 0:15/200:15;     % define the frequency in Hz for plotting
w = 2*pi*f;
[mag,phase] = bode(b,a,w);
```

The coefficients of the numerator and denominator are found to be given by `b = [0 0 0 31006]` and `a = [1 63 1974 31006]`, respectively. Hence the 5-Hz bandwidth filter is given by

$$H(s) = \frac{31{,}006}{s^3 + 63s^2 + 1974s + 31{,}006}$$

The resulting frequency response is plotted in Figure 9.41.

Example 9.26 *Highpass Filter Design*

To design a three-pole highpass filter with cutoff frequency $\omega = 4$ rad/sec, first design a three-pole Chebyshev or Butterworth filter with cutoff frequency of $\omega_c = 1$ rad/sec. Then transform it to a highpass filter. The commands are

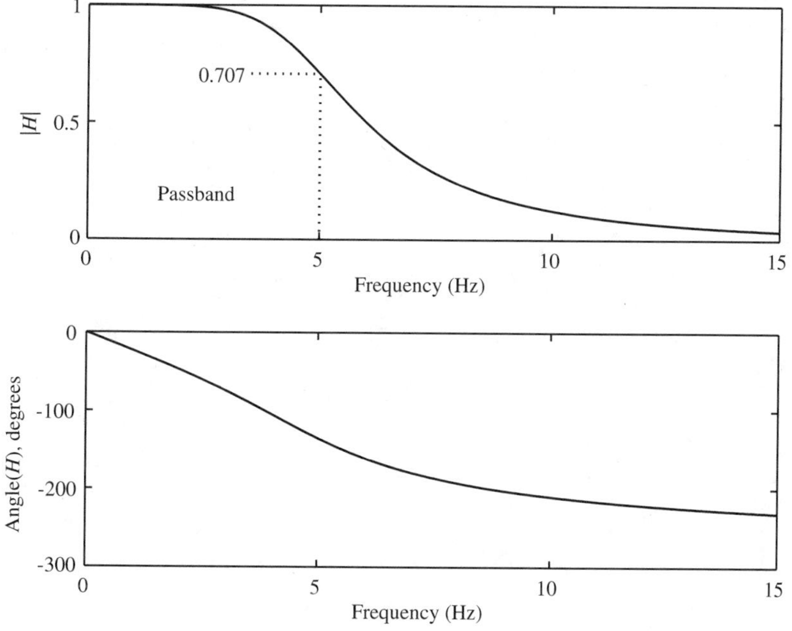

Figure 9.41 Frequency response curves for the lowpass filter in Example 9.25.

```
w0 = 4;                          % cutoff frequency
% create a 3 pole Chebyshev Type I filter with 3-dB passband
[z,p,k] = cheblap(3,3);
[b,a] = zp2tf(z,p,t); % convert to polynomials
% convert the lowpass filter to a highpass with
%   cutoff frequency w0
[b,a]=lp2hp(b,a,w0);
w = 0:0.025:10;
[mag,phase] = bode(b,a,w);
```

The coefficients of the numerator and denominator, located in the vectors `b` and `a`, respectively, are found to be `b = [1 0 0 0]` and `a = [1 14.8 39.1 255.4]`. Hence the resulting highpass filter is given by

$$H(s) = \frac{s^3}{s^3 + 14.8s^2 + 38.1s + 255.4}$$

The frequency response plot is shown in Figure 9.42.

Example 9.27 *Design Using MATLAB*

In Example 9.24, a three-pole Butterworth lowpass filter was transformed to a bandpass filter with the passband centered at $\omega = 4$ rad/sec and with the bandwidth equal to 2 rad/sec. To perform this conversion via MATLAB, use the following commands:

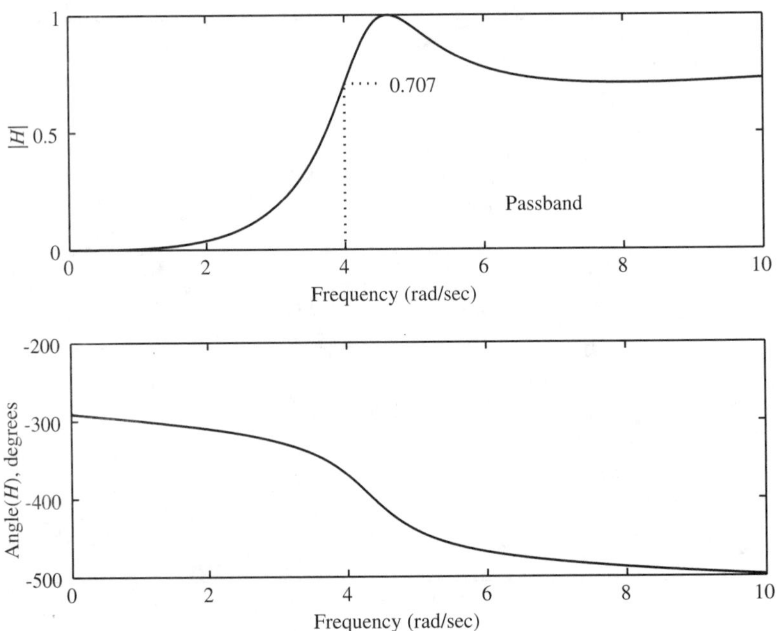

Figure 9.42 Frequency response of filter in Example 9.26.

```
w0 = 4;                        % center of band
wb = 2;                        % bandwidth
[z,p,k] = buttap(3);           % 3 pole Butterworth filter
[b,a] = zp2tf(z,p,k);          % convert to polynomials
% convert the lowpass filter to a bandpass
%      centered at w0 with bandwidth wb
[b,a]=lp2bp(b,a,w0,wb);
w = 0:0.05:10;
[mag,phase] = bode(b,a,w);
```

The numerator coefficients stored in the vector `b` match those obtained analytically in Example 9.21. However, the denominator coefficients calculated as `a = [1 4 56 136 896 1024 4096]` are different from those determined analytically. Although the coefficients have large differences, the corresponding poles are very close, resulting in very little error in the frequency response plot shown in Figure 9.43. The difference is due to roundoff errors in the computations.

Example 9.28 *Bandstop Filter*

To convert a three-pole Chebyshev lowpass filter to a bandstop filter with a stopband from $\omega = 4$ to $\omega = 6$, use the following commands:

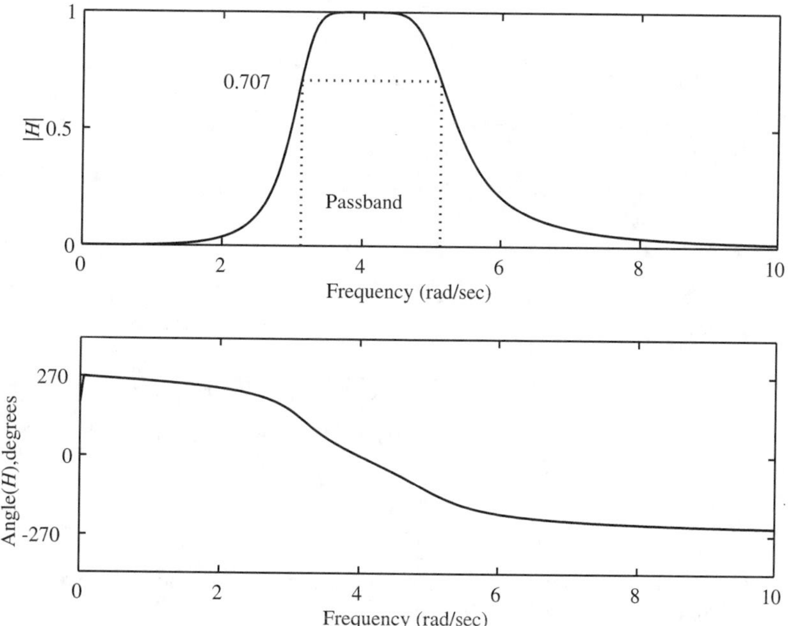

Figure 9.43 Response curves for filter in Example 9.27.

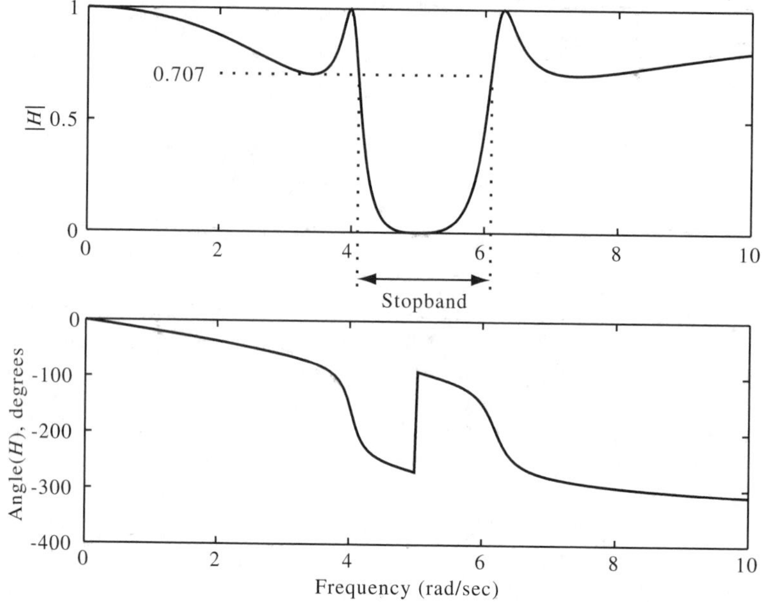

Figure 9.44 Frequency response of filter in Example 9.28.

```
w0 = 5;                          % center of stopband
wb = 2;                          % width of stopband
[z,p,k] = cheblap(3,3);          % 3 pole filter with 3-dB ripple
[b,a] = zp2tf(z,p,k);            % convert to polynomials
[b,a] = lp2bs(b,a,w0,wb);        % converts to a bandstop filter
w = 0:0.02:10;
[mag,phase] = bode(b,a,w);
```

The coefficients are calculated to be `b = [1 0 75 0 1875 0 15625]` and `a = [1 7 85 402 2113 4631 15625]`. Hence the bandstop filter is given by

$$H(s) = \frac{s^6 + 75s^4 + 1875s^2 + 15{,}625}{s^6 + 7s^5 + 85s^4 + 402s^3 + 2113s^2 + 4631s + 15{,}625}$$

The resulting frequency response plot is shown in Figure 9.44.

PROBLEMS

9.1. For the following linear time-invariant continuous-time systems, determine if the system is stable, marginally stable, or unstable.

(a) $H(s) = \dfrac{s - 4}{s^2 + 7s}$

(b) $H(s) = \dfrac{s+3}{s^2+3}$

(c) $H(s) = \dfrac{2s^2+3s+1}{s^3+2s^2+4}$

(d) $H(s) = \dfrac{3s^3-2s+6}{s^3+s^2+s+1}$

9.2. Consider the field-controlled dc motor given by the input/output differential equation

$$L_f I \frac{d^3y(t)}{dt^3} + (L_f k_d + R_f I)\frac{d^2y(t)}{dt^2} + R_f k_d \frac{dy(t)}{dt} = kx(t)$$

Assume that all the parameters I, L_f, k_d, R_f, and k are strictly positive (>0). Determine if the motor is stable, marginally stable, or unstable.

9.3. Consider the model for the ingestion and metabolism of a drug defined in Problem 1.19. Assuming that $k_1 > 0$ and $k_2 > 0$, determine if the system is stable, marginally stable, or unstable. What does your answer imply regarding the behavior of the system? Explain.

9.4. For the automobile suspension system defined in Example 2.2, assume that $M_1 = 100$, $M_2 = 1500$, $k_t = 1000$, $k_s = 200$, and $k_d = 10$. Is the system stable, marginally stable, or unstable?

9.5. Determine if the mass/spring system in Problem 2.15 is stable, marginally stable, or unstable. Assume that k_1, k_2, and k_3 are strictly positive (>0).

9.6. Determine if the small-signal model of the two connected pendulums in Problem 2.16 is stable, marginally stable, or unstable. Assume that all parameters are strictly positive (>0).

9.7. Consider the single-eye system studied in Problem 3.32. Assuming that $T_e > 0$, determine if the system is BIBO stable.

9.8. For the linear time-invariant continuous-time systems with impulse response $h(t)$ given below, determine if the system is BIBO stable.

(a) $h(t) = [2t^3 - 2t^2 + 3t - 2][u(t) - u(t-10)]$

(b) $h(t) = \dfrac{1}{t}$ for $t \geq 1$, $h(t) = 0$ for all $t < 1$

(c) $h(t) = \sin 2t$ for $t \geq 0$

(d) $h(t) = e^{-t}\sin 2t$ for $t \geq 0$

(e) $h(t) = e^{-t^2}$ for $t \geq 0$

9.9. Using the Routh–Hurwitz test, determine all values of the parameter k for which the following systems are stable.

(a) $H(s) = \dfrac{s^2+60s+800}{s^3+30s^2+(k+200)s+40k}$

(b) $H(s) = \dfrac{2s^3-3s+4}{s^4+s^3+ks^2+2s+3}$

(c) $H(s) = \dfrac{s^2+3s-2}{s^3+s^2+(k+3)s+3k-5}$

(d) $H(s) = \dfrac{s^4-3s^2+4s+6}{s^5+10s^4+(9+k)s^3+(90+2k)s^2+12ks+10k}$

9.10. Suppose that a system has the following transfer function:

$$H(s) = \frac{8}{s+4}$$

(a) Compute the system response to the following inputs. Identify the steady-state solution and the transient solution.

(i) $x(t) = u(t)$

(ii) $x(t) = tu(t)$

(iii) $x(t) = 2\,(\sin 2t)\,u(t)$

(iv) $x(t) = 2\,(\sin 10t)\,u(t)$

(b) Use MATLAB to compute the response numerically from $x(t)$ and $H(s)$. Plot the responses and compare them to the responses obtained analytically in part (a).

9.11. Consider three systems which have the following transfer functions:

Pole Positions and Step Response

(i) $H(s) = \dfrac{32}{s^2 + 4s + 16}$

(ii) $H(s) = \dfrac{32}{s^2 + 8s + 16}$

(iii) $H(s) = \dfrac{32}{s^2 + 10s + 16}$

For each system:

(a) Determine if the system is critically damped, underdamped, or overdamped.

(b) Calculate the step response of the system.

(c) Use MATLAB to compute the step response numerically. Plot the response and compare it to the plot of the response obtained analytically in part (b).

9.12. A first-order system has the step response shown in Figure P9.12. Determine the transfer function.

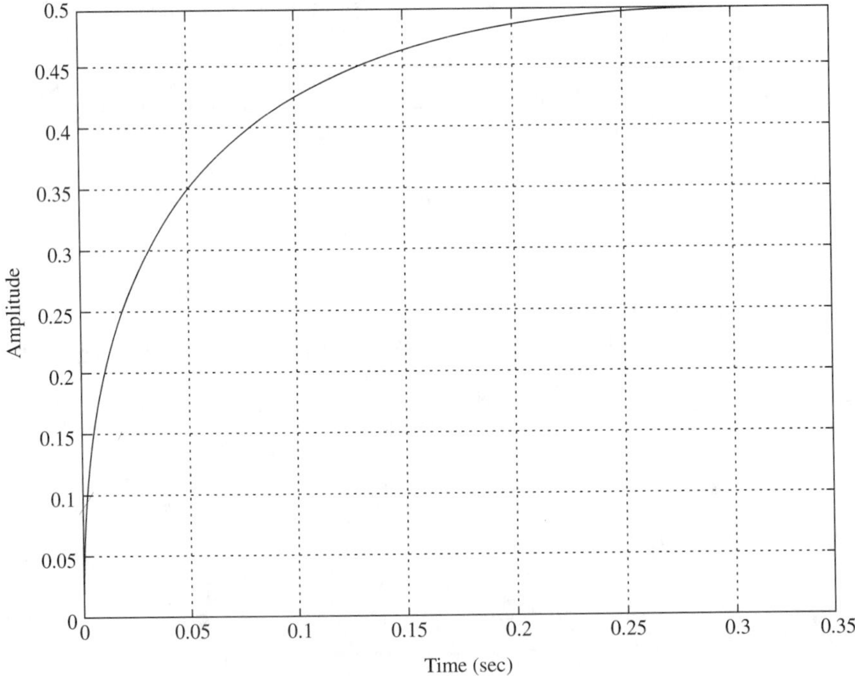

Figure P9.12

9.13. A second-order system has the step response shown in Figure P9.13. Determine the transfer function.

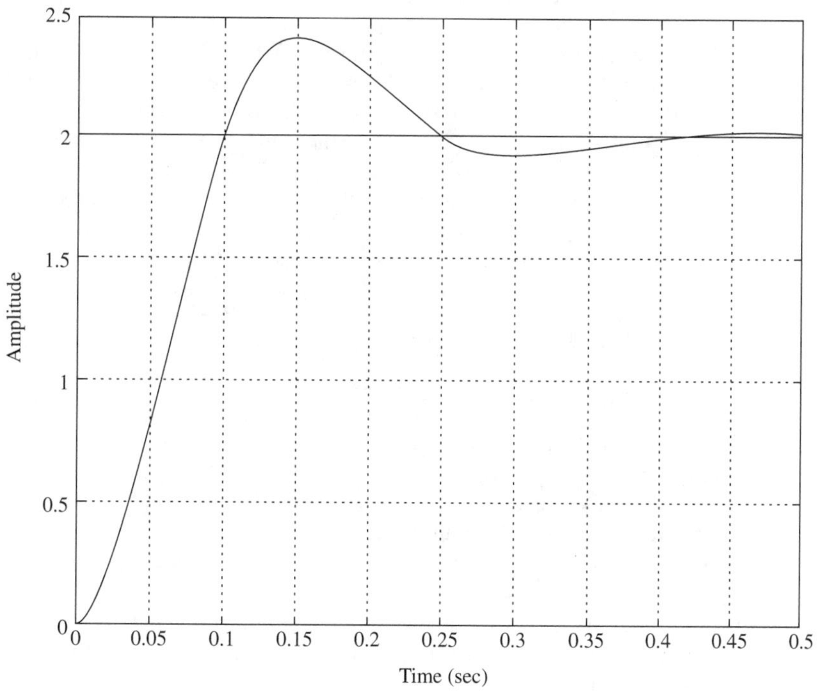

Figure P9.13

9.14. Consider the mass–spring–damper system with the input/output differential equation

$$M\frac{d^2y(t)}{dt^2} + D\frac{dy(t)}{dt} + Ky(t) = x(t)$$

where M is the mass, D is the damping constant, K is the stiffness constant, $x(t)$ is the force applied to the mass, and $y(t)$ is the displacement of the mass relative to the equilibrium position.

(a) Determine the pole locations for the cases **(i)** $M = 1, D = 50.4$, and $K = 3969$, and **(ii)** $M = 2, D = 50.4$, and $K = 3969$. Show the location of the poles on a pole-zero plot. Compute the natural frequency and the time constant for each of the cases. Which has the higher frequency of response? For which case does the transient response decay faster?

(b) Use MATLAB to compute the impulse response of the system for the two cases and compare your results with the predictions made in part (a).

(c) Repart parts (a) and (b) for the cases **(i)** $M = 1, D = 50.4$, and $K = 15{,}876$ and **(ii)** $M = 2, D = 50.4$, and $K = 15{,}876$.

Mass–Spring–Damper System

9.15. Again consider the mass-spring-damper system in Problem 9.14. Let $M = 1$, $D = 50.4$, and $K = 3969$.

(a) Compute the response to a unit step in the force.

(b) Compute the steady-state response to an input of $x(t) = 10\cos(20\pi t)u(t)$.

(c) Compute the steady-state response to an input of $x(t) = 10\cos(2\pi t)u(t)$.

(d) Use MATLAB to simulate the system with the inputs given in parts (a)–(c). Verify that your answers in parts (a)–(c) are correct by plotting them along with the corresponding results obtained from the simulation.

(e) Use the Mass–Spring–Damper demo available on the textbook Web page to simulate the system to the inputs given in parts (a)–(c) and compare the responses with those plotted in part (d). Change the damping parameter to $D = 127$, and use the applet to simulate the step response. Sketch the response.

9.16. Consider the two systems given by the following transfer functions:

(i) $H(s) = \dfrac{242.5(s + 8)}{(s + 2)[(s + 4)^2 + 81](s + 10)}$

(ii) $H(s) = \dfrac{115.5(s + 8)(s + 2.1)}{(s + 2)[(s + 4)^2 + 81](s + 10)}$

(a) Identify the poles and zeros of the system.

(b) Without computing the actual response, give the general form of the step response.

(c) Determine the steady-state value for the step response.

(d) Determine the dominant pole(s).

(e) Use MATLAB to compute and plot the step response of this system. Compare the plot to the answers expected in parts (b) to (d).

9.17. For each of the circuits in Figure P9.17, compute the steady-state response $y_{ss}(t)$ resulting from the inputs given below assuming that there is no initial energy at time $t = 0$.

(a) $x(t) = u(t)$

(b) $x(t) = (5\cos 2t)u(t)$

(c) $x(t) = [2\cos(3t + 45°)]u(t)$

9.18. Consider the mass–spring system in Problem 2.15. Assume that $M_1 = 1$, $M_2 = 10$, and $k_1 = k_2 = k_3 = 0.1$. Compute the steady-state response $y_{ss}(t)$ resulting from the inputs given below assuming no initial energy.

(a) $x(t) = u(t)$

(b) $x(t) = (10\cos t)u(t)$

(c) $x(t) = [\cos(5t - 30°)]u(t)$

9.19. For the automobile suspension system defined in Example 2.2, assume that $M_1 = 100$, $M_2 = 1500$, $k_t = 1000$, $k_s = 200$, and $k_d = 10$. Compute the steady-state response $y_{ss}(t)$ resulting from the inputs given below assuming no initial energy.

(a) $x(t) = u(t)$

(b) $x(t) = (\cos 0.1t)u(t)$

(c) $x(t) = (\cos t)u(t)$

9.20. A linear time-invariant continuous-time system has transfer function $H(s) = 2/(s + 1)$. Compute the transient response $y_{tr}(t)$ resulting from the input $x(t) = 3\cos 2t - 4\sin t$, $t \geq 0$, with no initial energy in the system.

9.21. A linear time-invariant continuous-time system has transfer function

$$H(s) = \frac{s^2 + 16}{s^2 + 7s + 12}$$

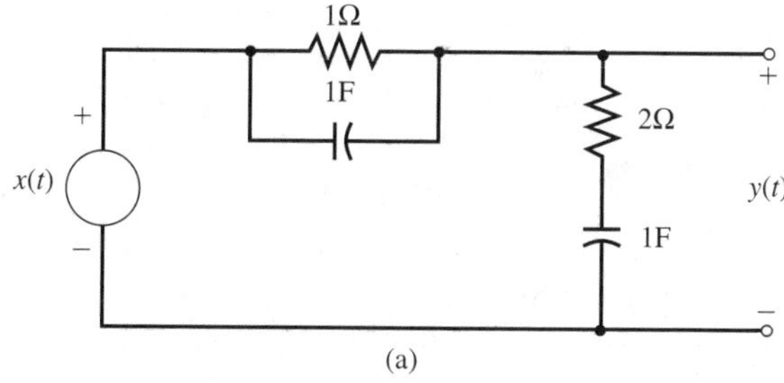

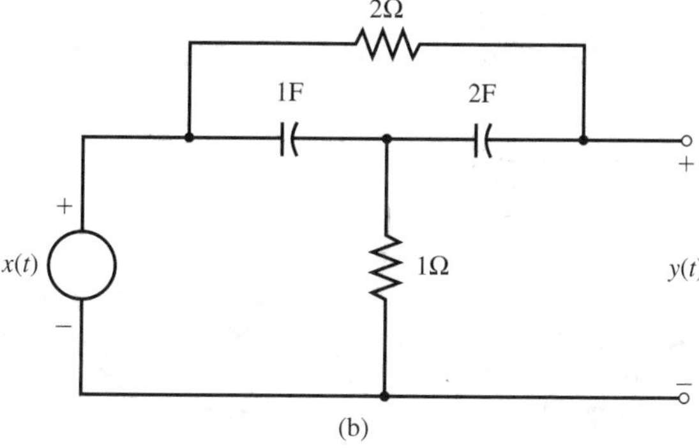

Figure P9.17

Compute the steady-state and transient responses resulting from the input $x(t) = 2\cos 4t$, $t \geq 0$, with no initial energy at time $t = 0$.

9.22. A linear time-invariant continuous-time system has transfer function

$$H(s) = \frac{s^2 + 1}{(s + 1)(s^2 + 2s + 17)}$$

Compute both the steady-state response $y_{ss}(t)$ and the transient response $y_{tr}(t)$ when the input $x(t)$ is

(a) $x(t) = u(t)$ with no initial energy

(b) $x(t) = \cos t$, $t \geq 0$, with no initial energy

(c) $x(t) = \cos 4t$, $t \geq 0$, with no initial energy

9.23. A linear time-invariant continuous-time system has transfer function

$$H(s) = \frac{s + 2}{(s + 1)^2 + 4}$$

The input $x(t) = C\cos(\omega_0 t + \theta)$ is applied to the system for $t \geq 0$ with zero initial energy at time $t = 0$. The resulting steady-state response $y_{ss}(t)$ is

$$y_{ss}(t) = 6\cos(t + 45°), \qquad t \geq 0$$

(a) Compute C, ω_0, and θ.

(b) Compute the Laplace transform $Y_{tr}(s)$ of the transient response $y_{tr}(t)$ resulting from this input.

9.24. A linear time-invariant continuous-time system has transfer function $H(s)$ with $H(0) = 3$. The transient response $y_{tr}(t)$ resulting from the step-function input $x(t) = u(t)$ with the system at rest at time $t = 0$ has been determined to be

$$y_{tr}(t) = -2e^{-t} + 4e^{-3t}, \qquad t \geq 0$$

(a) Compute the system's transfer function $H(s)$.

(b) Compute the steady-state response $y_{ss}(t)$ when the system's input $x(t)$ is equal to $2\cos(3t + 60°)$, $t \geq 0$, with the system at rest at time $t = 0$.

9.25. A linear time-invariant continuous-time system has transfer function $H(s)$. The input $x(t) = 3(\cos t + 2)\cos(2t - 30°)$, $t \geq 0$, produces the steady-state response $y_{ss}(t) = 6\cos(t - 45°) + 8\cos(2t - 90°)$, $t \geq 0$, with no initial energy. Compute $H(1)$ and $H(2)$.

9.26. Consider a second-order system in the form

$$H(s) = \frac{\omega_n^2}{s^2 + 2\zeta\omega_n s + \omega_n^2}$$

Let $s \to j\omega$ to obtain $H(\omega)$ and suppose that $0 < \zeta < 1$. Without factoring the denominator, find an expression for $|H(\omega)|$. To determine if a peak exists in $|H(\omega)|$, take the derivative of $|H(\omega)|$ with respect to ω. Show that a peak exists for $\omega \neq 0$ only if $\zeta \leq 1/\sqrt{2}$. Determine the height of the peak. What happens to the peak as $\zeta \to 0$?

9.27. Sketch the magnitude and phase plots for the following systems. In each case, compute $|H(\omega)|$ and $\angle H(\omega)$ for $\omega = 0$, $\omega =$ 3-dB points, $\omega = \omega_p$ and $\omega \to \infty$. Here ω_p is the value of ω for which $|H(\omega)|$ is maximum. Verify your calculations by plotting the frequency response using the M-file `bode`.

(a) $H(s) = \dfrac{10}{s + 5}$

(b) $H(s) = \dfrac{5(s + 1)}{s + 5}$

(c) $H(s) = \dfrac{s + 10}{s + 5}$

(d) $H(s) = \dfrac{4}{(s + 2)^2}$

(e) $H(s) = \dfrac{4s}{(s + 2)^2}$

(f) $H(s) = \dfrac{s^2 + 2}{(s + 2)^2}$

(g) $H(s) = \dfrac{4}{s^2 + \sqrt{2}(2s) + 4}$

9.28. Sketch the magnitude and phase plots for the circuits shown in Figure P9.28. In each case, compute $|H(\omega)|$ and $\angle H(\omega)$ for $\omega = 0$, $\omega =$ 3-dB points, and $\omega \to \infty$.

9.29. Repeat Problem 9.28 for the circuits in Figure P9.17.

9.30. Consider the RLC circuit shown in Figure P9.30. Choose values for R and L such that the damping ratio $\zeta = 1$ and the circuit is a lowpass filter with approximate 3-dB bandwidth equal to 20 rad/sec; that is, $|H(\omega)| \geq (0.707)|H(0)|$ for $0 \leq \omega \leq 20$.

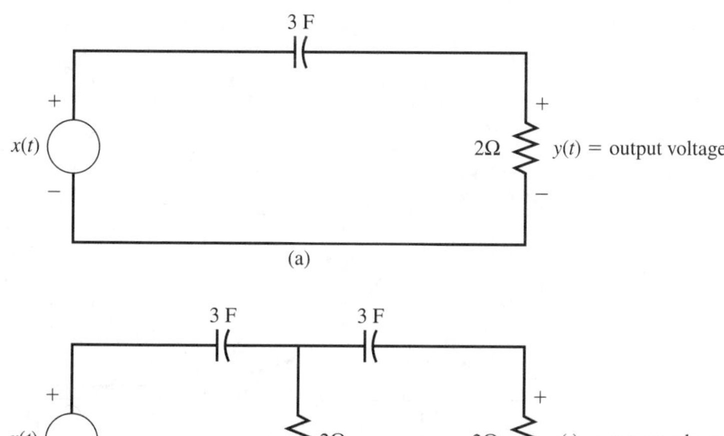

Figure P9.28

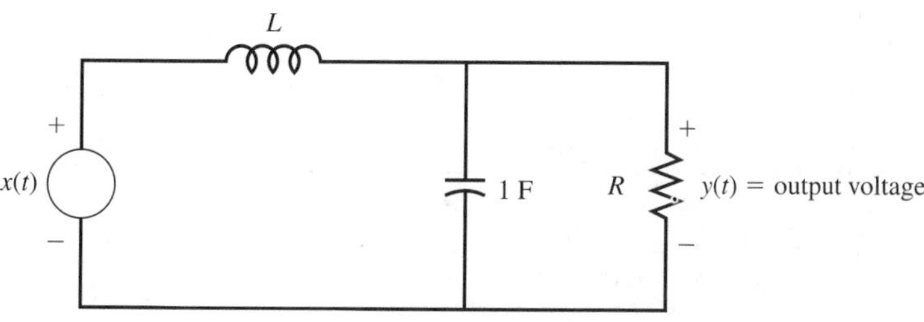

Figure P9.30

9.31. A linear time-invariant continuous-time system has transfer function $H(s)$. It is known that $H(0) = 1$ and that $H(s)$ has two poles and no zeros. In addition, the magnitude function $|H(\omega)|$ is shown in Figure P9.31. Determine $H(s)$.

9.32. A linear time-invariant continuous-time system has transfer function $H(s) = K/(s + a)$, where $K > 0$ and $a > 0$ are unknown. The steady-state response to $x(t) = 4 \cos t$, $t \geq 0$, is $y_{ss}(t) = 20 \cos(t + \phi_1)$, $t \geq 0$. The steady-state response to $x(t) = 5 \cos 4t$, $t \geq 0$, is $y_{ss}(t) = 10 \cos(4t + \phi_2)$, $t \geq 0$. Here ϕ_1, ϕ_2 are unmeasurable phase shifts. Find K and a.

9.33. Using the M-file `bode`, determine the frequency response curves for the mass–spring system in Problem 2.15. Take $M_1 = 1$, $M_2 = 10$, and $k_1 = k_2 = k_3 = 0.1$.

9.34. Repeat Problem 9.33 for the automobile suspension system defined in Example 2.2. Assume that $M_1 = 100$, $M_2 = 1500$, $k_t = 1000$, $k_s = 200$, and $k_d = 10$.

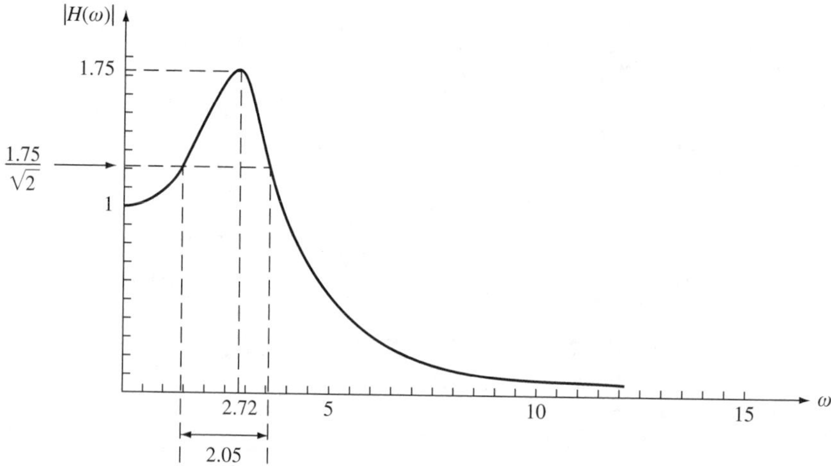

Figure P9.31

9.35. Draw the asymptotic Bode plots (both magnitude and phase plots) for the following systems. Compare your plots to the actual Bode plots obtained from MATLAB.

(a) $H(s) = \dfrac{16}{(s+1)(s+8)}$

(b) $H(s) = \dfrac{10(s+4)}{(s+1)(s+10)}$

(c) $H(s) = \dfrac{10}{s(s+6)}$

(d) $H(s) = \dfrac{10}{(s+1)(s^2+4s+16)}$

(e) $H(s) = \dfrac{10}{(s+1)(s^2+s+16)}$

(f) $H(s) = \dfrac{1000(s+1)}{(s+20)^2}$

9.36. A linear time-invariant continuous-time system has a rational transfer function $H(s)$ with two poles and two zeros. The frequency function $H(\omega)$ of the system is given by

$$H(\omega) = \frac{-\omega^2 + j3\omega}{8 + j12\omega - 4\omega^2}$$

Determine $H(s)$.

9.37. Consider the three-pole Butterworth filter given by the transfer function

$$H(s) = \frac{\omega_c^3}{s^3 + 2\omega_c s^2 + 2\omega_c^2 s + \omega_c^3}$$

(a) Derive an expression for the impulse response $h(t)$ in terms of the 3-dB bandwidth ω_c. Plot $h(t)$ when $\omega_c = 1$ rad/sec.

(b) Compare your result in part (a) with the impulse response of an ideal lowpass filter with frequency function $H(\omega) = p_2(\omega)$. Discuss the similarities and differences in the two impulse responses.

9.38. Again consider the three-pole Butterworth filter defined in Problem 9.37.

(a) For the case when $\omega_c = 2\pi$, compute the output response of the filter when the input is $x(t) = u(t) - u(t - 1)$ with no initial energy in the system.

(b) Repeat part (a) for the case when $\omega_c = 4\pi$.

(c) Using MATLAB, plot the responses found in parts (a) and (b).

(d) Are the results obtained in part (c) expected? Explain.

9.39. For the three-pole Butterworth filter with $\omega_c = 1$, compute the output response $y(t)$ when the input $x(t)$ is

(a) $x(t) = 1, -\infty < t < \infty$

(b) $x(t) = 2 \cos t, -\infty < t < \infty$

(c) $x(t) = \cos (10t + 30°), -\infty < t < \infty$

(d) $x(t) = 2(\cos t)(\sin t), -\infty < t < \infty$

9.40. Again consider the three-pole Butterworth filter with $\omega_c = 1$. The output response resulting from the input $x(t) = \cos 0.5t, -\infty < t < \infty$, can be expressed in the form $y(t) = B \cos[0.5(t - t_d)], -\infty < t < \infty$, where t_d is the time delay through the filter. The response resulting from the input $x(t) = \cos t, -\infty < t < \infty$, can be expressed in the form $y(t) = C \cos (t - t_d + \phi), -\infty < t < \infty$, where ϕ is the phase distortion resulting from the nonlinear phase characteristic of the filter. Compute t_d and ϕ.

9.41. Repeat Problem 9.38 for the three-pole Chebyshev filter with transfer function

$$H(s) = \frac{0.251\omega_c^3}{s^3 + 0.597\omega_c s^2 + 0.928\omega_c^2 s + 0.251\omega_c^3}$$

9.42. Repeat Problem 9.39 for the three-pole Chebyshev filter with $\omega_c = 1$.

9.43. The objective of this problem is to design both a highpass and a bandpass filter starting from the two-pole Butterworth filter with transfer function

$$H(s) = \frac{\omega_c^2}{s^2 + \sqrt{2}\omega_c s + \omega_c^2}$$

(a) Design the highpass filter so that the 3-dB bandwidth runs from $\omega = 10$ to $\omega = \infty$.

(b) Design the bandpass filter so that the 3-dB bandwidth runs from $\omega = 10$ to $\omega = 20$.

(c) Using the M-file `bode`, determine the frequency response curves of the filters constructed in parts (a) and (b).

9.44. Repeat Problem 9.43 for the two-pole Chebyshev filter with transfer function

$$H(s) = \frac{0.50\omega_c^2}{s^2 + 0.645\omega_c s + 0.708\omega_c^2}$$

9.45. Design a three-pole Butterworth stopband filter with a stopband from $\omega = 10$ to $\omega = 15$ rad/sec.

(a) Plot the frequency response curves for the resulting filter.

(b) From the magnitude curve plotted in part (a), determine the expected amplitude of the steady-state responses $y_{ss}(t)$ to the following signals: **(i)** $x(t) = \sin 5t$, **(ii)** $x(t) = \sin 12t$, and **(iii)** $x(t) = \sin 5t + \sin 12t$.

(c) Verify your prediction in part (b) by using MATLAB to compute and plot the response of the system to the signals defined in part (b). You may use `lsim` and integrate long enough for the response to reach steady state. [*Note:* When simulating a continuous-time system to find the response, computers approximate the system as being discrete time. Therefore, when defining the signals $x(t)$ for a time vector `t=0:T:tf`, make sure that the time increment `T` for which $x(t)$ is defined satisfies the Nyquist sampling theorem; that is $2\pi/T$ is at least twice the highest frequency in $x(t)$. See the comments in Problem 1.2 for further information.]

9.46. Design a three-pole Chebyshev bandpass filter with a passband from $\omega = 10$ to $\omega = 15$ rad/sec. Allow a 3-dB ripple in the passband.

(a) Plot the frequency response curves for the resulting filter.

(b) From the magnitude curve plotted in part (a), determine the expected amplitude of the steady-state responses $y_{ss}(t)$ to the following signals: **(i)** $x(t) = \sin 5t$, **(ii)** $x(t) = \sin 12t$, and **(iii)** $x(t) = \sin 5t + \sin 12t$.

(c) Verify your prediction in part (b) by using MATLAB to compute and plot the response of the system to the signals defined in part (b). (Consider the comment in Problem 9.45 regarding the selection of the time increment when using MATLAB.)

9.47. Design a lowpass Butterworth filter with a bandwidth of 10 rad/sec. Select an appropriate number of poles so that a 25-rad/sec signal is attenuated to a level that is no more than 5% of its input amplitude. Use MATLAB to compute and plot the response of the system to the following signals. (Consider the comments in Problem 9.45 regarding the selection of the time increment when using MATLAB.)

(a) $x(t) = \sin 5t$

(b) $x(t) = \sin 25t$

(c) $x(t) = \sin 5t + \sin 25t$

(d) $x(t) = w(t)$, where $w(t)$ is a random signal whose values are uniformly distributed between 0 and 1 (use `x = rand(201,1)` to generate the signal for the time vector `t = 0:.05:10`). Plot the random input $x(t)$ and compare it to the system response.

9.48. Design a highpass type 1 Chebyshev filter with a bandwidth of 10 rad/sec. Select an appropriate number of poles so that a 5-rad/sec signal is attenuated to a level which is no more than 10% of its input amplitude and there is at most a 3-dB ripple in the passband. Use MATLAB to compute and plot the response of the system to the following signals. (Consider the comment in Problem 9.45 regarding the selection of the time increment when using MATLAB.)

(a) $x(t) = \sin 5t$

(b) $x(t) = \sin 25t$

(c) $x(t) = \sin 5t + \sin 25t$

(d) $x(t) = w(t)$ where $w(t)$ is a random signal whose values are uniformly distributed between 0 and 1 (use `x = rand(201,1)` to generate the signal for the time vector `t = 0:.05:10`). Plot the random input $x(t)$ and compare it to the system response.

CHAPTER 10

Application to Control

One of the major applications of the transfer function framework is in the study of control. A very common type of control problem is forcing the output of a system to be equal to a desired reference signal, which is referred to as *tracking*. The tracking problem arises in a multitude of applications such as in industrial control and automation, where the objective is to control the position and/or velocity of a physical object. Examples are given in this chapter involving velocity control of a car and the control of the angular position of the shaft of a motor. The development begins in Section 10.1 with an introduction to the tracking problem, and then in Section 10.2 conditions are given for solving this problem in terms of a feedback control configuration. Here the focus is on the case when the reference is a constant signal, called a *set point*. In Section 10.3 the study of closed-loop system behavior as a function of a controller gain is given in terms of the root locus, and then in the last section of the chapter the root locus is applied to the problem of control system design.

10.1 INTRODUCTION TO CONTROL

Consider a causal linear time-invariant continuous-time system with input $x(t)$ and output $y(t)$. The system is given by its transfer function representation:

$$Y(s) = G_p(s)X(s) \tag{10.1}$$

where $Y(s)$ is the Laplace transform of the output $y(t)$, $X(s)$ is the Laplace transform of the input, and $G_p(s)$ is the transfer function of the system, which is often called the *plant.* Note the change in notation from $H(s)$ to $G_p(s)$ in denoting the transfer function. Throughout this chapter the transfer function of the given system will be denoted by $G_p(s)$, where the subscript "p" stands for "plant."

In many applications, the objective is to force the output $y(t)$ of the system to follow a desired signal $r(t)$, called the *reference signal.* This is called the *tracking problem;* that is, the objective is to find an input $x(t)$ so that the system output $y(t)$ is equal to (tracks) a desired reference signal $r(t)$. In this problem the input $x(t)$ is called a *control input.*

In many cases the reference $r(t)$ is a constant r_0, which is called a *set point.* Hence in *set-point control*, the objective is to find a control input $x(t)$ so that $y(t) = r_0$ for all t in some desired range of values. An example involving set-point control is given below.

Example 10.1 *Open Loop Control*

Consider a car moving on a level surface given by the input/output differential equation (see Example 9.1)

$$\frac{d^2y(t)}{dt^2} + \frac{k_f}{M}\frac{dy(t)}{dt} = \frac{1}{M}x(t) \tag{10.2}$$

where $x(t)$ is the drive or braking force applied to the car and $y(t)$ is the position of the car. In terms of the velocity $v(t) = dy(t)/dt$, the differential equation reduces to

$$\frac{dv(t)}{dt} + \frac{k_f}{M}v(t) = \frac{1}{M}x(t) \tag{10.3}$$

As noted in Section 1.4, the differential equation (10.3) specifies the velocity model of the car. From (10.3) the transfer function of the velocity model is

$$G_p(s) = \frac{1/M}{s + k_f/M} \tag{10.4}$$

Now with the output of the system defined to be the velocity $v(t)$, the objective in velocity control is to force $v(t)$ to be equal to a desired speed v_0. Hence in this problem the reference signal $r(t)$ is equal to the constant v_0 and v_0 is the set point. To solve the set point control problem [i.e., to force $v(t)$ to be equal to v_0], first suppose that the initial velocity $v(0)$ is zero and the goal is to find a control input $x(t)$ so that $v(t) = v_0$ for all $t > 0$. Then $V(s) = v_0/s$, and inserting this into the transfer function model $V(s) = G_p(s)X(s)$ and solving for $X(s)$ gives

$$X(s) = \frac{V(s)}{G_p(s)} = \frac{v_0/s}{(1/M)/(s + k_f/M)} = \frac{v_0M(s + k_f/M)}{s}$$

Taking the inverse transform results in the control input:

$$x(t) = v_0M\delta(t) + v_0k_f, \qquad t \geq 0 \tag{10.5}$$

where $\delta(t)$ is the impulse.

Obviously, the control (10.5) cannot be implemented since it contains an impulse. The presence of the impulse in the control input is a result of the requirement that

$v(t) = v_0$ for all $t > 0$ starting from $v(0) = 0$. In other words, the impulse is needed to change the velocity instantaneously from zero to the desired set point v_0. A nonimpulsive control input can be obtained by relaxing the tracking requirement so that $v(t) \to v_0$ as $t \to \infty$; that is, the velocity should converge to the desired velocity v_0 as t becomes large. In order for $v(t) \to v_0$, it is necessary that the transform $V(s)$ of $v(t)$ contain a pole at $s = 0$. Since the transfer function $G_p(s)$ of the velocity model does not have a pole at $s = 0$, in order for $V(s)$ to have a pole at zero, the transform $X(s)$ of the control input $x(t)$ must have a pole at 0. Hence the simplest possible form of $X(s)$ is A/s, where A is a (real) constant. The corresponding control input is $x(t) = Au(t)$, where $u(t)$ is the unit-step function. Then the question is whether or not there is a value of A for which the resulting response $v(t)$ converges to v_0. To answer this, first insert $X(s) = A/s$ into the transfer function representation, which gives

$$V(s) = G_p(s)X(s) = \frac{A/M}{s(s + k_f/M)} = \frac{A/k_f}{s} - \frac{A/k_f}{s + k_f/M}$$

Inverse transforming $V(s)$ yields

$$v(t) = \frac{A}{k_f}[1 - e^{-(k_f/M)t}], \qquad t \geq 0 \tag{10.6}$$

Then if A is set equal to $v_0 k_f$, the velocity $v(t)$ is given by

$$v(t) = v_0\,[1 - e^{-(k_f/M)t}], \qquad t \geq 0 \tag{10.7}$$

and since $k_f/M > 0$, it is seen that $v(t)$ converges to v_0 as $t \to \infty$. Therefore, with the input

$$x(t) = Au(t) = (v_0 k_f)u(t) \tag{10.8}$$

set point control is achieved in the limit as $t \to \infty$. Note that the implementation of the control (10.8) requires that the coefficient k_f in the velocity model (10.3) must be known. Also note that the control $x(t)$ given by (10.8) can be expressed in the form

$$x(t) = k_f r(t) \tag{10.9}$$

where $r(t)$ is the reference signal, which is equal to v_0 for $t \geq 0$. The control $x(t)$ given by (10.9) is referred to as an *open-loop control*, since it depends only on the reference signal $r(t)$ and not on the output $v(t)$.

From (10.7) it is seen that the rate at which $v(t)$ converges to v_0 depends directly on k_f/M; that is, the more positive k_f/M is, the faster $v(t)$ converges to v_0. Of course, k_f and M depend on the given system, and thus the ratio k_f/M is constrained by the system. The ratio k_f/M will be rather small since the mass of the car will be large and the coefficient k_f corresponding to the viscous friction will be relatively small. Hence the control $x(t) = k_f r(t)$ will not yield a suitably "fast response." It turns out that better performance can be achieved by generalizing the relationship between $r(t)$ and $x(t)$ as follows.

Let the Laplace transform $X(s)$ of the control signal $x(t)$ be given by

$$X(s) = \frac{B(s + k_f/M)}{s + C} R(s) \tag{10.10}$$

where B and C are real constants that are to be determined and $R(s)$ is the transform of the reference signal $r(t) = v_0$, $t \geq 0$. Inserting (10.10) into the transfer function representation $V(s) = G_p(s)X(s)$ and taking the inverse transform of $V(s)$ yields

$$v(t) = \frac{Bv_0}{CM}(1 - e^{-Ct}), \qquad t \geq 0$$

If $C > 0$, it is clear that the velocity $v(t)$ will converge to v_0 if $B = CM$. In addition, the rate at which $v(t)$ converges to v_0 can be made as fast as desired by choosing C to be a suitably large positive number. Hence, with $B = CM$ the control signal with transform $X(s)$ given by (10.10) achieves the objective of forcing the velocity $v(t)$ to converge to the set point v_0 with any desired rate of convergence. However, as discussed below, this type of control (i.e., open-loop control) is susceptible to influences on the system output that may result from unknown disturbances applied to the system. For example, the control defined by (10.10) may not work well when gravity acts on the car in going up or down hills.

Again consider a system or plant with input $x(t)$, output $y(t)$, transfer function $G_p(s)$, and where the goal is to track a reference signal $r(t)$. In open-loop control, $r(t)$ is applied to a controller or compensator with transfer function $G_c(s)$, and then the output of the controller is taken to be the control signal applied to the plant. A block diagram of open-loop control is given in Figure 10.1a, where $d(t)$ is a disturbance that may be applied to the system. Note that the transform $Y(s)$ of the plant output is given by

$$Y(s) = G_p(s)G_c(s)R(s) + G_p(s)D(s)$$

where $R(s)$ is the transform of the reference signal $r(t)$ and $D(s)$ is the transform of the disturbance $d(t)$.

As first observed in Example 10.1, the process shown in Figure 10.1a is referred to as *open-loop control* since the control signal $x(t)$ does not depend on the system output $y(t)$. Note that for the open-loop control defined by (10.10) in Example 10.1, the transfer function of the controller is

$$G_c(s) = \frac{B(s + k_f/M)}{s + C}$$

A major problem with open-loop control is that the output $y(t)$ of the plant will be "perturbed" by the disturbance input $d(t)$. However, since the control signal $x(t)$ does not depend on the plant output $y(t)$ in open-loop control, the control signal cannot compensate for the effect of the disturbance $d(t)$. To reduce the effect of disturbances, it is necessary that the control signal $x(t)$ depend directly on the plant output $y(t)$. This requires that $y(t)$ be measurable using some type of a sensor, in which case the measured output can be compared to the desired output $r(t)$. This results in the *tracking error* $e(t)$ given by

$$e(t) = r(t) - y(t)$$

which can be "fed back" to form the control signal $x(t)$. More precisely, in *feedback control* the error signal $e(t)$ is applied to the controller or compensator with transfer function $G_c(s)$ to yield the control signal $x(t)$ for the given system. A block diagram for the closed-loop control process is given in Figure 10.1b. The overall system shown

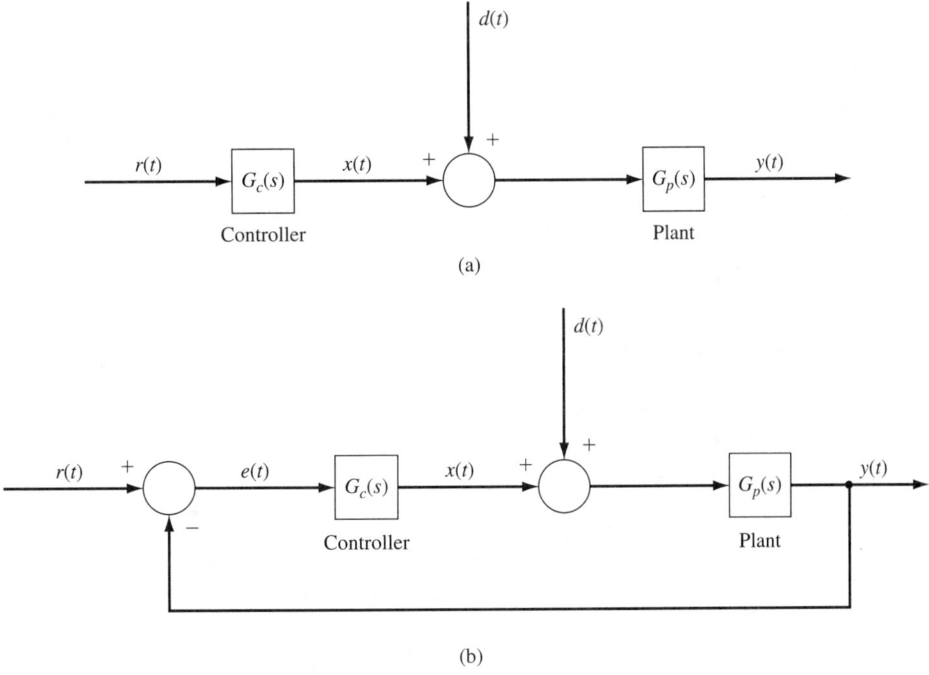

Figure 10.1 Block diagram of (a) open-loop control and (b) closed-loop control.

in Figure 10.1b is called the *closed-loop system*, since it is formed from the given system with transfer function $G_p(s)$ by "closing the loop" around $G_p(s)$ and $G_c(s)$. In the remainder of this chapter, the focus is on closed-loop control.

In the feedback control scheme shown in Figure 10.1b, the objective is to design the transfer function $G_c(s)$ of the controller so that the tracking error $e(t)$ converges to zero as $t \to \infty$, which is equivalent to requiring that the output $y(t)$ converges to the reference $r(t)$ as $t \to \infty$. This is sometimes referred to as *asymptotic tracking* since $y(t) = r(t)$ occurs in the limit as $t \to \infty$. The solution of the tracking problem using the configuration shown in Figure 10.1b is referred to as *output feedback control* since the system output $y(t)$ is fed back to the input; that is, the control signal $x(t)$ applied to the given system depends on the system output $y(t)$. The dependence of $x(t)$ on $y(t)$ is seen from the transform relationship

$$X(s) = G_c(s)E(s) = G_c(s)[R(s) - Y(s)] \tag{10.11}$$

which follows directly from the block diagram in Figure 10.1b. In (10.11), $R(s)$ is the transform of the reference $r(t)$ and $E(s)$ is the transform of the tracking error $e(t)$.

The simplest type of controller is the one with transfer function $G_c(s) = K_P$, where K_P is a (real) constant. In this case, (10.11) becomes

$$X(s) = K_P E(s) = K_P[R(s) - Y(s)]$$

and taking the inverse transform yields the control signal

$$x(t) = K_P e(t) = K_P[r(t) - y(t)] \tag{10.12}$$

The control given by (10.12) is called *proportional control* since the control signal $x(t)$ is directly proportional to the error signal $e(t)$. This explains the subscript "P" in K_P, which stands for "proportional."

Example 10.2 *Proportional Control*

Again consider the car on a level surface with the velocity model given by (10.3), and suppose that the goal is to force the velocity $v(t)$ to track a desired speed v_0, so that $r(t) = v_0 u(t)$. In this case the velocity $v(t)$ can be measured using a speedometer, and thus the tracking error

$$e(t) = r(t) - v(t) = v_0 - v(t)$$

can be computed. With proportional feedback control, the control signal $x(t)$ applied to the car is given by

$$x(t) = K_P[v_0 - v(t)] \tag{10.13}$$

With the control (10.13), the transform $V(s)$ of the resulting output $v(t)$ assuming that $v(0) = 0$ and $d(t) = 0$ (there is no disturbance) is

$$V(s) = G_p(s)X(s) = \frac{K_P/M}{s + k_f/M}\left[\frac{v_0}{s} - V(s)\right] \tag{10.14}$$

Solving (10.14) for $V(s)$ gives

$$V(s) = \frac{K_P v_0/M}{(s + k_f/M + K_P/M)s} \tag{10.15}$$

$$V(s) = \frac{-K_P v_0/(k_f + K_P)}{s + k_f/M + K_P/M} + \frac{K_P v_0/(k_f + K_P)}{s} \tag{10.16}$$

Inverse transforming (10.16) yields the response

$$v(t) = -\frac{K_P v_0}{k_f + K_P} e^{-[(k_f + K_P)/M]t} + \frac{K_P v_0}{k_f + K_P}, \quad t \geq 0 \tag{10.17}$$

From (10.17) it is seen that if $(k_f + K_P)/M > 0$, then $v(t)$ converges to $K_P v_0/(k_f + K_P)$. Since there is no finite value of K_P for which $K_P/(k_f + K_P) = 1$, the proportional controller will always result in a *steady-state tracking error* equal to

$$v_0 - \frac{K_P v_0}{k_f + K_P} = \left(1 - \frac{K_P}{k_f + K_P}\right) v_0 = \frac{k_f}{k_f + K_P} v_0 \tag{10.18}$$

However, the tracking error given by (10.18) can be made as small as desired by taking K_P to be suitably large compared to k_f. As will be seen from results given in the next section, it is possible to obtain a zero steady-state error by modifying the proportional controller.

From (10.17) it is seen that the rate at which $v(t)$ converges to the steady-state value can be made as fast as desired by again taking K_P to be suitably large. To see this, suppose that $k_f = 10$, $M = 1000$, and $v_0 = 60$. Then the transform $V(s)$ given by (10.15) becomes

$$V(s) = \frac{0.06K_P}{[s + 0.01(1 + 0.1K_P)]s}$$

The resulting velocity $v(t)$ can be computed using the MATLAB command `step(num,den)` with

```
num = 0.06*Kp;
den = [1 0.01+0.001*Kp];
```

Running the MATLAB software with $K_P = 100, 200$, and 500 results in the velocity responses shown in Figure 10.2. Note that the fastest response with the smallest steady-state error is achieved when $K_P = 500$, the largest value of K_P. For $K_P = 500$, the steady-state error is

$$\frac{k_f}{k_f + K_P} v_0 = \frac{10}{510} 60 = 1.176$$

Now suppose that a step disturbance force $d(t) = 50u(t)$ is applied to the car at time $t = 0$, where the disturbance may be a result of the car going down a long incline. If the open-loop control scheme shown in Figure 10.1a is used, the transform $V(s)$ of the velocity will be given by

$$V(s) = G_p(s)G_c(s)R(s) + G_p(s)\frac{50}{s}$$

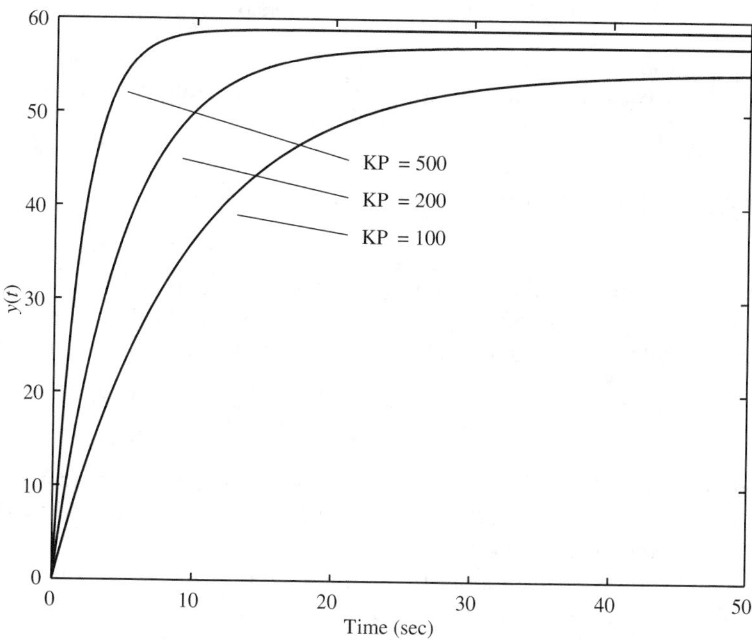

Figure 10.2 Velocity responses in Example 10.2 for $K_P = 100, 200$, and 500.

Thus the step disturbance will result in a perturbation in the velocity $v(t)$ of the car equal to the inverse transform of

$$G_p(s)\frac{50}{s} = \frac{(0.001)50}{(s + 0.01)s} = \frac{5}{s} - \frac{5}{s + 0.01}$$

which is equal to

$$5(1 - e^{-0.01t}), \quad t \geq 0$$

Hence the disturbance will result in a sizable error in achieving the desired set point of $v_0 = 60$. In contrast, if a step disturbance input $d(t)$ is applied to the car with the feedback control $x(t)$ given by (10.13), the transform $V(s)$ of the velocity is given by

$$V(s) = G_p(s)X(s) + G_p(s)\frac{50}{s} = K_P G_p(s)\left[\frac{v_0}{s} - V(s)\right] + G_p(s)\frac{50}{s} \qquad (10.19)$$

Solving (10.19) for $V(s)$ gives

$$V(s) = \frac{K_P G_p(s)}{1 + K_P G_p(s)}\frac{v_0}{s} + \frac{G_p(s)}{1 + K_P G_p(s)}\frac{50}{s}$$

and thus the perturbation of the velocity resulting from step disturbance is equal to the inverse transform of

$$\frac{G_p(s)}{1 + K_P G_p(s)}\frac{50}{s} = \frac{50/M}{(s + k_f/M + K_P/M)s}$$

For $k_f = 10$, $M = 1000$, and $K_P = 500$, the perturbation is

$$0.098(1 - e^{-0.51t}), \quad t \geq 0$$

Obviously, this term is much smaller than in the case of open-loop control, and thus in this example, closed-loop control is much more "robust" to a step disturbance than is open-loop control.

10.2 TRACKING CONTROL

Given a system with transfer function $G_p(s)$, a controller with transfer function $G_c(s)$, and a reference signal $r(t)$, again consider the feedback control configuration shown in Figure 10.1b. Throughout this section it is assumed that both the system and the controller are finite dimensional, and thus $G_p(s)$ and $G_c(s)$ are rational functions of s given by

$$G_p(s) = \frac{B_p(s)}{A_p(s)} \quad \text{and} \quad G_c(s) = \frac{B_c(s)}{A_c(s)} \qquad (10.20)$$

where $B_p(s)$, $A_p(s)$, $B_c(s)$, and $A_c(s)$ are polynomials in s with the degree of $A_p(s)$ equal to N and the degree of $A_c(s)$ is equal to q. Hence, the given system has N poles and the controller has q poles.

As discussed in Section 10.1, in tracking control the objective is to design the controller transfer function $G_c(s)$ so that the tracking error $e(t) = r(t) - y(t)$ converges to zero as $t \to \infty$. The solution to this problem involves the closed-loop poles; that is, the poles of the closed-loop system. The closed-loop poles are determined as follows.

First, from the block diagram in Figure 10.1b, when $d(t) = 0$ it is clear that the transform $Y(s)$ of the output $y(t)$ is given by

$$Y(s) = G_p(s)G_c(s)[R(s) - Y(s)] \tag{10.21}$$

where $R(s)$ is the transform of the reference input $r(t)$. Solving (10.21) for $Y(s)$ yields

$$Y(s) = \frac{G_p(s)G_c(s)}{1 + G_p(s)G_c(s)} R(s) \tag{10.22}$$

It is worth noting that the closed-loop transfer function representation (10.22) follows directly from the results in Section 8.6.

With $G_{\text{cl}}(s)$ defined to be the closed-loop transfer function, from (10.22)

$$G_{\text{cl}}(s) = \frac{G_p(s)G_c(s)}{1 + G_p(s)G_c(s)} \tag{10.23}$$

Then inserting (10.20) into (10.23) results in the following expression for the closed-loop transfer function:

$$G_{\text{cl}}(s) = \frac{[B_p(s)/A_p(s)][B_c(s)/A_c(s)]}{1 + [B_p(s)/A_p(s)][B_c(s)/A_c(s)]}$$

$$G_{\text{cl}}(s) = \frac{B_p(s)B_c(s)}{A_p(s)A_c(s) + B_p(s)B_c(s)} \tag{10.24}$$

From (10.24), it is seen that p is a pole of the closed-loop system if and only if

$$A_p(p)A_c(p) + B_p(p)B_c(p) = 0$$

Therefore, the closed-loop poles are the roots of the polynomial equation

$$A_p(s)A_c(s) + B_p(s)B_c(s) = 0 \tag{10.25}$$

Note that the degree of the polynomial in (10.25) is equal to $N + q$, where N is the degree of $A_p(s)$ and q is the degree of $A_c(s)$. Hence the number of closed-loop poles is equal to $N + q$, which is the sum of the number of poles of the given system and the number of poles of the controller.

Example 10.3 *Calculation of Closed-Loop Transfer Function*

Suppose that

$$G_p(s) = \frac{s + 2}{s^2 + 4s + 10}, \qquad G_c(s) = \frac{s + 1}{s(s + 10)}$$

Then both $G_p(s)$ and $G_c(s)$ have two poles which implies that the closed-loop system has four poles. Now

$$B_p(s) = s + 2, \qquad A_p(s) = s^2 + 4s + 10$$

$$B_c(s) = s + 1, \qquad A_c(s) = s(s + 10)$$

and inserting this into (10.24) gives the closed-loop transfer function

$$G_{cl}(s) = \frac{(s+2)(s+1)}{(s^2+4s+10)s(s+10)+(s+2)(s+1)}$$

$$G_{cl}(s) = \frac{s^2+3s+2}{s^4+14s^3+51s^2+103s+2}$$

The closed-loop poles are the roots of the equation

$$s^4+14s^3+51s^2+103s+2=0$$

Using the MATLAB command `roots` reveals that the closed-loop poles are -0.0196, -9.896, $-2.042 \pm j2.477$.

Again consider the tracking error $e(t) = r(t) - y(t)$. With $E(s)$ equal to the Laplace transform of the tracking error $e(t)$, from the block diagram in Figure 10.1b it is seen that

$$E(s) = R(s) - Y(s) \tag{10.26}$$

Inserting the expression (10.22) for $Y(s)$ into (10.26) yields

$$E(s) = R(s) - \frac{G_p(s)G_c(s)}{1+G_p(s)G_c(s)}R(s)$$

$$= \frac{1}{1+G_p(s)G_c(s)}R(s) \tag{10.27}$$

Then inserting (10.20) into (10.27) gives

$$E(s) = \frac{1}{1+[B_p(s)/A_p(s)][B_c(s)/A_c(s)]}R(s)$$

$$= \frac{A_p(s)A_c(s)}{A_p(s)A_c(s)+B_p(s)B_c(s)}R(s) \tag{10.28}$$

From the analysis given in Section 8.3, the tracking error $e(t)$ converges to zero as $t \to \infty$ if and only if all the poles of $E(s)$ are located in the open left-half plane. (This result also follows directly from the final value theorem.) From (10.28) it is seen that the poles of $E(s)$ include the closed-loop poles, that is, the values of s for which

$$A_p(s)A_c(s)+B_p(s)B_c(s)=0$$

As a result, a necessary condition for $e(t) \to 0$ is that the closed-loop system must be stable, and thus all the closed-loop poles must be located in the open left-half plane. It is important to stress that although stability of the closed-loop system is necessary for tracking, it is not sufficient for tracking. Additional conditions that guarantee tracking depend on the reference signal $r(t)$. This is investigated in detail below for the case of a step input.

Tracking a Step Reference

Suppose that the reference input $r(t)$ is equal to $r_0u(t)$, where r_0 is a real constant and $u(t)$ is the unit-step function. As discussed in Section 10.1, this case corresponds to set-point control, where the constant r_0 is the set point. Note that when $r(t) = r_0u(t)$, the resulting output response $y(t)$ with zero initial conditions is equal to r_0 times the step response of the closed-loop system. When $R(s) = r_0/s$, the expression (10.28) for the transform $E(s)$ of the tracking error becomes

$$E(s) = \frac{A_p(s)A_c(s)}{A_p(s)A_c(s) + B_p(s)B_c(s)}\frac{r_0}{s} \tag{10.29}$$

In this case, the poles of $E(s)$ are equal to the poles of the closed-loop system plus a pole at $s = 0$. Therefore, if the closed-loop system is stable, the conditions for applying the final value theorem to $E(s)$ are satisfied. Thus the limiting value of $e(t)$ as $t \to \infty$ can be computed using the final-value theorem, which gives

$$\lim_{t\to\infty} e(t) = \lim_{s\to 0} sE(s) = \frac{A_p(0)A_c(0)r_0}{A_p(0)A_c(0) + B_p(0)B_c(0)} \tag{10.30}$$

With the steady-state error e_{ss} defined by

$$e_{ss} = \lim_{t\to\infty} e(t)$$

from (10.30) it is seen that

$$e_{ss} = \frac{A_p(0)A_c(0)r_0}{A_p(0)A_c(0) + B_p(0)B_c(0)} \tag{10.31}$$

The steady-state error e_{ss} can be written in the form

$$e_{ss} = \frac{1}{1 + [B_p(0)/A_p(0)][B_c(0)/A_c(0)]} r_0 \tag{10.32}$$

and since

$$G_p(0) = \frac{B_p(0)}{A_p(0)} \quad \text{and} \quad G_c(0) = \frac{B_c(0)}{A_c(0)}$$

(10.32) can be expressed in the form

$$e_{ss} = \frac{1}{1 + G_p(0)G_c(0)} r_0 \tag{10.33}$$

From (10.33) it is clear that the steady-state error is zero if and only if

$$G_p(0)G_c(0) = \infty \tag{10.34}$$

which is the case if $G_p(s)G_c(s)$ has a pole at $s = 0$. The open-loop system defined by the transfer function $G_p(s)G_c(s)$ is said to be a *type 1 system* if $G_p(s)G_c(s)$ has a single pole at $s = 0$. Hence the system given by (10.22) will track the step input $r_0u(t)$

if $G_p(s)G_c(s)$ is a type 1 system. In addition, it follows from the results in Section 8.3 that the rate at which the error $e(t)$ approaches zero depends on the location of the closed-loop poles in the open left-half plane. In particular, the farther over in the left-half plane the closed-loop poles are, the faster the rate of convergence of $e(t)$ to zero. Also note that since the system output $y(t)$ is equal to $r_0 - e(t)$, the rate at which $y(t)$ converges to the set point r_0 is the same as the rate at which $e(t)$ converges to zero.

The open-loop system $G_p(s)G_c(s)$ is said to be a *type 0 system* if $G_p(s)G_c(s)$ does not have any poles at $s = 0$. This is equivalent to requiring that $G_p(0)G_c(0) \neq \infty$. Thus from the analysis above, it is clear that when the reference signal $r(t)$ is a step function and $G_p(s)G_c(s)$ is a type 0 system, the closed-loop system (10.22) will have a nonzero steady-state tracking error e_{ss} given by (10.33).

Suppose that the goal of the controller is to have zero tracking error for a step reference, but the original plant $G_p(s)$ does not have a pole at $s = 0$. To achieve the goal, the controller must have a pole at $s = 0$, so that the product $G_p(s)G_c(s)$ is type 1. A common type of controller used to achieve this goal is a *proportional plus integral* (PI) controller, which is given by

$$G_c(s) = K_P + \frac{K_I}{s} = \frac{K_P s + K_I}{s} \tag{10.35}$$

where K_P and K_I are real constants. In this case, the transform $X(s)$ of the control input $x(t)$ applied to the plant is given by

$$X(s) = G_c(s)E(s) = K_P E(s) + \frac{K_I}{s}E(s) \tag{10.36}$$

Inverse transforming (10.36) gives

$$x(t) = K_P e(t) + K_I \int_0^t e(\tau)d\tau \tag{10.37}$$

The first term on the right-hand side of (10.37) corresponds to proportional control (as discussed in Section 10.1), while the second term corresponds to integral control since this term is given in terms of the integral of the error $e(t)$. Thus, the subscript "I" in K_I stands for "integral." With the controller transfer function (10.35), the transform $E(s)$ of the error given by equation (10.28) becomes

$$E(s) = \frac{A_p(s)s}{A_p(s)s + B_p(s)(K_P s + K_I)} R(s)$$

Obviously, the coefficients of the denominator polynomial of $E(s)$ depend upon K_P and K_I. Thus, the poles of $E(s)$, which are the poles of the closed-loop system, can be modified through the selection of K_P and K_I. Hence, the objective of the control designer is to select values for K_P and K_I that result in closed loop poles that have an acceptable rate of convergence to zero of the error $e(t)$, or, equivalently, the rate of convergence of the output $y(t)$ to the reference r_0.

The results derived above are illustrated by the following examples.

Example 10.4 *Proportional Plus Integral Control*

Again consider velocity control of a car moving on a level surface as studied in Examples 10.1 and 10.2. The goal is to have the velocity $v(t)$ of the car converge to a desired velocity v_0 as $t \to \infty$. Recall that the transfer function $G_p(s)$ of the velocity model of the car is given by $G_p(s) = (1/M)/(s + k_f/M)$. In Example 10.2 proportional control was considered where $G_c(s) = K_P$. In this case

$$G_p(s)G_c(s) = \frac{K_P/M}{s + k_f/M}$$

Clearly, $G_p(s)G_c(s)$ does not have a pole at $s = 0$, and the open-loop system $G_p(s)G_c(s)$ is type 0. Hence, as first observed in Example 10.2, there is a steady-state error e_{ss} in tracking the step input $v_0u(t)$. From (10.33), e_{ss} is given by

$$e_{ss} = \frac{1}{1 + K_P/k_f}v_0 = \frac{k_f v_0}{k_f + K_P}$$

This checks with the result obtained in Example 10.2 [see (10.18)].

In order to meet the objective of zero steady-state error to a step reference, a PI controller of the form given in (10.35) is used. The transform of the error $E(s)$ is found from (10.27) to be

$$E(s) = \left[\frac{s(s + k_f/M)}{s(s + k_f/M) + (K_P s + K_I)(1/M)}\right]\frac{v_0}{s}$$

$$= \frac{(s + k_f/M)v_0}{s^2 + (k_f/M + K_P/M)s + K_I/M}$$

In this example, the design parameters K_P and K_I can be chosen to place the poles of $E(s)$ (equivalently, the closed loop poles) arbitrarily in the open left-half plane. Hence the rate of convergence to zero of the error $e(t)$, or equivalently, the rate of convergence of $v(t)$ to v_0, can be made as fast as desired by selecting values for K_P and K_I. To investigate this, suppose that $k_f = 10$, $M = 1000$, and $v_0 = 60$. Then the transform $V(s)$ of the velocity response is

$$V(s) = G_{cl}(s)\frac{v_0}{s} = \frac{G_p(s)G_c(s)}{1 + G_p(s)G_c(s)}\frac{60}{s} = \frac{0.06(K_P s + K_I)}{[s^2 + 0.01(1 + 0.1K_P)s + 0.001K_I]s}$$

When $K_I = 0$ so that there is no integral control, the response $v(t)$ was computed in Example 10.2 for three different values of K_P (see Figure 10.2). By taking K_I to be suitably large, the integral control action will eliminate the steady-steady state error seen in Figure 10.2. For instance, for $K_P = 500$ and $K_I = 1, 5$, and 10, the responses are plotted in Figure 10.3. Note that when $K_I = 1$, the integral action is not sufficiently strong to bring the velocity up to 60 during the 50-second time interval in the plot, although it is true that $v(t)$ is converging to 60 in the limit as $t \to \infty$. Note that for $K_I = 5$ the velocity reaches 60 in about 12 seconds, and increasing K_I to 10 results in $v(t)$ getting up to 60, but now there is some overshoot in the response. Clearly, $K_I = 5$ yields the best response and there is a reason for this; namely when $K_I = 5$, the controller transfer function $G_c(s) = (500s + 5)/s$ has a zero at $s = -5/500 = -0.01$, and this zero cancels the pole of $G_p(s)$ at $s = -0.01$. This results in a first-order closed-loop system with transfer function

$$G_{cl}(s) = \frac{\left(\frac{500s + 5}{s}\right)\left(\frac{0.001}{s + 0.01}\right)}{1 + \left(\frac{500s + 5}{s}\right)\left(\frac{0.001}{s + 0.01}\right)} = \frac{0.5}{s + 0.5}$$

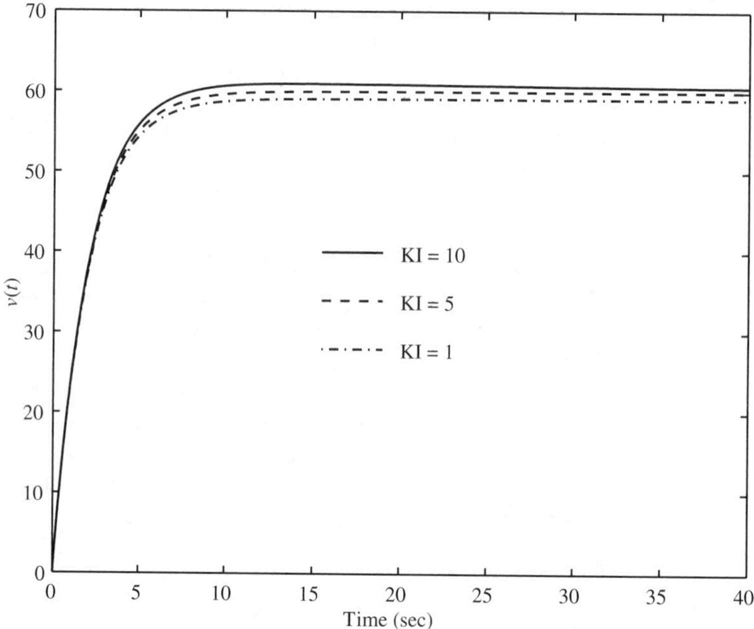

Figure 10.3 Velocity responses in Example 10.4.

Thus when $K_P = 500$ and $K_I = 5$, the expression for $V(s)$ above reduces to a first-order rational function; in other words, there is a pole–zero cancellation in the expression for $V(s)$. For these values of K_P and K_I,

$$V(s) = G_{cl}(s)\frac{v_0}{s} = \left(\frac{0.5}{s+0.5}\right)\frac{60}{s} = 60\left(\frac{1}{s} - \frac{1}{s+0.5}\right)$$

Then taking the inverse Laplace transform yields

$$v(t) = 60(1 - e^{-0.5t}), \qquad t \geq 0$$

Hence the transient $y_{tr}(t) = -60\exp(-0.5t)$ is a simple exponential that decays to zero with time constant $1/0.5 = 2$ seconds. Therefore, the velocity $v(t)$ converges to 60 with the rate of convergence corresponding to a time constant of 2 seconds. The key point here is that by choosing $G_c(s)$ so that it cancels the pole in $G_p(s)$, the closed-loop system becomes a first-order system which is much easier to deal with than a second-order system. It is common in practice to design the controller transfer function $G_c(s)$ so that it cancels one or more stable poles of the plant transfer function $G_p(s)$. Another example of this is given in Section 10.4.

In Example 10.4, the design parameters K_I and K_P of the PI controller could be chosen to place the closed loop poles arbitrarily in the open left-half plane. This was true with the car example because the plant was first order. When the plant is higher than first order, the closed loop poles cannot be made arbitrarily fast (that is, placed farther left in the s-plane) using a PI controller or a proportional controller, $G_c(s) = K_P$. In fact, a PI controller tends to slow down the response, that is, result in closed-loop poles that are to the right of the closed-loop poles achievable when using a simple proportional controller. If the goal is to speed up the closed loop response, even over that achievable with a proportional controller, a *proportional plus derivative controller* is often used; that is,

$$G_c(s) = K_P + K_D s \tag{10.38}$$

where K_P and K_D, are constants. In this case, the transform $X(s)$ of the control signal applied to the plant is given by

$$X(s) = G_c(s)E(s) = K_P E(s) + K_D s E(s) \tag{10.39}$$

Since multiplication by s in the s-domain corresponds to differentiation in the time domain, inverse transforming both sides of (10.39) results in the following expression for the control signal $x(t)$:

$$x(t) = K_P e(t) + K_D \frac{de(t)}{dt} \tag{10.40}$$

The first term on the right-hand side of (10.40) corresponds to proportional control, while the second term corresponds to derivative control, since this term is given in terms of the derivative of the tracking error $e(t)$. (Thus, the subscript "D" in K_D stands for "derivative.")

Example 10.5 *Proportional Plus Derivative Control*

A problem that arises in many applications is controlling the position of an object. An example is controlling the angular position of a valve in some chemical process, or controlling the angular position of a circular plate used in some manufacturing operation, such as drilling or component insertion. In such applications a fundamental problem is controlling the angular position of the shaft of a motor used to drive a specific mechanical structure (such as a valve or plate). Often, the motor used is a field-controlled dc (direct current) motor, which is illustrated in Figure 10.4. The load indicated in the figure is the structure (valve, plate, etc.) to which the motor shaft is connected. The input to the

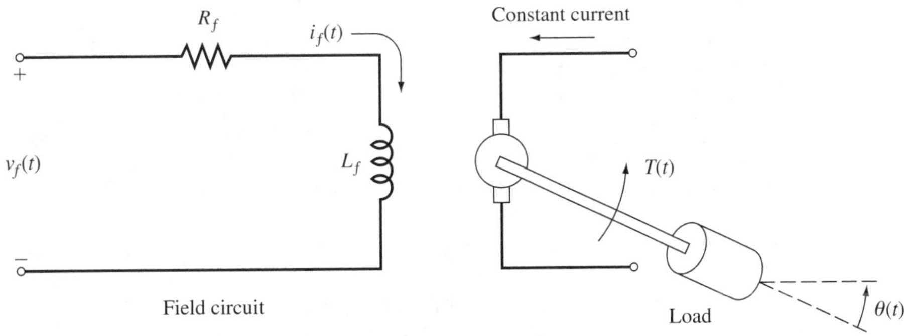

Figure 10.4 Field-controlled dc motor with load.

motor is the voltage $v_f(t)$ applied to the field circuit and the output is the angle $\theta(t)$ of the motor shaft. From the laws of mechanics, the torque $T(t)$ developed by the motor is related to the angle $\theta(t)$ by the differential equation

$$I\frac{d^2\theta(t)}{dt^2} + k_d\frac{d\theta(t)}{dt} = T(t) \qquad (10.41)$$

where I is the moment of inertia of the motor and load and k_d is the viscous friction coefficient of the motor and load. In the usual approximation of motor dynamics, it is assumed that the torque $T(t)$ is given approximately by

$$T(t) = k_m v_f(t) \qquad (10.42)$$

where k_m is the motor constant which is strictly positive ($k_m > 0$). Inserting (10.42) into (10.41) gives

$$I\frac{d^2\theta(t)}{dt^2} + k_d\frac{d\theta(t)}{dt} = k_m v_f(t) \qquad (10.43)$$

which is the input/output differential equation for the dc motor. Taking the Laplace transform of both sides of (10.40) results in the following transfer function representation for the dc motor:

$$\Theta(s) = \frac{k_m/I}{(s + k_d/I)s}V_f(s) \qquad (10.44)$$

where $\Theta(s)$ is the transform of $\theta(t)$ and $V_f(s)$ is the transform of $v_f(t)$. In this case the system (the dc motor with load) has a pole at $s = 0$, and thus for any controller transfer function $G_c(s) = B_c(s)/A_c(s)$, the open-loop system $G_p(s)G_c(s)$ will be type 1. Hence it is not necessary that $G_c(s)$ have a pole at $s = 0$ in order to track a step input $r(t) = \theta_0 u(t)$, where θ_0 is the desired angular position of the motor shaft. Choosing the simplest possible $G_c(s)$ results in the proportional controller given by $G_c(s) = K_P$ where K_P is a real constant. In this case, from (10.28) the transform $E(s)$ of the tracking error is

$$\begin{aligned} E(s) &= \left[\frac{(s + k_d/I)s}{(s + k_d/I)s + k_m K_P/I}\right]\frac{\theta_0}{s} \\ &= \frac{(s + k_d/I)\theta_0}{s^2 + (k_d/I)s + k_m K_P/I} \end{aligned} \qquad (10.45)$$

It follows from the Routh–Hurwitz stability test that the two poles of $E(s)$ are in the open left-half plane if and only if $k_d/I > 0$ and $k_m K_P/I > 0$. Since $k_d > 0$, $I > 0$, and $k_m > 0$, this condition is equivalent to $K_P > 0$. Therefore, for any value of $K_P > 0$ the tracking error $e(t) = \theta_0 - \theta(t)$ converges to zero, which implies that $\theta(t) \to \theta_0$.

Although the error $e(t)$ converges to zero for any $K_P > 0$, it is not possible to obtain an arbitrarily fast rate of convergence to zero by choosing K_P. In other words, it is not possible to place the poles of $E(s)$ arbitrarily far over in the left-half plane by choosing K_P. This follows directly from the expression (10.42) for $E(s)$, from which it is seen that the real parts of both of the two poles of $E(s)$ cannot be more negative than $-k_d/2I$. This can be verified by applying the quadratic formula to the polynomial $s^2 + (k_d/I)s + k_m K_P/I$ in the denominator of $E(s)$. A suitably fast rate of convergence of $e(t)$ to zero can be achieved by using a PD controller of the form (10.38)

With the PD controller given by (10.38), the transform $E(s)$ of the tracking error becomes

$$E(s) = \left[\frac{(s + k_d/I)s}{(s + k_d/I)s + (K_M/I)(K_P + K_D s)}\right]\frac{\theta_0}{s}$$
$$= \frac{(s + k_d/I)\theta_0}{s^2 + (k_d/I + K_M K_D/I)s + K_M K_P/I} \tag{10.46}$$

From (10.46) it is clear that the coefficients of the denominator polynomial of $E(s)$ can be chosen arbitrarily by choosing K_P and K_D, and thus the poles of $E(s)$ and the poles of the closed-loop system can be placed anywhere in the open left-half plane. Therefore, the rate of convergence of $\theta(t)$ to θ_0 can be made as fast as desired by choosing appropriate values of K_P and K_D. The form of the transient part of the response to a step input depends on the location of closed-loop poles. To investigate this, the output response $\theta(t)$ will be computed in the case when $I = 1$, $k_d = 0.1$, and $k_m = 10$. With these values for the system parameters, the closed-loop transfer function is

$$G_{cl}(s) = \frac{G_p(s)G_c(s)}{1 + G_p(s)G_c(s)}$$
$$= \frac{[10/s(s + 0.1)](K_P + K_D s)}{1 + [10/s(s + 0.1)](K_P + K_D s)}$$
$$= \frac{10(K_P + K_D s)}{s^2 + (0.1 + 10K_D)s + 10K_P}$$

Setting

$$s^2 + (0.1 + 10K_D)s + 10K_P = s^2 + 2\zeta\omega_n s + \omega_n^2 \tag{10.47}$$

results in the following form for $G_{cl}(s)$:

$$G_{cl}(s) = \frac{10(K_P + K_D s)}{s^2 + 2\zeta\omega_n s + \omega_n^2}$$

Except for the zero at $s = -K_P/K_D$, $G_{cl}(s)$ has the same form as the second-order transfer function studied in Section 9.3. If the effect of the zero is ignored, the analysis in Section 9.3 of the step response of the second-order case can be applied here. In particular, from the results in Section 9.3, it was shown that when the damping ratio ζ is between 0 and 1 the transient in the step response decays to zero at a rate corresponding to the exponential factor $\exp(-\zeta\omega_n t)$ [see (9.24)]. Selecting $\zeta\omega_n = 1$ and using (10.47) gives

$$\zeta\omega_n = 1 = \frac{0.1 + 10K_D}{2}$$

Solving for K_D yields $K_D = 0.19$. Now to avoid a large overshoot in the step response, the damping ratio ζ should not be smaller than $1/\sqrt{2}$. With $\zeta = 1/\sqrt{2}$ and $\zeta\omega_n = 1$, then $\omega_n = \sqrt{2}$ and using (10.47) gives

$$10K_P = \omega_n^2 = 2$$

and thus $K_P = 0.2$. The closed-loop transfer function $G_{cl}(s)$ is then

$$G_{cl}(s) = \frac{1.9s + 2}{s^2 + 2s + 2}$$

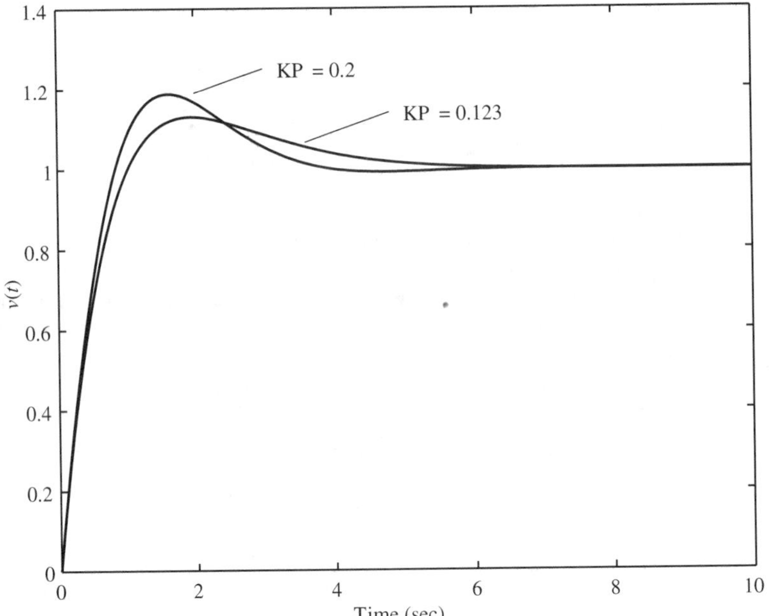

Figure 10.5 Step response with $K_D = 0.19$ and $K_P = 0.2$ and 0.123.

With $\theta_0 = 1$, the step response was then computed using the MATLAB command `step(num,den)` with

```
num = [1.9 2]
den = [1 2 2]
```

A plot of the response is given in Figure 10.5. Note that the overshoot is fairly pronounced. To reduce this, the damping ratio ζ should be increased. For example, setting $\zeta = 0.9$ but keeping $\zeta\omega_n = 1$ results in the values $K_P = 0.123$ and $K_D = 0.19$. The resulting step response is also shown in Figure 10.5. Note that the overshoot is smaller, but now the response is more "sluggish." To achieve a faster response, $\zeta\omega_n$ could be made larger than 1. The reader is invited to try this.

Consider the control signal $x(t)$ given in (10.40). For a step reference, $e(t) = r_0 - y(t)$. Therefore, (10.40) can be expressed in the form

$$x(t) = K_P e(t) - K_D \frac{dy(t)}{dt} \tag{10.48}$$

Hence to implement the control (10.48), it is necessary to measure the derivative $dy(t)/dt$ of the output $y(t)$. Unfortunately, this is often not possible to do in practice as a result of the presence of high-frequency noise in $y(t)$. For example, suppose that $y(t)$ contains a very small noise component equal to $10^{-4} \sin 10^6 t$. When this term is differentiated the result is $100 \sin 10^6 t$, which is not small in magnitude and can "swamp out" the signal terms. In practice, PD controllers are often implemented with an additional high-frequency filter to mitigate the effects of high-frequency noise.

10.3 ROOT LOCUS

Again consider the feedback control system with the transfer function representation

$$Y(s) = G_{cl}(s)R(s) \tag{10.49}$$

where the closed-loop transfer function $G_{cl}(s)$ is given by

$$G_{cl}(s) = \frac{G_p(s)G_c(s)}{1 + G_p(s)G_c(s)} \tag{10.50}$$

The closed-loop system is shown in Figure 10.6.

It is still assumed that the plant transfer function $G_p(s)$ has N poles and the controller transfer function $G_c(s)$ has q poles. Then if there are no pole–zero cancellations, the product $G_p(s)G_c(s)$ has $N + q$ poles, which are equal to the poles of the plant plus the poles of the controller. In addition, the zeros of $G_p(s)G_c(s)$ are equal to the zeros of the plant plus the zeros of the controller. With the zeros of $G_p(s)G_c(s)$ denoted by $z_1, z_2, \ldots, z_r$ and the poles denoted by $p_1, p_2, \ldots, p_{N+q}$, $G_p(s)G_c(s)$ can be expressed in factored form:

$$G_p(s)G_c(s) = K\frac{(s - z_1)(s - z_2)\cdots(s - z_r)}{(s - p_1)(s - p_2)\cdots(s - p_{N+q})} \tag{10.51}$$

In (10.51), K is a constant which contains the leading coefficients of the numerator and denominator polynomials of $G_p(s)G_c(s)$. Inserting (10.51) into (10.50) yields the following expression for the closed-loop transfer function:

$$G_{cl}(s) = \frac{K\dfrac{(s - z_1)(s - z_2)\cdots(s - z_r)}{(s - p_1)(s - p_2)\cdots(s - p_{N+q})}}{1 + K\dfrac{(s - z_1)(s - z_2)\cdots(s - z_r)}{(s - p_1)(s - p_2)\cdots(s - p_{N+q})}}$$

$$= \frac{K(s - z_1)(s - z_2)\cdots(s - z_r)}{(s - p_1)(s - p_2)\cdots(s - p_{N+q}) + K(s - z_1)(s - z_2)\cdots(s - z_r)} \tag{10.52}$$

From (10.52) it is seen that the closed-loop poles are the $N + q$ roots of the equation

$$(s - p_1)(s - p_2)\cdots(s - p_{N+q}) + K(s - z_1)(s - z_2)\cdots(s - z_r) = 0 \tag{10.53}$$

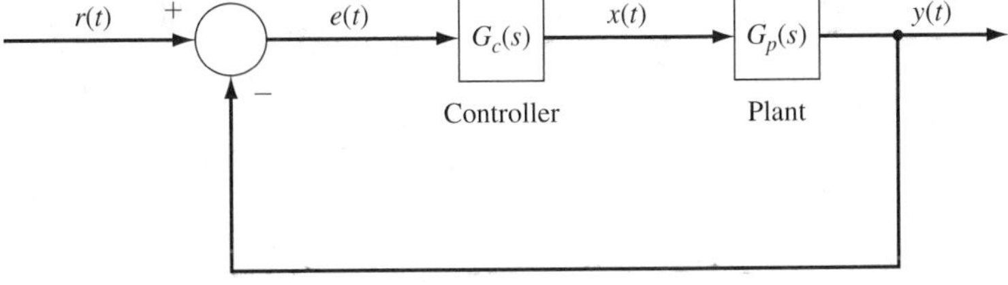

Figure 10.6 Feedback control system.

Obviously, the values of the $N + q$ closed-loop poles depend on the value of the constant K. In particular, note that when $K = 0$ the closed-loop poles are the same as the poles of $G_p(s)G_c(s)$.

Since the characteristics of the tracking error $e(t) = r(t) - y(t)$ depend directly on the values (or locations) of the closed-loop poles, in feedback control system design it is of major interest to know what possible closed-loop pole locations can be obtained by varying K. For example, K may correspond to a parameter (e.g., gain) in the controller that can be chosen by the designer, in which case the question arises as to whether or not there is a value of K that results in "good" pole locations. To answer this, it is first necessary to determine all closed-loop pole locations as K is varied over some range of values. This leads to the "180° root locus" (or the "$K > 0$ root locus"), which is the plot in the complex plane of the $N + q$ closed-loop poles as K is varied from 0 to ∞. Since only the $K > 0$ case is considered here, the 180° root locus or the $K > 0$ root locus will be referred to as the root locus. In the root locus construction, the constant K is called the *root-locus gain.*

Since there are $N + q$ closed-loop poles, the root locus has $N + q$ branches, where each branch corresponds to the movement in the complex plane of a closed-loop pole as K is varied from 0 to ∞. Since the closed-loop poles are the poles of $G_p(s)G_c(s)$ when $K = 0$, the root locus begins (when $K = 0$) at the poles of $G_p(s)G_c(s)$. As K is increased from zero, the branches of the root locus depart from the poles of $G_p(s)G_c(s)$, one branch per pole. As K approaches ∞, r of the branches move to the r zeros of $G_p(s)G_c(s)$, one branch per zero, and the other $N + q - r$ branches approach ∞.

A real or complex number p is on the root locus if and only if p is a root of (10.53) for some value of $K > 0$. That is, p is on the root locus if and only if for some $K > 0$,

$$(p - p_1)(p - p_2)\cdots(p - p_{N+q}) + K(p - z_1)(p - z_2)\cdots(p - z_r) = 0 \qquad (10.54)$$

Dividing both sides of (10.54) by $(p - p_1)(p - p_2)\cdots(p - p_{N+q})$ gives

$$1 + K\frac{(p - z_1)(p - z_2)\cdots(p - z_r)}{(p - p_1)(p - p_2)\cdots(p - p_{N+q})} = 0 \qquad (10.55)$$

Dividing both sides of (10.55) by K and rearranging terms yields

$$\frac{(p - z_1)(p - z_2)\cdots(p - z_r)}{(p - p_1)(p - p_2)\cdots(p - p_{N+q})} = -\frac{1}{K} \qquad (10.56)$$

Thus p is on the root locus if and only if (10.56) is satisfied for some $K > 0$.

Now if $P(s)$ is defined by

$$P(s) = \frac{(s - z_1)(s - z_2)\cdots(s - z_r)}{(s - p_1)(s - p_2)\cdots(s - p_{N+q})} \qquad (10.57)$$

then $KP(s) = G_p(s)G_c(s)$ and in terms of P, (10.56) becomes

$$P(p) = -\frac{1}{K} \qquad (10.58)$$

Thus p is on the root locus if and only if (10.58) is satisfied for some $K > 0$. Since $P(p)$ is a complex number in general, (10.58) is equivalent to the following two conditions:

$$|P(p)| = \frac{1}{K} \tag{10.59}$$

$$\angle P(p) = \pm 180^\circ \tag{10.60}$$

The condition (10.59) is called the *magnitude criterion* and the condition (10.60) is called the *angle criterion*. Any real or complex number p that satisfies the angle criterion (10.60) is on the root locus; that is, if (10.60) is satisfied, then (10.59) is also satisfied if

$$K = \frac{1}{|P(p)|} \tag{10.61}$$

In other words, for the value of K given by (10.61), p is on the root locus. This result shows that the root locus consists of all those real or complex numbers p such that the angle criterion (10.60) is satisfied. The use of the angle criterion is illustrated in the following example.

Example 10.6 *Root Locus for First-Order System*

Interactive Root Locus

Consider the closed-loop system with plant equal to the car with velocity model given by $G_p(s) = 0.001/(s + 0.01)$, and with proportional controller given by $G_c(s) = K_P$. Then

$$G_p(s)G_c(s) = \frac{0.001K_P}{s + 0.01} = K\frac{1}{s + 0.01} \tag{10.62}$$

where $K = 0.001\,K_P$. In this case, $G_p(s)G_c(s)$ has no zeros and one pole at $s = -0.01$, and thus $N + q = 1$ and $r = 0$. The root locus therefore has one branch that begins (when $K = 0$) at $s = -0.01$ and goes to ∞ as $K \to \infty$. From (10.62), $P(s)$ is

$$P(s) = \frac{1}{s + 0.01}$$

and thus

$$\angle P(p) = -\angle(p + 0.01)$$

Then $\angle P(p) = \pm 180^\circ$ if and only if p is a negative real number with $p < -0.01$. Hence the root locus consists of all negative real numbers p such that $p < -0.01$. The root locus is plotted in Figure 10.7. The arrow in Figure 10.7 shows the direction of movement of the closed-loop pole as $K \to \infty$. Note that in this case the closed-loop pole goes to ∞ by "moving out" on the negative real axis of the complex plane.

Given a negative real number $p < -0.01$, from (10.61) the value of K for which p is on the root locus is

$$K = \frac{1}{|1/(p + 0.01)|} = |p + 0.01|$$

For instance, for the closed-loop pole to be $p = -0.2$ the value of the root locus gain K is

$$K = |-0.2 + 0.01| = 0.19$$

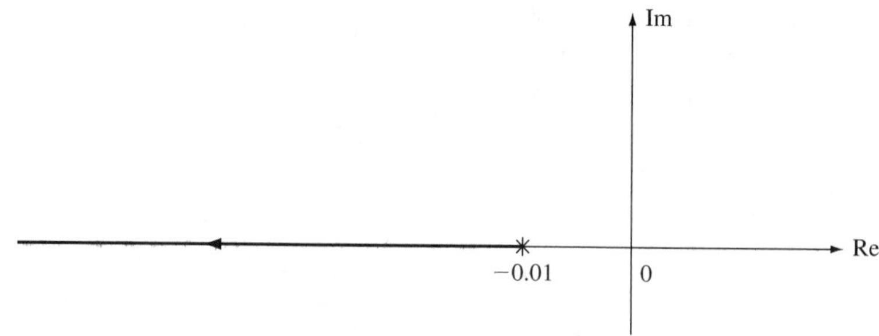

Figure 10.7 Root locus in Example 10.6.

Then since $K = 0.001K_P$ [see (10.62)], the gain K_P of the proportional controller must be

$$K_P = \frac{0.19}{0.001} = 190$$

This is the value of K_P that puts the closed-loop pole at $s = -0.2$. The reader is invited to check this result by computing the closed-loop transfer function with $G_c(s) = 190$.

Root-Locus Construction

Again consider the feedback control system in Figure 10.6 with $G_p(s)G_c(s)$ expressed in the factored form (10.51) and with $P(s)$ defined by (10.57) so that $KP(s) = G_p(s)G_c(s)$. Note that the zeros (respectively, poles) of $P(s)$ are the same as the zeros (respectively, poles) of $G_p(s)G_c(s)$. The closed-loop poles are the roots of equation (10.53) where the z_i are the zeros of $P(s)$ and the p_i are the poles of $P(s)$. A sketch of the root locus for $K > 0$ can be generated by numerically computing the poles of the closed-loop transfer function for specified values of K.

Example 10.7 *Root Locus for Second-Order System*

Interactive Root Locus

Consider the dc motor with transfer function $G_p(s) = 10/(s + 0.1)s$ and with proportional controller $G_c(s) = K_P$. Then

$$G_p(s)G_c(s) = \frac{10K_P}{(s + 0.1)s} = K\frac{1}{(s + 0.1)s}$$

and thus

$$K = 10K_P \quad \text{and} \quad P(s) = \frac{1}{(s + 0.1)s}$$

Since $P(s)$ has two poles at $p_1 = 0$ and $p_2 = -0.1$, there are two closed-loop poles and the root locus has two branches, which start at $s = 0$ and $s = -0.1$. The closed-loop transfer function is computed to be

$$G_{cl}(s) = \frac{KP(s)}{1 + KP(s)}$$
$$= \frac{K/s(s + 0.1)}{1 + K/s(s + 0.1)}$$
$$= K\frac{1}{s^2 + 0.1s + K}$$

The closed-loop poles are given by the roots of

$$s^2 + 0.1s + K = 0 \tag{10.63}$$

The root locus can be obtained numerically by substituting specific values for K into (10.63) and finding the roots of the resulting equation. In this case the quadratic formula can be used to solve for the roots:

$$s = -0.05 \pm 0.5\sqrt{0.01 - 4K}$$

For $K = 0$, the closed-loop poles are at $p_1 = 0$ and $p_2 = -0.1$, as expected. A computer program can be written that computes the roots of the equation in (10.63) starting at $K = 0$ and incrementing K by small amounts until a specified upper limit. In this particular example, it is obvious that for $0 < K < 0.0025$, the closed-loop poles are real and negative; for $K = 0.0025$, the closed-loop poles are real and equal. Finally, for $K > 0.0025$, the poles are complex and located at $p, \bar{p} = -0.05 \pm j0.5\sqrt{4K - 0.01}$. A plot of the resulting root locus is given in Figure 10.8.

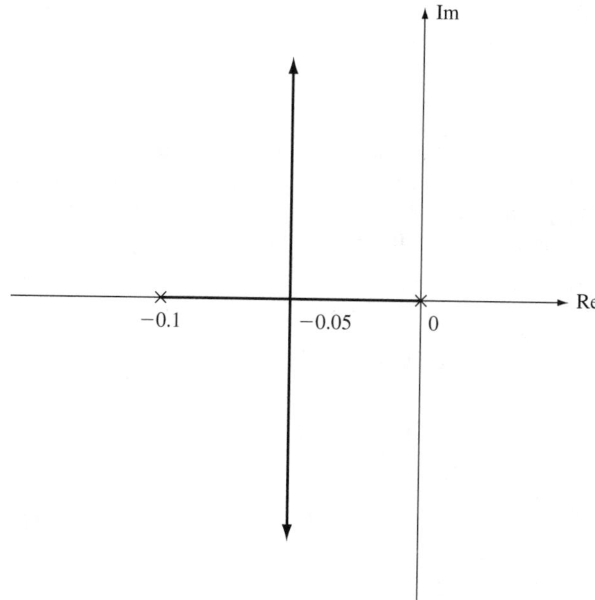

Figure 10.8 Root locus in Example 10.7.

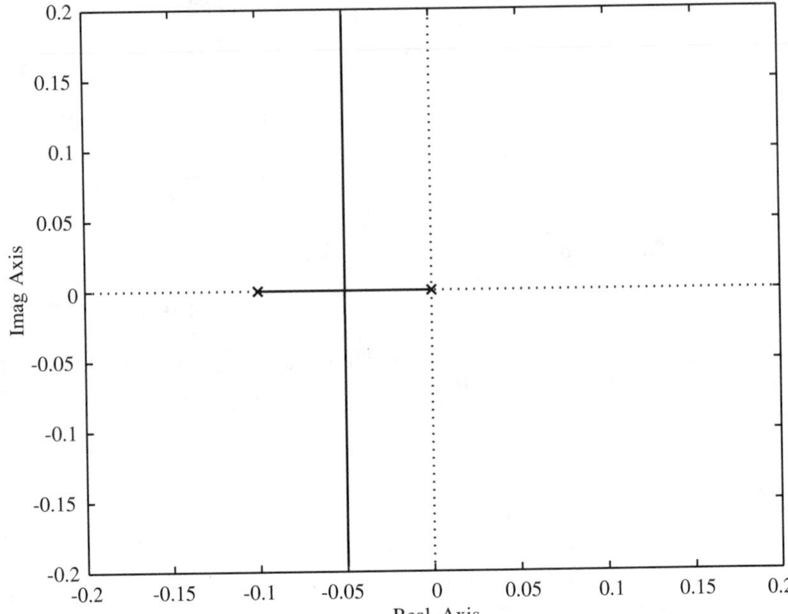

Figure10.9 MATLAB plot of root locus in Example 10.7.

The MATLAB M-file `rlocus` computes and plots the root locus given the numerator and denominator of $P(s)$. The commands to generate this plot are given by

```
num = 1;
den = [1 0.1 0];
rlocus(num,den)
```

where `num` contains the coefficients of the numerator of $P(s)$ and `den` contains the coefficients of the denominator of $P(s)$. The resulting plot is given in Figure 10.9. These commands automatically generate the values of K that will yield a good plot. To customize the plot, one can compute the root locus for gains specified in a vector K:

```
num = 1;
den = [1 0.1 0];
K = 0:.0005:.04;
p = rlocus(num,den,K);
plot(p)
```

Example 10.8 *Root Locus for Second-Order System with Zero*

Interactive Root Locus

Now suppose that

$$G_p(s) = \frac{2}{(s-1)(s^2+2s+5)} \quad \text{and} \quad G_c(s) = A(s+3)$$

where A is a real constant (a gain in the controller). Then

$$G_p(s)G_c(s) = \frac{2A(s+3)}{(s-1)(s^2+2s+5)} = K\frac{s+3}{(s-1)(s^2+2s+5)}$$

and thus

$$K = 2A \quad \text{and} \quad P(s) = \frac{s+3}{(s-1)(s^2+2s+5)} = \frac{s+3}{s^3+s^2+3s-5}$$

In this case, $P(s)$ has three poles at $p_1 = -1 + j2, p_2 = -1 - j2$, and $p_3 = 1$, and $P(s)$ has one zero at $z_1 = -3$. Therefore, the root locus has three branches beginning (when $K = 0$ or $A = 0$) at $-1 \pm j2$ and 1. A precise sketch of the root locus can be produced by using the MATLAB command `rlocus`. In this example, the following commands compute the root locus and then generate a plot:

```
num = [1 3];
den = [1 1 3 -5];
rlocus(num,den);
```

The resulting root locus is shown in Figure 10.10. The branches starting at the poles $-1 \pm j2$ go to infinity as $K \to \infty$, while the branch starting at $p_3 = 1$ goes to the zero at $z_1 = -3$ as $K \to \infty$.

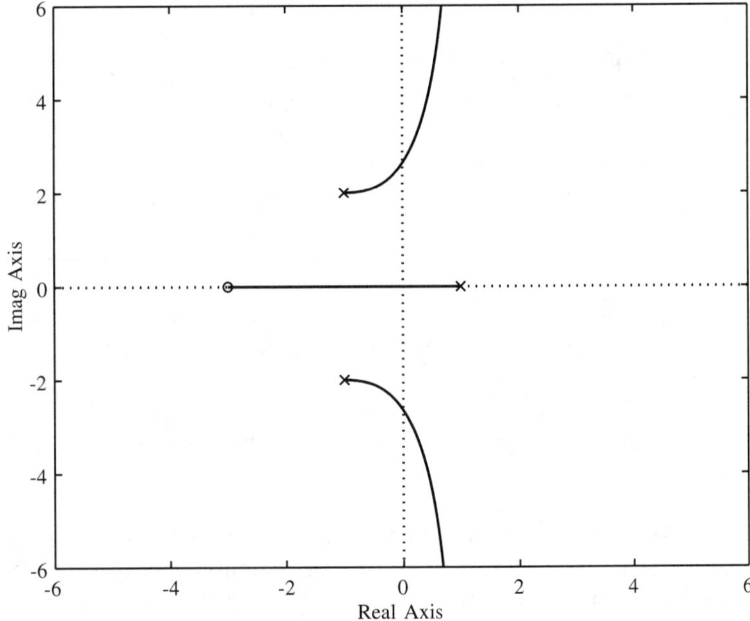

Figure 10.10 MATLAB plot of root locus.

The root locus shown in Figure 10.10 can be used to determine the range of values of K (or A) such that the closed-loop system is stable, that is, the range of values of K for which all three closed-loop poles are in the open left-half plane. First, note that since one of the branches starts at 1 when $K = 0$ and moves to the origin, the closed-loop system is not stable for $0 < K \leq c$, where c is the value of K for which there is a closed-loop pole at $s = 0$. The constant c can be determined by using the magnitude criterion (10.61), which gives

$$c = \frac{1}{|P(0)|} = \left|\frac{-5}{3}\right| = \frac{5}{3}$$

From the root locus in Figure 10.10 it is also clear that the closed-loop system is not stable for $K > b$, where b is the value of K for which the two complex poles are equal to $\pm j\omega_c$, where $\pm j\omega_c$ are the points on the imaginary axis where the two branches cross over into the right-half plane. From Figure 10.10 an approximate value of ω_c is 2.6. Then

$$b = \frac{1}{|P(j2.6)|} = \left|\frac{-j17.576 - 6.76 + j7.8 - 5}{j2.6 + 3}\right| = \sqrt{\frac{(11.76)^2 + (9.776)^2}{(2.6)^2 + 9}} = 3.85$$

Thus an approximate range for stability is $\frac{5}{3} < K < 3.85$, or since $K = 2A$, the range on A is $\frac{5}{6} < A < 1.925$. The exact range on K or A for stability can be determined by computing the exact value of K for which the two complex poles are on the imaginary axis. This can be carried out by using the Routh–Hurwitz test as follows. First, the closed-loop transfer function is

$$\begin{aligned} G_{cl}(s) &= \frac{KP(s)}{1 + KP(s)} \\ &= \frac{K(s + 3)}{s^3 + s^2 + (K + 3)s + (3K - 5)} \end{aligned}$$

and thus the Routh array is

s^3	1	$K + 3$
s^2	1	$3K - 5$
s^1	$\dfrac{(K + 3) - (3K - 5)}{1}$	
s^0	$3K - 5$	

From the results in Section 9.2, there are two poles on the $j\omega$-axis when the term in the first column of the Routh array indexed by s^1 is zero. Thus

$$K + 3 - (3K - 5) = 0$$

which gives $K = 4$. Therefore, the range for stability is $\frac{5}{3} < K < 4$, or in terms of A, $\frac{5}{6} < A < 2$.

An alternative means of sketching the root locus uses graphical construction rules that are derived from the angle and magnitude criteria (10.59) and (10.60). The graphical method provides insight into the effect of pole and zero locations on the shape of the root locus. Such insight is valuable for control design; however, only an introduction to control design using the root locus method is covered in this book.

For more information on the graphical construction rules and their use in control design, see Phillips and Harbor [1996].

10.4 APPLICATION TO CONTROL SYSTEM DESIGN

In this section the root-locus construction is applied to the problem of designing the controller transfer function $G_c(s)$ so that a desired performance is achieved in tracking a specific reference signal $r(t)$. In practice, performance is usually given in terms of an acceptable steady-state error and acceptable transient response. If the reference $r(t)$ is a step function (the case of set-point control), from the results in Section 10.2 the steady-state error will be zero if the open-loop system $G_p(s)G_c(s)$ is type 1. As noted before, if the plant transfer function $G_p(s)$ does not have a pole at $s = 0$, the controller transfer function $G_c(s)$ must have a pole at zero in order to have a type 1 system. Hence, in the case of set-point control, the best possible steady-state performance (i.e., zero steady-state error) is easily obtained by including (if necessary) a pole at zero in $G_c(s)$.

Achieving an acceptable transient response can be approached in terms of specifying desired pole locations for the dominant closed-loop poles. The behavior of the transient response can be characterized in terms of the poles of the closed-loop system as a result of the direct relationship between the form of the transient response and the system poles, as was shown to be the case in Chapter 9. For example, suppose that the controller transfer function $G_c(s)$ can be designed so that there are a pair of dominant closed-loop poles located at $p,\bar{p} = \alpha_0 \pm j\omega_0$, where $\alpha_0 < 0$. It follows directly from the analysis given in Section 9.3 that for there to be a pair of dominant closed-loop poles located at $p,\bar{p}$, there should not be any closed-loop zeros near $p,\bar{p}$, and all the other closed-loop poles must have real parts that are much less than the real part α_0 of the pair $p,\bar{p}$. If these conditions are satisfied, the pair $p,\bar{p}$ will be dominant, and as a result, the characteristics of the closed-loop system's transient response resulting from a reference input $r(t)$ will depend on the location of $p,\bar{p}$ in the complex plane. In particular, if $r(t)$ is a step, the analysis given in Section 9.3 can be applied to determine the region of the complex plane in which the values of $p,\bar{p}$ must lie for an acceptable transient. The root locus can then be plotted to determine if there is any value of K so that a closed-loop pole lies in the acceptable region of the complex plane. Generally, this procedure is first performed with a proportional controller, $G_c(s) = K_P$ since it is a simple controller to design and implement. If a dominant branch of the root locus does not lie in the desired region of the complex plane, a more complex controller $G_c(s)$ is used which reshapes the root locus.

Example 10.9 *Proportional Controller Design*

A controller is to be designed for the dc motor considered in Example 10.7, where

$$G_p(s) = \frac{10}{(s + 0.1)s}$$

The specifications for the closed-loop system are that the time constant τ be less than or equal to 25 seconds and the damping ratio ζ be greater than or equal to 0.4. Recall from Section 9.3 that $\tau \approx -1/\text{Re}\{p\}$, where p is the dominant pole, so in this case $\text{Re}\{p\} \leq -0.04$.

Also, recall that the damping ratio is defined by $\zeta = \cos\theta$, where θ is the angle of the pole position measured with respect to the negative real axis. These specifications can be transferred to the complex plane as shown in Figure 10.11. Any closed-loop pole that lies in the shaded region is acceptable. Next, examine a proportional controller, $G_c(s) = K_P$. Let $K = 10K_p$ and $P(s) = 1/(s + 0.1)s$. The root locus for the dc motor with a proportional controller is given in Figure 10.9 and is redrawn in Figure 10.12 along with the specifications. The pole designated as p_L in Figure 10.12 marks the point at which the root locus enters the desired region. The pole designated as p_H marks the point at which the root locus exits the desired region. The values of gain K which give closed-loop poles at p_L and p_H then specify the range of values of K for which the specifications are satisfied. To find the gain K_L that yields a closed-loop pole at p_L, use the magnitude criterion given in (10.61) for $p = p_L$, where $p_L = -0.04$, is obtained from the graph:

$$K_L = |(p + 0.1)p|_{p=-0.04} = 0.0024$$

Repeat the procedure to find the gain K_H that yields a closed-loop pole at p_H. Using the magnitude criterion on $p_H = -0.05 + j0.114$ yields $K_H = 0.0156$. Since the root-locus plot is continuous with respect to K, the range of K that satisfies the specifications is $0.0024 \leq K \leq 0.0156$ or $0.00024 \leq K_P \leq 0.00156$.

The step response of the closed-loop system for $K = 0.0024$ is obtained by using the following MATLAB commands:

```
K = 0.0024;
num = 1;
den = [1 .1 0];
[Ncl,Dcl] = cloop(num*K,den,-1);
step(Ncl,Dcl)
```

The command `cloop` computes the closed-loop transfer function from $P(s)$. The resulting step responses for $K = K_L$ and $K = K_H$ are shown in Figure 10.13. Note that the

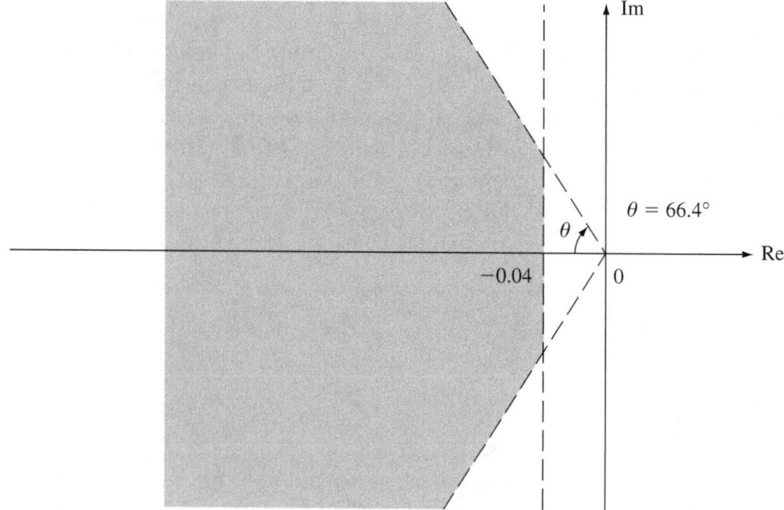

Figure 10.11 Complex plane showing region of acceptable pole positions.

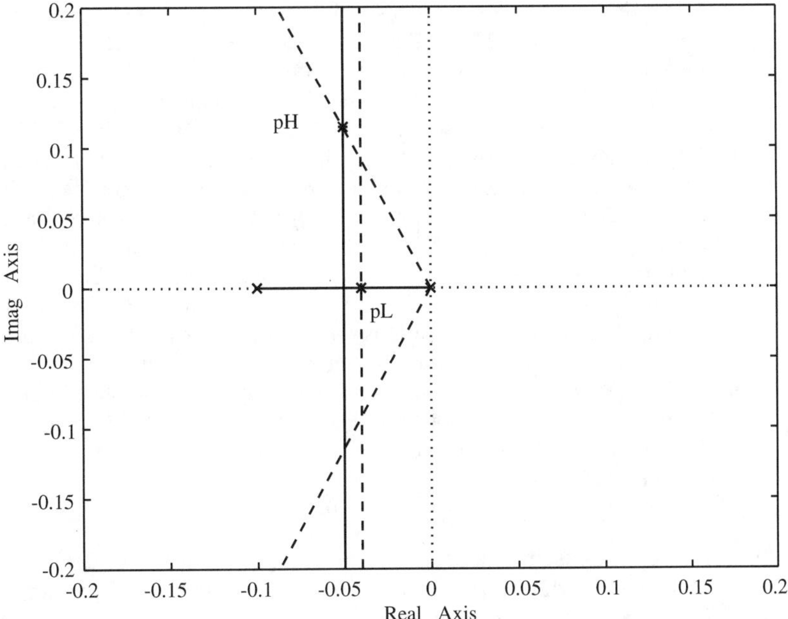

Figure 10.12 Root locus for Example 10.9 and region of acceptable pole positions.

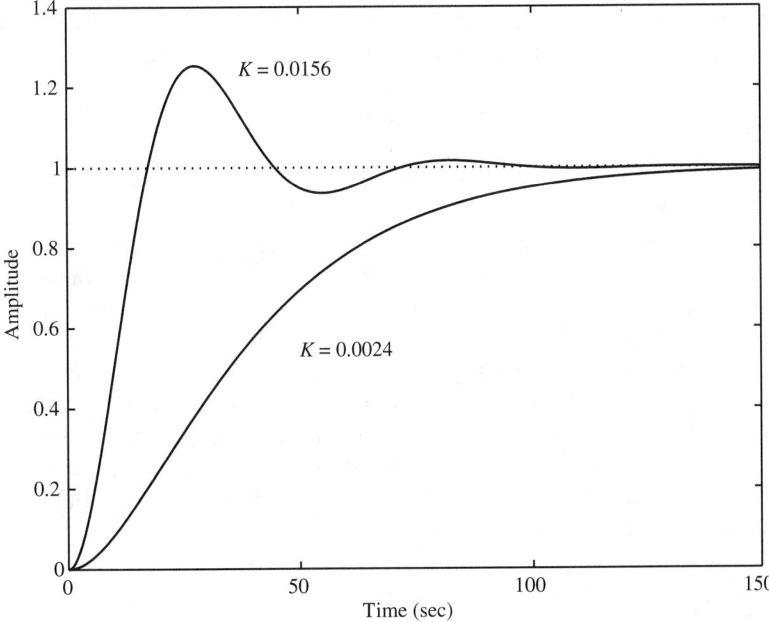

Figure 10.13 Step responses for closed-loop system with proportional control for $K = 0.0024$ and for $K = 0.0156$.

transient response for $K = K_L$ decays slower than might be expected for a system with a time constant of $\tau = 25$ sec. This is because the poles for $K = K_L$ are close to each other, one pole at $s = -0.04$ and the other at $s = -0.06$. In the approximation $\tau \approx -1/\text{Re}(p)$ it is assumed that p is the dominant pole and that the rest of the system poles are much farther left, so that their effect is negligible. The resulting transient response resembles a pure exponential with a decay rate of $\text{Re}(p)$. When the poles are close, as in this example, the resulting transient response is not nearly exponential, so that the approximation $\tau \approx 1/0.04$ is not very accurate. Approximations like the one for the time constant are often employed to obtain an initial control design that gives a reasonable time response, and iteration is then used to fine-tune the results. The reader is invited to investigate the step responses for other values of K between K_L and K_H.

Suppose that the specifications were modified so that $\tau \leq 10$ seconds and $\zeta \geq 0.4$. This requires that the real part of the dominant pole must lie to the left of -0.1. Examination of the root locus with a proportional controller shows that no gain exists so that a dominant branch of the root locus lies to the left of -0.1. Hence a proportional controller cannot satisfy the specifications. In this case, a more complex controller should be chosen as discussed below.

It is often the case that a proportional controller cannot satisfy the specifications. Since the root locus is defined by the poles and zeros of the transfer function $G_p(s)G_c(s)$, adding zeros and/or poles to $G_c(s)$ will change the shape of the root locus. An important part of control design is to determine where to put the controller zeros and poles to yield a desirable response. A further consideration for the control design is satisfying the specifications on steady-state errors. The discussion here is limited to PD, PI and PID controllers. The effect of each of these controllers on a root locus is examined below.

Consider the PD controller introduced in Example 10.5 and given by

$$\begin{aligned} G_c(s) &= K_P + K_D s \\ &= K_D\left(s + \frac{K_P}{K_D}\right) \end{aligned}$$

This type of controller contributes a zero to the rational function $P(s)$. The addition of this zero tends to pull the root locus to the left when compared to the root locus with a proportional gain. Thus the PD controller is generally used to speed up the transient response of a system over that obtainable with a proportional controller.

There are two design parameters for a PD controller, K_D and K_P. The ratio $-K_P/K_D$ defines the zero location of the controller. One rule of thumb is to choose the zero location to be in the left-hand plane and to the left of the rightmost pole. In some cases the zero may be chosen to cancel a stable real pole. Once the zero location has been determined, the regular root-locus design method illustrated in Example 10.9 can be used to select K_D.

Example 10.10 *PD Controller Design*

Consider the dc motor of Examples 10.5, 10.7, and 10.9. Now choose $G_c(s)$ as a PD controller and let $z = -K_P/K_D$. Then

$$G_p(s)G_c(s) = \frac{10K_D(s-z)}{(s+0.1)s} = K\frac{s-z}{(s+0.1)s}$$

and thus

$$K = 10K_D \quad \text{and} \quad P(s) = \frac{s-z}{(s+0.1)s}$$

Consider four different values for z: -0.05, -0.1, -0.2, and -1. The root locus for $z = -0.05$ is obtained using the MATLAB commands

```
num = [1 0.05];
den = [1 .1 0];
rlocus(num,den)
```

The root loci for the four different cases, $z = -0.05$, $z = -0.1$, $z = -0.2$, and $z = -1$, are given in Figure 10.14. For $z = -0.05$ there is a branch from $s = 0$ to $s = -0.05$. As K is increased, the closed-loop pole on this branch gets closer to the zero at -0.05. Recall from Section 9.3 that if a zero is very near a pole, the residue is small and the pole

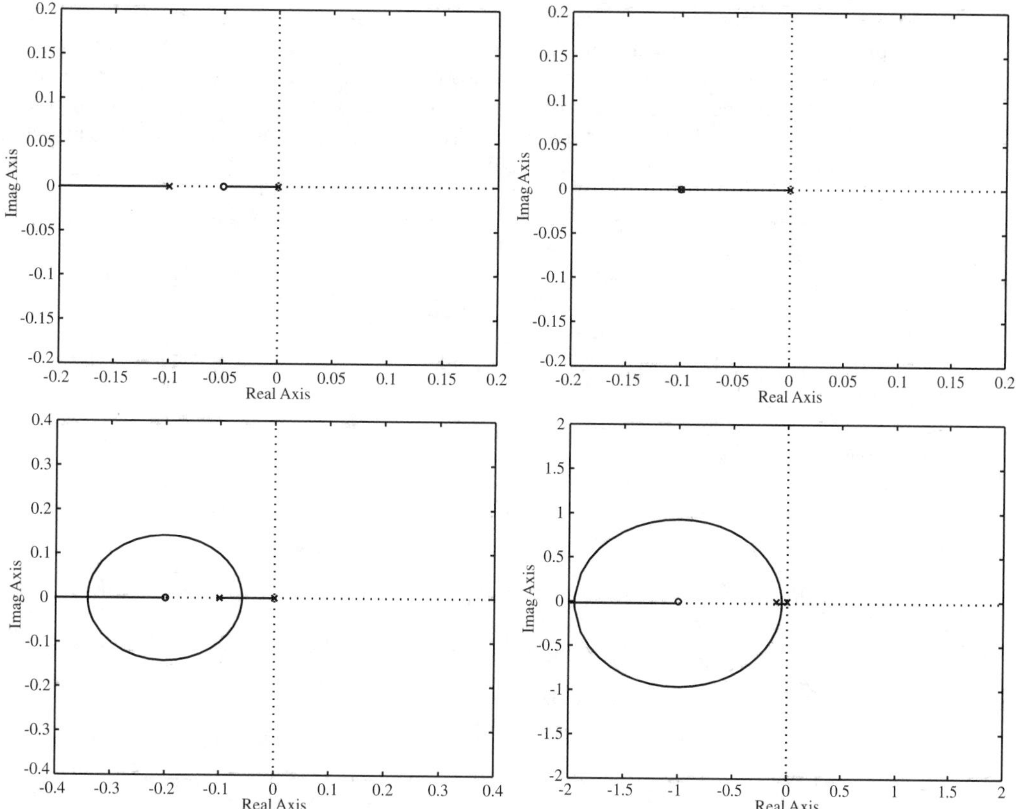

Figure 10.14 Root loci for PD controllers in Example 9.10: (a) $z = -0.05$; (b) $z = -0.1$; (c) $z = -0.2$; (d) $z = -1$.

is not dominant. Hence, as K is increased, the residue of the pole on this branch gets smaller, making it less dominant. The other branch goes to $-\infty$; therefore, the response can be made suitably fast by choosing a large value of K. For $z = -0.1$, the zero cancels a plant pole making the closed-loop system behave as a first-order response. The resulting branch starts at $s = 0$ and goes to $-\infty$ along the negative real axis as $K \to \infty$; hence the transient response can be made as fast as desired by increasing the value of K. For $z = -0.2$ the root locus has two branches starting at $s = 0$ and $s = -0.1$, which come together along the real axis to meet at $s = -0.06$, and then the branches split apart, forming a circular arc, which breaks into the real axis at $s = -0.34$; one branch then moves toward the zero at -0.2 and the other branch goes to $-\infty$ along the real axis. The circular arc means that this choice of zero location allows for an underdamped response. The dominant poles are farthest to the left when the closed-loop poles are both equal to -0.34. For $z = -1$, the circular arc has a large radius and crosses the real axis at a point that is farther left than for that obtained for $z = -0.2$. In this case the fastest response is obtained for the value of K that yields closed-loop poles at $s = -1.95$.

Now, let $z = -0.1$, which cancels a plant pole. Suppose that the specification requires that $\tau \leq 10$ seconds, so that the dominant pole must lie to the left of -0.1. Choose a desired closed-loop pole to be $p = -0.1$. Then use the magnitude criterion (10.61) to solve for the corresponding K, $K = 0.1$. The resulting control is $G_c(s) = 0.01(s + 0.1)$. Figure 10.15 shows the closed-loop step response of the dc motor with this PD controller. For comparison sake, the closed-loop dc motor with a proportional controller, $G_c(s) = 0.0005$, is also shown. The gain for the proportional control was chosen to give closed-loop poles at $-0.05 \pm j0.05$, which are as far left as possible with proportional control. The damping ratio of $\zeta = 0.707$ is large enough to give a reasonably small oscillation of the transient.

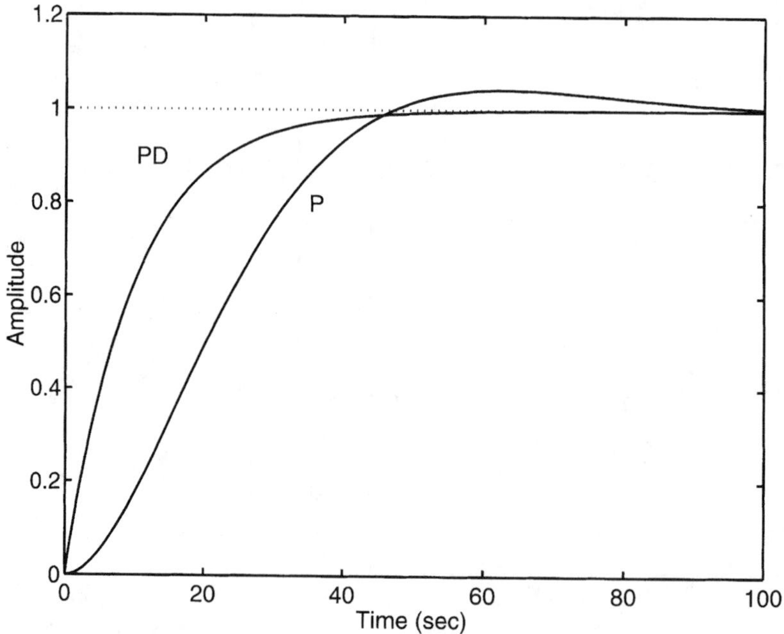

Figure 10.15 Closed-loop step response for PD and P controllers.

Consider the PI controller introduced in Example 10.5 and given by

$$G_c(s) = \frac{K_P s + K_I}{s} = K_P \frac{s + K_I/K_P}{s}$$

This controller increases the system type and is generally used to reduce the steady-state error, that is, increase the steady-state accuracy of the tracking between $y(t)$ and $r(t)$. The addition of a pole at the origin and a zero at $-K_I/K_P$ affects the shape of the root locus and therefore may affect the transient response. In general, the PI controller results in a root locus that is to the right of a root locus drawn for a proportional controller. Hence the transient response is generally slower than is possible with a proportional controller. The ratio of K_I/K_P is usually chosen so that the resulting zero is closer to the origin than any plant pole. The smaller the ratio, and hence the closer the zero is to the origin, the smaller the effect of the PI compensator on the root locus. Therefore, if a proportional controller can be found that gives desirable transient response but unacceptable steady-state error, a PI controller can be used to obtain nearly the same closed-loop pole location but much smaller steady-state errors.

Example 10.11 *PI Controller Design*

Consider a system with a transfer function

$$G_p(s) = \frac{1}{(s + 1)(s + 4)}$$

A controller is to be designed so that the output $y(t)$ tracks a reference input $r(t)$ with a small error. The root locus with a proportional controller is given in Figure 10.16a. To reduce the steady-state error, design a PI control of the form

$$G_c(s) = K_P + \frac{K_I}{s} = K_P \frac{s - z}{s}$$

where $z = -K_I/K_P$. In this case

$$K = K_P \quad \text{and} \quad P(s) = \frac{s - z}{s(s + 1)(s + 4)}$$

Consider three choices for z: -0.01, -1, and -3. The corresponding root loci are given in Figure 10.16b–d. Note that the root-locus branches for the proportional controller are farther left than the root loci for any of the PI controllers. Also notice that the closer the zero z is to the origin, the closer the PI root locus plot is to that of the P root locus.

A proportional plus integral plus derivative (PID) controller combines the benefits of a PI and a PD controller; that is, it increases the system type so that it decreases the steady-state error plus it improves the transient response by moving the root locus to the left. The general form of this controller is

$$\begin{aligned} G_c(s) &= K_P + K_D s + \frac{K_I}{s} \\ &= \frac{K_D s^2 + K_P s + K_I}{s} \end{aligned}$$

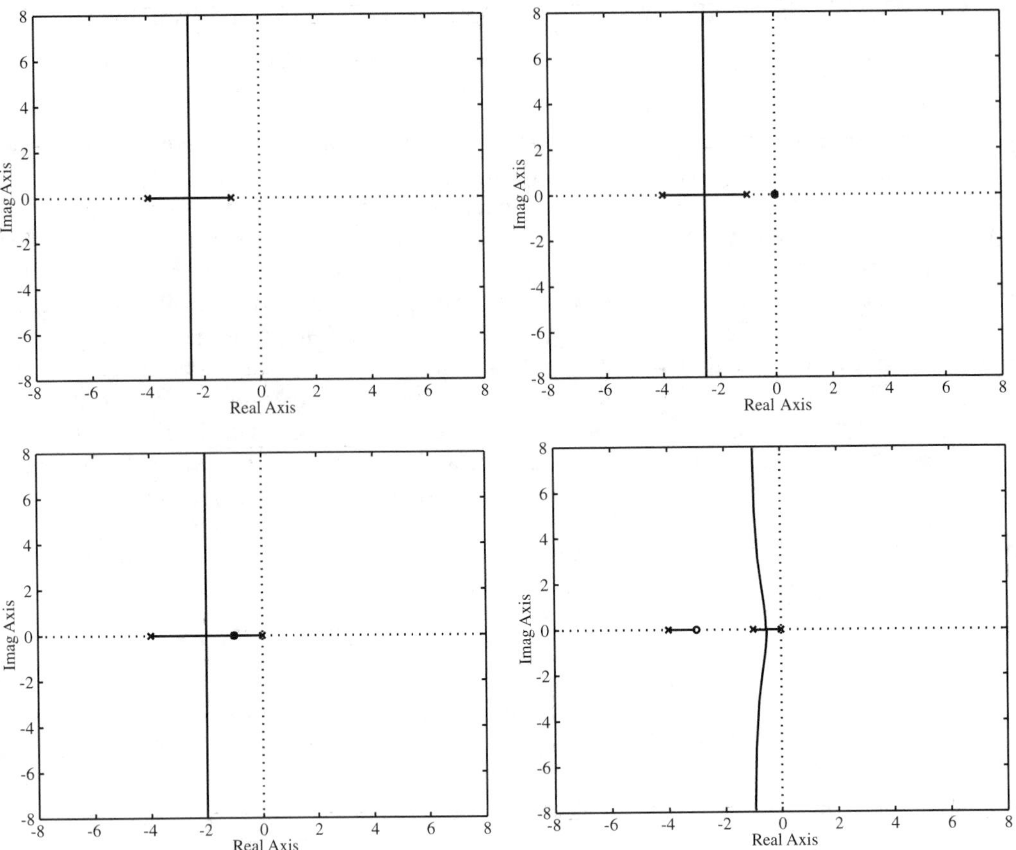

Figure 10.16 Root loci for Example 10.11: (a) P control; (b) PI control with $z = -0.01$; (c) PI control with $z = -1$; (d) PI control with $z = -3$.

This controller has one pole at the origin and two zeros. Denoting the zeros as z_1 and z_2, the controller has the general form

$$G_c(s) = K_D \frac{(s - z_1)(s - z_2)}{s}$$

Generally, one of the zeros is chosen to be near the origin (like that of a PI controller) and the other is chosen farther to the left (like a PD controller).

Example 10.12 *PID Controller Design*

Again consider the system given Example 10.11. A PI controller with $z_1 = -1$ was designed in Example 10.11. To this control, add a PD controller with a zero at $z_2 = -8$. Note that the zero of the PD controller can be chosen arbitrarily so that the root locus can be moved arbitrarily far to the left. The resulting PID controller has the form

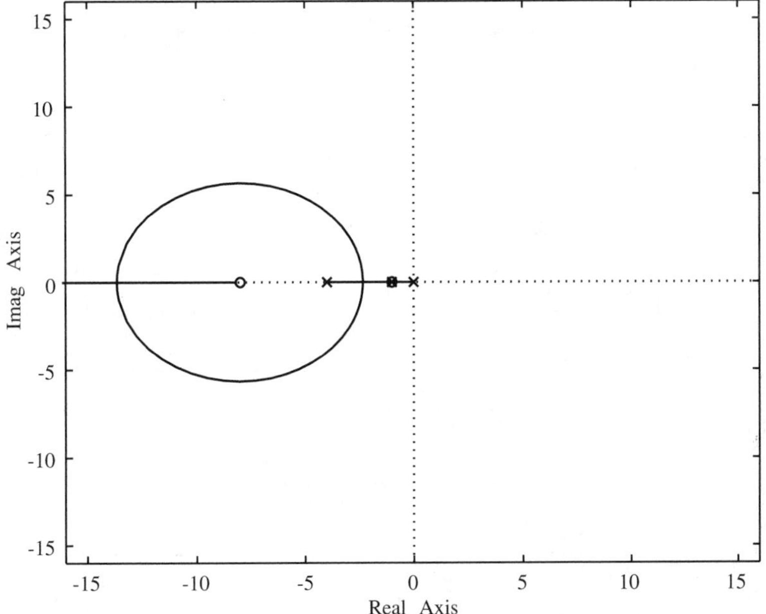

Figure 10.17 Root locus of PID controller.

$$G_c(s) = K_D \frac{(s+1)(s+8)}{s}$$

Then

$$K = K_D \quad \text{and} \quad P(s) = \frac{s+8}{(s+4)s}$$

The root locus, drawn in Figure 10.17, is farther to the left than any of the root loci drawn for a PI controller, as shown in Figure 10.16.

PROBLEMS

10.1. Consider the following system transfer function:

$$G_p(s) = \frac{1}{s+0.1}$$

(a) An open-loop control is shown in Figure P10.1a. Design the control, $G_c(s)$, so that the combined plant and controller $G_c(s)G_p(s)$ has a pole at $p = -2$, and the output $y(t)$ tracks a constant reference signal $r(t) = r_0 u(t)$ with zero steady-state error, where $e_{ss} = r_0 - y_{ss}$.

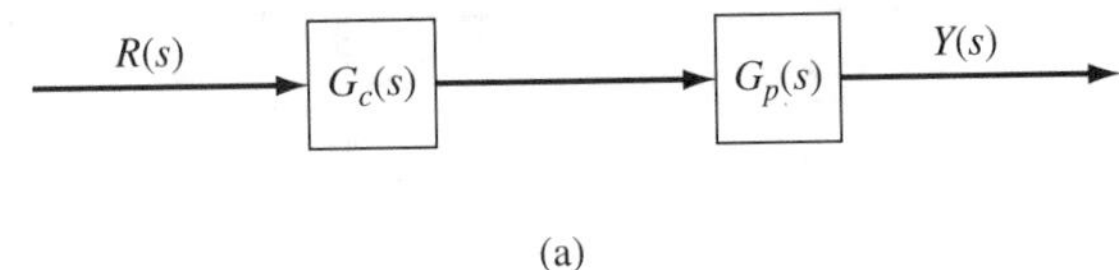

(a)

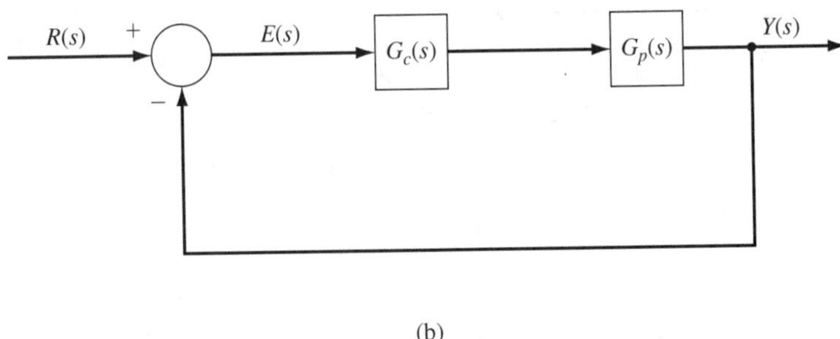

(b)

Figure P10.1

(b) Now suppose that the plant pole at $p = -0.1$ was modeled incorrectly and that the actual pole is $p = -0.2$. Apply the control designed in part (a) and the input $r(t) = r_0u(t)$ to the actual plant and compute the resulting steady-state error.

(c) A feedback controller $G_c(s) = 2(s + 0.1)/s$ is used in place of open-loop control as shown in Figure P10.1b. Verify that the closed-loop pole of the nominal system is at $p = -2$. (The nominal system has the plant pole at $p = -0.1$.) Let the input to the closed loop system be $r(t) = r_0u(t)$. Verify that the steady-state error $e_{ss} = r_0 - y_{ss}$ is zero.

(d) Compute the steady-state error of the actual closed-loop system (with plant pole at $p = -0.2$) when $r(t) = r_0u(t)$. Compare this error to that of the actual open-loop system computed in part (b).

(e) Simulate the responses of the systems in parts (a) to (d) when $r_0 = 1$. Explain the differences (and similarities) in responses.

10.2. Examine the effect of a disturbance on the performance of open- and closed-loop control systems by performing the following analysis.

(a) Consider an open-loop control system with a disturbance $D(s)$ as shown in Figure P10.2a. Define an error $E(s) = R(s) - Y(s)$ where $R(s)$ is a reference signal. Derive an expression for $E(s)$ in terms of $D(s)$, $X(s)$, and $R(s)$. Suppose that $D(s)$ is known. Can its effect be removed from $E(s)$ by proper choice of $X(s)$ and/or $G_c(s)$? Now suppose that $D(s)$ represents an unknown disturbance. Can its effect be removed (or reduced) from $E(s)$ by proper choice of $X(s)$ and/or $G_c(s)$? Justify your answers.

(b) Now consider the feedback system shown in Figure P10.2b. Derive an expression for $E(s)$ in terms of $D(s)$ and $R(s)$. Suppose that $D(s)$ represents an unknown disturbance and that $G_c(s) = K$. Can the effect of $D(s)$ be removed (or reduced) from $E(s)$ by proper choice of K? Justify your answer.

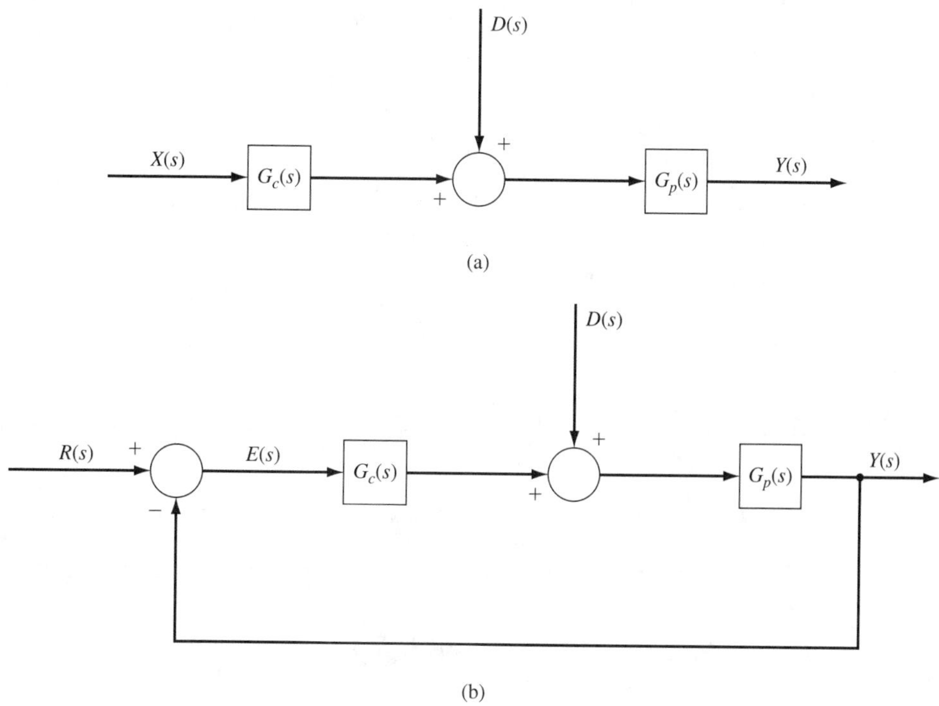

Figure P10.2

10.3. A rocket is drawn in Figure P10.3a, where $\theta(t)$ represents the angle between the rocket's orientation and its velocity, $\phi(t)$ represents the angle of the thrust engines, and $w(t)$ represents wind gusts which act as a disturbance to the rocket. The goal of the control design is to have the angle $\theta(t)$ track a reference angle $\theta_r(t)$. The angle of the thrust engines can be directly controlled by motors which position the engines; therefore, the plant output is $\theta(t)$ and the controlled input is $\phi(t)$. The system can be modelled by the following equation:

$$\Theta(s) = \frac{1}{s(s-1)}\Phi(s) + \frac{0.5}{s(s-1)}W(s)$$

(a) Consider an open-loop control $\Theta(s) = G_c(s)X(s)$, where $G_c(s)$, the controller transfer function, and $x(t)$, the command signal, can be chosen as desired. Is such a controller practical for having $\theta(t)$ track $\theta_r(t)$? Justify your answer.

(b) Now consider a feedback controller as shown in Figure P10.3b, where

$$G_c(s) = K(s+2)$$

Find an expression for the output $\Theta(s)$ of the closed-loop system in terms of $W(s)$ and $\Theta_r(s)$. Consider the part of the response due to $W(s)$; the lower this value, the better the disturbance rejection. How does the magnitude of this response depend on the magnitude of K?

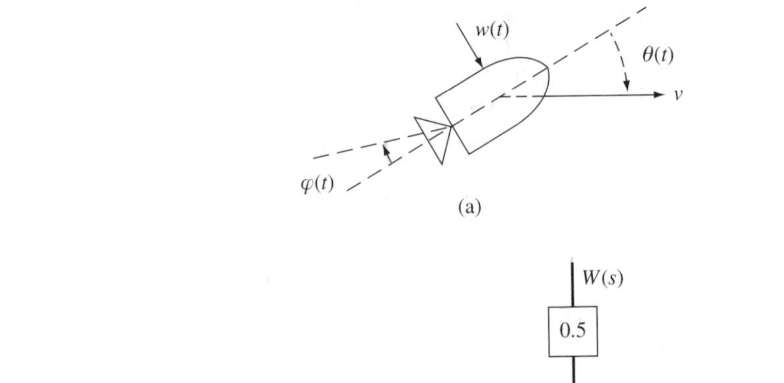

(a)

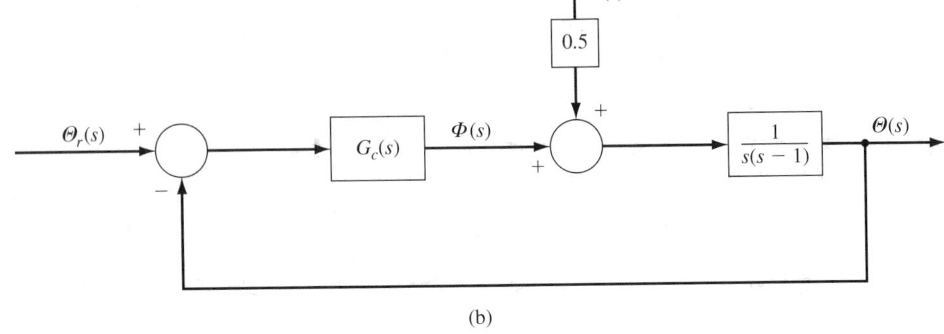

(b)

Figure P10.3

(c) Suppose that $\Theta_r(t) = 0$ and $w(t)$ is a random signal uniformly distributed between 0 and 1. Define a vector `w` in MATLAB as `w = rand(201,1)`, and define the time vector as `t=0:0.05:10`. Use `w` as the input to the closed-loop system and simulate the response for the time interval $0 \leq t \leq 10$. Perform the simulation for $K = 5$, 10, and 20 and plot the responses. Explain how the magnitude of the response is affected by the magnitude of K. Does this result match your prediction in part (b)?

10.4. Consider the feedback control system shown in Figure P10.4. Assume that there is no initial energy in the system at time $t = 0$.

(a) Derive an expression for $E(s)$ in terms of $D(s)$ and $R(s)$, where $E(s)$ is the Laplace transform of the error signal $e(t) = r(t) - y(t)$.

(b) Suppose that $r(t) = u(t)$ and $d(t) = 0$ for all t. Determine all (real) values of K so that $e(t) \to 0$ as $t \to \infty$.

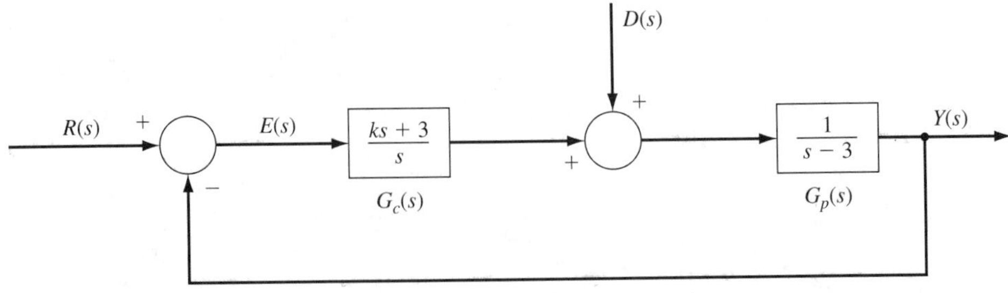

Figure P10.4

(c) Suppose that $r(t) = u(t)$ and $d(t) = u(t)$. Determine all (real) values of K so that $e(t) \to 0$ as $t \to \infty$.

(d) Suppose that $r(t) = u(t)$ and $d(t) = (\sin t)u(t)$. Determine all (real) values of K so that $e(t) \to 0$ as $t \to \infty$.

(e) Again suppose that $r(t) = u(t)$ and $d(t) = (\sin t)u(t)$. With the controller transfer function given by

$$G_c(s) = \frac{7s^3 + K_1 s + K_2}{s(s^2 + 1)}$$

determine all (real) values of K_1 and K_2 so that $e(t) \to 0$ as $t \to \infty$.

10.5. Consider a feedback connection as shown in Figure P10.1(b). The impulse response of the system with transfer function $G_p(s)$ is $h(t) = (\sin t)u(t)$.

(a) Determine the transfer function $G_c(s)$ so that the impulse response of the feedback connection is equal to $(\sin t)e^{-t}u(t)$.

(b) For $G_c(s)$ equal to your answer in part (a), compute the step response of the feedback connection.

10.6. Each of the following systems is to be controlled using feedback.

(i) $G_p(s) = \dfrac{s + 5}{s + 1}$

(ii) $G_p(s) = \dfrac{1}{s(s + 4)}$

For each system:

(a) Use the angle condition to determine which part of the real axis is on the root locus when $G_c(s) = K$. For the system in (ii), verify using the angle condition that $s = -2 + j\omega$ is on the root locus for all real ω.

(b) Calculate the closed-loop poles for specific values of $K > 0$, then use this information to plot the root locus.

(c) Verify the answers in parts (a) and (b) by using MATLAB to plot the root locus.

10.7. Use MATLAB to plot the root locus for each of the following systems.

(a) $G_p(s) = \dfrac{1}{(s + 1)(s + 10)}$; $\quad G_c(s) = K$

(b) $G_p(s) = \dfrac{1}{(s + 1)(s + 4)(s + 10)}$; $\quad G_c(s) = K$

(c) $G_p(s) = \dfrac{(s + 4)^2 + 4}{[(s + 2)^2 + 16](s + 8)}$; $\quad G_c(s) = K$

(d) $G_p(s) = \dfrac{s + 4}{(s + 6)^2 + 64}$; $\quad G_c(s) = K$

10.8. For each of the systems given in Problem 10.7, determine:

(a) The range of K that gives a stable response

(b) The value of K (if any) that gives a critically damped response

(c) The value(s) of K that gives the smallest time constant

10.9. For each of the closed-loop systems defined in Problems 10.7:

(a) Compute the steady-state error e_{ss} to a unit step input when $K = 100$.

(b) Verify your answer in part (a) by simulating the responses of the closed-loop systems to a step input.

Interactive Root Locus

10.10. Use MATLAB to plot the root locus for each of the following systems.

(a) $G_p(s) = \dfrac{1}{(s+1)(s+10)}$; $\quad G_c(s) = \dfrac{K(s+1.5)}{s}$

(b) $G(s) = \dfrac{1}{(s+1)(s+10)}$; $\quad G_c(s) = K(s+15)$

(c) $G_p(s) = \dfrac{1}{s(s-2)}$; $\quad G_c(s) = K(s+4)$

(d) $G_p(s) = \dfrac{1}{(s+2)^2+9}$; $\quad G_c(s) = \dfrac{K(s+4)}{s+10}$

(e) $G_p(s) = \dfrac{1}{(s+1)(s+3)}$; $\quad G_c(s) = \dfrac{K(s+6)}{s+10}$

10.11. Repeat Problem 10.9 for the closed-loop systems defined in Problem 10.10.

Interactive Root Locus

10.12. The transfer function of a dc motor is

$$G_p(s) = \frac{\Theta(s)}{V_a(s)} = \frac{60}{s(s+50)}$$

where $\theta(t)$ is the angle of the motor shaft and $v_a(t)$ is the input voltage to the armature. A closed-loop system is used to try to make the angle of the motor shaft track a desired motor angle $\theta_r(t)$. Unity feedback is used as shown in Figure P10.12, where $G_c(s) = K_P$ is the gain of an amplifier. Let $K = K_P(60)$.

(a) Plot the root locus for the system.

(b) Calculate the closed-loop transfer function for the following values of K: $K = 500$, 625, 5000, and 10,000. For each value of K, identify the corresponding closed-loop poles on the root locus plotted in part (a).

(c) Plot the step response for each value of K in part (b). For which value of K does the closed-loop response have the smallest time constant? The smallest overshoot?

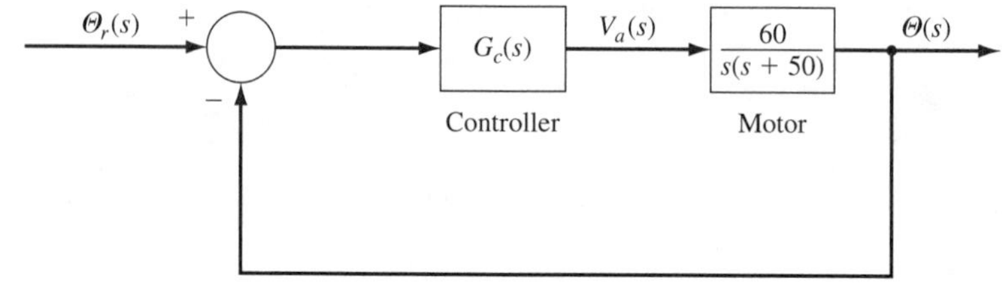

Figure P10.12

Interactive Root Locus

10.13. A system has the transfer function

$$G_p(s) = \frac{1}{(s+1)(s+7)}$$

(a) Sketch the root locus for a closed-loop system with a proportional controller, $G_c(s) = K_P$.

(b) Compute the closed-loop poles for $K_P = 5, 9, 73, 409$ and mark these pole positions on the root locus. Describe what type of closed-loop behavior you will expect for each of these selections of K_P. Calculate the steady-state error to a unit step function for each of these values of K_P.

(c) Verify the results of part (b) by using MATLAB to compute and plot the closed-loop step response for each value of K_P.

10.14. A system has the transfer function

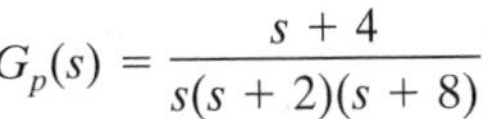

$$G_p(s) = \frac{s+4}{s(s+2)(s+8)}$$

Interactive Root Locus

(a) Sketch the root locus using MATLAB for a proportional controller, $G_c(s) = K_P$.

(b) Find a value of K_P that yields a closed-loop damping ratio of $\zeta = 0.707$ for the dominant poles. Give the corresponding closed-loop pole.

(c) Use MATLAB to compute and plot the closed-loop step response for the value of K_P found in part (b).

10.15. A third-order system has the transfer function

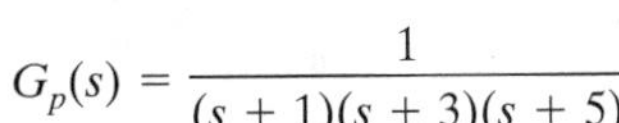

$$G_p(s) = \frac{1}{(s+1)(s+3)(s+5)}$$

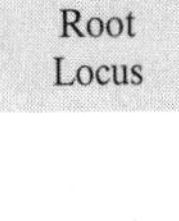

Interactive Root Locus

The performance specifications are that the dominant second-order poles have a damping ratio of $0.4 \le \zeta \le 0.707$ and $\zeta\omega_n > 1$.

(a) Plot the root locus for $G_c(s) = K_P$.

(b) From the root locus, find the value(s) of K_P that satisfy the criteria.

10.16. A system has the transfer function

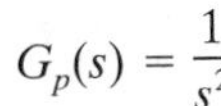

$$G_p(s) = \frac{1}{s^2}$$

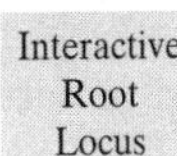

Interactive Root Locus

(a) Sketch the root locus for a closed-loop system with a proportional controller, $G_c(s) = K_P$. Describe what type of closed-loop response you will expect.

(b) Sketch the root locus for a PD controller of the form $G_c(s) = K_D s + K_P = K_D(s+2)$. Describe what type of closed-loop response you will expect as K_D is varied.

(c) Give the steady-state error of the closed-loop system with a PD controller for a step input when $K_D = 10$.

(d) Verify your result in part (c) by simulating the system.

10.17. The transfer function of the small-signal model of a simple pendulum is given by

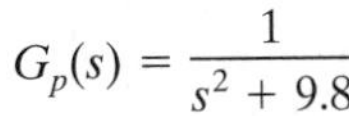

$$G_p(s) = \frac{1}{s^2 + 9.8}$$

Interactive Root Locus

The pendulum is placed in the closed-loop system illustrated in Figure P10.17.

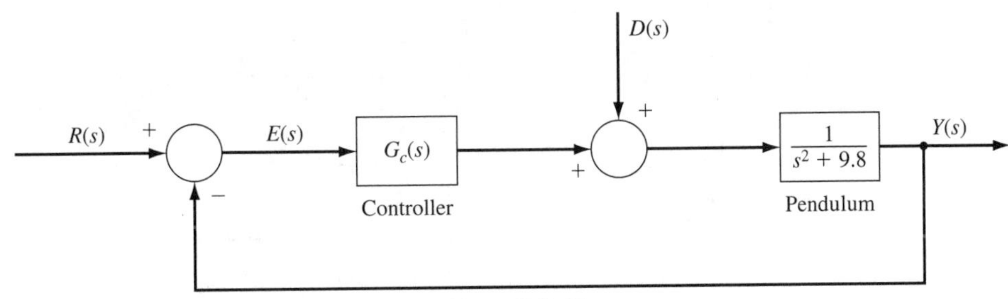

Figure P10.17

(a) Sketch the root locus for a closed loop system with a proportional controller, $G_c(s) = K_P$. Describe what type of closed-loop response you will expect for different ranges of values for K_P.

(b) Sketch the root locus for a PD controller of the form $G_c(s) = K_D s + K_P = K_D(s + 4)$. Describe what sort of closed-loop response you will expect as K_D is varied.

(c) Design the controller transfer function $G_c(s)$ so that when $r(t) = Au(t)$, where A is a constant, and $d(t) = 0$ for all t, $e(t) \to 0$ as $t \to \infty$.

Interactive Root Locus

10.18. A system has the transfer function

$$G_p(s) = \frac{s + 4}{(s + 1)(s + 2)}$$

(a) Sketch the root locus for a closed-loop system with a proportional controller, $G_c(s) = K_P$. Determine the value of K_P that will give closed loop poles with a time constant of $\tau = 0.5$ seconds.

(b) Compute the steady-state error of the step response for the value of K_P chosen in part (a).

(c) Design a PI controller so that the closed-loop system has a time constant of approximately 0.5 sec. For simplicity of design, select the zero of the controller to cancel the pole of the system. What is the expected steady-state error to a step input?

(d) Simulate the closed-loop system with two different controllers designed in parts (a) and (c) to verify the results of parts (b) and (c).

Interactive Root Locus

10.19. The system shown in Figure P10.19 is a temperature control system where the output temperature $T(t)$ should track a desired set-point temperature $r(t)$. The open-loop system has the transfer function

$$G_p(s) = \frac{0.05}{s + 0.05}$$

(a) Sketch the root locus for a closed-loop system with a proportional controller, $G_c(s) = K_P$. Suppose that the desired temperature is 70°F. Let $r(t) = 70u(t)$ and compute the gain required to yield a steady-state error of 2°. What is the resulting time constant of the closed-loop system?

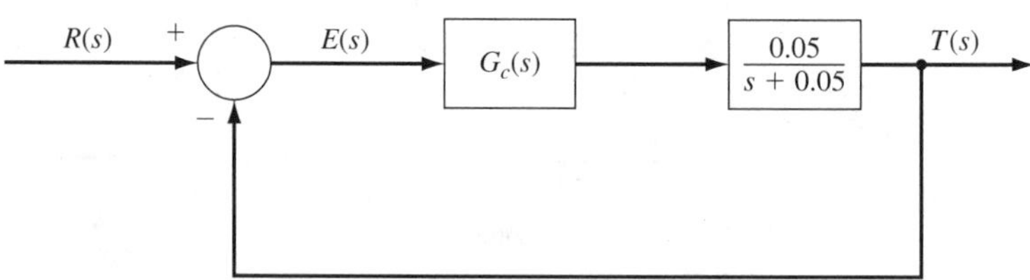

Figure P10.19

(b) Design a PI controller so that the closed-loop system has the same time constant as that computed in part (a). For simplicity of design, select the zero of the controller to cancel the pole of the system.

(c) To verify the results, simulate the response of the closed-loop system to $r(t) = 70u(t)$ for the two different controllers designed in parts (a) and (b).

10.20. A system is given by the transfer function

$$G_p = \frac{10}{s(s+1)}$$

Suppose that the desired closed-loop poles are at $-3 \pm j3$.

(a) Design a PD controller to obtain the desired poles. Use the angle criterion (10.60) evaluated at the desired closed-loop pole (i.e., $p = -3 + j3$) to determine the zero position.

(b) Simulate the step response of the closed-loop system.

10.21. A dc motor has the transfer function

$$\frac{\Omega(s)}{V_i(s)} \equiv G_p(s) = \frac{2}{(s+2)(s+10)}$$

where $\Omega(s)$ represents the motor speed and $V_i(s)$ represents the input voltage.

(a) Design a proportional controller to have a closed-loop damping ratio of $\zeta = 0.707$. For this value of K_P, determine the steady-state error for a unit step input.

(b) Design a PID controller so that the dominant closed-loop poles are at $-10 \pm 10j$. For simplicity, select one of the zeros of the controller to cancel the pole at -2. Then use the angle criterion (10.60) with $p = -10 + j10$ to determine the other zero position. What is the expected steady-state error to a step input?

(c) To verify your results, simulate the step response of the closed-loop system with the two different controllers designed in (a) and (b).

10.22. Consider the rocket described in Problem 10.3. A feedback loop measures the angle $\theta(t)$ and determines the corrections to the thrust engines.

(a) Design a PD controller to have the closed loop poles at $-0.5 \pm 0.5j$. (*Hint:* See the comment regarding the use of the angle criterion in Problem 10.20.)

(b) Simulate the response of the resulting closed-loop system to a unit impulse input.

Interactive Root Locus

10.23. An inverted pendulum shown in Figure P10.23 has the transfer function

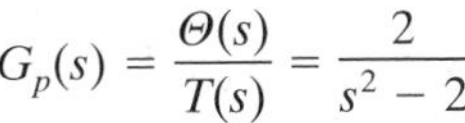

$$G_p(s) = \frac{\Theta(s)}{T(s)} = \frac{2}{s^2 - 2}$$

where $\Theta(s)$ represents the angle of the rod and $T(s)$ represents the torque applied by a motor at the base.

(a) Sketch the root locus for a proportional controller, $G_c(s) = K_P$. What type of closed-loop response would you expect for different values of K_P?

(b) Design a controller of the form

$$G_c(s) = K_L \frac{s - z_c}{s - p_c}$$

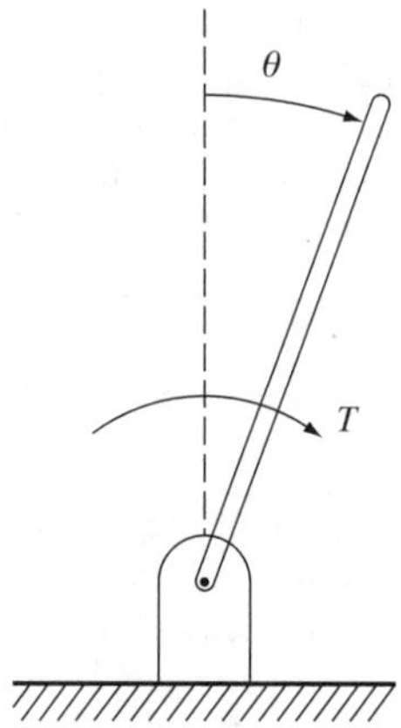

Figure P10.23

Choose $z_c = -3$ and solve for p_c from the angle criterion so that the dominant closed-loop poles are at $-3 \pm 3j$. (*Note:* The resulting controller is called a *lead controller.*) Draw the resulting root locus for this system and calculate the gain K_L that results in the desired closed-loop poles.

(c) Simulate the impulse response of the closed-loop system with the controller designed in part (b). (The impulse is equivalent to someone bumping the pendulum.)

Interactive Root Locus

10.24. A system has the transfer function

$$G_p(s) = \frac{1}{s(s + 2)}$$

(a) Sketch the root locus for a proportional controller, $G_c(s) = K_P$.

(b) Design a controller of the form

$$G_c(s) = K_L \frac{s - z_c}{s - p_c}$$

Select the zero of the controller to cancel the pole at -2. Solve for p_c from the angle criterion so that the dominant closed loop poles are at $-2 \pm 3j$. Draw the resulting root locus for this system and calculate the gain K_L that results in the desired closed-loop poles.

(c) Design another controller using the method described in part (b) except choose the zero to be at $z_c = -3$. Draw the resulting root locus for this system and calculate the gain K_L that results in the desired closed-loop poles.

(d) Compare the two controllers designed in parts (b) and (c) by simulating the step response of the two resulting closed-loop systems. Since both systems have the same dominant poles at $-2 \pm 3j$, speculate on the reason for the difference in the actual response.

10.25. Design a feedback controller that sets the position of a table tennis ball suspended in a plastic tube as illustrated in Figure P10.25. Here M is the mass of the ball, g the gravity constant, $y(t)$ the position of the ball at time t, and $x(t)$ the wind force on the ball due to the fan. The position $y(t)$ of the ball is continuously measured in real time using an ultrasonic sensor. The system is modeled by the differential equation

$$M\ddot{y}(t) = x(t) - Mg$$

The objective is to design the feedback controller so that $y(t) \to y_0$ as $t \to \infty$, where y_0 is the desired position (set point).

(a) Can the control objective be met using a proportional controller given by $G_c(s) = K_P$? Justify your answer.

(b) Can the control objective be met using a PI controller given by $G_c(s) = K_P + K_I/s$? Justify your answer.

(c) Design a PID controller that achieves the desired objective when $M = 1$ and $g = 9.8$.

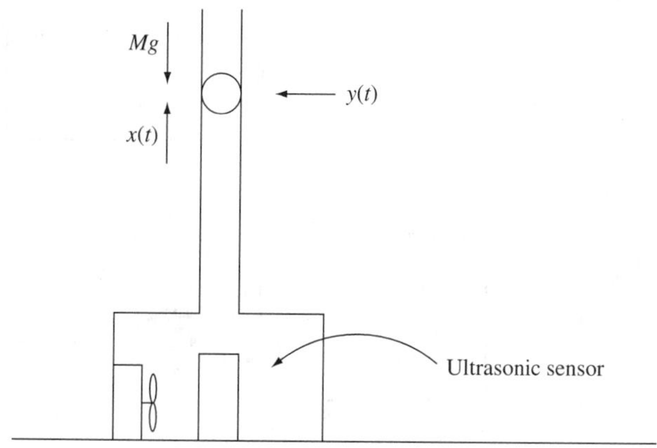

Figure P10.25

10.26. A proportional controller can be implemented using a simple amplifier. However, PD, PI, and PID controllers require a compensating network. Often this is achieved in analog with the use of operational amplifier (op amp) circuits. Consider the ideal op amp in Figure P10.26a. This op amp is an infinite impedance circuit element so that $v_a = 0$ and $i_a = 0$. These relationships also hold when the op amp is embedded in a circuit as shown in Figure P10.26b.

(a) Suppose that $R_1 = 1000\ \Omega$, $R_2 = 2000\ \Omega$, $C_1 = C_2 = 0$ in Figure 10.26b. Compute the transfer function between the input v_1 and the output v_2. (This circuit is known as an inverting circuit.)

(b) Suppose that $R_1 = 10\,\text{k}\Omega, R_2 = 20\,\text{k}\Omega, C_1 = 10\,\mu\text{F}$, and $C_2 = 0$ in Figure 10.26b. The resulting circuit is a PD controller. Compute the transfer function of the circuit.

(c) Suppose that $R_1 = 10\,\text{k}\Omega, R_2 = \infty$ (removed from circuit), $C_1 = 200\,\mu\text{F}$, and $C_2 = 10\,\mu\text{F}$ in Figure 10.26b. The resulting circuit is a PI controller. Compute the transfer function of the circuit.

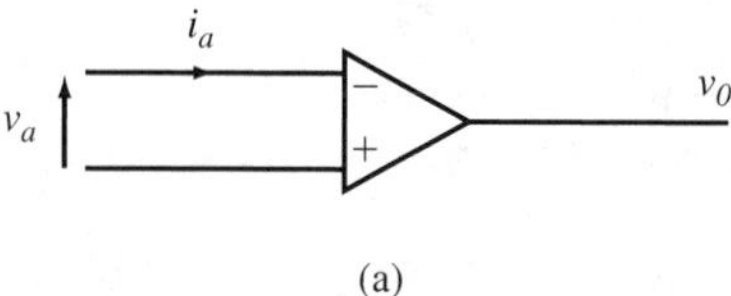

(a)

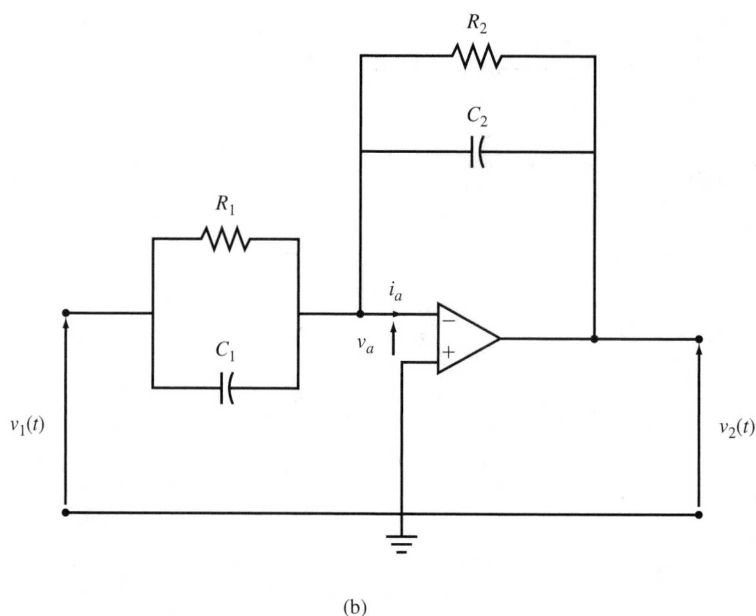

(b)

Figure P10.26

◆

CHAPTER 11

The z-Transform and Discrete-Time Systems

This chapter deals with the z-transform, which is the discrete-time counterpart of the Laplace transform. The z-transform operates on a discrete-time signal $x[n]$ in contrast to the Laplace transform, which operates on a continuous-time or analog signal $x(t)$. In Sections 11.1 and 11.2, the z-transform of a discrete-time signal $x[n]$ is defined, and then the basic properties of the z-transform are studied. In Section 11.3 the computation of the inverse z-transform is considered, and then in Section 11.4 the z-transform is applied to the study of causal linear time-invariant discrete-time systems. The development begins in Section 11.4 with the generation of the z-domain representation from the input/output difference equation, and then the transfer function representation is generated by applying the z-transform to the input/output convolution sum representation of a system. The transfer function of interconnections containing unit-delay elements and interconnections of blocks is also studied in Section 11.4. In the last two sections of the chapter, the transfer function representation is utilized in the study of stability and in the study of the frequency response behavior of a discrete-time system.

The theory of the z-transform and its application to causal linear time-invariant discrete-time systems closely resembles the theory of the Laplace transform and its application to causal linear time-invariant continuous-time systems. In particular, results and techniques in this chapter closely parallel the results and techniques given in Chapters 8 and 9 on the Laplace transform. However, there are some differences between the transform theory in the continuous-time case and the transform theory in the discrete-time case, although for the most part, these differences are minor. In reading this chapter, the reader should look for the similarities and differences in the two cases.

11.1 *z*-TRANSFORM OF A DISCRETE-TIME SIGNAL

Given the discrete-time signal $x[n]$, in Chapter 7 the discrete-time Fourier transform (DTFT) was defined by

$$X(\Omega) = \sum_{n=-\infty}^{\infty} x[n]e^{-j\Omega n} \tag{11.1}$$

Recall that $X(\Omega)$ is in general a complex-valued function of the frequency variable Ω.

The z-transform of the signal $x[n]$ is generated by adding the factor ρ^{-n} to the summation in (11.1), where ρ is a real number. The factor ρ^{-n} plays the same role as the exponential factor $e^{-\sigma t}$ that was added to the Fourier transform to generate the Laplace transform in the continuous-time case. Inserting ρ^{-n} in (11.1) gives

$$X(\Omega) = \sum_{n=-\infty}^{\infty} x[n]\rho^{-n}e^{-j\Omega n} \tag{11.2}$$

which can be rewritten as

$$X(\Omega) = \sum_{n=-\infty}^{\infty} x[n](\rho e^{j\Omega})^{-n} \tag{11.3}$$

The function $X(\Omega)$ given by (11.3) is now a function of the complex number

$$z = \rho e^{j\Omega}$$

so X should be written as a function of z, which gives

$$X(z) = \sum_{n=-\infty}^{\infty} x[n]z^{-n} \tag{11.4}$$

The function $X(z)$ given by (11.4) is the *two-sided* z-transform of the discrete-time signal $x[n]$. The *one-sided* z-transform of $x[n]$, also denoted by $X(z)$, is defined by

$$X(z) = \sum_{n=0}^{\infty} x[n]z^{-n} \tag{11.5}$$

As seen from (11.5) the one-sided z-transform is a power series in z^{-1} whose coefficients are the values of the signal $x[n]$.

Note that if $x[n] = 0$ for $n = -1, -2, \ldots$, the one- and two-sided z-transforms are identical. The one-sided z-transform can be applied to signals $x[n]$ that are nonzero for $n = -1, -2, \ldots$, but any nonzero values of $x[n]$ for $n < 0$ cannot be recovered from the one-sided z-transform. In this book, only the one-sided z-transform is pursued, which will be referred to as the z-transform.

Given a discrete-time signal $x[n]$ with z-transform $X(z)$, the set of all complex numbers z such that the summation on the right-hand side of (11.5) converges (i.e., exists) is called the *region of convergence* of the z-transform $X(s)$. The z-transform $X(z)$ exists (is well defined) for any value of z belonging to the region of convergence.

Example 11.1 *z-Transform of Unit Pulse*

Let $\delta[n]$ denote the unit pulse concentrated at $n = 0$ given by

$$\delta[n] = \begin{cases} 1, & n = 0 \\ 0, & n \neq 0 \end{cases}$$

Since $\delta[n]$ is zero for all n except $n = 0$, the z-transform is

$$\sum_{n=0}^{\infty} \delta[n]z^{-n} = \delta[0]z^{-0} = 1 \tag{11.6}$$

Thus the z-transform of the unit pulse $\delta[n]$ is equal to 1. In addition, it is obvious that the summation in (11.6) exists for any value of z, and thus the region of convergence of the z-transform of the unit pulse is the set of all complex numbers.

Note that the unit pulse $\delta[n]$ is the discrete-time counterpart to the unit impulse $\delta(t)$ in the sense that the z-transform of $\delta[n]$ is equal to 1 and the Laplace transform of $\delta(t)$ is also equal to 1. However, as noted in Section 1.3 the pulse $\delta[n]$ is not a sampled version of $\delta(t)$.

Example 11.2 *z-Transform of Shifted Pulse*

Given a positive integer q, consider the unit pulse $\delta[n - q]$ located at $n = q$. For example, when $q = 2$, $\delta[n - 2]$ is the pulse shown in Figure 11.1. For any positive integer value of q, the z-transform of $\delta[n - q]$ is

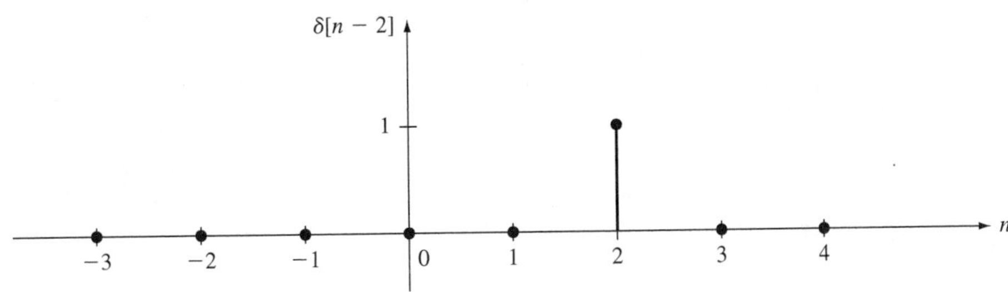

Figure 11.1 Unit pulse $\delta[n - 2]$ located at $n = 2$.

$$\sum_{n=0}^{\infty} \delta[n-q]z^{-n} = \delta[0]z^{-q} = z^{-q} = \frac{1}{z^q}$$

The region of convergence is the set of all complex numbers z such that $|z| > 0$.

Example 11.3 *Unit-Step Function*

Consider the discrete-time unit-step function $u[n]$ given by

$$u[n] = \begin{cases} 1, & n = 0, 1, 2, \ldots \\ 0, & n = -1, -2, \ldots \end{cases}$$

The z-transform $U(z)$ is

$$\begin{aligned} U(z) &= \sum_{n=0}^{\infty} u[n]z^{-n} \\ &= \sum_{n=0}^{\infty} z^{-n} \\ &= 1 + z^{-1} + z^{-2} + z^{-3} + \cdots \end{aligned} \tag{11.7}$$

The transform $U(z)$ can be expressed as a rational function of z: Multiplying both sides of (11.7) by $z - 1$ gives

$$\begin{aligned} (z-1)U(z) &= (z + 1 + z^{-1} + z^{-2} + \cdots) - (1 + z^{-1} + z^{-2} + \cdots) \\ &= z \end{aligned} \tag{11.8}$$

Dividing both sides of (11.8) by $z - 1$ yields

$$U(z) = \frac{z}{z-1} = \frac{1}{1 - z^{-1}} \tag{11.9}$$

Note that the form of the z-transform $U(z)$ of the discrete-time unit-step function $u[n]$ is different from the form of the Laplace transform $1/s$ of the continuous-time unit-step function $u(t)$.

The region of convergence for the z-transform $U(z)$ given by (11.9) includes the set of all complex numbers z such that $|z| > 1$. This follows from the result that

$$\left| \sum_{n=0}^{\infty} (1)z^{-n} \right| < \infty \qquad \text{if } |z| > 1 \tag{11.10}$$

To prove (11.10), first note that using (7.5) (see Problem 7.2), for any positive integer q,

$$\sum_{n=0}^{q} z^{-n} = \frac{(1/z)^{q+1} - 1}{(1/z) - 1} \tag{11.11}$$

where it is assumed that $z \neq 1$. Then using (11.11)

$$\left| \sum_{n=0}^{q} z^{-n} \right| = \left| \frac{(1/z)^{q+1} - 1}{(1/z) - 1} \right| \leq \frac{(1/|z|)^{q+1} + 1}{|(1/z) - 1|} \tag{11.12}$$

and using (11.12), if $|z| > 1$,

$$\left|\sum_{n=0}^{\infty} z^{-n}\right| = \lim_{q\to\infty}\left|\sum_{n=0}^{q} z^{-n}\right| \le \frac{1}{|(1/z) - 1|} < \infty$$

Thus (11.10) is verified.

Example 11.4 *z-Transform of $a^n u[n]$*

Given a real or complex number a, let $x[n] = a^n u[n]$. The z-transform $X(z)$ of $x[n]$ is given by

$$\begin{aligned} X(z) &= \sum_{n=0}^{\infty} a^n z^{-n} \\ &= 1 + az^{-1} + a^2 z^{-2} + \cdots \end{aligned} \tag{11.13}$$

This transform can also be written as a rational function in z: Multiplying both sides of (11.13) by $z - a$ gives

$$\begin{aligned} (z - a)X(z) &= (z + a + a^2 z^{-1} + a^3 z^{-2} + \cdots) - (a + a^2 z^{-1} + a^3 z^{-2} + \cdots) \\ &= z \end{aligned}$$

Hence

$$X(z) = \frac{z}{z - a} = \frac{1}{1 - az^{-1}} \tag{11.14}$$

Note that if $a = 1$, (11.14) is the same as (11.9). The region of convergence for the transform $X(z) = z/(z - a)$ includes the set of all complex numbers z such that $|z| > |a|$. This follows by using an argument similar to that given in Example 11.3. The details are left to the reader.

Relationship between the DTFT and the z-Transform

As shown above, the z-transform $X(z)$ of a discrete-time signal $x[n]$ can be viewed as a generalization of the discrete-time Fourier transform (DTFT) $X(\Omega)$. In fact, from (11.5) it appears that

$$X(\Omega) = X(z)\big|_{z=e^{j\Omega}} \tag{11.15}$$

However, (11.15) is *not valid* in general unless the region of convergence of $X(z)$ includes all complex numbers z such that $|z| = 1$. If this is the case, the DTFT $X(\Omega)$ of $x[n]$ is given by (11.15). For example, suppose that $x[n] = a^n u[n]$, where a is a real or complex number. In Example 11.4 it was shown that the z-transform of $x[n]$ is $X(z) = z/(z - a)$ and that the region of convergence of the z-transform includes the set of all complex numbers z such that $|z| > |a|$. Thus if $|a| < 1$, the DTFT of $x[n]$ exists (in the ordinary sense) and is given by

$$X(\Omega) = X(z)\big|_{z=e^{j\Omega}} = \frac{e^{j\Omega}}{e^{j\Omega} - a} = \frac{1}{1 - ae^{-j\Omega}} \tag{11.16}$$

TABLE 11.1 BASIC z-TRANSFORM PAIRS

$\delta[n] \leftrightarrow 1$
$\delta[n-q] \leftrightarrow \frac{1}{z^q} = z^{-q}$
$u[n] \leftrightarrow \frac{z}{z-1} = \frac{1}{1-z^{-1}}$
$a^n u[n] \leftrightarrow \frac{z}{z-a} = \frac{1}{1-az^{-1}}$, a real or complex

Given a signal $x[n]$ with z-transform $X(z)$, the transform pair notation

$$x[n] \leftrightarrow X(z)$$

will sometimes be used to denote the fact that $X(z)$ is the z-transform of $x[n]$, and conversely, that $x[n]$ is the inverse z-transform of $X(z)$. The transform pairs derived in the examples given above are shown in Table 11.1.

As seen in Table 11.1, the z-transform can sometimes be expressed as a ratio of polynomials in z or z^{-1}. In this book, preference will be given to expressing z-transforms in terms of positive powers of z, as opposed to negative powers of z.

11.2 PROPERTIES OF THE z-TRANSFORM

The z-transform possesses a number of properties that are useful in deriving transform pairs and in the application of the transform to the study of linear causal time-invariant discrete-time systems. These properties are very similar to the properties of the Laplace transform that were given in Section 8.2. In this section the properties of the z-transform are stated and proved. As an illustration of the use of the properties, a collection of common transform pairs is generated from the basic set of pairs given in Table 11.1.

Linearity

The z-transform is a linear operation, as is the Laplace transform. Hence if $x[n] \leftrightarrow X(z)$ and $v[n] \leftrightarrow V(z)$, then for any real or complex scalars a, b,

$$ax[n] + bv[n] \leftrightarrow aX(z) + bV(z) \tag{11.17}$$

The proof of (11.17) follows directly from the definition of the z-transform. The details are omitted.

Example 11.5 *Linearity*

Let $x[n] = u[n]$ and $v[n] = a^n u[n]$, where $a \neq 1$. From Table 11.1,

$$u[n] \leftrightarrow \frac{z}{z-1} \quad \text{and} \quad a^n u[n] \leftrightarrow \frac{z}{z-a}$$

Hence by linearity

$$u[n] + a^n u[n] \leftrightarrow \frac{z}{z-1} + \frac{z}{z-a} = \frac{2z^2 - (1+a)z}{(z-1)(z-a)}$$

Right Shift of $x[n]u[n]$

Suppose that $x[n] \leftrightarrow X(z)$. Given a positive integer q, consider the discrete-time signal $x[n-q]u[n-q]$, which is the q-step right shift of $x[n]u[n]$. Then

$$x[n-q]u[n-q] \leftrightarrow z^{-q}X(z) \tag{11.18}$$

To prove this property, first note that by definition of the z-transform

$$x[n-q]u[n-q] \leftrightarrow \sum_{n=0}^{\infty} x[n-q]u[n-q]z^{-n}$$

Then since $u[n-q] = 1$ for $n \geq q$ and $u[n-q] = 0$ for $n < q$,

$$x[n-q]u[n-q] \leftrightarrow \sum_{n=q}^{\infty} x[n-q]z^{-n} \tag{11.19}$$

Consider a change of index in the summation in (11.19): With $\bar{n} = n - q$ so that $n = \bar{n} + q$, then $\bar{n} = 0$ when $n = q$ and $\bar{n} = \infty$ when $n = \infty$. Hence

$$\begin{aligned}\sum_{n=q}^{\infty} x[n-q]z^{-n} &= \sum_{\bar{n}=0}^{\infty} x[\bar{n}]z^{-(\bar{n}+q)} \\ &= z^{-q}\sum_{\bar{n}=0}^{\infty} x[\bar{n}]z^{-\bar{n}} \\ &= z^{-q}X(z)\end{aligned}$$

Therefore, combining this with (11.19) yields the transform pair (11.18).

Example 11.6 *z-Transform of a Pulse*

Given a positive integer q, the objective is to determine the z-transform of the pulse $p[n]$ defined by

$$p[n] = \begin{cases} 1, & n = 0, 1, 2, \ldots, q-1 \\ 0, & \text{all other } n \end{cases}$$

Writing $p[n]$ in terms of the unit-step function $u[n]$ gives

$$p[n] = u[n] - u[n-q]$$

From Table 11.1, the z-transform of $u[n]$ is $z/(z-1)$ and thus by the right-shift property (11.18), the z-transform of $u[n-q]$ is equal to

$$z^{-q}\frac{z}{z-1} = \frac{z^{-q+1}}{z-1}$$

Thus, by linearity the z-transform of the pulse $p[n]$ is

$$\frac{z}{z-1} - \frac{z^{-q+1}}{z-1} = \frac{z(1-z^{-q})}{z-1} = \frac{z^{q}-1}{z^{q-1}(z-1)}$$

Right Shift of $x[n]$

Suppose that $x[n] \leftrightarrow X(z)$. Then

$$x[n-1] \leftrightarrow z^{-1}X(z) + x[-1] \tag{11.20}$$

$$x[n-2] \leftrightarrow z^{-2}X(z) + x[-2] + z^{-1}x[-1] \tag{11.21}$$

$$\vdots$$

$$x[n-q] \leftrightarrow z^{-q}X(z) + x[-q] + z^{-1}x[-q+1] + \cdots + z^{-q+1}x[-1] \tag{11.22}$$

Note that if $x[n] = 0$ for $n = -1, -2, \ldots, -q$, the transform pair (11.22) reduces to

$$x[n-q] \leftrightarrow z^{-q}X(z) \tag{11.23}$$

which is identical to the transform pair (11.18).

To prove the transform pair (11.20), first note that by definition of the z-transform

$$x[n-1] \leftrightarrow \sum_{n=0}^{\infty} x[n-1]z^{-n} \tag{11.24}$$

Defining the change of index $\bar{n} = n - 1$ in the summation in (11.24) gives

$$x[n-1] \leftrightarrow \sum_{\bar{n}=-1}^{\infty} x[\bar{n}]z^{-(\bar{n}+1)} = \sum_{\bar{n}=0}^{\infty} x[\bar{n}]z^{-(\bar{n}+1)} + x[-1]$$

$$\leftrightarrow z^{-1}\sum_{\bar{n}=0}^{\infty} x[\bar{n}]z^{-\bar{n}} + x[-1]$$

$$\leftrightarrow z^{-1}X(z) + x[-1]$$

Thus the transform pair (11.20) is verified. The verification of (11.21) and (11.22) for $q > 2$ can be demonstrated in a similar manner. The details are left to the interested reader.

Left Shift in Time

In contrast to the Laplace transform, the z-transform does have a left-shift property as follows. Given the discrete-time signal $x[n]$ and a positive integer q, the q-step left shift of $x[n]$ is the signal $x[n+q]$. Now suppose that $x[n] \leftrightarrow X(z)$. Then

$$x[n+1] \leftrightarrow zX(z) - x[0]z \tag{11.25}$$

$$x[n+2] \leftrightarrow z^2X(z) - x[0]z^2 - x[1]z \tag{11.26}$$

$$\vdots$$

$$x[n+q] \leftrightarrow z^qX(z) - x[0]z^q - x[1]z^{q-1} - \cdots - x[q-1]z \tag{11.27}$$

To prove (11.25), first observe that

$$x[n+1] \leftrightarrow \sum_{n=0}^{\infty} x[n+1]z^{-n} \tag{11.28}$$

Defining the change of index $\bar{n} = n + 1$ in the summation in (11.28) gives

$$\begin{aligned} x[n+1] &\leftrightarrow \sum_{\bar{n}=1}^{\infty} x[\bar{n}]z^{-(\bar{n}-1)} \\ &\leftrightarrow z\sum_{\bar{n}=1}^{\infty} x[\bar{n}]z^{-\bar{n}} = z\left[\sum_{\bar{n}=0}^{\infty} x[\bar{n}]z^{-\bar{n}} - x[0]\right] \\ &\leftrightarrow z[X(z) - x[0]] \end{aligned}$$

Hence (11.25) is verified.

Example 11.7 *Left Shift of Unit-Step Function*

Consider the one-step left shift $u[n+1]$ of the discrete-time unit-step function $u[n]$. By the left-shift property (11.25), the z-transform of $u[n+1]$ is equal to

$$zU(z) - u[0]z = \frac{z^2}{z-1} - z = \frac{z^2 - z(z-1)}{z-1} = \frac{z}{z-1}$$

Hence the z-transform of $u[n+1]$ is equal to the z-transform of $u[n]$. This result is not unexpected since $u[n+1] = u[n]$ for $n = 0, 1, 2, \ldots$

Multiplication by n and n^2

If $x[n] \leftrightarrow X(z)$, then

$$nx[n] \leftrightarrow -z\frac{d}{dz}X(z) \tag{11.29}$$

and

$$n^2x[n] \leftrightarrow z\frac{d}{dz}X(z) + z^2\frac{d^2}{dz^2}X(z) \tag{11.30}$$

To prove (11.29), first recall the definition of the z-transform:

$$X(z) = \sum_{n=0}^{\infty} x[n]z^{-n} \tag{11.31}$$

Taking the derivative with respect to z of both sides of (11.31) yields

$$\frac{d}{dz}X(z) = \sum_{n=0}^{\infty}(-n)x[n]z^{-n-1}$$

$$= -z^{-1}\sum_{n=0}^{\infty} nx[n]z^{-n} \tag{11.32}$$

Thus

$$-z\frac{d}{dz}X(z) = \sum_{n=0}^{\infty} nx[n]z^{-n} \tag{11.33}$$

Now the right-hand side of (11.33) is equal to the z-transform of the signal $nx[n]$, and thus (11.29) is verified. The proof of (11.30) follows by taking the second derivative of $X(z)$ with respect to z. The details are left to the reader.

Example 11.8 *z-Transform of $na^nu[n]$*

Let $x[n] = a^nu[n]$, where a is any nonzero real or complex number. From Table 11.1,

$$X(z) = \frac{z}{z-a}$$

Then

$$z\frac{d}{dz}X(z) = z\left[\frac{-z}{(z-a)^2} + \frac{1}{z-a}\right] = \frac{-az}{(z-a)^2}$$

which gives the transform pair

$$na^nu[n] \leftrightarrow \frac{az}{(z-a)^2} \tag{11.34}$$

Note that when $a = 1$, (11.34) becomes

$$nu[n] \leftrightarrow \frac{z}{(z-1)^2} \tag{11.35}$$

Example 11.9 *z-Transform of* $n^2a^nu[n]$

To compute the z-transform of the signal $n^2a^nu[n]$, first set $x[n] = a^nu[n]$, so that $X(z) = z/(z - a)$. Then

$$\frac{d^2}{dz^2}X(z) = \frac{2a}{(z-a)^3}$$

and thus, using the results in Example 11.8 and the transform pair (11.30) gives

$$n^2a^nu[n] \leftrightarrow \frac{-az}{(z-a)^2} + \frac{2az^2}{(z-a)^3}$$

$$\leftrightarrow \frac{az(z+a)}{(z-a)^3} \tag{11.36}$$

Setting a = 1 in (11.36) results in the transform pair

$$n^2u[n] \leftrightarrow \frac{z(z+1)}{(z-1)^3} \tag{11.37}$$

Multiplication by a^n

If $x[n] \leftrightarrow X(z)$, then for any nonzero real or complex number a,

$$a^nx[n] \leftrightarrow X\left(\frac{z}{a}\right) \tag{11.38}$$

By (11.38), multiplication by a^n in the time domain corresponds to scaling of the z variable in the transform domain. To prove (11.38), observe that

$$a^nx[n] \leftrightarrow \sum_{n=0}^{\infty} a^nx[n]z^{-n}$$

$$\leftrightarrow \sum_{n=0}^{\infty} x[n]\left(\frac{z}{a}\right)^{-n} = X\left(\frac{z}{a}\right)$$

Example 11.10 *z-Transform of* $a^np[n]$

Let $p[n]$ denote the pulse defined by $p[n] = u[n] - u[n - q]$, where q is a positive integer. From Example 11.6, the z-transform of the pulse is

$$\frac{z(1 - z^{-q})}{z - 1}$$

Then using (11.38) results in the transform pair

$$a^np[n] \leftrightarrow \frac{(z/a)[1 - (z/a)^{-q}]}{(z/a) - 1}$$

$$\leftrightarrow \frac{z(1 - a^qz^{-q})}{z - a}$$

Multiplication by cos Ωn and sin Ωn

If $x[n] \leftrightarrow X(z)$, then for any positive real number Ω,

$$(\cos \Omega n)x[n] \leftrightarrow \frac{1}{2}[X(e^{j\Omega}z) + X(e^{-j\Omega}z)] \tag{11.39}$$

and

$$(\sin \Omega n)x[n] \leftrightarrow \frac{j}{2}[X(e^{j\Omega}z) - X(e^{-j\Omega}z)] \tag{11.40}$$

To prove (11.39) and (11.40), first note that using Euler's formula yields

$$(\cos \Omega n)x[n] = \frac{1}{2}[e^{-j\Omega n}x[n] + e^{j\Omega n}x[n]] \tag{11.41}$$

$$(\sin \Omega n)x[n] = \frac{j}{2}[e^{-j\Omega n}x[n] - e^{j\Omega n}x[n]] \tag{11.42}$$

By (11.38),

$$e^{-j\Omega n}x[n] \leftrightarrow X(e^{j\Omega}z) \quad \text{and} \quad e^{j\Omega n}x[n] \leftrightarrow X(e^{-j\Omega}z) \tag{11.43}$$

Then using (11.43) with (11.41) and (11.42) yields (11.39) and (11.40).

Example 11.11 *z-Transform of Sinusoids*

Let $v[n] = (\cos \Omega n)u[n]$. With $x[n]$ set equal to the unit step $u[n]$, $X(z) = z/(z - 1)$, and using (11.39) gives

$$\begin{aligned}(\cos \Omega n)u[n] &\leftrightarrow \frac{1}{2}\left(\frac{e^{j\Omega}z}{e^{j\Omega}z - 1} + \frac{e^{-j\Omega}z}{e^{-j\Omega}z - 1}\right)\\ &\leftrightarrow \frac{1}{2}\left[\frac{e^{j\Omega}z(e^{-j\Omega}z - 1) + e^{-j\Omega}z(e^{j\Omega} - 1)}{(e^{j\Omega}z - 1)(e^{-j\Omega}z - 1)}\right]\\ &\leftrightarrow \frac{1}{2}\left[\frac{z^2 - e^{j\Omega}z + z^2 - e^{-j\Omega}z}{z^2 - (e^{j\Omega} + e^{-j\Omega})z + 1}\right]\\ &\leftrightarrow \frac{z^2 - (\cos \Omega)z}{z^2 - (2\cos \Omega)z + 1}\end{aligned} \tag{11.44}$$

Similarly, using (11.40) results in the transform pair

$$(\sin \Omega n)u[n] \leftrightarrow \frac{(\sin \Omega)z}{z^2 - (2\cos \Omega) + 1} \tag{11.45}$$

Example 11.12 *a^n Times a Sinusoid*

Now let $v[n] = a^n(\cos \Omega n)u[n]$. The z-transform of $v[n]$ can be computed by setting $x[n] = a^n u[n]$ and using the multiplication by cos Ωn property. However, it is easier to set $x[n]$ equal to $(\cos \Omega n)u[n]$ and then apply the multiplication by a^n property. Using (11.38) and the transform pair (11.44) gives

$$a^n(\cos \Omega n)u[n] \leftrightarrow \frac{(z/a)^2 - (\cos \Omega)(z/a)}{(z/a)^2 - (2\cos \Omega)(z/a) + 1}$$

$$\leftrightarrow \frac{z^2 - (a\cos \Omega)z}{z^2 - (2a\cos \Omega)z + a^2} \tag{11.46}$$

Using (11.38) and the transform pair (11.45) results in the transform pair

$$a^n(\sin \Omega n)u[n] \leftrightarrow \frac{(a\sin \Omega)z}{z^2 - (2a\cos \Omega)z + a^2} \tag{11.47}$$

Summation

Given the discrete-time signal $x[n]$ with $x[n] = 0$ for $n = -1, -2, \ldots$, let $v[n]$ denote the sum of $x[n]$, defined by

$$v[n] = \sum_{i=0}^{n} x[i] \tag{11.48}$$

To derive an expression for the z-transform of $v[n]$, first note that $v[n]$ can be expressed in the form

$$v[n] = \sum_{i=0}^{n-1} x[i] + x[n]$$

and using the definition (11.48) of $v[n]$ gives

$$v[n] = v[n-1] + x[n] \tag{11.49}$$

Then taking the z-transform of both sides of (11.49) and using the right-shift property yields

$$V(z) = z^{-1}V(z) + X(z)$$

and solving for $V(z)$ gives

$$V(z) = \frac{1}{1 - z^{-1}}X(z)$$

$$= \frac{z}{z-1}X(z) \tag{11.50}$$

Hence the z-transform of the sum of a signal $x[n]$ is equal to $z/(z-1)$ times the z-transform of the signal.

Example 11.13 *z-Transform of $(n+1)u[n]$*

Let $x[n] = u[n]$. Then the sum is

$$v[n] = \sum_{i=0}^{n} u[i] = (n+1)u[n]$$

and thus the sum of the step is a ramp. By (11.50), the transform of the sum is

$$V(z) = \frac{z}{z-1}X(z) = \frac{z^2}{(z-1)^2}$$

This yields the transform pair

$$(n+1)u[n] \leftrightarrow \frac{z^2}{(z-1)^2} \tag{11.51}$$

Convolution

Given two discrete-time signals $x[n]$ and $v[n]$ with both signals equal to zero for $n = -1, -2, \ldots$, in Chapter 3 the convolution of $x[n]$ and $v[n]$ was defined by

$$x[n] * v[n] = \sum_{i=0}^{n} x[i]v[n-i]$$

Note that since $v[n] = 0$ for $n = -1, -2, \ldots$, the convolution sum can be taken from $i = 0$ to $i = \infty$; that is, the convolution operation is given by

$$x[n] * v[n] = \sum_{i=0}^{\infty} x[i]v[n-i] \tag{11.52}$$

Taking the z-transform of both sides of (11.52) yields the transform pair

$$\begin{aligned} x[n] * v[n] &\leftrightarrow \sum_{n=0}^{\infty}\left[\sum_{i=0}^{\infty} x[i]v[n-i]\right]z^{-n} \\ &\leftrightarrow \sum_{i=0}^{\infty} x[i]\left[\sum_{n=0}^{\infty} v[n-i]z^{-n}\right] \end{aligned} \tag{11.53}$$

Using the change of index $\bar{n} = n - i$ in the second summation of (11.53) gives

$$\begin{aligned} x[n] * v[n] &\leftrightarrow \sum_{i=0}^{\infty} x[i]\left[\sum_{\bar{n}=-i}^{\infty} v[\bar{n}]z^{-\bar{n}-i}\right] \\ &\leftrightarrow \sum_{i=0}^{\infty} x[i]\left[\sum_{\bar{n}=0}^{\infty} v[\bar{n}]z^{-\bar{n}-i}\right], \text{ since } v[\bar{n}] = 0 \text{ for } \bar{n} < 0 \\ &\leftrightarrow \left[\sum_{i=0}^{\infty} x[i]z^{-i}\right]\left[\sum_{\bar{n}=0}^{\infty} v[\bar{n}]z^{-\bar{n}}\right] \\ &\leftrightarrow X(z)V(z) \end{aligned} \tag{11.54}$$

From (11.54), it is seen that the z-transform of the convolution $x * v$ is equal to the product $X(z)V(z)$, where $X(z)$ and $V(z)$ are the z-transforms of $x[n]$ and $v[n]$, respectively. Therefore, convolution in the discrete-time domain corresponds to a product in the z-transform domain. This result is obviously analogous to the result in the continuous-time framework where convolution corresponds to multiplication in the s-domain. Examples of the use of the transform pair (11.54) will be given in Section 11.4 when the transfer function representation is developed for linear time-invariant discrete-time systems.

Initial-Value Theorem

If $x[n] \leftrightarrow X(z)$, the initial values of $x[n]$ can be computed directly from $X(z)$ by using the relationships

$$x[0] = \lim_{z\to\infty} X(z) \tag{11.55}$$

$$x[1] = \lim_{z\to\infty} [zX(z) - zx[0]]$$

$$\vdots$$

$$x[q] = \lim_{z\to\infty} [z^q X(z) - z^q x[0] - z^{q-1}x[1] - \cdots - zx[q-1]] \tag{11.56}$$

To prove (11.55), first note that

$$z^{-n} \to 0 \qquad \text{as } z \to \infty \text{ for all } n \geq 1$$

and thus

$$x[n]z^{-n} \to 0 \qquad \text{as } z \to \infty \text{ for all } n \geq 1$$

Thus, taking the limit as $z \to \infty$ of both sides of

$$X(z) = \sum_{n=0}^{\infty} x[n]z^{-n}$$

yields (11.55).

In the next section it will be shown that if the transform $X(z)$ is a rational function of z, the initial values of $x[n]$ can be calculated by a long-division operation.

Final-Value Theorem

Given a discrete-time signal $x[n]$ with z-transform $X(z)$, suppose that $x[n]$ has a limit as $n \to \infty$. Then the final-value theorem states that

$$\lim_{n\to\infty} x[n] = \lim_{z\to 1} (z - 1)X(z) \tag{11.57}$$

The proof of (11.57) is analogous to the proof that was given of the final-value theorem in the continuous-time case. The details are not pursued here.

As in the continuous-time case, care must be exercised in using the final-value theorem since the limit on the right-hand side of (11.57) may exist even though $x[n]$ does not have a limit as $n \to \infty$. Existence of the limit of $x[n]$ as $n \to \infty$ can readily be checked if the transform $X(z)$ is rational in z; that is, $X(z)$ can be written in the form $X(z) = B(z)/A(z)$, where $B(z)$ and $A(z)$ are polynomials in z with real coefficients. Here it is assumed that $B(z)$ and $A(z)$ do not have any common factors; if there are common factors, they should be canceled.

Now letting $p_1, p_2, \ldots, p_N$ denote the poles of $X(s) = B(z)/A(z)$ [i.e., the roots of $A(z) = 0$], $x[n]$ has a limit as $n \to \infty$ if and only if the magnitudes $|p_1|$, $|p_2|, \ldots, |p_N|$ are all strictly less than 1, except that one of the p_i's may be equal to 1. This is equivalent to the condition that all the poles of $(z - 1)X(z)$ have magnitudes strictly less than 1. The proof that the pole condition on $(z - 1)X(z)$ is necessary and sufficient for the existence of the limit follows from the results given in the next section. If this condition is satisfied, the limit of $x[n]$ as $n \to \infty$ is given by

$$\lim_{n\to\infty} x[n] = [(z - 1)X(z)]_{z=1} \tag{11.58}$$

As in the continuous-time case, the relationship (11.58) makes it possible to determine the limiting value of a time signal directly from the transform of the signal (without having to compute the inverse transform).

Example 11.14 *Limiting Value*

Suppose that

$$X(z) = \frac{3z^2 - 2z + 4}{z^3 - 2z^2 + 1.5z - 0.5}$$

In this case $X(z)$ has a pole at $z = 1$, and thus there is a pole–zero cancellation in $(z - 1)X(z)$. Performing the cancellation gives

$$(z - 1)X(z) = \frac{3z^2 - 2z + 4}{z^2 - z + 0.5}$$

Using the MATLAB command `roots` reveals that the poles of $(z - 1)X(z)$ are $z = 0.5 \pm j0.5$. The magnitude of both these poles is equal to 0.707, and therefore $x[n]$ has a limit as $n \to \infty$. From (11.58), the limit is

$$\lim_{n\to\infty} x[n] = [(z - 1)X(z)]_{z=1} = \left[\frac{3z^2 - 2z + 4}{z^2 - z + 0.5}\right]_{z=1} = \frac{5}{0.5} = 10$$

The properties of the z-transform given above are summarized in Table 11.2. In Table 11.3, a collection of common z-transform pairs is given, which includes the transform pairs that were derived above using the properties of the z-transform.

TABLE 11.2 PROPERTIES OF THE z-TRANSFORM

Property	Transform Pair/Property
Linearity	$ax[n] + bv[n] \leftrightarrow aX(z) + bV(z)$
Right shift of $x[n]u[n]$	$x[n-q]u[n-q] \leftrightarrow z^{-q}X(z)$
Right shift of $x[n]$	$x[n-1] \leftrightarrow z^{-1}X(z) + x[-1]$
	$x[n-2] \leftrightarrow z^{-2}X(z) + x[-2] + z^{-1}x[-1]$
	$\vdots$
	$x[n-q] \leftrightarrow z^{-q}X(z) + x[-q] + z^{-1}x[-q+1] + \cdots + z^{-q+1}x[-1]$
Left shift in time	$x[n+1] \leftrightarrow zX(z) - x[0]z$
	$x[n+2] \leftrightarrow z^2X(z) - x[0]z^2 - x[1]z$
	$x[n+q] \leftrightarrow z^qX(z) - x[0]z^q - x[1]z^{q-1} - \cdots - x[q-1]z$
Multiplication by n	$nx[n] \leftrightarrow -z\dfrac{d}{dz}X(z)$
Multiplication by n^2	$n^2x[n] \leftrightarrow z\dfrac{d}{dz}X(z) + z^2\dfrac{d^2}{dz^2}X(z)$
Multiplication by a^n	$a^nx[n] \leftrightarrow X\left(\dfrac{z}{a}\right)$
Multiplication by $\cos \Omega n$	$(\cos \Omega n)x[n] \leftrightarrow \dfrac{1}{2}[X(e^{j\Omega}z) + X(e^{-j\Omega}z)]$
Multiplication by $\sin \Omega n$	$(\sin \Omega n)x[n] \leftrightarrow \dfrac{j}{2}[X(e^{j\Omega}z) - X(e^{-j\Omega}z)]$
Summation	$\sum_{i=0}^{n} x[i] \leftrightarrow \dfrac{z}{z-1}X(z)$
Convolution	$x[n] * v[n] \leftrightarrow X(z)V(z)$
Initial-value theorem	$x[0] = \lim_{z\to\infty} X(z)$
	$x[1] = \lim_{z\to\infty} [zX(z) - zX[0]]$
	$\vdots$
	$x[q] = \lim_{z\to\infty} [z^qX(z) - z^qx[0] - z^{q-1}x[1] - \cdots - zx[q-1]]$
Final-value theorem	If $X(z)$ is rational and the poles of $(z-1)X(z)$ have magnitudes < 1, then $\lim_{n\to\infty} x[n] = [(z-1)X(z)]_{z=1}$

11.3 COMPUTATION OF THE INVERSE z-TRANSFORM

If $X(z)$ is the z-transform of the discrete-time signal $x[n]$, the signal can be computed from $X(z)$ by taking the inverse z-transform of $X(z)$ given by

$$x[n] = \frac{1}{j2\pi}\int X(z)z^{k-1}\,dz \tag{11.59}$$

The integral in (11.59) is evaluated by integrating along a counterclockwise closed circular contour that is contained in the region of convergence of $X(z)$.

TABLE 11.3 COMMON z-TRANSFORM PAIRS

$$\delta[n] \leftrightarrow 1$$

$$\delta[n-q] \leftrightarrow \frac{1}{z^q}, \quad q = 1, 2, \ldots$$

$$u[n] \leftrightarrow \frac{z}{z-1}$$

$$u[n] + u[n-q] \leftrightarrow \frac{z^q - 1}{z^{q-1}(z-1)}, \quad q = 1, 2, \ldots$$

$$a^n u[n] \leftrightarrow \frac{z}{z-a}, \; a \text{ real or complex}$$

$$nu[n] \leftrightarrow \frac{z}{(z-1)^2}$$

$$(n+1)u[n] \leftrightarrow \frac{z^2}{(z-1)^2}$$

$$n^2 u[n] \leftrightarrow \frac{z(z+1)}{(z-1)^3}$$

$$na^n u[n] \leftrightarrow \frac{az}{(z-a)^2}$$

$$n^2 a^n u[n] \leftrightarrow \frac{az(z+a)}{(z-a)^3}$$

$$n(n+1)a^n u[n] \leftrightarrow \frac{2az^2}{(z-a)^3}$$

$$(\cos \Omega n)u[n] \leftrightarrow \frac{z^2 - (\cos \Omega)z}{z^2 - (2\cos \Omega)z + 1}$$

$$(\sin \Omega n)u[n] \leftrightarrow \frac{(\sin \Omega)z}{z^2 - (2\cos \Omega)z + 1}$$

$$a^n(\cos \Omega n)u[n] \leftrightarrow \frac{z^2 - (a\cos \Omega)z}{z^2 - (2a\cos \Omega)z + a^2}$$

$$a^n(\sin \Omega n)u[n] \leftrightarrow \frac{(a\sin \Omega)z}{z^2 - (2a\cos \Omega)z + a^2}$$

When the transform $X(z)$ is a rational function of z, the inverse z-transform can be computed (thank goodness!) without having to evaluate the integral in (11.59). The computation of $x[n]$ from a rational $X(z)$ is considered in this section. When $X(z)$ is rational, $x[n]$ can be computed by expanding $X(z)$ into a power series in z^{-1} or by expanding $X(z)$ into partial fractions. The following development begins with the power-series expansion approach.

Expansion by Long Division

Let $X(z)$ be given in the rational form $X(z) = B(z)/A(z)$ with the polynomials $B(z)$ and $A(z)$ written in descending powers of z. To compute the inverse z-transform $x[n]$ for a finite range of values of n, $X(z)$ can be expanded into a power series in z^{-1} by dividing $A(z)$ into $B(z)$ using long division. The values of the signal $x[n]$ are then "read off" from the coefficients of the power-series expansion. The process is illustrated by the following example.

Example 11.15 *Inverse z-Transform via Long Division*

Suppose that

$$X(z) = \frac{z^2 - 1}{z^3 + 2z + 4}$$

Dividing $A(z)$ into $B(z)$ gives

$$\begin{array}{r l}
 & z^{-1} + 0z^{-2} - 3z^{-3} - 4z^{-4} + \cdots \\
z^3 + 2z + 4 \,\big)\!\! & z^2 \quad - 1 \\
 & \underline{z^3 \quad + 2 \quad + 4z^{-1}} \\
 & \quad - 3 \quad - 4z^{-1} \\
 & \quad \underline{- 3 \qquad\quad - 6z^{-2} - 12z^{-3}} \\
 & \qquad - 4z^{-1} + 6z^{-2} + 12z^{-3} \\
 & \qquad \underline{- 4z^{-1} \qquad - 8z^{-3} - 16z^{-4}} \\
 & \qquad\qquad 6z^{-2} + 20z^{-3} + 16z^{-4} \\
 & \qquad\qquad\qquad \vdots
\end{array}$$

Thus

$$X(z) = z^{-1} - 3z^{-3} - 4z^{-4} \quad \cdots \tag{11.60}$$

By definition of the z-transform,

$$X(z) = x[0] + x[1]z^{-1} + x[2]z^{-2} + \cdots \tag{11.61}$$

Equating (11.60) and (11.61) yields the following values for $x[n]$:

$$x[0] = 0, \quad x[1] = 1, \quad x[2] = 0, \quad x[3] = -3, \quad x[4] = -4, \quad \cdots$$

From the results in Example 11.15, it is seen that the initial values $x[0], x[1], x[2], \ldots$ of a signal $x[n]$ can be computed by carrying out the first few steps of the expansion of $X(z) = B(z)/A(z)$ using long division. In particular, note that the initial value $x[0]$ is nonzero if and only if the degree of $B(z)$ is equal to the degree of $A(z)$. If the degree of $B(z)$ is strictly less than the degree of $A(z)$ minus 1, both $x[0]$ and $x[1]$ are zero, and so on.

Instead of carrying out the long division by hand, MATLAB can be used to compute $x[n]$ for any desired range of values of n. This is accomplished using the command `dimpulse` that arises in the transfer function formulation (considered in the next section). Consider a z-transform of the form

$$X(z) = \frac{b_M z^M + b_{M-1} z^{M-1} + \cdots + b_0}{a_N z^N + a_{N-1} z^{N-1} + \cdots + a_0}$$

where $M \leq N$. To solve for the signal values $x[n]$ for $n = 0$ to $n = q - 1$, use the commands

```
num = [bM bM-1 . . . b0];
den = [aN aN-1 . . . a0];
x = dimpulse(num,den,q);
```

The computation of $x[n]$ from $X(z)$ using MATLAB is illustrated by the following example.

Example 11.16 *Inverse z-Transform Using MATLAB*

Consider the z-transform given in Example 11.15:

$$X(z) = \frac{z^2 - 1}{z^3 + 2z + 4}$$

To evaluate $x[n]$ for $n = 0$ to $n = 19$ the MATLAB commands are

```
num = [1 0 -1];
den = [1 0 2 4];
x = dimpulse(num,den,20);
```

Running the program results in the following vector (the first element of which is $x[0]$):

```
x =

     0
     1
     0
    -3
    -4
     6
    20
     4
   -64
   -88
   112
   432
   128
 -1312
 -1984
  2112
  9216
  3712
-26880
-44288
```

Note that the values of $|x[n]|$ appear to be growing without bound as n increases. As will be seen from the development given below, the unbounded growth of the magnitude of $x[n]$ is a result of $X(z)$ having a pole with magnitude > 1.

Inversion via Partial Fraction Expansion

Using long division as described above, the inverse z-transform $x[n]$ of $X(z)$ can be computed for any finite range of integer values of n. However, if an analytical expression for $x[n]$ is desired that is valid for all $n \geq 0$, it is necessary to use partial fraction expansion as was done in the Laplace transform theory. The steps are as follows.

Again suppose that $X(z)$ is given in the rational form $X(z) = B(z)/A(z)$. If the degree of $B(z)$ is equal to $A(z)$, the partial fraction expansion in Section 8.3 cannot be applied directly to $X(z)$. However, dividing $A(z)$ into $B(z)$ yields the following form for $X(z)$:

$$X(z) = x[0] + \frac{R(z)}{A(z)}$$

where $x[0]$ is the initial value of the signal $x[n]$ at time $n = 0$ and $R(z)$ is a polynomial in z whose degree is strictly less than that of $A(z)$. The rational function $R(z)/A(z)$ can then be expanded by partial fractions.

There is another approach that avoids having to divide $A(z)$ into $B(z)$; namely, first expand

$$\frac{X(z)}{z} = \frac{B(z)}{zA(z)}$$

The rational function $X(z)/z$ can be expanded into partial fractions since the degree of $B(z)$ is strictly less than the degree of $zA(z)$ in the case when $B(z)$ and $A(z)$ have the same degrees. After $X(z)/z$ has been expanded, the result can be multiplied by z to yield an expansion for $X(z)$. The inverse z-transform of $X(z)$ can then be computed term by term. There are two cases to consider.

Distinct Poles. Suppose that the poles $p_1, p_2, \ldots, p_N$ of $X(z)$ are distinct and are all nonzero. Then $X(z)/z$ has the partial fraction expansion

$$\frac{X(z)}{z} = \frac{c_0}{z} + \frac{c_1}{z - p_1} + \frac{c_2}{z - p_2} + \cdots + \frac{c_N}{z - p_N} \tag{11.62}$$

where c_0 is the real number given by

$$c_0 = \left[z\frac{X(z)}{z}\right]_{z=0} = X(0) \tag{11.63}$$

and the other residues in (11.62) are real or complex numbers given by

$$c_i = \left[(z - p_i)\frac{X(z)}{z}\right]_{z=p_i}, \qquad i = 1, 2, \ldots, N \tag{11.64}$$

Multiplying both sides of (11.62) by z yields the following expansion for $X(z)$:

$$X(z) = c_0 + \frac{c_1 z}{z - p_1} + \frac{c_2 z}{z - p_2} + \cdots + \frac{c_N z}{z - p_N} \tag{11.65}$$

Then taking the inverse z-transform of each term in (11.65) and using Table 11.3 gives

$$x[n] = c_0\delta[n] + c_1 p_1^n + c_2 p_2^n + \cdots + c_N p_N^n, \qquad n = 0, 1, 2, \ldots \tag{11.66}$$

From (11.66), it is clear that the form of the time variation of the signal $x[n]$ is determined by the poles $p_1, p_2, \ldots, p_N$ of the rational function $X(z)$. Hence the poles of

$X(z)$ determine the characteristics of the time variation of $x[n]$. The reader will recall that this is also the case in the Laplace transform theory, except that here the terms comprising $x[n]$ are of the form cp^n; whereas in the continuous-time case, the terms are of the form $c \exp(pt)$.

If all the poles of $X(z)$ are real, the terms comprising the signal defined by (11.66) are all real. However, if two or more of the poles are complex, the corresponding terms in (11.66) will be complex. Such terms can be combined to yield a real form. To see this, suppose that the pole $p_1 = a + jb$ is complex, so that $b \neq 0$. Then one of the other poles of $X(z)$ must be equal to the complex conjugate $\bar{p}_1$ of p_1. Suppose that $p_2 = \bar{p}_1$; then in (11.66) it must be true that $c_2 = \bar{c}_1$. Hence the second and third terms of the right-hand side of (11.66) are equal to

$$c_1 p_1^n + \bar{c}_1 \bar{p}_1^n \tag{11.67}$$

This term can be expressed in the form

$$2|c_1|\, \sigma^n \cos(\Omega n + \angle c_1) \tag{11.68}$$

where

$$\sigma = |p_1| = \text{magnitude of the pole } p_1$$

and

$$\Omega = \angle p_1 = \text{angle of } p_1$$

The verification that (11.67) and (11.68) are equivalent is left to the homework problems (see Problem 11.7). Using (11.68) in (11.66) results in the following expression for $x[n]$:

$$x[n] = c_0\delta[n] + 2|p_1|\, \sigma^n \cos(\Omega n + \angle c_1) + c_3 p_3^n + \dots + c_n p_n^n, \quad n = 0, 1, 2, \dots \tag{11.69}$$

The expression (11.69) shows that if $X(z)$ has a pair of complex poles p_1, p_2 with magnitude σ and angle $\pm\Omega$, the signal $x[n]$ contains a term of the form

$$2|c|\sigma^n \cos(\Omega n + \angle c)$$

The computation of the inverse z-transform using the procedure above is illustrated in the following example.

Example 11.17 *Complex Pole Case*

Suppose that

$$X(z) = \frac{z^3 + 1}{z^3 - z^2 - z - 2}$$

Here

$$A(z) = z^3 - z^2 - z - 2$$

Using the MATLAB command `roots` reveals that the roots of $A(z)$ are

$$p_1 = -0.5 - j0.866$$

$$p_2 = -0.5 + j0.866$$

$$p_3 = 2$$

Then expanding $X(z)/z$ gives

$$\frac{X(z)}{z} = \frac{c_0}{z} + \frac{c_1}{z + 0.5 + j0.866} + \frac{\bar{c}_1}{z + 0.5 - j0.866} + \frac{c_3}{z - 2}$$

where

$$c_0 = X(0) = \frac{1}{-2} = -0.5$$

$$c_1 = \left[(z + 0.5 + j0.866)\frac{X(z)}{z}\right]_{z=-0.5-j0.866} = 0.429 + j0.0825$$

$$c_3 = \left[(z - 2)\frac{X(z)}{z}\right]_{z=2} = 0.643$$

From (11.66), the inverse z-transform is

$$x[n] = -0.5\delta[n] + c_1(-0.5 - j0.866)^n + \bar{c}_1(-0.5 + j0.866)^n + 0.643(2)^n, \quad n = 0, 1, 2, \ldots$$

The second and third terms in $x[n]$ can be written in real form by using the form (11.68). Here the magnitude and angle of p_1 are given by

$$|p_1| = \sqrt{(0.5)^2 + (0.866)^2} = 1$$

$$\angle p_1 = \pi + \tan^{-1}\frac{0.866}{0.5} = \frac{4\pi}{3} \text{ rad}$$

and the magnitude and angle of c_1 are given by

$$|c_1| = \sqrt{(0.429)^2 + (0.0825)^2} = 0.436$$

$$\angle c_1 = \tan^{-1}\frac{0.0825}{0.429} = 10.89°$$

Then rewriting $x[n]$ in the form (11.69) yields

$$x[n] = -0.5\delta[n] + 0.873\cos\left(\frac{4\pi}{3}n + 10.89°\right) + 0.643(2)^n, \quad n = 0, 1, 2, \ldots$$

It should be noted that the poles and the associated residues (the c_i) for $X(z)/z$ can be computed using the MATLAB commands

```
num = [1 0 0 1];
den = [1 -1 -1 -2 0];
[r,p] = residue(num,den)
```

Running the program yields

```
r =                              p =

    0.6429                          2.0000
    0.4286 - 0.0825i               -0.5000 + 0.8660i
    0.4286 + 0.0825i               -0.5000 - 0.8660i
   -0.5000                               0
```

which matches with the poles and residues computed above.

The inverse z-transform of $X(z)$ can also be computed using the numerical method given in Example 11.16: The following commands compute $x[n]$ for $n = 0$ to $n = 19$.

```
num = [1 0 0 1];
den = [1 -1 -1 -2];
x = dimpulse(num,den,20);
```

The reader is invited to plot this response and compare it to the response calculated analytically.

Repeated poles. Again let $p_1, p_2, \ldots, p_N$ denote the poles of $X(z) = B(z)/A(z)$ and assume that all the p_i are nonzero. Suppose that the pole p_1 is repeated r times and that the other $N - r$ poles are distinct. Then $X(z)/z$ has the partial fraction expansion

$$\frac{X(z)}{z} = \frac{c_0}{z} + \frac{c_1}{z - p_1} + \frac{c_2}{(z - p_1)^2} + \cdots + \frac{c_r}{(z - p_1)^r} + \frac{c_{r+1}}{z - p_{r+1}} + \cdots + \frac{c_N}{z - p_N} \tag{11.70}$$

In (11.70), $c_0 = X(0)$ and the residues $c_{r+1}, c_{r+2}, \ldots, c_n$ are computed in the same way as in the distinct pole case [see (11.64)]. The constants $c_r, c_{r-1}, \ldots, c_1$ are given by

$$c_r = \left[(z - p_1)^r \frac{X(z)}{z}\right]_{z=p_1}$$

$$c_{r-1} = \left[\frac{d}{dz}(z - p_1)^r \frac{X(z)}{z}\right]_{z=p_1}$$

$$c_{r-2} = \frac{1}{2!}\left[\frac{d^2}{dz^2}(z - p_1)^r \frac{X(z)}{z}\right]_{z=p_1}$$

$$\vdots$$

$$c_{r-i} = \frac{1}{i!}\left[\frac{d^i}{dz^i}(z - p_1)^r \frac{X(z)}{z}\right]_{z=p_1}$$

Then multiplying both sides of (11.70) by z gives

$$X(z) = c_0 + \frac{c_1 z}{z - p_1} + \frac{c_2 z}{(z - p_2)^2} + \cdots + \frac{c_r z}{(z - p_1)^r} + \frac{c_{r+1} z}{z - p_{r+1}} + \cdots + \frac{c_N z}{z - p_N} \tag{11.71}$$

The inverse z-transform of the terms

$$\frac{c_i z}{(z - p_1)^i} \tag{11.72}$$

can be computed for $i = 2$ and 3 by using the transform pairs in Table 11.2. This results in the transform pairs

$$c_2 n(p_1)^{n-1} u[n] \leftrightarrow \frac{c_2 z}{(z - p_1)^2} \tag{11.73}$$

$$\frac{1}{2} c_3 n(n-1)(p_1)^{n-2} u[n] \leftrightarrow \frac{c_3 z}{(z - p_1)^3} \tag{11.74}$$

The inverse transform of (11.72) for $i = 4, 5, \ldots$ can be computed by repeatedly using the multiplication by n property of the z-transform. This results in the transform pair

$$\frac{c_i}{(i-1)!} n(n-1)\cdots(n-i+2)(p_1)^{n-i-1} u[n-i+2] \leftrightarrow \frac{c_i z}{(z - p_1)^i}, \qquad i = 4, 5, \ldots$$

Example 11.18 *Repeated Pole Case*

Suppose that

$$X(z) = \frac{6z^3 + 2z^2 - z}{z^3 - z^2 - z + 1}$$

Then

$$\frac{X(z)}{z} = \frac{6z^2 + 2z - 1}{z^3 - z^2 - z + 1} = \frac{6z^2 + 2z - 1}{(z-1)^2(z+1)}$$

Note that the common factor of z in the numerator and denominator of $X(z)/z$ has been canceled. To eliminate unnecessary computations, any common factors in $X(z)/z$ should be canceled before performing a partial fraction expansion.

Now the poles of $X(z)/z$ are $p_1 = 1$, $p_2 = 1$, $p_3 = -1$, and thus the expansion has the form

$$\frac{X(z)}{z} = \frac{c_1}{z-1} + \frac{c_2}{(z-1)^2} + \frac{c_3}{z+1}$$

where

$$c_2 = \left[(z-1)^2 \frac{X(z)}{z}\right]_{z=1} = \frac{6+2-1}{2} = 3.5$$

$$c_1 = \left[\frac{d}{dz}(z-1)^2 \frac{X(z)}{z}\right]_{z=1} = \left[\frac{d}{dz}\frac{6z^2+2z-1}{z+1}\right]_{z=1}$$

$$= \left.\frac{(z+1)(12z+2)-(6z^2+2z-1)(1)}{(z+1)^2}\right|_{z=1}$$

$$= \frac{2(14)-7}{4} = 5.25$$

and

$$c_3 = \left[(z+1)\frac{X(z)}{z}\right]_{z=-1} = \frac{6-2-1}{(-2)^2} = 0.75$$

Hence

$$X(z) = \frac{5.25z}{z-1} + \frac{3.5z}{(z-1)^2} + \frac{0.75z}{z+1}$$

Using the transform pair (11.73) results in the following inverse transform:

$$x[n] = 5.25(1)^n + 3.5n(1)^{n-1} + 0.75(-1)^n, \qquad n = 0, 1, 2, \ldots$$

$$= 5.25 + 3.5n + 0.75(-1)^n, \qquad n = 0, 1, 2, \ldots$$

As a check on the results above, the poles and residues of $X(z)/z$ can be computed using the commands

```
num = [6 2 -1];
den = [1 -1 -1 1];
[r,p] = residue(num,den)
```

Running the program yields

```
r =                    p =

    5.2500                 1.0000
    3.5000                 1.0000
    0.7500                -1.0000
```

This matches with the poles and residues computed above. Note that the residue 5.25 for the first occurrence (in the list above) of the pole at $p = 1$ corresponds to the term $c_1/(z-1)$ in the expansion of $X(z)/z$, and the constant 3.5 for the second occurrence of $p = 1$ corresponds to the term $c_2/(z-1)^2$.

As in Example 11.17, the inverse z-transform can be computed using the following MATLAB commands:

```
num = [6 2 -1 0];
den = [1 -1 -1 1];
x = dimpulse(num,den,20);
```

Pole locations and the form of a signal. Given a discrete-time signal $x[n]$ with rational z-transform $X(z) = B(z)/A(z)$, by the results above it is seen that there is a direct relationship between the poles of $X(z)$ and the form of the time variation of the signal $x[n]$. In particular, if $X(z)$ has a nonrepeated real pole p, then $x[n]$ contains a term of the form $c(p)^n$ for some constant c, and if the pole p is repeated twice, $x[n]$ contains the terms $c_1(p)^n$ and $c_2n(p)^n$ for some constants c_1 and c_2. If $X(z)$ has a nonrepeated pair $a \pm jb$ of complex poles with magnitude σ and angles $\pm\,\Omega$, then $x[n]$ contains a term of the form $c\sigma^n \cos(\Omega n + \theta)$ for some constants c and θ. If the complex pair $a \pm jb$ is repeated twice, $x[n]$ contains the terms $c_1\sigma^n \cos(\Omega n + \theta_1) + c_2n\sigma^n \cos(\Omega n + \theta_2)$ for some constants $c_1, c_2, \theta_1, \theta_2$.

Note that these relationships between signal terms and poles are analogous to those in the Laplace transform theory of continuous-time signals. In fact, as was the case for continuous-time signals, the behavior of a discrete-time signal as $n \to \infty$ can be determined directly from the poles of $X(z)$. In particular, it follows from the results above that $x[n]$ converges to 0 as $n \to \infty$ if and only if all the poles $p_1, p_2, \ldots, p_N$ of $X(z)$ have magnitudes that are strictly less than 1; that is,

$$|p_i| < 1 \qquad \text{for } i = 1, 2, \ldots, N \tag{11.75}$$

In addition, it follows that $x[n]$ converges (as $n \to \infty$) to a finite constant if and only if (11.75) is satisfied, except that one of the p_i may be equal to 1. If this is the case, $x[n]$ converges to the value of the residue in the expansion of $X(z)/z$ corresponding to the pole at 1. In other words,

$$\lim_{n\to\infty} x[n] = \left[(z-1)\frac{X(z)}{z}\right]_{z=1} = [(z-1)X(z)]_{z=1}$$

Note that this result is consistent with the final-value theorem given in Section 11.2.

11.4 TRANSFER FUNCTION REPRESENTATION

In this section the transfer function representation is generated for the class of causal linear time-invariant discrete-time systems. The development begins with discrete-time systems defined by an input/output difference equation. Systems given by a first-order input/output difference equation are considered first.

Pole Positions and Impulse Response

First-Order Case

Consider the linear time-invariant discrete-time system given by the first-order input/output difference equation

$$y[n] + ay[n-1] = bx[n] \tag{11.76}$$

where a and b are real numbers, $y[n]$ is the output, and $x[n]$ is the input. Taking the z-transform of both sides of (11.76) and using the right-shift property (11.20) gives

$$Y(z) + a[z^{-1}Y(z) + y[-1]] = bX(z) \tag{11.77}$$

where $Y(z)$ is the z-transform of the output response $y[n]$ and $X(z)$ is the z-transform of the input $x[n]$. Solving (11.77) for $Y(z)$ yields

$$Y(z) = -\frac{ay[-1]}{1 + az^{-1}} + \frac{b}{1 + az^{-1}}X(z) \tag{11.78}$$

and multiplying the terms on the right-hand side of (11.78) by z/z gives

$$Y(z) = -\frac{ay[-1]z}{z + a} + \frac{bz}{z + a}X(z) \tag{11.79}$$

Equation (11.79) is the *z-domain representation* of the discrete-time system defined by the input/output difference equation (11.76). The first term on the right-hand side of (11.79) is the z-transform of the part of the output response resulting from the initial condition $y[-1]$, and the second term on the right-hand side of (11.79) is the z-transform of the part of the output response resulting from the input $x[n]$ applied for $n = 0, 1, 2, \ldots$.

The system given by (11.76) has no initial energy at time $n = 0$ if $y[-1] = 0$, in which case (11.79) reduces to

$$Y(z) = \frac{bz}{z + a}X(z) \tag{11.80}$$

Defining

$$H(z) = \frac{bz}{z + a}$$

(11.80) becomes

$$Y(z) = H(z)X(z) \tag{11.81}$$

The function $H(z)$ is called the *transfer function* of the system since it specifies the transfer from the input to the output in the z-domain assuming no initial energy ($y[-1] = 0$). Equation (11.81) is the *transfer function representation* of the system.

For any initial condition $y[-1]$ and any input $x[n]$ with rational z-transform $X(z)$, the output $y[n]$ can be computed by taking the inverse z-transform of $Y(z)$ given by (11.79). The procedure is illustrated by the following example.

Example 11.19 *Step Response*

For the system given by (11.76), suppose that $a \neq -1$ and $x[n]$ is equal to the unit-step function $u[n]$. Then $X(z) = z/(z - 1)$, and from (11.79) the z-transform of the output response is

$$Y(z) = -\frac{ay[-1]z}{z+a} + \frac{bz}{z+a}\left(\frac{z}{z-1}\right)$$
$$= -\frac{ay[-1]z}{z+a} + \frac{bz^2}{(z+a)(z-1)} \tag{11.82}$$

Expanding

$$\frac{bz^2}{(z+a)(z-1)}\frac{1}{z} = \frac{ab/(a+1)}{z+a} + \frac{b/(a+1)}{z-1}$$

and taking the inverse z-transform of both sides of (11.82) gives

$$y[n] = -ay[-1](-a)^n + \frac{b}{a+1}[a(-a)^n + (1)^n]$$
$$= -ay[-1](-a)^n + \frac{b}{a+1}[-(-a)^{n+1} + 1], \qquad n = 0, 1, 2, \ldots \tag{11.83}$$

If the initial condition $y[-1]$ is zero, (11.83) reduces to

$$y[n] = \frac{b}{a+1}[-(-a)^n + 1], \qquad n = 0, 1, 2, \ldots \tag{11.84}$$

The output $y[n]$ given by (11.84) is called the *step response* since it is the output response when the input $x[n]$ is the unit step $u[n]$ with no initial energy prior to the application of $u[n]$.

Second-Order Case

Pole Positions and Impulse Response

Now consider the discrete-time system given by the second-order input/output difference equation

$$y[n] + a_1y[n-1] + a_2y[n-2] = b_0x[n] + b_1x[n-1] \tag{11.85}$$

Taking the z-transform of both sides of (11.85) and using the right-shift properties (11.20) and (11.21) gives (assuming that $x[-1] = 0$)

$$Y(z) + a_1[z^{-1}Y(z) + y[-1]] + a_2[z^{-2}Y(z) + z^{-1}y[-1] + y[-2]]$$
$$= b_0X(z) + b_1z^{-1}X(z)$$

Solving for $Y(z)$ gives

$$Y(z) = \frac{-a_1y[-1] - a_2y[-1]z^{-1} - a_2y[-2]}{1 + a_1z^{-1} + a_2z^{-2}} + \frac{b_0 + b_1z^{-1}}{1 + a_1z^{-1} + a_2z^{-2}}X(z) \tag{11.86}$$

Multiplying both sides of (11.86) by z^2/z^2 yields

$$Y(z) = \frac{-(a_1y[-1] + a_2y[-2])z^2 - a_2y[-1]z}{z^2 + a_1z + a_2} + \frac{b_0z^2 + b_1z}{z^2 + a_1z + a_2}X(z) \tag{11.87}$$

Equation (11.87) is the z-domain representation of the discrete-time system given by the second-order input/output difference equation (11.85). The first term on the right-hand side of (11.87) is the z-transform of the part of the output response resulting from the initial conditions $y[-1]$ and $y[-2]$, and the second term on the right-hand side of (11.87) is the z-transform of the part of the output response resulting from the input $x[n]$ applied for $n \geq 0$.

There is no initial energy in the system at $n = 0$ if $y[-1] = y[-2] = 0$, in which case (11.87) reduces to the transfer function representation

$$Y(z) = \frac{b_0 z^2 + b_1 z}{z^2 + a_1 z + a_2} X(z) \tag{11.88}$$

From (11.88) it is seen that the transfer function $H(z)$ of the system is

$$H(z) = \frac{b_0 z^2 + b_1 z}{z^2 + a_1 z + a_2} \tag{11.89}$$

Note that $H(z)$ is a second-order rational function of z.

Example 11.20 *Second-Order System*

Consider the discrete-time system given by the input/output difference equation

$$y[n] + 1.5y[n-1] + 0.5y[n-2] = x[n] - x[n-1]$$

By (11.89), the transfer function of the system is

$$H(z) = \frac{z^2 - z}{z^2 + 1.5z + 0.5}$$

Suppose that the goal is to compute the output response $y[n]$ when $y[-1] = 2, y[-2] = 1$, and the input $x[n]$ is the unit step $u[n]$. Then by (11.87), the z-transform of the response is

$$\begin{aligned}
Y(z) &= \frac{-[(1.5)(2) + (0.5)(1)]z^2 - (0.5)(2)z}{z^2 + 1.5z + 0.5} + \frac{z^2 - z}{z^2 + 1.5z + 0.5}\left(\frac{z}{z-1}\right) \\
&= \frac{-3.5z^2 - z}{z^2 + 1.5z + 0.5} + \frac{z^2}{z^2 + 1.5z + 0.5} \\
&= \frac{-2.5z^2 - z}{z^2 + 1.5z + 0.5} \\
&= \frac{0.5z}{z + 0.5} - \frac{3z}{z + 1}
\end{aligned}$$

Then taking the inverse z-transform gives

$$y[n] = 0.5(-0.5)^n - 3(-1)^n, \qquad n = 0, 1, 2, \ldots$$

The response to a step input when $y[-1] = 2$ and $y[-2] = 1$ can be obtained using the `filter` command. The initial conditions required by this command are related to but not equal to $y[-1]$ and $y[-2]$. For the general second-order difference equation given in (11.85), define an initial condition vector to be `zi=[-a1*y[-1]-a2*y[-2], -a2*y[-1]]` if $x = 0$ for $n < 0$. The use of the `filter` command is demonstrated below:

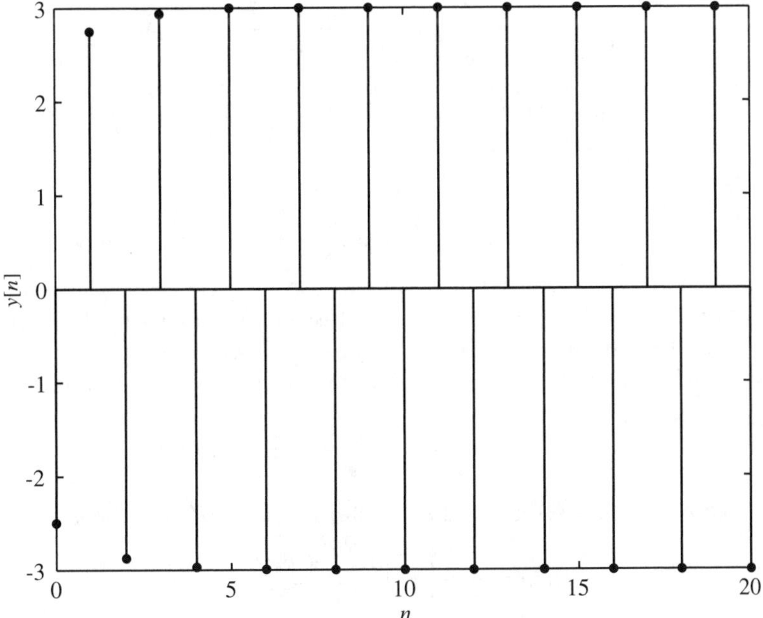

Figure 11.2 Step response in Example 11.20.

```
num = [1 -1 0];
den = [1 1.5 .5];
n = 0:20;
x = ones(1,length(n));
zi = [-1.5*2-0.5*1,-0.5*2];
y = filter(num,den,x,zi);
```

The response of this system is shown in Figure 11.2. Note that this response matches the result obtained analytically, where the term $0.5(0.5)^n$ decays to zero quickly and the term $-3(-1)^n$ simply oscillates between -3 and 3.

*N*th-Order Case

Now suppose that the discrete-time system under study is specified by the Nth-order input/output difference equation

$$y[n] + \sum_{i=1}^{N} a_i y[n - i] = \sum_{i=1}^{M} b_i x[n - i] \tag{11.90}$$

Taking the z-transform of both sides of (11.90) results in the z-domain representation of the system:

$$Y(z) = \frac{C(z)}{A(z)} + \frac{B(z)}{A(z)} X(z) \tag{11.91}$$

where

$$B(z) = b_0 z^N + b_1 z^{N-1} + \cdots + b_M z^{N-M}$$

and

$$A(z) = z^N + a_1 z^{N-1} + \cdots + a_{N-1} z + a_N$$

and where $C(z)$ is a polynomial in z whose coefficients are determined by the initial conditions $y[-1], y[-2], \ldots, y[-N]$ (assuming that $x[n] = 0$ for $n = -1, -2, \ldots, -M$). The system defined by (11.90) has no initial energy at time $n = 0$ if $C(z) = 0$, in which case (11.91) reduces to the transfer function representation

$$Y(z) = \frac{B(z)}{A(z)} X(z) \tag{11.92}$$

From (11.92) it is seen that the transfer function $H(z)$ of the system is

$$H(z) = \frac{B(z)}{A(z)} = \frac{b_0 z^N + b_1 z^{N-1} + \cdots + b_M z^{N-M}}{z^N + a_1 z^{N-1} + \cdots + a_{N-1} z + a_N} \tag{11.93}$$

Transform of the Input/Output Convolution Sum

Suppose that the discrete-time system is given by the input/output convolution relationship

$$y[n] = h[n] * x[n] = \sum_{i=0}^{\infty} h[i]x[n-i], \qquad n = 0, 1, 2, \ldots \tag{11.94}$$

where $h[n]$ is the unit-pulse response of the system (see Section 3.1). Recall that $y[n]$ given by (11.94) is the output response resulting from input $x[n]$ with no initial energy in the system prior to the application of $x[n]$.

If the input $x[n]$ is zero for $n = -1, -2, \ldots$, the z-transform can be applied to both sides of (11.94), which results in the transfer function representation

$$Y(z) = H(z)X(z) \tag{11.95}$$

In (11.95), $H(z)$ is the transfer function of the system, which is equal to the z-transform of the unit-pulse response $h[n]$. Note that if $x[n]$ is the unit pulse $\delta[n]$, then $X(z) = 1$ and (11.95) reduces to $Y(z) = H(z)$. The relationship between the unit-pulse response $h[n]$ and the transfer function $H(z)$ can be expressed in terms of the transform pair notation

$$h[n] \leftrightarrow H(z) \tag{11.96}$$

The transform pair (11.96) is analogous to the relationship generated in Section 8.5 between the impulse response and the transfer function in the continuous-time case. As in the continuous-time theory, (11.96) provides a major link between the time domain and the transform domain in the study of discrete-time systems.

As shown above, if the discrete-time system is given by an input/output difference equation, the transfer function is a rational function of z. It turns out that the converse is also true; that is, if $H(z)$ is rational, the system can be described by an input/output difference equation. To see this, suppose that $H(z)$ can be expressed in the rational form (11.93). Then multiplying both sides of (11.92) by $A(z)$ gives

$$A(z)Y(z) = B(z)X(z)$$

and taking the inverse z-transform of this results in the difference equation (11.90). Since only finite-dimensional discrete-time systems can be described by an input/output difference equation, the conclusion is that a linear time-invariant (and causal) discrete-time system is finite dimensional if and only if its transfer function $H(z)$ is rational in z.

If a given discrete-time system is finite dimensional, so that $H(z)$ is rational, the poles and zeros of the system are defined to be the poles and zeros of $H(z)$. It is also possible to define the pole–zero diagram of a discrete-time system as was done in the continuous-time case (see Section 8.5).

It is important to observe that if the input $x[n]$ is nonzero for at least one positive value of n, so that $X(z) \neq 0$, then both sides of the transfer function representation (11.95) can be divided by $X(z)$, which gives

$$H(z) = \frac{Y(z)}{X(z)} \tag{11.97}$$

Thus the transfer function $H(z)$ is equal to the ratio of the z-transforms of the output and input. Note that since $H(z)$ is unique, the ratio $Y(z)/X(z)$ cannot change as the input $x[n]$ ranges over some collection of input signals. From (11.97) it is also seen that $H(z)$ can be determined from the output response to any input that is not identically zero for $n \geq 0$.

Example 11.21 *Computation of Transfer Function*

A linear time-invariant discrete-time system has unit-pulse response

$$h[n] = 3(2^{-n})\cos\left(\frac{\pi n}{6} + \frac{\pi}{12}\right), \qquad n = 0, 1, 2, \ldots \tag{11.98}$$

where the argument of the cosine in (11.98) is in radians. The transfer function $H(z)$ of the system is equal to the z-transform of $h[n]$ given by (11.98). To compute the transform of $h[n]$, first expand the cosine in (11.98) using the trigonometric identity

$$\cos(a + b) = (\cos a)(\cos b) - (\sin a)(\sin b) \tag{11.99}$$

Applying (11.99) to (11.98) gives

$$h[n] = 3(2^{-n})\left[\cos\left(\frac{\pi n}{6}\right)\cos\left(\frac{\pi}{12}\right) - \sin\left(\frac{\pi n}{6}\right)\sin\left(\frac{\pi}{12}\right)\right], \qquad n \geq 0$$

$$= 2.898\left(\frac{1}{2}\right)^n \cos\left(\frac{\pi n}{6}\right) - 0.776\left(\frac{1}{2}\right)^n \sin\left(\frac{\pi n}{6}\right), \qquad n \geq 0 \qquad (11.100)$$

Taking the z-transform of (11.100) yields

$$H(z) = 2.898\frac{z^2 - [0.5\cos(\pi/6)]z}{z^2 - [\cos(\pi/6)]z + 0.25} - 0.776\frac{[0.5\sin(\pi/6)]z}{z^2 - [\cos(\pi/6)]z + 0.25}$$

$$= \frac{2.898z^2 - 1.449z}{z^2 - 0.866z + 0.25}$$

If both $H(z)$ and $X(z)$ are rational functions of z, the output response can be computed by first expanding the product $H(z)X(z)$ (or $H(z)X(z)/z$) by partial fractions. The process is illustrated by the following example.

Example 11.22 *Computation of Step Response*

Suppose that the objective is to compute the step response of the system in Example 11.21. Then inserting $X(z) = z/(z - 1)$ and $H(z)$ into $Y(z) = H(z)X(z)$ gives

$$Y(z) = \frac{2.898z^3 - 1.449z^2}{(z - 1)(z^2 - 0.866z + 0.25)}$$

Since the zeros of $z^2 - 0.866z + 0.25$ are complex, to avoid complex arithmetic, $Y(z)/z$ can be expanded into the form

$$\frac{Y(z)}{z} = \frac{cz + d}{z^2 - 0.866z + 0.25} + \frac{c_3}{z - 1}$$

where

$$c_3 = \left[(z - 1)\frac{Y(z)}{z}\right]_{z=1} = \frac{2.898 - 1.449}{1 - 0.866 + 0.25} = 3.773$$

Hence

$$Y(z) = \frac{cz^2 + dz}{z^2 - 0.866z + 0.25} + \frac{3.773z}{z - 1} \qquad (11.101)$$

Putting the right-hand side of (11.101) over a common denominator and equating coefficients gives

$$c + 3.773 = 2.898$$

$$d - c - (0.866)(3.773) = -1.449$$

Solving for c and d yields

$$c = -0.875 \quad \text{and} \quad d = 0.943$$

Now to determine the inverse z-transform of the first term on the right-hand side of (11.101), set

$$z^2 + 0.866z + 0.25 = z^2 - (2a\cos\Omega)z + a^2$$

Then

$$a = \sqrt{0.25} = 0.5$$

$$\Omega = \cos^{-1}\left(\frac{0.866}{2a}\right) = \frac{\pi}{6}\text{ rad}$$

and thus

$$\frac{cz^2 + dz}{z^2 - 0.866z + 0.25} = \frac{-0.875z^2 + 0.943z}{z^2 - (\cos\pi/6)z + 0.25} \tag{11.102}$$

Expressing the right-hand side of (11.102) in the form

$$\frac{\alpha(z^2 - 0.5(\cos\pi/6)z)}{z^2 - (\cos\pi/6)z + 0.25} + \frac{\beta(\sin\pi/6)z}{z^2 - (\cos\pi/6)z + 0.25}$$

results in $\alpha = -0.875$ and

$$-0.5\alpha\cos\frac{\pi}{6} + \beta\sin\frac{\pi}{6} = 0.943 \tag{11.103}$$

Solving (11.103) for β yields $\beta = 1.128$. Thus

$$Y(z) = \frac{-0.875(z^2 - 0.5(\cos\pi/6)z)}{z^2 - (\cos\pi/6)z + 0.25} + \frac{1.128(\sin\pi/6)z}{z^2 - (\cos\pi/6)z + 0.25} + \frac{3.773z}{z - 1}$$

Finally, using Table 11.3 gives

$$y[n] = -0.875\left(\frac{1}{2}\right)^n\cos\frac{\pi n}{6} + 2.26\left(\frac{1}{2}\right)^n\sin\frac{\pi n}{6} + 3.773, \qquad n = 0, 1, 2, \ldots$$

The step response also can be computed numerically using the MATLAB command `dstep`. For this example, the following commands compute and plot the step response:

```
num = [2.898 -1.449 0];
den = [1 -.866 .25];
n = 0:20;
y = dstep(num,den,21);
stem(n,y)
```

The resulting plot of the step response is given in Figure 11.3. Note that the response settles to a steady-state value of 3.773, which matches the limiting value of $y[n]$ obtained by examining the expression above for $y[n]$.

Note that the poles of $(z - 1)Y(z)$ are $0.433 \pm j0.25$, which have a magnitude of 0.5. Then since the poles of $(z - 1)Y(z)$ have magnitude less than 1, the final-value theorem can be applied to compute the limiting value of $y[n]$. The result is

$$\lim_{n\to\infty} y[n] = [(z-1)Y(z)]_{z=1} = \left[\frac{2.898z^3 - 1.449z^2}{z^2 - 0.866z + 0.25}\right]_{z=1} = \frac{1.449}{0.384} = 3.773$$

which checks with the value above.

Transfer Function of Interconnections

The transfer function of a linear time-invariant discrete-time system can be computed directly from a signal-flow diagram of the system. This approach is pursued here for systems given by an interconnection of unit-delay elements and an interconnection of blocks. The following results are directly analogous to those derived in Sections 8.5 and 8.6 for linear time-invariant continuous-time systems.

Interconnections of unit-delay elements. The *unit-delay element* is a system whose input/output relationship is given by

$$y[n] = x[n-1] \tag{11.104}$$

The system given by (11.104) is referred to as the *unit-delay element* since the output $y[n]$ is equal to a one-step time delay of the input $x[n]$. As shown in Figure 11.4, the unit-delay element is represented by a box with a D that stands for "delay."

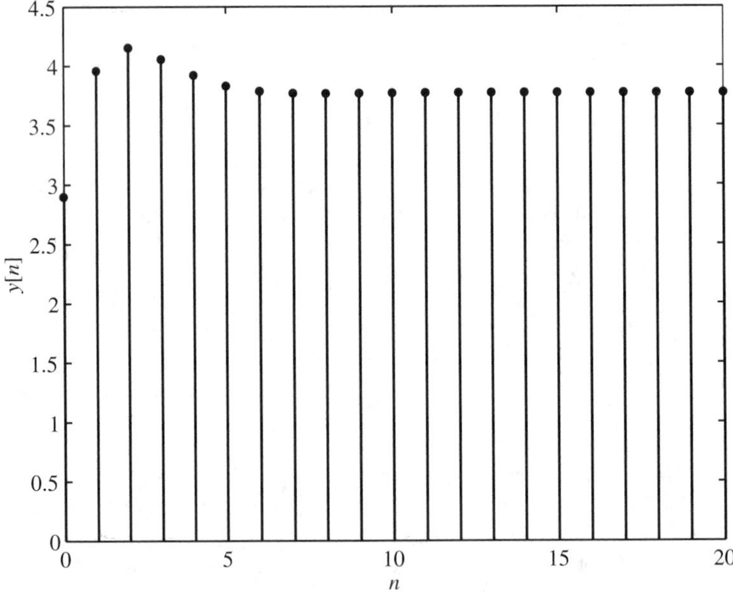

Figure 11.3 Step response in Example 11.22.

Figure 11.4 Unit-delay element.

Taking the z-transform of both sides of (11.104) with $x[-1] = 0$ yields the transfer function representation of the unit-delay element:

$$Y(z) = z^{-1}X(z) \tag{11.105}$$

From (11.105) it is seen that the transfer function of the unit-delay element is equal to $1/z$. Note that the unit delay is the discrete-time counterpart to the integrator in the sense that the unit delay has transfer function $1/z$ and the integrator has transfer function $1/s$.

Now suppose that a discrete-time system is given by an interconnection of unit delays, adders, subtracters, and scalar multipliers. The transfer function for any such interconnection can be computed by working in the z-domain with the unit delays represented by their transfer function $1/z$. The procedure is very similar to that considered in Section 8.5 for continuous-time systems consisting of interconnections of integrators.

Example 11.23 *Computation of Transfer Function*

Consider the discrete-time system given by the interconnection in Figure 11.5. Note that the outputs of the two unit-delay elements in Figure 11.4 are denoted by $q_1[n]$ and $q_2[n]$. The z-domain representation of the system with all initial conditions equal to zero is shown in Figure 11.6. From this figure it is clear that

$$zQ_1(z) = Q_2(z) + X(z) \tag{11.106}$$

$$zQ_2(z) = Q_1(z) - 3Y(z) \tag{11.107}$$

$$Y(z) = 2Q_1(z) + Q_2(z) \tag{11.108}$$

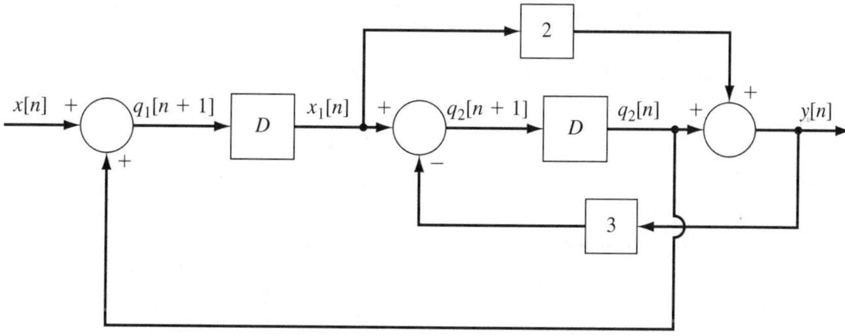

Figure 11.5 System in Example 11.23.

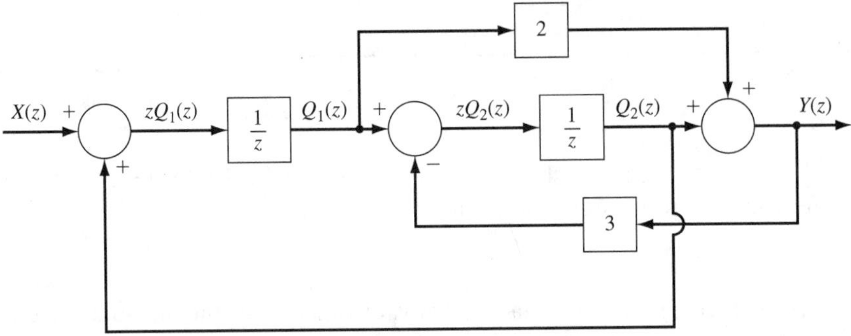

Figure 11.6 Representation of system in the z-domain.

Solving (11.106) for $Q_1(z)$ and inserting the result into (11.107) and (11.108) yields

$$zQ_2(z) = z^{-1}Q_2(z) + z^{-1}X(z) - 3Y(z) \tag{11.109}$$

$$Y(z) = 2z^{-1}Q_2(z) + z^{-1}X(z) + Q_2(z) \tag{11.110}$$

Solving (11.109) for $Q_2(z)$ and inserting the result into (11.110) gives

$$Y(z) = \frac{2z^{-1} + 1}{z - z^{-1}}[z^{-1}X(z) - 3Y(z)] + 2z^{-1}X(z)$$

Then

$$\left[1 + \frac{3(2z^{-1} + 1)}{z - z^{-1}}\right]Y(z) = \frac{z^{-1}(2z^{-1} + 1)}{z - z^{-1}}X(z) + 2z^{-1}X(z)$$

$$\frac{z + 5z^{-1} + 3}{z - z^{-1}}Y(z) = \left[\frac{z^{-1}(2z^{-1} + 1)}{z - z^{-1}} + 2z^{-1}\right]X(z)$$

$$= \frac{z^{-1} + 2}{z - z^{-1}}X(z)$$

Thus

$$Y(z) = \frac{z^{-1} + 2}{z + 5z^{-1} + 3}X(z) = \frac{2z + 1}{z^2 + 3z + 5}X(z)$$

so the transfer function is

$$H(z) = \frac{2z + 1}{z^2 + 3z + 5}$$

Transfer function of basic interconnections. The transfer function of series, parallel, and feedback connections have exactly the same form as in the continuous-time case. The results are displayed in Figure 11.7.

The transfer function equivalences for moving pick-off points and adders or subtracters also have exactly the same form as in the continuous-time case. In par-

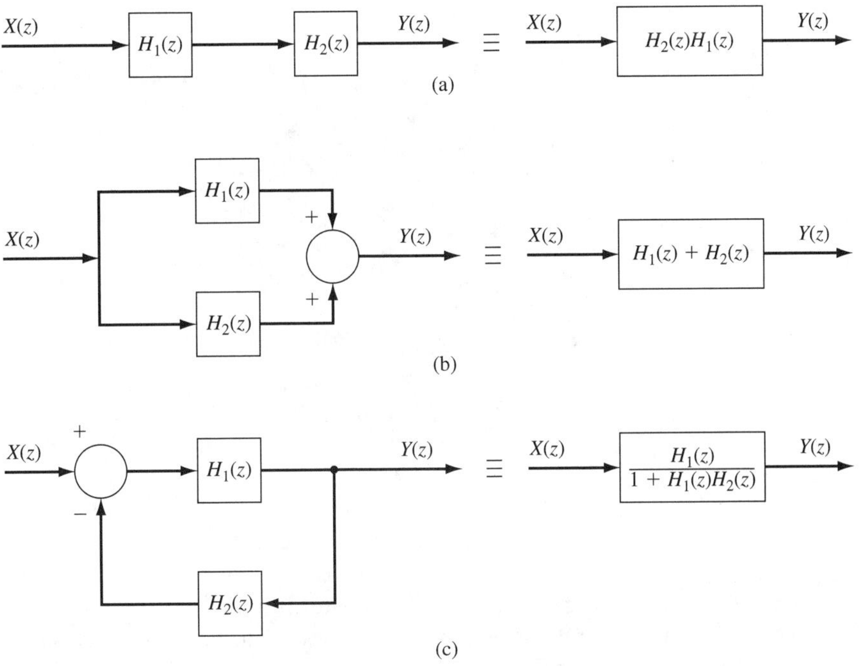

Figure 11.7 Transfer functions of basic interconnections: (a) series connection; (b) parallel connection; (c) feedback connection.

ticular, simply replace s by z in Figures 8.20 and 8.21 to obtain the equivalences in the discrete-time case.

Block-diagram reduction. By using the equivalences in Figure 11.7 and the equivalences for moving pick-off points and adders/subtracters, it is possible to reduce a given block diagram in order to compute the transfer function of the system. Again, the process is directly analogous to block-diagram reduction in the continuous-time case (see Section 8.6). The transfer function of a block diagram can also be determined by using Mason's theorem. The procedure is identical to the continuous-time case discussed in Section 8.6.

11.5 STABILITY OF DISCRETE-TIME SYSTEMS

Consider the linear time-invariant discrete-time system given by the transfer function

$$H(z) = \frac{B(z)}{A(z)} = \frac{b_M z^M + b_{M-1} z^{M-1} + \cdots + b_1 z + b_0}{a_N z^N + a_{N-1} z^{N-1} + \cdots + a_1 z + a_0} \tag{11.111}$$

The system is assumed to be causal, and thus $M \leq N$. It is also assumed that the polynomials $B(z)$ and $A(z)$ do not have any common factors. If there are common factors, they should be canceled.

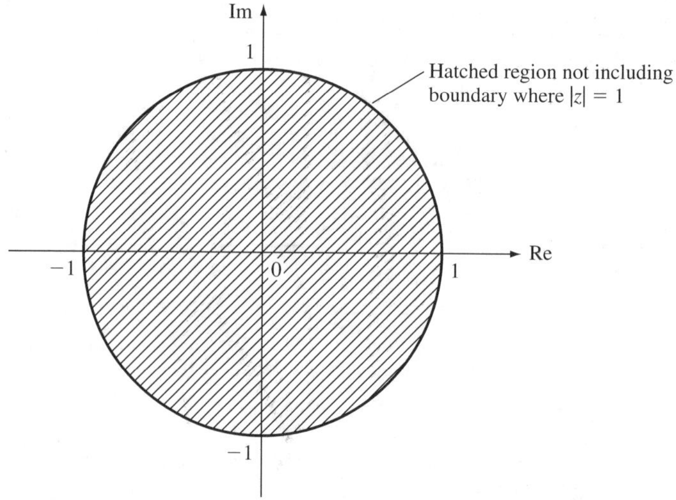

Figure 11.8 Open unit disk.

As observed in Section 11.4, the transfer function $H(z)$ is the z-transform of the system's unit-pulse response $h[n]$. From the development in Section 11.3, it follows that the form of the time variation of $h[n]$ is directly determined by the poles of the system, which are the roots of $A(z) = 0$. In particular, if $H(z)$ has a real pole p, then $h[n]$ contains a term of the form $c(p)^n$, and if $H(z)$ has a complex pair $a \pm jb$ of poles with magnitude σ and angles $\pm\Omega$, then $h[n]$ contains a term of the form $c(\sigma)^n \cos(\Omega n + \theta)$. If $H(z)$ has repeated poles, it will contain terms of the form $cn(n+1)\cdots(n+i)(p)^n$ and/or $cn(n+1)\cdots(n+i)\sigma^n \cos(\Omega n + \theta)$.

It follows from the relationship between the form of $h[n]$ and the poles of $H(z)$ that $h[n]$ converges to zero as $n \to \infty$ if and only if

$$|p_i| < 1 \qquad \text{for } i = 1, 2, \ldots, N \tag{11.112}$$

where $p_1, p_2, \ldots, p_N$ are the poles of $H(z)$. The condition (11.112) is equivalent to requiring that all the poles be located in the *open unit disk* of the complex plane. The open unit disk is that part of the complex plane consisting of all complex numbers whose magnitude is strictly less than 1. The open unit disk is the hatched region shown in Figure 11.8.

A discrete-time system with transfer function $H(z)$ given by (11.111) is said to be *stable* if its unit-pulse response $h[n]$ converges to zero as $n \to \infty$. Thus stability is equivalent to requiring that all the poles of the system lie in the open unit disk of the complex plane. The system is marginally stable if the unit-pulse response $h[n]$ is bounded; that is,

$$|h[n]| < c \qquad \text{for all } n \tag{11.113}$$

where c is a finite positive constant. It also follows from the relationship between the form of $h[n]$ and the poles of $H(z)$ that a system is marginally stable if and only if

$|p_i| \le 1$ for all nonrepeated poles of $H(z)$, and $|p_i| < 1$ for all repeated poles. This is equivalent to requiring that all poles lie in the open unit disk, except that nonrepeated poles can be located on the *unit circle* (i.e., all those complex numbers z such that $|z| = 1$).

Finally, a system is unstable if the magnitude of $h[n]$ grows without bound as $n \to \infty$; that is,

$$|h[n]| \to \infty \qquad \text{as } n \to \infty \tag{11.114}$$

The instability condition (11.114) is equivalent to having one or more poles located outside the closed unit disk (all complex numbers z such that $|z| > 1$) or having one or more repeated poles located on the unit circle.

From the results above, it is seen that the stability boundary in the discrete-time case is the unit circle of the complex plane. In contrast, in the continuous-time case the stability boundary is the imaginary axis ($j\omega$-axis) of the complex plane. Thus the stability boundary in the discrete-time case is quite different from the stability boundary in the continuous-time case. In Chapter 12 it is shown that there is a mapping between the stability region in the continuous-time case and the stability region in the discrete-time case.

The convergence of $h[n]$ to zero turns out to be equivalent (when $H(z)$ is rational) to absolute summability of $h[n]$; that is,

$$\sum_{n=0}^{\infty} |h[n]| < \infty$$

Thus stability is equivalent to absolute summability of the unit-pulse response $h[n]$. In addition, convergence of $h[n]$ to zero is equivalent to bounded-input bounded-output (BIBO) stability, which implies that the output $y[n]$ is a bounded signal whenever the input $x[n]$ is a bounded signal (assuming no initial energy in the system prior to the application of $x[n]$). In mathematical terms, BIBO stability means that whenever $|x[n]| \le c_1$ for all n and for some finite positive constant c_1, then $|y[n]| \le c_2$ for all n and for some finite positive constant c_2, where $y[n]$ is the response to $x[n]$ with no initial energy in the system. Hence stability of a system (i.e., convergence of $h[n]$ to zero) is equivalent to BIBO stability.

Jury Stability Test

Again consider the discrete-time system with transfer function $H(z) = B(z)/A(z)$ with

$$A(z) = a_N z^N + a_{N-1} z^{N-1} + \cdots + a_1 z + a_0, \qquad a_N > 0$$

By using the Routh–Hurwitz test as discussed in Section 9.2, it is possible to test for stability of a continuous-time system without having to compute the poles of the system. There is a discrete-time counterpart to the Routh–Hurwitz test, which is referred to as the *Jury test.* The Jury test is based on the Jury table, which is displayed in Table 11.4. Here it is assumed that the degree of $A(z)$ is greater than or equal to 2.

TABLE 11.4 JURY TABLE

Row	z^0	z^1	z^2	$\cdots$	z^{N-2}	z^{N-1}	z^N
1	a_0	a_1	a_2	$\cdots$	a_{N-2}	a_{N-1}	a_N
2	a_N	a_{N-1}	a_{N-2}	$\cdots$	a_2	a_1	a_0
3	b_0	b_1	b_2	$\cdots$	b_{N-2}	b_{N-1}	
4	b_{N-1}	b_{N-2}	b_{N-3}	$\cdots$	b_1	b_0	
5	c_0	c_1	c_2	$\cdots$	c_{N-2}		
6	c_{N-2}	c_{N-3}	c_{N-4}	$\cdots$	c_0		
$\vdots$	$\vdots$	$\vdots$	$\vdots$				
$2N-5$	d_0	d_1	d_2	d_3			
$2N-4$	d_3	d_2	d_1	d_0			
$2N-3$	e_0	e_1	e_2				

As in the Routh array, the first two rows of the Jury table are filled by the coefficients of $A(z)$. However, the manner in which the coefficients are distributed in the first two rows of the Jury table is different from that in the Routh array. The coefficients in the third and fourth rows of the Jury table are given by

$$b_i = a_0 a_i - a_{N-i} a_N, \qquad i = 0, 1, 2, \ldots, N-1$$

The coefficients in the fifth and sixth rows are given by

$$c_i = b_0 b_i - b_{N-i-1bN-1}, \qquad i = 0, 1, 2, \ldots, N-2$$

The process continues down to the $(2N-3)$th row, whose last element e_2 is given by

$$e_2 = d_0 d_2 - d_1 d_3$$

Having computed the Jury table, by the Jury test the system is stable (all poles are in the open unit disk) if and only if all the following conditions are satisfied:

$$A(1) > 0 \quad \text{and} \quad (-1)^n A(-1) > 0$$

$$a_N > |a_0|$$

$$|b_0| > |b_{N-1}|$$

$$|c_0| > |c_{N-2}|$$

$$\vdots$$

$$|e_0| > |e_2|$$

Example 11.24 *Jury Test*

Suppose that

$$A(z) = z^2 + a_1 z + a_0$$

The Jury table is shown in Table 11.5.

TABLE 11.5 JURY TABLE FOR EXAMPLE 11.24

Row	z^0	z^1	z^2
1	a_0	a_1	1

From the conditions given above, the roots of $A(z) = 0$ are in the open unit disk if and only if

$$A(1) = 1 + a_1 + a_0 > 0$$

$$(-1)^2 A(-1) = 1 - a_1 + a_0 > 0$$

$$1 > |a_0|$$

The first two conditions are equivalent to

$$-a_1 < 1 + a_0 \quad \text{and} \quad a_1 < 1 + a_0$$

which in turn is equivalent to

$$|a_1| < 1 + a_0$$

Hence the roots of $A(z) = 0$ are in the open unit disc if and only if

$$|a_1| < 1 + a_0 \quad \text{and} \quad |a_0| < 1 \tag{11.115}$$

This is an interesting result, since it is not obvious that (11.115) is a necessary and sufficient condition for both roots of a second-degree polynomial to be in the open unit disc.

11.6 FREQUENCY RESPONSE OF DISCRETE-TIME SYSTEMS

Again consider the linear time-invariant finite-dimensional discrete-time system with the transfer function

$$H(z) = \frac{B(z)}{A(z)} = \frac{b_M z^M + b_{M-1} z^{M-1} + \cdots + b_1 z + b_0}{a_N z^N + a_{N-1} z^{N-1} + \cdots + a_1 z + a_0} \tag{11.116}$$

Throughout this section it is assumed that the system is stable, and thus all the poles of $H(z)$ are located in the open unit disk of the complex plane. As in the continuous-time case (see Section 9.4), the frequency response characteristics of the system can be determined by examining the response to a sinusoidal input. In particular, let the system input $x[n]$ be given by

$$x[n] = C\cos(\Omega_0 n), \qquad n = 0, 1, 2, \ldots \tag{11.117}$$

where C and Ω_0 are real numbers. From Table 11.3 the z-transform of the sinusoidal input (11.117) is

$$X(z) = \frac{C[z^2 - (\cos \Omega_0)z]}{z^2 - (2\cos \Omega_0)z + 1}$$

If there is no initial energy in the system at time $n = 0$, the z-transform of the resulting output response is

$$Y(z) = \frac{CB(z)[z^2 - (\cos \Omega_0)z]}{A(z)[z^2 - (2\cos \Omega_0)z + 1]}$$

Now

$$\begin{aligned} z^2 - (2\cos\Omega_0)z + 1 &= (z - \cos\Omega_0 - j\sin\Omega_0)(z - \cos\Omega_0 + j\sin\Omega_0) \\ &= (z - e^{j\Omega_0})(z - e^{-j\Omega_0}) \end{aligned}$$

and thus

$$Y(z) = \frac{CB(z)[z^2 - (\cos \Omega_0)z]}{A(z)(z - e^{j\Omega_0})(z - e^{-j\Omega_0})}$$

Dividing $Y(z)$ by z gives

$$\frac{Y(z)}{z} = \frac{CB(z)(z - \cos \Omega_0)}{A(z)(z - e^{j\Omega_0})(z - e^{-j\Omega_0})}$$

Pulling out the terms $z - e^{j\Omega_0}$ and $z - e^{-j\Omega_0}$ yields

$$\frac{Y(z)}{z} = \frac{\eta(z)}{A(z)} + \frac{c}{z - e^{j\Omega_0}} + \frac{\bar{c}}{z - e^{-j\Omega_0}}$$

where $\eta(z)$ is a polynomial in z with the degree of $\eta(z)$ less than N. The constant c is given by

$$\begin{aligned} c = \left[(z - e^{j\Omega_0})\frac{Y(z)}{z}\right]_{z=e^{j\Omega_0}} &= \left[\frac{CB(z)(z - \cos\Omega_0)}{A(z)(z - e^{-j\Omega_0})}\right]_{z=e^{j\Omega_0}} \\ &= \frac{CB(e^{j\Omega_0})(e^{j\Omega_0} - \cos\Omega_0)}{A(e^{j\Omega_0})(e^{j\Omega_0} - e^{-j\Omega_0})} \\ &= \frac{CB(e^{j\Omega_0})(j\sin\Omega_0)}{A(e^{j\Omega_0})(j2\sin\Omega_0)} \\ &= \frac{CB(e^{j\Omega_0})}{2A(e^{j\Omega_0})} = \frac{C}{2}H(e^{j\Omega_0}) \end{aligned}$$

Multiplying the expression for $Y(z)/z$ by z gives

$$Y(z) = \frac{z\eta(z)}{A(z)} + \frac{(C/2)H(e^{j\Omega_0})z}{z - e^{j\Omega_0}} + \frac{(C/2)\overline{H(e^{j\Omega_0})}z}{z - e^{-j\Omega_0}} \tag{11.118}$$

Let $y_{tr}[n]$ denote the inverse z-transform of $z\eta(z)/A(z)$. Since the system is stable, the roots of $A(z) = 0$ are within the open unit disk of the complex plane, and thus

$y_{tr}[n]$ converges to zero as $n \to \infty$. Hence $y_{tr}[n]$ is the transient part of the output response resulting from the sinusoidal input $x[n] = C\cos(\Omega_0 n)$.

Now let $y_{ss}[n]$ denote the inverse z-transform of the second and third terms on the right-hand side of (11.118). By using the trigonometric identity (8.71), $y_{ss}[n]$ can be written in the form

$$y_{ss}[n] = C|H(e^{j\Omega_0})|\cos[\Omega_0 n + \angle H(e^{j\Omega_0})], \qquad n = 0, 1, 2, \ldots \qquad (11.119)$$

The response $y_{ss}[n]$ clearly does not converge to zero as $n \to \infty$, and thus it is the steady-state part of the output response resulting from the input $x[n] = C\cos(\Omega_0 n)$.

Note that the steady-state response to a sinusoidal input is also a sinusoid with the same frequency, but is amplitude scaled by the amount $|H(e^{j\Omega_0})|$ and is phase shifted by the amount $\angle H(e^{j\Omega_0})$. This result corresponds to the development given in Chapter 7 in terms of the discrete-time Fourier transform (DTFT). More precisely, it follows directly from the formulation in Section 7.4 that the output response y[n] resulting from the input

$$x[n] = C\cos(\Omega_0 n), \qquad n = 0, \pm 1, \pm 2, \ldots$$

is given by

$$y[n] = C|H(\Omega_0)|\cos[\Omega_0 n + \angle H(\Omega_0)], \qquad n = 0, \pm 1, \pm 2, \ldots \qquad (11.120)$$

where $H(\Omega_0)$ is the value at $\Omega = \Omega_0$ of the DTFT $H(\Omega)$ of the unit-pulse response $h[n]$; that is,

$$H(\Omega_0) = H(e^{j\Omega_0}) = \sum_{n=0}^{\infty} h[n]e^{-j\Omega_0 n}$$

Now since the system is stable, the unit-pulse response $h[n]$ is absolutely summable, and thus the DTFT of $h[n]$ is equal to the transfer function $H(z)$ evaluated at $z = e^{j\Omega}$; that is,

$$H(\Omega) = H(z)\big|_{z=e^{j\Omega}} \qquad (11.121)$$

It then follows that the expressions (11.119) and (11.120) for the output response $y[n]$ are identical for $n \geq 0$, and thus the transfer function analysis given above directly corresponds to the Fourier analysis given in Chapter 7.

As first defined in Section 7.4, the DTFT $H(\Omega)$ of $h[n]$ is the frequency response function of the system, and the plots of $|H(\Omega)|$ and $\angle H(\Omega)$ versus Ω are the magnitude and phase plots of the system. Since

$$|H(e^{j\Omega_0})| = |H(\Omega)|_{\Omega=\Omega_0} \quad \text{and} \quad \angle H(e^{j\Omega_0}) = \angle H(\Omega)\big|_{\Omega=\Omega_0}$$

from (11.119) it is seen that the steady-state response resulting from the sinusoidal input $x[n] = C\cos(\Omega_0 n)$ can be determined directly from the magnitude and phase plots. Note that this fundamental result is directly analogous to the development in the continuous-time case given in Section 9.4.

Computation of the Frequency Response Curves

The MATLAB command `freqz` can be used to compute the frequency response function $H(\Omega)$ for any desired set of values of the frequency variable Ω. Once $H(\Omega)$ is determined, the magnitude and phase plots can be computed and plotted. To use the command `freqz`, the transfer function $H(z)$ must have $M \le N$ and be defined in terms of polynomials in z^{-1}. Given a transfer function in the form (11.116), $H(z)$ can be put into the proper form by multiplying $H(z)$ by z^{-N}/z^{-N}. This yields

$$H(z) = \frac{d_N + d_{N-1}z^{-1} + \cdots + d_0 z^{-N}}{a_N + a_{N-1}z^{-1} + \cdots + a_0 z^{-N}}$$

Note that if $M = N$, the coefficients d_i in the numerator of $H(z)$ are equal to the b_i in (11.116). Otherwise, if $M < N$, then $d_i = b_i$ for $i = 0, 1, 2, \ldots, M$ and $d_i = 0$ for $i > M$. Now the MATLAB commands that will generate the frequency response curves are

```
num = [dN dN-1 ... d0];
den = [aN aN-1 ... a0];
OMEGA = -pi:pi/150:pi;
H = freqz(num,den,OMEGA);
mag = abs(H);
phase = 180/pi*unwrap(angle(H));
```

It should be noted that the use of the command `unwrap` is to smooth out the phase plot since the `angle` command may result in jumps of $\pm 2\pi$. The use of the MATLAB software is illustrated in the following example.

Example 11.25 *Frequency Response Curves*

Consider the discrete-time system with transfer function

$$H(z) = \frac{1}{z - 0.5}$$

To use MATLAB for computation of the frequency response curves, first rewrite $H(z)$ as

$$H(z) = \frac{z^{-1}}{1 - 0.5z^{-1}}$$

Then the MATLAB commands to obtain the frequency response are

```
num = [0 1];
den = [1 -.5];
OMEGA = -pi:pi/150;pi;
H = freqz(num,den,OMEGA);
subplot(211),plot(OMEGA,abs(H));
subplot(212),plot(OMEGA,180/pi*unwrap(angle(H)));
```

The MATLAB-generated plots are shown in Figure 11.9.

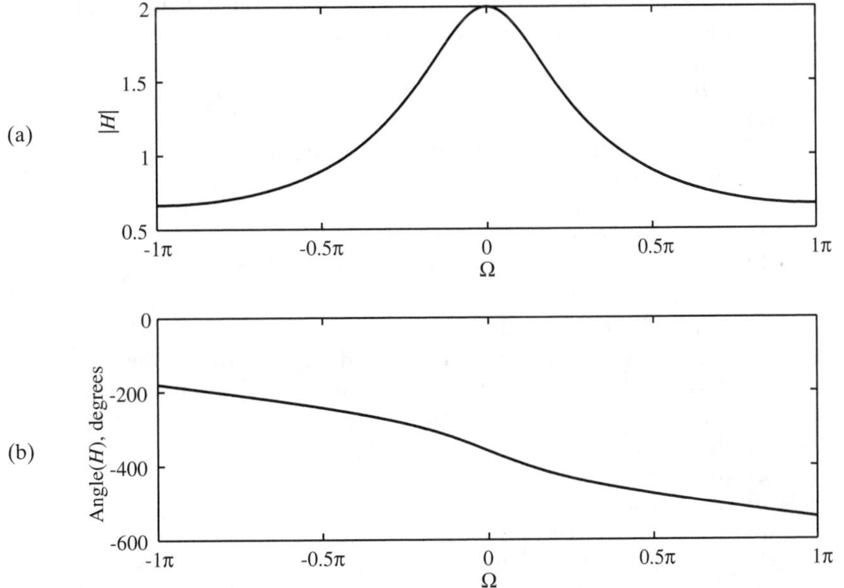

Figure 11.9 Magnitude (a) and phase (b) plots for the system in Example 11.25.

PROBLEMS

11.1. Consider the discrete-time signal $x[n]$, where

$$x[n] = \begin{cases} b^n, & \text{for } n = 0, 1, 2, \ldots, N-1 \\ 0, & \text{for all other } N \end{cases}$$

Here b is an arbitrary real number and N is a positive integer.

(a) For what real values of b does $x[n]$ possess a z-transform?

(b) For the values of b determined in part (a), find the z-transform of $x[n]$. Express your answer as a ratio of polynomials.

11.2. A discrete-time signal $x[n]$ has z-transform

$$X(z) = \frac{z}{8z^2 - 2z - 1}$$

Determine the z-transform $V(z)$ of the following signals.

(a) $v[n] = x[n-4]u[n-4]$

(b) $v[n] = x[n+2]u[n+2]$

(c) $v[n] = \cos(2n)x[n]$

(d) $v[n] = e^{3n}x[n]$

(e) $v[n] = n^2x[n]$

(f) $v[n] = x[n]*x[n]$

(g) $v[n] = x[0] + x[1] + x[2] + \cdots + x[n]$

11.3. Compute the z-transform of the following discrete-time signals. Express your answer as a ratio of polynomials in z whenever possible.

(a) $x[n] = \delta[n] + 2\delta[n-1]$

(b) $x[n] = 1$ for $n = 0,1$ and $x[n] = 2$ for all $n \geq 2$ (i.e., $n = 2, 3, \ldots$)

(c) $x[n] = e^{0.5n}u[n] + u[n-2]$

(d) $x[n] = e^{0.5n}$ for $n = 0,1$, and $x[n] = 1$ for all $n \geq 2$

(e) $x[n] = \sin(\pi n/2)u[n-2]$

(f) $x[n] = (0.5)^n nu[n]$

(g) $x[n] = u[n] - nu[n-1] + (1/3)^n u[n-2]$

(h) $x[n] = n$ for $n = 0,1,2$ and $x[n] = -n$ for all $n \geq 3$

(i) $x[n] = (n-1)u[n] - nu[n-3]$

(j) $x[n] = (0.25)^{-n}u[n-2]$

11.4. Using the transform pairs in Table 11.3 and the properties of the z-transform in Table 11.2, determine the z-transform of the following discrete-time signals.

(a) $x[n] = (\cos^2 \omega n)u[n]$

(b) $x[n] = (\sin^2 \omega n)u[n]$

(c) $x[n] = n(\cos \omega n)u[n]$

(d) $x[n] = n(\sin \omega n)u[n]$

(e) $x[n] = ne^{-bn}(\cos \omega n)u[n]$

(f) $x[n] = ne^{-bn}(\sin \omega n)u[n]$

(g) $x[n] = e^{-bn}(\cos^2 \omega n)u[n]$

(h) $x[n] = e^{-bn}(\sin^2 \omega n)u[n]$

11.5. Let $x[n]$ be a discrete-time signal with $x[n] = 0$ for $n = -1, -2, \ldots$. The signal $x[n]$ is said to be summable if

$$\sum_{n=0}^{\infty} x[n] < \infty$$

If $x[n]$ is summable, the sum x_{sum} of $x[n]$ is defined by

$$x_{\text{sum}} = \sum_{n=0}^{\infty} x[n]$$

Now suppose that the z-transform $X(z)$ of $x[n]$ can be expressed in the form

$$X(z) = \frac{B(z)}{a_N(z-p_1)(z-p_2)\cdots(z-p_N)}$$

where $B(z)$ is a polynomial in z. By using the final-value theorem, show that if $|p_i| < 1$ for $i = 1, 2, \ldots, N$, $x[n]$ is summable and

$$x_{\text{sum}} = \lim_{z \to 1} X(z)$$

11.6. Using the results of Problem 11.5, compute x_{sum} for the following signals. In each case, assume that $x[n] = 0$ for all $n < 0$.

(a) $x[n] = a^n$, $|a| < 1$

(b) $x[n] = n(a^n)$, $|a| < 1$

(c) $x[n] = a^n \cos \pi n$, $|a| < 1$

(d) $x[n] = a^n \sin(\pi n/2)$, $|a| < 1$

11.7. Let p and c be complex numbers defined in polar coordinates as $p = \sigma e^{j\Omega}$ and $c = |c|e^{j\angle c}$. Prove the following relationship.

$$cp^n + \bar{c}\,\bar{p}^n = 2\,|c|\sigma^n \cos(\Omega n + \angle c)$$

11.8. A discrete-time signal $x[n]$ has z-transform

$$X(z) = \frac{z+1}{z(z-1)}$$

Compute $x[0], x[1]$, and $x[10{,}000]$.

11.9. Compute the inverse z-transform $x[n]$ of the following transforms. Determine $x[n]$ for all integers $n \geq 0$.

(a) $X(z) = \dfrac{z}{z^2+1}$

(b) $X(z) = \dfrac{z^2}{z^2+1}$

(c) $X(z) = \dfrac{1}{z^2+1} + \dfrac{1}{z^2-1}$

(d) $X(z) = \dfrac{z^2}{z^2+1} + \dfrac{z}{z^2-1}$

(e) $X(z) = \dfrac{z^2-1}{z^2+1}$

(f) $X(z) = \dfrac{z+2}{(z-1)(z^2+1)}$

(g) $X(z) = \dfrac{z^2+2}{(z-1)(z^2+1)}$

(h) $X(z) = \ln\left(\dfrac{2z-1}{2z}\right)$

11.10. For the transforms given in Problem 11.9 (a) to (g), compute the inverse z-transform numerically using `dimpulse`. Compare these results with the answers obtained analytically for $n = 0$ to $n = 5$.

11.11. Find the inverse z-transform $x[n]$ of the following transforms. Determine $x[n]$ for all n.

(a) $X(z) = \dfrac{z+0.3}{z^2+0.75z+0.125}$

(b) $X(z) = \dfrac{5z+1}{4z^2+4z+1}$

(c) $X(z) = \dfrac{4z+1}{z^2-z+0.5}$

(d) $X(z) = \dfrac{z}{16z^2+1}$

(e) $X(z) = \dfrac{2z+1}{z(10z^2-z-2)}$

(f) $X(z) = \dfrac{z+1}{(z-0.5)(z^2-0.5z+0.25)}$

(g) $X(z) = \dfrac{z^3+1}{(z-0.5)(z^2-0.5z+0.25)}$

(h) $X(z) = \dfrac{z+1}{z(z-0.5)(z^2-0.5z+0.25)}$

11.12. For each of the transforms given in Problem 11.11, compute the inverse z-transform numerically using `dimpulse`. Compare these results with the answers obtained analytically.

11.13. By using the z-transform, compute the convolution $x[n]*v[n]$ for all $n \geq 0$, where

(a) $x[n] = u[n] + 3\delta(n-1], v[n] = u[n-2]$

(b) $x[n] = u[n], v[n] = nu[n]$

(c) $x[n] = \sin(\pi n/2)u[n], v[n] = e^{-n}u[n-2]$

(d) $x[n] = u[n-1] + \delta[n], v[n] = e^{-n}u[n] - 2e^{-2n}u[n-2]$

11.14. A linear time-invariant discrete-time system has unit-pulse response

$$h[n] = \begin{cases} \dfrac{1}{n}, & \text{for } n = 1, 2, 3 \\ n - 2, & \text{for } n = 4, 5 \\ 0, & \text{for all other } n \end{cases}$$

Compute the transfer function $H(z)$.

11.15. The input $x[n] = (-1)^n u[n]$ is applied to a linear time-invariant discrete-time system. The resulting output response $y[n]$ with no initial energy in the system is given by

$$y[n] = \begin{cases} 0, & \text{for } n < 0 \\ n + 1, & \text{for } n = 0, 1, 2, 3 \\ 0, & \text{for } n \geq 4 \end{cases}$$

Determine the transfer function $H(z)$ of the system.

11.16. For the system defined in Problem 11.15, compute the output response $y[n]$ resulting from the input $x[n] = (1/n)(u[n-1] - u[n-3])$ with no initial energy in the system.

11.17. A system is described by the difference equation

$$y[n] + 0.7y[n-1] = u[n]; \qquad y[-1] = 1$$

(a) Find an analytical expression for $y[n]$.

(b) Verify your result by simulating the system using MATLAB.

11.18. Repeat Problem 11.17 for the system described by the following difference equation

$$y[n] - 0.2y[n-1] - 0.8y[n-2] = 0; \qquad y[-1] = 1, y[-2] = 1$$

11.19. A linear time-invariant discrete-time system is described by the input/output difference equation

$$y[n+2] + y[n] = 2x[n+1] - x[n]$$

(a) Compute the unit-pulse response $h[n]$.

(b) Compute the step response $y[n]$.

(c) Compute $y[n]$ for all $n \geq 0$ when $x[n] = 2^n u[n]$ with $y[-1] = 3$ and $y[-2] = 2$.

(d) An input $x[n]$ with $x[-2] = x[-1] = 0$ produces the output response $y[n] = (\sin \pi n)u[n]$ with no initial energy at time $n = 0$. Determine $x[n]$.

(e) An input $x[n]$ with $x[-2] = x[-1] = 0$ produces the output response $y[n] = \delta[n-1]$. Compute $x[n]$.

(f) Verify the results of parts (a) to (e) via computer simulation.

11.20. A linear time-invariant discrete-time system is given by the input/output difference equation

$$y[n] + y[n-1] - 2y[n-2] = 2x[n] - x[n-1]$$

Find an input $x[n]$ with $x[n] = 0$ for $n < 0$ that gives the output response $y[n] = 2(u[n] - u[n-3])$ with initial conditions $y[-2] = 2, y[-1] = 0$.

11.21. The input $x[n] = u[n] - 2u[n-2] + u[n-4]$ is applied to a linear time-invariant discrete-time system. The resulting response with no initial energy is $y[n] = nu[n] - nu[n-4]$. Compute the transfer function $H(z)$.

11.22. A system has the transfer function

$$H(z) = \frac{-0.4z^{-1} - 0.5z^{-2}}{(1 - 0.5z^{-1})(1 - 0.8z^{-1})}$$

(a) Compute an analytical expression for the step response.
(b) Verify your result by simulating the step response using MATLAB.

11.23. Repeat Problem 11.22 for the system with the transfer function

$$H(z) = \frac{z^2 - 0.1}{z^2 - 0.6484z + 0.36}$$

11.24. A linear time-invariant discrete-time system has transfer function

$$H(z) = \frac{z^2 - z - 2}{z^2 + 1.5z - 1}$$

(a) Compute the unit-pulse response $h[n]$ for all $n \geq 0$.
(b) Compute the step response $y[n]$ for all $n \geq 0$.
(c) Compute the output values $y[0]$, $y[1]$, $y[2]$ resulting from the input $x[n] = 2^n \sin(\pi n/4) + \tan(\pi n/3), n = 0, 1, 2, \ldots$, with the system at rest at time $n = 0$.
(d) If possible, find an input $x[n]$ with $x[n] = 0$ for all $n < 0$ and such that the output response $y[n]$ resulting from $x[n]$ is given by $y[0] = 2, y[1] = -3, y[n] = 0$ for all $n \geq 2$. Assume that the system is at rest at $n = 0$.
(e) Verify the results of parts (a) to (d) by computer simulation.

11.25. A linear time-invariant discrete-time system has transfer function

$$H(z) = \frac{z}{(z - 0.5)^2(z^2 + 0.25)}$$

(a) Find the unit-pulse response $h[n]$ for all $n \geq 0$.
(b) Simulate the unit-pulse response using MATLAB and compare this result to the result for $h[n]$ obtained analytically in part (a).

11.26. Consider a car on a level surface given by the input/output differential equation

$$\frac{d^2y(t)}{dt^2} + \frac{k_f}{M}\frac{dy(t)}{dt} = \frac{1}{M}x(t)$$

Recall that $y(t)$ is the position of the car at time t and $x(t)$ is the drive or braking force applied to the car. By using the Euler approximation of the derivatives as shown in equations (2.62) to (2.64), obtain the discrete-time simulation of the car. Determine the transfer function $H_d(z)$ of the discrete-time simulation. Here d stands for "discretized."

11.27. The input $x[n] = (0.5)^n u[n]$ is applied to a linear time-invariant discrete-time system with the initial conditions $y[-1] = 8$ and $y[-2] = 4$. The resulting output response is

$$y[n] = 4(0.5)^n u[n] - n(0.5)^n u[n] - (-0.5)^n u[n]$$

Find the transfer function $H(z)$.

11.28. A linear time-invariant discrete-time system has transfer function

$$H(z) = \frac{3z}{(z + 0.5)(z - 0.5)}$$

The output response resulting from the input $x[n] = u[n]$ and initial conditions $y[-1]$ and $y[-2]$ is

$$y[n] = [(0.5)^n - 3(-0.5)^n + 4]u[n]$$

Determine the initial conditions $y[-1]$, $y[-2]$, and the part of the output response due to the initial conditions.

11.29. A linear time-invariant discrete-time system has unit-pulse response $h[n]$ equal to the Fibonacci sequence; that is, $h[0] = 0$, $h[1] = 1$, and $h[n] = h[n - 2] + h[n - 1]$ for $n \geq 2$. Show that the system's transfer function $H(z)$ is rational in z. Express $H(z)$ as a ratio of polynomials in positive powers of z.

11.30. For each of the transfer functions

(i) $H(z) = z/(z - 0.5)$
(ii) $H(z) = z/(z + 0.5)$
(iii) $H(z) = z/(z - 1)$
(iv) $H(z) = z/(z + 1)$
(v) $H(z) = z/(z - 2)$
(vi) $H(z) = z/(z + 2)$

(a) Compute the pole. From this pole position, describe the type of behavior that you would expect in the transient response.
(b) Verify your prediction in part (a) by determining an analytical expression for the unit-pulse response.
(c) Simulate the unit-pulse response and compare to the answer obtained analytically in part (b).

11.31. For the transfer functions

(i) $H(z) = (z^2 - 0.75z)/(z^2 - 1.5z + 2.25)$
(ii) $H(z) = (z^2 - 0.5z)/(z^2 - z + 1)$
(iii) $H(z) = (z^2 - 0.25z)/(z^2 - 0.5z + 0.25)$

(a) Compute the pole positions. From knowledge of the pole positions, describe the type of behavior you would expect in the transient response.
(b) Without computing the actual response, give a general expression for the step response.
(c) Verify your prediction by simulating the system for a step input.

11.32. By using the z-domain representation, determine the transfer functions of the discrete-time systems shown in Figure P11.32.

11.33. Use block diagram reduction to determine the transfer function of the discrete-time system shown in Figure P11.33.

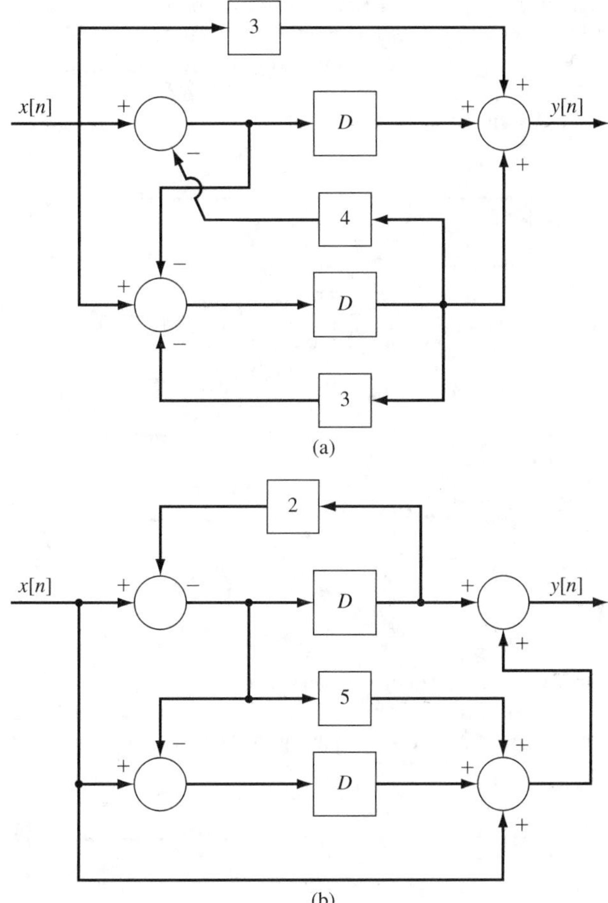

Figure P11.32

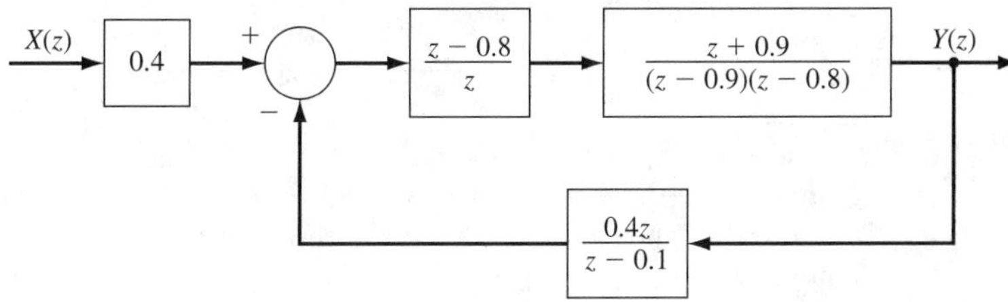

Figure P11.33

11.34. Consider the discrete-time system shown in Figure P11.34.

(a) Determine the transfer function $H(z)$ of the system.

(b) Determine the system's input/output difference equation.

(c) Compute the output response $y[n]$ when $x[n] = 4u[n]$ with no initial energy in the system.

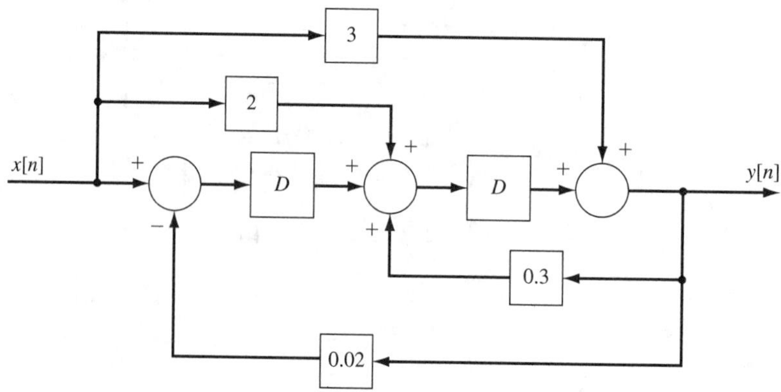

Figure P11.34

11.35. Consider the cascade connection shown in Figure P11.35. Determine the unit-pulse response $h_2[n]$ of the system with transfer function $H_2(z)$ so that when $x[n] = \delta[n]$ with no initial energy, the response $y[n]$ is equal to $\delta[n]$.

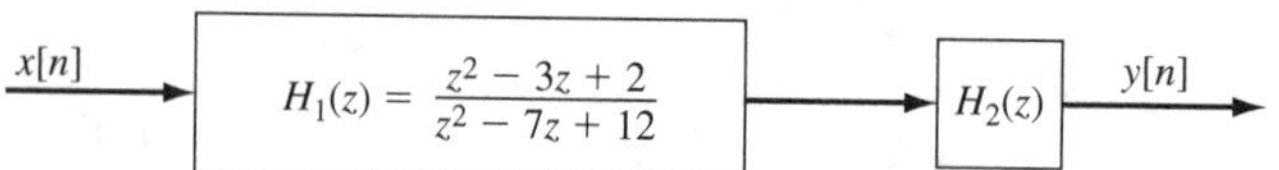

Figure P11.35

11.36. A linear time-invariant discrete-time system is given by the feedback connection shown in Figure P11.36. In Figure P11.36, $X(z)$ is the z-transform of the system's input $x[n]$, $Y(z)$ is the z-transform of the system's output $y[n]$, and $H_1(z)$, $H_2(z)$ are the transfer functions of the subsystems given by

$$H_1(z) = \frac{z}{z+1}, \qquad H_2(z) = \frac{9}{z-8}$$

(a) Determine the unit-pulse response of the overall system.

(b) Compute the step response of the overall system.

(c) Compute $y[n]$ when $x[n] = (0.5)^n u[n]$ with $y[-1] = -3, y[-2] = 4$.

(d) Compute $y[n]$ when $x[n] = (0.5)^n u[n]$ with $y[-2] = 1, w[-1] = 2$, where $w[n]$ is the output of the feedback system in Figure P11.36.

(e) Verify the results of parts (a) to (d) via computer simulation.

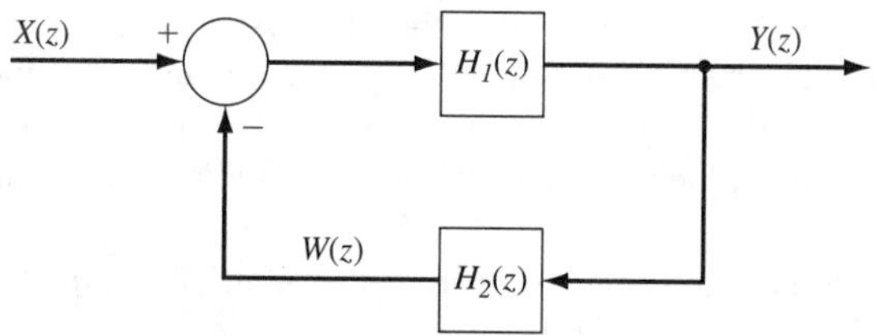

Figure P11.36

11.37. A model for the generation of echoes is shown in Figure P11.37. As shown, each successive echo is represented by a delayed and scaled version of the output, which is fed back to the input.

(a) Determine the transfer function $H(z)$ of the echo system.

(b) Suppose that we would like to recover the original signal $x[n]$ from the output $y[n]$ by using a system with transfer function $W(z)$ [and with input $y[n]$ and output $x[n]$]. Determine $W(z)$.

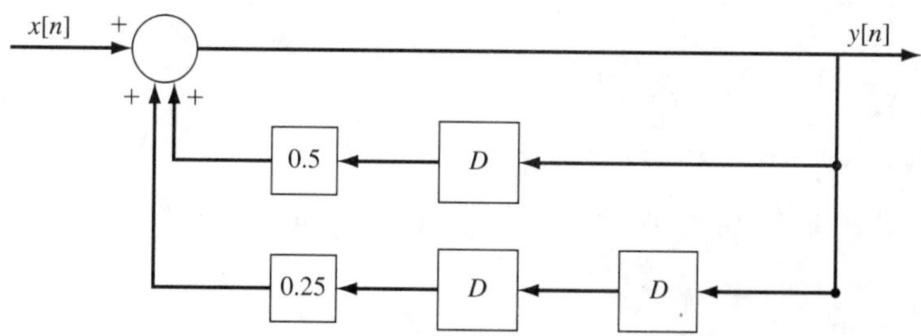

Figure P11.37

11.38. A linear time-invariant discrete-time system is given by the cascade connection shown in Figure P11.38.

(a) Compute the unit-pulse response of the overall system.

(b) Compute the input/output difference equation of the overall system.

(c) Compute the step response of the overall system.

(d) Compute $y[n]$ when $x[n] = u[n]$ with $y[-1] = 3, q[-1] = 2$.

(e) Compute $y[n]$ when $x[n] = (0.5)^n u[n]$ with $y[-2] = 2, q[-2] = 3$.

(f) Verify the results of parts (a) and (c) to (e) via computer simulation.

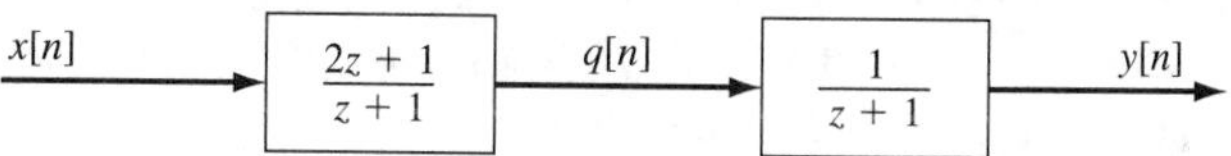

Figure P11.38

11.39. A linear time-invariant discrete-time system is excited by the input $x[n] = \delta[n] + 2u[n-1]$. The resulting output response with no initial energy is $y[n] = (0.5)^n u[n]$. Determine if the system is stable, marginally stable, or unstable. Justify your answer.

11.40. Determine if the system in Problem 11.34 is stable, marginally stable, or unstable. Justify your answer.

11.41. A discrete-time system is given by the input/output difference equation

$$y[n+2] - y[n+1] + y[n] = x[n+2] - x[n+1]$$

Is the system stable, marginally stable, or unstable? Justify your answer.

11.42. For the linear time-invariant discrete-time systems with unit-pulse response $h[n]$ given below, determine if the system is BIBO stable.

(a) $h[n] = \sin(\pi n/6)(u[n] - u[n-10])$
(b) $h[n] = (1/n)u[n-1]$
(c) $h[n] = (1/n^2)u[n-1]$
(d) $h[n] = e^{-n}\sin(\pi n/6)u[n]$

11.43. Using the Jury test, determine whether or not the following linear time-invariant discrete-time systems are stable.

(a) $H(z) = \dfrac{z-4}{z^2 + 1.5z + 0.5}$

(b) $H(z) = \dfrac{z^2 - 3z + 1}{z^3 + z^2 - 0.5z + 0.5}$

(c) $H(z) = \dfrac{1}{z^3 + 0.5z + 0.1}$

11.44. A linear time-invariant discrete-time system has transfer function

$$H(z) = \frac{z}{z + 0.5}$$

(a) Find the transient response and steady-state response resulting from the input $x[n] = 5\cos 3n$, $n = 0, 1, 2, \ldots$, with no initial energy in the system at time $n = 0$.
(b) Sketch the frequency response curves.
(c) Plot the actual frequency response curves using MATLAB.

11.45. A linear time-invariant discrete-time system has transfer function

$$H(z) = \frac{-az + 1}{z - a}$$

where $-1 < a < 1$.

(a) Compute the transient and steady-state responses when the input $x[n] = \cos(\pi n/2)u[n]$ with no initial energy in the system.
(b) When $a = 0.5$, sketch the frequency response curves.
(c) Show that $|H(\Omega)| = C$ for $0 \le \Omega \le 2\pi$, where C is a constant. Derive an expression for C in terms of a. Since $|H(\Omega)|$ is constant, the system is called an *allpass filter.*

11.46. Consider the discrete-time system shown in Figure P11.46. Compute the steady-state output response $y_{ss}[n]$ and the transient output response $y_{tr}[n]$ when $y[-1] = x[-1] = 0$ and $x[n] = 2\cos(\pi n/2)u[n]$.

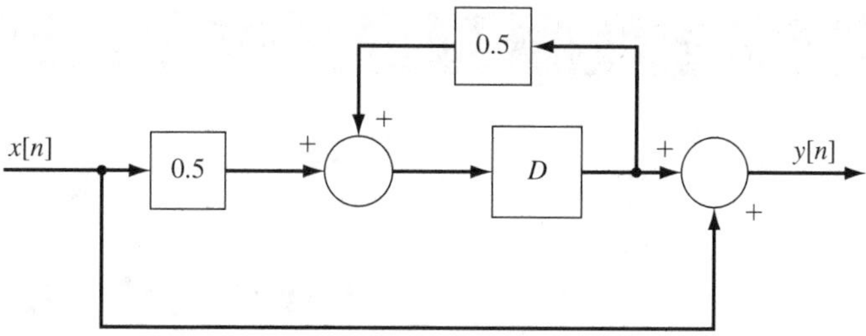

Figure P11.46

11.47. A discrete-time system is given by the input/output difference equation

$$y[n + 2] + 0.3y[n + 1] + 0.02y[n] = x[n + 1] + 3x[n]$$

(a) Compute the transient and steady-state output responses when $x[n] = (\cos \pi n)u[n]$ with no initial energy.

(b) Use MATLAB to plot the frequency response curves of the system.

11.48. A linear time-invariant discrete-time system has transfer function

$$H(z) = \frac{z}{(z^2 + 0.25)(z - 0.5)^2}$$

(a) Assuming that the system is at rest at time $n = 0$, find the transient and steady-state response to $x[n] = 12\cos(\pi n/2)u[n]$.

(b) Use MATLAB to plot the frequency response curves of the system.

11.49. Consider the discrete-time system given by the input/output difference equation

$$y[n] = \frac{x[n] + x[n - 1]}{2}$$

This system is called an *averager.* Sketch the frequency response curves of the averager.

11.50. The *differencer* is the discrete-time system with the input/output difference equation

$$y[n] = x[n] - x[n - 1]$$

Sketch the frequency response curves of the differencer.

◆

CHAPTER 12

Design of Digital Filters and Controllers

In this chapter the continuous- and discrete-time techniques developed in Chapters 9 to 11 are utilized to design digital filters and digital controllers. The development begins in Section 12.1 with the study of the discretization of continuous-time signals and systems. Here a Fourier analysis is given for analog signal discretization, which is then utilized to generate a frequency-domain condition for the discretization of continuous-time systems. In Section 12.2 the design of digital filters is pursued in terms of the discretization of analog prototype filters. As with analog filters discussed in Chapter 9, digital filters can be lowpass, highpass, bandpass, and bandstop, or can take on an arbitrary frequency response function characteristic. Digital filters can have an infinite impulse (i.e. unit-pulse) response or the impulse response may decay to zero in a finite number of steps. Correspondingly, filters are classified as infinite impulse response (IIR) or finite impulse response (FIR). In Section 12.2 the design of IIR filters is developed in terms of analog prototypes that are then mapped into digital filters using the bilinear transformation. The use of MATLAB to carry out this design process is considered in Section 12.3. Then in Section 12.4 the design of FIR filters is developed by truncating or windowing the impulse response of an IIR filter.

The mapping concept discussed in Section 12.2 for transforming analog filters to digital filters can also be used to map continuous-time controllers to discrete-time (digital) controllers. This is discussed in Section 12.5 along with a brief development of the response matching technique. Here part of the emphasis is on step response matching, which is commonly used in digital control. The mapping of an analog controller into a digital controller is illustrated in the dc motor application considered in Chapter 10.

12.1 DISCRETIZATION

Let $x(t)$ be a continuous-time signal that is to be sampled, and let $X(\omega)$ denote the Fourier transform of $x(t)$. As discussed in Section 5.3, the plots of the magnitude $|X(\omega)|$ and the angle $\angle X(\omega)$ versus ω display the amplitude and phase spectra of the signal $x(t)$. Now with the sampling interval equal to T, as discussed in Section 5.5 the sampled signal $x_s(t)$ can be represented by multiplying the signal $x(t)$ by the impulse train $p(t)$, that is $x_s(t) = x(t)p(t)$, where

$$p(t) = \sum_{n=-\infty}^{\infty} \delta(t - nT)$$

The sampling operation is illustrated in Figure 12.1a. Recall that the sampling frequency ω_s is equal to $2\pi/T$.

In Section 5.5, it was shown that the Fourier transform $X_s(\omega)$ of the sampled signal $x_s(t)$ is given by

$$X_s(\omega) = \sum_{n=-\infty}^{\infty} \frac{1}{T} X(\omega - n\omega_s) \tag{12.1}$$

Note that $X_s(\omega)$ consists of scaled replicas of $X(\omega)$ shifted in frequency by multiples of the sampling frequency ω_s. Also note that $X_s(\omega)$ is a periodic function of ω with period equal to ω_s.

Physically, sampling is accomplished through the use of an analog-to-digital (A/D) converter which first samples the signal to obtain $x_s(t)$ and then converts it into a string of pulses with amplitude 0 or 1. The process of sampling an analog signal $x(t)$ and then converting the sampled signal $x_s(t)$ into a binary-amplitude signal is also called *pulse-code modulation.* The binary-amplitude signal is constructed by

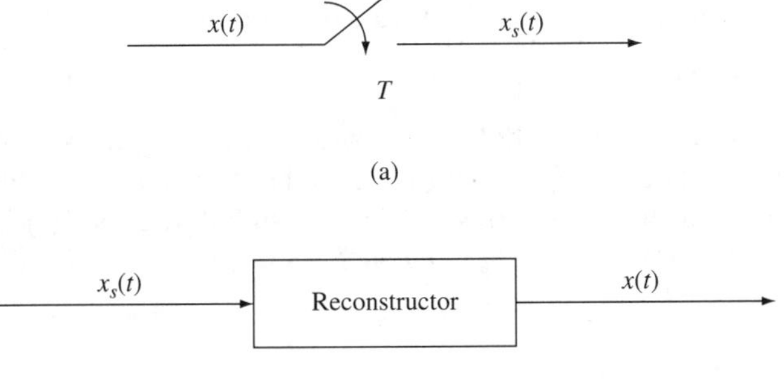

Figure 12.1 Signal sampler (a) and signal reconstructor (b).

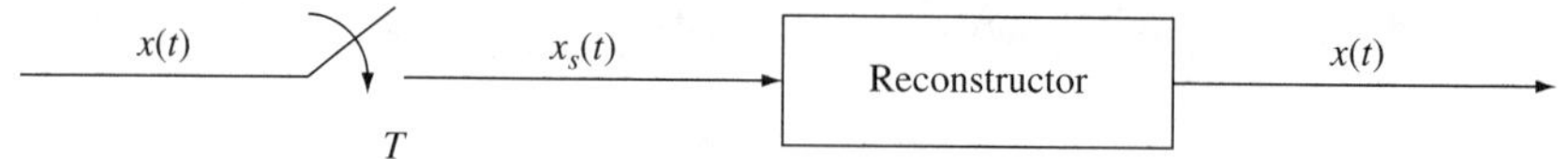

Figure 12.2 Cascade of a signal sampler and reconstructor.

quantizing and then encoding the sampled signal $x_s(t)$. A detailed description of the process is left to a more advanced treatment of sampling.

The continuous-time signal $x(t)$ can be regenerated from the sampled signal $x_s(t) = x(t)p(t)$ by using a *signal reconstructor*, illustrated in Figure 12.1b. As shown in the figure, the output of the signal reconstructor is exactly equal to $x(t)$, in which case the signal reconstructor is said to be "ideal." Note that if an (ideal) signal reconstructor is put in cascade with a sampler, the result is the analog signal $x(t)$. More precisely, for the cascade connection shown in Figure 12.2, if the input is $x(t)$, the output is $x(t)$. This shows that signal reconstruction is the inverse of signal sampling, and conversely.

It was shown in Section 6.4 that if the signal $x(t)$ is bandlimited with bandwidth B, that is, $|X(\omega)| = 0$ for $\omega > B$, and if the sampling frequency ω_s is greater than or equal to $2B$, then $x(t)$ can be exactly recovered from the sampled signal $x_s(t)$ by applying $x_s(t)$ to an ideal lowpass filter with bandwidth B. In most practical situations, signals are not perfectly bandlimited, and as a result aliasing may occur. However, as discussed in Section 6.4 the distortion due to aliasing can be reduced significantly if the signal is lowpass filtered before it is sampled. In particular, if the frequency content of the signal is not very small for frequencies larger than half of the sampling frequency ω_s, the effect of aliasing can be reduced by first lowpass filtering with a filter, often called an *anti-aliasing filter,* having a bandwidth less than or equal to $\omega_s/2$.

Hold Operation

Instead of lowpass filtering, there are other methods for reconstructing a continuous-time signal $x(t)$ from the sampled signal $x_s(t)$. One of these is the hold operation illustrated in Figure 12.3. The output $\tilde{x}(t)$ of the hold device is given by

$$\tilde{x}(t) = x(nT), \qquad nT \le t < nT + T \tag{12.2}$$

From (12.2) it is seen that the hold operation "holds" the value of the sampled signal at time nT until it receives the next value of the sampled signal at time $nT + T$. The output $\tilde{x}(t)$ of the hold device is a piecewise-constant analog signal; that is, $\tilde{x}(t)$ is constant over each T-second interval $nT \le t < nT + T$. Since the amplitude

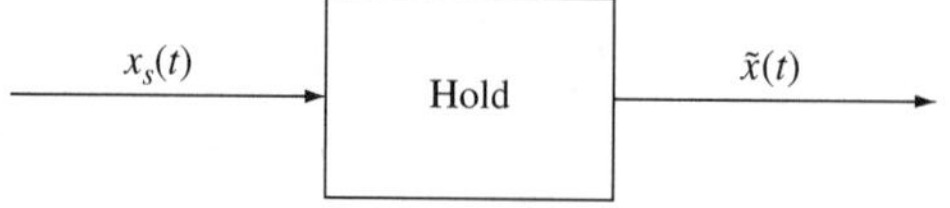

Figure 12.3 Hold operation.

of $\tilde{x}(t)$ is constant over each T-second interval, the device is sometimes called a *zero-order hold.*

It turns out that the hold device corresponds to a type of lowpass filter. To see this, the frequency response function of the hold device will be computed. First, the Fourier transform $X_s(\omega)$ of the sampled signal $x_s(t) = x(t)p(t)$ will be expressed in a form different from that given in (12.1): Using the definition of the Fourier transform gives

$$
\begin{aligned}
X_s(\omega) &= \int_{-\infty}^{\infty} x_s(t)e^{-j\omega t}\,dt \\
&= \int_{-\infty}^{\infty} \sum_{n=-\infty}^{\infty} x(t)\,\delta(t-nT)e^{-j\omega t}\,dt \\
&= \sum_{n=-\infty}^{\infty} \int_{-\infty}^{\infty} x(t)\,\delta(t-nT)e^{-j\omega t}\,dt \\
&= \sum_{n=-\infty}^{\infty} x(nT)e^{-j\omega nT}
\end{aligned}
\tag{12.3}
$$

Now the Fourier transform $\tilde{X}(\omega)$ of the output $\tilde{x}(t)$ of the hold device is given by

$$
\begin{aligned}
\tilde{X}(\omega) &= \int_{-\infty}^{\infty} \tilde{x}(t)e^{-j\omega t}\,dt \\
&= \sum_{n=-\infty}^{\infty} \int_{nT}^{nT+T} x(nT)e^{-j\omega t}\,dt \\
&= \sum_{n=-\infty}^{\infty} \left[\int_{nT}^{nT+T} e^{-j\omega t}\,dt\right] x(nT) \\
&= \sum_{n=-\infty}^{\infty} \left[-\frac{1}{j\omega}e^{-j\omega t}\right]_{t=nT}^{t=nT+T} x(nT) \\
&= \frac{1-e^{-j\omega T}}{j\omega} \sum_{n=-\infty}^{\infty} e^{-j\omega nT}x(nT)
\end{aligned}
\tag{12.4}
$$

Then using (12.3) in (12.4) yields

$$
\tilde{X}(\omega) = \frac{1-e^{-j\omega T}}{j\omega} X_s(\omega) \tag{12.5}
$$

From (12.5) it is seen that the frequency function $H_{\text{hd}}(\omega)$ of the hold device is given by

$$
H_{\text{hd}}(\omega) = \frac{1-e^{-j\omega T}}{j\omega}
$$

A plot of the magnitude function $|H_{\text{hd}}(\omega)|$ is given in Figure 12.4 for the case when $T = 0.2$. As seen from the plot, the frequency response function of the hold device

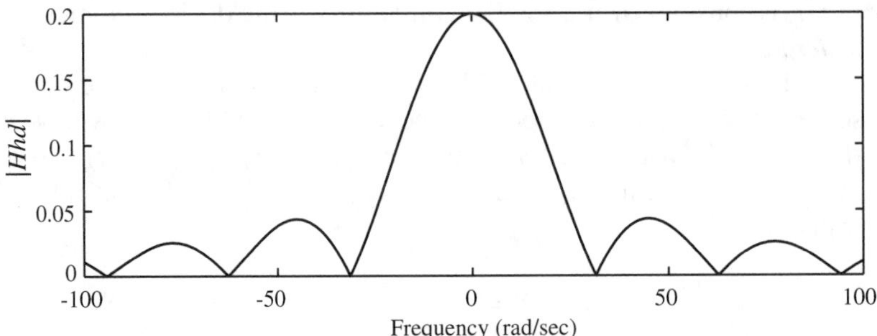

Figure 12.4 Magnitude of the frequency response function of the hold device with $T = 0.2$.

does correspond to that of a lowpass filter. The hold device is often used in digital control, which is considered briefly in Section 12.5.

System Discretization

Now consider a linear time-invariant continuous-time system with transfer function $H(s)$. As illustrated in Figure 12.5a, $y(t)$ is the output response of the system resulting from input $x(t)$ with no initial energy in the system prior to the application of the input. Now given $T > 0$, let $x[n]$ and $y[n]$ denote the discrete-time signals formed from the values of $x(t)$ and $y(t)$ at the times $t = nT$; that is,

$$x[n] = x(t)\big|_{t=nT} \quad \text{and} \quad y[n] = y(t)\big|_{t=nT}$$

A discretization of the given continuous-time system with transfer function $H(s)$ is a linear time-invariant discrete-time system with input $u[n]$, output $\hat{y}[n]$, and transfer

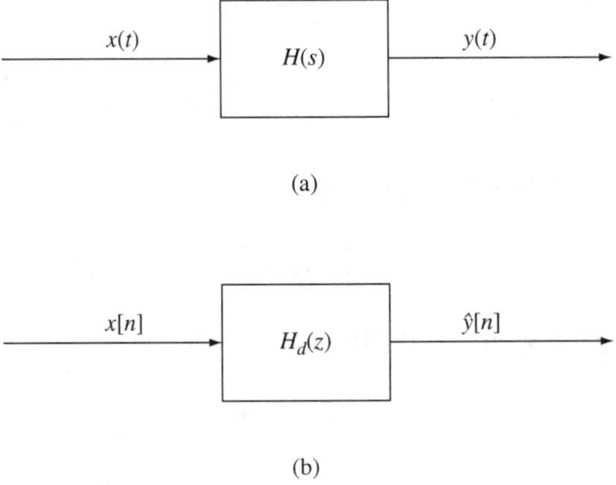

Figure 12.5 Given continuous-time system (a) and discretization (b).

function $H_d(z)$, where the subscript "d" stands for "discrete." The discrete-time system is illustrated in Figure 12.5b. To be a discretization, the behavior of the discrete-time system must correspond to that of the continuous-time system at the sample times $t = nT$. One way to specify this correspondence is to require that the output $\hat{y}[n]$ of the discrete-time system satisfy the condition

$$\hat{y}[n] = y[n] = y(nT), \qquad n = 0, \pm 1, \pm 2, \ldots \tag{12.6}$$

In other words, the output of the discretization should be exactly equal to (i.e., matches) the output values of the continuous-time system at the sample times $t = nT$. As shown below, this matching condition can be expressed in the frequency domain.

Let $\hat{Y}(\Omega)$ denote the discrete-time Fourier transform (DTFT) of $\hat{y}[n]$ given by

$$\hat{Y}(\Omega) = \sum_{n=-\infty}^{\infty} \hat{y}[n]e^{-j\Omega n}$$

Then for (12.6) to be satisfied, it must be true that

$$\hat{Y}(\Omega) = \sum_{n=-\infty}^{\infty} y(nT)e^{-j\Omega n} \tag{12.7}$$

Now let $Y_s(\omega)$ denote the Fourier transform of the sampled output signal $y_s(t) = y(t)p(t)$. Then replacing x by y in (12.3) yields

$$Y_s(\omega) = \sum_{n=-\infty}^{\infty} y(nT)e^{-j\omega nT} \tag{12.8}$$

and setting $\omega = \Omega/T$ in (12.8) gives

$$Y_s(\omega)\big|_{\omega=\Omega/T} = \sum_{n=-\infty}^{\infty} y(nT)e^{-j\Omega n} \tag{12.9}$$

But the right-hand sides of (12.7) and (12.9) are identical, and thus for (12.6) to be satisfied it must be true that

$$\hat{Y}(\Omega) = Y_s(\omega)\big|_{\omega=\Omega/T} = Y_s\left(\frac{\Omega}{T}\right) \tag{12.10}$$

Since $\hat{Y}(\Omega)$ is periodic in Ω with period 2π, it is only necessary to consider (12.10) for $-\pi \le \Omega \le \pi$, that is,

$$\hat{Y}(\Omega) = Y_s(\omega)\big|_{\omega=\Omega/T} = Y_s\left(\frac{\Omega}{T}\right), \qquad -\pi \le \Omega \le \pi \tag{12.11}$$

From (12.11) it is seen that the DTFT $\hat{Y}(\Omega)$ of the discrete-time system output $\hat{y}[n]$ must equal the Fourier transform $Y_s(\omega)$ of the sampled output $y_s(t)$ with $Y_s(\omega)$ evaluated at $\omega = \Omega/T$. This is the key "matching condition" in the frequency domain. Unfortunately, in general it is not possible to satisfy (12.11) exactly; however, (12.11) can be satisfied approximately, and there is more than one way to carry this out, depending on the application under study. This is pursued in the next two sections in the application to digital filtering and then in Section 12.5 in the application to digital control.

12.2 DESIGN OF IIR FILTERS

If a digital filter is to be used to filter continuous-time signals, the design specifications are usually given in terms of the continuous-time frequency spectrum, for example bandwidth and passband ripple. Design methods are well established to meet these specifications using prototype lowpass, highpass,bandpass or bandstop analog filters. Two examples of these are the Butterworth and Chebyshev filters discussed in Section 9.6.

It is a reasonable strategy, therefore, first to design an analog prototype filter with frequency response function $H(\omega)$, and then to select a digital filter that best approximates the behavior of the desired analog filter. To obtain a good approximation, the frequency response function $H_d(\Omega)$ of the digital filter should be designed so that the matching condition (12.11) is satisfied. As shown below, this results in a transformation that maps the analog prototype filter given by $H(\omega)$ to a digital filter given by $H_d(\Omega)$. The resulting digital filter will have an infinite impulse response (IIR), and thus this design approach yields an IIR filter. There are other methods for designing IIR filters; however, only the analog-to-digital transformation approach is considered here. See Oppenheim and Schafer [1989] for a discussion of other design methods, especially numerical methods.

To design $H_d(\Omega)$, first note that in the ω domain

$$Y(\omega) = H(\omega)X(\omega) \tag{12.12}$$

where $H(\omega)$ is the desired frequency response function (the analog prototype). Throughout this section it is assumed that the analog filter is stable so that $H(\omega)$ is equal to the transfer function $H(s)$ with $s = j\omega$. It is also assumed that

$$|H(\omega)| \approx 0 \qquad \text{for } \omega > \omega_s/2 \tag{12.13}$$

and

$$|X(\omega)| \approx 0 \qquad \text{for } \omega > \omega_s/2 \tag{12.14}$$

where again $\omega_s = 2\pi/T$ is the sampling frequency.

With the conditions (12.13) and (12.14), it follows that

$$Y_s(\omega) = H(\omega)X_s(\omega) \qquad \text{for } -\omega_s/2 \le \omega \le \omega_s/2 \tag{12.15}$$

where $Y_s(\omega)$ is the Fourier transform of the sampled output $y_s(t)$ and $X_s(\omega)$ is the Fourier transform of the sampled input $x_s(t)$. Actually, (12.15) holds only approximately since $|H(\omega)|$ and $|X(\omega)|$ are only approximately zero for $\omega > \omega_s/2$.

Combining the frequency matching condition (12.11) and (12.15) reveals that the DTFT $\hat{Y}(\Omega)$ of the discrete-time system output $\hat{y}[n]$ is given by

$$\begin{aligned} \hat{Y}(\Omega) &= [H(\omega)X_s(\omega)]_{\omega=\Omega/T}, \qquad -\pi \le \Omega \le \pi \\ &= H\left(\frac{\Omega}{T}\right)X_s\left(\frac{\Omega}{T}\right), \qquad -\pi \le \Omega \le \pi \end{aligned} \tag{12.16}$$

But

$$X_s\left(\frac{\Omega}{T}\right) = X_d(\Omega) \tag{12.17}$$

where $X_d(\Omega)$ is the DTFT of $x[n]$. The proof of (12.17) follows by setting $\omega = \Omega/T$ in (12.3).

Then inserting (12.17) into (12.16) yields

$$\hat{Y}(\Omega) = H\left(\frac{\Omega}{T}\right)X_d(\Omega), \qquad -\pi \le \Omega \le \pi \tag{12.18}$$

But from the results of Chapter 7 [see (7.50)],

$$\hat{Y}(\Omega) = H_d(\Omega)X_d(\Omega), \qquad -\pi \le \Omega \le \pi \tag{12.19}$$

where

$$H_d(\Omega) = H_d(z)\big|_{z=e^{j\Omega}}$$

is the frequency response function of the discrete-time system. Finally, comparing (12.18) and (12.19) reveals that

$$H_d(\Omega) = H\left(\frac{\Omega}{T}\right), \qquad -\pi \le \Omega \le \pi \tag{12.20}$$

This is the fundamental design requirement specified in the frequency domain. Note that (12.20) defines a transformation that maps the continuous-time filter with frequency function $H(\omega)$ into the digital filter with frequency function $H_d(\Omega)$. More precisely, as seen from (12.20), $H_d(\Omega)$ is constructed simply by setting $\omega = \Omega/T$ in $H(\omega)$.

The transformation given by (12.20) can be expressed in terms of the Laplace transform variable s and the z-transform variable z as follows:

$$H_d(z) = H(s)\big|_{s=(1/T)\ln z} \tag{12.21}$$

where $\ln z$ is the natural logarithm of the complex variable z. To verify that (12.21) does imply (12.20), simply set $z = e^{j\Omega}$ on both sides of (12.21). The relationship (12.21) shows how the continuous-time filter with transfer function $H(s)$ can be transformed into the digital filter with transfer function $H_d(z)$. Unfortunately, due to the nature of the function $(1/T)\ln z$, it is not possible to use (12.21) to derive an expression for $H_d(z)$ which is rational in z (i.e., a ratio of polynomials in z). However, the log function $(1/T)\ln z$ can be approximated by

$$\frac{1}{T}\ln z \approx \frac{2}{T}\frac{z-1}{z+1} \tag{12.22}$$

This leads to the following transformation from the z-domain to the s-domain:

$$s = \frac{2}{T}\frac{z-1}{z+1} \tag{12.23}$$

Solving (12.23) for s in terms of z yields the inverse transformation given by

$$z = \frac{1 + (T/2)s}{1 - (T/2)s} \tag{12.24}$$

The transformation defined by (12.24) is called the *bilinear transformation* from the complex plane into itself. The term *bilinear* means that the relationship in (12.24) is a bilinear function of z and s. For a detailed development of bilinear transformations in complex function theory, see Churchill et al. [1976]. For a derivation of the bilinear transformation using the trapezoidal integration approximation, see Problem 12.3.

The bilinear transformation (12.24) has the property that it maps the open left-half plane into the open unit disk. In other words, if $\text{Re}\, s < 0$, then z given by (12.24) is located in the open unit disk of the complex plane; that is, $|z| < 1$. In addition, the transformation maps the $j\omega$-axis of the complex plane onto the unit circle of the complex plane.

Now using the approximation (12.22) in (12.21) results in

$$H_d(z) = H\left(\frac{2}{T}\frac{z - 1}{z + 1}\right) \tag{12.25}$$

By (12.25) the transfer function of the digital filter is (approximately) equal to the transfer function of the given continuous-time system, with s replaced by $(2/T)[(z - 1)/(z + 1)]$.

Note that since the bilinear transformation maps the open left-half plane into the open unit disk, the poles of $H_d(z)$ given by (12.25) are in the open unit disk if and only if the given continuous-time system is stable (which is assumed above). Thus the digital filter is stable, which shows that the bilinear transformation preserves stability. This is a very desirable property for any discretization process.

For the digital filter with transfer function $H_d(z)$ given by (12.25), in general the matching condition (12.21) is not satisfied since the bilinear transformation is an approximation to $s = (1/T) \ln z$. As a result, if $H(s)$ is the transfer function of an analog lowpass filter with cutoff frequency ω_c, in general the corresponding cutoff frequency of the discretization $H_d(z)$ will not be equal to $\omega_c T$. The cutoff frequency of the discretization is said to be *warped* from the desired value $\omega_c T$, which results in an error in the digital filter realization of the given analog filter. The amount of warping can be computed as follows.

Setting $z = e^{j\Omega}$ in (12.25) gives

$$H_d(\Omega) = H_d(z)\Big|_{z=e^{j\Omega}} = H\left(\frac{2}{T}\frac{e^{j\Omega} - 1}{e^{j\Omega} + 1}\right) \tag{12.26}$$

Now since the inverse of the bilinear transformation maps the unit circle onto the $j\omega$-axis, the point $(2/T)[(e^{j\Omega} - 1)/(e^{j\Omega} + 1)]$ must be equal to some point on the $j\omega$-axis; that is,

$$j\omega = \frac{2}{T}\frac{e^{j\Omega} - 1}{e^{j\Omega} + 1} \tag{12.27}$$

for some value of ω. Hence

$$\begin{aligned}\omega &= \frac{2}{T}\frac{(1/j)(e^{j\Omega}-1)}{e^{j\Omega}+1} \\ &= \frac{2}{T}\frac{(1/j2)(e^{j(\Omega/2)}-e^{-j(\Omega/2)})}{(1/2)(e^{j(\Omega/2)}+e^{-j(\Omega/2)})} \\ &= \frac{2}{T}\tan\frac{\Omega}{2} \qquad (12.28)\end{aligned}$$

The inverse relationship is

$$\Omega = 2\tan^{-1}\frac{\omega T}{2} \qquad (12.29)$$

Combining (12.26) and (12.27) yields

$$H_d(\Omega) = H(\omega)$$

where Ω is given by (12.29).

Therefore, if ω_c is the cutoff frequency of the given analog filter [with transfer function $H(s)$], the corresponding cutoff frequency Ω_c of the discretization $H_d(z)$ is given by

$$\Omega_c = 2\tan^{-1}\frac{\omega_c T}{2}$$

The amount of warping from the desired value $\Omega_c = \omega_c T$ depends on the magnitude of $(\omega_c T)/2$. If $(\omega_c T)/2$ is small so that $\tan^{-1}[(\omega_c T)/2)] \approx (\omega_c T)/2$, then

$$\Omega_c \approx 2\left(\frac{\omega_c T}{2}\right) = \omega_c T$$

Thus in this case the warping is small.

Example 12.1 *Two-Pole Butterworth Filter*

Consider the two-pole Butterworth filter with transfer function

$$H(s) = \frac{\omega_c^2}{s^2 + \sqrt{2}\,\omega_c s + \omega_c^2}$$

Constructing the discretization $H_d(z)$ defined by (12.25) yields

$$\begin{aligned}H_d(z) &= H\left(\frac{2}{T}\frac{z-1}{z+1}\right) \\ &= \frac{\omega_c^2}{\left(\frac{2}{T}\frac{z-1}{z+1}\right)^2 + \sqrt{2}\,\omega_c\left(\frac{2}{T}\frac{z-1}{z+1}\right) + \omega_c^2}\end{aligned}$$

$$= \frac{(T^2/4)(z+1)^2\,\omega_c^2}{(z-1)^2 + (T/\sqrt{2})\omega_c(z+1)(z-1) + (T^2/4)(z+1)^2\,\omega_c^2}$$

$$= \frac{(T^2\omega_c^2/4)(z^2+2z+1)}{\left(1 + \frac{\omega_c T}{\sqrt{2}} + \frac{T^2\omega_c^2}{4}\right)z^2 + \left(\frac{T^2\omega_c^2}{2} - 2\right)z + \left(1 - \frac{\omega_c T}{\sqrt{2}} + \frac{T^2\omega_c^2}{4}\right)}$$

For $\omega_c = 2$ and $T = 0.2$, the transfer function is given by

$$H_d(z) = \frac{0.0302(z^2 + 2z + 1)}{z^2 - 1.4514z + 0.5724}$$

The magnitude function $|H_d(\Omega)|$ plotted in Figure 12.6 is obtained by using the following MATLAB commands:

```
numd = 0.0302*[1 2 1];
dend = [1 -1.4514 0.5724];
w = -pi:2*pi/300:pi;
H = freqz(numd,dend,w);
plot(w,abs(H));
```

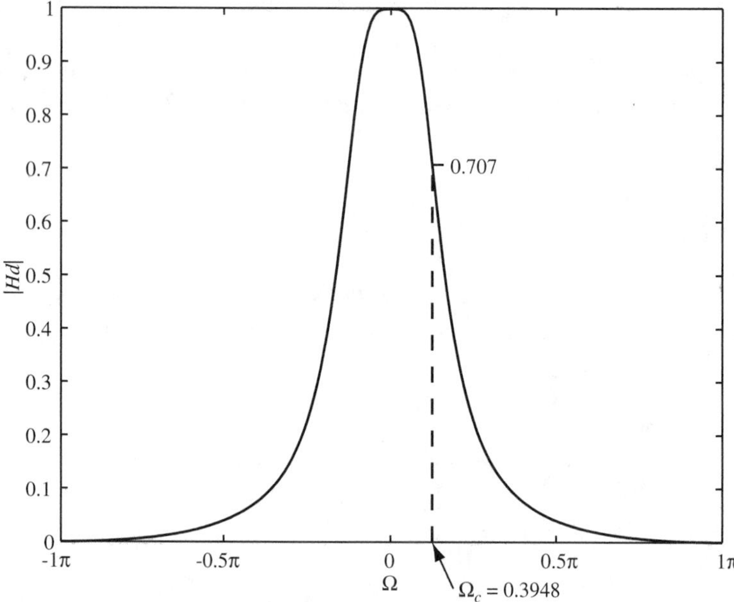

Figure 12.6 Magnitude function of discretization.

In this case, the cutoff frequency of the discretization is

$$\Omega_c = 2\tan^{-1}\frac{(2)(0.2)}{2} = 0.3948$$

which is very close to the desired value $\omega_c T = 0.4$.

The effect of warping can be eliminated by *prewarping* the analog filter prior to applying the bilinear transformation. In this process the cutoff frequency of the analog filter is designed so that the corresponding cutoff frequency Ω_c of the digital filter is equal to $\omega_c T$, where ω_c is the desired value of the analog filter cutoff frequency. The relationship $\Omega_c = \omega_c T$ follows directly from the desired matching condition (12.20). Hence, from (12.29) it is seen that the prewarped analog cutoff frequency, denoted by ω_p, should be selected as

$$\omega_p = \frac{2}{T}\tan\frac{\Omega_c}{2} \tag{12.30}$$

Thus the analog filter should be designed to have the analog cutoff frequency ω_p (instead of ω_c) so that the distortion introduced by the bilinear transformation will be canceled by the prewarping. The procedure is illustrated by the following example.

Example 12.2 *Prewarping*

The two-pole lowpass filter designed in Example 12.1 can be redesigned so that the cutoff frequencies match by prewarping the analog frequency as discussed above. For a desired analog cutoff frequency of $\omega_c = 2$ and $T = 0.2$, the desired digital cutoff frequency is $\Omega_c = 0.4$. The prewarped analog frequency is calculated from (12.30) to be $\omega_p = 2.027$. Substituting ω_p for ω_c in the transfer function of the digital filter constructed in Example 12.1 yields the redesigned filter with transfer function

$$H_d(z) = \frac{0.0309(z^2 + 2z + 1)}{z^2 - 1.444z + 0.5682}$$

12.3 DESIGN OF IIR FILTERS USING MATLAB

In this section it is shown that MATLAB can be used to design a digital filter using the Butterworth and Chebyshev analog prototypes discussed in Section 9.6. It should be noted that there are other analog prototype filters that are available in MATLAB, but these are not considered in this book.

To design an IIR filter via MATLAB, the first step is to use MATLAB to design an analog filter that meets the desired criteria, and then the analog filter is mapped to a discrete-time (digital) filter using the bilinear transformation. Recall from Chapter 9 that the design of an analog filter using MATLAB begins with the design of an N-pole lowpass filter with a bandwidth normalized to 1 rad/sec. If the analog filter is a Butterworth, the command used is `buttap`, while the command for a Chebyshev filter is `cheb1ap`. Then the filter is transformed via frequency transformations into a lowpass filter with a different bandwidth or into a highpass, bandpass, or bandstop filter with the desired frequency requirements. In MATLAB, the resulting filter analog transfer function is stored with the numerator and denominator coefficients in vectors. This transfer function can be mapped to a digital-filter transfer function using the command `bilinear`.

Example 12.3 *MATLAB Design of Butterworth Filter*

Digital Filtering of Continuous-Time Signals

The two-pole lowpass Butterworth filter with $\omega_c = 2$ and $T = 0.2$ designed in Example 12.1 can be found using the following commands:

```
[z,p,n] = buttap(2);    % creates a 2-pole filter
% convert z,p,n to a transfer function form
[num,den] = zp2tf(z,p,n);
% need to transform the cutoff frequency to 2
wc = 2;
[num,den] = lp2lp(num,den,wc);
T = 0.2;
[numd,dend] = bilinear(num,den.1/T)
```

The program designs a two-pole lowpass Butterworth filter with cutoff frequency of 1 rad/sec and then transforms it to a lowpass filter with $\omega_c = 2$. Recall from Section 9.6 that the frequency transformation is performed by the command lp2lp. Then the bilinear transformation is used to map the filter to the z-domain. The resulting vectors containing the coefficients of the digital filter are given by

```
numd = [0.0302   0.0605   0.0302]
dend = [1   -1.4514   0.5724]
```

This result corresponds exactly to the filter generated in Example 12.1.

MATLAB includes the M-files `butter` and `cheby1` that already contain all the steps needed for design based on analog prototypes. These commands first design the appropriate analog filter, then transform it to discrete time using the bilinear transformation. The available filter types are lowpass, highpass, and bandstop. The M-files require that the number of poles be determined and the digital cutoff frequencies be specified. Recall that the continuous-time frequency ω is related to the discrete-time frequency Ω by $\Omega = \omega T$. Hence the digital cutoff frequency is $\Omega_c = \omega_c T$, where ω_c is the desired analog cutoff frequency. The M-files also require that the cutoff frequency be normalized by π.

Example 12.4 *Alternate Design*

Consider the lowpass filter design in Example 12.1. This design can be accomplished using the following commands:

```
N = 2;            % number of poles
T = 0.2;          % sampling time
wc = 2;           % analog cutoff frequency
Omegac = wc*T/pi;   % normalized digital cutoff frequency
[numd,dend] = butter(N,Omegac)
```

The resulting filter is defined by the numerator and denominator coefficients

```
numd = [0.0309 0.0619 0.0309];
dend = [1 -1.444 0.5682];
```

This matches the results found in Example 12.2, which uses the prewarping method.

Example 12.5 *Chebyshev Type 1 Highpass Filter*

Digital Filtering of Continuous-Time Signals

Now the objective is to design a Chebyshev type 1 highpass filter with an analog cutoff frequency of $\omega_c = 2$ rad/sec and sampling interval $T = 0.2$, and a passband ripple of 3 dB. The MATLAB commands are

```
N = 2;     % number of poles
Rp = 3;    % passband ripple
T = .2;    % sampling period
wc = 2;    % analog cutoff frequency
Omegac = wc*T/pi; % normalized digital cutoff frequency
[numd,dend] = cheby1(N,Rp,Omegac,'high')
```

The filter is given by

```
numd = [0.5697 -1.1394 0.5697]
dend = [1 -1.516 0.7028]
```

The frequency response of the digital filter is given in Figure 12.7.

Example 12.6 *Filtering Specific Frequencies*

Digital Filtering of Continuous-Time Signals

For the signal

$$x(t) = 1 + \cos t + \cos 5t$$

the objective is to remove the cos $5t$ component by using a two-pole digital lowpass Butterworth filter. Since the highest-frequency component of $x(t)$ is 5 rad/sec, to avoid aliasing the sampling frequency should be at least 10 rad/sec. Thus a sampling period of $T = 0.2$ is sufficiently small to avoid aliasing (equivalently, $\omega_s = 2\pi/T = 10\pi$ rad/sec). In addition, a filter cutoff frequency ω_c of 2 rad/sec should result in attenuation of the component cos $5t$ with little attenuation of $1 + \cos t$. Hence the filter designed in Examples 12.3 and 12.4 should be adequate for the filtering task.

The following commands create the sampled version of $x(t)$ and then filter it using the command `filter`:

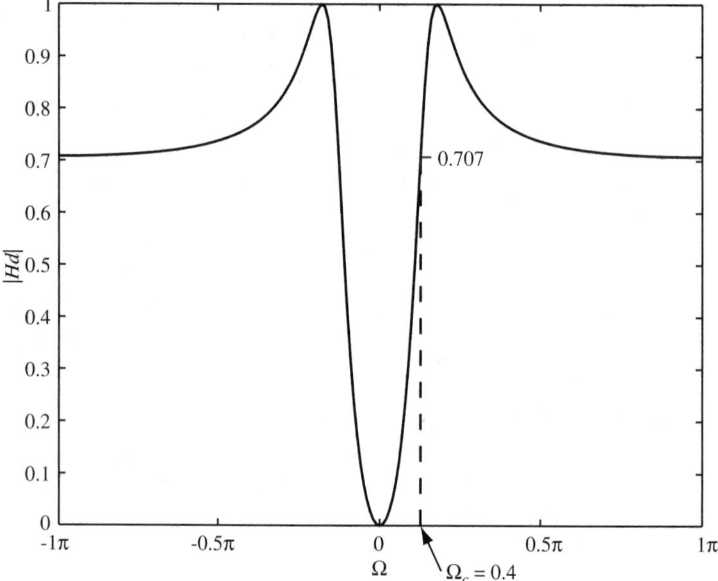

Figure 12.7 Frequency response function of digital filter in Example 12.5.

```
numd = [0.0309 0.0619 0.0309]; % define digital filter
dend = [1 -1.444 0.5682];
n = 0:80;
T = 0.2;
x = 1 + cos(T*n) + cos(T*5*n);
y = filter(numd,dend,x);
% plot x(t) with more resolution
t = 0:0.06:15;
xa = 1 + cos(t) + cos(5*t);
subplot(211),plot(t,xa);  % analog input, x(t)
subplot(212),plot(n*T,y);  % analog output, y(t)
pause
subplot(211),stem(n*T,x);   % sampled input, x[n]
subplot(212),stem(n*T,y);   % output of filter, y[n]
```

The discrete-time signal $x[\text{n}]$ and the digital filter output $y[n]$ are plotted in Figure 12.8. In the program above, the analog output $y(t)$ is generated by using the command `plot`, which approximates the output of an ideal reconstructor. The plots of the analog input $x(t)$ and analog output $y(t)$ are shown in Figure 12.9. Note that there is an initial transient in $y(t)$ due to the effect of the initial conditions, and as t increases, the analog output $y(t)$ quickly settles to steady-state behavior since the poles of the digital filter are inside the unit circle and thus the filter is stable. From the plot of $y(t)$ in Figure 12.9b, it is clear that the frequency component $\cos 5t$ has been significantly attenuated; however,

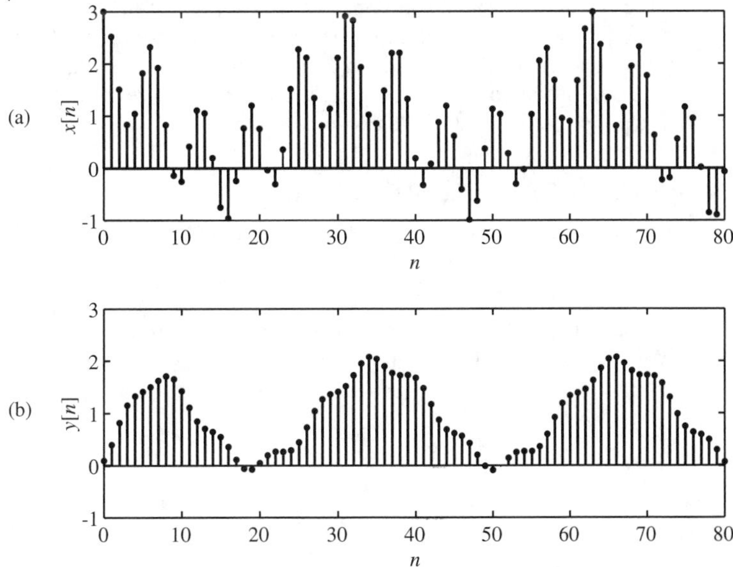

Figure 12.8 Plot of (a) discrete-time signal $x[n]$ and (b) output $y[n]$ of the digital filter.

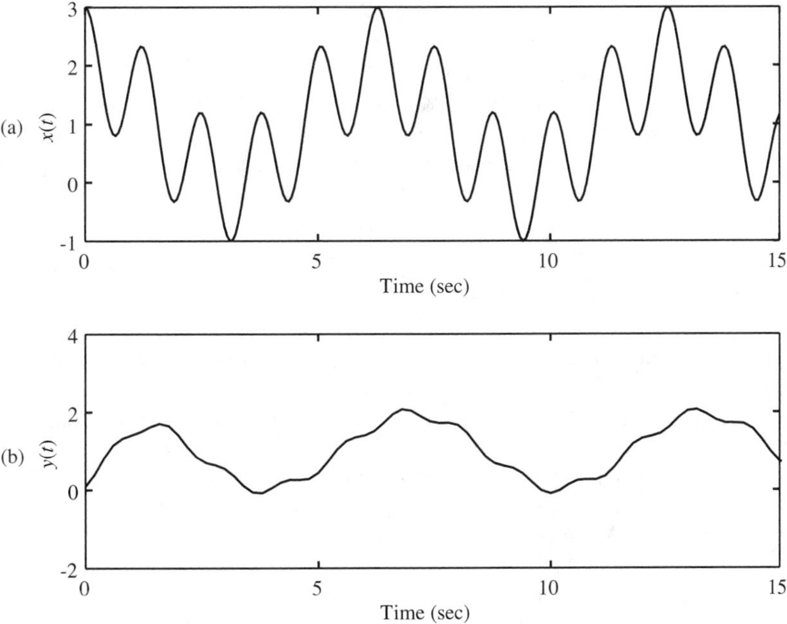

Figure 12.9 Plot of (a) analog signal $x(t)$ and (b) analog output $y(t)$.

there is still some small component of cos $5t$ present in $y(t)$. A better result could have been achieved by using a larger-order filter since additional poles can yield a sharper transition between the passband and the stopband (see Chapter 9). To achieve better rejection of the cos $5t$ term, the reader is invited to rewrite the above program with a five-pole filter instead of the two-pole filter.

Example 12.7 *Removing Signal Components*

Again consider the signal $x(t)$ defined in Example 12.6:

$$x(t) = 1 + \cos t + \cos 5t$$

In this example it is desired to remove the component $1 + \cos t$ by using the digital highpass filter designed in Example 12.5. The following commands implement the digital filter and plot the results:

```
numd = [0.5697  -1.1394 0.5697];  % define the digital filter
dend = [1 -1.516 0.7028];
n = 0:75;
T = 0.2;
x = 1 + cos(T*n) + cos(T*5*n);
y = filter(numd,dend,x);
% plot x(t) with more resolution
t = 0:0.1:15;
xa = 1 + cos(t) + cos(5*t);
subplot(211), plot(t,xa);  % analog input, x(t)
subplot(212), plot(n*T,y);  % analog output, y(t)
```

The plots of $x(t)$ and $y(t)$ are shown in Figure 12.10. In the plot of $y(t)$, notice that the dc component of $x(t)$ has been filtered out and the cos t component has been reduced by about 85% while the cos $5t$ component is left intact. Had the filter been designed with an analog cutoff frequency of a larger value, say $\omega_c = 3$, the component cos t would be further reduced along with some attenuation of cos $5t$. The reader is invited to try this as an exercise.

Example 12.8 *Filtering Random Signals*

Digital Filtering of Continuous-Time Signals

Consider the random continuous-time signal $x(t)$ shown in Figure 12.11a. It is assumed that the signal is bandlimited to 5π rad/sec, so that a sampling time of $T = 0.2$ is acceptable. The sampled signal is first sent through the digital lowpass filter designed in Example 12.4. The resulting analog output $y(t)$ is shown in Figure 12.11b. Notice that the filtered signal $y(t)$ is smoother than the input signal $x(t)$, which is a result of the removal of the higher-frequency components in $x(t)$.

The sampled signal is then sent through the digital highpass filter designed in Example 12.4 and the resulting analog output is shown in Figure 12.12b. Notice that the dc component has been removed while the peak-to-peak amplitude of the signal $x(t)$ remains approximately equal to 1.

To generate the filtered output $y(t)$, replace the definition of $x(t)$ in the M-files for Examples 12.6 and 12.7 with the following command:

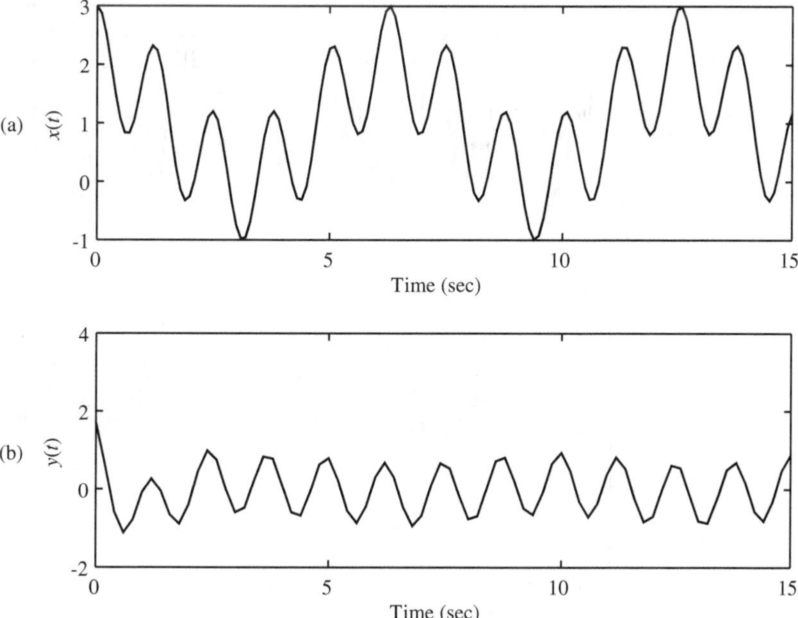

Figure 12.10 Plot of (a) analog signal $x(t)$ and (b) analog output $y(t)$ in Example 12.7.

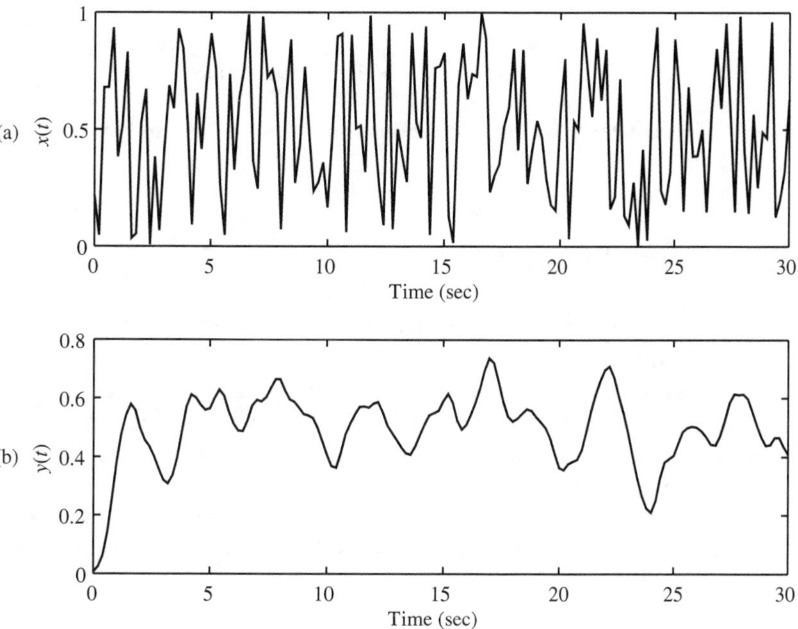

Figure 12.11 Plots of (a) the signal $x(t)$ and (b) the lowpass filtered analog output $y(t)$ in Example 12.8.

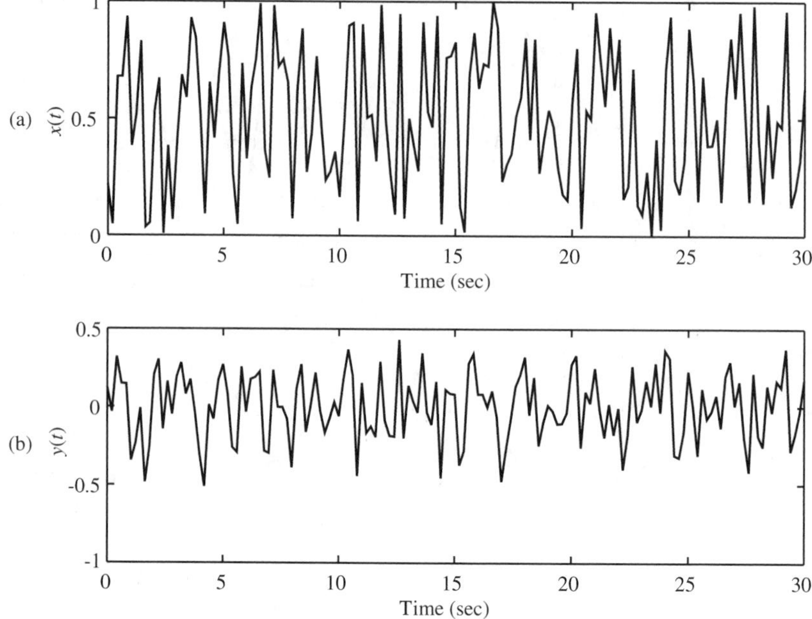

Figure 12.12 Plots of (a) the signal $x(t)$ and (b) the highpass filtered analog output $y(t)$ in Example 12.8.

```
x = rand(1,length(n));
```

This creates a vector with random numbers that are uniformly distributed between 0 and 1.

12.4 DESIGN OF FIR FILTERS

In contrast to an IIR filter, an FIR filter is a digital filter where the impulse response (i.e., the unit-pulse response) $h[n]$ is zero for all $n \geq N$. Such filters are useful in practice because they can be designed to have linear phase; that is, $\angle H_d(\Omega) = -c\Omega$ for some positive number c. Nonlinear phase (which is common with IIR filters) may cause some distortion in the time signal, while linear phase causes only a time delay of the signal being processed.

An FIR filter can be designed by truncating the impulse response of an IIR filter. In particular, let $H(\Omega)$ represent a desired IIR filter with impulse response $h[n]$. A corresponding FIR filter is given by

$$h_d[n] = \begin{cases} h[n], & 0 \leq n \leq N - 1 \\ 0, & \text{otherwise} \end{cases}$$

where N is the length of the filter. The transfer function of the FIR filter is given by

$$H_d(z) = \sum_{n=0}^{N-1} h_d[n]z^{-n}$$

The corresponding frequency response can be calculated directly from the definition of the DTFT:

$$H_d(\Omega) = \sum_{n=0}^{N-1} h_d[n]e^{-jn\Omega} \tag{12.31}$$

Ideally, $H_d(\Omega)$ should be a close approximation to the DTFT of $h[n]$ (the desired IIR filter); that is, $H_d(\Omega) \approx H(\Omega)$. However, the truncation of $h[n]$ introduces some errors in the frequency response. One of the errors is a ripple in the magnitude plot, which is caused by the Gibbs phenomenon, discussed in Section 5.2.

Analytically, the truncation of the infinite impulse response can be expressed as a multiplication by a signal $w[n]$ called a window:

$$h_d[n] = w[n]h[n] \tag{12.32}$$

where

$$w[n] = \begin{cases} 1 & \text{if } 0 \le n \le N-1 \\ 0, & \text{otherwise} \end{cases} \tag{12.33}$$

Recall from Chapter 7 that multiplication in the time domain corresponds to a convolution in the frequency domain. Hence, taking the DTFT of both sides of (12.32) yields

$$H_d(\Omega) = \frac{1}{2\pi}\int_{-\pi}^{\pi} H(\Omega - \lambda)W(\lambda)\, d\lambda \tag{12.34}$$

where the DTFT $W(\Omega)$ of $w[n]$ is given by

$$W(\Omega) = \frac{\sin(\Omega N/2)}{\sin(\Omega/2)}e^{-j\Omega(N-1)/2}$$

The plot of $|W(\Omega)|$ is shown in Figure 12.13 for for $N = 10$. Notice that there is a main lobe and sidelobes with regularly spaced zero crossings at $\Omega = 2\pi m/N$ for $m = 0, \pm 1, \pm 2, \ldots$.

To achieve a perfect match between $H_d(\Omega)$ and $H(\Omega)$, from (12.34) it is seen that $W(\Omega)$ would have to be equal to the impulse $2\pi\delta(\Omega)$. This corresponds to $w[n] = 1$ for all n, and thus for such a $w[n]$, there is no signal truncation and the filter given by (12.32) would be IIR.

However, it is possible to have $W(\Omega)$ be a close approximation to the impulse function; in particular, the narrower the main lobe and the smaller the size of the sidelobes of $|W(\Omega)|$, the closer $W(\Omega)$ approximates an impulse function. As N is increased, the width of the main lobe of $|W(\Omega)|$ becomes narrower. Since N is the length of the filter, this means that the larger the value of N, the closer $W(\Omega)$ approximates an impulse, and hence the closer $H_d(\Omega)$ approximates $H(\Omega)$. From a practical standpoint, however, a filter with a very large length would be too complex to implement.

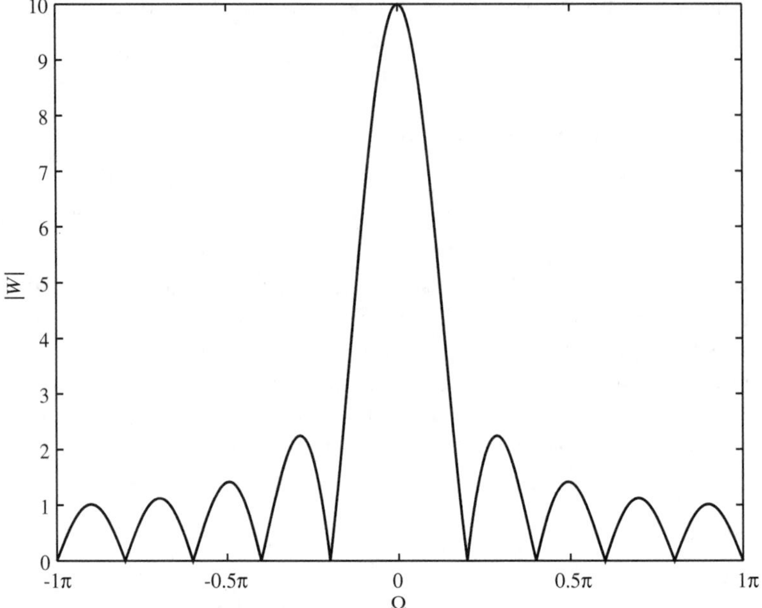

Figure 12.13 Plot of the magnitude $|W(\Omega)|$ of the DTFT of $w[n]$.

Another issue is designing the FIR filter so that it has linear phase. As mentioned above, linear phase corresponds to a time delay in the filter output, whereas a nonlinear phase causes some distortion in the signal. A filter $H_d(\Omega)$ has linear phase if the impulse response has even symmetry, that is, the filter has the following property: $h_d[n] = h_d[2m - n]$ where m is an integer or an integer divided by 2. If the length N of the filter is set equal to $2m + 1$, there is symmetry in the impulse response as it is reflected about $n = m$. This is illustrated in Figure 12.14 for the case $m = 7/2, N = 8$ and for $m = 3, N = 7$. Note that a filter can have symmetry if N is even or odd valued, as seen in Figure 12.14.

To prove that a filter with even symmetry has linear phase, substitute $h_d[n] = h_d[2m - n]$ into the frequency response function $H_d(\Omega)$ given by (12.31):

$$H_d(\Omega) = \sum_{n=0}^{N-1} h_d[2m - n]e^{-j\Omega n} \tag{12.35}$$

Let $k = 2m - n$ and rewrite (12.35) as

$$\begin{aligned} H_d(\Omega) &= \sum_{k=2m}^{0} h_d[k]e^{-j\Omega(2m-k)} \\ &= e^{-j2m\Omega}\sum_{k=0}^{2m} h_d[k]e^{j\Omega k} \\ &= e^{-j2m\Omega}\overline{H_d(\Omega)} \end{aligned} \tag{12.36}$$

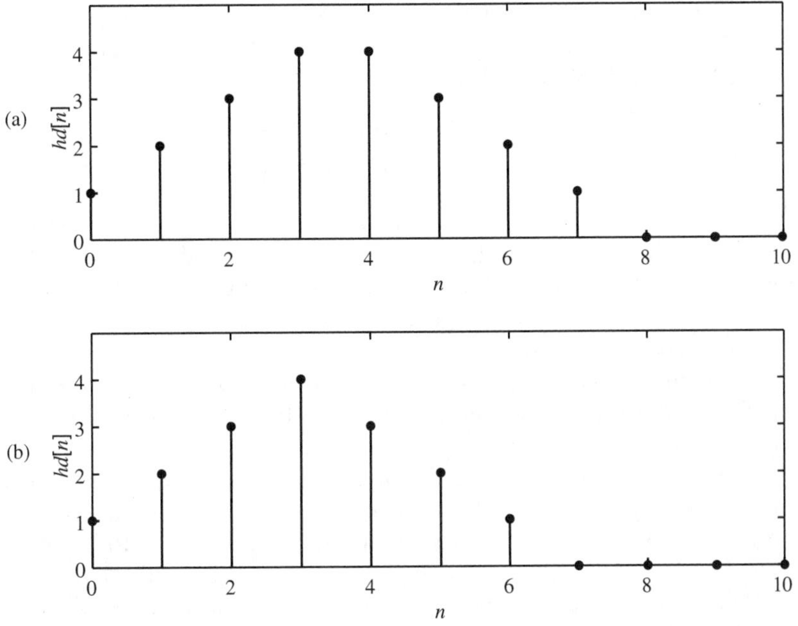

Figure 12.14 FIR filters with even symmetry with (a) $m = 7/2, N = 8$ and (b) $m = 3$, $N = 7$.

The angles of the expressions on both sides of (12.36) must match; that is,

$$\begin{aligned}\angle H_d(\Omega) &= -2m\Omega + \angle \overline{H}_d(\Omega)\\ &= -2m\Omega - \angle H_d(\Omega)\end{aligned}$$

So $\angle H_d(\Omega) = -m\Omega$ which shows that the phase is linear.

An FIR filter that has linear phase can be designed by first selecting an ideal filter with the desired frequency characteristics. If $H_i(\Omega)$ denotes the frequency response function of the ideal filter, the filter has zero phase and thus $H_i(\Omega)$ is real valued. From the results in Chapter 7, taking the inverse DTFT of a real-valued frequency function $H_i(\Omega)$ will produce a time function $h_i[n]$ with the property that $h_i[n] = h_i[-n]$. Hence $h_i[n]$ must have nonzero values for $n < 0$, and therefore the ideal filter is noncausal. To produce a casual filter, the impulse response $h_i[n]$ must be delayed a sufficient number of samples so that the important characteristics of the delayed $h_i[n]$ occur for $n \geq 0$. The delayed filter with impulse response $h_i[n - m]$ can be truncated for $n < 0$ without a loss of important information. Delaying and truncating the impulse response $h_i[n]$ also preserves the symmetry in $h_i[n]$, so that the resulting impulse response $h_d[n]$ has the property for linear phase (i.e., $h_d[n] = h_d[2m - n]$).

One design procedure for a linear phase FIR filter is summarized in the following steps: First select an ideal filter $H_i(\Omega)$ with the desired frequency characteristics. Calculate the inverse DTFT $h_i[n]$ to find the impulse response. Select the value of the delay m so that the delayed impulse response $h_i[n - m]$ is sufficiently small for $n < 0$. Truncate $h_i[n - m]$ for $n < 0$ and for $n > N - 1$, where $N = 2m + 1$ to yield the FIR filter with impulse response:

$$h_d[n] = \begin{cases} h_i[n - m] & \text{for } 0 \le n \le N - 1 \\ 0, & \text{otherwise} \end{cases} \qquad (12.37)$$

In the development above, it was assumed that the delay m is an integer that results in an odd-length filter (i.e., $N = 2m + 1$ is odd). An even-length FIR filter with linear phase can also be designed by selecting m to be half of an integer. The filter would still be defined as in (12.37), but it would no longer be a delayed and truncated version of the ideal filter.

A more standard but equivalent design procedure is to reorder the steps in the procedure above as follows. Since shifting in the time domain by m samples is equivalent to multiplying by $e^{-j\Omega m}$ in the frequency domain, the shift in time can be performed prior to taking the inverse DTFT. The difference in the approaches relates to a preference between working in the time domain or in the frequency domain. This design procedure can be summarized in the following steps: Select an ideal filter with frequency function $H_i(\Omega)$, and then multiply by $e^{-j\Omega m}$, where m is either an integer or an integer divided by 2. The inverse DTFT of the product $e^{-j\Omega m}H_i(\Omega)$ is then computed and the resulting sequence is truncated for $n < 0$ and $n > N - 1$, where $N = 2m + 1$ to obtain the definition of the FIR filter as given in (12.37).

Example 12.9 *FIR Lowpass Filter*

Consider the ideal lowpass filter with frequency response function $H_i(\Omega)$ shown in Figure 12.15. Note that the cutoff frequency is Ω_c. To make the filter causal, introduce a

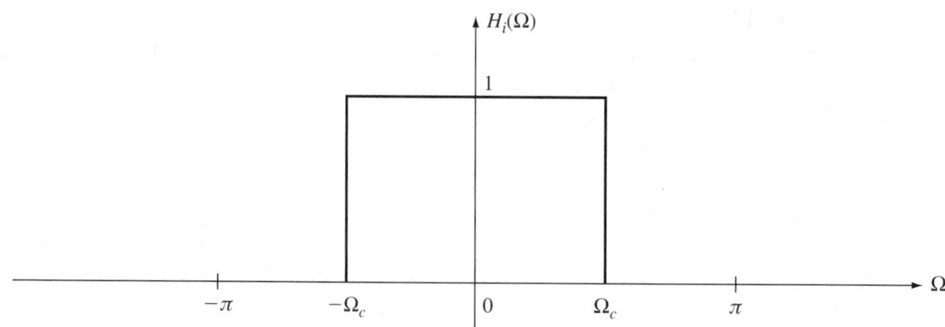

Figure 12.15 Frequency response function $H_i(\Omega)$ of the ideal lowpass filter in Example 12.9.

phase shift of $e^{-j\Omega m}$ in $H_i(\Omega)$. The frequency response function $H(\Omega)$ of the resulting filter is then given by $H(\Omega) = H_i(\Omega)e^{-j\Omega m}$. From the definition of $H_i(\Omega)$, $H(\Omega)$ can be written in the form

$$H(\Omega) = \begin{cases} e^{-j\Omega m} & \text{if } |\Omega| \leq \Omega_c \\ 0 & \text{if } |\Omega| > \Omega_c \end{cases} \tag{12.38}$$

The impulse response $h[n]$ of this filter can be computed by taking the inverse DTFT of (12.38) using the results in Chapter 7. This yields

$$h[n] = \frac{\sin(\Omega_c(n - m))}{\pi(n - m)} = \frac{\Omega_c}{\pi}\,\text{sinc}\left[\frac{\Omega_c(n - m)}{\pi}\right]$$

The FIR filter is obtained by truncating the response $h[n]$ for $n < 0$ and for $n > N - 1 = 2m$, which gives

$$h_d[n] = \begin{cases} \dfrac{\Omega_c}{\pi}\,\text{sinc}\left[\dfrac{\Omega_c(n - m)}{\pi}\right] & \text{for } 0 \leq n \leq N - 1 \\ 0, & \text{otherwise} \end{cases} \tag{12.39}$$

For $\Omega_c = 0.4$, the impulse response of the ideal filter with a zero phase shift (i.e., in the case $m = 0$) is shown in Figure 12.16. Note the nonzero values of the impulse response for $n < 0$, which results from the noncausal nature of this filter. Shown in Figure 12.17 are the impulse responses of the resulting FIR filter defined by (12.39) for the cases when $m = 10$ and $m = 21/2$. Notice that the FIR filter lengths are $N = 21$ and $N = 22$, respectively, and that both filters have linear phase since the impulse responses have symmetry about the point m.

The frequency response function $H_d(\Omega)$ of the FIR filter with impulse response $h_d[n]$ is found by direct computation from the definition given by (12.31). This yields the magnitude function $|H_d(\Omega)|$ shown in Figure 12.18a for the case when $N = 21$. Comparing Figures 12.15 and 12.18a reveals that in contrast to the frequency function $H(\Omega)$

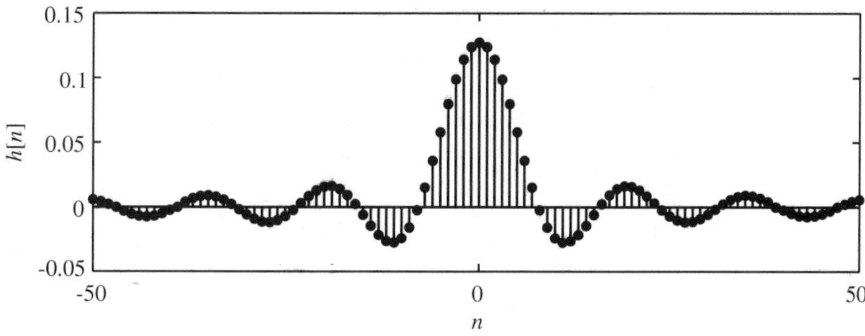

Figure 12.16 Impulse response of the ideal filter in Example 12.9.

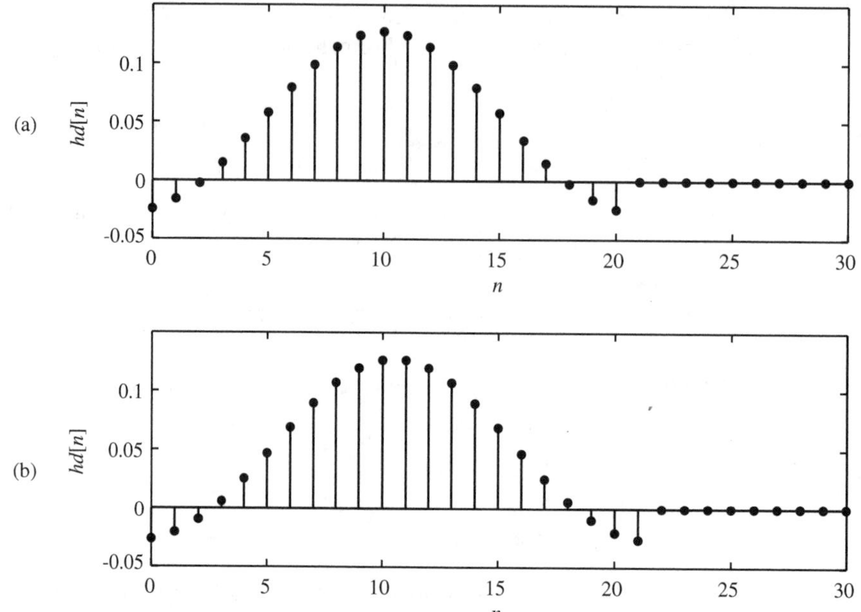

Figure 12.17 Impulse responses of FIR filter for (a) $N = 21$ and (b) $N = 22$.

of the ideal IIR filter, the frequency function $H_d(\Omega)$ of the FIR filter does not have a sharp transition between the passband and the stopband. Also, there is some ripple in the magnitude plot of $H_d(\Omega)$, which is a result of the truncation process. For $N = 41$, the magnitude of $H_d(\Omega)$ is plotted in Figure 12.18b. Note that the ripple has approximately the same magnitude as in the case when $N = 21$, but now the transition is much sharper.

The MATLAB commands to compute the frequency response function of the FIR filter for $N = 21$ are

```
Omegac = .4; % digital cutoff frequency
N = 21; % filter length
m = (N-1)/2; % phase shift
n = 0:2*m+10; % define points for plot
h = Omegac/pi*sinc(Omegac*(n-m)/pi); % delayed ideal filter
w = [ones(1,N) zeros(1,length(n)-N)]; % window
hd = h.*w;
Omega = -pi:2*pi/300:pi; % plot the frequency response
Hd = freqz(hd,1,Omega);
plot(Omega,abs(Hd));
```

To conclude this example, consider the analog signal $x(t)$ shown in Figure 12.19a. Recall that this signal was used in Examples 12.6 and 12.7. The lowpass FIR filter designed for $N = 21$ above can be used to filter the signal to remove the cos $5t$ component.

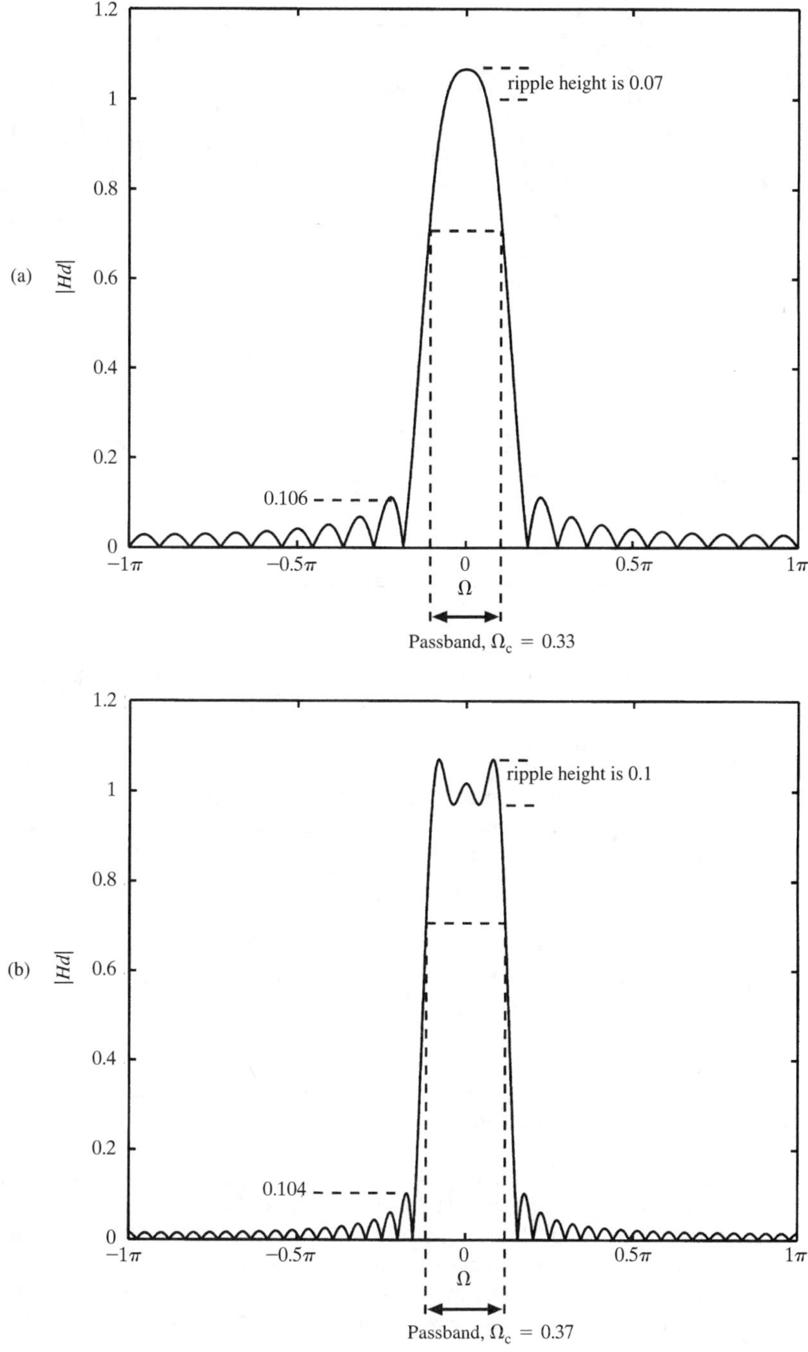

Figure 12.18 FIR filter magnitude function $|H_d(\Omega)|$ for the case (a) $N = 21$ and (b) $N = 41$.

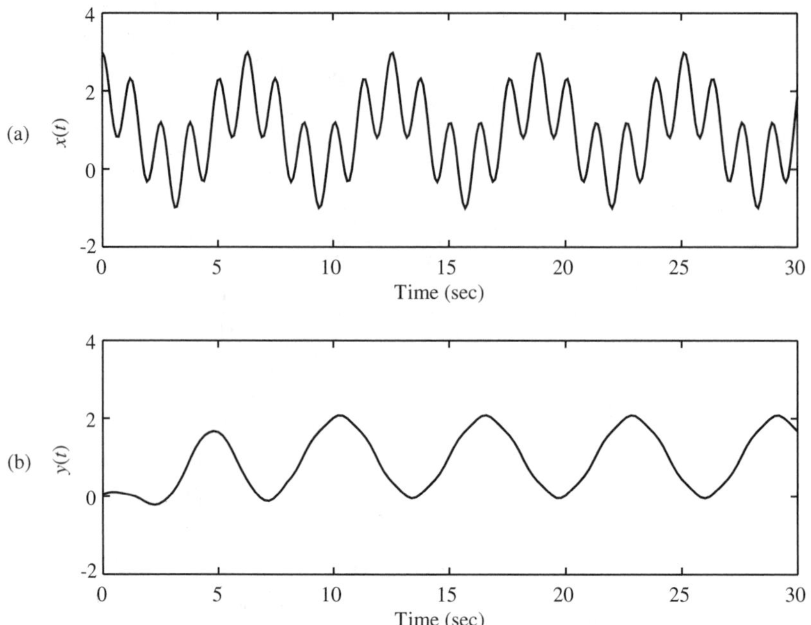

Figure 12.19 Plots of (a) the signal $x(t)$ and (b) the analog output $y(t)$.

The lowpass filter constructed above was designed with $\Omega_c = 0.4$, which is the same cutoff frequency as for the IIR filter designed in Example 12.6. The following MATLAB commands show how the signal $x(t)$ is filtered using the FIR filter designed above.

```
n = 0:150;
T = .2;
x = 1+cos(n*T)+cos(5*T*n); % sampled input, x[n]
y = filter(hd,1,x); % sampled output, y[n]
t = 0:.1:30; % plot x(t) with more resolution
x = 1 + cos(t) + cos(5*t);
subplot(211), plot(t,x)   % input,  x(t)
subplot(212), plot(n*T,y) % output, y(t)
```

The resulting analog output $y(t)$ is shown in Figure 12.19b. Notice that the cos $5t$ component of $x(t)$ is filtered and the dc value and low-frequency component are passed without attenuation. Had a larger value of N been used in the FIR filter design given above, the analog response $y(t)$ would be even smoother than that seen in Figure 12.19b since more of the high-frequency component would be filtered. The reader is invited to repeat the problem of filtering $x(t)$ using an FIR filter having length $N = 41$.

Windows

As discussed above, in FIR filter design the infinite-length impulse response $h[n]$ is multiplied by a window $w[n]$ to yield the truncated impulse response given by $h_d[n] = w[n]h[n]$. The particular window $w[n]$ defined by (12.33) is referred to as the *rectangular window* since it produces an abrupt truncation of $h[n]$. It turns out that the ripple in $H_d(\Omega)$ resulting from the use of the rectangular window (e.g., see Figure 12.18) can be reduced by using a window that tapers off gradually. There are several types of windows that have a gradual transition, each producing a different effect on the resulting FIR filter. Two examples are the Hanning and Hamming windows, which are defined below along with the rectangular window.

Rectangular:

$$w[n] = 1, \qquad 0 \leq n \leq N-1$$

Hanning:

$$w[n] = \frac{1}{2}\left(1 - \cos\frac{2\pi n}{N-1}\right), \qquad 0 \leq n \leq N-1$$

Hamming:

$$w[n] = 0.54 - 0.46\cos\frac{2\pi n}{N-1}, \qquad 0 \leq n \leq N-1$$

All three of the windows above are plotted in Figure 12.20 for $N = 21$ (stem plotting has been suppressed so that the comparisons between the functions are more apparent).

The log of the magnitude of the DTFT of the window function $w[n]$ is plotted in decibels for the rectangular, Hanning, and Hamming windows in Figure 12.21 for $N = 21$. As discussed in the first part of this section, the frequency response function $H_d(\Omega)$ of an FIR filter with impulse response $h_d[n] = h[n]w[n]$ is a better approximation to the desired frequency response $H(\Omega)$ when the main lobe of $|W(\Omega)|$ is narrow and the sidelobes are small in value. The type of window function $w[n]$ that is used is based on these criteria. In particular, the nonrectangular windows have much smaller sidelobes than the rectangular window, and as a result, there is much less ripple in the frequency response function of the FIR filter. However, for nonrectangular windows the main lobes are wider, which means that the transition region between the passband and stopband of the FIR filter is more gradual. A more sophisticated window called a *Kaiser window* is generally used for design of practical filters since it allows the designer the freedom to trade off the sharpness of the pass-to-stopband transitions with the magnitude of the ripples. See Oppenheim and Schafer [1989] for a more detailed discussion of windows.

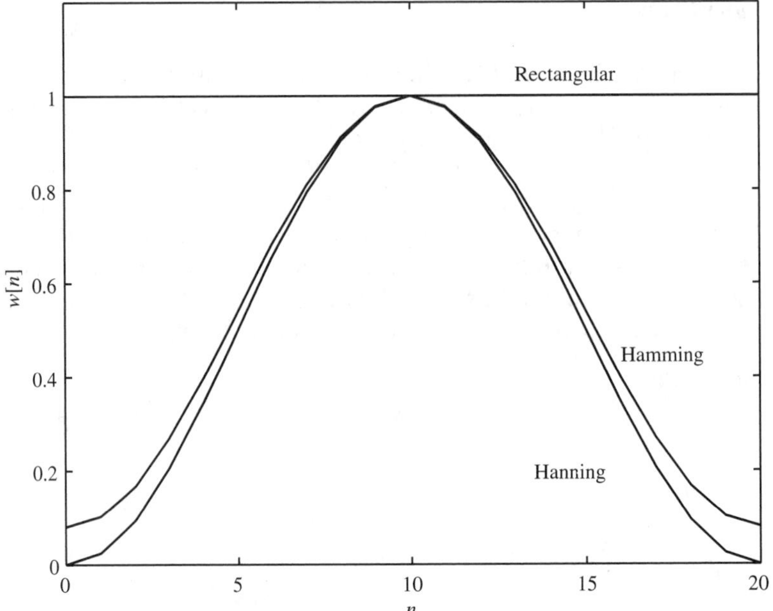

Figure 12.20 The rectangular, Hanning, and Hamming window functions for $N = 21$.

Example 12.10 *Lowpass Filtering Using Hanning and Hamming Windows*

Consider the lowpass filter designed in Example 12.9. Instead of using a rectangular window to truncate the infinite impulse response, Hanning and Hamming windows will be used. The MATLAB commands given in Example 12.9 can be rerun where the definition of w is replaced by the following statement for the Hanning window:

```
w = [0 hanning(N-2)' zeros(1,length(n)-N+1)];
```

and the following statement for the Hamming window:

```
w = [hamming(N)' zeros(1,length(n)-N)];
```

With $N = 41$ ($m = 20$), the impulse response for the FIR filter designed using the rectangular window is shown in Figure 12.22a, while Figure 12.22b shows the corresponding frequency function. Figures 12.23 and 12.24 show the impulse response and the frequency response of the FIR filter designed using the Hanning and Hamming windows, respectively. Note that the ripple in the frequency response designed using the rectangular window is very noticeable, while the ripple in the other filter frequency responses is negligible. Also note that the transition region between the passband and stopband is more gradual for the nonrectangular windows.

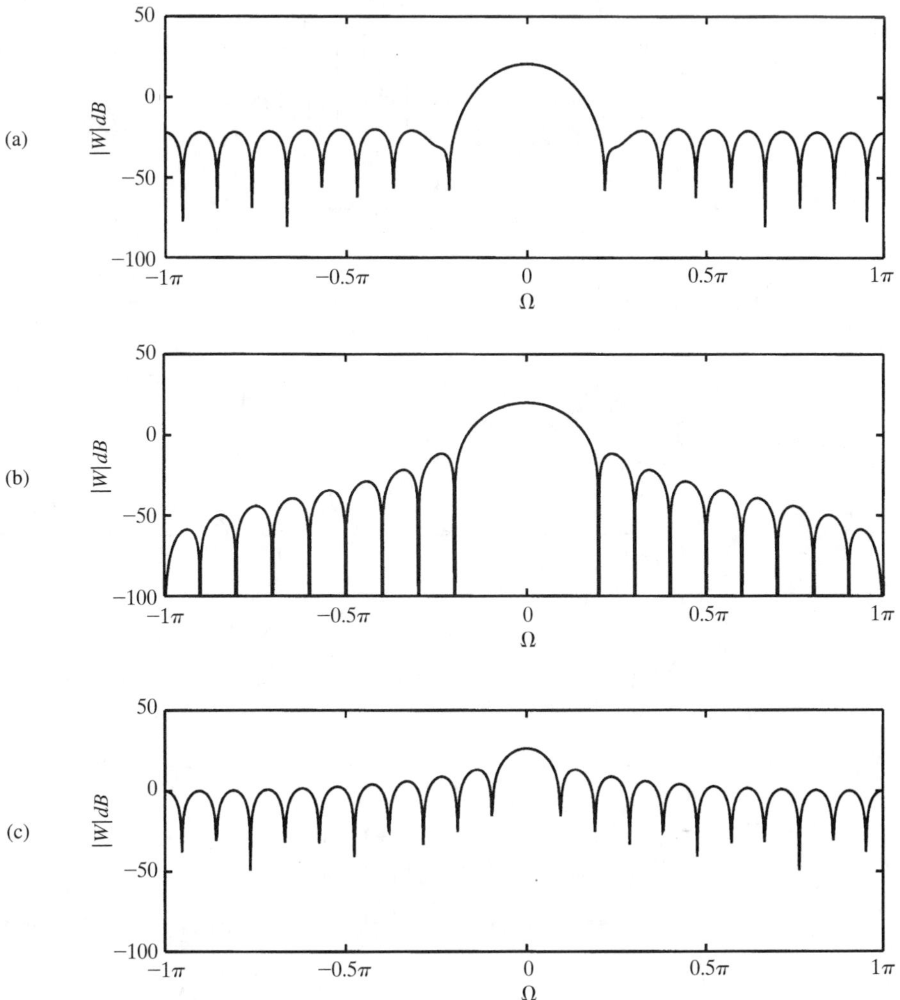

Figure 12.21 Log magnitude of the DTFT of the (a) rectangular, (b) Hanning, and (c) Hamming window functions for $N = 21$.

The Signal Processing Toolbox contains the command `fir1`, which automatically performs the commands required in Examples 12.9 and 12.10. See the tutorial that is available on the Web site.

An alternative way of designing FIR filters is to use numerical techniques to derive the filter coefficients to match arbitrary frequency characteristics. For

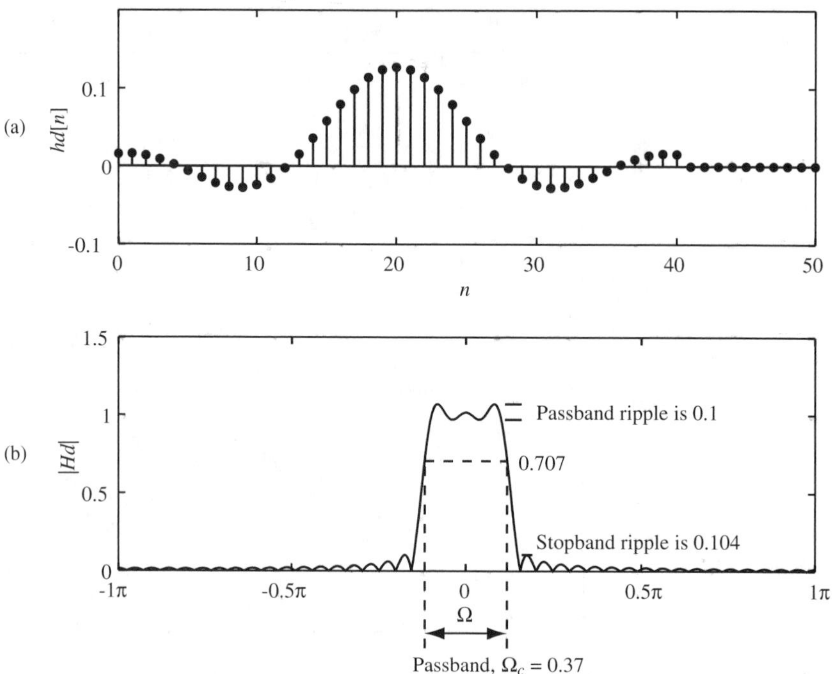

Figure 12.22 Plot of (a) impulse response and (b) filter frequency response for the rectangular window.

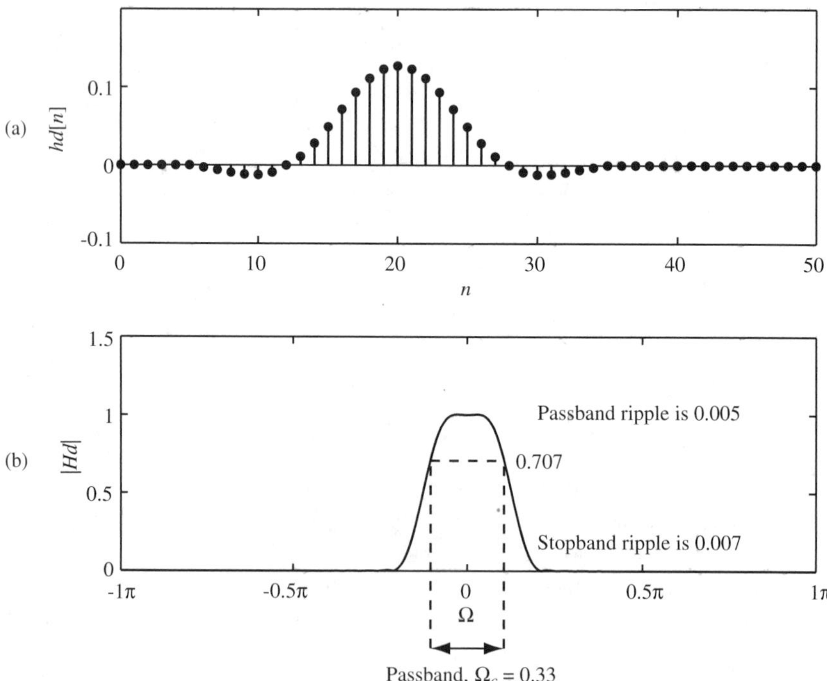

Figure 12.23 Plot of (a) impulse response and (b) filter frequency response for the Hanning window.

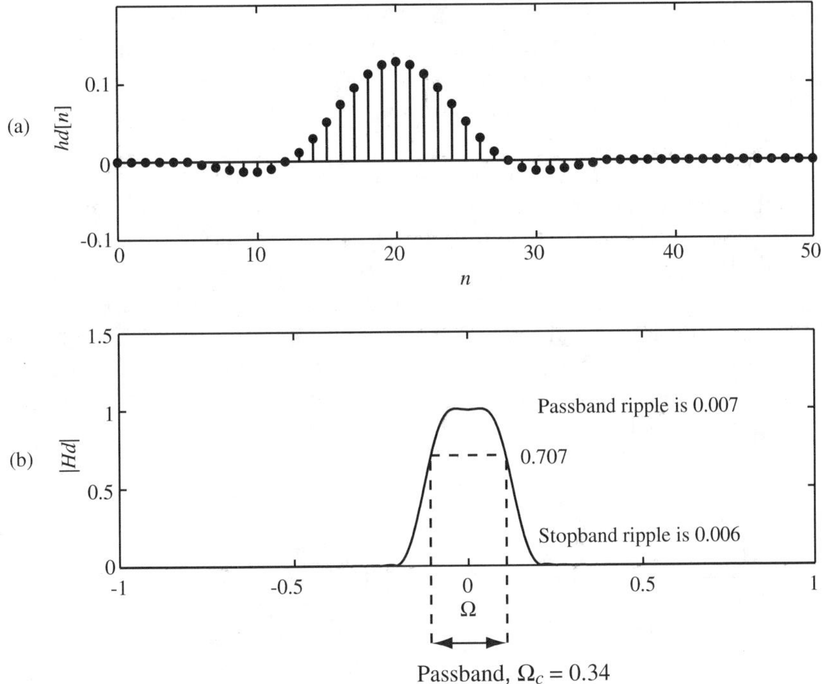

Figure 12.24 Plot of (a) impulse response and (b) filter frequency response for the Hamming window.

example, MATLAB includes a command that utilizes the Parks–McClellan algorithm, but this topic is beyond the scope of this book. See Oppenheim and Schafer [1989] for a discussion of various algorithmic techniques.

12.5 DESIGN OF DIGITAL CONTROLLERS

Digital control of a continuous-time system has become very standard in recent years as computer chips have become smaller, cheaper, and more powerful. Very complicated control structures can be implemented easily using a digital signal processing chip, whereas an equivalent analog controller may require very complex hardware. Applications where digital control has been used are the engine controllers in many automobiles, flight controls on aircraft, equipment control in manufacturing systems, climate control in buildings, and process controllers in chemical plants.

Digital control started becoming commonplace in the 1970s and early 1980s as computers were becoming cheaper and more compact. The theory for continuous-time control design was already mature at that time, so the first method for digital controller design was based on discretizing a standard continuous-time controller and then implementing the discretization using a sampler-and-hold circuit. A continuous-time system (or plant) with digital controller is illustrated in Figure 12.25b, while Figure 12.25a shows the standard analog controller configuration that was studied in Chapter 10.

The method of digital controller design where a continuous-time controller is discretized (i.e., mapped to a discrete-time controller) is often referred to as *analog emulation.* More recently, direct design methods have become more established which entail mapping the continuous-time plant into the discrete-time domain and then designing the controller using the discrete-time counterparts to the root-locus method discussed in Chapter 10, as well as frequency-domain design techniques based on Bode plots. The discussion in this section is limited to the analog emulation method of design. For details on direct design methods, see Franklin et al. [1990].

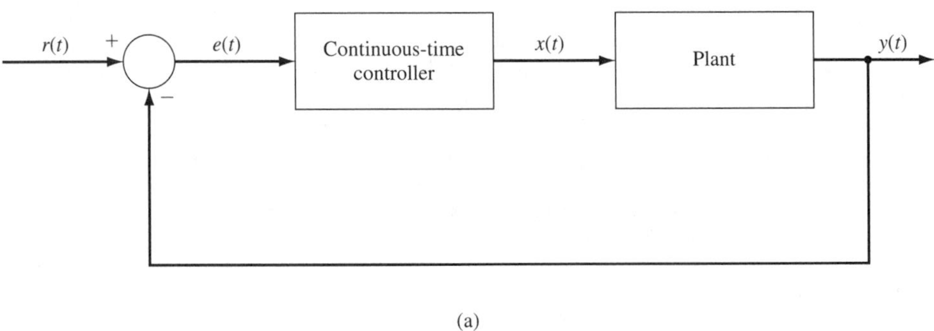

(a)

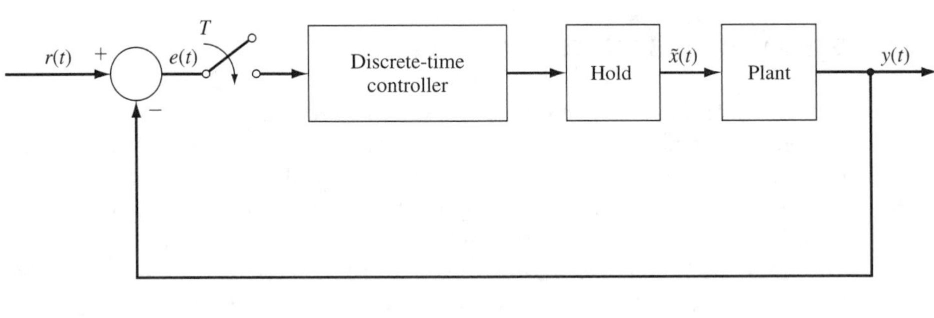

(b)

Figure 12.25 Block diagram of control system with (a) analog controller and (b) digital controller.

The analog emulation method for designing digital controllers is very similar to the design of digital filters by using analog prototypes. In fact, in both methods a continuous-time system (either a filter or a controller) is first designed and then is mapped to a discrete-time transfer function. For instance, the bilinear transformation that was developed in Section 12.2 can be used to map an analog controller with transfer function $G_c(s)$ to a digital controller with transfer function $G_d(z)$. The implementation of the digital controller is achieved using a computer or digital signal processing chip, whose output is converted to a continuous-time signal using a D/A converter. This is the same process that is used for the implementation of a digital filter.

In addition to the bilinear transformation developed in Section 12.2, in applications involving digital control there are a number of other techniques that are often used to transform a continuous-time transfer function to a discrete-time function. One such method, termed *response matching,* involves matching the output of a continuous-time system to the output of a discrete-time system when the input is a specific function $x(t)$. In particular, consider a continuous-time system with transfer function $G(s)$, and let $y(t)$ be the output resulting from a specific input $x(t)$ with no initial energy in the system. In response matching, the objective is to construct a discrete-time system with transfer function $G_d(z)$ such that when the input $x[n]$ to the discrete-time system is $x[n] = x(nT) = x(t)|_{t=nT}$, the output $y[n]$ of the discrete-time system is $y[n] = y(nT) = y(t)|_{t=nT}$, where T is the sampling interval. In other words, for the specific input under consideration, the output $y(t)$ of the continuous-time system matches the output $y[n]$ of the discrete-time system at the sample times $t = nT$. Clearly, the transfer function $G_d(z)$ of the desired discrete-time system is given by

$$G_d(z) = \frac{Y(z)}{X(z)}$$

where $X(z)$ and $Y(z)$ are the z-transforms of the discretized input $x(nT)$ and output $y(nT)$, respectively.

Of particular interest in digital control is step-response matching, where the input $x(t)$ is a step function. In this case, the output $y(t)$ is computed by taking the inverse Laplace transform of $Y(s) = G(s)/s$, where $G(s)$ is the transfer function of the given continuous-time system. To determine the corresponding discrete-time system, $y(t)$ is discretized to obtain $y[n] = y(nT)$, and then the z-transform $Y(z)$ of $y[n]$ is computed. The transfer function $G_d(z)$ of the discrete-time system is given by

$$G_d(z) = Y(z)\frac{z-1}{z} \qquad (12.40)$$

The process is illustrated by the following example.

Example 12.11 *Step-Response Matching*

Consider the continuous-time system with transfer function

$$G(s) = 0.2\frac{s+0.1}{s+2}$$

The transform of the step response of this system is

$$Y(s) = 0.2\frac{s+0.1}{s(s+2)}$$

$$= \frac{0.01}{s} + \frac{0.19}{s+2} \tag{12.41}$$

and taking the inverse Laplace transform of (12.41) gives the step response:

$$y(t) = 0.01 + 0.19e^{-2t}, \qquad t \geq 0$$

The discretized version of $y(t)$ is

$$y[n] = 0.01 + 0.19e^{-2nT}, \qquad n \geq 0 \tag{12.42}$$

and taking the z-transform of (12.42) gives

$$Y(z) = 0.01\frac{z}{z-1} + 0.19\frac{z}{z-e^{-2T}}$$

$$= \frac{0.2z^2 - (0.01e^{-2T} + 0.19)z}{(z-1)(z-e^{-2T})}$$

Hence using (12.40) yields the following transfer function for the corresponding discrete-time system:

$$G_d(z) = \frac{0.2z - (0.01e^{-2T} + 0.19)}{z - e^{-2T}}$$

MATLAB can be used to perform step-response matching. For instance, for the system in Example 12.11, the coefficients of the transfer function $G(s)$ are stored in num and den, a value is defined for the sampling time T, and the command c2dm is used:

```
num = .2*[1 .1];
den = [1 2];
[numd,dend] = c2dm(num,den,T,'zoh');
```

The reader is invited to run this and check the results against those obtained in Example 12.11. It is worth noting that in the command c2dm, if the option 'zoh' is replaced by 'tustin', the resulting computer computation uses the bilinear transformation to obtain the discrete-time system.

The main differences between the design of digital filters and the design of digital controllers are that in digital control, the determination of the sampling period must take into account the effect of the feedback, and a delay introduced by a digital controller may affect stability in the feedback loop. These considerations are discussed in more detail below.

Given a continuous-time plant with transfer function $G_p(s)$, suppose that an analog controller with transfer function $G_c(s)$ has been designed using a method such as the root-locus technique discussed in Section 10.3. To map $G_c(s)$ to a digital equivalent $G_d(z)$ using the bilinear transformation or the response matching technique, it is necessary to determine the appropriate sampling period T. Generally, the smaller the sampling time, the better the matching between the desired continuous-time controller and the digital controller that is implemented. Improving the efficiency of the computer program that performs the discrete-time calculations can reduce the sampling time; however, a reduction in sampling time generally requires the use of faster (and more expensive) A/D converters and digital signal processors. Thus it is important to compute the maximum sampling time that yields a good approximation in using the digital controller in place of the analog controller. An appropriate sampling frequency can be determined via the following analysis.

In the block diagram shown in Figure 12.25b, note that the signal to be sampled is the error $e(t) = r(t) - y(t)$, where $r(t)$ is the reference and $y(t)$ is the measured output signal. The reference signal has a frequency content that is usually known; however, the frequency content of $y(t)$ depends on the controller given by the transfer function $G_d(z)$. In general, $y(t)$ is not strictly bandlimited, but higher frequencies are attenuated sufficiently so that sampling will not cause substantial aliasing errors. To determine an appropriate sampling frequency, the frequency content of $y(t)$ for the system with the analog controller can first be found by using the ω-domain relationship:

$$Y(\omega) = G_{\text{cl}}(\omega)R(\omega) \tag{12.43}$$

where the closed-loop frequency response function $G_{cl}(\omega)$ is given by

$$G_{\text{cl}}(\omega) = \frac{G_c(\omega)G_p(\omega)}{1 + G_c(\omega)G_p(\omega)}$$

Thus, from (12.43) it is seen that the frequency content of $y(t)$ depends on the frequency response characteristics of the closed-loop system and on the frequency content of the reference input $r(t)$. Now for sufficiently small sampling times, the digital implementation of the controller closely approximates the designed analog controller so that the frequency content of the measured output $y(t)$ (when the digital controller is utilized) is closely approximated by (12.43). Hence the sampling frequency should be chosen to be higher than twice the largest significant frequency component of $Y(\omega)$ given by (12.43), and it should be sufficiently large so that (12.43) gives a good approximation of the frequency content of the output when the digital controller is used. One rule of thumb given in Franklin et al. [1990] is to choose the sampling frequency to be greater than 20 times the bandwidth of the analog closed-loop system.

Once a specific sampling interval T is chosen, the continuous-time controller $G_c(s)$ can be mapped to a discrete-time equivalent using the step-response matching method or the bilinear transformation. As mentioned previously, other methods of mapping are also available. Most will yield nearly the same answer if the sampling period is small with respect to the system's natural frequencies. The discretization process is illustrated below.

Example 12.12 *Digital Control of dc Motor*

A digital controller implementation will be given for an analog controller for the dc motor defined in Example 10.5. Here the plant (the dc motor) is given by

$$G_p(s) = \frac{10}{(s + 0.1)s}$$

and the analog controller has transfer function

$$G_c(s) = 0.2\frac{s + 0.1}{s + 2}$$

The corresponding closed-loop transfer function is given by

$$G_{\text{cl}}(s) = \frac{5}{s^2 + 2s + 5}$$

The natural frequency of the closed-loop system is $\omega_n = \sqrt{5} \approx 2.24$ rad/sec. Since this is a second-order system with no zeros, the natural frequency is approximately the same as the bandwidth. Thus, to get good performance from the digital controller, the sampling frequency is chosen to be $\omega_s \approx 20\omega_n$ or $T = 0.14$ second.

From the results in Example 12.11, the digital controller found using step-response matching for $T = 0.14$ is

$$G_d(z) = 0.2\frac{z - 0.988}{z - 0.756} \tag{12.44}$$

To use the bilinear transformation to find the digital controller, substitute $s = 2(z - 1)/T(z + 1)$ in $G_c(s)$ for s:

$$G_d(z) = G_c(s)\big|_{s=2(z-1)/T(z+1)} = \frac{0.177(z - 0.986)}{z - 0.754} \tag{12.45}$$

Due to the small sampling time, the digital controllers in (12.44) and (12.45) found from the two different mapping methods are nearly identical.

To simulate the response of the plant with digital control, the M-file `hybrid` can be used. This M-file is available from the Web site. The commands to derive $G_d(z)$ from $G_c(s)$ using the bilinear transformation and to simulate the response to a step input are

```
T = 0.14;
Nc = .2*[1 .1]; % analog controller
Dc = [1 2];
[Nd,Dd] = bilinear(Nc,Dc,1/T); % digital controller
t = 0:.5*T:10;
u = ones(1,length(t));     % step input
Np = 10; % plant
Dp = [1 .1 0];
[theta,uc] = hybrid(Np,Dp,Nd,Dd,T,t,u);
```

The responses to a step reference are shown in Figure 12.26 for three different digital controllers that are implemented with different sampling frequencies. Also shown in

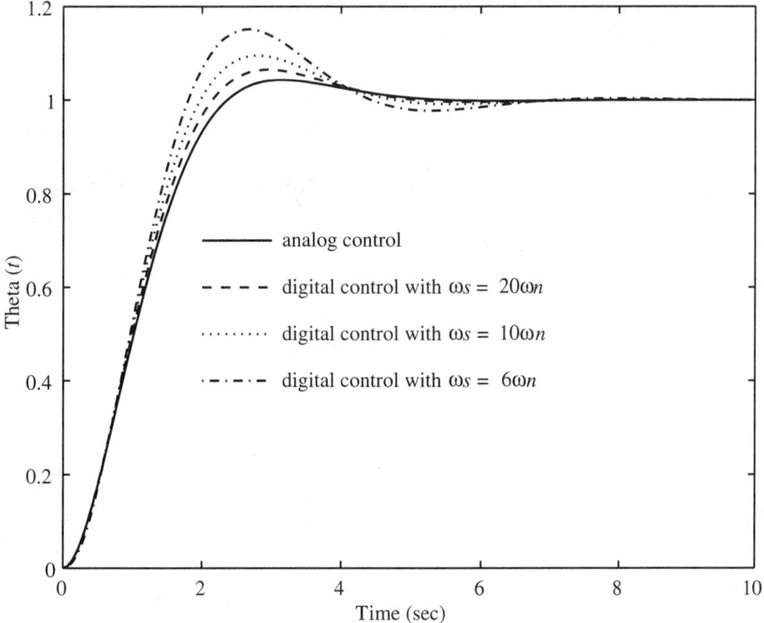

Figure 12.26 Response to a step reference signal in Example 12.12.

Figure 12.26 is the step response found using the analog controller. Notice that there is very little degradation in the response, due to the digital implementation when $\omega_s = 20\omega_n$, but the degradation is more apparent with a smaller sampling frequency such as $\omega_s = 10\omega_n$ and $\omega_s = 6\omega_n$.

PROBLEMS

12.1. Determine an appropriate sampling frequency for the following signals that avoids aliasing:

(a) $x(t) = 3\,\mathrm{sinc}^2(t/2\pi), -\infty < t < \infty$

(b) $x(t) = 4\,\mathrm{sinc}(t/\pi)\cos 2t, -\infty < t < \infty$

(c) $x(t) = e^{-5t}u(t)$

12.2. Digitize the following systems using the bilinear transformation. Assume that $T = 0.2$ second.

(i) $H(s) = 2/(s + 2)$

(ii) $H(s) = 4(s + 1)/(s^2 + 4s + 4)$

(iii) $H(s) = 2s/(s^2 + 1.4s + 1)$

(a) For each continuous-time system, simulate the step response using `step`

(b) For each discrete-time system derived in part (a), simulate the step response using `dstep`. Compare these responses, $y[n] = y(nT)$, to the corresponding responses $y(t)$ obtained in part (a) by plotting the results.

12.3. The bilinear transformation was introduced in (12.23) as a means of approximating the exact mapping $s = (1/T)\ln(z)$. An alternative derivation for the bilinear transformation involves the trapezoidal approximation of an integral:

$$\int_{nT}^{(n+1)T} f(t)\,dt \approx \frac{T}{2}[f([n+1]T) + f(nT)]$$

where the right-hand side of the expression represents the area of the trapezoid that best fits under the curve $f(t)$ from $t = nT$ to $t = (n+1)T$ (see Figure P12.3). Now consider a first-order continuous-time system:

$$H(s) = \frac{Y(s)}{X(s)} = \frac{a}{s+a}$$

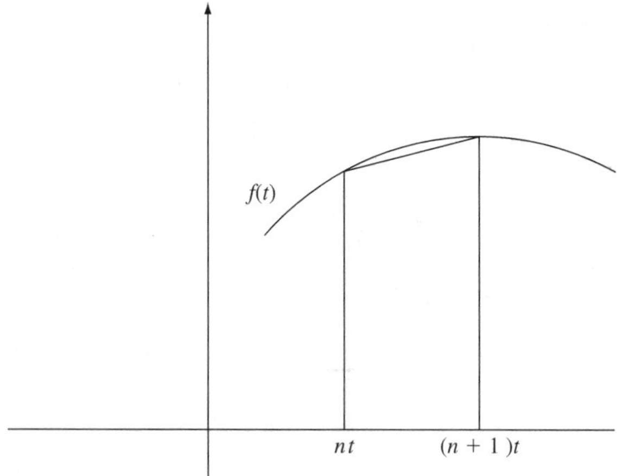

Figure P12.3

To derive the bilinear transformation, perform the following steps: (1) find the differential equation that relates the input $x(t)$ to the output $y(t)$; (2) integrate both sides of the differential equation from $t = nT$ to $t = (n+1)T$ using the trapezoidal approximation where appropriate; (3) obtain a corresponding difference equation by letting $y[n] = y(nT)$ and $x[n] = x(nT)$; (4) compute the digital transfer function $H_d(z)$ of the difference equation; and (5) find a relationship between s and z so that $H(s) = H_d(z)$.

12.4. Consider the hybrid system shown in Figure P12.4. Compute exact values for $y(1), y(2)$, and $y(3)$ when:

(a) $w(0) = -1, y(0) = 1$, and $x(t) = \delta(t) =$ unit impulse

(b) $w(0) = y(0) = 0$ and $x(t) = e^{-2t}u(t)$

(c) $w(0) = 2, y(0) = -1$, and $x(t) = u(t)$

12.5. Consider the one-pole lowpass filter given by the transfer function

$$H(s) = \frac{B}{s+B}$$

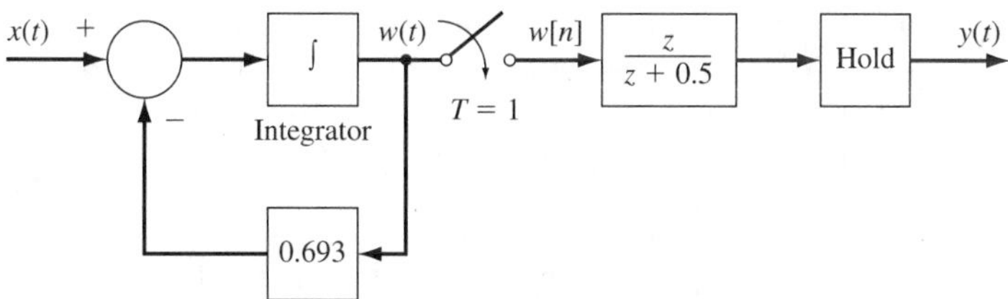

Figure P12.4

(a) Design a discrete-time system that realizes this filter using the bilinear transformation for $\omega_s = B$, $\omega_s = 2B$, and $\omega_s = 5B$.

(b) Sketch the frequency response of the continuous-time system for $B = 10$ and mark the three different sampling frequencies. Which should give a more accurate discrete-time realization?

(c) Use MATLAB to compute the step responses of the three discrete-time systems obtained in part (a) with $B = 10$. Plot the responses versus $t = nT$, where T is the sampling period. On the same graph, plot the step response of the original continuous-time system and compare the accuracy of the three discretizations.

12.6. Suppose that the sampled sinusoid $x(nT) = \cos \omega_0 nT$, $n = 0, \pm 1, \pm 2, \ldots$, is applied (with $\omega_s = 2B$) to the discretization constructed in Problem 12.5. Determine the range of values of ω_0 for which the peak magnitude of the resulting output response is greater than or equal to 0.707. In other words, determine the "effective" 3-dB bandwidth of the discretization. Compare your results with that obtained in part (b) of Problem 12.5.

12.7. A two-pole bandpass filter is given by the transfer function

$$H(s) = \frac{100}{s^2 + 2s + 101}$$

(a) Determine the 3-dB points of the filter.

(b) Digitize the filter using the bilinear transformation for an arbitrary sampling interval T.

(c) Plot the frequency response function of the digital filter $H_d(z)$ obtained in part (b). Take $T = 0.02$ second.

12.8. Consider the two-pole highpass filter given by the transfer function

$$H(s) = \frac{s^2}{s^2 + \sqrt{2}s + 1}$$

(a) Determine the 3-dB point of the filter.

(b) Discretize the filter using the bilinear transformation.

(c) With the sampling interval $T = 0.1$, plot the frequency response of the digital filter obtained in part (b).

12.9. Consider the two-pole Chebyshev lowpass filter given by

$$H(s) = \frac{0.5\omega_c^2}{s^2 + 0.645\omega_c s + 0.708\omega_c^2}$$

where ω_c is the 3-dB point.

(a) Discretize the filter using the bilinear transformation.

(b) Determine the output response $y(nT)$ of the discretized filter to the inputs $x(nT)$ given below. Take $\omega_c = 6\pi$ and $T = 0.01$. Plot the resulting analog signal $y(t)$ generated from an ideal reconstructor [i.e., plot $y(nT)$ using the `plot` command].

- **(i)** $x(nT) = p_1(nT)$, where $p_1(t) = u(t + 1/2) - u(t - 1/2)$
- **(ii)** $x(nT) = p_1(nT) + 0.5w(nT)$, where $w(nT)$ is a noise signal whose values are random numbers between 0 and 1 (use `rand` in MATLAB to generate the signal)
- **(iii)** $x(nT) = p_1(nT) + w(nT)$
- **(iv)** $x(nT) = (1 - 2|nT|)p_1(nT)$
- **(v)** $x(nT) = (1 - 2|nT|)p_1(nT) + 0.5w(nT)$

12.10. Design a three-pole lowpass IIR filter to have an analog cutoff frequency of $\omega_c = 15$. The sampling interval is $T = 0.1$. Perform the design twice, once without prewarping the frequency and once with prewarping the frequency. Plot the frequency curves of the two filters and compare the actual digital cutoff frequencies. Base your design on a Butterworth analog prototype filter.

12.11. Use analytical methods to design a three-pole lowpass IIR digital filter that has an analog cutoff frequency of $\omega_c = 10$ rad/sec and assume a sampling period of $T = 0.1$ second. Base your design on a Chebyshev analog prototype filter with a passband ripple of 3 dB.

(a) Specify the desired digital cutoff frequency, Ω_c. Also, give the largest frequency component of an input $x(t)$ that would be allowed to avoid aliasing.

(b) Verify your analytical design by using MATLAB to design the filter numerically. Plot the frequency response function for the resulting digital filter. Measure the actual digital cutoff frequency.

(c) Use MATLAB to simulate the response to the following sampled signal: $x(nT) = 1 + \sin \pi nT + \sin 6\pi nT$. Plot the analog input and outputs, $x(t)$ and $y(t)$, and the digital filter inputs and outputs, $x[n]$ and $y[n]$.

12.12. Use analytical methods to design a three-pole highpass IIR digital filter that has an analog cutoff frequency of $\omega_c = 10$ rad/sec and assume a sampling period of $T = 0.1$ second. Base your design on a Butterworth analog prototype filter.

(a) Specify the desired digital cutoff frequency, Ω_c. Also, give the largest frequency component of $x(t)$ that would be allowed to avoid aliasing.

(b) Verify your analytical design by using MATLAB to design the filter numerically. Plot the frequency response function for the resulting digital filter. Measure the actual digital cutoff frequency.

(c) Use MATLAB to simulate the response to the following sampled signal: $x(nT) = 1 + \sin \pi nT + \sin 6\pi nT$. Plot the analog input and outputs, $x(t)$ and $y(t)$, and the digital filter inputs and outputs, $x[n]$ and $y[n]$.

12.13. Use MATLAB to design a three-pole bandpass IIR digital filter that has an analog passband of $\omega_c = 1$ to $\omega_c = 5$ rad/sec; assume a sampling period of $T = 0.1$ second. Base your design on a Butterworth analog prototype filter. Specify the filter in terms of $H_d(z)$.

(a) Plot the frequency response function for the resulting digital filter.

(b) From the frequency response curve plotted in part (a), estimate the amplitude of the response to the following input signals: $x(t) = 1$, $x(t) = \sin \pi t$, and $x(t) = \sin 6\pi t$.

(c) Simulate the response of the filter to the sampled signal $x(nT) = 1 + \sin \pi nT + \sin 6\pi nT$. Compare the response to the expected amplitudes derived in part (b).

(d) Simulate the response of the filter to the sampled random signal $x(nT)$ generated from the MATLAB command `x = rand(1,200)`. Plot the input and corresponding response in continuous time.

12.14. Design a lowpass FIR filter of length $N = 30$ which has a cutoff frequency of $\Omega_c = \pi/3$. Use the rectangular window.

(a) Plot the impulse response of the filter.

(b) Plot the magnitude of the frequency response curve and determine the amount of the ripple in the passband.

(c) Compute and plot the response of the filter to an input of $x[n] = 1 + 2\cos(\pi n/6) + 2\cos(2\pi n/3)$. Discuss the filtering effect on the various components of $x[n]$.

12.15. Design a highpass FIR filter of length $N = 15$ that has a cutoff frequency of $\Omega_c = \pi/2$. Use the rectangular window.

(a) Plot the impulse response of the filter.

(b) Plot the magnitude of the frequency response curve and determine the amount of ripple in the passband.

(c) Repeat the design for length $N = 31$ and compare the two filters in terms of ripple and transition region.

(d) Compute and plot the response of each filter to an input of $x[n] = 2 + 2\cos(\pi n/3) + 2\cos(2\pi n/3)$. Compare the filters in terms of their effect on the various components of $x[n]$.

12.16. Design a lowpass FIR filter of length $N = 10$ that has a cutoff frequency of $\Omega_c = \pi/3$. Perform your design using **(i)** a rectangular window, **(ii)** a Hamming window, and **(iii)** a Hanning window.

(a) Plot the impulse responses of each filter.

(b) Plot the magnitude of the frequency response for each filter and compare the filters in terms of the stopband and passband ripples.

(c) Compute and plot the response of each filter to an input of $x[n] = 2 + 2\cos(\pi n/4) + 2\cos(\pi n/2)$. Compare the filters in terms of their effect on the various components of $x[n]$.

12.17. Consider the single-car system described in Problem 11.26 and given by the input/output differential equation

$$\frac{d^2y(t)}{dt^2} + \frac{k_f}{M}\frac{dy(t)}{dt} = \frac{1}{M}x(t)$$

where $y(t)$ is the position of the car at time t. Suppose that $M = 1$ and $k_f = 0.1$.

(a) Discretize the system using the bilinear transformation for $T = 5$ seconds.

(b) Discretize the system using step-response matching for $T = 5$ seconds.

(c) Consider the discrete-time approximation given in Problem 11.26 resulting from the Euler approximation. Obtain a transfer function $H_d(z)$ for this approximation for $T = 5$.

(d) For each digitization obtained in parts (a) to (c), plot the analog output $y(nT)$ of the discrete-time step response. Compare these results with the step response for the original continuous-time system. Plot all results for $t = 0$ to $t = 50$ seconds.

(e) Repeat parts (a) to (d) for $T = 1$ second.

12.18. Each continuous-time system given below represents a desired closed-loop transfer function that will be achieved by the use of a digital feedback controller. Determine an appropriate sampling frequency ω_s for each system.

(a) $G_{cl}(s) = 10/(s + 10)$

(b) $G_{cl}(s) = 4/(s^2 + 2.83s + 4)$

(c) $G_{cl}(s) = 9(s + 1)/(s^2 + 5s + 9)$

12.19. Consider the following continuous-time system:

$$G_p(s) = \frac{2}{s + 2}$$

Digital control is to be applied to this system so that the resulting closed-loop system has a pole at $s = -4$.

(a) Design a continuous-time feedback controller $G_c(s)$ that achieves the desired closed-loop pole.

(b) Obtain a digital controller from part (a), where $T = 0.25$ second. Using this control, simulate the step response of the hybrid system.

(c) Repeat part (b) with $T = 0.1$ second. Compare the resulting response with that obtained in part (b) and determine which is closer to the desired closed-loop response.

12.20. Consider the following continuous-time system:

$$G_p(s) = \frac{1}{(s + 1)(s + 2)}$$

The following lead controller is designed that gives closed-loop poles at $s = -4 \pm 8j$:

$$G_c(s) = \frac{73(s + 2)}{s + 5}$$

(a) Digitize the controller using step-response matching for $T = 0.1$ second. Using this control, simulate the step response of the hybrid system.

(b) Repeat part (a) with $T = 0.05$ second. Compare the resulting response with that obtained in part (a) and determine which is closer to the desired closed-loop response.

CHAPTER 13

State Representation

The models that have been considered so far in this book are mathematical representations of the input/output behavior of the system under study. In this chapter a new type of model is defined, which is specified in terms of a collection of variables that describe the internal behavior of the system. These variables are called the state variables of the system. The model that is defined in terms of the state variables is called the state or state-variable representation. The objective of this chapter is to define the state model and to study the basic properties of this model for both continuous-time and discrete-time systems. An in-depth development of the state approach to systems can be found in a number of textbooks. For example, the reader may want to refer to Kailath [1980], Brogan [1991], or Rugh [1996].

The state model is given in terms of a matrix equation, and thus the reader should be familiar with matrix algebra. A brief review is given in Appendix B. As a result of the matrix form of the state model, it is easily implemented on a computer. A number of commercially available software packages, such as MATLAB, can be used to carry out the matrix operations arising in the state model. In particular, MATLAB is built around operations involving matrices and vectors and thus is well suited for the study of the state model. MATLAB's definition of operations in terms of matrices and vectors results in state model computations being as easy as standard calculator operations.

The development of the state model begins in Section 13.1 with the notion of state and the definition of the state equations for a continuous-time system. The construction of state models from input/output differential equations is considered in

Section 13.2. The solution of the state equations is studied in Section 13.3. Then in Section 13.4 the discrete-time version of the state model is presented. In Section 13.5 the notion of equivalent state representations is studied, and in Section 13.6 the discretization of continuous-time state models is pursued.

13.1 STATE MODEL

Consider a single-input single-output causal continuous-time system with input $v(t)$ and output $y(t)$. Throughout this chapter the input will be denoted by $v(t)$, rather than $x(t)$ since the symbol "$x(t)$" will used to denote the system state as defined below.

Given a value t_1 of the time variable t, in general it is not possible to compute the output response $y(t)$ for $t \geq t_1$ from only knowledge of the input $v(t)$ for $t \geq t_1$. The reason for this is that the application of the input $v(t)$ for $t < t_1$ may put energy into the system that affects the output response for $t \geq t_1$. For example, a voltage or current applied to an RLC circuit for $t < t_1$ may result in voltages on the capacitors and/or currents in the inductors at time t_1. These voltages and currents at time t_1 can then affect the output of the RLC circuit for $t \geq t_1$.

Given a system with input $v(t)$ and output $y(t)$, for any time point t_1 the state $x(t_1)$ of the system at time $t = t_1$ is defined to be that portion of the past history ($t \leq t_1$) of the system required to determine the output response $y(t)$ for all $t \geq t_1$ given the input $v(t)$ for $t \geq t_1$. A nonzero state $x(t_1)$ at time t_1 indicates the presence of energy in the system at time t_1. In particular, the system is in the zero state at time t_1 if and only if there is no energy in the system at time t_1. If the system is in the zero state at time t_1, the response $y(t)$ for $t \geq t_1$ can be computed from knowledge of the input $v(t)$ for $t \geq t_1$. If the state at time t_1 is not zero, knowledge of the state is necessary to be able to compute the output response for $t \geq t_1$.

If the given system is finite dimensional, the state $x(t)$ of the system at time t is an N-element column vector given by

$$x(t) = \begin{bmatrix} x_1(t) \\ x_2(t) \\ \vdots \\ x_N(t) \end{bmatrix}$$

The components $x_1(t), x_2(t), \ldots, x_N(t)$ are called the *state variables* of the system. For example, suppose that the given system is an RLC circuit. From circuit theory, any energy in the circuit at time t is completely characterized by the voltages across the capacitors at time t and the currents in the inductors at time t. Thus the state of the circuit at time t can be defined to be a vector whose components are the voltages across the capacitors at time t and the currents in the inductors at time t. If the number of capacitors in the circuit is equal to N_C and the number of inductors in the circuit is equal to N_L, the total number of state variables is equal to $N_C + N_L$.

Now suppose that the system is an integrator with the input/output relationship

$$y(t) = \int_{t_0}^{t} v(\lambda)\, d\lambda, \qquad t > t_0 \tag{13.1}$$

where $y(t_0) = 0$. It will be shown that the state $x(t)$ of the integrator can be chosen to be the output $y(t)$ of the integrator at time t. To see this, let t_1 be an arbitrary value of time with $t_1 > t_0$. Then rewriting (13.1) yields

$$y(t) = \int_{t_0}^{t_1} v(\lambda)\, d\lambda + \int_{t_1}^{t} v(\lambda)\, d\lambda, \qquad t \geq t_1 \tag{13.2}$$

From (13.2) it is seen that the first term on the right-hand side of (13.2) is equal to $y(t_1)$. Therefore,

$$y(t) = y(t_1) + \int_{t_1}^{t} v(\lambda)\, d\lambda, \qquad t \geq t_1 \tag{13.3}$$

The relationship (13.3) shows that $y(t_1)$ characterizes the energy in the system at time t_1. More precisely, from (13.3) it is seen that $y(t)$ can be computed for all $t \geq t_1$ from knowledge of $v(t)$ for $t \geq t_1$ and knowledge of $y(t_1)$. Thus the state at time t_1 can be taken to be $y(t_1)$.

Now consider an interconnection of integrators, adders, subtracters, and scalar multipliers. Since adders, subtracters, and scalar multipliers are memoryless devices, the energy in the interconnection is completely characterized by the values of the outputs of the integrators. Thus the state at time t can be defined to be a vector whose components are the outputs of the integrators at time t.

State Equations

Consider a single-input single-output finite-dimensional continuous-time system with state $x(t)$ given by

$$x(t) = \begin{bmatrix} x_1(t) \\ x_2(t) \\ \vdots \\ x_N(t) \end{bmatrix}$$

The state $x(t)$ is a vector-valued function of time t. In other words, for any particular value of t, $x(t)$ is an N-element column vector. The vector-valued function $x(t)$ is called the *state trajectory of the system*.

The system with state $x(t)$ can be modeled by the state equations given by

$$\dot{x}(t) = f(x(t), v(t), t) \tag{13.4}$$

$$y(t) = g(x(t), v(t), t) \tag{13.5}$$

In (13.4), $\dot{x}(t)$ is the derivative of the state vector with the derivative taken component by component; that is,

$$\dot{x}(t) = \begin{bmatrix} \dot{x}_1(t) \\ \dot{x}_2(t) \\ \vdots \\ \dot{x}_N(t) \end{bmatrix}$$

On the right-hand side of (13.4), f is a function of the state $x(t)$ at time t, the input $v(t)$ at time t, and time t. Hence, by (13.4), the derivative of the state at time t is a function of time t and the state and input at time t.

The function f in (13.4) is a vector-valued function of several variables. That is, if values for t, $v(t)$, and the components of $x(t)$ are inserted in $f(x(t), v(t), t)$, the result is an N-element column vector that is equal to $\dot{x}(t)$. Since $x(t)$ and $\dot{x}(t)$ are N-element column vectors, (13.4) is a vector differential equation. In particular, (13.4) is a first-order vector differential equation.

In the second equation (13.5), g is another function of the state $x(t)$ at time t, the input $v(t)$ at time t, and time t. So the output $y(t)$ of the system at time t is a function of $x(t)$ and $v(t)$ at time t and time t. Equation (13.5) is called the *output equation* of the system.

Equations (13.4) and (13.5) comprise the state model of the system. This representation is a time-domain model of the system since the equations are in terms of functions of time. Note that the state model is specified in two parts; (13.4) describes the state response resulting from the application of an input $v(t)$ with initial state $x(t_0) = x_0$, while (13.5) gives the output response as a function of the state and input. The two parts of the state model correspond to a cascade decomposition of the system as illustrated in Figure 13.1. The double line for $x(t)$ in Figure 13.1 indicates that $x(t)$ is a vector signal.

From Figure 13.1 it is seen that the system state $x(t)$ is an "internal" vector variable of the system; that is, the state variables [the components of $x(t)$] are signals within the system. Since the state model is specified in terms of the internal vector variable $x(t)$, the representation is an internal model of the system. The form of this model is quite different from that of the external or input/output models studied in Chapters 2 and 3.

The functions f and g in the state equations (13.4) and (13.5) may be nonlinear functions of their arguments, in which case the given system is nonlinear. The system

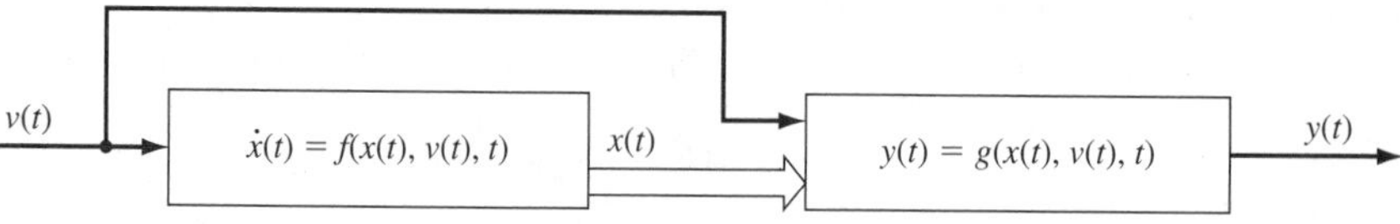

Figure 13.1 Cascade structure corresponding to state model.

is linear if and only if the functions f and g are both linear. If f and g are linear, the state equations can be written in the form

$$\dot{x}(t) = A(t)x(t) + b(t)v(t) \tag{13.6}$$

$$y(t) = c(t)x(t) + d(t)v(t) \tag{13.7}$$

In (13.6), $A(t)$ is a $N \times N$ matrix whose entries are functions of time t, and $b(t)$ is an N-element column vector whose components are functions of t. In (13.7), $c(t)$ is a N-element row vector with time-varying components and $d(t)$ is a real-valued function of time. The number N of state variables is called the *dimension* of the state model (or system).

The system is time invariant if and only if $A(t), b(t), c(t)$, and $d(t)$ are constant, that is, they do not vary with t. If this is the case, the state model is given by

$$\dot{x}(t) = Ax(t) + bv(t) \tag{13.8}$$

$$y(t) = cx(t) + dv(t) \tag{13.9}$$

Here the components of the matrix A and the vectors b and c are real numbers. The element d in (13.9) is a real number. The term $dv(t)$ is a "direct feed" between the input $v(t)$ and output $y(t)$. If $d = 0$, there is no direct connection between $v(t)$ and $y(t)$.

With a_{ij} equal to the ij entry of A and b_i equal to the ith component of b, (13.8) can be written in the expanded form

$$\begin{aligned}
\dot{x}_1(t) &= a_{11}x_1(t) + a_{12}x_2(t) + \cdots + a_{1N}x_N(t) + b_1v(t) \\
\dot{x}_2(t) &= a_{21}x_1(t) + a_{22}x_2(t) + \cdots + a_{2N}x_N(t) + b_2v(t) \\
&\vdots \\
\dot{x}_N(t) &= a_{N1}x_1(t) + a_{N2}x_2(t) + \cdots + a_{NN}x_N(t) + b_Nv(t)
\end{aligned}$$

With $c = [c_1 \quad c_2 \quad \cdots \quad c_N]$, the expanded form of (13.9) is

$$y(t) = c_1x_1(t) + c_2x_2(t) + \cdots + c_Nx_N(t) + dv(t)$$

From the expanded form of the state equations, it is seen that the derivative $\dot{x}_i(t)$ of the ith state variable and the output $y(t)$ are equal to linear combinations of all the state variables and the input.

13.2 CONSTRUCTION OF STATE MODELS

In the first part of this section it is shown how to construct a state model from the input/output differential equation of the system. The development begins with the first-order case.

Consider the single-input single-output continuous-time system given by the first-order input/output differential equation

$$\dot{y}(t) = f(y(t), v(t), t) \tag{13.10}$$

Defining the state $x(t)$ of the system to be equal to $y(t)$ results in the state model

$$\dot{x}(t) = f(x(t), v(t), t)$$
$$y(t) = x(t)$$

Thus it is easy to construct a state model from a first-order input/output differential equation. If the given system is linear and time invariant so that

$$\dot{y}(t) = -ay(t) + bv(t)$$

for some constants a and b, then the state model is

$$\dot{x}(t) = -ax(t) + bv(t)$$
$$y(t) = x(t)$$

In terms of the notation of (13.8) and (13.9), the coefficients A, b, c, and d of this state model are

$$A = -a, \qquad b = b, \qquad c = 1, \qquad d = 0$$

Now suppose that the given system has the second-order input/output differential equation

$$\ddot{y}(t) = f(y(t), \dot{y}(t), v(t), t) \tag{13.11}$$

Defining the state variables by

$$x_1(t) = y(t), \qquad x_2(t) = \dot{y}(t)$$

yields the state equations

$$\dot{x}_1(t) = x_2(t)$$
$$\dot{x}_2(t) = f(x_1(t), x_2(t), v(t), t)$$
$$y(t) = x_1(t)$$

If the system with input/output differential equation (13.11) is linear and time invariant, (13.11) can be written in the form

$$\ddot{y}(t) = -a_1\dot{y}(t) - a_0y(t) + b_0v(t)$$

Then with $x_1(t) = y(t)$ and $x_2(t) = \dot{y}(t)$, the state model is

$$\begin{bmatrix} \dot{x}_1(t) \\ \dot{x}_2(t) \end{bmatrix} = \begin{bmatrix} 0 & 1 \\ -a_0 & -a_1 \end{bmatrix} \begin{bmatrix} x_1(t) \\ x_2(t) \end{bmatrix} + \begin{bmatrix} 0 \\ b_0 \end{bmatrix} v(t)$$

$$y(t) = \begin{bmatrix} 1 & 0 \end{bmatrix} \begin{bmatrix} x_1(t) \\ x_2(t) \end{bmatrix}$$

The definition of state variables in terms of the output and derivatives of the output extends to any system given by the Nth-order input/output differential equation

$$y^{(N)}(t) = f(y(t), y^{(1)}(t), \ldots, y^{(N-1)}(t), v(t), t) \tag{13.12}$$

With the state variables defined by

$$x_i(t) = y^{(i-1)}(t), \qquad i = 1, 2, \ldots, N$$

the resulting state equations are

$$\begin{aligned}
\dot{x}_1(t) &= x_2(t) \\
\dot{x}_2(t) &= x_3(t) \\
&\vdots \\
\dot{x}_{N-1}(t) &= x_N(t) \\
\dot{x}_N(t) &= f(x_1(t), x_2(t), \ldots, x_N(t), v(t), t) \\
y(t) &= x_1(t)
\end{aligned}$$

From the constructions above, it is tempting to conclude that the state variables of a system can always be defined to be equal to the output $y(t)$ and derivatives of $y(t)$. Unfortunately, this is not the case. For instance, suppose that the system is given by the linear second-order input/output differential equation

$$\ddot{y}(t) + a_1\dot{y}(t) + a_0 y(t) = b_1\dot{v}(t) + b_0 v(t) \tag{13.13}$$

where $b_1 \neq 0$. Note that (13.13) is not a special case of (13.12) since $\ddot{y}(t)$ depends on $\dot{v}(t)$.

If $x_1(t) = y(t)$ and $x_2(t) = \dot{y}(t)$, it is not possible to eliminate the term $b_1\dot{v}(t)$ in (13.13). Thus there is no state model with respect to this definition of state variables. But the system does have the following state model:

$$\begin{bmatrix} \dot{x}_1(t) \\ \dot{x}_2(t) \end{bmatrix} = \begin{bmatrix} 0 & 1 \\ -a_0 & -a_1 \end{bmatrix} \begin{bmatrix} x_1(t) \\ x_2(t) \end{bmatrix} + \begin{bmatrix} 0 \\ 1 \end{bmatrix} v(t), \qquad y(t) = [b_0 \quad b_1] \begin{bmatrix} x_1(t) \\ x_2(t) \end{bmatrix} \tag{13.14}$$

To verify that (13.14) is a state model, it must be shown that the input/output differential equation corresponding to (13.14) is the same as (13.13). Expanding (13.14) gives

$$\dot{x}_1(t) = x_2(t) \tag{13.15}$$

$$\dot{x}_2(t) = -a_0 x_1(t) - a_1 x_2(t) + v(t) \tag{13.16}$$

$$y(t) = b_0 x_1(t) + b_1 x_2(t) \tag{13.17}$$

Differentiating both sides of (13.17) and using (13.15) and (13.16) yields

$$\begin{aligned}
\dot{y}(t) &= b_0 x_2(t) + b_1[-a_0 x_1(t) - a_1 x_2(t) + v(t)] \\
&= -a_1 y(t) + (a_1 b_0 - a_0 b_1)x_1(t) + b_0 x_2(t) + b_1 v(t)
\end{aligned} \tag{13.18}$$

Differentiating both sides of (13.18) and again using (13.15) and (13.16) gives

$$\begin{aligned}\ddot{y}(t) &= -a_1\dot{y}(t) + (a_1b_0 - a_0b_1)x_2(t)\\ &\quad + b_0[-a_0x_1(t) - a_1x_2(t) + v(t)] + b_1\dot{v}(t)\\ &= -a_1\dot{y}(t) - a_0y(t) + b_0v(t) + b_1\dot{v}(t)\end{aligned}$$

This is the same as the input/output differential equation (13.13) of the given system, and thus (13.14) is a state model.

The state variables $x_1(t)$ and $x_2(t)$ in the state model (13.14) can be expressed in terms of $v(t)$, $y(t)$, and $\dot{y}(t)$. The derivation of these expressions is left to the interested reader (see Problem 13.3).

Now consider the linear time-invariant system given by the Nth-order input/output differential equation

$$y^{(N)}(t) + \sum_{i=0}^{N-1} a_i y^{(i)}(t) = \sum_{i=0}^{N-1} b_i v^{(i)}(t)$$

This system has the N-dimensional state model $\dot{x}(t) = Ax(t) + bv(t)$, $y(t) = cx(t)$, where

$$A = \begin{bmatrix} 0 & 1 & 0 & \cdots & 0 \\ 0 & 0 & 1 & & 0 \\ \vdots & \vdots & & \ddots & \vdots \\ 0 & 0 & 0 & & 1 \\ -a_0 & -a_1 & -a_2 & \cdots & -a_{N-1} \end{bmatrix}, \quad b = \begin{bmatrix} 0 \\ 0 \\ \vdots \\ 0 \\ 1 \end{bmatrix}, \quad c = [b_0 b_1 + b_{N-1}]$$

The verification that this is a state model is omitted.

Integrator Realizations

Any linear time-invariant system given by the N-dimensional state model

$$\dot{x}(t) = Ax(t) + bv(t)$$

$$y(t) = c(t) + dv(t)$$

can be realized by an interconnection of N integrators and combinations of adders, subtracters, and scalar multipliers. The steps of the realization process are as follows.

Step 1. For each state variable $x_i(t)$, construct an integrator and define the output of the integrator to be $x_i(t)$. The input to the ith integrator will then be equal to $\dot{x}_i(t)$. Note that if there are N state variables, the integrator realization will contain N integrators.

Step 2. Put an adder/subtracter in front of each integrator. Feed into the adders/subtracters scalar multiples of the state variables and input according to the vector equation $\dot{x}(t) = Ax(t) + bv(t)$.

Step 3. Put scalar multiples of the state variables and input into an adder/subtracter to realize the output $y(t)$ in accordance with the equation $y(t) = cx(t) + dv(t)$.

Example 13.1 *Integrator Realization*

Consider a two-dimensional state model with arbitrary coefficients; that is,

$$\begin{bmatrix} \dot{x}_1(t) \\ \dot{x}_2(t) \end{bmatrix} = \begin{bmatrix} a_{11} & a_{12} \\ a_{21} & a_{22} \end{bmatrix} \begin{bmatrix} x_1(t) \\ x_2(t) \end{bmatrix} + \begin{bmatrix} b_1 \\ b_2 \end{bmatrix} v(t)$$

$$y(t) = [c_1 \quad c_2] \begin{bmatrix} x_1(t) \\ x_2(t) \end{bmatrix}$$

Following the steps above results in the realization shown in Figure 13.2.

There is a converse to the result that any linear time-invariant system given by a state model has an integrator realization. Namely, any system specified by an interconnection consisting of N integrators and combinations of adders, subtracters, and scalar multipliers has a state model of dimension N. A state model can be computed directly from the interconnection by employing the following steps.

Step 1. Define the output of each integrator in the interconnection to be a state variable. Then if the output of the ith integrator is $x_i(t)$, the input to this integrator is $\dot{x}_i(t)$.

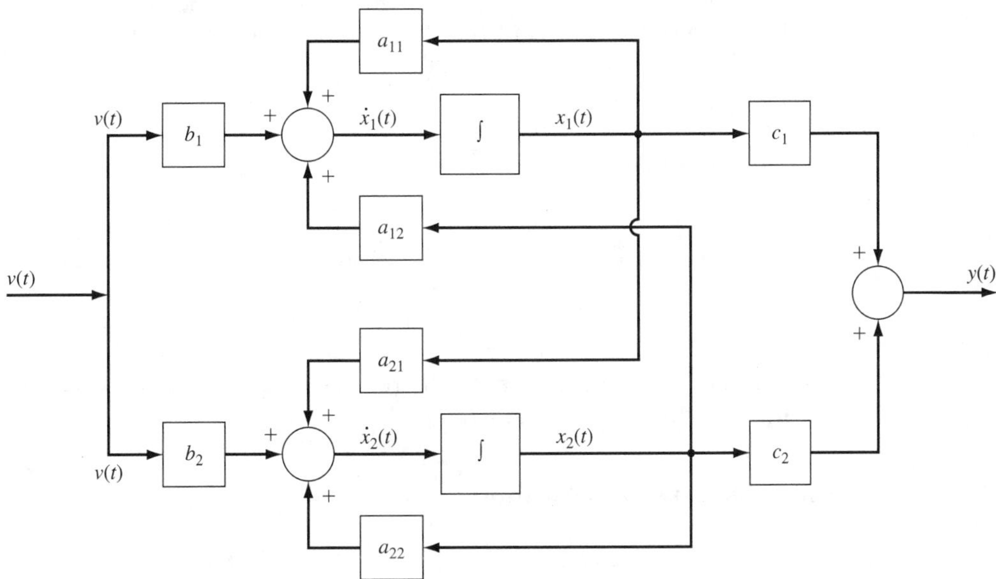

Figure 13.2 Realization in Example 13.1.

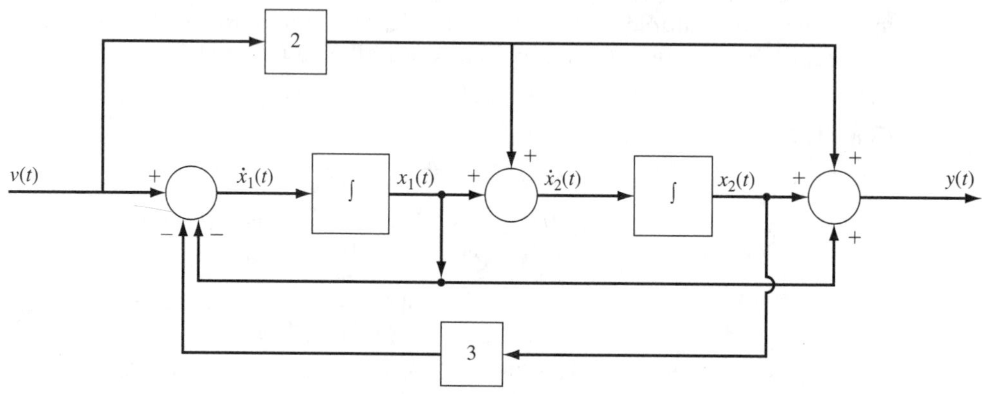

Figure 13.3 System in Example 13.2.

Step 2. By looking at the interconnection, express each $\dot{x}_i(t)$ in terms of a sum of scalar multiples of the state variables and input. Writing these relationships in matrix form yields the vector equation $\dot{x}(t) = Ax(t) + bv(t)$.

Step 3. Again looking at the interconnection, express the output $y(t)$ in terms of scalar multiples of the state variables and input. Writing this in vector form yields the output equation $y(t) = cx(t) + dv(t)$.

Example 13.2 *State Equations from an Integrator Realization*

Consider the system shown in Figure 13.3. With the output of the first integrator denoted by $x_1(t)$ and the output of the second integrator denoted by $x_2(t)$, from Figure 13.3

$$\dot{x}_1(t) = -x_1(t) - 3x_2(t) + v(t)$$

$$\dot{x}_2(t) = x_1(t) + 2v(t)$$

Also from Figure 13.3,

$$y(t) = x_1(t) + x_2(t) + 2v(t)$$

Thus the coefficient matrices of the state model are

$$A = \begin{bmatrix} -1 & -3 \\ 1 & 0 \end{bmatrix}, \quad b = \begin{bmatrix} 1 \\ 2 \end{bmatrix}, \quad c = [1 \quad 1], \quad d = 2$$

From the above results it is seen that there is a one-to-one correspondence between integrator realizations and state models.

Multi-Input Multi-Output Systems

The state model generalizes very easily to multi-input multi-output systems. In particular, the state model of a p-input r-output linear time-invariant finite-dimensional continuous-time system is given by

$$\dot{x}(t) = Ax(t) + Bv(t)$$

$$y(t) = Cx(t) + Dv(t)$$

where now B is a $N \times p$ matrix of real numbers, C is a $r \times N$ matrix of real numbers, and D is a $r \times p$ matrix of real numbers. The matrix A is still $N \times N$, as in the single-input single-output case.

If a p-input r-output system is specified by a collection of coupled input/output differential equations, a state model can be constructed by generalizing the procedure given above in the single-input single-output case. The process is illustrated by the following example.

Example 13.3 *Two-Car System*

Consider two cars moving along a level surface as shown in Figure 13.4. It is assumed that the mass of both cars is equal to M and that the coefficient corresponding to viscous friction is the same for both cars and is equal to k_f. As illustrated, $d_1(t)$ is the position of the first car at time t, $d_2(t)$ is the position of the second car at time t, $f_1(t)$ is the drive or braking force applied to the first car, and $f_2(t)$ is the drive or braking force applied to the second car. Thus the motion of the two cars is given by the differential equations

$$\ddot{d}_1(t) + \frac{k_f}{M}\dot{d}_1(t) = \frac{1}{M}f_1(t)$$

$$\ddot{d}_2(t) + \frac{k_f}{M}\dot{d}_2(t) = \frac{1}{M}f_2(t)$$

The first car also has a radar, which gives a measurement of the distance

$$w(t) = d_2(t) - d_1(t)$$

between the two cars at time t.

The inputs of the two-car system are defined to be $f_1(t)$ and $f_2(t)$, and thus the system is a two-input system. The output is defined to be

$$y(t) = \begin{bmatrix} \dot{d}_1(t) \\ \dot{d}_2(t) \\ w(t) \end{bmatrix}$$

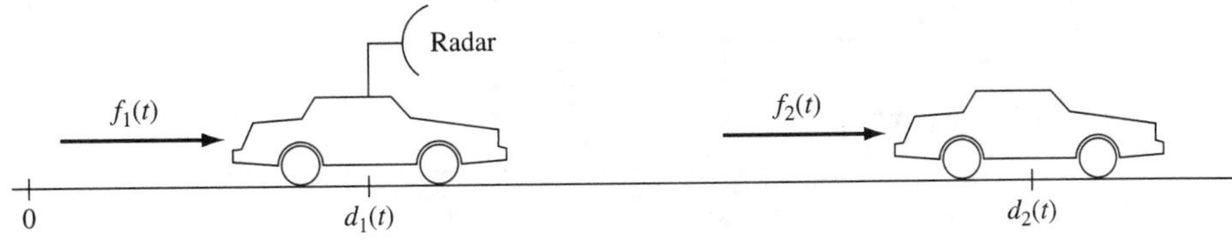

Figure 13.4 Two-car system.

With this definition of the output, the system is a three-output system. Now if the state variables are defined by

$$x_1(t) = \dot{d}_1(t)$$
$$x_2(t) = \dot{d}_2(t)$$
$$x_3(t) = w(t)$$

then the state model of the system is

$$\begin{bmatrix} \dot{x}_1(t) \\ \dot{x}_2(t) \\ \dot{x}_3(t) \end{bmatrix} = \begin{bmatrix} \frac{-k_f}{M} & 0 & 0 \\ 0 & \frac{-k_f}{M} & 0 \\ -1 & 1 & 0 \end{bmatrix} \begin{bmatrix} x_1(t) \\ x_2(t) \\ x_3(t) \end{bmatrix} + \begin{bmatrix} \frac{1}{M} & 0 \\ 0 & \frac{1}{M} \\ 0 & 0 \end{bmatrix} \begin{bmatrix} f_1(t) \\ f_2(t) \end{bmatrix}$$

$$y(t) = \begin{bmatrix} 1 & 0 & 0 \\ 0 & 1 & 0 \\ 0 & 0 & 1 \end{bmatrix} \begin{bmatrix} x_1(t) \\ x_2(t) \\ x_3(t) \end{bmatrix}$$

If a p-input r-output system is given by an interconnection of integrators, adders, subtracters, and scalar multipliers, a state model can be constructed directly from the interconnection. The process is very similar to the steps given above in the single-input single-output case.

Example 13.4 *Two-Input Two-Output System*

Consider the two-input two-output system shown in Figure 13.5. From the figure,

$$\dot{x}_1(t) = -3y_1(t) + v_1(t)$$
$$\dot{x}_2(t) = v_2(t)$$
$$y_1(t) = x_1(t) + x_2(t)$$
$$y_2(t) = x_2(t)$$

Inserting the expression for $y_1(t)$ into the expression for $\dot{x}_1(t)$ gives

$$\dot{x}_1(t) = -3[x_1(t) + x_2(t)] + v_1(t)$$

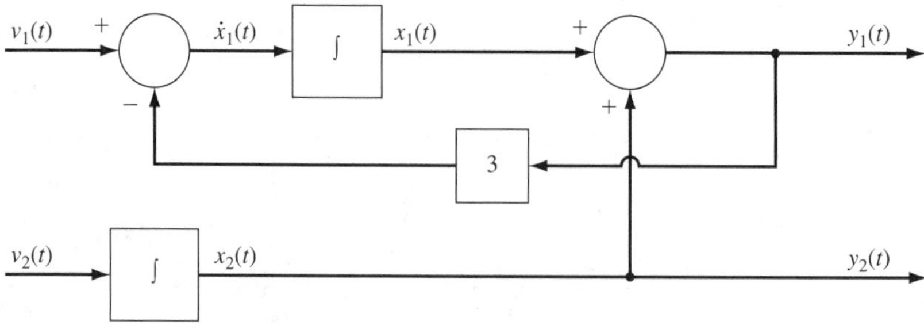

Figure 13.5 System in Example 13.4.

Putting these equations in matrix form results in the state model

$$\begin{bmatrix} \dot{x}_1(t) \\ \dot{x}_2(t) \end{bmatrix} = \begin{bmatrix} -3 & -3 \\ 0 & 0 \end{bmatrix} \begin{bmatrix} x_1(t) \\ x_2(t) \end{bmatrix} + \begin{bmatrix} 1 & 0 \\ 0 & 1 \end{bmatrix} \begin{bmatrix} v_1(t) \\ v_2(t) \end{bmatrix}$$

$$\begin{bmatrix} y_1(t) \\ y_2(t) \end{bmatrix} = \begin{bmatrix} 1 & 1 \\ 0 & 1 \end{bmatrix} \begin{bmatrix} x_1(t) \\ x_2(t) \end{bmatrix}$$

13.3 SOLUTION OF STATE EQUATIONS

Consider the p-input r-output linear time-invariant continuous-time system given by the state model

$$\dot{x}(t) = Ax(t) + Bv(t) \tag{13.19}$$

$$y(t) = Cx(t) + Dv(t) \tag{13.20}$$

Recall that the matrix A is $N \times N$, B is $N \times p$, C is $r \times N$, and D is $r \times p$. Given an initial state $x(0)$ at initial time $t = 0$ and an input $v(t)$, $t \geq 0$, in this section an analytical expression is derived for the solution $x(t)$ of (13.19). Then from this, (13.20) will be used to derive an expression for the output response $y(t)$. The numerical solution of (13.19) and (13.20) using MATLAB will be demonstrated in examples given below. The numerical solution procedure used by MATLAB is discussed in Section 13.6.

The development begins by considering the free (unforced) vector differential equation

$$\dot{x}(t) = Ax(t), \qquad t > 0 \tag{13.21}$$

with initial state $x(0)$. To solve (13.21), it is necessary to define the *matrix exponential* e^{At}, which is a generalization of the scalar exponential e^{at}. For each real value of t, e^{At} is defined by the matrix power series

$$e^{At} = I + At + \frac{A^2t^2}{2!} + \frac{A^3t^3}{3!} + \frac{A^4t^4}{4!} + \cdots \tag{13.22}$$

where I is the $N \times N$ identity matrix. The matrix exponential e^{At} is a $N \times N$ matrix of time functions. Later it will be shown how the elements of e^{At} can be computed by using the Laplace transform.

There are a couple of properties of e^{At} that are needed: First, for any real numbers t and λ,

$$e^{A(t+\lambda)} = e^{At}e^{A\lambda} \tag{13.23}$$

The relationship (13.23) can be checked by setting $t = t + \lambda$ in (13.22).

Taking $\lambda = -t$ in (13.23) gives

$$e^{At}e^{-At} = e^{A(t-t)} = I_N \tag{13.24}$$

The relationship (13.24) shows that the matrix e^{At} always has an inverse, which is equal to the matrix e^{-At}.

The derivative $d/dt(e^{At})$ of the matrix exponential e^{At} is defined to be the matrix formed by taking the derivative of the components of e^{At}. The derivative $d/dt\,(e^{At})$ can be computed by taking the derivative of the terms comprising the matrix power series in (13.22). The result is

$$\begin{aligned}\frac{d}{dt}e^{At} &= A + A^2t + \frac{A^3t^2}{2!} + \frac{A^4t^3}{3!} + \cdots \\ &= A\left(I + At + \frac{A^2t^2}{2!} + \frac{A^3t^3}{3!} + \cdots\right) \\ &= Ae^{At} = e^{At}A \end{aligned} \tag{13.25}$$

Again consider the problem of solving (13.21). It is claimed that the solution is

$$x(t) = e^{At}x(0), \qquad t \geq 0 \tag{13.26}$$

To verify that the expression (13.26) for $x(t)$ is the solution, take the derivative of both sides of (13.26). This gives

$$\begin{aligned}\frac{d}{dt}x(t) &= \frac{d}{dt}[e^{At}x(0)] \\ &= \left[\frac{d}{dt}e^{At}\right]x(0) \\ &= Ae^{At}x(0) = Ax(t)\end{aligned}$$

Thus, the expression (13.26) for $x(t)$ does satisfy the vector differential equation (13.21).

From (13.26) it is seen that the state $x(t)$ at time t resulting from state $x(0)$ at time $t = 0$ with no input applied for $t \geq 0$ can be computed by multiplying $x(0)$ by the matrix e^{At}. As a result of this property, the matrix e^{At} is called the *state-transition matrix* of the system.

Solution to Forced Equation

An expression for the solution to the forced equation (13.19) will now be derived. The solution can be computed by using a matrix version of the integrating factor method. In Chapter 1 this solution technique was used to solve a first-order scalar differential equation. The following steps are simply a matrix version of the derivation given in Section 1.4.

Multiplying both sides of (13.19) on the left by e^{-At} and rearranging terms yields

$$e^{-At}[\dot{x}(t) - Ax(t)] = e^{-At}Bv(t) \tag{13.27}$$

From (13.25), the left-hand side of (13.27) is equal to the derivative of $e^{-At}x(t)$. Thus

$$\frac{d}{dt}[e^{-At}x(t)] = e^{-At}Bv(t) \tag{13.28}$$

Integrating both sides of (13.28) gives

$$e^{-At}x(t) = x(0) + \int_0^t e^{-A\lambda}\, Bv(\lambda)\, d\lambda$$

Finally, multiplying both sides on the left by e^{At} gives

$$x(t) = e^{At}x(0) + \int_0^t e^{A(t-\lambda)}Bv(\lambda)\, d\lambda, \qquad t \geq 0 \tag{13.29}$$

Equation (13.29) is the complete solution of (13.19) resulting from initial state $x(0)$ and input $v(t)$ applied for $t \geq 0$. This equation is a generalization of the expression that was given for the solution of a first-order scalar differential equation [see (2.5)].

Output Response

Inserting (13.29) into (13.20) results in the following expression for the complete output response $y(t)$ resulting from initial state $x(0)$ and input $v(t)$:

$$y(t) = Ce^{At}x(0) + \int_0^t Ce^{A(t-\lambda)}Bv(\lambda)\, d\lambda + Dv(t), \qquad t \geq 0 \tag{13.30}$$

From the definition of the unit impulse $\delta(t)$, (13.30) can be rewritten in the form

$$y(t) = Ce^{At}x(0) + \int_0^t \{Ce^{A(t-\lambda)}Bv(\lambda) + D\delta(t-\lambda)v(\lambda)\}\, d\lambda, \qquad t \geq 0 \tag{13.31}$$

Then defining

$$y_{zi}(t) = Ce^{At}x(0) \tag{13.32}$$

and

$$y_{zs}(t) = \int_0^t \{Ce^{A(t-\lambda)}Bv(\lambda) + D\delta(t-\lambda)v(\lambda)\}\, d\lambda \tag{13.33}$$

from (13.31)

$$y(t) = y_{zi}(t) + y_{zs}(t)$$

The term $y_{zi}(t)$ is called the *zero-input response* since it is the complete output response when the input $v(t)$ is zero. The term $y_{zs}(t)$ is called the *zero-state response,* since it is the complete output response when the initial state $x(0)$ is zero.

The zero-state response $y_{zs}(t)$ is the same as the response to input $v(t)$ with no initial energy in the system at time $t = 0$. In the single-input single-output case, from the results in Chapter 3

$$y_{zs}(t) = h(t) * v(t) = \int_0^t h(t-\lambda)v(\lambda)\,d\lambda, \qquad t \geq 0 \tag{13.34}$$

where $h(t)$ is the impulse response of the system. Equating the right-hand sides of (13.33) and (13.34) yields

$$\int_0^t \{Ce^{A(t-\lambda)}Bv(\lambda) + D\delta(t-\lambda)v(\lambda)\}\,d\lambda = \int_0^t h(t-\lambda)v(\lambda)\,d\lambda \tag{13.35}$$

For (13.35) to hold for all inputs $v(t)$, it must be true that

$$h(t-\lambda) = Ce^{A(t-\lambda)}B + D\delta(t-\lambda), \qquad t \geq \lambda$$

or

$$h(t) = Ce^{At}B + D\delta(t), \qquad t \geq 0 \tag{13.36}$$

From the relationship (13.36), it is possible to compute the impulse response directly from the coefficient matrices of the state model of the system.

Solution via the Laplace Transform

Again consider a p-input r-output N-dimensional system given by the state model (13.19) and (13.20). To compute the state and output responses resulting from initial state $x(0)$ and input $v(t)$ applied for $t \geq 0$, the expressions (13.29) and (13.30) can be used. To avoid having to evaluate the integrals in (13.29) and (13.30), the Laplace transform can be used to compute the state and output responses of the system. The transform approach is a matrix version of the procedure that was given for solving a scalar first-order differential equation. The steps are as follows.

Taking the Laplace transform of the state equation (13.19) gives

$$sX(s) - x(0) = AX(s) + BV(s) \tag{13.37}$$

where $X(s)$ is the Laplace transform of the state vector $x(t)$ with the transform taken component by component; that is,

$$X(s) = \begin{bmatrix} X_1(s) \\ X_2(s) \\ \vdots \\ X_N(s) \end{bmatrix}$$

where $X_i(s)$ is the Laplace transform of $x_i(t)$. The term $V(s)$ in (13.37) is the Laplace transform of the input $V(t)$, where again the transform is taken component by component.

Now (13.37) can be rewritten in the form

$$(sI - A)X(s) = x(0) + BV(s) \tag{13.38}$$

where I is the $N \times N$ identity matrix. Note that in factoring out s from $sX(s)$, it is necessary to multiply s by I. The reason for this is that A cannot be subtracted from the scalar s, since A is a $N \times N$ matrix. However, A can be subtracted from the diagonal matrix sI. By definition of the identity matrix, the product $(sI - A)X(s)$ is equal to $sX(s) - AX(s)$.

It turns out that the matrix $sI - A$ in (13.38) always has an inverse $(sI - A)^{-1}$. (It will be seen later why this is true.) Thus both sides of (13.38) can be multiplied on the left by $(sI - A)^{-1}$. This gives

$$X(s) = (sI - A)^{-1}x(0) + (sI - A)^{-1}BV(s) \tag{13.39}$$

The right-hand side of (13.39) is the Laplace transform of the state response resulting from initial state $x(0)$ and input $v(t)$ applied for $t \geq 0$. The state response $x(t)$ can then be computed by taking the inverse Laplace transform of the right-hand side of (13.39).

Comparing (13.39) with (13.29) reveals that $(sI - A)^{-1}$ is the Laplace transform of the state-transition matrix e^{At}. Since e^{At} is a well-defined function of t, this shows that the inverse $(sI - A)^{-1}$ must exist. Also note that

$$e^{At} = \text{inverse Laplace transform of } (sI - A)^{-1} \tag{13.40}$$

The relationship (13.40) is very useful for computing e^{At}. This will be illustrated in an example given below.

Taking the Laplace transform of the output equation (13.20) yields

$$Y(s) = CX(s) + DV(s) \tag{13.41}$$

Inserting (13.39) into (13.41) gives

$$Y(s) = C(sI - A)^{-1}x(0) + [C(sI - A)^{-1}B + D]V(s) \tag{13.42}$$

The right-hand side of (13.42) is the Laplace transform of the complete output response resulting from initial state $x(0)$ and input $v(t)$.

If $x(0) = 0$ (no initial energy in the system) (13.42) reduces to

$$Y(s) = Y_{zs}(s) = H(s)V(s) \tag{13.43}$$

where $H(s)$ is the transfer function matrix of the system given by

$$H(s) = C(sI - A)^{-1}B + D \tag{13.44}$$

So the transfer function matrix can be computed directly from the coefficient matrices of the state model of the system.

Example 13.5 *Two-Dimensional System*

Consider the two-input three-output two-dimensional system with state model $\dot{x}(t) = Ax(t) + Bv(t)$, $y(t) = Cx(t)$, where

$$A = \begin{bmatrix} -3 & 1 \\ -2 & -1 \end{bmatrix}, \quad B = \begin{bmatrix} 3 & 2 \\ 2 & 1 \end{bmatrix}, \quad C = \begin{bmatrix} 1 & 2 \\ -2 & 2 \\ 1 & -1 \end{bmatrix}$$

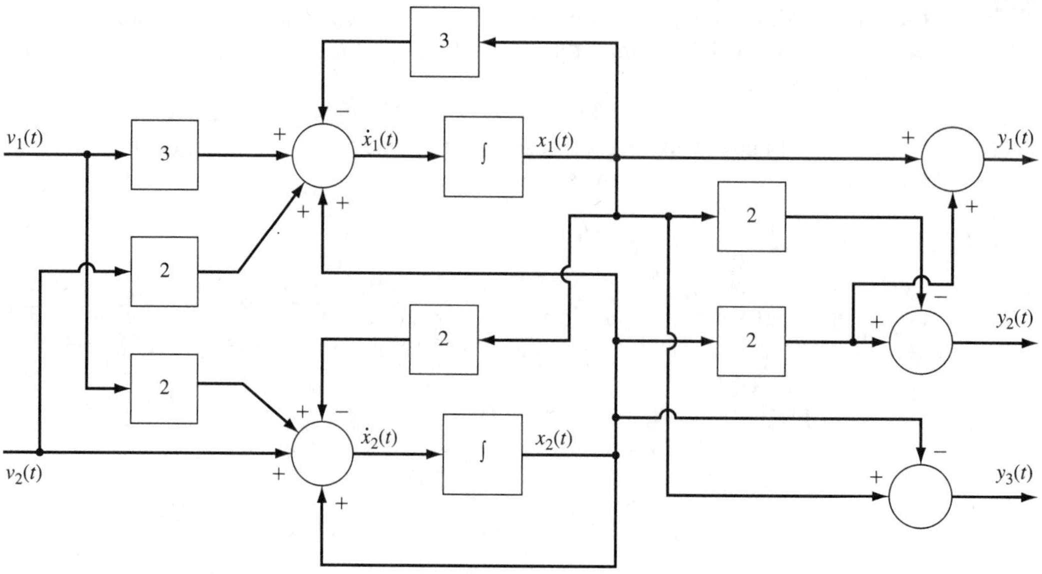

Figure 13.6 Realization of system in Example 13.5.

An integrator realization of this system is given in Figure 13.6.

The state-transition matrix e^{At} of the system will be computed first. For the A matrix given above,

$$(sI - A)^{-1} = \begin{bmatrix} s+3 & -1 \\ 2 & s+1 \end{bmatrix}^{-1} = \frac{1}{s^2 + 4s + 5}\begin{bmatrix} s+1 & 1 \\ -2 & s+3 \end{bmatrix}$$

$$= \frac{1}{(s+2)^2 + 1}\begin{bmatrix} s+1 & 1 \\ -2 & s+3 \end{bmatrix}$$

By (13.40) e^{At} is equal to the inverse Laplace transform of $(sI - A)^{-1}$. Using table lookup gives

$$e^{At} = e^{-2t}\begin{bmatrix} \cos t - \sin t & \sin t \\ -2\sin t & \cos t + \sin t \end{bmatrix}$$

Now the state response $x(t)$ resulting from any initial state $x(0)$ with zero input is given by

$$x(t) = e^{At}x(0), \qquad t \geq 0$$

For example, if

$$x(0) = \begin{bmatrix} 1 \\ 1 \end{bmatrix}$$

then

$$x(t) = e^{-2t}\begin{bmatrix} \cos t - \sin t & \sin t \\ -2\sin t & \cos t + \sin t \end{bmatrix}\begin{bmatrix} 1 \\ 1 \end{bmatrix}$$

$$= e^{-2t}\begin{bmatrix} \cos t \\ \cos t - \sin t \end{bmatrix}, \qquad t \geq 0$$

Now the state response $x(t)$ resulting from the input

$$v(t) = \begin{bmatrix} u(t) \\ e^{-t} \end{bmatrix}, \qquad t \geq 0$$

will be computed. Then

$$V(s) = \begin{bmatrix} \dfrac{1}{s} \\ \dfrac{1}{s+1} \end{bmatrix}$$

and using (13.39) with $x(0) = 0$ yields

$$X(s) = \begin{bmatrix} s+3 & -1 \\ 2 & s+1 \end{bmatrix}^{-1}\begin{bmatrix} 3 & 2 \\ 2 & 1 \end{bmatrix}\begin{bmatrix} \dfrac{1}{s} \\ \dfrac{1}{s+1} \end{bmatrix}$$

$$= \frac{1}{s^2 + 4s + 5}\begin{bmatrix} s+1 & 1 \\ -2 & s+3 \end{bmatrix}\begin{bmatrix} \dfrac{5s+3}{s(s+1)} \\ \dfrac{3s+2}{s(s+1)} \end{bmatrix}$$

$$= \frac{1}{[(s+2)^2 + 1]s(s+1)}\begin{bmatrix} 5s^2 + 11s + 5 \\ 3s^2 + s \end{bmatrix}$$

Expanding the components of $X(s)$ gives

$$X(s) = \begin{bmatrix} \dfrac{-1.5s - 0.5}{(s+2)^2 + 1} + \dfrac{1}{s} + \dfrac{0.5}{s+1} \\ \dfrac{s+6}{(s+2)^2 + 1} - \dfrac{1}{s+1} \end{bmatrix}$$

Taking the inverse Laplace transform of $X(s)$ yields

$$x(t) = \begin{bmatrix} e^{-2t}(-1.5\cos t + 2.5\sin t) + 1 + 0.5e^{-t} \\ e^{-2t}(\cos t + 4\sin t) - e^{-t} \end{bmatrix}, \qquad t \geq 0$$

The output response is

$$y(t) = Cx(t) = \begin{bmatrix} e^{-2t}(0.5\cos t + 10.5\sin t) + 1 - 1.5e^{-t} \\ e^{-2t}(5\cos t + 3\sin t) - 2 - 3e^{-t} \\ e^{-2t}(-2.5\cos t - 1.5\sin t) + 1 + 1.5e^{-t} \end{bmatrix}, \qquad t \geq 0$$

MATLAB can be used to solve for the state and output responses of the system resulting from any initial state $x(0)$ with zero input. For example, when $x_1(0) = 1$ and $x_2(0) = 1$, the resulting responses can be computed using the following commands:

```
A = [-3 1;-2 -1]; B = [3 2;2 1];
C = [1 2;-2 2;1 -1]; D = zeros(3,2);
t = 0:.04:8; % simulate for 0<t<8 sec
x0 = [1 1]'; % define the I.C.
v = zeros(length(t),2); % zeros the input vector
[y,x] = lsim(A,B,C,D,v,t,x0);
subplot(211),
plot(t,x(:,1),'-',t,x(:,2),'--')
subplot(212),
plot(t,y(:,1),'-',t,y(:,2),'--',t,y(:,3),'-.')
```

Running the program produces the plots of $x_1(t)$ and $x_2(t)$ shown in Figure 13.7a and the plots of $y_1(t)$, $y_2(t)$, and $y_3(t)$ shown in Figure 13.7b.

The response of the system resulting from the inputs $v_1(t) = u(t)$ and $v_2(t) = e^{-t}$ with zero initial state can be solved using the commands

```
v(:,1) = ones(length(t),1);
v(:,2) = exp(-t)';
x0 = [0 0]';
[y,x] = lsim(A,B,C,D,v,t,x0);
```

The corresponding plots of the state $x(t)$ and the output $y(t)$ are given in Figure 13.8. When $x(0) = 0$, the output response $y(t)$ may be computed by taking the inverse Laplace transform of the transfer function representation $Y(s) = H(s)V(s)$. In this example, the transfer function is

$$\begin{aligned} H(s) &= C(sI - A)^{-1}B \\ &= \begin{bmatrix} 1 & 2 \\ -2 & 2 \\ 1 & -1 \end{bmatrix} \begin{bmatrix} s+3 & -1 \\ 2 & s+1 \end{bmatrix}^{-1} \begin{bmatrix} 3 & 2 \\ 2 & 1 \end{bmatrix} \\ &= \frac{1}{s^2 + 4s + 5} \begin{bmatrix} 1 & 2 \\ -2 & 2 \\ 1 & -1 \end{bmatrix} \begin{bmatrix} s+1 & 1 \\ -2 & s+3 \end{bmatrix} \begin{bmatrix} 3 & 2 \\ 2 & 1 \end{bmatrix} \\ &= \frac{1}{s^2 + 4s + 5} \begin{bmatrix} s-3 & 2s+7 \\ -2s-6 & 2s+4 \\ s+3 & -s-2 \end{bmatrix} \begin{bmatrix} 3 & 2 \\ 2 & 1 \end{bmatrix} \end{aligned}$$

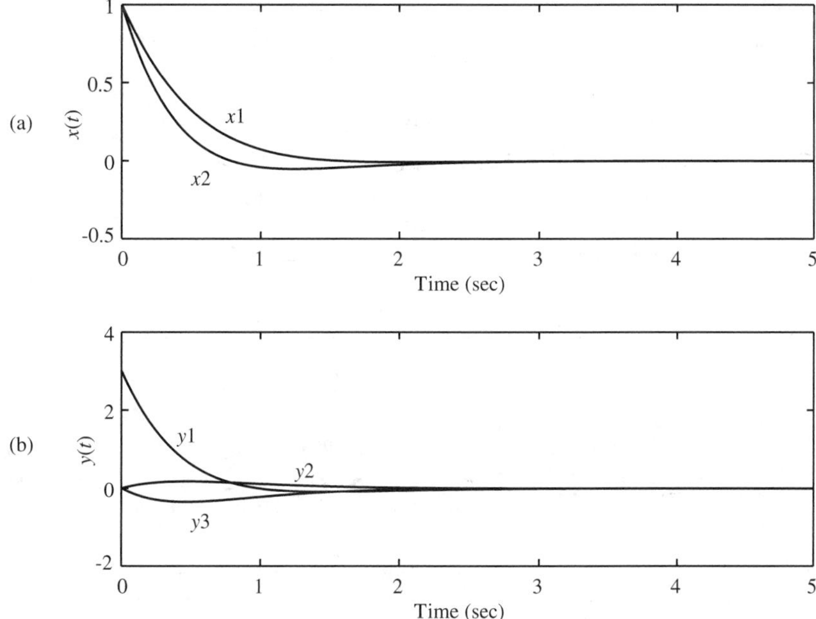

Figure 13.7 Plots of the state responses (a) and output responses (b) resulting from initial states $x_1(0) = 1$ and $x_2(0) = 1$.

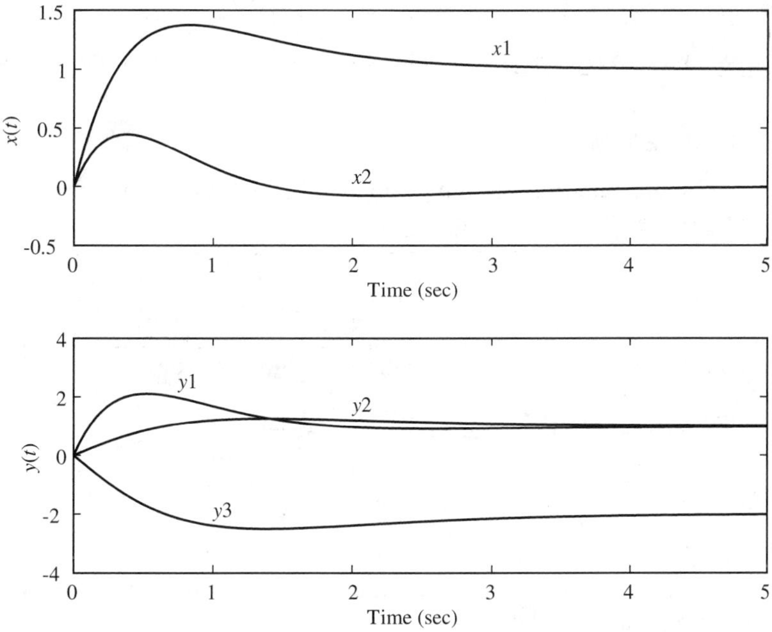

Figure 13.8 Plots of the state responses (a) and output responses (b) resulting from inputs $v_1(t) = u(t)$ and $v_2(t) = e^{-t}$.

$$H(s) = \frac{1}{s^2 + 4s + 5}\begin{bmatrix} 7s + 5 & 4s + 1 \\ -2s - 10 & -2s - 8 \\ s + 5 & s + 4 \end{bmatrix}$$

MATLAB can be used to calculate the transfer function matrix $H(s)$: The first column of $H(s)$ is found by typing

```
[num1,den1] = ss2tf(A,B,C,D,1)
```

Using this command for the system above results in

```
num1 =                  and den1 =
         7    5                 1   4    5
        -2  -10
         1    5
```

Comparing this to the result obtained above analytically shows that `num1` contains the numerator coefficients in the first column of $H(s)$ and `den1` contains the denominator coefficients. The second column of the transfer function matrix $H(s)$ is found by using

```
[num2,den2] = ss2tf(A,B,C,D,2)
```

which results in

```
num2 =                  and den2 =
         4   1                     1    4    5
        -2  -8
         1   4
```

where `num2` contains the numerator coefficients in the second column of $H(s)$.

13.4 DISCRETE-TIME SYSTEMS

A p-input r-output finite-dimensional linear time-invariant discrete-time system can be modeled by the state equations

$$x[n + 1] = Ax[n] + Bv[n] \tag{13.45}$$

$$y[n] = Cx[n] + Dv[n] \tag{13.46}$$

The state vector $x[n]$ is the N-element column vector

$$x[n] = \begin{bmatrix} x_1[n] \\ x_2[n] \\ \vdots \\ x_N[n] \end{bmatrix}$$

As in the continuous-time case, the state $x[n]$ at time n represents the past history (before time n) of the system.

The input $v[n]$ and output $y[n]$ are the column vectors

$$v[n] = \begin{bmatrix} v_1[n] \\ v_2[n] \\ \vdots \\ v_p[n] \end{bmatrix}, \qquad y[n] = \begin{bmatrix} y_1[n] \\ y_2[n] \\ \vdots \\ y_r[n] \end{bmatrix}$$

The matrices A, B, C, and D in (13.45) and (13.46) are $N \times N, N \times p, r \times N$, and $r \times p$, respectively. Equation (13.45) is a first-order vector difference equation. Equation (13.46) is the output equation of the system.

Construction of State Models

Consider a single-input single-output linear time-invariant discrete-time system with the input/output difference equation

$$y[n + N] + \sum_{i=0}^{N-1} a_i y[n + i] = bv[n] \tag{13.47}$$

Defining the state variables

$$x_{i+1}[n] = y[n + i], \qquad i = 0, 1, 2, \ldots, N - 1 \tag{13.48}$$

results in a state model of the form (13.45) and (13.46) with

$$A = \begin{bmatrix} 0 & 1 & 0 & \cdots & 0 \\ 0 & 0 & 1 & & 0 \\ \vdots & \vdots & & \ddots & \vdots \\ 0 & 0 & 0 & \cdots & 1 \\ -a_0 & -a_1 & -a_2 & \cdots & -a_{N-1} \end{bmatrix}, \qquad B = \begin{bmatrix} 0 \\ 0 \\ \vdots \\ 0 \\ b \end{bmatrix}$$

$$C = [1 \quad 0 \quad 0 \quad \cdots \quad 0], \qquad D = 0$$

If the right-hand side of (13.47) is modified so that it is in the more general form

$$\sum_{i=0}^{N-1} b_i x[n + i]$$

the state model above is still a state model, except that B and C must be modified so that

$$B = \begin{bmatrix} 0 \\ 0 \\ \vdots \\ 0 \\ 1 \end{bmatrix}$$

$$C = [b_0 \quad b_1 \quad \cdots \quad b_{N-1}]$$

In this case, the state variables $x_i[n]$ are functions of $y[n]$, $v[n]$ and left shifts of $y[n]$ and $x[n]$; that is, $x_i[n]$ is no longer given by (13.48). This state model is the discrete-time counterpart of the state model that was generated from a linear constant-coefficient input/output differential equation in the continuous-time case (see Section 13.2).

Realizations Using Unit-Delay Elements

Given a discrete-time system with the N-dimensional state model (13.45) and (13.46), a realization of the system can be constructed consisting of an interconnection of N unit delay elements and combinations of adders, subtracters, and scalar multipliers. Conversely, if a discrete-time system is specified by an interconnection of unit delays, adders, subtracters, and scalar multipliers, a state model of the form (13.45) and (13.46) can be generated directly from the interconnection. The procedure for going from interconnection diagrams to state models (and conversely) is analogous to that given in the continuous-time case.

Example 13.6 *Interconnection of Unit Delay Elements*

Consider the three-input two-output three-dimensional discrete-time system given by the interconnection in Figure 13.9. From the diagram

$$x_1[n+1] = -x_2[n] + v_1[n] + v_3[n]$$

$$x_2[n+1] = x_1[n] + v_2[n]$$

$$x_3[n+1] = x_2[n] + v_3[n]$$

$$y_1[n] = x_2[n]$$

$$y_2[n] = x_1[n] + x_3[n] + v_2[n]$$

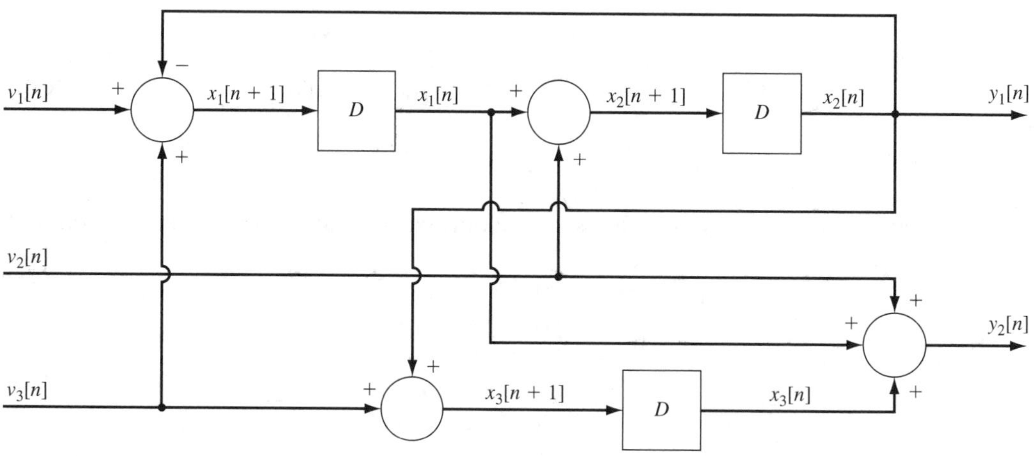

Figure 13.9 System in Example 13.6.

Writing these equations in matrix form results in the state model

$$\begin{bmatrix} x_1[n+1] \\ x_2[n+1] \\ x_3[n+1] \end{bmatrix} = \begin{bmatrix} 0 & -1 & 0 \\ 1 & 0 & 0 \\ 0 & 1 & 0 \end{bmatrix} \begin{bmatrix} x_1[n] \\ x_2[n] \\ x_3[n] \end{bmatrix} + \begin{bmatrix} 1 & 0 & 1 \\ 0 & 1 & 0 \\ 0 & 0 & 1 \end{bmatrix} \begin{bmatrix} v_1[n] \\ v_2[n] \\ v_3[n] \end{bmatrix}$$

$$\begin{bmatrix} y_1[n] \\ y_2[n] \end{bmatrix} = \begin{bmatrix} 0 & 1 & 0 \\ 1 & 0 & 1 \end{bmatrix} \begin{bmatrix} x_1[n] \\ x_2[n] \\ x_3[n] \end{bmatrix} + \begin{bmatrix} 0 & 0 & 0 \\ 0 & 1 & 0 \end{bmatrix} \begin{bmatrix} v_1[n] \\ v_2[n] \\ v_3[n] \end{bmatrix}$$

Solution of State Equations

Again consider the p-input r-output discrete-time system with the state model

$$x[n+1] = Ax[n] + Bv[n] \tag{13.49}$$

$$y[n] = Cx[n] + Dv[n] \tag{13.50}$$

The vector difference equation (13.49) can be solved by using a matrix version of recursion. The process is a straightforward generalization of the recursive procedure for solving a first-order scalar difference equation (see Chapter 2). The steps are as follows.

It is assumed that the initial state of the system is the state $x[0]$ at initial time $n = 0$. Then setting $n = 0$ in (13.49) gives

$$x[1] = Ax[0] + Bv[0] \tag{13.51}$$

Setting $n = 1$ (13.49) and using (13.51) yields

$$\begin{aligned} x[2] &= Ax[1] + Bv[1] \\ &= A[Ax[0] + Bv[0]] + Bv[1] \\ &= A^2x[0] + ABv[0] + Bv[1] \end{aligned}$$

If this process is continued, for any integer value of $n \geq 1$

$$x[n] = A^n x[0] + \sum_{i=0}^{n-1} A^{n-i-1}Bv[i], \qquad n \geq 1 \tag{13.52}$$

The right-hand side of (13.52) is the state response resulting from initial state $x[0]$ and input $v[n]$ applied for $n \geq 0$. Note that if $v[n] = 0$ for $n \geq 0$, then

$$x[n] = A^n x[0], \qquad n \geq 0 \tag{13.53}$$

From (13.53) it is seen that the state transition from initial state $x[0]$ to state $x[n]$ at time n (with no input applied) is equal to $x[0]$ times the matrix A^n. Therefore, in the discrete-time case the state-transition matrix is the matrix A^n.

Inserting (13.52) into the output equation (13.50) gives

$$y[n] = CA^n x[0] + \sum_{i=0}^{n-1} CA^{n-i-1}Bv[i] + Dv[n], \qquad n \geq 1 \tag{13.54}$$

The right-hand side of (13.54) is the complete output response resulting from initial state $x[0]$ and input $v[n]$. The term

$$y_{zi}[n] = CA^n x[0], \qquad n \geq 0$$

is the zero-input response, and the term

$$\begin{aligned} y_{zs}[n] &= \sum_{i=0}^{n-1} CA^{n-i-1}Bv[i] + Dv[n], \qquad n \geq 1 \\ &= \sum_{i=0}^{n} [CA^{n-i-1}Bu[n-i-1] + \delta[n-i]D]v[i], \qquad n \geq 1 \end{aligned} \tag{13.55}$$

is the zero-state response (where $\delta[n]$ = unit pulse located at $n = 0$).

In the single-input single-output case,

$$y_{zs}[n] = \sum_{i=0}^{n} h[n-i]v[i] \tag{13.56}$$

where $h[n]$ is the unit-pulse response of the system. Comparing (13.55) and (13.56) reveals that

$$\begin{aligned} h[n-i] &= CA^{n-i-1}Bu[n-i-1] + \delta[n-i]D \\ h[n] &= CA^{n-1}Bu[n-1] + \delta[n]D \end{aligned}$$

or

$$h[n] = \begin{cases} D, & n = 0 \\ CA^{n-1}B, & n \geq 1 \end{cases}$$

Solution via the *z*-Transform

Taking the z-transform of the vector difference equation (13.49) gives

$$zX(z) - zx[0] = AX(z) + BV(z) \tag{13.57}$$

where $X(z)$ and $V(z)$ are the z-transforms of $x[n]$ and $v[n]$, respectively, with transforms taken component by component. Solving (13.57) for $X(z)$ gives

$$X(z) = (zI - A)^{-1}zx[0] + (zI - A)^{-1}BV(z) \tag{13.58}$$

The right-hand side of (13.58) is the z-transform of the state response $x[n]$ resulting from initial state $x[0]$ and input $v[n]$.

Comparing (13.58) and (13.52) shows that $(zI - A)^{-1}z$ is the z-transform of the state-transition matrix A^n. Thus

$$A^n = \text{inverse } z\text{-transform of } (zI - A)^{-1}z \tag{13.59}$$

Taking the z-transform of (13.50) and using (13.58) yields

$$Y(z) = C(zI - A)^{-1}zx[0] + [C(zI - A)^{-1}B + D]V(z) \tag{13.60}$$

The right-hand side of (13.60) is the z-transform of the complete output response resulting from initial state $x[0]$ and input $v[n]$.

If $x[0] = 0$,

$$Y(z) = Y_{zs}(z) = [C(zI - A)^{-1}B + D]V(z) \tag{13.61}$$

Since

$$Y_{zs}(z) = H(z)V(z)$$

where $H(z)$ is the transfer function matrix, by (13.61)

$$H(z) = C(zI - A)^{-1}B + D \tag{13.62}$$

Example 13.7 *Computation of Transfer Function Matrix*

Again consider the system in Example 13.6. The state-transition matrix A^n will be computed first.

$$(zI - A)^{-1} = \begin{bmatrix} z & 1 & 0 \\ -1 & z & 0 \\ 0 & -1 & z \end{bmatrix}^{-1} = \frac{1}{(z^2 + 1)z}\begin{bmatrix} z^2 & -z & 0 \\ z & z^2 & 0 \\ 1 & z & z^2 + 1 \end{bmatrix}$$

Expanding the components of $(zI - A)^{-1}z$ and using table lookup gives

$$A^n = \begin{bmatrix} +\cos\frac{\pi}{2}n & -\sin\frac{\pi}{2}n & 0 \\ \sin\frac{\pi}{2}n & +\cos\frac{\pi}{2}n & 0 \\ -\left(\cos\frac{\pi}{2}n\right)u[n-1] & -\sin\frac{\pi}{2}n & \delta[n] \end{bmatrix}$$

The state response $x[n]$ resulting from any initial state $x[0]$ (with no input applied) can be computed by using (13.53). For example, if

$$x[0] = \begin{bmatrix} 1 \\ 1 \\ 0 \end{bmatrix},$$

then

$$x[n] = A^n x[0] = \begin{bmatrix} \cos\frac{\pi}{2}n - \sin\frac{\pi}{2}n \\ \sin\frac{\pi}{2}n + \cos\frac{\pi}{2}n \\ -\left(\cos\frac{\pi}{2}n\right)u[n-1] + \sin\frac{\pi}{2}n \end{bmatrix}, \qquad n \geq 0$$

MATLAB contains the command `dlsim`, which can be used to compute the state response $x[n]$. For example, to compute $x[n]$ for the system above with $x_1[0] = x_2[0] = 1$ and $x_3[0] = 0$, the commands are

```
A = [0 -1 0;1 0 0;0 1 0]; B = [1 0 1;0 1 0;0 0 1];
C = [0 1 0;1 0 1]; D = [0 0 0;0 1 0];
x0 = [1 1 0]';                          % define the I.C.
n = 0:1:10;
v = zeros(length(n),3);
[y,x] = dlsim(A,B,C,D,v,x0);
```

Running the program results in the values of $x[n]$ for $n = 1$ to 5 given in the table below.

	$x[0]$	$x[1]$	$x[2]$	$x[3]$	$x[4]$	$x[5]$
x_1	1	−1	−1	1	1	−1
x_2	1	1	−1	−1	1	1
x_3	0	1	1	−1	−1	1

The reader is invited to verify that these values are an exact match with the values obtained from the analytical solution given above.

To conclude the example, the transfer function matrix of the system will be computed. From (13.62),

$$\begin{aligned}
H(z) &= C(zI - A)^{-1}B + D \\
&= \frac{1}{(z^2+1)z}\begin{bmatrix} 0 & 1 & 0 \\ 1 & 0 & 1 \end{bmatrix}\begin{bmatrix} z^2 & -z & 0 \\ z & z^2 & 0 \\ 1 & z & z^2+1 \end{bmatrix}\begin{bmatrix} 1 & 0 & 1 \\ 0 & 1 & 0 \\ 0 & 0 & 1 \end{bmatrix} + \begin{bmatrix} 0 & 0 & 0 \\ 0 & 1 & 0 \end{bmatrix} \\
&= \frac{1}{(z^2+1)z}\begin{bmatrix} z & z^2 & z \\ z^2+1 & (z^2+1)z & 2(z^2+1) \end{bmatrix}
\end{aligned}$$

To use MATLAB to calculate the transfer function matrix, type

```
[num1,den] = ss2tf(A,B,C,D,1);    % for column 1 coefficients
[num2,den] = ss2tf(A,B,C,D,2);    % for column 2 coefficients
[num3,den] = ss2tf(A,B,C,D,3);    % for column 3 coefficients
```

This yields `den = [1 0 1 0]`, which contains the denominator coefficients and

```
num1 =                              num2 =

        0  0  1  0                          0  1  0  0
        0  1  0  1                          1  0  1  0

num3 =

        0  0  1  0
        0  2  0  2
```

The reader will note that these coefficients match those found above analytically.

13.5 EQUIVALENT STATE REPRESENTATIONS

Unlike the transfer function model, the state model of a system is not unique. The relationship between state models of the same system is considered in this section. The following analysis is developed in terms of continuous-time systems. The theory in the discrete-time case is very similar, so attention will be restricted to the continuous-time case.

Consider a p-input r-output N-dimensional linear time-invariant continuous-time system given by the state model

$$\dot{x}(t) = Ax(t) + Bv(t) \tag{13.63}$$

$$y(t) = Cx(t) + Dv(t) \tag{13.64}$$

Let P denote a fixed $N \times N$ matrix with entries that are real numbers. It is required that P be invertible, so the determinant $|P|$ of P is nonzero (see Appendix B). The inverse of P is denoted by P^{-1}.

In terms of the matrix P, a new state vector $\bar{x}(t)$ can be defined for the given system, where

$$\bar{x}(t) = Px(t) \tag{13.65}$$

The relationship (13.65) is called a *coordinate transformation* since it is a mapping from the original state coordinates to the new state coordinates.

Multiplying both sides of (13.65) on the left by the inverse P^{-1} of P gives

$$P^{-1}\bar{x}(t) = P^{-1}Px(t)$$

By definition of the inverse, $P^{-1}P = I$, where I is the $N \times N$ identity matrix. Hence

$$P^{-1}\bar{x}(t) = Ix(t) = x(t)$$

or

$$x(t) = P^{-1}\bar{x}(t) \tag{13.66}$$

Via (13.66) it is possible to go from the new state vector $\bar{x}(t)$ back to the original state vector $x(t)$. Note that if P were not invertible, it would not be possible to go back to $x(t)$ from $\bar{x}(t)$. This is the reason P must be invertible.

In terms of the new state vector $\bar{x}(t)$, it is possible to generate a new state-equation model of the given system. The steps are as follows.

Taking the derivative of both sides of (13.65) and using (13.63) yields

$$\begin{aligned}\dot{\bar{x}}(t) &= P\dot{x}(t) = P[Ax(t) + Bv(t)]\\ &= PAx(t) + PBv(t)\end{aligned} \tag{13.67}$$

Inserting the expression (13.66) for $x(t)$ into (13.67) and (13.64) gives

$$\dot{\bar{x}}(t) = PA(P^{-1})\bar{x}(t) + PBv(t) \tag{13.68}$$

$$y(t) = C(P^{-1})\bar{x}(t) + Dv(t) \tag{13.69}$$

Defining the matrices

$$\overline{A} = PA(P^{-1}), \qquad \overline{B} = PB, \qquad \overline{C} = C(P^{-1}), \qquad \overline{D} = D, \tag{13.70}$$

(13.68) and (13.69) can be written in the form

$$\dot{\bar{x}}(t) = \overline{A}\bar{x}(t) + \overline{B}v(t) \tag{13.71}$$

$$y(t) = \overline{C}\bar{x}(t) + \overline{D}v(t) \tag{13.72}$$

Equations (13.71) and (13.72) are the state equations of the given system in terms of the new state vector $\bar{x}(t)$. Thus it is possible to generate a new N-dimensional state model from the original N-dimensional state model. Since the above construction can be carried out for any invertible $N \times N$ matrix P, and there are an infinite number of such matrices, it is possible to generate an infinite number of new state models from a given state model.

Let the original state model be denoted by the quadruple (A, B, C, D) and the new state model be denoted by the quadruple $(\overline{A}, \overline{B}, \overline{C}, \overline{D})$. The state models (A, B, C, D) and $(\overline{A}, \overline{B}, \overline{C}, \overline{D})$ are said to be related by the coordinate transformation P since the state vector $\bar{x}(t)$ of the latter is related to the state vector $x(t)$ of the former by the relationship $\bar{x}(t) = Px(t)$. Any two such state models are said to be *equivalent*. The only difference between two equivalent state models is in the labeling of states. More precisely, the states of $(\overline{A}, \overline{B}, \overline{C}, \overline{D})$ are linear combinations [given by $\bar{x}(t) = Px(t)$] of the states of (A, B, C, D).

It should be stressed that the notion of equivalent state models applies only to state models having the same dimension. State models with different dimensions cannot be related by a coordinate transformation.

Any two equivalent state models have exactly the same input/output relationship. In particular, the transfer function matrices corresponding to any two equivalent models are the same. To prove this, let $H(s)$ and $\overline{H}(s)$ denote the transfer function matrices associated with (A, B, C, D) and $(\overline{A}, \overline{B}, \overline{C}, \overline{D})$, respectively; that is,

$$H(s) = C(sI - A)^{-1}B + D \tag{13.73}$$

$$\overline{H}(s) = \overline{C}(sI - \overline{A})^{-1}\overline{B} + \overline{D} \tag{13.74}$$

Inserting the expressions (13.70) for $\overline{A}, \overline{B}, \overline{C}, \overline{D}$ into (13.74) yields

$$\overline{H}(s) = C(P^{-1})[sI - PA(P^{-1})]^{-1}PB + D$$

Now

$$sI - PA(P^{-1}) = P(sI - A)(P^{-1})$$

In addition, for any $N \times N$ invertible matrices X, Y, W,

$$(WXY)^{-1} = (Y^{-1})(X^{-1})(W^{-1})$$

Thus

$$\begin{aligned}\overline{H}(s) &= C(P^{-1})P(sI - A)^{-1}(P^{-1})PB + D \\ &= C(sI - A)^{-1}B + D \\ &= H(s)\end{aligned}$$

So the transfer function matrices are the same.

Example 13.8 *Equivalent State Models*

Consider the system given by the second-order differential equation

$$\ddot{y}(t) + 2\dot{y}(t) + 3y(t) = \dot{v}(t) + 2v(t)$$

This equation is in the form of (13.13), and thus a state model for this system is obtained by using (13.14). This results in the state model

$$\begin{bmatrix}\dot{x}_1(t)\\ \dot{x}_2(t)\end{bmatrix} = \begin{bmatrix}0 & 1\\ -3 & -2\end{bmatrix}\begin{bmatrix}x_1(t)\\ x_2(t)\end{bmatrix} + \begin{bmatrix}0\\ 1\end{bmatrix}v(t), \qquad y(t) = [2 \quad 1]\begin{bmatrix}x_1(t)\\ x_2(t)\end{bmatrix}$$

The state variables $x_1(t)$ and $x_2(t)$ in the state model above have no physical meaning, and thus another set of state variables might be just as suitable. For example, let $\bar{x}(t) = Px(t)$, where

$$P = \begin{bmatrix}1 & 1\\ 0 & 1\end{bmatrix}$$

It is easy to show that $|P|$ is not zero, so this is a valid transformation matrix. In terms of $\bar{x}(t)$, the new state model is given by (13.71) and (13.72) where the quadruple $(\overline{A}, \overline{B}, \overline{C}, \overline{D})$ can be found from (13.70). The new coefficient matrices can be determined using the following MATLAB commands:

```
A = [0 1;-3 -2]; B = [0 1]'; C = [2 1]; D = 0;
P = [1 1;0 1];
Abar = P*A*inv(P);
Bbar = P*B;
Cbar = C*inv(P);
Dbar = D;
```

MATLAB contains these commands in the following M-file:

```
[Abar,Bbar,Cbar,Dbar] = ss2ss(A,B,C,D,P);
```

Running MATLAB results in the matrices

$$\overline{A} = \begin{bmatrix}-3 & 2\\ -3 & 1\end{bmatrix}; \qquad \overline{B} = \begin{bmatrix}1\\ 1\end{bmatrix}; \qquad \overline{C} = [2 \quad -1]; \qquad \overline{D} = 0$$

The transfer function matrices for both of the state models above can be calculated using the command

```
[num,den] = ss2tf(A,B,C,D,1)
```

Inserting the values for A, B, C, and D results in the transfer function

$$C(sI - A)^{-1}B + D = \frac{s + 2}{s^2 + 2s + 3}$$

Using the same commands on the second state model shows that the transfer functions are indeed identical.

By considering a coordinate transformation, it is sometimes possible to go from one state model (A, B, C, D) to another state model $(\overline{A}, \overline{B}, \overline{C}, \overline{D})$ for which one or more of the coefficients matrices $\overline{A}, \overline{B}, \overline{C}$ have a special form. Such models are called canonical models or canonical forms. Examples are the diagonal form, control-canonical form, and the observer-canonical form. Due to the special structure of these canonical forms, they can result in a significant simplification in the solution to certain classes of problems. For example, the control-canonical form is very useful in the study of state feedback. For an in-depth development of the various canonical forms, the reader is referred to Kailath [1980], Brogan [1991], or Rugh [1996].

Example of the Diagonal Form

Consider the *RLC* series circuit shown in Figure 13.10. To determine a state model of the circuit, the state variables can be defined to be the current $i(t)$ in the inductor and the voltage $v_c(t)$ across the capacitor, that is,

$$x_1(t) = i(t)$$

$$x_2(t) = v_c(t)$$

Summing the voltages around the loop gives

$$Ri(t) + L\frac{di(t)}{dt} + v_c(t) = v(t)$$

Hence

$$\dot{x}_1(t) = -\frac{R}{L}x_1(t) - \frac{1}{L}x_2(t) + \frac{1}{L}v(t)$$

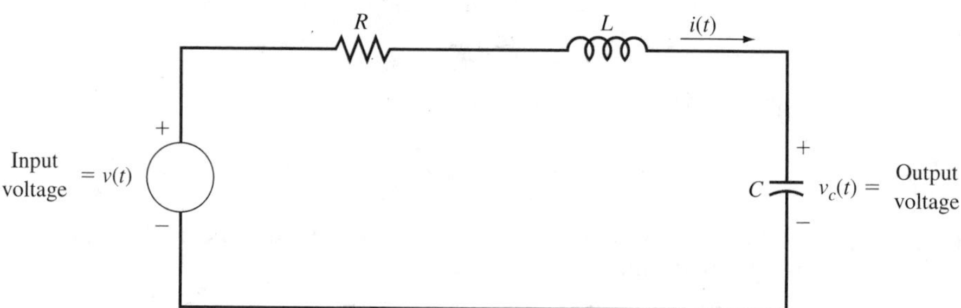

Figure 13.10 Series *RLC* circuit.

Also,

$$\dot{x}_2(t) = \frac{1}{C}x_1(t)$$

$$y(t) = x_2(t)$$

So the circuit has the state model $\dot{x}(t) = Ax(t) + Bv(t), y(t) = Cx(t)$, where

$$A = \begin{bmatrix} -\dfrac{R}{L} & -\dfrac{1}{L} \\ \dfrac{1}{C} & 0 \end{bmatrix}, \qquad B = \begin{bmatrix} \dfrac{1}{L} \\ 0 \end{bmatrix}, \qquad C = [0 \quad 1]$$

Now the question is whether or not there is a coordinate transformation $\bar{x}(t) = Px(t)$ such that $\overline{A} = PA(P^{-1})$ is in the diagonal form

$$\overline{A} = \begin{bmatrix} \bar{a}_1 & 0 \\ 0 & \bar{a}_2 \end{bmatrix}$$

Part of the interest in the diagonal form is the simplification that results from this form. For instance, if $\overline{A}$ is in the diagonal form given below, the state transition matrix has the simple form

$$e^{\overline{A}t} = \begin{bmatrix} e^{\bar{a}_1 t} & 0 \\ 0 & e^{\bar{a}_2 t} \end{bmatrix}$$

Not every matrix A can be put into a diagonal form $\overline{A}$ by a coordinate transformation $\bar{x}(t) = Px(t)$. There are systematic methods for studying the existence and computation of diagonal forms. In the following development a direct procedure is given for determining the existence of a diagonal form for the series RLC circuit.

First, with det $(sI - A)$ equal to the determinant of $sI - A$ (see Appendix B), inserting the above A into det $(sI - A)$ gives

$$\det(sI - A) = \det\begin{bmatrix} s + \dfrac{R}{L} & \dfrac{1}{L} \\ -\dfrac{1}{C} & s \end{bmatrix} = s^2 + \frac{R}{L}s + \frac{1}{LC} \tag{13.75}$$

With $\overline{A}$ equal to the diagonal form given above,

$$\det(sI - \overline{A}) = \det\begin{bmatrix} s - \bar{a}_1 & 0 \\ 0 & s - \bar{a}_2 \end{bmatrix} = (s - \bar{a}_1)(s - \bar{a}_2) \tag{13.76}$$

From results in matrix algebra, it can be shown that

$$\det(sI - A) = \det(sI - \overline{A})$$

Then equating (13.75) and (13.76), it is seen that $\bar{a}_1$ and $\bar{a}_2$ must be the zeros of $s^2 + (R/L)s + (1/LC)$. Thus

$$\bar{a}_1^2 + \frac{R}{L}\bar{a}_1 + \frac{1}{LC} = 0$$

$$\bar{a}_2^2 + \frac{R}{L}\bar{a}_2 + \frac{1}{LC} = 0$$

Now since $\overline{A} = PA(P^{-1}), \overline{A}P = PA$. Setting $\overline{A}P = PA$ gives

$$\begin{bmatrix} \bar{a}_1 & 0 \\ 0 & \bar{a}_2 \end{bmatrix} \begin{bmatrix} p_1 & p_2 \\ p_3 & p_4 \end{bmatrix} = \begin{bmatrix} p_1 & p_2 \\ p_3 & p_4 \end{bmatrix} \begin{bmatrix} -\frac{R}{L} & -\frac{1}{L} \\ \frac{1}{C} & 0 \end{bmatrix}$$

Equating the entries of $\overline{A}P$ and PA results in the equations

$$\bar{a}_1 p_1 = -\frac{Rp_1}{L} + \frac{p_2}{C} \tag{13.77}$$

$$\bar{a}_1 p_2 = -\frac{p_1}{L} \tag{13.78}$$

$$\bar{a}_2 p_3 = -\frac{Rp_3}{L} + \frac{p_4}{C} \tag{13.79}$$

$$\bar{a}_2 p_4 = -\frac{p_3}{L} \tag{13.80}$$

Equations (13.77) and (13.78) reduce to the single constraint

$$p_2 = -\frac{p_1}{L\bar{a}_1}$$

Equations (13.79) and (13.80) reduce to the single constraint

$$p_4 = -\frac{p_3}{L\bar{a}_2}$$

Therefore,

$$P = \begin{bmatrix} p_1 & -\frac{p_1}{L\bar{a}_1} \\ p_3 & -\frac{p_3}{L\bar{a}_2} \end{bmatrix}$$

Finally, since P must be invertible, $\det P \neq 0$, and thus

$$-\frac{p_1 p_3}{L\bar{a}_2} + \frac{p_1 p_3}{L\bar{a}_1} \neq 0 \tag{13.81}$$

Equation (13.81) is satisfied if and only if $\bar{a}_1 \neq \bar{a}_2$. Hence the diagonal form exists if and only if the zeros of $s^2 + (R/L)s + (1/LC)$ are distinct. Note that the transformation matrix P given above is not unique.

13.6 DISCRETIZATION OF STATE MODEL

Again consider a p-input r-output N-dimensional continuous-time system given by the state model

$$\dot{x}(t) = Ax(t) + Bv(t) \tag{13.82}$$

$$y(t) = Cx(t) + Dv(t) \tag{13.83}$$

In this section it is shown that the state representation can be discretized in time, which results in a discrete-time simulation of the given continuous-time system.

From the results in Section 13.3, the state response $x(t)$ resulting from initial state $x(0)$ and input $v(t)$ is given by

$$x(t) = e^{At}x(0) + \int_0^t e^{A(t-\lambda)}Bv(\lambda)\, d\lambda, \qquad t > 0$$

Now suppose that the initial time is changed from $t = 0$ to $t = \tau$, where τ is any real number. Solving the vector difference equation (13.82) with the initial time $t = \tau$ gives

$$x(t) = e^{A(t-\tau)}x(\tau) + \int_\tau^t e^{A(t-\lambda)}Bv(\lambda)\, d\lambda, \qquad t > \tau \tag{13.84}$$

Let T be a fixed positive number. Then setting $\tau = nT$ and $t = nT + T$ in (13.84), where n is the discrete-time index yields

$$x(nT + T) = e^{AT}x(nT) + \int_{nT}^{nT+T} e^{A(nT+T-\lambda)}Bv(\lambda)\, d\lambda \tag{13.85}$$

Equation (13.85) looks like a state equation for a discrete-time system except that the second term on the right-hand side of (13.85) is not in the form of a matrix times $v(nT)$. This term can be expressed in such a form if the input $v(t)$ is constant over the T-second intervals $nT \leq t \leq nT + T$; that is,

$$v(t) = v(nT), \qquad nT \leq t < nT + T \tag{13.86}$$

If $v(t)$ satisfies (13.86), (13.85) can be written in the form

$$x(nT + T) = e^{AT}x(nT) + \left\{\int_{nT}^{nT+T} e^{A(nT+T-\lambda)}B\, d\lambda\right\}v(nT) \tag{13.87}$$

Let B_d denote the $N \times p$ matrix defined by

$$B_d = \int_{nT}^{nT+T} e^{A(nT+T-\lambda)}B\, d\lambda \tag{13.88}$$

Carrying out the change of variables $\bar{\lambda} = nT + T - \lambda$ in the integral in (13.88) gives

$$B_d = \int_T^0 e^{A\bar{\lambda}} B(-d\bar{\lambda}) = \int_0^T e^{A\lambda} B \, d\lambda$$

From this expression it is seen that B_d is independent of the time index n.

Now let A_d denote the $N \times N$ matrix defined by

$$A_d = e^{AT}$$

Then in terms of A_d and B_d, the difference equation (13.87) can be written in the form

$$x(nT + T) = A_d x(nT) + B_d v(nT) \tag{13.89}$$

Setting $t = nT$ in both sides of (13.83) results in the discretized output equation

$$y(nT) = Cx(nT) + Dv(nT) \tag{13.90}$$

Equations (13.89) and (13.90) are the state equations of a linear time-invariant N-dimensional discrete-time system. This discrete-time system is a discretization in time of the given continuous-time system. If the input $v(t)$ is constant over the T-second intervals $nT \leq t < nT + T$, the values of the state response $x(t)$ and output response $v(t)$ for $t = nT$ can be computed exactly by solving the state equations (13.89) and (13.90). Since (13.89) can be solved recursively, the discretization process yields a numerical method for solving the state equation of a linear time-invariant continuous-time system.

If $v(t)$ is not constant over the T-second intervals $nT \leq t < nT + T$, the solution of (13.89) and (13.90) will yield approximate values for $x(nT)$ and $y(nT)$. In general, the accuracy of the approximate values will improve as T is made smaller. So even if $v(t)$ is not piecewise constant, the representation (13.89) and (13.90) serves as a discrete-time simulation of the given continuous-time system.

Since the step function $u(t)$ is constant for all $t > 0$, in the single-input single-output case the step response of the discretization (13.89) and (13.90) will match the values of the step response of the given continuous-time system. Hence the above discretization process is simply a state version of step-response matching (see Section 12.5).

The discretization in time given above can be used to discretize any system given by a Nth-order linear constant-coefficient input/output differential equation. In particular, a state model of the system can be constructed by using the realization given in Section 13.2, and then the coefficient matrices of this state model can be discretized, which yields a discretization in time of the given continuous-time system.

Example 13.9 *Car on a Level Surface*

Consider the car on a level surface with state model

$$\begin{bmatrix} \dot{x}_1(t) \\ \dot{x}_2(t) \end{bmatrix} = \begin{bmatrix} 0 & 1 \\ 0 & \dfrac{-k_f}{M} \end{bmatrix} \begin{bmatrix} x_1(t) \\ x_2(t) \end{bmatrix} + \begin{bmatrix} 0 \\ \dfrac{1}{M} \end{bmatrix} f(t)$$

$$y(t) = [1 \quad 0] \begin{bmatrix} x_1(t) \\ x_2(t) \end{bmatrix}$$

where $x_1(t)$ is the position, $x_2(t)$ is the velocity of the car and $f(t)$ is the drive or braking force applied to the car. To compute the discretized matrices A_d and B_d for this system, it is first necessary to calculate the state transition matrix e^{At}. For the A above,

$$(sI - A)^{-1} = \begin{bmatrix} s & -1 \\ 0 & s + \dfrac{k_f}{M} \end{bmatrix}^{-1} = \frac{1}{s(s + k_f/M)} \begin{bmatrix} s + \dfrac{k_f}{M} & 1 \\ 0 & s \end{bmatrix}$$

Thus

$$e^{At} = \begin{bmatrix} 1 & \dfrac{M}{k_f}\left[1 - \exp\left(-\dfrac{k_f}{M}t\right)\right] \\ 0 & \exp\left(-\dfrac{k_f}{M}t\right) \end{bmatrix}$$

and

$$e^{At} = \begin{bmatrix} 1 & \dfrac{M}{k_f}\left[1 - \exp\left(-\dfrac{k_fT}{M}\right)\right] \\ 0 & \exp\left(-\dfrac{k_fT}{M}\right) \end{bmatrix}$$

$$B_d = \int_0^T e^{A\lambda}B\, d\lambda = \begin{bmatrix} \displaystyle\int_0^T \frac{1}{k_f}\left[1 - \exp\left(-\frac{k_f}{M}\lambda\right)\right] d\lambda \\ \displaystyle\int_0^T \frac{1}{M}\exp\left(-\frac{k_f}{M}\lambda\right) d\lambda \end{bmatrix}$$

$$= \begin{bmatrix} \dfrac{T}{k_f} - \dfrac{M}{k_f^2}\left[1 - \exp\left(-\dfrac{k_fT}{M}\right)\right] \\ \dfrac{1}{k_f}\left[1 - \exp\left(-\dfrac{k_fT}{M}\right)\right] \end{bmatrix}$$

When $M = 1, k_f = 0.1$, and $T = 0.1$, the discretized matrices are

$$A_d = \begin{bmatrix} 1 & 0.09950166 \\ 0 & 0.99004983 \end{bmatrix}, \qquad B_d = \begin{bmatrix} 0.00498344 \\ 0.09950166 \end{bmatrix}$$

The state model of the resulting discrete-time simulation is

$$\begin{bmatrix} x_1(0.1n + 0.1) \\ x_2(0.1n + 0.1) \end{bmatrix} = \begin{bmatrix} 1 & 0.09950166 \\ 0 & 0.99004983 \end{bmatrix} \begin{bmatrix} x_1(0.1n) \\ x_2(0.1n) \end{bmatrix} + \begin{bmatrix} 0.00498344 \\ 0.09950166 \end{bmatrix} f(0.1n)$$

where

$$x_1(0.1n) = y(0.1n) \quad \text{and} \quad x_2(0.1n) = \dot{y}(0.1n)$$

An approximation for A_d and for B_d can be found numerically by first writing e^{AT} in series form using (13.22):

$$e^{AT} = \text{I} + AT + \frac{A^2T^2}{2!} + \frac{A^3T^3}{3!} + \cdots$$

The matrix A_d can be found by truncating the series above after a few terms, and the matrix B_d can be found by substituting the truncated series into equation (13.88) and evaluating, which yields

$$B_d = \sum_{k=0}^{N} \frac{A^k T^{k+1}}{(k+1)!} B$$

MATLAB performs a similar computation of A_d and B_d by using the command `c2d`. For instance, in Example 13.9 the MATLAB commands are

```
kf = .1; m=1;
A = [0 1;0 -kf/m]; B = [0 1/m]; C = [1 0];
T = 0.1;
[Ad,Bd] = c2d(A,B,T)
```

This programs yields

$$A_d = \begin{bmatrix} 1.0 & 0.09950166 \\ 0 & 0.99004983 \end{bmatrix} \quad \text{and} \quad B_d = \begin{bmatrix} 0.004983375 \\ 0.099501663 \end{bmatrix}$$

which is consistent with the result obtained in Example 13.9.

PROBLEMS

13.1. For the circuit in Figure P13.1, find the state model with the state variables defined to be $x_1(t) = i_L(t), x_2(t) = v_C(t)$, and with the output defined as $y(t) = i_L(t) + v_C(t)$.

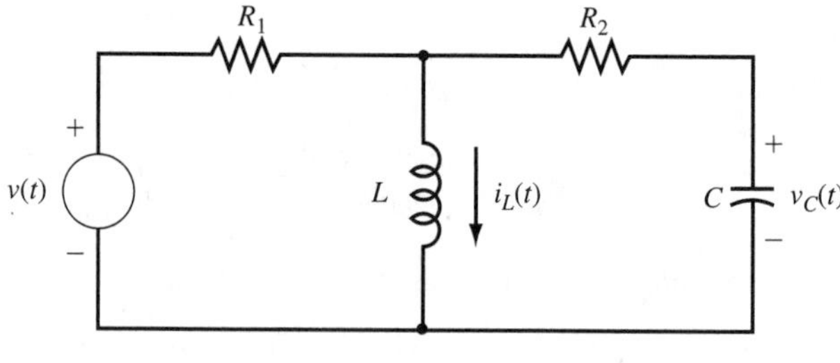

Figure P13.1

13.2. For the circuits in Figure P13.2, find the state model with the state variables as defined in the circuit diagrams.

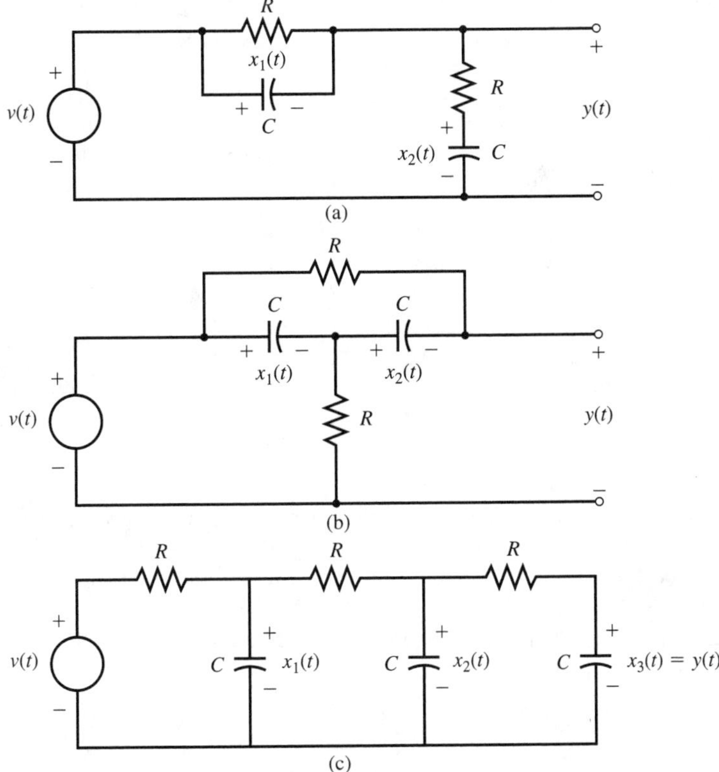

Figure P13.2

13.3. A system is described by the input/output differential equation

$$\ddot{y} + a_1\dot{y} + a_0 y = b_1\dot{v} + b_0 v$$

The following state model for this system was given in Section 13.2:

$$\begin{bmatrix} \dot{x}_1 \\ \dot{x}_2 \end{bmatrix} = \begin{bmatrix} 0 & 1 \\ -a_0 & -a_1 \end{bmatrix}\begin{bmatrix} x_1 \\ x_2 \end{bmatrix} + \begin{bmatrix} 0 \\ 1 \end{bmatrix} v$$

$$y = [b_0 \quad b_1]\begin{bmatrix} x_1 \\ x_2 \end{bmatrix}$$

Find expressions for x_1 and x_2 in terms of v, y, and $\dot{y}$.

13.4. When the input $v(t) = \cos t$, $t \geq 0$, is applied to a linear time-invariant continuous-time system, the resulting output response (with no initial energy in the system) is

$$y(t) = 2 - e^{-5t} + 3\cos t, \quad t \geq 0$$

Find a state model of the system with the smallest possible number of state variables. Verify the model by simulating the response to $v(t) = \cos t$.

13.5. A linear time-invariant continuous-time system has the transfer function

$$H(s) = \frac{s^2 - 2s + 2}{s^2 + 3s + 1}$$

Find a state model of the system with the smallest possible number of state variables.

13.6. Consider the two-input two-output linear time-invariant continuous-time system shown in Figure P13.6.

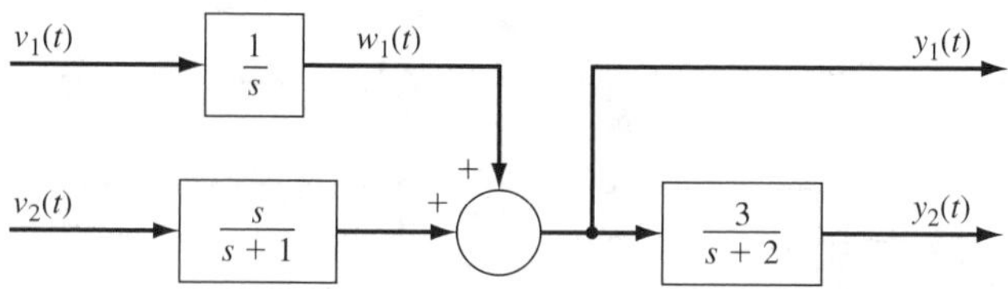

Figure P13.6

(a) Find the state model of the system with the state variables defined to be $x_1(t) = v(t), x_2(t) = y_1(t) - w(t) - v_2(t)$, and $x_3(t) = y_2(t)$.

(b) Find the state model of the system with the state variables defined to be $x_1(t) = y_1(t) - v_2(t), x_2(t) = y_2(t)$, and $x_3(t) = y_1(t) - w(t) - v_2(t)$.

13.7. For the system shown in Figure P13.7, find a state model with the number of state variables equal to 2.

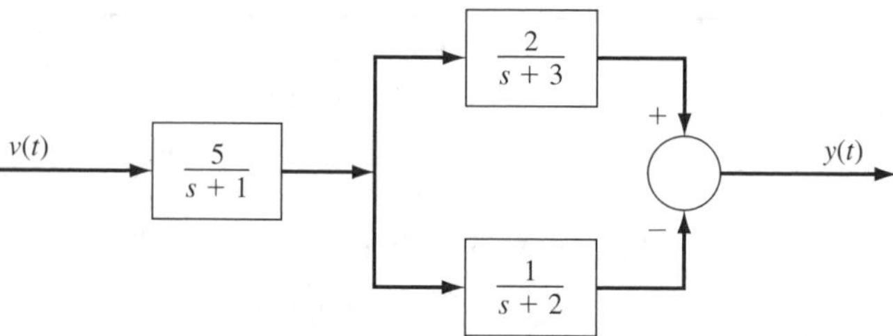

Figure P13.7

13.8. A linear time-invariant continuous-time system has the transfer function

$$H(s) = \frac{bs}{s^2 + a_1 s + a_0}$$

Find the state model of the system with the state variables defined to be

$$x_1(t) = y(t) \quad \text{and} \quad x_2(t) = \int_{-\infty}^{t} y(\lambda)\, d\lambda$$

13.9. For the two-input two-output system shown in Figure P13.9, find a state model with the smallest possible number of state variables.

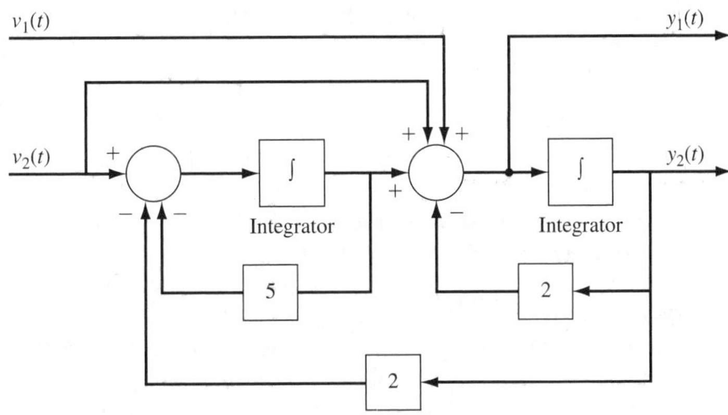

Figure P13.9

13.10. The input $x[n] = -2 + 2^n, n = 0, 1, 2, \ldots$, is applied to a linear time-invariant discrete-time system. The resulting response is $y[n] = 3^n - 4(2^n), n = 0, 1, 2, \ldots$, with no initial energy in the system. Find a state model of the system with the smallest possible number of state variables. Verify the model by simulating its response to $x[n] = -2 + 2^n, n \geq 0$.

13.11. A linear time-invariant continuous-time system has the state model

$$\begin{bmatrix} \dot{x}_1(t) \\ \dot{x}_2(t) \end{bmatrix} = \begin{bmatrix} 0 & 1 \\ -1 & 2 \end{bmatrix} \begin{bmatrix} x_1(t) \\ x_2(t) \end{bmatrix} + \begin{bmatrix} 0 \\ 1 \end{bmatrix} v(t)$$

$$y(t) = [1 \quad 2] \begin{bmatrix} x_1(t) \\ x_2(t) \end{bmatrix}$$

Derive an expression for $x_1(t)$ and $x_2(t)$ in terms of $v(t)$, $y(t)$, and (if necessary) the derivatives of $v(t)$ and $y(t)$.

13.12. Consider the system consisting of two masses and three springs shown in Figure P13.12. The masses are on wheels which are assumed to be frictionless. The input $v(t)$ to the system is the force $v(t)$ applied to the first mass. The position of the first mass is $w(t)$ and the position of the second mass is the output $y(t)$, where both $w(t)$ and $y(t)$ are defined with respect to some equilibrium position. With the state variables defined by $x_1(t) = w(t), x_2(t) = \dot{w}(t), x_3(t) = y(t)$, and $x_4(t) = \dot{y}(t)$, find the state model of the system.

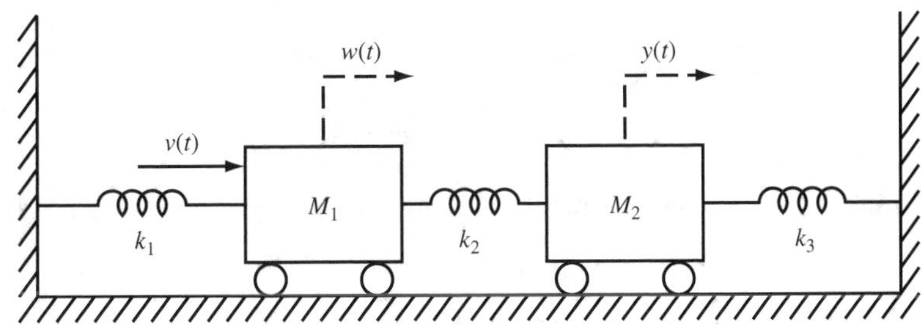

Figure P13.12

13.13. An automobile suspension system is modeled by the equations

$$M_1 \frac{d^2w(t)}{dt^2} + k_t[w(t) - v(t)] = k_s[y(t) - w(t)] + k_d\left[\frac{dy(t)}{dt} - \frac{dw(t)}{dt}\right]$$

$$M_2 \frac{d^2y(t)}{dt^2} + k_s[y(t) - w(t)] + k_d\left(\frac{dy(t)}{dt} - \frac{dw(t)}{dt}\right) = 0$$

where $w(t)$ is the position of the wheel and the output $y(t)$ is the position of the frame relative to an equilibrium position. The input $v(t)$ is the ground position. [See Figure 2.5 with the notation change: $q(t) = w(t)$.] With the state variables defined by $x_1(t) = w(t)$, $x_2(t) = \dot{w}(t)$, $x_3(t) = y(t)$, and $x_4(t) = \dot{y}(t)$, find the state model of the system.

13.14. Consider the "dueling pendulums" shown in Figure P13.14. Each pendulum has mass M and length L, the angular position of the pendulum on the left is the output $y(t)$, and the angular position of the one on the right is $\theta(t)$. Mg is the force on each pendulum due to gravity, where g is the gravity constant. The distance between the two pendulums is $d(t)$, which is assumed to be nonnegative for all t. As shown, a spring is attached between the two pendulums. The force on each mass due to the spring depends on the amount the spring is stretched. The input to the two-pendulum system is a force $f(t)$ applied to the mass on the right with the force applied tangential to the motion of the mass. *Assuming that the angles $\theta(t)$ and $y(t)$ are small for all t*, derive the state model for the system with the state variables defined by $x_1(t) = \theta(t)$, $x_2(t) = \dot{\theta}(t)$, $x_3(t) = y(t)$, $x_4(t) = \dot{y}(t)$, and $x_5(t) = d(t) - d_0$, where d_0 is the distance between the pendulums when they are in the vertical resting position.

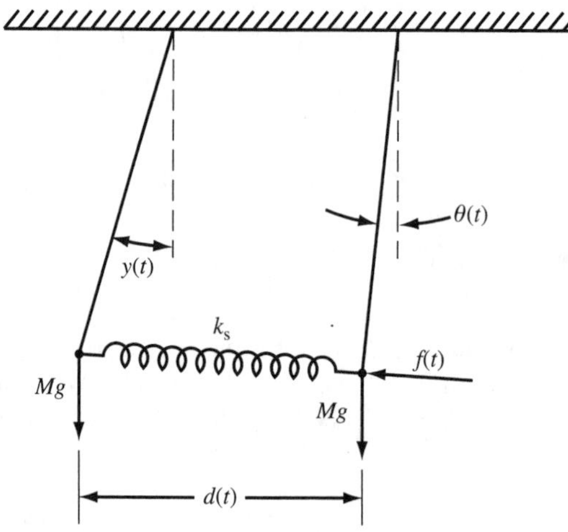

Figure P13.14

13.15. A two-input two-output linear time-invariant continuous-time system has the transfer function matrix

$$H(s) = \begin{bmatrix} \dfrac{1}{s+1} & 0 \\ \dfrac{1}{s+1} & \dfrac{1}{s+2} \end{bmatrix}$$

Find the state model of the system with the state variables defined to be $x_1(t) = y_1(t)$, $x_2(t) = y_2(t)$, where $y_1(t)$ is the first system output and $y_2(t)$ is the second system output.

13.16. A linear time-invariant continuous-time system is given by the state model $\dot{x}(t) = Ax(t) + Bv(t), y(t) = Cx(t)$, where

$$A = \begin{bmatrix} 0 & 2 \\ 0 & -1 \end{bmatrix}, \quad B = \begin{bmatrix} 1 & 0 \\ 1 & -1 \end{bmatrix}, \quad C = [1 \quad 3]$$

(a) Compute the state transition matrix e^{At}.
(b) Compute the transfer function matrix $H(s)$.
(c) Compute the impulse response function matrix $H(t)$.
(d) Compute the state response $x(t)$ for $t > 0$ resulting from initial state $x(0) = [-1 \quad 1]'$ (prime denotes transpose) and input $v(t) = [u(t) \quad u(t)]'$.
(e) If possible, find a nonimpulsive input $v(t)$ with $v(t) = 0$ for $t < 0$ such that the state response $x(t)$ resulting from initial state $x(0) = [1 \quad -1]'$ and input $v(t)$ is given by $x(t) = [u(t) \quad -u(t)]'$.

13.17. Consider a single car moving on a level surface given by the input/output differential equation

$$\frac{d^2y(t)}{dt^2} + \frac{k_f}{M}\frac{dy(t)}{dt} = \frac{1}{M}v(t)$$

where $y(t)$ is the position of the car at time t. With state variables $x_1(t) = y(t), x_2(t) = \dot{y}(t)$, the car has the state model $\dot{x}(t) = Ax(t) + bv(t), y(t) = cx(t)$, where

$$A = \begin{bmatrix} 0 & 1 \\ 0 & \dfrac{-k_f}{M} \end{bmatrix}, \quad b = \begin{bmatrix} 0 \\ \dfrac{1}{M} \end{bmatrix}, \quad c = [1 \quad 0]$$

In the following independent parts, take $M = 1$, and $k_f = 0.1$.
(a) Using the state model, derive an expression for the state response $x(t)$ resulting from initial conditions $y(0) = y_0, \dot{y}(0) = v_0$, with $v(t) = 0$, all t.
(b) With $v(t) = 0$ for $0 \le t \le 10$, it is known that $y(10) = 0, \dot{y}(10) = 55$. Compute $y(0)$ and $\dot{y}(0)$.
(c) The force $v(t) = 1$ is applied to the car for $0 \le t \le 10$. The state $x(5)$ at time $t = 5$ is known to be $x(5) = [50 \quad 20]'$. Compute the initial state $x(0)$ at time $t = 0$.
(d) Now suppose that $v(t) = 0$ for $10 \le t \le 20$, and $y(10) = 5, y(20) = 50$. Compute the state $x(10)$ at time $t = 10$.
(e) Verify the answers for parts (b) to (d) by simulating the response of the state model.

13.18. A two-input two-output linear time-invariant continuous-time system is given by the state model $\dot{x}(t) = Ax(t) + Bv(t), y(t) = Cx(t)$, where

$$A = \begin{bmatrix} 1 & -1 \\ 0 & 2 \end{bmatrix}, \quad B = \begin{bmatrix} 1 & 1 \\ 0 & 0 \end{bmatrix}, \quad C = \begin{bmatrix} 1 & -1 \\ 1 & -1 \end{bmatrix}$$

(a) The output response $y(t)$ resulting from some initial state $x(0)$ with $v(t) = 0$ for all $t \ge 0$ is given by

$$y(t) = \begin{bmatrix} 2e^{2t} \\ 2e^{2t} \end{bmatrix}, \quad t \ge 0$$

Compute $x(0)$.

(b) The output response $y(t)$ resulting from some initial state $x(0)$ and input $v(t) = [u(t) \quad u(t)]'$ is given by

$$y(t) = \begin{bmatrix} 4e^t - 2e^{2t} - 2 \\ 4e^t - 2e^{2t} - 2 \end{bmatrix}, \qquad t \geq 0$$

Compute $x(0)$.

13.19. A linear time-invariant continuous-time system has state model $\dot{x}(t) = Ax(t) + Bv(t)$, $y(t) = Cx(t)$, where

$$A = \begin{bmatrix} -8 & -4 \\ 12 & 6 \end{bmatrix}, \qquad B = \begin{bmatrix} 1 & 1 \\ 2 & 2 \end{bmatrix}, \qquad C = [1 \quad -2]$$

The following parts are independent.

(a) Suppose that $y(2) = 3$ and $\dot{y}(2) = 5$. Compute $x(2)$.

(b) Suppose that $v(t) = 0$ for $0 \leq t \leq 1$ and that $x(1) = [1 \quad 1]'$. Compute $x(0)$.

(c) Suppose that $x(0) = [1 \quad 1]'$. If possible, find an input $v(t)$ such that the output response resulting from $x(0)$ and $v(t)$ is zero, that is, $y(t) = 0, t > 0$.

13.20. The ingestion and metabolism of a drug in a human in modeled by the equations

$$\frac{dw(t)}{dt} = -k_1 w(t) + v(t)$$

$$\frac{dy(t)}{dt} = k_1 w((t) - k_2 y(t)$$

where the input $v(t)$ is the ingestion rate of the drug, the output $y(t)$ is the mass of the drug in the bloodstream, and $w(t)$ is the mass of the drug in the gastrointestinal tract. In the following parts, assume that $k_1 \neq k_2$.

(a) With the state variables defined to be $x_1(t) = w(t)$ and $x_2(t) = y(t)$, find the state model of the system.

(b) With A equal to the system matrix found in part (a), compute the state transition matrix e^{At}.

(c) Compute the inverse of e^{At}.

(d) Using your answer in part (c), compute the state $x(t)$ for all $t > 0$ when $v(t) = 0$ for $t \geq 0$ and the initial state is $x(0) = [M_1 \quad M_2]'$.

(e) Using the state model found in part (a), compute the state response $x(t)$ for all $t > 0$ when $v(t) = e^{at}, t \geq 0$, and $x(0) = [M_1 \quad M_2]'$. Assume that $a \neq k_1, \neq k_2$.

13.21. Consider an inverted pendulum on a motor-driven cart as illustrated in Figure P13.21. Here $\theta(t)$ is the angle of the pendulum from the vertical position, $d(t)$ is the position of the cart at time t, $v(t)$ is the drive or braking force applied to the cart, and M is the mass of the cart. The mass of the pendulum is m. From the laws of mechanics (see Section 2.2), the process is described by the differential equations

$$(J + mL^2)\ddot{\theta} - mgL \sin \theta(t) + mL\ddot{d}(t) \cos \theta(t) = 0$$

$$(M + m)\ddot{d}(t) + mL\ddot{\theta}(t) = v(t)$$

where J is the moment of inertia of the inverted pendulum about the center of mass, g is the gravity constant, and L is one-half the length of the pendulum. We assume that

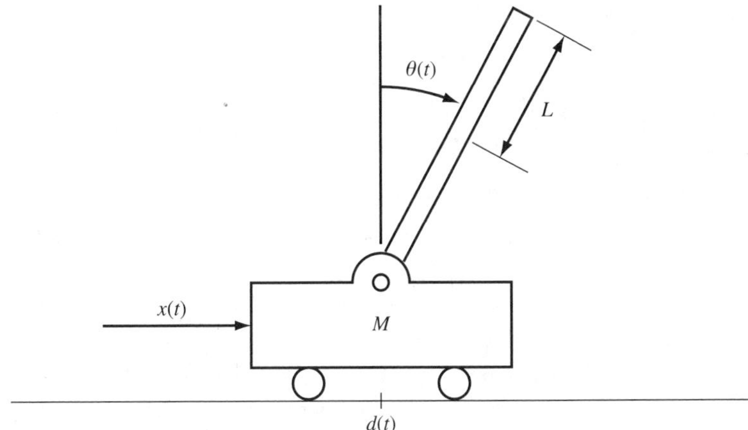

Figure P13.21

the angle $\theta(t)$ is small, and therefore cos $\theta(t) \approx 1$ and sin $\theta(t) \approx \theta(t)$. In the following parts, take $J = 1, L = 1, g = 9.8, M = 1$, and $m = 0.1$.

(a) With the state variables defined to be $x_1 = \theta(t), x_2 = \dot{\theta}(t), x_3 = d(t)$, and $x_4 = \dot{d}(t)$, and the output defined to be $\theta(t)$, find the state model of the inverted pendulum.

(b) With A equal to the system matrix found in part (a), compute the state transition matrix e^{At}.

(c) Compute the inverse $(e^{At})^{-1}$ of the state transition matrix.

(d) Using your answer in part (c), compute the state $x(5)$ at time $t = 5$ assuming that $x(10) = [10° \quad 0 \quad 5 \quad 2]'$ and $v(t) = 0$ for $5 \leq t \leq 10$.

(e) Using the state model, compute the state response $x(t)$ for all $t > 0$ when $\theta(0) = 10°, d(0) = 0, \dot{\theta}(0) = 0, \dot{d}(0) = 0$, and $v(t) = 0$ for $t \geq 0$.

(f) Repeat part (e) with $\theta(0) = 0, \dot{\theta}(0) = 1, d(0) = 0, \dot{d}(0) = 0$, and $v(t) = 0$ for $t \geq 0$.

(g) Repeat part (e) with $\theta(0) = 0, \dot{\theta}(0) = 0, d(0) = 0, \dot{d}(0) = 1$, and $v(t) = 0$ for $t \geq 0$.

(h) Verify the results of parts (d) to (g) by simulating the response of the state model.

13.22. A linear time-invariant discrete-time system is given by the state model $x[n + 1] = Ax[n] + Bv[n], y[n] = Cx[n]$, where

$$A = \begin{bmatrix} -1 & 1 \\ -1 & -2 \end{bmatrix}, \quad B = \begin{bmatrix} 0.5 & 1 \\ -1 & -0.5 \end{bmatrix}, \quad C = \begin{bmatrix} 2 & 1 \\ -1 & -2 \end{bmatrix}$$

(a) Compute $x[1], x[2]$, and $x[3]$ when $x(0) = [1 \quad 1]'$ and $v[n] = [n \quad n]'$.

(b) Compute the transfer function matrix $H(z)$.

(c) Suppose that $x[0] = [0 \quad 0]'$. Find an input $v[n]$ that sets up the state $x[2] = [-1 \quad 2]'$; that is, the state $x[2]$ of the system at time $n = 2$ resulting from input $v[n]$ is equal to $[-1 \quad 2]'$.

(d) Now suppose that $x[0] = [1 \quad -2]'$. Find an input $v[n]$ that drives the system to the zero state at time $n = 2$; that is, $x[2] = [0 \quad 0]'$.

(e) Verify the results of parts (a) to (d) by simulating the response of the state model.

13.23. Consider the discrete-time system with state model $x[n+1] = Ax[n] + Bv[n]$, $y[n] = cx[n]$, where

$$A = \begin{bmatrix} 1 & 0 \\ 0.5 & 1 \end{bmatrix}, \quad B = \begin{bmatrix} 2 \\ 1 \end{bmatrix}, \quad C = \begin{bmatrix} 2 & -1 \\ 1 & 0 \\ 1 & -1 \end{bmatrix}$$

The following parts are independent.

(a) Compute $y[0]$, $y[1]$, and $y[2]$ when $x[0] = [-1 \quad 2]'$ and the input is $v[n] = \sin(\pi/2)n$.

(b) Suppose that $x[3] = [1 \quad -1]'$. Compute $x[0]$ assuming that $v[n] = 0$ for $n = 0, 1, 2, \ldots$.

(c) Suppose that $y[3] = [1 \quad 2 \quad -1]'$. Compute $x[3]$.

(d) Verify the results of parts (a) to (c) by simulating the response of the state model.

13.24. A discrete-time system has the state model $x[n+1] = Ax[n] + Bv[n]$, $y[n] = Cx[n]$, where

$$CB = \begin{bmatrix} 6 \\ 3 \end{bmatrix} \quad \text{and} \quad CAB = \begin{bmatrix} 22 \\ 11 \end{bmatrix}$$

When $x[0] = 0$, it is known that $y[1] = [6 \quad 3]'$ and $y[2] = [4 \quad 2]'$. Compute $v[0]$ and $v[1]$.

13.25. A continuous-time system has state model $\dot{x}(t) = Ax(t) + bv(t)$, $y(t) = Cx(t)$, where

$$A = \begin{bmatrix} 3 & -2 \\ 9 & -6 \end{bmatrix}, \quad b = \begin{bmatrix} 1 \\ 2 \end{bmatrix}$$

(a) Determine if there is a coordinate transformation $\bar{x}(t) = Px(t)$ such that $\overline{A}$ is in diagonal form. If such a transformation exists, give P and $\overline{A}$.

(b) Verify the results in part (a) by using MATLAB to compute $\overline{A}$ and $\overline{b}$.

13.26. We are given two continuous-time systems with state models

$$\dot{x}(t) = A_1 x(t) + b_1 v(t), \; y(t) = C_1 x(t)$$

$$\dot{\bar{x}}(t) = A_2 \bar{x}(t) + b_2 v(t), \; y(t) = C_2 \bar{x}(t)$$

where

$$A_1 = \begin{bmatrix} 1 & 1 \\ 0 & 2 \end{bmatrix}, \quad A_2 = \begin{bmatrix} 4 & 2 \\ -3 & -1 \end{bmatrix}, \quad b_1 = \begin{bmatrix} 1 \\ 1 \end{bmatrix},$$

$$b_2 = \begin{bmatrix} 5 \\ 3 \end{bmatrix}, \quad C_1 = \begin{bmatrix} 1 & 2 \\ 1 & 2 \end{bmatrix}, \quad C_2 = \begin{bmatrix} 0 & 1 \\ 0 & 1 \end{bmatrix}$$

Determine if there is a coordinate transformation $\bar{x}(t) = Px(t)$ between the two systems. Determine P if it exists.

13.27. A linear time-invariant continuous-time system has state model $\dot{x}(t) = Ax(t) + bv(t)$, $y(t) = cv(t)$. It is known that there is a coordinate transformation $\bar{x}(t) = P_1 x(t)$ such that

$$\overline{A} = \begin{bmatrix} -2 & 0 & 0 \\ 0 & -1 & 0 \\ 0 & 0 & 1 \end{bmatrix}, \quad \overline{b} = \begin{bmatrix} 1 \\ 1 \\ 1 \end{bmatrix}, \quad \overline{c} = [-1 \quad 1 \quad 1]$$